# Cash Crops

P. M. Priyadarshan • S. Mohan Jain
Editors

# Cash Crops

## Genetic Diversity, Erosion, Conservation and Utilization

Springer

*Editors*
P. M. Priyadarshan
(erstwhile with the Rubber Research
Institute of India)
Thiruvananthapuram, Kerala, India

S. Mohan Jain
Department of Agricultural Sciences
University of Helsinki
Helsinki, Finland

ISBN 978-3-030-74925-5 ISBN 978-3-030-74926-2 (eBook)
https://doi.org/10.1007/978-3-030-74926-2



This Springer imprint is published by the registered company Springer Nature Switzerland AG
The registered company address is: Gewerbestrasse 11, 6330 Cham, Switzerland

# Preface

Breeding crop plants in the current century is a daunting challenge. Plant breeders have already made much effort in developing new high yielding plant varieties. These efforts are further augmented with the advent of new stresses related to climate change. Any successful breeding programme solely depends on the availability of genetic diversity. Genetic diversity and genetic variation are to be separately dealt with. While genetic diversity is the range of genetic characteristics in a crop, the genetic variation refers to the genetic differences among individuals in a specific trait, which are based on DNA sequence.

Staple or food crops like cereals, wheat, rice, and maize are absolutely essential for sustenance. However, there are other crops like coffee, tea, cocoa, cotton, oilseeds, sugar cane, oil palm, date palm, and rubber that are grown only for raising income to the farmers. A differentiation of food with cash crops suggests that most food crops are considered not to be traded much through markets. But this definition does not hold good for developing countries where most of the crops are grown for revenue. However, the debate on the differentiation between food crops and cash crops still continues.

A looming threat to the world in the current century is population explosion where food and nutrition security, climate change, and ever-growing human population growth have become major global issues. Estimates suggest that the world population will reach close to 10 billion by 2050, with a food gap of 70% between the crop calories available in 2006 and projected calorie demand in 2050. A sustained average annual growth in agricultural production of 44 million metric tons per year, for 30 years, is essential to meet this demand. Breeding activities have achieved a linear increase at an average rate of 32 million metric tons per year. Creating and utilizing new genetic diversity is the only option before the plant breeders to make the production of new varieties more sustainable. Breeding new cash crop varieties will ensure increased food and nutrition security. This challenge is further made intriguing since most cash crops are long term in nature.

Crop diversity around the world is declining, posing a challenge to both environment and food security. With the advent of Green Revolution, staple crops like wheat, rice, soybean, and corn have become dominant in arable lands worldwide.

The lack of genetic diversity and increasing uniformity of any crop may render the same unsustainable. Hence, the FAO in 2018 had set out a new set of conservation guidelines titled, "Voluntary Guidelines for the Conservation and Sustainable Use of Farmers' Varieties." This publication highlights how crop genetic diversity enables resistance to environmental and climatic shocks (https://reliefweb.int/sites/reliefweb.int/files/resources/ca5601en.pdf) and functions as a tool for researchers and policymakers. The FAO notes that almost all countries report the "genetic erosion" of crops, with maize, wheat, and rice accounting for 51% of all plant-based food.

The loss of crop genetic diversity poses a "particularly severe" threat to global food security and nutrition—one that risks the achievement of sustainable development goals (SDG 2 by FAO = zero hunger) on eradicating hunger and malnutrition by 2030. It entails the necessity to provide the farmers and local communities with the information and support in relation to crop conservation and sustainable utilization. Yet, almost 690 million people remain chronically undernourished, amid signs of diminishing momentum towards reaching Zero Hunger.

There are several upcoming innovative technologies that are being implemented in breeding new plants. The use of "CRISPR single base editors" is one such tool that brings much promise in deriving new genetic diversity. Single base editing can produce genotypes with broad-spectrum disease resistance that ensures higher yield under climatic stressful conditions. Speed breeding by creating artificial climatic conditions to breed plants through raising more generations per year is another innovative option. The applications of most promising futuristic technologies are breeding for C4 photosynthesis, high-throughput phenotyping, RNA-seq (whole transcriptome shotgun *sequencing*), metabolomic profiling, phenomics, genomic selection, and climate resilient breeding with artificial intelligence. However, these technologies must go a long way before successfully being implemented in commercial breeding programmes.

Beginning with a chapter on introduction to cash crops, this book contains 14 other chapters on cardamom, cocoa, coconut, cashew nut, coffee, date palm, ground nut, jute, lentil, mustard, oil palm, rubber, sugar cane, and tomato authored by well-known scientists on their respective crops. This book is expected to assist students, researchers, scientists, and faculties as a guide to genetic diversity of cash crops.

We gratefully acknowledge the guidance and assistance rendered by all the reviewers who have done a meticulous job with respective chapters. Finally, we profusely thank Springer Nature for publishing this book.

Thiruvananthapuram, Kerala, India — P. M. Priyadarshan
Helsinki, Finland — S. Mohan Jain

# Contents

# About the Editors

**P. M. Priyadarshan** is a prominent *Hevea* rubber breeder. He began his research career by breeding triticale and wheat while doing MPhil and PhD degrees at Chaudhary Charan Singh University, Meerut, under the guidance of Prof. P.K. Gupta. During the 1980s he focused on the in vitro culture of spices. He joined the Rubber Research Institute of India (Rubber Board, Ministry of Commerce, Govt. of India) as a Plant Breeder in 1990 and specialized in breeding *Hevea* rubber for sub-optimal environments. He was a visiting scientist at the Hardwood Tree Improvement and Regeneration Centre, Purdue University, USA. In 2009, he became the Institute's Deputy Director and managed its Central Experiment Station until 2016. He has been involved in breeding cereals, spices, and *Hevea* rubber for 32 years. He is a reviewer of several international journals like *Industrial Crops and Products* and *Tree Genetics and Genomes*. He has published several research papers and chapters in journals and books of international repute like *Advances in Agronomy*, *Advances in Genetics*, and *Plant Breeding Reviews*. He has edited books on *Breeding Plantation Tree Crops* (Springer), *Breeding Major Food Staples* (Blackwell-Wiley), and the *Genomics of Tree Crops* (Springer). He also authored books on the *Biology of Hevea Rubber* (Springer) and on *Plant Breeding: Classical to Modern* (Springer).

**S. Mohan Jain** received his MPhil, 1973, and PhD, 1978, from Jawaharlal Nehru University, New Delhi, India. He was a postdoctoral fellow in Israel and USA and visiting scientist/professor in Japan, Malaysia, Germany, and Italy. He worked in both academia and industry and served as a Technical Officer, Plant Breeding and Genetics, International Atomic Energy Agency (IAEA), Vienna, Austria, 1999–2005. Dr. Jain is on the editorial board of *Euphytica, In Vitro, Propagation of Ornamental Plants* and reviewer of *Plant Cell and Organ Culture*, *Plant Cell Reports*, and few others. His has more than 165 publications in peer-reviewed journals, book chapters, and conference proceedings and has edited 65 books. He was an invited speaker and acted as a Chairperson in several international conferences worldwide. He was awarded Nobel Peace Prize certificate, in commemoration of the awarding to IAEA of the Nobel Peace Prize for 2005. He was also former consultant to IAEA, the European Union, the Government of Grenada, Iranian Date Palm Company, and the Egyptian Government. Currently, he is Adjunct Professor, Department of Agricultural Sciences, University of Helsinki, Helsinki, Finland.

# Cash Crops: An Introduction

**P. M. Priyadarshan and S. Mohan Jain**

**Abstract** Cash crops are grown for cash generation rather than for sustenance. The commodities are produced and consumed as fruits, flowers, foliage, stems, roots, latex, or any plant parts that are consumed directly or processed products as fiber, rubber, sugar, beverages, and biofuel. An array of crops like wheat, corn, oats, potatoes, cherries, apples, strawberries, soybean, bananas, cotton, jute, oranges, jojoba, jatropha, groundnut, and brassicas come under the umbrella of cash crops. Breeding activities have achieved a linear increase at an average rate of 32 million metric tons per year against an expected turnover of 44 million metric tons per year. The ensuing climate change, low agricultural diversity, and high intensity of agricultural inputs can exacerbate food insecurity and instability. Conserving crop biodiversity is an urgent undertaking. The Inter-governmental Panel on Climate Change (IPCC) predicts that 25–30% of plant species will be extinct or endangered in the next century. As such, cash crops are utilized by humans to earn money, and such crops are prone to climate changes. Hence, safe guarding the existing diversity and producing new diversity in cash crops are of utmost importance.

Molecular marker techniques like SSRs, DArTseq, and SNP genotyping assays in cash crops are on the progress. In many crops, data obtained through PCR analysis of DNA fragments from ancient DNA samples have shown evolutionary changes within the gene pool over a long time. During the last three decades, under the auspices of FAO, the international community has strived to developing and maintaining a global system on plant genetic resources for food and agriculture. CGIAR centers make significant contributions to both the global system and SDG Target 2.5 through international PGRFA collections. Case situations on cacao, coffee, sugarcane, and *Hevea* rubber are explained.

**Keywords** Cash crops · Genetic diversity · Molecular tools · Genomics · Climate change · Biotechnology

---

P. M. Priyadarshan (✉)
Erstwhile with the Rubber Research Institute of India, Kottayam, Kerala, India

S. M. Jain
Department of Agricultural Sciences, University of Helsinki, Helsinki, Finland
e-mail: mohan.jain@helsinki.fi



P. M. Priyadarshan, S. M. Jain (eds.), *Cash Crops*,
https://doi.org/10.1007/978-3-030-74926-2_1

## 1 Introduction

Cash crops are grown for cash generation rather than for sustenance. The commodities are produced and consumed as fruits, flowers, foliage, stems, roots, latex, or any plant parts that are consumed directly or processed products as fiber, rubber, sugar, beverages, and biofuel. An array of crops like wheat, corn, oats, potatoes, cherries, apples, strawberries, soybean, bananas, cotton, jute, oranges, jojoba, jatropha, groundnut, and brassicas come under the umbrella of cash crops. However, some of the major crops are cocoa, coffee, oil palm, sugar cane, and rubber.

Current estimates are that nearly 690 million people are hungry (8.9% the world population) (FAO 2020). The world is not on track to achieve Zero Hunger by 2030. If recent trends continue, the number of people affected by hunger would surpass 840 million by 2030. Hence, the world requires a dramatic increment in food production in the next 30 years. The current human population of 7.7 billion is expected to reach 8.6 billion in 2030 and 10 billion by 2050 (Pourkheirandish et al. 2020). Concurrently, the effects of climate change pose a looming threat to agriculture via drought and salinity that limit agricultural land and water use (Godfray et al. 2010).

The history of agriculture shows that wild grains were collected and consumed starting from 20,000 BC onwards. The eight Neolithic founder crops viz. emmer wheat, einkorn wheat, hulled barley, peas, lentils, chickpeas, bitter vetch, and flax were cultivated from 9500 BC in the eastern Mediterranean area. Rice was domesticated in China some 10,000 years ago (Gross and Zhao 2014). The Anthropocene epoch is believed to have made the man-made changes in Earth's biodiversity and biogeography (Martin et al. 2019). Anthropocene epoch is an unofficial unit of geologic time, used to describe the most recent period in Earth's history when human activity started to have a significant impact on the planet's climate and ecosystems. Such changes in the human-mediated spread of crops beyond their regions of domestication into other parts of the world are central to these arguments (Khoury et al. 2016). Analyses made after 1950s indicate that there have been major influxes (and subsequent domination) of crops into the food supplies, diets, agricultural economies, and farmlands in many parts of the world (Nelson et al. 2016).

Agricultural trade liberalization in the 1980s encouraged the production and export of a few select crops or genotypes at the regional- or country-level (Li 2015). Such programs only ensured that patterns of crop diversification over the past 50 years have differed drastically among regions. At the same time, climatic limitations to growing certain crops in higher latitudes also have likely led to less drastic or immediate shifts in agricultural diversity in these regions, as compared to lower latitudes (Ramankutty et al. 2002). Furthermore, existing global evaluations of crop diversity have primarily focused on taxonomic diversity measured as crop species richness across global scales (Khoury et al. 2016; Martin et al. 2019).

By the year 2050, the world population may reach nine billion and the demand for food may grow by 70% (Dhankher and Foyer 2018). Breeding activities have achieved a linear increase at an average rate of 32 million metric tons per year (Alston et al. 2009) against an expected turnover of 44 million metric tons per year

(Tester and Langridge 2010). The ensuing climate change, low agricultural diversity, and high intensity of agricultural inputs can exacerbate food insecurity and instability (Dhankher and Foyer 2018). For instance, current estimates indicate that an increase of 1 °C might cause a 10–20% reduction in the world's production of maize (Tito et al. 2018). In fact, meta-analyses of climate change and its impact suggest that by 2030 crop yields may be decreased by 50% (Campbell et al. 2018).

The livelihood of millions of smallholder farmers and the survival of several national economies in Africa, Latin America, and Asia depend on crops usually considered commodities, or minor or orphan crops like coffee (*Coffea arabica* L.) and cacao (*Theobroma cacao* L.) (Davis et al. 2019; Farrell et al. 2018; Schroth, et al. 2016). For coffee, rapid deforestation and climate change may lead to the extinction of many wild African species (Davis et al. 2019). Such crops are at risk due to climate change (Davis et al. 2019; Schroth et al. 2016). For cacao, mostly grown in West Africa, the maximum temperature tolerated (38 °C) could be exceeded during hot and dry *El Niño* years (Schroth et al. 2016). The rising temperatures, longer droughts, and excessive rainfall have reduced coffee yields by 30% in Colombia since 2008 (Van der Vossen et al. 2015).

## 2 Genetic Diversity Utilization and Exchange of Germplasm

Conserving crop biodiversity is an urgent undertaking. The Inter-governmental Panel on Climate Change (IPCC) predicts that 25–30% of plant species will be extinct or endangered in the next century (Foden et al. 2018; Sintayehu 2018). As such, cash crops are utilized by humans to earn money, and such crops are prone to climate changes. Hence, safe guarding the existing diversity and producing new diversity in cash crops are of utmost importance. Crop diversity around the world is declining posing a challenge to both environment and food security. With the advent of Green Revolution, staple crops like wheat, rice, soybean, and corn have become dominant in arable lands of the world. The option to keep the environment intact is to conserve and to produce newer genetic diversity of crops. Under these circumstances, cash crops ensure food, nutrition, and income to the farmers in addition to conserving genetic diversity.

Genetic diversity can be defined as any quantitative measure of the variability of a population, which reflects the equilibrium between mutation and the loss of genetic variation (Carvalho 2019). Molecular markers for plants were initially isoenzymes (Tanksley 1983). In due course, DNA markers could be used to detect allelic variation in the genes underlying the target characteristics (Collard and Mackill 2007). The popularization and modernization of genetic markers led to primers developed to detect variability in accessions (Camacho et al. 2017).

New technologies, like next-generation sequencing (NGS), have permitted the accumulation of large quantities of molecular data through the analysis of marker regions (Goodwin et al. 2016). Such studies make estimates of genetic diversity extremely robust.

## *Molecular Tools and Genome Editing*

Molecular marker techniques are employed in the evaluation of genetic diversity and construction of genetic and physical maps. Physical mapping of linked markers helps in relating genetic distances to physical distances (see Nadeem et al. 2017 for a review). Large amounts of sequence data and genomic and cDNA libraries shall be made available over time. They could be used to draw genes coding for potentially useful traits. SSR, DArTseq, and SNP genotyping assays in cash crops are on the progress (De Wever 2019; Spinoso-Castillo et al. 2020; Ali et al. 2019; Pootakham et al. 2020). In many crops, data obtained through PCR analysis of DNA fragments from ancient DNA samples have shown evolutionary changes within the gene pool over long time (Utge et al. 2020). However, it is very unlikely that such data can replace the germplasm conservation of whole organisms but can give definite leads to genetic diversity (Rasmussen 2020).

Genome editing is another new area that stems promise towards creating genetic variability at will (Xu et al. 2019). The clustered regularly interspaced short palindromic repeats (CRISPR) is one such technique that is largely being used to manipulate genomes as desired (Zhu et al. 2020).CRISPR-Cas genome editing technology has altered plant molecular biology beyond all expectations that allows precise genetic manipulation and provides the opportunity to create germplasm with beneficial traits. CRISPR-Cas can be combined with double-haploid (DH) production (Dwivedi et al. 2015) and speed breeding (Watson et al. 2018) to create more desirable genetic variation. DH lines are derived by crossing a genotype with an inducer line whose haploid chromosome set is lost in the zygote. This is followed by doubling the remaining haploid chromosome compliment to achieve a completely homozygous line. This is raised in a single generation (Wolter et al. 2019). The genotype subjected to multiplex genome editing can be used for DH production for immediate homozygous fixation of the edited alleles. In this way, multiple generations of selfing can be avoided to fix homozygosity (Jacquier et al. 2020).

The base editing system enables precise C»T or A»G editing in a specified sequence range by fusion of Cas9 nickase with cytidine or adenine deaminase (Gaudelli et al. 2017). Zong et al. (2018) demonstrated the usefulness of base editing for generating new transcriptional alleles in wheat. Using enhanced base editor, three regulatory elements in the *TaVRN1-A1* promoter in wheat protoplasts were targeted that are involved in the regulation of vernalization. A variety of mutations in all three targeted regulatory elements were identified through deep-sequencing. The EvolvR system is yet another elegant way by which site-specific genetic diversity can be generated (Halperin et al. 2018). It relies on the fusion between an engineered error-prone polymerase domain to a Cas9 nickase. It enables the diversification of all nucleotides at a specific site and within a tuneable window length of up to 350 bp. In this window, the mutation rate can be elevated to more than seven million times (Wolter et al. 2019). If a large collection of random mutations is required at a specific locus, EvolvR has an advantage over base editing in terms of a larger diversity of mutations.CRISPR enables exploitation of wild relatives of crops as a valuable source of allele mining and expansion of crop germplasm. This should address issues related to genetic impoverishment of many crops and the resistance of wild plants against a broad range of stresses (Østerberg et al. 2017). Unfortunately, the development of new crop varieties by genome editing is hampered by strict GMO (genetically modified organism) regulation, like the European Union, where the authorization of new varieties developed by genome editing techniques is subjected to time and cost-intensive admission procedures. But most other countries are not facing such an impediment. However, on a global scale, CRISPR-Cas will continue to revolutionize plant breeding through creation of new germplasm.

The continuous deforestation and excessive utilization of fossil fuels has resulted in escalation of concentration of $CO_2$ from 280 to 400μmol$^{-1}$. $CO_2$ concentration may further elevate to twofold, i.e., up to 800μmol$^{-1}$ by the end this century. Emission of dangerous gases, especially $CO_2$, is the main factor for the greenhouse effect and warmer average global temperatures (Vaughan et al. 2018). From 1990 to 2016, the climatic events that occurred are manifold (Fig. 1). Derivation of new climate-smart crop varieties is the answer for guaranteed food security (Wheeler and Von Braun 2013; FAO 2015). Molecular markers are also used to assess plant response to climate change. The microgeographical genetic differences, and various molecular markers, indicated that although phenotypic plasticity buffers against environmental changes when the climate events become more extreme, such resistance weakens (Jump and Peñuelas 2005). The use of molecular markers is crucial as it indicates adaptability to climate changes. There are a range of strategies available like high-throughput single-nucleotide polymorphism (SNP) genotyping, genomic selection, and trait mapping to enhance sustainable crop production and resilience to climate change (Pourkheirandish et al. 2020). Various genomic approaches for the development of climate change resilient crops are given in Table 1. The "pyramid" approach to introducing favorable alleles and gene combinations is also prime (Nutan et al. 2020).

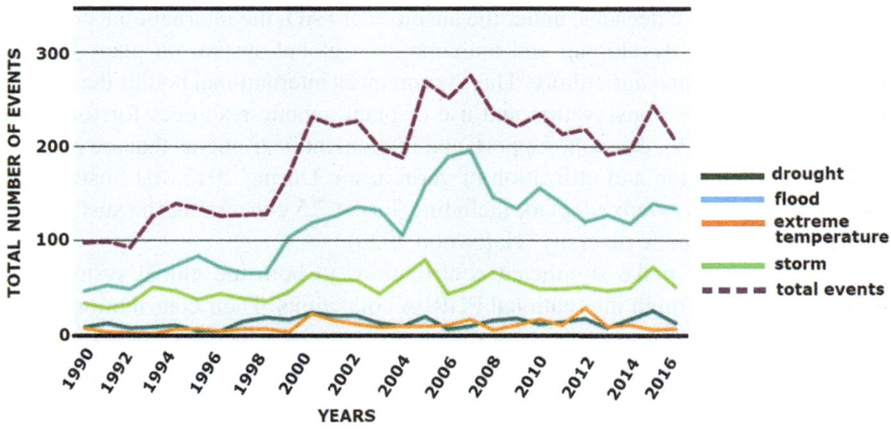


**Fig. 1** An increasing number of extreme climate-related events occurred during 1990–2016. (Source: Food and Agriculture Organization (FAO) based on data from the Emergency Events Database (EMDAT) (https://www.emdat.be/)

**Table 1** Summary of different approaches, which can be used to improve crop diversity and resilience

| Approach | Desired outcome |
|---|---|
| *Using genomics to improve crop plant diversity and resilience* | |
| Accessing genetic diversity of crop wild relatives (CWRs) | Diversification of the existing breeding resources |
| De novo crop domestication | Domestication of completely new crops using wild species |
| Engineering polyploidy | Controlled genome duplication or bridging the genomes of two related species |
| Harnessing plant-microbe interactions | Optimal choice of suitable crops for the specific soil type and geographic location |
| The challenge of climate change and plant diseases | Prediction of pathogen evolution and prevalence and deployment of suitable protective measures ahead of time |
| Genome editing for nutritionally enhanced crops | Editing of target genes to improve crop nutritional value |
| *Accessing new breeding targets using genomic technologies* | |
| Third-generation sequencing | Use of long sequencing reads for higher quality reference genome construction |
| Accurate gene prediction and functional annotation | Precise candidate gene identification |
| Analysis of the non-coding part of genome | Identification of new functional genomic sequences and breeding targets |
| Pangenome as a reference sequence | Inclusion of species-wide genomic variation in the analysis |
| *Pairing genomics with other emerging technologies* | |
| Machine learning and crop plant genomics | Use of artificial intelligence for crop genotype and phenotype prediction |
| Speed breeding | Shortening the breeding cycle |
| High-throughput phenotyping | Increased resolution, accuracy and speed of plant phenotyping |

Adopted from: Pourkheirandish M. et al. 2020; courtesy: Frontiers in Plant Science

## *CGIAR Gene Banks*

During the last three decades, under the auspices of FAO, the international community has strived to developing and maintaining a global system on plant genetic resources for food and agriculture. This system owes international bodies that monitor the status of the conservation and use of plant genetic resources for food and agriculture (PGRFA), develop, support, and implement instruments that are responsible for conservation and utilization of germplasm. During 2015, the Sustainable Development Goals were adopted, including Target 2.5 concerning the sustainable management of genetic diversity (Halewood 2020).

CGIAR centers make significant contributions to both the global system and SDG Target 2.5 through international PGRFA collections. Their contributions are:

- Assembling and conserving PGRFA,
- Adding value to those materials through extensive characterization, evaluation, and documentation,
- Health testing, and,

- Supplying samples that are free of pests and diseases to researchers, plant breeders, farmers, national and community gene banks, and seed companies around the world.

CGIAR collections include over 760,000 accessions of crops, forages, and trees, originally from 207 countries, as well as pre-bred materials. Over the last 10 years, the CGIAR Centers' gene banks have distributed more than 1.1 million PGRFA samples to recipients in 163 countries (Halewood 2020). Such transfers amount to nearly 23% of all PGRFA following the norms of multilateral system of access and benefit sharing created by the International Treaty on Plant Genetic Resources for Food and Agriculture. Even though CGIAR never addresses the diversity of cash crops directly, many genetic resources are scattered in several of its institutions and other academic universities and bodies like USDA. A common platform to coordinate the conservation and use of genetic resources of cash crops is an urgent need. As case examples, cocoa, coffee, sugar cane, and rubber are explained in a little detail here.

## *Cacao*

Cacao (*Theobroma cacao* L.), "the food of Gods" is next to tea and coffee as a beverage. Native to Amazon basin, cocoa was domesticated and distributed to different regions by the natives Mayas, Aztecs, and Pipil-Nicaraos (Bartley 2005; Malhotra and Apshara 2017). There has been a phenomenal increase in the geographical expansion of cocoa genetic resources over the last 30 years (Bartley 2005). Currently, 40–50 million people depend on cocoa for their livelihood, and the production is 3.97 mt contributed by Africa (71%), Latin America (14%), and Asia and Oceania (14%) (Anga 2013). Cóte d'Ivoire is the major cocoa producer followed by Ghana, Cameroon, Nigeria, and Brazil (Pipitone 2016; Malhotra and Apshara 2017).

The first introduction of cacao was by Alonso Pinzón in 1510 in southern Yucatan (Bartley 2005). In the sixteenth century, the Europeans started to cultivate cacao in Asia and Africa where the Criollo, Amelonado, and the Trinitario hybrids started their route of dispersal from the Americas to the old world. Cacao was domesticated to southeast Asia (to Indonesia) in 1560 by the Dutch (Van Hall 1932). Around 1770, the Dutch introduced cacao to Peninsular Malaysia (Thong et al. 1992). In 1798, the British took cacao to Madras, India, from the island of Amboina, and it was introduced into Ceylon (now Sri Lanka) from Trinidad at about the same time (Ratnam 1961; Wood 1991). Remnants of the ancient Criollo, Amelonado, and Trinitario populations can still be found in Asia and Pacific regions, such as Indonesia and South Pacific (Fiji and Samoa) (Susilo et al. 2011).

The "Upper Amazon" stretching from Marañón River in Peru to the frontier of Brazil is the location of cacao's primary gene pool. Here, a series of major river systems in Peru, Ecuador, Colombia, and Brazil flow into the Marañón and Amazon rivers. Wild cacao populations are found in these river basins in both spontaneous (without human interference) and subspontaneous forms (wild cacao trees exploited by man) prior to European occupation (Almeida 2001; Bartley 2005). It is hypothesized that gene flow in cacao is limited, and mating is likely confined within patches (Chapman and Soria 1983).

During late 1920s, there was an outbreak of "witches' broom" disease (WBD) in Trinidad. This necessitated a search for genetic resistance in the Upper Amazonian. During the 1930s and 1940s, wild germplasm was collected from the Upper Amazon basin of Ecuador and Peru (Pound 1945; Wood and Lass 2008). The Pound collection (named after the collector F. J. Pound) is the foundation to the modern cacao breeding programs. The tributaries of Rio Ucayali, Rio Morona, and Rio Marañón contributed to the pound collection (Pound 1945; Bartley 2005; Zhang et al. 2009). Thus "Pound collection" was established in Iquitos, Peru. Approximately 80% of the world's chocolate production comes from the Forastero type of cacao. This is a variety favored over the Criollo for its disease resistance and high yield (Rusconi and Conti 2010). The third genetic group, Trinitario, is a hybrid of Criollo and Forastero derived in Trinidad. Trinitario is cultivated in many parts of South and Central America, Africa, South-East Asia, and Oceania for its aroma, productivity, and disease-resistance property (Wickramasuriya and Dunwell 2018).

A considerable portion of the global cacao diversity is in situ. This includes two international collections maintained at the Cocoa Research Centre of the University of the West Indies (CRC/UWI), Trinidad and Tobago, and at the Centro Agronómico Tropical de Investigación y Enseñanza (CATIE), Costa Rica (Laliberté et al. 2018). Laliberté et al. (2018) further enumerated the following strategies for cacao genetic conservation:

- Cacao gene pool is conserved in situ and ex situ for the long term by a global network of partners.
- Global system for the safe exchange of cacao germplasm is strengthened.
- Use of cacao genetic diversity is optimized.
- Effectiveness of global efforts to conserve and use cacao.
- genetic resources is assured.

CATIE (Centro Agronómico Tropical de Investigación y Enseñanza, Cost Rica) has international collection of cocoa, coffee, and other cash crops. The collections were initiated in the decade 1940 and were put in 2004 under the article 15 of the International Treaty for Plant Genetic Resources for Food and Agriculture (ITPGRFA) and thus in the public domain, which means that every accession is available for everyone for investigation, experimentation, or training.

The International Cacao Collection at CATIE (IC3) is one of the two most diverse and important gene banks worldwide, in particular, the ten currently defined genetic groups are represented by accessions. IC3 was founded in Turrialba, Costa Rica, in 1944 and declared as an international gene bank by IPGRI (now The Alliance Biodiversity International) in 1978. Currently, the collection preserves 1235 wild and cultivated genotypes from countries of origin of cacao, but also from producing countries in America, Africa, and Asia. Genetic resources are conserved in the field in the form of six trees per accession, over 10 ha at two replications located in La

Montaña Farm in Turrialba 602 m.a.s.l. and in La Lola Farm in Matina, the Atlantic Coast at 40 m.a.s.l. The conservation in two places assures the accurate and safety preservation in case of lose any accessions and the better evaluation of the accessions for production, resistance against pests and diseases, and quality (see: https://www.catie.ac.cr/en/products-and-services/collections-and-germplasm-banks/international-cocoa-collection) (Dominique Dessauw and William Solano, CATIE, personal communication).

Global Network for Cacao Genetic Resources (CacaoNet) could finalize the Global Strategy in 2012 (Laliberté 2012). The Global Strategy provides a clear framework to secure funding to ensure conservation of cacao diversity to directly benefit millions of small-scale cacao farmers. The International Cocoa Organization, Côte d'Ivoire and International Cocoa Germplasm Database (ICGD), University of Reading, are the two other active genetic resources.

## *Coffee*

Coffee is an important agricultural export commodity in many Asian, African, and Latin American countries, providing a livelihood to more than 125 million people worldwide. The genus *Coffea* comprises more than 124 species of which only two species (*Coffea arabica*—arabica coffee and *C. canephora*—robusta coffee) are commercially cultivated for beverage production (Misra 2019). These 124 species are distributed in Africa, Madagascar, the Comoros Islands, the Mascarene Islands (La Réunion and Mauritius), tropical Asia, and Australia (Krishnan et al. 2015). Climate change poses unprecedented challenges to sustainable coffee cultivation. World coffee production is estimated at 169.34 million bags (one bag with 60 kg) in 2019/2020 (International Coffee Organization).

Possibly around the sixth century, from its center of origin in Ethiopia, coffee made its way to Yemen, with the first record of consumption as a beverage by practitioners of Sufism around 1450 (Weinberg and Bealer 2001). From there, coffee spread to Cairo, Damascus, and Istanbul, leading to the birth of the coffeehouse. Following this, coffeehouses opened in Europe, the first one in Venice in 1645 and in Oxford in 1650. Dutch East India Company commenced coffee cultivation in Java using seeds obtained from Mocha in Yemen in the 1690s (Vega et al. 2008).

In the 1960s, FAO took the initiative to prevent loss of genetic resources and enlarge the genetic base through collecting mainly *C. arabica* germplasm along with noncultivated species (Engelmann et al. 2007; Krishnan 2013; Vega et al. 2008). In addition to these international collecting missions, local researchers within origin countries have performed their own collecting missions, such as in Ethiopia, Madagascar, and Cote d'Ivoire (Labouisse et al. 2008). The FAO report, *State of the World's Plant Genetic Resources*, released in 1998, documented 21,087 coffee accessions conserved worldwide (Anthony et al. 2007). The FAO World Information and Early Warning System (WIEWS) Coffea *Germplasm Report* (2009–2011) is the most comprehensive inventory of coffee germplasm held in living collections. In

2016, the Global Crop Diversity Trust, in partnership with World Coffee Research, led the development of the Global Conservation Strategy for Coffee Genetic Resources, which was scheduled for completion in early 2017 (Engelmann et al. 2017; Bramel et al. 2017: https://cdn.croptrust.org/wp/wp-content/uploads/2017/07/Coffee-Strategy_Mid_Res.pdf).

CATIE has one of the most important coffee collections in the world, with nearly 2000 introductions. The collection began in 1949 with materials introduced from Brazil, Guatemala, and El Salvador, and later, in the 1960s, it was strengthened with wild *Coffea arabica* materials from Ethiopia and Yemen, including the original materials collected by FAO, ORSTOM (now IRD), and IPGRI. This is considered the fourth largest collection of Coffea spp. in the world and the most important collection of *C. arabica* in the American continent, due to the number of introductions and the genetic diversity conserved. It is considered one of the four Origin collections in the world, the only one outside the African continent, according to the Global Strategy for the Conservation of Coffee Genetic Resources developed by the Global Crop Diversity Trust and World Coffee Research (WCR). The agreement signed with the International Treaty on Plant Genetic Resources for Food and Agriculture (FAO) defines it as the most important "International Collection" of *C. arabica* under public domain.

The collection has different types of genetic material such as:

- Wild Arabica coffee genotypes collected by FAO and ORSTOM in Ethiopia, IPGRI in Yemen, as well as diploid species;
- Varieties, mutants, and selections with resistance to coffee rust;
- Inter- and intraspecific hybrids; and,
- Research material.

These genetic resources are considered as the genetic heritage of the region, in relation to their potential to be able to respond quickly to current and future problems of Latin American coffee growing. For more than 60 years, this collection has contributed to genetic improvement programs around the world. The accessions with resistance to rust allowed the generation of some F1 hybrids. CATIE plays a leading role in asexual reproduction via somatic embryogenesis and by micro-cuttings (William Solano, CATIE, personal communication; Mata-Quirós et al. 2017).

## *Sugar Cane*

As a perennial $C_4$ grass crop, sugarcane is cultivated worldwide in tropical and subtropical regions as a major source of sucrose (Park et al. 2015), with a global crop value of $61 billion a year (FAOSTAT 2016). Sugarcane accounts for 75% of global sucrose production (Kandel et al. 2018). In addition to its importance as a food crop, it has one of the highest solar energy conversion efficiency and highest biomass yield among the known crops (Lam et al. 2009; Henry 2010; Byrt et al. 2011).

Modern sugarcane cultivars are derived from interspecific hybridizations between the domestic and wild species *Saccharum officinarum* ($2n = 80$) and *Saccharum spontaneum* ($2n = 40$–$128$). Both species are highly polyploid (Aitken et al. 2005), with a basic chromosome number of $x = 10$ for *S. officinarum* and $x = 8$ for *S. spontaneum* (D'Hont et al. 1998). The genome of current sugarcane cultivars is composed of approximately 70–80% *S. officinarum* and 10–20% *S. spontaneum*, with only 10% from the recombination of these two species (D'Hont et al. 1996).

Modern cultivars have limited genetic variation (Fickett et al. 2020). Almost 20 *S. officinarum* clones are involved in the genealogy of sugarcane cultivars with only a few being used extensively (Raboin et al. 2008). Basic crosses are made with clones of *S. spontaneum, S. robustum,* and species of other genera within the *Saccharum* complex to broaden the genetic base (Moore et al. 2013; Ming et al. 2010).

A "World Collection of Sugarcane and Related Grasses" (WCSRG) is maintained at the National Germplasm Repository of the USDA-ARS Subtropical Horticulture Research Station, Miami, FL, USA. A total of 342 *S. spontaneum* clones were assessed using stratified random sampling over geographical origins and principal component cluster groups to select a 75-clone core collection (Tai and Miller 2001). By and large, this is one of the largest sugarcane germplasm collections in the world. There is yet another base broadening (basic breeding) program at the USDA-ARS Sugarcane Research Unit in Houma, Louisiana (Fickett et al. 2020).

## *Rubber*

Rubber is an industrial raw material indispensable to humans with more than 55,000 vivid products made from it. The *para* rubber tree, the *Hevea brasiliensis* (Willd. *ex* A. Juss.) Müll. Arg., (Euphorbiaceae), is the chief contributor to natural rubber production worldwide. *Hevea* rubber is native of Amazon basin (Priyadarshan and Clément-Demange 2004). *Hevea* has eleven species, all native to Amazon that are inter-crossable (Priyadarshan and Goncalves 2003). All *Hevea* species have $2n = 36$ chromosomes, with the exception of one triploid clone of *H. guianensis* ($2n = 54$) and the existence of one genotype of *H. pauciflora* with $2n = 18$ (Baldwin 1947; Majumder 1964). Although *Hevea* behaves as a diploid, it is believed to be an amphidiploid ($2n = 36$; $x = 9$) that stabilized during the course of evolution (Clément-Demange et al. 2000).

It is difficult to evaluate how narrow the genetic base initially was for what has now become the "Wickham" domesticated population (Dean 1987; Baulkwill 1989). Much importance was conferred to a small number of 22 seedlings disseminated from Singapore to Malaysia after 1876. A significant part of the Wickham seedlings which germinated in Kew Gardens was then sent to Ceylon (now Sri Lanka), raised, and disseminated to different countries, especially India. However, it is significant that the original Wickham stock was collected in only one Brazilian site, Boïm, on the Western banks of the Tapajoz river, not far from Santarém. All these contentions are debatable (Thomas 2001). Directional selection applied to such populations for more than a century, and the low fruit-set in *Hevea* probably further contributed to the narrow genetic base (Priyadarshan 2016). Genetic diversity can now be compared to that of the available wild Amazonian populations by use of molecular genetic markers (Priyadarshan 2017a, b).

Since the introduction of rubber to southeast Asian countries by Wickham and Cross in 1877 through Kew Botanic Gardens, there have been attempts to enrich the germplasm with new genetic diversity (Simmonds 1989). *Hevea* rubber was evolved because of evolution spanning over thousands of years. Today, the cultivation of *Hevea* rubber spans to several continents and new environments that are both optimal and suboptimal (Priyadarshan 2003). With the initiatives taken up by the IRRDB (International Rubber Research and Development Board), 64,734 seeds, 1413 m of budwood from 194 high yielding trees and 1160 seedlings were collected during 1981 from Acre, Rondonia, and Mato-Grosso states of Brazil, from 60 different locations spread to 16 districts (Nouy 1982; Tan 1987; Simmonds 1989). Of this, 37.5% of the seeds went to Malaysia and 12.5% to Côte d'Ivoire while half of the collections were retained in Brazil. The clonal selections were brought to Malaysia and Côte d'Ivoire after quarantine measures (of 1 year in Guadalupe Island) for South American Leaf Blight (SALB—*Microcyclus ulei*). After the establishment of two IRRDB Germplasm Centers in Malaysia and Côte d'Ivoire, other IRRDB member countries were supplied with budwood from this material according to their request. Malaysia alone established 8900 seedlings and 109 clones from this exploration (Pushparajah 2001). Crosses between Wickham and Amazonian accessions could introduce more variation. Breeding at IRCA (Institut de Recherches sur le Caoutchouc en Afrique), Côte d'Ivoire, under the auspices of CIRAD (Centre de coopération internationale en recherche agronomique pour le développement), involve utilization of Amazonian accessions (Clement-Demange et al. 1998).

Identification of all Wickham clones could be done with 13 probes associated with restriction enzyme *EcoRI* (Besse et al. 1993). The cultivated clones are genetically close to the Mato-Grosso genotypes. A comprehensive map of genomic variation across rubber tree plastomes (plastid genomes) exhibited higher genomic variants in wild rubber than that of cultivated ones (Feng et al. 2020). mtDNA of Wickham clones has lesser variation because their female progenitors are all primary clones (Supriya and Priyadarshan, 2019). The cytoplasmic donors of most of the improved clones are either PB 56 or Tjir 1. While the cytoplasm of PB 56 was transferred through PB 5/51, the cytoplasm of Tjir 1 was through RRII 105, RRIM 600, and RRIM 605 (Priyadarshan 2017a, b).

The original 22 seedlings of Wickham collection from which the day-to-day *Hevea* clones were evolved had been genetically narrowed to enrich the *Hevea* gene pool. Also, these populations were subjected to several rounds of controlled crossing that further narrowed the diversity. Moreover, the strategy followed by the breeders to select only the desirable genotypes and to reject the unwanted ones (without precisely assessing yield and other secondary attributes) is the main reason that reduced diversity. Concerted efforts to infuse the Amazonian germplasm through controlled crossings never met with enriching the diversity as desired and as expected. This is because selection was and is always been in favor of higher yield. Of late, it is noteworthy that effrots are being incurred to breed *Hevea* at molecular level.

## 3 Future Perspectives

The conservation of crop genetic resources, including their wild relatives, is of utmost importance for the future of mankind. Most crops be stored in seed gene banks. But for crops that produce recalcitrant (non-storable) seeds such as cacao, coffee, and coconut, in vitro and cryopreserved collections provide an alternative (Panis et al. 2020). International gene banks have a collection of over 760,000 conserved accessions of various plants, most of these accessions are within the multilateral system governed by the International Treaty on Plant Genetic Resources for Food and Agriculture (ITPGRFA) (Lee et al. 2020). However, in spite of the success in collection and conservation, only a small portion of the genetic diversity has been used in crop breeding programs. Genetically diverse plant germplasms stored in ex situ gene banks are excellent resources for breeding new high yielding and sustainable crop varieties. Some gene banks do follow cost-effective genotyping technologies. However, the adoption of modern phenotyping is lagging. These phenotyping tools and high-throughput genotyping must accelerate the derivation of more resilient food crops for the future (Nguyen and Norton 2020).

Li et al. (2018) pointed out two major limitations preventing the exploitation of gene bank genetic resources for breeding programs:

- Time and available resources for thorough characterization of accessions at a large scale.
- Identifying and introducing the allelic variance into elite breeding materials.

This missing characterization data makes searching of an accession with specific desirable agronomic trait a daunting task (Wambugu et al. 2018). The gene banks are increasingly required *to move beyond providing basic passport data* that defines only the identity and origin to thoroughly catalogue agronomic, physiological, and genetic traits (Anglin et al. 2018). A three-step strategy that combines genomics and phenomics to effectively mine gene bank genetic resources has been proposed (McCouch et al. 2013):

- Obtain a sample of sequence information from the genomes of all non-duplicate plant samples in the world's gene banks that are available under the terms and conditions of the ITPGRFA. This may perhaps come up to two million. This "fingerprint" for each plant will serve as the basis for assessing genetic relationships which can be used to systematically select subsets of material for in-depth investigations.
- Analyze the phenotypes of gene bank accessions to evaluate their traits and overall performance. This is rather costly, intellectually challenging, complex, and

time-consuming process. All gene bank accessions cannot be evaluated under all relevant environments, even with high-throughput phenotyping technologies. The genetics of high-performing offspring can be traced back to DNA inherited from wild or landrace donors. Such donors must be quite often less productive.

- Create an internationally accessible informatics infrastructure to catalogue the world's seed/genetic diversity. Details of passport, genomic, and phenotypic information can be directly linked to seeds and genetic stocks. This can augment plant improvement programs. This is a task that requires unprecedented effort. Currently, seed data are recorded and managed by different people like gene bank curators, agronomists, and breeders following database systems. This calls for a coordinated system of data management and benefit sharing.

When we consider coffee as an example, *C. arabica* is an allotetraploid ($2n = 4x = 44$) originating from a hybridization event of the two diploid species *C. canephora* and *C. eugenioides* ($2n = 2x = 22$). Interestingly, these progenitor species harbor a greater level of genetic variability and are an important source of genes to broaden the narrow Arabica genetic base. Merot-L'anthoene et al. (2019) selected 8580 unique and informative SNPs from *C. canephora* and *C. arabica* sequencing data, with 40% of the SNP located in annotated genes. A panel of *C. canephora* accessions was successfully discriminated, and over 70% of the SNP markers were transferable across the three species. Furthermore, the canephora-derived subgenome of *C. arabica* was shown to be more closely related to *C. canephora* accessions from northern Uganda. This study is a clear indication of the wide range of variability available in cash crops.Gene bank phenomics still lag in evaluating available plant genetic resources despite the development of DNA technology (Philipp et al. 2018). High-throughput phenotyping (HTP) using sensors and imagers is a promising, efficient, and cost-effective approach to collect phenotypic data for multiple traits across large-scale trials that can then be used together with genomic data for accurate selection in breeding (Araus and Kefauver 2018). With a comprehensive phenomics approach combining pedigree, genomic, and phenotyping data (McCouch et al. 2013), the true value of gene bank genetic resources is evident. Such strategic efforts are prime to double our current rate of genetic gain to feed the growing world population under the anticipated dynamic climate changes.

**Acknowledgments** The authors are thankful to Dr. William Solano and Dr. Rolando H. Cerda of CATIE, Costa Rica and Dominique Dessauw of CATIE/CIRAD, for providing details of cacao and coffee germplasm collections. Special thanks to Dr. Sarada Krishnan, Director (Hort.), Denver Botanic Gardens for suggestions on coffee.

## References

Aitken KS et al (2005) A combination of AFLP and SSR markers provides extensive map coverage and identification of homo(eo)logous linkage groups in a sugarcane cultivar. Theor Appl Genet 110:789–801. https://doi.org/10.1007/s00122-004-1813-7

Ali A et al (2019) Genetic diversity and population structure analysis of *Saccharum* and *Erianthus* genera using microsatellite (SSR) markers. Sci Rep 9:395. https://doi.org/10.1038/s41598-018-36630-7

Almeida CMVC (2001) Ecology of natural populations. In: Dias LAS (ed) Genetic improvement of cacao, Chap. 4 (EcoPort version by P. Griffee). FAO, Rome. http://ecoport.org/ep?SearchType¼earticleView&earticleId¼197&page¼-2#section2690

Alston JM et al (2009) Agricultural research, productivity, and food prices in the long run. Science 325:1209–1210

Anga JM (2013) World cocoa economy as from ICCO 2013, present and future prospects. In: Malaysian international cocoa conference (MICC 2013), 7–8 October 2013, Kuala Lumpur

Anglin NL et al (2018) A case of need: linking traits to genebank accessions. Biopreserv Biobank 16:337–349

Anthony F, Dussert S, Dulloo E (2007) Coffee genetic resources. In: Engelmann F, Dulloo ME, Astorga C, Dussert S, Anthony F (eds) Conserving coffee genetic resources: complementary strategies for ex situ conservation of coffee (Coffea arabica L.) genetic resources. A case study in CATIE, Costa Rica. Bioversity International, Rome, pp 12–22

Araus JL, Kefauver SC (2018) Breeding to adapt agriculture to climate change: affordable phenotyping solutions. Curr Opin Plant Biol 45:237–247

Baldwin JJT (1947) Hevea: a first interpretation. A cytogenetic survey of a controversial genus, with a discussion of its implications to taxonomy and to rubber production. J Hered 38:54–64

Bartley BGD (2005) The genetic diversity of cacao and its utilization. CABI Publishing, Wallingford

Baulkwill WJ (1989) The history of natural rubber production. In: Webster CC, Baulkwill WJ (eds) Rubber. Longman Scientific and Technical, Harlow, pp 1–56

Besse P, Lebrun P, Seguin M, Lanaud C (1993) DNA fingerprints in *Hevea brasiliensis* (rubber tree) using human minisatellite probes. Heredity 70:237–244

Bramel P et al (2017) Global conservation strategy for coffee genetic resources. Development of the global strategy for conservation of coffee genetic resources. Crop trust. https://cdn.croptrust.org/wp/wp-content/uploads/2017/07/Coffee-Strategy_Mid_Res.pdf

Byrt CS et al (2011) C4 plants as biofuel feedstocks: optimising biomass production and feedstock quality from a lignocellulosic perspective. J Integr Plant Biol 53:120–135. https://doi.org/10.1111/j.1744-7909.2010.01023.x

Camacho LMD et al (2017) Development, characterization and cross-amplification of microsatellite markers for *Chrysolaena obovata*, an important Asteraceae from Brazilian Cerrado. J Genet 96:47–53

Campbell BM et al (2018) Urgent action to combat climate change and its impacts (SDG 13): transforming agriculture and food systems. Curr Opin Environ Sustain 34:13–20

Carvalho YGS (2019) Recent trends in research on the genetic diversity of plants: implications for conservation. Diversity 11:62. https://doi.org/10.3390/d11040062

Chapman RK, Soria SJ (1983) Comparative Forcipomyia (Diptera, Ceratopogonidae) pollination of cacao in Central America and Southern Mexico. Revista Theobroma (Brasil) 13:129–139

Clément-Demange A, Legnaté H, Chapuset T, Pinard F, Seguin M (1998) Characterization and use of the IRRDB germplasm in Ivory Coast and French Guyana: status in 1997. p. 71–88. In: Cronin ME (ed) Proc. IRRDB Symp. Natural Rubber in Vietnam, 13–15 Oct. 1997, Vol. 1. Int. Rubb. Res. and Develop. Board (IRRDB), Hertford, UK

Clément-Demange A et al (2000) Rubber tree. In: Charrier A, Jacquot M, Hamon S, Nicolas D (eds) Tropical plant breeding. Collection Reperes, CIRAD-ORSTOM, Montpellier, pp 455–480

Collard BCY, Mackill DJ (2007) Marker-assisted selection: an approach for precision plant breeding in the twenty-first century. Philos Trans R Soc B 363:557–572

D'Hont A, Grivet L, Feldmann P, Rao PS, Berding N, Glaszmann JC (1996) Characterisation of the double genome structure of modern sugarcane cultivars (*Saccharum* spp.) by molecular cytogenetics. Mol Gen Genet 250(4):405–413. https://doi.org/10.1007/bf02174028

D'Hont A, Ison D, Alix K, Roux C, Glaszmann JC (1998) Determination of basic chromosome numbers in the genus *Saccharum* by physical mapping of ribosomal RNA genes. Genome 41(2):221–225

Davis AP et al (2019) High extinction risk for wild coffee species and implications for coffee sector sustainability. Sci Adv 5:eaav3473. https://doi.org/10.1126/sciadv.aav3473
De Wever J (2019) The development of a novel SNP genotyping assay to differentiate cacao clones. Sci Rep 9:9512. https://doi.org/10.1038/s41598-019-45884-8
Dean W (1987) Brazil and the struggle for rubber. Cambridge University Press, Cambridge
Dhankher OP, Foyer CH (2018) Climate resilient crops for improving global food security and safety. Plant Cell Environ 41:877–884
Dwivedi SL et al (2015) Haploids: constraints and opportunities in plant breeding. Biotechnol Adv 33:812–829. https://doi.org/10.1016/j.biotechadv.2015.07.001
Engelmann F, Dulloo ME, Astorga C, Dussert S, Anthony F (eds) (2007) Complementary strategies for ex situ conservation of coffee (Coffea arabica L.) genetic resources. A case study in CATIE, Costa Rica. Topical reviews in agricultural biodiversity. Bioversity International, Rome
Engelmann F et al (2017) Conserving coffee genetic resources. Bioversity International, Rome
FAO (2015) Coping with climate change—the roles of genetic resources for food and agriculture. FAO, Rome. 110p
FAO (2020) The state of food security and nutrition in the world. Food and Agriculture Organization, Rome. 320 pages. https://doi.org/10.4060/ca9692en
FAOSTAT (2016) FAOSTAT. https://faostat3.fao.org/home
Farrell AD et al (2018) Climate adaptation in a minor crop species: is the cocoa breeding network prepared for climate change? Agroecol Sustain Food Syst 42:812–833
Feng L-Y et al (2020) Higher genomic variation in wild than cultivated rubber trees, *Hevea brasiliensis*, revealed by comparative analyses of chloroplast genomes. Front Ecol Evol 8:237. https://doi.org/10.3389/fevo.2020.00237
Fickett ND et al (2020) An enriched sugarcane diversity panel for utilization in genetic improvement of sugarcane. Nat Sci Rep 10:13390. https://doi.org/10.1038/s41598-020-70292-8
Foden WB et al (2018) Climate change vulnerability assessment of species. WIREs Clim Change 2018:e551. https://doi.org/10.1002/wcc
Gaudelli NM et al (2017) Programmable base editing of A•T to G•C in genomic DNA without DNA cleavage. Nature 551:464–471. https://doi.org/10.1038/nature24644
Godfray H et al (2010) The future of the global food system. Philos Trans R Soc Lond Ser B Biol Sci 365:2769–2777. https://doi.org/10.1098/rstb.2010.0180
Goodwin S, McPherson JD, McCombie WR (2016) Coming of age: ten years of next-generation sequencing technologies. Nat Rev Genet 17:333
Gross BL, Zhao Z (2014) Archaeological and genetic insights into the origins of domesticated rice. PNAS 111(17):6190–6197. https://doi.org/10.1073/pnas.1308942110; first published April 21, 2014
Halewood M (2020) Germplasm acquisition and distribution by CGIAR genebanks. Plants 9:1296. https://doi.org/10.3390/plants9101296
Halperin SO et al (2018) CRISPR-guided DNA polymerases enable diversification of all nucleotides in a tunable window. Nature 560:248. https://doi.org/10.1038/s41586-018-0384-8
Henry RJ (2010) Evaluation of plant biomass resources available for replacement of fossil oil. Plant Biotechnol J 8:288–293. https://doi.org/10.1111/j.1467-7652.2009.00482.x
Jacquier NMA et al (2020) Puzzling out plant reproduction by haploid induction for innovations in plant breeding. Nat Plants 6:610–619. https://doi.org/10.1038/s41477-020-0664-9
Jump AS, Peñuelas J (2005) Running to stand still: adaptation and the response of plants to rapid climate change. Ecol Lett 8:1010–1020
Kandel R et al (2018) Potentials, challenges, and genetic and genomic resources for sugarcane biomass improvement. Front Plant Sci 9:151. https://doi.org/10.3389/fpls.2018.00151
Khoury CK et al (2016) Origins of food crops connect countries worldwide. Proc Roy Soc B Biol Sci 283(1832):20160792. https://doi.org/10.1098/rspb.2016.0792. WOS:000378318700021
Krishnan S (2013) Current status of coffee genetic resources and implications for conservation. CAB Rev 8(16):1–9
Krishnan S et al (2015) Current status of coffee genetic resources: implications for conservation—case study in Madagascar. Acta Hortic 1101. ISHS 2015; Proc. XXIX IHC–IV

international symposium on plant genetic resources, pp 15–20. https://doi.org/10.17660/ActaHortic.2015.1101.3
Labouisse J-P, Bellachew B, Kotecha S, Bertrans B (2008) Current status of coffee (Coffea arabica L.) genetic resources in Ethiopia: implications for conservation. Genet Resour Crop Evol 55:1079–1093
Laliberté B (2012) Global strategy for the conservation and use of cacao genetic resources. Bioversity International, Rome
Laliberté B et al (2018) Conserving and exploiting cocoa genetic resources: the key challenges. In: Umaharan P (ed) Achieving sustainable cultivation of cocoa. Burleigh Dodds Science Publishing, Cambridge
Lam E et al (2009) Improving sugarcane for biofuel: engineering for an even better feedstock. GCB Bioenergy 1:251–255
Lee J et al (2020) Advanced strategic research to promote the use of rice genetic resources. Agronomy 10:1629. https://doi.org/10.3390/agronomy10111629
Li TM (2015) Can there be food sovereignty here? J Peasant Stud 42:205–211
Li H et al (2018) Fast-forwarding genetic gain. Trends Plant Sci 23:184–186
Majumder SK (1964) Chromosome studies of some species of *Hevea*. J Rubber Res Inst Malaysia 18:269–273
Malhotra S, Apshara SE (2017) Genetic resources of cocoa (*Theobroma cacao* L.) and their utilization—an appraisal. Indian J Genet 77:199–213
Martin AR et al (2019) Regional and global shifts in crop diversity through the anthropocene. PLoS One 14(2):e0209788. https://doi.org/10.1371/journal.pone.0209788
Mata-Quirós A et al (2017) Understanding the genetic structure and parentage of the clonal series of cacao UF, CC, and ARF preserved in the international cacao collection at CATIE (IC3). In: International symposium on cocoa research (ISCR), Lima, Peru, 13-17 November 2017
McCouch S et al (2013) Feeding the future. Nature 499:23–24. https://www.nature.com/articles/499023a#supplementary-information
Merot-L'anthoene V et al (2019) Development and evaluation of a genome-wide Coffee 8.5K SNP array and its application for high-density genetic mapping and for investigating the origin of *Coffea arabica* L. Plant Biotechnol J 17:1418–1430. https://doi.org/10.1111/pbi.13066
Ming R et al (2010) Sugarcane improvement through breeding and biotechnology. In: Janick J (ed) Plant breeding reviews, vol 27. Wiley, New York, pp 15–118
Misra MK (2019) Genetic resources and breeding of coffee (*coffea* spp.). In: Al-Khayri JM, Jain SM, Johnson D (eds) Advances in plant breeding strategies: nut and beverage crops. Springer Nature, Berlin, pp 475–515
Moore PH, Paterson AH, Tew T (2013) Sugarcane: the crop, the plant, and domestication. In: Moore PH, Botha FC (eds) Sugarcane: physiology, biochemistry and functional biology. Wiley, New York, pp 1–15
Nadeem MA et al (2017) DNA molecular markers in plant breeding: current status and recent advancements in genomic selection and genome editing. Biotechnol Biotechnol Equip 32:261–285
Nelson EJ et al (2016) Commercial plant production and consumption still follow the latitudinal gradient in species diversity despite economic globalization. PLoS One 11:e0163002. https://doi.org/10.1371/journal.pone.0163002
Nguyen GN, Norton SL (2020) Gene bank phenomics: a strategic approach to enhance value and utilization of crop germplasm. Plants 9:817. https://doi.org/10.3390/plants9070817
Nouy B (1982) Status report on new Hevea germplasm collected from Brasil. IRRDB African Germplasm Centre. Status: Sept. 1982
Nutan KK et al (2020) Integrating dynamics of yield traits in rice in response to environmental changes. J Exp Bot 71:490–506
Østerberg JT, Xiang W, Olsen LI, Edenbrandt AK, Vedel SE, Christiansen A et al (2017) Accelerating the domestication of new crops: feasibility and approaches. Trends Plant Sci 22:373–384. https://doi.org/10.1016/j.tplants.2017.01.004

Panis B et al (2020) Challenges and prospects for the conservation of crop genetic resources in field gene banks, in in vitro collections and/or in liquid nitrogen. Plan Theory 9:1634. https://doi.org/10.3390/plants9121634

Park J-W et al (2015) Cold responsive gene expression profiling of sugarcane and *Saccharum spontaneum* with functional analysis of a cold inducible *Saccharum* homolog of NOD26-Like intrinsic protein to salt and water stress. PLoS One 10:e0125810

Philipp N et al (2018) Leveraging the use of historical data gathered during seed regeneration of an ex-situ genebank collection of wheat. Front Plant Sci 9:609

Pipitone L (2016) Overview of cocoa supply and demand. In: ICCO cocoa. Market outlook conference September 2016. International Cocoa. Organization (ICCO), London. https://www.icco.org/about-us/international-cocoaagreements/cat_view/294-cocoa-market-outlook-conference-september-2016.html

Pootakham W et al (2020) Development of molecular markers in *Hevea brasiliensis* for marker-assisted breeding. In: Matsui M, Chow KS (eds) The rubber tree genome. Compendium of plant genomes. Springer, Cham. https://doi.org/10.1007/978-3-030-42258-5_5

Pound FJ (1945) A note on the cacao population of South America. In: Report and proceedings of the cocoa research conference held at Colonial Office, May–June 1945. His Majesty's Stationery Office, London, pp 131–133

Pourkheirandish M et al (2020) Global role of crop genomics in the face of climate change. Front Plant Sci 11:922. https://doi.org/10.3389/fpls.2020.00922

Priyadarshan PM (2003) Breeding *Hevea brasiliensis* for environmental constraints. Adv Agron 79:351–400. (Review Article—Academic Press)

Priyadarshan PM (2016) Genetic diversity and erosion in *Hevea* rubber. In: Ahuja MR, Jain SM (eds) Genetic diversity and erosion in plants, Sustainable development and biodiversity, vol 8. Springer, Berlin. https://doi.org/10.1007/978-3-319-25954-3_6

Priyadarshan PM (2017a) Biology of *Hevea* rubber. Springer Nature, Berlin

Priyadarshan PM (2017b) Refinements to *Hevea* rubber breeding. Tree Genet Genomes 13:1–17. https://doi.org/10.1007/s11295-017-1101-8

Priyadarshan PM, Clément-Demange A (2004) Breeding *Hevea* rubber: formal and molecular genetics. Adv Genet 52:51–115

Priyadarshan PM, Goncalves P d S (2003) *Hevea* gene pool for breeding. Genet Resour Crop Evol 50:101–114

Pushparajah E (2001) Natural rubber. In: Last FT (ed) Tree crop ecosystems, Ecosytems of the world series, vol 19. Elsevier Science, Amsterdam, pp 379–407

Raboin LM et al (2008) Analysis of genome-wide linkage disequilibrium in the highly polyploid sugarcane. Theor Appl Genet 116:701–714

Ramankutty N et al (2002) The global distribution of cultivable lands: current patterns and sensitivity to possible climate change. Glob Ecol Biogeogr 11:377–392

Rasmussen SK (2020) Molecular genetics, genomics, and biotechnology in crop plant breeding. Agronomy 10:439. https://doi.org/10.3390/agronomy10030439

Ratnam R (1961) Introduction of *Criollo cacao* into Madras state. South Indian Horticult 9:24–29

Rusconi M, Conti A (2010) Theobroma cacao L., the food of the gods: a scientific approach beyond myths and claims. Pharmacol Res 61:5–13

Schroth G et al (2016) Vulnerability to climate change of cocoa in West Africa: patterns, opportunities and limits to adaptation. *Sci Total Environ* 556:231–241

Simmonds NW (1989) Rubber breeding. In: Webster CC, Baulkwill WJ (eds) Rubber. Longmann Scientific and Technical, Essex, pp 85–124

Sintayehu DW (2018) Impact of climate change on biodiversity and associated key ecosystem services in Africa: a systematic review. Ecosyst Health Sustain 4(9):225–239. https://doi.org/10.1080/20964129.2018.1530054

Spinoso-Castillo JL et al (2020) Genetic diversity of coffee (*Coffea* spp.) in Mexico evaluated by using DArTseq and SNP markers. Genet Resour Crop Evol 67:1795–1806

Supriya R, Priyadarshan PM (2019) Genomic technologies for *Hevea* breeding. Adv Genet 104:1–73. https://doi.org/10.1016/bs.adgen.2019.04.001
Susilo A et al (2011) Assessing genetic diversity in java fine-flavor cocoa (*Theobroma cacao* L.) germplasm by simple sequence repeat (SSR) markers. Trop Agric 55:84–92
Tai PYP, Miller JD (2001) A core collection for *Saccharum spontaneum* L. from the world collection of sugarcane. Crop Sci 41:879–885
Tan H (1987) Strategies in rubber tree breeding. In: Abbott AJ, Atkin RK (eds) Improving vegetatively propagated crops. Academic Press, London, pp 28–54
Tanksley SD (1983) Molecular markers in plant breeding. Plant Mol Biol Rep 1:3–8
Tester M, Langridge P (2010) Breeding technologies to increase crop production in a changing world. Science 327:818–822
Thomas KK (2001) Role of Clement Robert Markham in the introduction of *Hevea* rubber into the British India. Planter 77:287–292
Thong KC et al (1992) Cocoa in Peninsular Malaysia I: the early history. Cocoa Growers Bull 45:7–25
Tito R et al (2018) Global climate change increases risk of crop yield losses and food insecurity in the tropical Andes. Glob Chang Biol 24:e592–e602
Utge J et al (2020) A mobile laboratory for ancient DNA analysis. PLoS One 15:e0230496. https://doi.org/10.1371/journal.pone.0230496
Van der Vossen H et al (2015) Next generation variety development for sustainable production of Arabica coffee (*Coffea Arabica*, L.): a review. Euphytica 204:243–256
Van Hall CJJ (1932) Cacao, 2nd edn. Macmillan, London
Vaughan MM et al (2018) The effects of climate change associated abiotic stresses on maize phytochemical defenses. Phytochem Rev 17:37–49
Vega FE, Ebert AW, Ming R (2008) Coffee germplasm resources, genomics, and breeding. In: Janick J (ed) Plant breeding review, vol 30. Wiley, New York, pp 415–447
Wambugu PW et al (2018) Role of genomics in promoting the utilization of plant genetic resources in genebanks. Brief Funct Genome 17:198–206
Watson A et al (2018) Speed breeding is a powerful tool to accelerate crop research and breeding. Nat Plants 4:23. https://doi.org/10.1038/s41477-017-0083-8
Weinberg BA, Bealer BK (2001) The world of caffeine: the science and culture of the world's most popular drug. Routledge, London, 512 p
Wheeler T, Von Braun J (2013) Climate change impacts on global food security. Science 341:508–513
Wickramasuriya AM, Dunwell JM (2018) Cacao biotechnology: current status and future prospects. Plant Biotechnol J 16:4–17
Wolter et al (2019) Plant breeding at the speed of light: the power of CRISPR/Cas to generate directed genetic diversity at multiple sites. BMC Plant Biol 19:176. https://doi.org/10.1186/s12870-019-1775-1
Wood GAR (1991) A history of early cocoa introductions. Cocoa Growers Bull 44:7–12
Wood GAR, Lass RA (eds) (2008) Cocoa, 4th edn. Wiley Online Library, Hoboken. https://doi.org/10.1002/9780470698983. http://onlinelibrary.wiley.com/book/10.1002/9780470698983
Xu J et al (2019) Genome editing for horticultural crop improvement. Horticult Res 6:113. https://doi.org/10.1038/s41438-019-0196-5
Zhang DP et al (2009) Molecular characterization of an earliest cacao (Theobroma cacao L) collection from Peruvian Amazon using microsatellite DNA markers. Tree Genet Genomes 5:595–607. https://doi.org/10.1007/s11295-009-0212-2
Zhu H et al (2020) Applications of CRISPR–Cas in agriculture and plant biotechnology. Nat Rev Mol Cell Biol 21:661–677
Zong Y et al (2018) Efficient C-to-T base editing in plants using a fusion of nCas9 and human APOBEC3A. Nat Biotechnol 36:950. https://doi.org/10.1038/nbt.4261

# Part I
# Beverages

# Cacao

**Andrew Daymond and Frances Bekele**

**Abstract** Cacao originates from the rainforests of South America, the centre of diversity being the Upper Amazon region. However, present-day cultivation extends through the humid tropics with the largest proportion being in West Africa. Ex situ conservation of cacao is particularly important due to the incremental loss of native (wild) habitats and replacement of traditional varieties on farms. It is estimated that over 24,000 cacao accessions are conserved in genebanks, many of which are associated with national research institutes. There are currently two international genebanks for cacao (in Trinidad and in Costa Rica) that have been put under the auspices of the International Treaty on Plant Genetic Resources for Food and Agriculture. Breeding approaches have tended to focus on improving yields and resistance to pests and diseases and have often utilised a relatively narrow genetic base. In order to meet on-going and future challenges to cocoa production, including the need for greater climate resilience and more easily managed, productive cacao farms, it is necessary to utilise a broader range of genetic resources. Advances in genomic technology may accelerate breeding efforts and can be used to identify off-types and redundancies in collections. Underlying breeding efforts, there is a critical need for on-going conservation, characterisation, evaluation and utilisation of cacao genetic resources.

**Keywords** *Theobroma* cacao · Genetic diversity · Conservation · Characterisation · Breeding

---

A. Daymond (✉)
School of Agriculture, Policy and Development, The University of Reading, Reading, UK
e-mail: a.j.daymond@reading.ac.uk

F. Bekele
Cocoa Research Centre, The University of the West Indies,
St. Augustine, Trinidad and Tobago
e-mail: frances.bekele@sta.uwi.edu



P. M. Priyadarshan, S. M. Jain (eds.), *Cash Crops*,
https://doi.org/10.1007/978-3-030-74926-2_2

# 1 Origin and Botany

The cacao[1] tree, *Theobroma cacao* L., is a member of the *Malvaceae* family (Alverson et al. 1999) (formerly classified in the *Sterculiaceae* family) and is endemic to the understorey of the rainforests of South and Central America. The greatest diversity of cacao and its wild relatives is found in the Upper Amazon tributaries region (Cheesman 1944; Motamayor et al. 2002). This perennial crop is cultivated throughout the humid tropics for its seed (colloquially and subsequently referred to as "beans"), which are utilised extensively by the confectionary industry in the production of chocolate products. Cocoa (see Footnote 1) butter (fat extracted from the beans) is also sometimes used by the cosmetics industry.

Cultivated cacao is typically maintained to a height of between 3 and 5 m (sometimes shorter when grafted), although wild cacao can be much taller. Cacao seedlings are orthotropic in nature and produce leaves in a spiral pattern. The leaves are produced periodically through a series of "flushes" after which the terminal bud remains dormant until the next flush event (Fig. 1a). Newly formed flush leaves are soft and delicate and their colour ranges from light green to deep red depending on the genotype. Expansion and hardening of the flush leaves coincides with the development of photosynthetic apparatus (Baker and Hardwick 1973).

Growth of the juvenile plant continues vertically until it reaches a height of between 1 and 2 m, at which point it forms a "jorquette", whereby usually three to five branches are formed at the apex. The jorquette branches then display plagiotropic growth, and pairs of leaves are formed opposite each other. Jorquette formation essentially marks the transition from the vegetative to the reproductive phase of the tree. Seedlings come into bearing 3–4 years after planting, although there is some genotypic variation for this trait. Grafted trees come into bearing earlier, often less than 2 years after grafting.

Cacao is semi-cauliflorous in nature in that flowers and fruits are usually formed on both the branches and on the trunk. The flowers are small (1–1.5 cm in diameter; Fig. 1b) and if left without pollination fall off the tree 1–2 days after opening. Pollination of flowers is by various insects, of which midges belonging to the genus *Forcipomyia* have been cited as being particularly important (Toxopeus 2001a). Cacao flowers are hermaphrodite. Whilst some genotypes are self-compatible, others are incompatible, requiring the pollen from another tree to achieve successful pollination (Cope 1962). Typically, less than 5% of flowers form fruits (colloquially and subsequently referred to as "pods"). The time from pod set to ripening is usually between 5 and 6 months, the duration being influenced by both genotype and environmental factors, particularly temperature (Hadley et al. 1994; Daymond and Hadley 2008). Not all pods that set reach maturity, a proportion are lost through a fruit thinning process known as cherelle wilt, which arises as a result of competition

[1] The term "cacao" refers to the plant, whereas the term "cocoa" strictly refers to the end product (dried bean). This convention is used here. In the literature "cocoa" has also been used interchangeably with "cacao" when referring to the plant.

**Fig. 1** Characteristics of the cacao tree. (**a**) Flushing seedlings, (**b**) close-up of a flower, (**c**) a mature cacao tree (Ghana) bearing flowers, immature and ripe pods. (**d**) Harvested pods

for assimilates (Nichols 1964). Cherelle wilt is characterised by the initial cessation of growth of the young pod (cherelle), after which it turns yellow and then blackens and shrivels, but remains on the tree (McKelvie 1956).

The mature pod varies considerably in size and form with the majority being between 15 and 25 cm in length (Fig. 1c, d). It consists of a hard over-casing (pericarp), the husk, in which is contained typically between 20 and 50 beans that are covered with a mucilaginous pulp. In most, although not all, regions in which cacao is cultivated the harvested beans are fermented before being dried. In much of West Africa, this is achieved via the heap method, whereby the contents of harvested pods are piled onto banana or plantain leaves and covered over with more leaves. The beans are left to ferment typically for 5–7 days and are regularly turned over (usually every 48 h). An alternative method, often practised in South America, is to use wooden boxes with slits in them, out of which the pulp is able to drain. The fermentation process is important in improving the flavour of the final product, and the amount of fermentation needed can vary amongst varieties. After the beans are fermented, they are dried before going to market and usually the final product from the farmer's perspective.

## 2 History and Modern Cultivation

Historical cultivation of cacao in Mesoamerica, from as long ago as 3900 years BP (Zarrillo et al. 2018), and the importance of the crop in these cultures is well documented (de la Cruz et al. 1995). The cocoa beans were ground and mixed with other ingredients to make various drinks. Cocoa was economically and socially important amongst the Maya, Olmec and Aztec civilisations and was used as a form of currency. Cocoa-based drinks were used in various rituals. Recently, evidence has been found for the cultivation of cacao approximately 5300 years BP in what is now south-east Ecuador (Zarrillo et al. 2018). The findings were based on three classes of archaeological evidence recovered from stone and ceramic artefacts, these being starch grains from cacao, absorbed theobromine residues and ancient DNA.

The first European contact with cocoa was during the conquest of Mexico in the sixteenth century. Subsequently, with the opening up of a European market for cocoa drinks, cacao cultivation began to spread to other parts of South America and the Caribbean, most noticeably to Venezuela, Trinidad and Jamaica. Criollo cacao was taken to the Philippines circa 1600 and from there was introduced into parts of Indonesia, namely Sulawesi and Java (Wood 2001a). The first cultivation in the north-eastern Brazilian state of Bahia (where today the largest proportion of Brazilian cacao is grown) is thought to have taken place in 1746 where seed of the Amelonado type is believed to have been introduced from the state of Pará. The resulting variety came to be known as "*Comum*" in Bahia. Cacao of the Amelonado type was transferred from Bahia to São Tomé in 1822. In 1855, it was introduced from São Tomé to what was then the island of Fernando Pó (which today is Bioko, a part of Equatorial Guinea) and later in the century to Ghana and Nigeria (Wood 2001a).

Currently, cacao cultivation is spread throughout the humid tropical regions of West Africa, Central and South America and South East Asia. Its cultivation is limited by climate, in that cacao cannot withstand low temperatures or prolonged dry periods, unless irrigation is used (Lahive et al. 2019). According to the International

Cocoa Organisation (ICCO 2019), the proportion of cocoa beans produced in West Africa, Central and South America and Asia was around 77%, 17% and 6%, respectively, for the year 2018–2019. The six largest cocoa producers in the year 2018–2019 were: Côte d'Ivoire, Ghana, Ecuador, Nigeria, Cameroon and Indonesia. Other significant producing countries in West Africa are: Sierra Leonne, Togo and Uganda; in South America are: Brazil, Peru, Dominican Republic, Colombia and Mexico and in Asia are: Papua New Guinea and India. The majority of the estimated five to six million cacao farms worldwide are smallholdings with a size of less than 5 ha (Daymond et al. 2018), although larger plantations also exist, e.g. in parts of South America.

In West Africa, cacao is propagated by seed, with seed gardens often providing bi-parental crosses ("hybrids") to farmers (although farmers may also grow from seed from their own farms). In parts of South East Asia, the Americas and the Caribbean, clonal cacao is cultivated as well as seed-raised cultivars. Clonal propagation can be by grafting, budding or rooted cuttings, the latter being particularly practised in the Caribbean. Somatic embryogenesis-derived plants have been distributed to farmers in parts of Indonesia and Ecuador, although at present represent a small proportion of cacao cultivated globally.

## 3 Genetic Diversity of Cacao

The genetic diversity of cacao manifests itself in various phenotypic characteristics, e.g. leaf size, flower characteristics and notably pod shape, size, degree of wartiness and colour (Fig. 2). Cacao has traditionally been divided into two main groups, Criollo and Forastero. A third group, Trinitario, represents crosses between Criollo and Forastero types; the first of these crosses are believed to have taken place in

**Fig. 2** Genetic diversity in cacao pod morphology

Trinidad, which gave rise to this name. The Criollo group is understood to be genetically similar to those cultivated by the Mayan cultures (Wood 2001a). It is characterised by having plump, white bean cotyledons and has light green leaf flushes. The chocolate produced from Criollo beans has a more subtle flavour, and these beans generally attract a premium. Today few plantations cultivate Criollo types, for example on some generally older farms in Venezuela, Central America, Madagascar, Sri Lanka and Samoa (Fowler 2009). The term Forastero means ‘foreigner’ or ‘outsider’ in Spanish, implying types distinct to Criollo, and usually have smaller, flatter beans with purple cotyledons. Forastero types are native to the Amazon basin and are sometimes subdivided into Upper Amazon and Lower Amazon (Bhattacharjee 2018). Distinct subgroups or cultivars have been recognised within the Forastero umbrella, these include Nacional (known for its fine flavour attributes and particularly cultivated in Ecuador; Enríquez 1993), Refractarios, which were collected from trees showing an absence of witches’ broom disease in Ecuador and Amelonado. The latter formed the basis of early cultivation in West Africa.

Whilst the groupings of Criollo, Forastero and Trinitario are still used by the cocoa trade, they do not reflect the full genetic diversity of the species. In particular, the Forastero grouping contains considerable genetic diversity, causing some to question how meaningful this term is (e.g. Bartley 2005). An alternate classification was proposed by Motamayor et al. (2008) to reflect better the genetic diversity of the crop. The classification was based on a cluster analysis of microsatellite data from 1241 accessions covering a large geographical area in Central and South America. Ten groupings or clusters were derived from this analysis, these being: Marañon, Curaray, Criollo, Iquitos, Nanay, Contamana, Amelonado, Purús, Nacional and Guiana. The authors suggested that these groupings should be the starting point for collection and curation activities as well as being a basis for designing new mating schemes in breeding programmes (Fig. 3).

## *Wild Relatives of Cacao*

There are 22 accepted species within the genus *Theobroma*, of which *T. cacao* is by far the most widely cultivated (Cuatrecasas 1964; Table 1). Other species have been proposed within the genus *Theobroma,* but are not yet fully accepted (Bhattacharjee 2018). Amongst the *Theobroma* species, 15 are considered to be edible, but only three, apart from *T. cacao*, are cultivated, these being *T. bicolor*, *T. angustifolium* and *T. grandiflorum*. The latter is cultivated in the Americas, notably in Brazil (known locally as *cupuaçu,* or *cupuassu* in Spanish-speaking countries) where its pulp is used in preparing beverages. There are also 17 species of the genus *Herrania*, a wild relative of *Theobroma*.

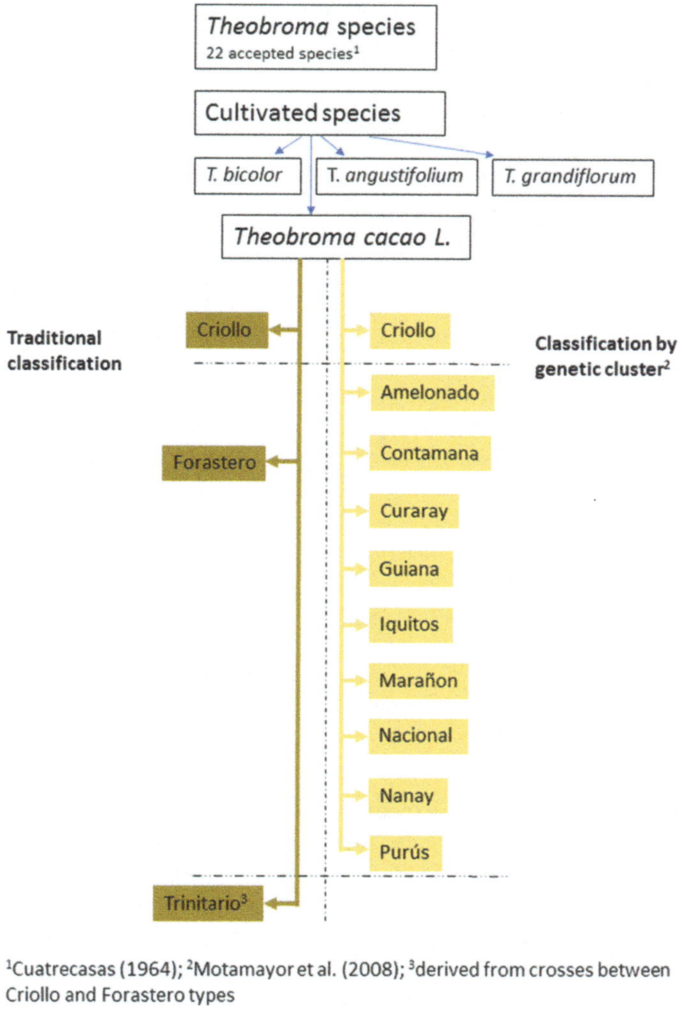


**Fig. 3** Diversity within *Theobroma* species. Diversity within cacao (*T. cacao*) is illustrated through its traditional classification and through the genetic cluster analysis of Motamayor et al. (2008)

## 4 Conservation of Wild Cacao and Landraces

The genetic diversity of cacao can be said to be represented by wild cacao within the Amazon basin, relics of ancient cultivation and trees present on modern farms (CacaoNet 2012). Since cultivated cacao is usually only one or two generations removed from wild germplasm, wild cacao can be incorporated into breeding programmes, in contrast to many field crops (Zhang and Motilal 2016).

**Table 1** Species of *Theobroma* (grouped according to Cuatrecasas 1964)

| Section | Species | Authority |
|---|---|---|
| Andropetalum | *T. mammosum* | Cuatrec. & León |
| Glossopetalum | *T. angustifolium* | Moçiño & Sessé |
| | *T. canumanense* | Pires & Fróes |
| | *T. chocoense* | Cuatrec. |
| | *T. cirmolinae* | Cuatrec. |
| | *T. grandiflorum* | (Willd. ex Spreng.) Schum |
| | *T. hylaeum* | Cuatrec. |
| | *T. nemorale* | Cuatrec. |
| | *T. obovatum* | Klotzsch ex Bernoulli |
| | *T. simiarum* | Donn. Smith. |
| | *T. sinuosum* | Pavón ex Hubber |
| | *T. stipulatum* | Cuatrec. |
| | *T. subincanum* | Mart. |
| Oreanthes | *T. bernouillii* | Pittier |
| | *T. glaucum* | Karst |
| | *T. speciosum* | Willd. |
| | *T. sylvestre* | Mart. |
| | *T. velutinum* | Benoist |
| Rhytidocarpus | *T. bicolor* | Humb. & Bonpl. |
| Telmatocarpus | *T. gileri* | Cuatrec. |
| | *T. microcarpum* | Mart. |
| Theobroma | *T. cacao* | Carl Linnaeus (L.) |

The first major collecting expeditions of wild germplasm were made by F.J. Pound in the 1930s and 1940s, which were stimulated by the outbreak of witches' broom disease in Trinidad. Collections of pods in Peru over the period 1938–1943 included tributaries of the Rio Ucayali, Rio Morona and Rio Marañon (Pound 1938, 1945; Zhang and Motilal 2016). They provided five germplasm groups: IMC, NA, PA, MO and SCA. In a subsequent expedition of the same areas, Pound collected budwood, which formed the basis of the 'POUND' clones. They are sometimes referred to as 'P' clones, but note that 'P' has been used for other germplasm (Turnbull and Hadley 2019). In addition to the Peruvian collections, Pound also collected material from western Ecuador that yielded Ecuadorian Refractario germplasm. They are stored at the International Cocoa Genebank in Trinidad (ICGT) (Lockwood and End 1993). The Pound collection forms a large part of the Trinidad genebank and, to date, has supplied the most widely utilised materials in cacao breeding (Zhang and Motilal 2016).

The threat of genetic erosion, primarily due deforestation in the Amazon region, combined with a perceived need to broaden the genetic base of cacao preserved ex situ, led to a number of subsequent collections within the Amazon in Brazil, Ecuador, Bolivia, Colombia, Peru, French Guiana and Guyana (see Lockwood and End 1993; Zhang and Motilal 2016;). The collection expedition of Allen during

1980s was the first to collect in a systematic manner to maximise genetic diversity (Allen 1987). Passport and preliminary characterisation data, including pod and bean weights and dimensions, were collected from these materials. An expedition in the 1990s in Belize was designed to collect Criollo types, which are under-represented in ex situ collections (Mooleedhar et al. 1995).

Recent collections in the Peruvian Amazon (Loreto, San Martin, Ucayali and Madre de Dios) have resulted in the establishment of 441 wild cacao accessions at ICT (*Instituto de Cultivos Tropicales*), Tarapoto, where they have been characterised (Arévalo-Gardini et al. 2018). A project to collect wild cacao from the likely area of origin of the Nacional variety was established between INIAP (*Instituto Nacional de Investigaciones Agricolas*) and CIRAD (*Centre de Coopération Internationale en Recherche Agronomique pour le Développement*) and resulted in the collection from the Zamora Chinchipe province of 83 wild accessions in 2010 and 48 in 2013. The areas of collection had experienced rapid genetic erosion due to tree felling and opening of land for cattle rearing (Loor-Solórzano et al. 2015).

It has been suggested by Motilal (2018) that future collecting should be focused around unexplored areas of the centre of diversity of cacao (i.e. the upper Amazon basin), where there is a higher chance of encountering new or rare alleles. Motilal (2018) proposes that such collecting should be stratified sampling of river basins reflecting the distribution of wild cacao.

When it comes to landraces that are conserved ex situ, a notable group is the Imperial College Selections (ICS) that were selected from farms in Trinidad. A survey was initiated by Pound in 1931 (Pound 1932), initially of 50,000 trees. A preliminary shortlist of 1000 trees was then selected by eliminating trees with small pods or pods with thick husks. This was then reduced further to 100 clones, by 1936, on the basis of high yield (in this case assessed as numbers of pods per tree) and low pod index (numbers of pods required to produce 1 kg of dry beans). The clones were named ICS 1 to ICS 100. Subsequent testing of these clones across locations indicated that some of the clones had more stable yield across soil types than others (Toxopeus 2001b). A number of ICS clones have been utilised in breeding programmes (Bekele et al. 2020), and some are used as cultivars (e.g. ICS 60 and ICS 95).

In situ conservation of cacao germplasm is difficult to achieve due to the loss of the natural habitat of cacao and replacement of traditional varieties on farm with hybrids. For this reason, most cacao conservation is as ex situ genebanks. Since the seed of cacao is recalcitrant, genebanks comprise living collections of trees. Cryopreservation techniques involving a vitrification process have been developed for somatic embryos (Adu-Gyamfi and Wetten 2012), and so this may represent a future means of conserving or backing up germplasm.

Between the records in the ICGD and www.cacaonet.org, a total of 54 genebanks are listed across 39 different countries. It is estimated that over 24,000 accessions are conserved ex situ (CacaoNet 2012). The country with the largest number of named accessions in genebanks is Brazil, where CEPLAC (*Comissão Executiva do Plano da Lavoura Cacaueira*) has collections in Bahia, Belém and Rondônia. Other large national collections in South America include those of INIAP (*Instituto*

*Nacional de Investigaciones Agropecuarias*) in Ecuador, ICT (*Instituto de Cultivos Tropicales*) in Peru, INIA (*Instituto Nacional de Investigaciones Agricoles*) in Venezuela and Agrosavia in Colombia. Away from the centre of origin, significant germplasm collections can be found in Ghana (Cocoa Research Institute of Ghana—CRIG), Côte d'Ivoire (*Centre National de Recherche Agronomique*—CNRA) and Malaysia (Malaysian Cocoa Board—MCB). CacaoNet (2012) highlighted inadequacies in funding of many ex situ collections and, in some cases, the need for safety duplication of cacao germplasm.

There are two international genebanks for cacao (i.e. collections in the public domain). The International Cocoa Genebank, Trinidad (ICGT) is managed by the Cocoa Research Centre of The University of the West Indies. It was founded in 1982 and as of January 2019 has approximately 2200 accessions, which are routinely characterised for traits of economic interest (Bekele et al. 2020). The International Cacao Collection at CATIE (IC3) in Costa Rica is managed by the *Centro Agronómico Tropical de Investigación y Enseñanza* (CATIE). It was established in 1944 and as of October 2019 has 1235 accessions (A. Mata pers. comm 2020). Both institutes have signed agreements with the International Treaty on Plant Genetic Resources for Food and Agriculture (ITPGRFA) to maintain their germplasm within the public domain. As such, material from these collections is supplied with the Standard Material Transfer Agreement for plant genetic resources (Correa 2006), which effectively ensures that onward distribution of material is with a commitment to maintain it within the public domain. Given the risk of transferring pests and diseases of cacao, international movement of cacao germplasm must be via intermediate quarantine. The principal route for international distribution of cacao germplasm is via the International Cocoa Quarantine Centre at the University of Reading (ICQC, R), where imported material undergoes a 2-year quarantine period that includes indexing for viruses. Material from plants that have completed quarantine at ICQC, R are distributed in response to requests from research institutes and growers, usually in the form of budwood (Daymond 2018). The USDA-ARS cacao quarantine station in Miami distributes material regionally within the Americas and sends material to ICQC, R for onward distribution in W Africa and SE Asia.

## 5 Crop Improvement: Overview of Traits of Importance to Breeders

A range of traits important to breeders are summarised in Table 2. The International Cocoa Germplasm Database (ICGD) (www.icgd.reading.ac.uk) acts as online repository for data that has utility for cacao breeders (See Box 1). It is notable that some accessions have been characterised in far more detail than others. Data on

**Table 2** Traits of importance to cacao breeders

| Trait | Expressed as | Factors influencing the expression of the trait |
|---|---|---|
| Yield | kg dry beans $ha^{-1}$; pods $tree^{-1}$ | Planting density, local climate, age of the crop, shade intensity |
| Precocity | Years to bearing | |
| Yield efficiency | Ratio of yield to vegetative growth | May be affected by local climate |
| Compatibility | Self-compatible, self-incompatible | |
| Bean size | g dry weight | Local climate, season |
| Pod value (index) | Number of pods to produce 1 kg of dry beans | Local climate, season |
| Bean lipid content | Percentage | Local climate, season |
| Bean quality/ flavour | Flavour descriptors | Local climate, soil type, post-harvest processing |
| Resistance to diseases | Descriptors, e.g. 'resistant', 'moderately resistant', 'susceptible' | Background disease pressure. Whether or not there is a dry season may impact the severity of fungal diseases. The phenology of flowering/fruiting of a given clone in relation to the local climate may allow disease 'escape' |
| Resistance to pests | Descriptors as above | Background pest pressure |
| Cadmium uptake | Bean cadmium concentration | Soil pH |

germplasm needs to be seen in the context of the situation in which the data were collected, in that the expression of a particular trait may vary according to, for example, environmental and husbandry conditions. This is particularly the case for yield, whereby the extent to which maximum yield potential will be expressed depends on factors such as the local climate (e.g. the duration of the dry season), the age of the crop, the intensity of any shade present (from overhead trees) and the planting density of the crop. The latter is particularly important given that different cacao genotypes vary in their vigour and yield efficiency (Daymond et al. 2002), and therefore the optimal planting density will vary according to variety (Lockwood and Pang 1996). In the case of diseases and pests, their impact on the performance of a genotype, in terms of growth and yield, will be influenced by factors such as the background pressure of the disease/pest in the location of assessment and whether 'disease escape' is occurring, i.e. whether peaks of cropping are occurring at times when the background disease pressure is lower which may, in turn, be influenced by environmental conditions.

**Box 1**
**The International Cocoa Germplasm Database (ICGD)**
**www.icgd.reading.ac.uk**

The International Cocoa Germplasm Database (ICGD), which is based at the University of Reading, is an information service for the cocoa research community, particularly researchers and breeders. The online database contains information on the origins, characteristics, distribution and availability of cocoa germplasm. Searches on specific clones provide information on which research institutes hold the clone in question and data relating to that clone (including pod morphology, bean size and lipid content, compatibility and resistance to pests and diseases). Pedigree and collection details of clones are also provided. Where available, molecular fingerprinting data is also provided for clones in the form of either microsatellite and/or SNP data. This permits users to cross check the identity of a clone, which they hold against references in the database.

The ICGD also allows searches to be made for clones that have particular traits and this can be in combination with one or more locations at which the germplasm is held (Turnbull and Hadley 2019).

**The International Cocoa Germplasm Database (ICGD)**

**www.icgd.reading.ac.uk**

The International Cocoa Germplasm Database (ICGD), which is based at the University of Reading, is an information service for the cocoa research community, particularly researchers and breeders. The online database contains information on the origins, characteristics, distribution and availability of cocoa germplasm. Searches on specific clones provide information on which research institutes hold the clone in question and data relating to that clone (including pod morphology, bean size and lipid content, compatibility and resistance to pests and diseases). Pedigree and collection details of clones are also provided. Where available, molecular fingerprinting data is also provided for clones in the form of either microsatellite and/or SNP data. This permits users to cross check the identity of a clone, which they hold against references in the database.

The ICGD also allows searches to be made for clones that have particular traits and this can be in combination with one or more locations at which the germplasm is held (Turnbull and Hadley, 2019).

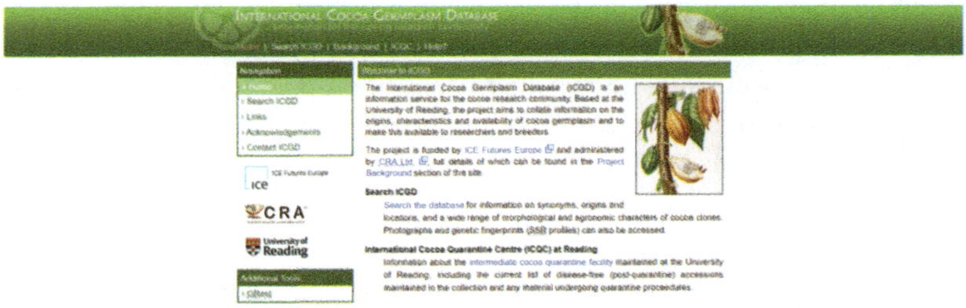

An issue of increasing importance is cadmium bioaccumulation in cocoa beans (Ramtahal et al. 2015), particularly in light of European Union directives that lowered the amount of permissible cadmium in food products and addressed cocoa specifically through regulations implemented in January 2019. Cadmium accumulation can particularly be an issue where cacao is cultivated on volcanic soils. Genotypic variation has been found both in the amount of cadmium accumulated and that partitioned between vegetative and reproductive components (Lewis et al. 2018) and this has the potential to be utilised in cultivar development.

## 6 Climate Resilience

Since cacao is grown throughout the humid tropics, cultivated cacao will often experience a greater range of environmental stresses than in its natural habitat. In W Africa, there is a dry season that typically extends from early December until the end of February, although the precise timing and duration can vary from year to year. The dry season is usually more marked in *El Nino* years. With climate change, in some areas, the dry season may be longer and there is likely to be more year-to-year variation in its intensity. Based on an ensemble of models, Dunning et al. (2018) predict a delay in the onset of the wet season in West Africa and in increase in the mean daily rainfall total during the wet season.

Juvenile cacao is particularly sensitive to water deficit and seedlings will not recover under severe drought conditions. For this reason, water deficit presents a particular challenge when it comes to establishment. Prolonged periods of soil water limitation can also have a consequent impact on yields. Conversely, flooding can have a negative impact on cacao performance, particularly impacting on photosynthetic rates (Sena Gomes and Kozlowski 1986; de Almeida et al. 2016; da Silva Branco et al. 2017).

There is a genotypic variation in responses to water deficit. For example, Ofori et al. (2015) examined the response of different families of cacao that had recently been established in the field to withholding water. They found variation amongst families in terms of vigour and survival in response to water deficit. A number of researchers have shown variation in various physiological and morphological parameters amongst clonal cacao that have been subjected to drought (Balasimha and Daniel 1988; Balasimha et al. 1991; Antwi et al. 1994; Araque et al. 2012; dos Santos et al. 2014). Balasimha et al. (2013) screened 52 international cacao clones, growing in the field in India, to water deficit by withholding irrigation during the dry season. They observed differential responses of germplasm to drought in terms of their rate of stomatal conductance and chlorophyll fluorescence.

Limited studies to date have examined tolerance to flooding. Bertolde et al. (2010) observed genotypic variation in this trait amongst juvenile cacao clones and suggested that the lower levels of chlorosis found in more tolerant clones could be used as selection trait.

Temperature is a major factor that limits where cacao can be cultivated. Generally, cacao is not cultivated where night temperatures regularly fall below 15 °C, although cacao can tolerate occasional nights where the temperature falls within the range of 10–15 °C, which sometimes occurs at night during the dry season in West Africa. In Bahia, Brazil, the cooler period between May and July has an impact on flowering and pod setting, and a consequent effect on cropping cycle. In most cacao-growing regions, maximum monthly temperatures range from 26 to 34 °C, although slightly higher temperatures have been recorded in Nigeria and Southern India (Lahive et al. 2019). However, daily maximum temperatures can be higher than monthly values (Asare et al. 2017). Excessively high temperatures can have a negative impact on photosynthetic rate. Balasimha et al. (1991) measured a decline in photosynthetic rate for field-grown cacao in India when monthly temperatures were greater than 34 °C. Furthermore, greater pod losses through cherelle wilt have been reported at higher temperatures (Daymond and Hadley 2008). Climate change predictions are that temperatures will increase in cacao-growing regions. Exploitation of germplasm that is more resilient to higher temperatures could be part of a strategy to adapt to climate change. To date, genotypic variation in vegetative and pod growth has been observed amongst a limited number of clones in response to differences in current temperature regimes (Daymond and Hadley 2004, 2008). However, there is limited information on genotypic variation in high temperature tolerance (Daymond and Hadley 2009) and thus a need for more research within this area. There appears to be no information on genotypic variation in low temperature tolerance.

## 7 Disease and Pest Resistance

Challenges from diseases are present to a greater or lesser extent in most cacao-growing areas. Some diseases and pests are fairly ubiquitous, most noticeably *Phytophthora palmivora*, the causal agent of *Phytophthora* pod rot, is encountered in most cacao-growing areas. Other diseases and pests are confined to particular geographical regions. For example, Cacao swollen shoot virus disease is only encountered in West Africa, witches' broom disease is only found in South America and parts of the Caribbean and the pest Cocoa Pod Borer is only found in South East Asia. Table 3 provides a summary of the major diseases of cacao.

Different levels of resistance to the major diseases and pests of cacao have been recorded in germplasm collections. A complete resistance in a cultivar is rare; however, cultivation of partially resistant germplasm in combination with good agricultural practices can result in a marked reduction in losses due to pests and diseases. It is important for breeders and growers to have access to a range of resistant/tolerant material since resistance may break down over time. This is illustrated by the case of the search for germplasm resistant to witches' broom disease. Amongst the material collected by F.J. Pound in the 1930s and 1940s, the clones SCA 6 and SCA 12 were found to have a high level of resistance in Trinidad. The two clones were widely used as parents in Trinidad for a number of 'TSH' (Trinidad Selected

**Table 3** Distribution of key pests and diseases of cacao (adapted from End et al. 2017)

| Disease | Known areas encountered |
|---|---|
| Cacao swollen shoot virus (CSSV) | Benin, Côte d'Ivoire, Ghana, Liberia, Nigeria, Sierra Leone and Togo. Reports in Sri Lanka |
| Witches' broom disease | Brazil (Bahia, Espirito Santo, Amazonian regions), Bolivia, Colombia, Ecuador, French Guiana, Grenada, Guyana, Panama, Peru, St Vincent, Suriname, Trinidad and Tobago and Venezuela |
| *Moniliophthora* pod rot (frosty pod rot or moniliasis disease) | Belize, Bolivia, Colombia, Costa Rica, Ecuador, El Salvador, Guatemala, Honduras, Jamaica, Mexico, Nicaragua, Panama, Peru and western Venezuela |
| *Phytophthora* pod rot (black pod) | *Phytophthora palmivora* is found in most cacao-growing countries worldwide<br>*P. megakarya*: Bioko (Fernando Pó), Cameroon, Côte d'Ivoire, Gabon, Ghana, Nigeria, São Tomé and Principe and Togo |
| Vascular streak dieback (*Ceratobasidium theobromae*) | Most cacao-growing areas in South and South East Asia: China (Hainan Island), India, Indonesia, West Malaysia and Sabah, Myanmar, PNG (islands of New Guinea, New Britain, New Ireland), southern Philippines, Thailand and Vietnam |
| *Verticillium* wilt of cacao | Worldwide, especially Brazil, Colombia and Uganda |
| *Ceratocystis* wilt | Brazil, Cameroon, Colombia, Costa Rica, Ecuador, French Guiana, Trinidad and Venezuela |
| *Pests* | |
| Cocoa pod borer (*Conopomorpha cramerella*) | South East Asia including Malaysia, Indonesia, the Philippines and Papua New Guinea |
| Mirids and other Heteropterous plant sucking bugs | All cacao-growing regions except the Caribbean |

Hybrids) clones. However, when witches' broom disease arrived in Bahia, Brazil and spread in the 1990s, the TSH material showed limited resistance over time, due to the presence of different strains of the disease (Ahnert and Eskes 2018).

*Phytophthora* pod rot or black pod is the most widespread disease of cacao. *Phytophthora palmivora* is encountered in all the cacao-growing areas, whilst the more aggressive *P. megakarya* is encountered in parts of West Africa. A range of resistance to black pod amongst germplasm has been reported by various researchers (e.g. Iwaro et al. 1997; Tahi et al. 2000; Pokou et al. 2008), and resistance to black pod has been a major focus of breeding programmes.

Frosty pod rot is caused by a fungus, *Moniliophthora roreri*, which is a particularly severe disease that is found in parts of South and Central America and in Jamaica. It represents a notable threat to other cacao-growing regions due in part to the fact that the fungal spores of *M. roreri* are long lived, making it more easily spread by human activity. In Costa Rica, an extensive screening exercise has been conducted within CATIE's international genebank (IC3). This process initially identified 278 clones to be 'resistant' or 'moderately resistant' and one clone, CATIE R-6, as being highly resistant. A process of re-evaluation of this material is currently underway at CATIE under uniform conditions to confirm their resistance level (Phillips-Mora et al. 2018). Based on screening for frosty pod rod resistance combined with high yields, six clones are being distributed in Central America (Phillips-Mora et al. 2013).

In the case of vascular streak dieback, material that exhibits resistance includes some clones developed by the Malaysian Cocoa Board (Turnbull and Hadley 2019), e.g. PBC 123, which is also grown extensively in Indonesia (where it is known as 'Sulawesi 1'). In Papua New Guinea, some Trinitario clones (e.g. KA2–101) have shown resistance over a period of time dating back to the 1960s (McMahon and Purwantara 2016).

Regarding pests, mirids as a group are economically significant, particularly *Sahlbergella singularis* and *Distantiella theobroma* in West Africa and *Helopeltis* spp. in South-East Asia (Campbell 2017). Sounigo et al. (2003) demonstrated genotypic differences in resistance to mirids, amongst a progeny trial. They found the most resistant parents amongst Upper Amazon Forastero, whereas germplasm from the Trinitario group appeared to be more susceptible.

## 8 Bean Quality

Key quality traits for cocoa beans include a high fat (or lipid) content, low shell percentage (the 'shell' being the dried testa) and bean weight. The importance of bean weight lies in the fact that smaller beans generally have a high shell percentage and so require more processing. Beans with a dry weight of less than 1 g or "bean count" (the number of beans needed to produce 100 g) of greater than 100 are considered inferior and often attract a lower price (Wood 2001b). Environmental parameters are known to impact on bean size, specifically higher temperatures and water deficit have been shown to be correlated with a reduction in bean size (Toxopeus and Wessel 1970; End 1990; Daymond and Hadley 2008; Handley 2016). Thus, it is important to consider bean quality data within the context of the environment in which the data were collected. Furthermore, there is some evidence from a limited range of germplasm that bean size is more stable for some genotypes than others in relation to changes in environmental conditions (Daymond and Hadley 2008). Stability of bean size might therefore be a selection criterion in relation to breeding for a changing environment.

Screens of germplasm within genebanks have found a significant amount of variation in bean fat content. Khan (1998) reported a range of 46.1–59.8% cotyledon fat content amongst 247 accessions within the International Cocoa Genebank in Trinidad. Furthermore, Khan et al. (2008) observed a range of 44.4–59.8% amongst a further 323 accessions, consisting mainly of Upper Amazon Forasteros and Refractarios. The latter study recommended various accessions for utilisation in breeding for improved bean quality. Other genebanks in which bean fat content has been screened include the *Centro de Pesquisa do Cacao* collection near Ilhéus, Bahia, Brazil where a range of 45.4–60.3% across 490 accessions was reported by Pires et al. (1998). They found a weak negative correlation between dry bean weight and fat content when comparing across genotypes.

## 9 Traditional Breeding Approaches

Breeding programmes of perennial tree crops, such as cacao, are often long-term endeavours requiring a considerable amount of land and resources. Progress in cacao breeding has also been hindered by insufficient knowledge of the genetic diversity of cacao. There is still much untapped genetic diversity conserved ex situ that may be utilised in future breeding programmes. Traditional approaches can be augmented by modern genomic approaches, based on recent progress achieved in the latter.

The consensus of opinion is that sustainable progress in cacao breeding for improved production and yield requires systematic screening for genetically superior individuals in segregating populations (Eskes et al. 1996; Pires et al. 1999; Lockwood et al. 2007; Monteiro et al. 2009). Hybrids between heterozygous parents, including double crosses and triple crosses, where the parents are derived from different source populations, have been recognised as the best potential sources of superior selections (Bekele and Phillips-Mora 2019). Iwaro et al. (2009) described the genetic gain achieved, in Trinidad, through hybrid screening for black pod disease resistance along with favourable yield potential. This genetic gain is currently being harnessed to further improve the Trinidad and Tobago Ministry of Agriculture's Trinidad Selected Hybrid breeding programme (Iwaro et al. 2010; Maharaj et al. 2011; Bekele and Phillips-Mora 2019) through gene pyramiding. The latter is designed to assemble several desirable genes for specific traits from multiple parents into a single genotype, as explained by Servin et al. (2004).

Conventional breeding methods, such as recurrent selection and backcrossing, which spanned decades over several generations to produce superior genotypes, are no longer the only means available to accomplish gene pyramiding. With the advent of genotyping by sequencing, the identification of molecular markers, which tag traits of interest, can expedite the selection of genotypes that harbour the associated genes that encode favourable traits. Introgression of multiple favourable traits in progeny can thus now be accelerated by marker-assisted selection and breeding with genomics (Iwata et al. 2016; Varshney et al. 2014). The process of gene pyramiding is affected by the number of genes to be transferred, the distance between the target genes and flanking markers, the number of genotypes selected in each breeding generation and the pedigree of the genotypes involved (Joshi and Nayak 2010).

Bi-parental, diallel crosses schemes (complete and partial), factorial mating designs (North Carolina II model where each member of a group of males is mated to each member of a group of females) and recurrent selection schemes have been the predominant traditional breeding approaches utilised in cacao improvement (Table 4). Ahnert (2009) advocated ideotype breeding to increase production potential through selection of individual traits in accordance with a specified model. Traits included precocity (early bearing), plant vigour and growth rate, growth habit, durable disease resistance, bean size, shape and lipid content, pollination compatibility status and yield. A recurrent scheme with $F_1$ crosses and self-pollinated generations, in a circular partial diallel mating design, was proposed.

**Table 4** Examples of traditional breeding approaches applied for cacao improvement (adapted from Bekele and Phillips-Mora 2019)

| Country | Mating design | Breeding objectives |
|---|---|---|
| Brazil | Recurrent breeding scheme | Multi-trait selection for resistance to witches' broom (WB), black pod (BP), frosty pod rot (FP), high yield, high bean fat content and other traits |
| | Bi-parental crosses; complete diallel design | Improvement in yield (number of pods per plant), resistance to WB and *Ceratocystis* wilt |
| Cameroon | 6 × 6 diallel crossing scheme | Multi-trait selection for improved yield, yield potential and resistance to (BP) diseases |
| Colombia | Recurrent Selection programme | 160 full-sib progenies established to accumulate favourable alleles for yield and resistance to four diseases (FP, WB, BP and *Ceratocystis* wilt) |
| Costa Rica | Bi-parental crosses | Multi-trait selection for disease resistance: FP (mainly) and BP, high-yield and good quality (flavour). Pyramiding of FP resistant genes |
| Côte d'Ivoire | Reciprocal Recurrent Selection Scheme | Selection for adaptation and for resistance to BP, good yield potential, self-compatibility, large bean size and good flavour attributes |
| | 5 × 5 diallel and in a 4 × 2NC II factorial design | Resistance to BP (laboratory and field) |
| | Reciprocal Recurrent Selection Scheme | Resistance to BP and yield |
| Ghana | Bi-parental crosses | Improvement in yield, establishment capabilities, resistance to Cacao swollen shoot virus (CSSV) disease resistance and *P. megakarya* resistance |
| | Diallel or factorial | Field resistance to CSSV disease, combining ability analysis of BP disease incidence |
| India | Bi-parental crosses | Stable yield (over time and environment), bean size/weight, compatibility, resistance to BP, *Ceratocystis wilt* and vascular streak dieback (VSD) (laboratory and field) and drought tolerance (Apshara 2019) |
| Malaysia | Reciprocal Recurrent Selection Scheme | Resistance to cacao diseases, particularly VSD |
| Nigeria | Bi-parental crosses, progeny trials; diallel crossing scheme | Establishment ability, vegetative vigour, precocity, yield, resistance to BP and CSSV |
| Papua New Guinea | Bi-parental and poly crosses<br>Reciprocal Recurrent Selection Scheme | Durability of field resistance to BP disease, yield, vigour and compact architecture |
| | | Resistance to VSD |
| Trinidad and Tobago | Reciprocal Recurrent Selection Scheme | Resistance to *Ceratocystis* wilt and WB, large bean size and weight and high yield |

Progress in cacao breeding, combined with long-term improvement, has been achieved through reciprocal recurrent selection schemes (RRSS). RRSS utilise the products of recombination from controlled breeding programmes to generate new clonal cultivars on a continuing basis as improvement progresses. Population breeding and recurrent selection have been utilised in cacao breeding in Africa and the Americas since the 1990s (Ahnert and Eskes 2018). RRSS have produced favourable results in several countries, including Brazil, Colombia, Côte d'Ivoire, Malaysia, Papua New Guinea and Trinidad and Tobago (Table 4). Heterosis (hybrid vigour) (Stuber 1994) has been exploited in these hybrid breeding programmes, as recommended by Ahnert and Eskes (2018). This has enabled the expression of superior traits in progeny compared to the parental genotypes. Hybridisation of genetically diverse parental types and non-additive gene action tend to favour heterosis.

Currently, a large-scale mass selection breeding programme is in progress in Ghana, entitled the Mabang Megakarya Selection Programme (MMSP). It was established in 2005 as a public–private partnership and has been supported by the Ghana Cocoa Board (COCOBOD), the Dutch government, the Ghana Cocoa Growers Research Association Ltd. (GCGRA—a UK-based scientific research association managed by representatives from the chocolate industry and the University of Ghana), Cocoa Research (UK) Ltd. and the chocolate industry (Mars-Wrigley and Mondelez). The MMSP is located at a Cocoa Research Institute of Ghana (CRIG) substation, close to the northern border of the Ashanti region.

The MMSP involves two strands, one involving seedlings and the other clonal trials. Currently, seed gardens provide mixed 'hybrids' to farmers resulting from bi-parental crosses in seed gardens. The seedling trials at MMSP were established with the aim of identifying crosses that outperform current seed garden materials. The performance of populations derived from crosses with over 30 different pollen parents is currently being evaluated.

The clonal trials involve large-scale evaluation of a broad range of clones (about 3500 to date; R. Lockwood pers. comm.) including international material, clones identified in previous trials and farmer selections. The promising individuals are then taken to a second phase where they are replicated in larger numbers and evaluated for yield and disease tolerance. Ultimately, the best candidate clones will be planted in multisite trials, which will include farmer field trials. This multilocational, farmer participatory approach has been adopted to optimise the results of the MMSP and should serve as a model in traditional and modern cacao breeding.

## 10 Genomic Resources

Molecular (genetic) markers, such as single-nucleotide polymorphisms (SNPs), have proven indispensable for elucidating genetic identities, diversity and phylogenetics within cacao collections, mapping candidate genes and paving the way for marker-assisted selection and breeding with genomics in cacao (Brown et al. 2008;

Guiltinan et al. 2008; Allegre et al. 2012; Lanaud et al. 2017). Application of molecular methods and research on cacao genomics culminated in the sequencing of the genome, in 2011, by two independent research teams. The first team, led by USDA in collaboration with Mars and IBM, produced a sequence that covers 92% of the genome with approximately 29,000 genes. The sequencing was done using the highly homozygous cultivar, MATINA 1–6 (Saski et al. 2011; Motamayor et al. 2013). The second team, the International Cacao Genome Sequencing (ICGS) consortium, led by CIRAD, in collaboration with other institutes in France, the USA, Brazil, Côte d'Ivoire, Korea and Venezuela, sequenced an ancient Mayan cacao variety, a homozygous Criollo, B97–61/B2, which was domesticated in Belize (Argout et al. 2011, 2017). Version 2.0 of the Criollo genome of Argout et al. (2017) has 96.7% of the assembly anchored to the 10 chromosomes of the *T. cacao* L. genome. The two genomes produced were very similar apart from having 12 small regions located on different chromosomes in the two cacao varieties. The latter phenomenon could be attributed to transposon (jumping gene) activity (Guiltinan and Maximova 2015). Re-sequencing of 200 cacao accessions was conducted by Cornejo et al. (2018) using the MATINA 1–6 sequence as a reference.

Advances in cacao genomics have the potential to shorten the breeding cycle of this perennial crop thereby impacting progress on yield improvement, abiotic (drought and other climatic constraints) and biotic (disease and pest) resistance and assessment of cocoa quality (including flavour) (Ahnert and Eskes 2018; Bekele and Phillips-Mora 2019). Polygenic traits, which are significantly affected by environment, are more difficult to score without a tool for identifying and following the major genes influencing the phenotype. The inheritance of resistance to black pod, frosty pod and witches' broom diseases and yield in cacao has been found to be polygenic (Simmonds 1993). Recent research has facilitated the tagging of some of these traits with molecular markers to identify Quantitative Trait Loci (QTLs) to expedite genetic improvement (Bekele and Phillips-Mora 2019). Devy et al. (2018) were able to differentiate bulk and fine cacao in Indonesia using MYB nucleotide sequence fragments. Building on the notable achievements in cacao genomics made thus far, as recounted by Guiltinan and Maximova (2015), new discoveries will further contribute to the application of genomic resources to better manage and utilise cacao genetic diversity conserved ex situ as well as in situ. The full potential impact of the application of genomic resources and genomic innovations in cacao breeding is explained by Varshney et al. (2014). The latter authors detailed the benefits of applying genome sequencing technologies in crop breeding.

A particular application of SNP genotyping assays is addressing mislabelling, which is often encountered in cacao collections and can impede effective utilisation. SNP analyses can also be used to identify redundancy in collections, for example, two accessions with different names but the same molecular fingerprint. Mahabir et al. (2019) recently reported on a 'robust set' of 96 SNPs to be used as a 'minimal core SNP panel' for DNA fingerprinting of cacao. Other SNP sets and assays have been recently proposed by, for example, DuVal et al. (2017). Currently, a Reference Genotype Working Group is working on designating reference accessions against which the fidelity of particular clones can be verified; these references will be highlighted within the International Cocoa Germplasm Database (ICGD) (Turnbull et al. 2018). The group will also propose a core set of markers, based on those already being utilised by several research groups.

## 11 Modern Breeding Approaches

The advent of the mapping of the cacao genome and advances in genomics has enabled the molecular approach to cacao breeding. Modern breeding approaches such as genomic selection and marker-assisted selection are fairly new tools at the disposal of the cacao breeder (Guiltinan et al. 2008; Rodriguez-Medina et al. 2019). Some of the relevant genomic innovations include QTL mapping (Cervantes-Martinez and Brown 2004), genome-wide association mapping (Marcano et al. 2007, 2009; Motilal et al. 2016; Da Silva et al. 2016), identification of candidate genes and haplotype inferencing (Cervantes-Martinez and Brown 2004; Fouet et al. 2011; Utro et al. 2013; Gutiérrez-López et al. 2016). Cacao genetic diversity can now be fully exploited using genome-wide association studies and other gene-mapping techniques coupled with phenotyping approaches.

Modern breeding approaches can facilitate rapid selection of new genetic combinations in superior varieties through techniques such as QTL mapping and bulk segregant analysis (Wilkinson 2003). Progress in improvement of multiple traits simultaneously may be achieved via marker-assisted gene pyramiding (Servin et al. 2004). It is left to be seen whether other genomic innovations such as targeted editing of genomes using CRISPR-Cas technology (Fister et al. 2018) will also be adopted in cacao.

### *Association Mapping*

Association mapping involves searching for genotype–phenotype correlations in unrelated individuals (Bradbury et al. 2007) and is based on non-random association of alleles (gene variants) in a population or linkage disequilibrium (Stack et al. 2015). It is an alternative to classical QTL (family) mapping studies, which are constrained by the paucity of large productive progenies from known parental origin for perennial crops such as cacao.

Linkage maps with genetic markers covering the whole genome facilitate the elucidation of genes or chromosomal regions influencing traits of interest (Botstein et al. 1980; Allegre et al. 2012). Since the cacao genome has been fully mapped and there is tight coverage with molecular markers (Argout et al. 2011, 2017; Saski et al. 2011), association mapping and admixture mapping studies (Stack et al. 2015) are now facilitated. Marcano et al. (2009) conducted a genome-wide mapping study for yield factors and morphological traits in cacao. Royaert et al. (2011, 2016) identified marker-trait associations for self-compatibility and resistance to witches' broom disease, respectively, in a segregating mapping population of cacao. Sounigo et al. (2012) found several associations related to SSRs and yield in Cameroon.

Motamayor et al. (2013) identified candidate genes regulating pod colour in cacao. Da Silva et al. (2016) performed genome-wide association studies (GWAS) to locate markers for self-incompatibility in a mapping population involving 295 trees. Putative markers were located on chromosome 4 within 196 kb, suggesting the location of a major gene in this region. Motilal et al. (2016) employed GWAS to identify putative genes linked to bean number and disease traits. Romero Navarro et al. (2017) compared the results of GWAS and genomic prediction to identify associated markers and develop predictive models for frosty pod rot and black pod diseases, as well as yield traits in a population of improved clones. Osorio-Guarín et al. (2020) performed GWAS, using data from 102 cacao accessions, and detected two genes putatively associated with productivity (number of healthy pods) and seven encoding frosty pod disease resistance.

### *Genomic Selection*

The relatively slow progress realised in cacao breeding to date has the potential to be accelerated by the advancement made towards genomic selection in *T. cacao* (Ribeyre et al. 2017; McElroy et al. 2018). Genomic selection (GS) predicts genotypic performance from genome-wide marker data. It is a form of genomic-assisted or marker-assisted selection (MAS), dubbed GS-MAS, in which genetic markers covering the whole genome are used so that all QTL are in linkage disequilibrium with at least one marker.

Genomic selection requires that a test or *training* population, which represents the genetic diversity of a larger breeding population, which is thoroughly genotyped and phenotyped. This then facilitates the estimation of breeding values (GEBVs) that predict phenotypic performance on the basis of genetic marker frequencies. The larger breeding population must also be genotyped, and the GEBVs are used to predict the phenotypes of genotypes in that population. GS-MAS has been found useful in other crops for complex traits controlled by many QTLs (such as yield) and with low heritability ($h^2$) (Bernardo and Yu 2007), once the markers with the most significant associations with traits are close to functional genes. Consequently, this process has the potential to reduce the time and expense associated with large-scale phenotyping of progeny and, if successfully deployed, could thus significantly accelerate the process of achieving genetic gain in cacao.

## 12 Phenotyping Approaches

As a perennial tree crop, cacao does not easily lend itself to phenotyping via some of the platforms that have been developed primarily for short stature crops. Andrade-Sanchez (2014) described such a system for cotton. Nevertheless, a number of approaches have been proposed and applied to screen germplasm for resistance to both biotic and abiotic stresses. Such tests need to be conducted using standardised approaches in order to be reproducible. It is important that approaches developed for laboratory and greenhouse level screening are validated at a field level if they are to be applied with confidence.

The developments of phenotyping methods for resilience to water deficit, flooding and high temperatures are vital in the context of climate change. Tolerance to water deficit is a complex issue to screen for, in part due to the likely need to select for multiple traits. Furthermore, the objective of selecting for resilience will vary according to the age of the crop and the severity of stress encountered. Under severe stress, limiting water loss from the crop, e.g. through stomatal closure, may be the priority in order to avoid loss of trees. However, under more mild stress events, maintaining yields might be the priority. A number of studies have suggested traits associated with resilience to water deficit (Lahive et al. 2019). Balasimha and Daniel (1988) developed a screening method by measuring leaf water potential on detached leaves. The genotypes showing slower rate of decline in leaf water potential were considered more drought tolerant. Nevertheless, there remains a need to correlate a broader range of physiological and morphological responses to water deficit with yield, growth and survival.

In the case of tolerance to the major diseases of cacao, a number of screening methods have been proposed. A method to assess the level of resistance of germplasm to frosty pod rot has been developed and extensively employed in CATIE, Costa Rica. The method consists of spraying 2–3-month-old cacao pods with a spore suspension, this being the age when pods are most susceptible to frosty pod rot infection. These pods are then covered for 2 days with a transparent polythene bag, which contains paper towel and water in order to maintain high humidity. After 9 weeks, the pods are sectioned longitudinally, and percentage of internal necrosis is scored. The level of resistance of the clone is then classified according to the percentage of necrosis noted (categories: 'susceptible', 'moderately susceptible', 'moderately resistant', 'resistant' and 'highly resistant') (Phillips-Mora et al. 2018). A rapid screening method for resistance to *Phytophthora* pod rot based on inoculation of leaf discs was proposed by Nyassé et al. (1995). The test uses a five-point scale in order to categorise the severity of infection. Tahi et al. (2000) concluded that a good correlation can be found between the leaf disc method and field observations for *P. palmivora* as long as, critically, the test is carried out using a standardised procedure and under optimal conditions. Barreto et al. (2015) also found a positive correlation between data obtained from the leaf disc method and collected field data, in this case for three *Phytophthora* species (*P. palmivora*, *P. citrophthora* and *P. capsica*).

Laboratory-based tests for mirid resistance include feeding preference tests whereby contained mirids are presented with stems from different germplasm and after a period of feeding the number of lesions are recorded. This test has, however, produced contradictory results when the same germplasm has been tested in different countries (N'Guessan et al. 2008; Anike et al. 2009). Awudzi et al. (2020) has suggested that the growth environment of cacao trees can impact on the feeding preference of mirids, and therefore comparative tests of germplasm should be conducted on trees grown under similar environmental conditions.

## 13 Conclusions and Prospects

Cacao is a major commodity crop, and demand for cocoa beans is continuing to grow. Historically, increased demand for cocoa beans has been met through expansion of growing areas. This trend is not sustainable. Moving forward, there is a need to increase farm yields $ha^{-1}$, to improve farmer livelihoods and to make cacao farming generally more attractive to future generations. To achieve these ends, new varieties need to be developed in combination with optimisation of agronomic and soil conservation practices. To date, in most cases, cacao breeding activities have been focused on improving yields and resistance to diseases. There is a need to widen the focus of breeding to include resilience to abiotic stresses (particularly in relation to climate change), bean quality and low cadmium accumulation. Furthermore, selection for lower stature (compact) trees with high-yield efficiency, an approach used for a range of other trees crops, would allow better ease of management of the crop. Breeding will need to utilise a broader range of genetic resources than has previously been the case. Thus, there is a critical need for on-going conservation, characterisation, evaluation and utilisation of cacao genetic resources.

## References

Adu-Gyamfi R, Wetten A (2012) Cryopreservation of cocoa (*Theobroma cacao* L.) somatic embryos by vitrification. CryoLetters 33:494–505

Ahnert D (2009) Ideotype breeding in cocoa. In: Bekele F, End M, Eskes AB (eds) Proceedings of the international workshop on cocoa breeding for farmers' needs, 15-17th October 2006, San José, Costa Rica, pp 157–166

Ahnert D, Eskes AB (2018) Developments in cacao breeding programmes in Africa and the Americas. In: Umaharan P (ed) Achieving sustainable cultivation of cocoa. Burleigh Dodds Science Publishing, Cambridge, pp 111–150

Allegre M, Argout X, Boccara M et al (2012) Discovery and mapping of a new expressed sequence tag-single nucleotide polymorphism and simple sequence repeat panel for large-scale genetic studies and breeding of *Theobroma cacao* L. DNA Res 19:23–35

Allen JB (1987) London cocoa trade amazon project. Final report phase 2. Cocoa growers' bulletin (United Kingdom), 39, 94 pp

Alverson WS, Whitlock BA, Nyffeler R, Bayer C, Baum DA (1999) Phylogeny of the core Malvales: evidence from ndhF sequence data. Am J Bot 86:1474–1486

Andrade-Sanchez P, Gore MA, Heun JT, Thorp KR, Carmo-Silva E, French AN, Salvucci ME, White JW (2014) Development and evaluation of a field-based high throughput phenotyping platform. Funct Plant Biol 41:68–79

Anike JC, Omoloye AA, Aikpokpodion PO, Okelana FA, Eskes AB (2009) Evaluation of resistance in selected cocoa genotypes to the brown cocoa mirid, *Sahlbergella singularis* Haglund in Nigeria. Crop Prot 28:350–355

Antwi A, Lopez FB, McDavid CR, Spence J (1994). Leaf morphology and drought tolerance in cocoa. In: Proceedings of the 11th international cocoa research conference, 18-24th July 1993, Yamoussoukro, Côte d'Ivoire, pp 617–624

Apshara SE (2019) Cocoa genetic resources and their utilization in palm-based cropping systems of India. In: Aikpokpodion P (ed) *Theobroma cacao*—science for sustainability of global cocoa economy. IntechOpen, London. https://doi.org/10.5772/intechopen.82077

Araque O, Jaimez RE, Tezara W, Coronel I, Urich R, Espinoza W (2012) Comparative photosynthesis, water relations, growth and survival rates in juvenile Criollo cacao cultivars (*Theobroma cacao*) during dry and wet seasons. Exp Agric 48:513–522

Arévalo-Gardini E, Zhang D, Arevalo-Gardini J, Pinedo AF, Baligar VC (2018). Nuevas Colecciones de germoplasma de cacaos silvestres y domesticados del Perú y sus implicancias en el mejoramiento genético. In: Proceedings of the 2017 international symposium on cocoa research (ISCR), 13-17th November 2017, Lima, Peru. https://www.icco.org/about-us/icco-news/388-proceedings-of-the-international-symposium-on-cocoa-research-2017.html

Argout X, Salse J, Aury J-M, Guiltinan MJ et al (2011) The genome of *Theobroma cacao*. Nat Genet 43(2):101–109

Argout X, Martin G, Droc G, Fouet O, Labadie K, Rivals E, Aury JM, Lanaud C (2017) The cacao Criollo genome v2.0: an improved version of the genome for genetic and functional genomic studies. BMC Genomics 18(1):730–739

Asare R, Asare RA, Asante W, Markussen B, Raebild A (2017) Influences of shading and fertilization on on-farm yields of cocoa in Ghana. Exp Agric 53:416–431

Awudzi GK, Hadley P, Hatcher PE, Daymond AJ (2020) Mirid feeding preference as influenced by light and temperature mediated changes in plant nutrient concentration in cocoa. Ann Appl Biol 177:395–403

Baker NR, Hardwick K (1973) Biochemical and physiological aspects of leaf development in cocoa (*Theobroma cacao*). I. Development of chlorophyll and photosynthetic activity. New Phytol 72:1315–1324

Balasimha D, Daniel EV (1988) A screening method for drought tolerance in cocoa. Curr Sci 57(7):395

Balasimha D, Daniel EV, Bhat PG (1991) Influence of environmental factors on photosynthesis in cocoa trees. Agric For Meteorol 55:15–21

Balasimha D, Apshara SA, Jose CT (2013) Genotypic variations in chlorophyll fluorescence and stomatal conductance of cocoa in relation to drought tolerance. J Plant Crop 41:40–45

Barreto MA, Santos JCS, Corrêa RX, Luz EDMN, Marelli J, Souza AP (2015) Detection of genetic resistance to cocoa black pod disease caused by three *Phytophthora* species. Euphytica 206:677–687

Bartley BGD (2005) The genetic diversity of cacao and its diversification. CABI, Wallingford

Bekele F, Phillips-Mora W (2019) Cacao (*Theobroma cacao* L.) breeding. In: Al-Khayri J, Jain S, Johnson D (eds) Advances in plant breeding strategies: industrial and food crops, vol 6. Springer-Verlag, New York, pp 409–487

Bekele, F.L., Bidaisee, G.G., Singh, H. et al. (2020) Morphological characterisation and evaluation of cacao (*Theobroma cacao* L.) in Trinidad to facilitate utilisation of Trinitario cacao globally. Genet Resour Crop Evol 67, 621–643. https://doi.org/10.1007/s10722-019-00793-7

Bernardo R, Yu J (2007) Prospects for genomewide selection for quantitative traits in maize. Crop Sci 47(3):1082–1090

Bertolde FZ, de Almeida A-AF, Corrêa RX, Gomes FP, Gaiotto FA, Baligar VC, Loguercio LL (2010) Molecular, physiological and morphological analysis of waterlogging tolerance in clonal genotypes of *Theobroma cacao* L. Tree Physiol 30(1):56–67

Bhattacharjee R (2018) Taxonomy and classification of cacao. In: Umaharan P (ed) Achieving sustainable cultivation of cocoa. Burleigh Dodds Science Publishing Limited, Cambridge, pp 3–18

Botstein D, White RL, Skolnick M, Davis RW (1980) Construction of a genetic linkage map in man using restriction fragment length polymorphisms. Am J Hum Genet 32:314–331

Bradbury PJ, Zhang Z, Kroon DE, Cassteven TM, Ramdoss Y, Buckler ES (2007) TASSEL: software for association mapping of complex traits in diverse samples. Bioinformatics 23(19):2633–2635

Brown JS, Sautter RT, Tondo CT et al (2008) A composite linkage map from the combination of three crosses made from commercial clones of cacao, *T cacao* L. Trop Plant Biol 1(2):120–130

CacaoNet (2012) A global strategy for the conservation and use of cacao genetic resources, as the foundation for a sustainable cocoa economy (B. Laliberté, compiler). Bioversity International, Montpellier, 175 pp

Campbell C (2017) Insect pests. In: End MJ, Daymond AJ, Hadley P (eds) Technical guidelines for the safe movement of cacao germplasm (revised from the FAO/IPGRI technical guidelines no. 20) (second update, November 2017). Global Cacao Genetic Resources Network (CacaoNet). Bioversity International, Rome

Cervantes-Martinez C, Brown JS (2004) A haplotype-based method for QTL mapping of F1 populations in outbred plant species. Crop Sci 44(5):1572–1583

Cheesman EE (1944) Notes on the nomenclature, classification and possible relationships of cocoa populations. Trop Agric 21:144–159

Cope FW (1962) The mechanism of pollen incompatibility in *Theobroma cacao* L. Heredity 17:157–182

Cornejo OE, Yee M, Dominguez V et al (2018) Population genomic analyses of the chocolate tree, *Theobroma cacao* L., provide insights into its domestication process. Commun Biol 1:167

Correa CM (2006) Considerations on the standard material transfer agreement under the FAO treaty on plant genetic resources for food and agriculture. J World Intellect Property 9:137–165

Cuatrecasas J (1964) Cacao and its allies: a taxonomic revision of the genus *Theobroma*. Contrib US Herbarium 35:379–614

da Silva Branco MC, de Almeida A-AF, Dalmolin AC, Ahnert D, Baligar VC (2017) Influence of low light intensity and soil flooding on cacao physiology. Sci Hortic 217:243–257

da Silva MR, Clément D, Gramacho KP, Monteiro WR, Argout X, Lanaud C, Lopes U (2016) Genome-wide association mapping of sexual incompatibility genes in cacao (*Theobroma cacao* L.). Tree Genet Genomes 12(3):62

Daymond AJ (2018) Safe handling and movement of cocoa germplasm for breeding. In: Umaharan P (ed) Achieving sustainable cultivation of cocoa. Burleigh Dodds Science Publishing Limited, Cambridge, pp 101–110

Daymond AJ, Hadley P (2004) The effects of temperature and light integral on chlorophyll and chlorophyll fluorescence of four contrasting genotypes of cacao (*Theobroma cacao*). Ann Appl Biol 145:257–262

Daymond AJ, Hadley P (2008) Differential effects of temperature on fruit development and bean quality of contrasting genotypes of cacao (*Theobroma cacao* L.). Ann Appl Biol 153:175–185

Daymond AJ, Hadley P (2009) Screening of cocoa germplasm for high temperature tolerance. In: Proceedings 15th international cocoa research conference, San José, Costa Rica. Cocoa Producers' Alliance, Lagos, pp 87–94

Daymond AJ, Hadley P, Machado RCR, Ng E (2002) Genetic variation in partitioning to the yield component of cacao (*Theobroma cacao* L.). HortScience 37:799–801

Daymond AJ, Acheampong K, Prawoto A, Abdoellah et al (2018) Mapping COCOA PROductivity in Ghana, Indonesia and Côte d'Ivoire. In: Proceedings of the 2017 international symposium on cocoa research (ISCR), 13-17th November 2017. Lima, Peru. https://www.icco.org/about-us/icco-news/388-proceedings-of-the-international-symposium-on-cocoa-research-2017.html

de Almeida J, Tezara W, Herrera A (2016) Physiological responses to drought and experimental water deficit and waterlogging of four clones of cacao (*Theobroma cacao* L.) selected for cultivation in Venezuela. Agric Water Manage 171:80–88

de la Cruz M, Whitkus R, Gomez-Pompa A, Mota-Bravo L (1995) Origins of cacao cultivation. Nature 375:542–543

Devy L, Anita-Sari I, Saputra TI, Susilo AW, Wachjar A (2018) Identification of molecular marker based on MYB transcription factor for the selection of Indonesian Fine Cacao (*Theobroma cacao* L.). Pelita Perkebunan 34(2):59–68

Dos Santos EA, de Almeida A-AF, Ahnert D, da Conceição AS, Pirovani CP, Pires JL, Valle RR, Baligar VC (2014) Molecular, physiological and biochemical responses of *Theobroma cacao* L. genotypes to soil water deficit. PLoS One 9(12):e115746

Dunning CM, Black E, P., & Allan, R. P. (2018) Later wet seasons with more intense rainfall over Africa under future climate change. J Clim 31:9719–9738

DuVal A, Gezan SA, Mustiga G, Stack C, Marelli J-P, Chaparro J, Livingstone DIII, Royaert S, Motamayor JC (2017) Genetic parameters and the impact of off-types for *Theobroma cacao* L in a breeding program in Brazil. Front Plant Sci 8:2059

End MJ (1990) A study of the effects of the photo-thermal environment on fruit and seed growth and development in *Theobroma cacao* L. PhD thesis, The University of Reading, Reading

End MJ, Daymond AJ, Hadley P (2017) Technical guidelines for the safe movement of cacao germplasm (revised from the FAO/IPGRI technical guidelines no. 20) (second update, November 2017).Global Cacao Genetic Resources Network (CacaoNet). Bioversity International, Rome

Enríquez GA (1993) Characteristics of cacao "Nacional" of Ecuador. In: Proceedings of the international workshop on conservation, characterization and utilization of cocoa genetic resources in the 21st century, 13-17th September 1992, pp 269–278. Port-of-Spain, Trinidad & Tobago. Cocoa Research Unit, St. Augustine

Eskes AB, Paulin D, Clement D et al (1996) Selection methods applied and genetic knowledge generated in cocoa breeding in Côte d'Ivoire and Cameroon. In: End MJ, Eskes AB, Lee MT, Lockwood G (eds) Proceedings of the international workshop on cocoa breeding strategies 18–19th October 1994, Kuala Lumpur, Malaysia. INGENIC and Malaysian Cocoa Board, London, pp 41–56

Fister AS, Landherr L, Maximova SN, Guiltinan MJ (2018) Transient expression of CRISPR/Cas9 machinery targeting TcNPR3 enhances defense response in *Theobroma cacao*. Front Plant Sci 9:268

Fouet O, Allegre M, Argout X et al (2011) Structural characterization and mapping of functional EST-SSR markers in Theobroma cacao. Tree Genet Genomes 7(4):799–817

Fowler MS (2009) Cocoa beans: from tree to factory. In: Beckett ST (ed) Industrial chocolate manufacture and use, 4th edn. Blackwell Publishing, Oxford, pp 10–48

Guiltinan MJ, Maximova SN (2015) Applications of genomics to the improvement of cacao. In: Wilson PK, Hurst WJ (eds) Chocolate and health: chemistry, nutrition and therapy. The Royal Society of Chemistry, Cambridge, pp 67–81

Guiltinan MJ, Verica J, Zhang D, Figueira A (2008) Genomics of *Theobroma cacao*, 'the Food of the Gods'. In: Moore PH, Ming R (eds) Genomics of tropical crop plants. Springer, New York, pp 145–170

Gutiérrez-López N, Ovando-Medina I, Salvador-Figueroa M, Molina-Freaner F, Avendano-Arrazate CH, Vazquez-Ovando A (2016) Unique haplotypes of cacao trees as revealed by trnH-psbA chloroplast DNA. Peer J 4:e1855

Hadley P, Acheampong K, Pearson S, End MJ, Weis E (1994) The effects of environmental factors on cherelle wilt in cocoa grown in controlled environments. In: Proceedings 11th international cocoa research conference, Yamoussoukro, Côte d'Ivoire. Cocoa Producers' Alliance, Lagos, pp 661–666

Handley LR (2016) The effects of climate change on the reproductive development of *Theobroma cacao* L. PhD thesis, The University of Reading

ICCO (2019) Production of cocoa beans. www.icco.org. Accessed October 2019

Iwaro AD, Sreenivasan TN, Umaharan P (1997) Foliar resistance to *Phytophthora palmivora* as an indicator of pod resistance in Theobroma cacao. Plant Dis 81:619–624

Iwaro AD, Bharath S, Bekele FL, Butler DR (2009) Assessment of genetic gain in a germplasm enhancement programme for resistance to black pod disease. In: Proceedings of the 15th international cocoa research conference, San José, Costa Rica. Cocoa Producers' Alliance, Lagos, pp 33–39

Iwaro AD, Bekele FL, Butler DR et al (2010) Recent progress in breeding for specific traits in cocoa to meet challenges to production. In: Proceedings of the international congress on tropical agriculture: overcoming challenges to developing sustainable agri-food systems in the tropics 30th November–5th December 2008, Port of Spain. The University of the West Indies, St. Augustine, pp 43–52

Iwata H, Minamikawa MF, Kajiya-Kanegae H, Ishimori M, Hayashi T (2016) Genomics-assisted breeding in fruit trees. Breed Sci 66(1):100–115

Joshi RK, Nayak S (2010) Gene pyramiding—a broad spectrum technique for developing durable stress resistance in crops. Biotechnol Mol Biol Rev 5(3):51–60

Khan N (1998) The CRU/ACRI cocoa butterfat project. In: Report for 1997, Cocoa Research Unit. The University of the West Indies, Trinidad, pp 53–61

Khan N, Motilal LA, Sukha DA, Bekele FL, Iwaro AD, Bidaisee GG, Umaharan P, Grierson LH, Zhang D (2008) Variability of butterfat content in cacao (*Theobroma cacao* L.): combination and correlation with other seed-derived traits at the International Cocoa Genebank, Trinidad. Plant Genet Resour Char Utiliz 6:175–186

Lahive F, Hadley P, Daymond AJ (2019) The physiological responses of cacao to the environment and the implications for climate change resilience. A review. Agron Sustain Dev 39:5

Lanaud C, Fouet O, Legavre T et al (2017) Deciphering the *Theobroma cacao* self-incompatibility system: from genomics to diagnostic markers for self-compatibility. J Exp Bot 68(17):4775–4790

Lewis C, Lennon AM, Umaharan P (2018) Genetic variation in bioaccumulation and partitioning of cadmium in Theobroma cacao L. Sci Total Environ 640–641:696–703

Lockwood G, End MJ (1993) History, technique and future needs for cacao collection. In: Proceedings of international workshop on conservation and utilization of cocoa genetic resources in the 21st century, 13-19th September, 1992. Cocoa Research Unit, The University of the West Indies, Port of Spain, pp 1–14

Lockwood G, Pang JTY (1996) Yields of cocoa clones in response to planting density in Malaysia. Exp Agric 32:41–47

Lockwood G, Owusu-Ansah F, Adu-Ampomah Y (2007) Heritability of single plant yield and incidence of black pod disease in cocoa. Exp Agric 43(4):455–462

Loor-Solórzano RG, Lachenaud P, Fouet O, Argout X, Pena G, Macias JC, Amores-Puyutaxi FM, Valdez F, Hurtado J, Lanaud C (2015) Rescue of cacao genetic resources related to the Nacional variety: surveys in the Ecuadorian Amazon (2010-2013). Espamcienca 6(E):7–15

Mahabir A, Motilal LA, Gopaulchan D, Ramkissoon S, Sankar A, Umaharan P (2019) Development of a core SNP panel for cacao (*Theobroma cacao* L.) identity analysis. Genome 63:103–114

Maharaj K, Maharaj P, Bekele FL, Ramnath D, Bidaisee GG, Bekele I, Jennings K, Sankar R (2011) Trinidad selected hybrids: an investigation of the phenotypic and agro-economic traits of 20 selected cacao cultivars. Trop Agric 88(4):175–185

Marcano M, Pugh T, Cros E et al (2007) Adding value to cocoa (*Theobroma cacao* L.) germplasm information with domestication history and admixture mapping. Theor Appl Genet 114(5):877–884

Marcano M, Morales S, Hoyer MT et al (2009) A genomewide admixture mapping study for yield factors and morphological traits in a cultivated cocoa (*Theobroma cacao* L.) population. Tree Genet Genomes 5(2):329–337

McElroy MS, Navarro AJ, Mustiga G et al (2018) Prediction of cacao (*Theobroma cacao*) resistance to *Moniliophthora* spp. diseases via genome-wide association analysis and genomic selection. Front Plant Sci 9:A343

McKelvie AD (1956) Cherelle wilt of cacao I. Pod development and its relation to wilt. J Exp Bot 7:252–263

McMahon P, Purwantara A (2016) Vascular streak dieback (*Ceratobasidium Theobroma*): history and biology. In: Bailey BA, Meinhardt LW (eds) Cacao diseases. A history of old enemies and new encounters. Springer International Publishing, Basel, pp 307–335

Monteiro WR, Lopes UV, Clement D (2009) Genetic improvement in cocoa. In: Jain SM, Priyadarshan PM (eds) Breeding plantation tree crops: tropical species. Springer, New York, pp 589–626

Mooleedhar V, Maharaj W, O'Brien H (1995) The collection of Criollo cocoa germplasm in Belize. Cocoa Growers' Bull 49:26–40

Motamayor JC, Risterucci AM, Lopez PA, Ortiz CF, Moreno A, Lanaud C (2002) Cacao domestication I: the origin of the cacao cultivated by the Mayas. Heredity 89:380–386

Motamayor JC, Lachenaud P, da Silva e Mota JW, Loor R, Kuhn DN, Brown S, Schnell RJ (2008) Geographic and genetic population differentiation of the Amazonian chocolate tree (*Theobroma cacao* L). PLoS One 3(10):e3311

Motamayor JC, Mockaitis K, Schmutz J, Haiminen N, Livingstone D III, Cornejo O, Findley SD, Zheng P, Utro F, Royaert S, Saski C (2013) The genome sequence of the most widely cultivated cacao type and its use to identify candidate genes regulating pod color. Genome Biol 14(6):r53

Motilal LA (2018) The role of genebanks in preserving the genetic diversity of cacao. In: Umaharan P (ed) Achieving sustainable cultivation of cocoa. Burleigh Dodds Science Publishing Limited, Cambridge, pp 47–99

Motilal LA, Zhang D, Mischke S, Meinhardt LW, Boccara M, Fouet O, Lanaud C, Umaharan P (2016) Association mapping of seed and disease resistance traits in *Theobroma cacao* L. Planta 244(6):1265–1276

N'Guessan KF, N'Goran JAK, Eskes AB (2008) Resistance of cacao (*Theobroma cacao* L.) to *Sahlbergella singularis* (Hemiptera: Miridae): investigation of antixenosis, antibiosis and tolerance. Int J Trop Insect Sci 28:201–210

Nichols R (1964) Studies of fruit development of cacao (*Theobroma cacao*) in relation to cherelle wilt. I Development of the pericarp. Ann Bot 28:619–635

Nyassé S, Cilas C, Herail C, Blaha G (1995) Leaf inoculation as an early screening test cocoa (Theobroma cocoa L.) resistance to Phytophthora black pod disease. Crop Prot 14:657–663

Ofori A, Padi FK, Acheampong K, Lowor S (2015) Genetic variation and relationships of traits related to drought tolerance in cocoa (*Theobroma cacao* L.) under shade and no shade condition in Ghana. Euphytica 201:411–421

Osorio-Guarín JA, Berdugo-Cely JA, Coronado-Silva RA, Baez E, Jaimes Y, Yockteng R (2020) Genome-wide association study reveals novel candidate genes associated with productivity and disease resistance to *Moniliophthora* spp. in cacao (*Theobroma cacao* L.). G3 Genes Genomes Genet 10(5):1713–1725

Phillips-Mora W, Arciniegas-Leal A, Mata-Quiros A, Motamayor JC (2013) Catalogue of cacao clones selected by CATIE for commercial plantings. Technical series. Technical manual 105. CATIE, Costa Rica. 68 p

Phillips-Mora W, Mata-Quiros A, Arciniegas-Leal A (2018) Generation of cacao clones with durable resistant against frosty pod rot. In: Proceedings of the 2017 international symposium on cocoa research (ISCR), 13-17th November 2017, Lima, Peru. https://www.icco.org/about-us/icco-news/388-proceedings-of-the-international-symposium-on-cocoa-research-2017.html

Pires JL, Cascardo JCM, Lambert SV, Figueira A (1998) Increasing cocoa butter yield through genetic improvement of Theobroma cacao L.: seed fat content variability, inheritance, and association with seed yield. Euphytica 103:115–121

Pires JL, Monteiro WR, Luz ED et al (1999) Cocoa breeding for witches' broom resistance at CEPEC, Bahia, Brazil. In: Proceedings of the international workshop on the contribution of disease resistance to cocoa variety improvement, Salvador, 24–26th November 1996. INGENIC, London, pp 91–101

Pokou ND, N'Goran JAK, Kébé I, Eskes A, Tahi M, Sangaré A (2008) Levels of resistance to phytophthora pod rot in cocoa accessions selected on farm in Côte d'Ivoire. Crop Prot 27:302–309

Pound FJ (1932) The genetic constitution of the cacao crop. In: First annual report cacao research, Trinidad, 1932, pp 27–29

Pound FJ (1938) Cacao and Witches' Broom disease (*Marasmius perniciosus*) of South America with notes on other species of *Theobroma*. Yuille's Printery, Port-of-Spain

Pound FJ (1945) A note on the cocoa population of South America. In: Report and proceedings of the cocoa research conference held at The Colonial Office, London, May-June 1945, Colonial No. 192, His Majesty's Stationery Office, London (pp 95–97). Reprinted in 1982. In: Archives of cocoa research, vol 1, pp 93–97

Ramtahal G, Chang Yen I, Ahmad N, Bekele I, Bekele F, Maharaj K, Wilson L, Harrynanan L (2015) Prediction of soil cadmium bioavailability to cacao (*Theobroma cacao* L.) using single-step extraction procedures. Commun Soil Sci Plant Anal 46(20):2585–2594

Ribeyre F, Sounigo O, Argout X, Cilas C, Efombagn MI, Denis M, Bouvet JM, Fouet O, Lanaud C (2017) The genomic selection of *Theobroma cacao* L: a new strategy of marker assisted selection to improve breeding efficiency and predict useful traits in new populations. In: Proceedings of the 2017 international symposium on cocoa research (ISCR), 13-17th November 2017, Lima, Peru. https://www.icco.org/about-us/icco-news/388-proceedings-of-the-international-symposium-on-cocoa-research-2017.html

Rodriguez-Medina C, Arana AC, Sounigo O, Argout X, Alvarado GA, Yockteng R (2019) Cacao breeding in Colombia, past, present and future. Breed Sci 69(3):373–382

Romero Navarro JA, Phillips-Mora W, Arciniegas-Leal A et al (2017) Application of genome wide association and genomic prediction for improvement of cacao productivity and resistance to black and frosty pod diseases. Front Plant Sci 8:1905

Royaert S, Phillips-Mora W, Leal AM, Cariaga K, Brown JS, Kuhn DN, Schnell RJ, Motamayor JC (2011) Identification of marker-trait associations for self-compatibility in a segregating mapping population of Theobroma cacao L. Tree Genet Genomes 7(6):1159–1168

Royaert S, Jansen J, da Silva DV, de Jesus Branco SM, Livingstone DS, Mustiga G, Marelli JP, Araújo IS, Corrêa RX, Motamayor JC (2016) Identification of candidate genes involved in Witches' broom disease resistance in a segregating mapping population of *Theobroma cacao* L in Brazil. BMC Genomics 17:107

Saski CA, Feltus FA, Staton ME, Blackmon BP, Ficklin SP, Kuhn DN, Schnell RJ, Shapiro H, Motamayor JC (2011) A genetically anchored physical framework for *Theobroma cacao* cv. Matina 1-6. BMC Genomics 12(1):413

Sena Gomes AR, Kozlowski T (1986) The effects of flooding on water relations and growth of Theobroma cacao var. catongo seedlings. J Horticult Sci 61:265–276

Servin B, Martin OC, Mézard M (2004) Toward a theory of marker-assisted gene pyramiding. Genetics 168(1):513–523

Simmonds NW (1993) The breeding of perennial crops. In: International workshop on conservation, characterisation and utilisation of cocoa genetic resources in the 21st century, Port-of-Spain (Trinidad and Tobago) 13-17th Sept 1992. Cocoa Research Centre, Port-of-Spain

Sounigo O, Coulibaly N, Bruna L, N'Goran J, Cilas C, Eskes AB (2003) Evaluation of resistance of *Theobroma cacao* L. to mirids in Côte d'Ivoire: results of comparative progeny trials. Crop Prot 22:615–621

Sounigo O, Efombagn B, Lemainque A et al (2012) Association mapping on cocoa: a way to identify functional SSR markers linked to yield, tolerance to black pod and mirids assessed in Cameroon and develop a marker assisted breeding programme. In: Proceedings of the 16th international cocoa research conference, 16–21st November 2009, Bali. Cocoa Producers' Alliance, Lagos, pp 153–158

Stack JC, Royaert S, Gutiérrez O, Nagai C, Holanda ISA, Schnell R, Motamayor JC (2015) Assessing microsatellite linkage disequilibrium in wild, cultivated, and mapping populations of Theobroma cacao L. and its impact on association mapping. Tree Genet Genomes 11(2):19

Stuber CW (1994) Heterosis in plant breeding. Plant Breed Rev 12:227–251

Tahi M, Kébé I, Eskes AB, Ouattara S, Sangare A, Mondeil F (2000) Rapid screening of cacao genotypes for field resistance to *Phytophthora palmivora* using leaves, twigs and roots. Eur J Plant Pathol 106:87–94

Toxopeus H (2001a) Botany, types and populations. In: Wood GAR, Lass RA (eds) Cocoa. Blackwell Science, Oxford, pp 11–37

Toxopeus H (2001b) Planting material. In: Wood GAR, Lass RA (eds) Cocoa. Blackwell Science, Oxford, pp 80–92

Toxopeus H, Wessel M (1970) Studies in pod and bean values in Theobroma cacao in Nigeria. I Environmental effects on West African Amelonado with particular attention to annual rainfall. Netherlands J Agric Sci 18:132–139

Turnbull CJ, Hadley P (2019) International Cocoa Germplasm Database (ICGD). [Online Database]. CRA Ltd./ICE Futures Europe/University of Reading, Reading. http://www.icgd.reading.ac.uk

Turnbull CJ, Daymond AJ, Gutierrez O, Hadley P, Livingstone D, Motamayor JC, Phillips W, Umaharan P, Zhang D (2018) Adopting reference genotypes to identify off-types in cacao collections. In: Proceedings of the 2017 international symposium on cocoa research (ISCR), 13-17th November 2017, Lima, Peru. https://www.icco.org/about-us/icco-news/388-proceedings-of-the-international-symposium-on-cocoa-research-2017.html

Utro F, Haiminen N, Livingstone D, Cornejo OE, Royaert S, Schnell RJ, Motamayor JC, Kuhn DN, Parida L (2013) iXora: exact haplotype inferencing and trait association. BMC Genet 14(1):48

Varshney RK, Terauchi R, McCouch SR (2014) Harvesting the promising fruits of genomics: applying genome sequencing technologies to crop breeding. PLoS Biol 12(6):e1001883

Wilkinson MJ (2003) The application and constraints of new technologies in plant breeding. In: Proceedings of the international workshop on new technologies in cocoa breeding, Kota Kinabalu, 16–17 October 2000. INGENIC, London, pp 12–24

Wood GAR (2001a) History and development. In: Wood GAR, Lass RA (eds) Cocoa. Blackwell Science, Oxford, pp 1–10

Wood GAR (2001b) Quality and inspection. In: Wood GAR, Lass RA (eds) Cocoa. Blackwell Science, Oxford, pp 505–527

Zarrillo S, Gaikwad N, Lanaud C et al (2018) The use and domestication of *Theobroma cacao* during the mid-Holocene in the upper Amazon. Nat Ecol Evol 2:1879–1888

Zhang D, Motilal L (2016) Origin, dispersal, and current global distribution of cacao genetic diversity. In: Bailey BA, Meinhardt LW (eds) Cacao diseases. A history of old enemies and new encounters. Springer International Publishing, Basel, pp 3–32

# Coffee: Genetic Diversity, Erosion, Conservation, and Utilization

**Sarada Krishnan**

**Abstract** The predominantly cultivated coffee, *Coffea arabica,* is characterized by low genetic diversity making it highly vulnerable to biotic and abiotic stresses. The key to sustaining the coffee crop and the industry that depends on the crop will depend on the genetic diversity of the crop. In addition to the two cultivated species, *C. arabica* and *C. canephora*, there are over 100 additional species with geographic distribution in Africa, Madagascar, Indian Ocean Islands, and Australasia. Almost 60% of *Coffea* species are in some form of threat in their natural habitats, and genetic erosion is a challenge in ex situ collections. The conservation and utilization of the full extent of these coffee genetic resources will lead to the development of improved varieties that can cope with biotic and abiotic stresses that will be heightened by climate change. The development of the global conservation strategy for coffee genetic resources provides a roadmap for the long-term conservation of coffee genetic resources. The utilization of advances made in molecular technology will be critical in developing resilient varieties that will be crucial in sustaining the future of the crop and the industry that depends on the crop. This chapter provides an overview of the origin and spread of coffee, the economic importance of the crop, advances in genetic diversity studies, conservation of genetic resources both in situ and ex situ, some of the major pests and diseases, coffee quality attributes, breeding approaches, and threats such as climate change. The sustainable future of the crop depends not just on the producing countries but also on the consuming countries.

**Keywords** Coffee genetic resources · In situ conservation · Ex situ conservation · Climate change · Coffee quality · Crop improvement

---

S. Krishnan (✉)
Denver Botanic Gardens, Denver, CO, USA



P. M. Priyadarshan, S. M. Jain (eds.), *Cash Crops*,
https://doi.org/10.1007/978-3-030-74926-2_3

# 1 Introduction: Origin and Spread of Coffee

The genus *Coffea* belonging to the *Rubiaceae* family consists of 124 species distributed in tropical Africa, Madagascar, the Comoros and the Mascarenes (Mauritius and Reunion), tropical Asia, and Australia (Davis et al. 2006, 2011). Of the 124 species, only two are economically important for the production of the beverage coffee, *C. arabica* L. (Arabica coffee) and *C. canephora* A. Froehner (robusta coffee). Higher beverage quality is associated with *C. arabica* which accounted for 58.8% of total production in 2018, whereas *C. canephora* accounted for 41.2% of total production (ICO 2020). In recent years, *C. liberica* Hiern is gaining some attention in the specialty market with special cupping events in Borneo organized by coffee professionals from Malaysia and Germany (Christensen 2019). This chapter deals with the genetic diversity and production of Arabica coffee with some references to robusta coffee and other crop wild relatives.

The highlands of southwestern Ethiopia and the Boma plateau of South Sudan are the primary centers of origin of *C. arabica* (Meyer 1965; Thomas 1942) (Fig. 1). *Coffea arabica* is an allotetraploid ($2n = 4x = 44$) that originated from two different diploid ($2n = 2x = 22$) wild ancestors, *C. eugenoides* S. Moore as the female parent and *C. canephora* or ecotypes related to these two species (Lashermes et al. 1999). This is the only tetraploid species in this genus, all others being diploid. Due to the

**Fig. 1** *Coffea arabica* 'Typica' in cultivation in Jamaica. (Photo credit: Sarada Krishnan)

nature of its origin, reproductive biology and evolution, and due to the narrow genepool of the founding population from which they spread around the world, Arabica coffee has very low genetic diversity (Anthony et al. 2002; Lashermes et al. 1999; Vega et al. 2008). It is self-compatible and mostly reproduces by self-fertilization. Meyer (1965) reported self-fertilization occurring in 83–95% of population of *C. arabica* grown in Brazil, and Fazuoli et al. (2000) reported self-fertility occurring in about 90% of the flowers. Meyer (1965) reported that self-fertilization occurs only in about 40–60% of the wild coffee plants grown at the Jimma Agricultural Research Center in Ethiopia alluding to the heterozygous nature of the plants expected in its center of origin.

The total global Arabica coffee production in 2019–2020 was 5.63 million tons. The 2020–2021 production is predicted to increase to 6.11 million tons (USDA FAS 2020). Brazil is the highest producer of Arabica coffee followed by Colombia, Ethiopia, and Honduras. The top 10 producers account for about 88% of the total global coffee production and about 86% of the total exports. Among the top 10 producers, Brazil, Vietnam, and Colombia together produce and export about 60% of the total world production and export (Table 1).

*C. canephora* (Fig. 2) has a much wider distribution, spreading from West to East Africa in Ghana, Guinea, Guinea Bissau, Cote d'Ivoire, Liberia, Nigeria, Cabinda, Cameroon, Congo, Central African Republic, Democratic Republic of Congo, Gabon, Sudan, South Sudan, Tanzania, and Uganda to the south to Angola (Davis et al. 2006). Vietnam is the highest producer of robusta coffee with a production of 1.81 million tons during the 2019–2020 season followed by Brazil and Indonesia with a production of 1.1 million tons and 567,000 tons, respectively (USDA FAS 2020).

**Table 1** World's top ten coffee producers—production, export, and proportions of world production and export during 2018–2019 (ICO 2020)

| Country | Production 000 60 kg bags | % of world production | Exports 000 60 kg bags | % of world exports |
|---|---|---|---|---|
| 1. Brazil | 62,925 | 36.81 | 37,614 | 29.15 |
| 2. Vietnam | 31,174 | 18.24 | 26,474 | 20.52 |
| 3. Colombia | 13,858 | 8.11 | 13,502 | 10.46 |
| 4. Indonesia | 9419 | 5.51 | 4718 | 3.66 |
| 5. Ethiopia | 7776 | 4.55 | 3801 | 2.95 |
| 6. Honduras | 7328 | 4.29 | 6812 | 5.28 |
| 7. India | 5302 | 3.10 | 6022 | 4.67 |
| 8. Uganda | 4704 | 2.75 | 4454 | 3.45 |
| 9. Peru | 4263 | 2.50 | 4001 | 3.10 |
| 10. Guatemala | 4007 | 2.34 | 3607 | 2.80 |
| Total world production and export | 170,937 | 88.20 | 129,028 | 86.04 |

**Fig. 2** *Coffea canephora* in cultivation in India. (Photo credit: Sarada Krishnan)

The cultivated phase of *C. arabica* was restricted to Yemen for a long time before it spread to other parts of the world (Meyer 1965). Seeds from the South Western Ethiopian forests were taken to Yemen in the fourteenth century where it was cultivated for market demands in Mocha and Cairo in the late fifteenth century (Scalabrin et al. 2020). From Mocha in Yemen, two significant dispersions are documented, the first one leading to the development of the "Typica" line and the second one to the "Bourbon" line. In 1670, seeds were smuggled to India by Baba Budan, from which the Dutch East India Company took seeds to Java in 1696 and 1699 (Scalabrin et al. 2020). From Java, plants were taken the Amsterdam Botanical Garden in 1706. From this, a plant was taken to France in 1713, which was used by Antoine de Jussieu in first describing coffee (Krishnan 2017). He named the plant as *Jasminum arabicanum,* which was later classified by Linnaeus under the genus *Coffea* in 1737 and as *Coffea arabica* in 1953 (Charrier and Berthaud 1985). In 1720, one plant was taken to the French colony of Martinique in the Caribbean from France. From Martinique, coffee spread throughout the Caribbean islands: Haiti (1725), Guadeloupe (1726), Jamaica (1730), Cuba (1748), and Puerto Rico (1755). The Dutch around this same time introduced plants from Amsterdam to their South American colony in Suriname in 1718, from where coffee was introduced to French Guiana in 1719 and Brazil in 1727. This was the basis of the "Typica" genetic line of coffee. The second dispersion event that led to the "Bourbon" genetic line originated from coffee trees introduced from Mocha in Yemen to Bourbon (Reunion) Islands in 1715 and 1718 (Anthony et al. 2002; Vega et al. 2008), which reached the Americas and East Africa in the mid-nineteenth century (Scalabrin et al. 2020). Coffee cultivation was introduced in Ceylon (Sri Lanka) in 1740 by the French. Ceylon become a major producer of coffee until 1869 when Ceylon's thriving coffee industry was devastated by the fugal disease, the coffee leaf rust (*Hemileia vastatrix* Berkeley and Broome), leading to the replacement of coffee by tea by the 1900s (Damania 2003).

## 2 Genetic Diversity

Cultivated Arabica coffee is characterized by low genetic diversity whereas higher diversity exists in the wild germplasm of the Ethiopian forests. When *C. arabica* cultivated in Yemen was compared with Ethiopian germplasm collected in the southwest highlands of Illubabor and Kaffa provinces in Ethiopia, a clear genetic separation was observed, supporting the hypothesis of varied Ethiopian origins of the plants that were transferred to and selected in Yemen for cultivation (Charrier et al. 2009).

Using 736 *C. arabica* accessions, Scalabrin et al. (2020) analyzed the genetic diversity in the species and its relationship with historical records and geographic distribution using a genome-wide SNP assay. The study concluded that the severity of the bottleneck effect and the unsubstantiated carryover of ancestral diversity by the most likely parental *Coffea* species into *C. arabica* population suggest *a single hybridization event* that led to the origin of the tetraploid *C. arabica* about 10,000 or 20,000 years ago. The available data and in silico simulations supported the recent origin of *C. arabica* accessions from a single individual after the polyploidization event (Scalabrin et al. 2020). The distribution of private alleles among the three species, *C. arabica*, *C. canephora,* and *C. eugenioides*, showed that the majority of the SNPs in *C. arabica* was not shared with either of the parental species confirming that the present-day variation in *C. arabica* arose after the polyploidization event. Scalabrin et al. (2020) also argued that the presence of genetic differentiation in wild *C. arabica* in the Southwestern most range of distribution in the rainforests of Ethiopia and that the cultivated varieties used worldwide were genetically similar to the germplasm used in Eastern Ethiopian and Yemeni plantations from which the Typica and Bourbon lineages arose. This concurs with the historical accounts of the global spread of cultivated coffee out of Yemen.

Several genetic diversity studies have been conducted in coffee. A sample representation of some of the research areas and authors is presented in Table 2.

One of the challenges in coffee-producing countries is the lack of a national system for distributing improved varieties to farmers with the exception of Colombia and Honduras for *C. arabica* and Vietnam for *C. canephora.* Since the 1990s, the Centro Agronomico Tropical de Investigacion Ensenanza (CATIE) genebank in Costa Rica has developed hybrids using Latin American varieties and wild genotypes from their collections. Five of the most productive hybrids are multiplied by tissue culture utilizing somatic embryogenesis for distribution to local farmers (Bramel et al. 2017) (Fig. 3). Many regions still grow varieties selected in the 1950s such as Mundo Novo, Caturra, or Catuai. This has led to a weak coffee seed sector that is poorly organized in many coffee-growing countries, mostly functioning through informal seed exchange. In the context of climate change, disease

**Table 2** Listing of genetic diversity studies in coffee

| Research topic | Authors |
|---|---|
| Identification of molecular markers and characterizing genetic diversity | Anagbogu et al. (2019)<br>Anthony et al. (2001)<br>Baruah et al. (2003)<br>Combes et al. (2000)<br>Cubry et al. (2008)<br>Lashermes et al. (1996)<br>Maluf et al. (2005)<br>Mishra et al. (2014)<br>Missio et al. (2009, 2010, 2011)<br>Moncada and McCouch (2004)<br>Musloi et al. (2009)<br>Poncet et al. (2004), Poncey et al. (2007)<br>Rovelli et al. (2000)<br>da Silva et al. (2019)<br>Silvestrini et al. (2007, 2008)<br>Teressa et al. (2010)<br>Tesfaye et al. (2007, 2014)<br>Tessema et al. (2011)<br>Yan et al. (2019) |
| Genetic diversity studies in wild species of coffee | Andrianasolo et al. (2013)<br>Krishnan et al. (2013a, b)<br>Razafinarivo et al. (2012, 2013) |
| Phylogenetic relationships among *Coffea* species | Cros et al. (1998)<br>Davis et al. (2007)<br>Lashermes et al. (1997)<br>Maurin et al. (2007) |
| Genetic inheritance and introgression of traits | Herrera et al. (2002, 2004, 2009)<br>Lashermes et al. (2000a)<br>Leroy et al. (2011)<br>Pearl et al. (2004)<br>Prakash et al. (2002, 2004)<br>Priyono and Sumirat (2012)<br>Van der Vossen and Walyaro (2009) |
| Marker-assisted selection | Alkimim et al. (2017)<br>Krishnan (2014)<br>Lashermes et al. (2000b)<br>Prakash et al. (2011) |
| Utilizing gene editing technology | Breitler et al. (2018) |

pressures, and growing demand for high-quality coffee, interest in improved coffee varieties has increased in recent years. In order to ensure that the farmers are getting the true-to-type proclaimed variety and the benefits of improved varieties, tools to monitor the genetic conformity of the material are needed. Vieira et al. (2010) used 127 SSR markers to analyze genetic similarity in 25 *C. arabica* varieties composed of 19 commercially important Brazilian varieties and 6 interspecific hybrids of *C. arabica, C. canephora,* and *C. liberica.* The DNA fingerprint data generated

**Fig. 3** Clonal propagation of coffee by tissue culture for distribution to farmers at the Centro Agronomico Tropical de Investigacion y Ensenanza (CATIE) genebank in Costa Rica. (Photo credit: Sarada Krishnan)

from this study could serve as a reference database for identification of varieties, though this was restricted to Brazilian varieties. Using 8 SSR markers representing a total of 95 alleles, Pruvot-Woehl et al. (2020) have developed a robust DNA fingerprinting method to authenticate *C. arabica* varieties utilizing 2533 Arabica samples representing all possible Arabica genetic background from germplasm collections and those cultivated around the world. These eight SSR markers were powerful in discriminating *C. arabica* varieties. The study also revealed that several generations of uncontrolled pollination have resulted in a significant portion of the cultivated Arabica plants in the field to be admixed as in the case of Catimors and Sarchimors in Central America. Due to the lack of plot isolation for seed harvesting, cross pollination has led to genetic drift of an originally pure variety, the result of which has been the cultivation of varieties with mixed genetics in farmers' fields.

## 3 In Situ Conservation

Genetic diversity is essential for the continued evolution of a species to confer adaptations to the changing environment. When population sizes get small, this leads to loss of genetic diversity termed as genetic erosion, which will restrict gene flow that is needed to maintain genetic diversity. To prevent genetic erosion, conserving crops and their wild relatives in situ (in their natural habitats) allows for continued evolution, conferring adaptive traits. This allows gene flow, thereby maintaining genetic diversity while preventing genetic drift (Bramel and Krishnan 2019). In situ conservation has its own inherent threats caused by human population pressures such as deforestation and conversion of land to agriculture (Krishnan 2013). Climate change leading to loss of suitable habitat is also gaining much attention (Davis et al. 2012, 2019; Moat et al. 2019). In addition, access of germplasm by researchers and breeders becomes challenging due to remote locations of in situ sites.

Even though *C. arabica* has low genetic diversity, the natural forests in Southwestern Ethiopia reveal genetic differentiation representing a yet-to-be tapped reservoir of novel genetic variation that could be utilized for crop improvement (Scalabrin et al. 2020). To conserve the entire genepool of coffee genetic resources, in situ conservation is critical, though it is highly neglected (Krishnan 2018). Deforestation caused by human population pressures is one of the key factors causing erosion of coffee genetic resources in Ethiopia (Krishnan 2013; Labouisse et al. 2008).

In 2010, as part of the Man and Biosphere (MAB) reserve program of the United Nation's Educational, Scientific and Cultural Organization (UNESCO), the Yayu Biosphere Reserve, and the Kafa Biosphere Reserve (Fig. 4) were formed in the last remaining coffee forests of Ethiopia (Bramel et al. 2017). Occupying about 167,021 ha, the biosphere is partitioned into three zones: (1) core zone which is strictly protected without any human interaction except for research and monitoring, (2) buffer zone which is slightly disturbed forest where collection of non-timber forest products, honey, and harvesting of wild coffee is allowed without manipulating the forest canopy and ground vegetation, and (3) transition zone where coffee is grown in the human settlements using appropriate agricultural practices and local landraces (Bramel et al. 2017; OEFCCA and OFWE 2018). The main aim of the Man and Biosphere project is to promote interdisciplinary research, training, and

**Fig. 4** *Coffea arabica* growing in situ in the natural forests of Ethiopia in the Kafa Biosphere Reserve. (Photo credit: Sarada Krishnan)

communications in the field of ecosystem conservation and the rational use of natural resources (OEFCCA and OFWE 2018).

In 2017, the Oromia Environment, Forest and Climate Change Authority (OEFCCA) and the Oromia Forest and Wildlife Enterprise (OFWE) of Ethiopia engaged in a consultancy with numerous stakeholders to develop the Yayu Coffee Forest Biosphere Reserve (YCFBR) Management Plan. The purpose of the plan is to conserve the biodiversity and the natural resources of the YCFBR while ensuring the sustainable and equitable sharing of benefits and responsibilities through four specific objectives (OEFCCA and OFWE 2018):

- Promote conservation and protection of YCFBR
- Enhance sustainable economic development
- Establish and strengthen public–private partnerships
- Develop communication and public relationship strategies

By applying the IUCN Red List of Threatened Species criteria, Davis et al. (2019) conducted a gap analysis to identify the conservation status, both ex situ and in situ, of all 124 species of *Coffea*. Their results showed that about 72% of the species occur in at least one protected area, and 28% are not known to occur in any protected area. Of the 124 species, 60% (75 species) were assessed as threatened with extinction with 13 species falling under the category of critically endangered (CR), 40 species as endangered (EN), and 22 species as vulnerable (VU). Much work needs to be done with regard to in situ conservation, and all schemes should engage multiple stakeholders taking into consideration the socioeconomic conditions of the local communities, policies, and political will of the governing authorities, and scientific understanding of the biology of the species and optimum population size.

## 4 Ex Situ Collections

Coffee seeds were first classified as recalcitrant and then later designated as having intermediate storage behavior since they can survive desiccation of water content less than 0.20 g $H_2O$/g dry weight. Despite this, they cannot be put into long-term storage since they do not survive the combination of low temperatures and complete desiccation. Additionally, the different species of *Coffea* exhibit different levels of sensitivity to desiccation (Eira et al. 2006). Considering that *Coffea* spp. cannot be stored as seeds, the best alternative for conserving coffee germplasm ex situ is in field genebanks (Fig. 5). Though these ex situ collections are easily accessible to users such as breeders, there are many challenges associated with field genebanks (Bramel et al. 2017; Krishnan 2013). They are typically grown in ecological conditions unsuitable for their growth and are exposed to infestation by pests and diseases leading to genetic erosion due to strong selection pressure (Anthony et al. 2007a, b; Dulloo et al. 1998; Vega et al. 2008). Another challenge associated with field

**Fig. 5** Kenya Coffee Research Institute's *ex situ* field genebank. (Photo credit: Sarada Krishnan)

collections is the risk of transfer of pests and diseases through germplasm exchange (Dulloo et al. 2009).

Awareness of the threats to coffee genetic resources due to lack of genetic variability in cultivated coffee and increasing deforestation of forests in which coffee occurs naturally increased during the second half of the twentieth century (Anthony et al. 2007a). This led to the undertaking of extensive germplasm collecting expeditions by the Food and Agriculture Organization (FAO) of the United Nations, various French organizations such as Office de La Recherche Outre-Mer (ORSTOM; now known as Institute de Recherche pour le Developpement—IRD), Centre de Cooperation Internationale en Recherche Agronomique pour le Developpement (CIRAD) and the Museum of Natural History, Paris, and International Plant Genetic Resources Institute (IPGRI; now known as Bioversity International) in the 1960s, 1970s, and 1980s (Charrier and Berthaud 1985). During these expeditions, the emphasis was on collecting *C. arabica* germplasm, but a number of wild relatives were also collected, especially from Madagascar. Germplasms from these collecting missions were distributed to several coffee genebanks throughout the world—Ethiopia, Brazil, Cameroon, Central African Republic, Congo, Costa Rica, Cote d'Ivoire, Guinea, India, Kenya, Madagascar, Peru, Tanzania and Yemen (Bramel et al. 2017).

Inventories of coffee collections in ex situ genebanks have been done periodically. The most current one was done in 2016 during the development of the Global Conservation Strategy for Coffee Genetic Resources (Bramel et al. 2017) by the Crop Trust and World Coffee Research. To get a better understanding of current holdings of coffee collections in ex situ genebanks, a survey was developed to document what is conserved, how it is conserved, how the collections are evaluated, characterized, and utilized, and the security of the collections (Krishnan et al. 2018).

Based on literature available on previous inventories and that listed in FAO WIEWS (2009–2011) database, surveys were sent to 32 genebanks around the world. Of these, 16 genebanks responded to the survey and collectively these genebanks hold 21,026 *Coffea* accessions, a majority of them holding *C. arabica* accessions (57.2%) followed by *C. canephora* at 3.1% (Bramel et al. 2017). In addition to the survey, site visits to seven genebanks were conducted for the development of the strategy. A costing study of the collections held at Centro Agronomico Tropical de Investigacion y Ensenanza (CATIE) in Costa Rica was also undertaken. This collection was selected because this is the only genebank holding coffee that has signed the Article 15 agreement with the Governing Body of the International Treaty on Plant Genetic Resources for Food and Agriculture (ITPGRFA), making the germplasm available for distribution globally for the diversification and improvement of coffee (Bramel et al. 2017).

The Global Conservation Strategy for Coffee Genetic Resources concluded, based on the survey and site visits, that the current coffee conservation system is not sustainable, secure, cost effective, or rational. In order to make the system effective, there needs to be a global system to secure unique accessions both ex situ and in situ for use by future generations. To achieve this, the basic principles that should be adhered to are:

- Sustainable management, conservation, and international distribution by key collection holders of the genetic diversity of the genepool for the long term.
- Management of all global collections at international genebank standards through development of a quality management system, safety duplication, standard operating procedures, and comprehensive risk management strategy.
- Minimize redundancies in collections.
- Facilitation of germplasm exchange within the framework of an Access and Benefit Sharing (ABS) system.
- Development of a comprehensive, integrated accession-level information system to allow users to search for, select, and order germplasm from the collections.
- Linking ex situ conservation with in situ and on-farm conservation efforts.
- Promotion to enhance conservation and use of collections through global collaborations on breeding, research, and conservation efforts.

- Other complementary conservation strategies to support conservation in field genebanks include cryopreservation and in vitro slow growth techniques. Since 1997, the Institut de Recherche pour le Developpement (IRD) in Montpellier, France, has been making considerable efforts to develop coffee cryopreservation protocols. Progress has been made in understanding the mechanisms of coffee seed sensitivity to desiccation and exposure to liquid nitrogen as well as developing rewarming and rehydration protocols to achieve 100% survival of frozen seeds (Dussert et al. 2007). Using explants such as shoot apex, orthotropic nodes, and zygotic embryos, slow growth in vitro techniques has been achieved for *C. arabica* at 20 °C and for *C. canephora* at 23 °C to facilitate medium-term conservation (Santana-Buzzy et al. 2007).

# 5 Major Pests and Diseases

## *Coffee Berry Borer:* Hypothenemus hampei *(Ferrari)*

The most serious pest of coffee in many of the major coffee-producing countries in the world is the coffee berry borer, *Hypothenemus hampei* (Coleoptera: Curculionidae: Scolytinae), an insect endemic to Africa (Vega et al. 2009, 2012). It invaded coffee plantations throughout South and Central America, Mexico, and the Caribbean following its accidental introduced into Brazil in 1913 (Infante et al. 2012). The coffee berry borer has been transported around the world, most probably through seeds containing the borer. There are very few coffee-producing countries that are still free of this pest with Papua New Guinea and Nepal still remaining free of the pest. An incursion prevention program was mounted in Papua New Guinea in 2007 to prevent invasion from Papua Province in Indonesia (CABI 2020). The first discovery of this pest in Hawaii was in 2010, first in the Big Island, followed by Oahu in 2014.

About 8 weeks after flowering, the adult female cuts a characteristic hole at the blossom end of large green berries. The eggs are deposited in internal galleries and upon hatching, the larvae, feed on the seed. A single berry may be infested with up to 20 larvae. Many berries fall immaturely. In heavy infestations, they have been known to attack 100% of berries. The insect remains inside the berry most of its life, making it difficult to control. The total life cycle is 25–35 days. Yield and quality of the marketable product are significantly reduced (CABI 2020; Crowe 2009; Vega et al. 2009, 2012).

There is no simple and cheap method of control of this insect. Cultural control measures are recommended with chemical control used as a supplement to cultural measures. Between fruiting, breeding in overripe berries either on the tree or fallen to the ground will cause the infestation to be carried over. Cultural measures that can be adopted to reduce infestations include: reducing heavy shade, keeping the coffee bush open by pruning, picking coffee at least once a week during main harvest season, stripping the trees of any remnant berries once harvesting is done, ensuring that berries are not left on the ground, and destroying all infested berries by burning or rapid sun drying (Crowe 2009).

## *Coffee Leaf Miner:* Leucoptera *spp.*

Twenty species of leaf miners of the genus *Leucoptera* (Lepidoptera: Lyonetiidae) have been described, infesting 65 host species. Of these, four species are known to infest *Coffea* species: *L. coffeella, L. meyricki, L. coma,* and *L. caffeina* (Filho et al. 1999; Filho 2006). The leaf miner is a moth whose larvae feed inside the leaf tissue and consume the palisade parenchyma. An introduced pest from Africa, *L. coffeella* (Guérin-Méneville) is one of the most serious pests in Brazil on *Coffea arabica* with crop losses of up to 50% possible. In eastern Africa from Ethiopia to South Africa, *L. caffeina* and *L. meyricki* are major pests of Arabica coffee that have also been recorded as attacking the indigenous wild coffee, *C. eugenioides* and other shrubs in the *Rubiaceae* family (Crowe 2009).

The coffee leaf miner, *L. coffeella*, a monophagous pest attacking only coffee plants, was first introduced to Brazil around 1851 probably on nursery stock imported from the Antilles and Bourbon Island (as cited in Filho 2006). Infested coffee has large, irregular, brown spots on the upper surface of the leaf caused by larvae that bore through the epidermis and feed in the palisade tissue (Crowe 2009). This reduces photosynthetic leaf area. Rubbing or exposing the spots reveal fresh mines and the small whitish caterpillars. Mined leaves shed prematurely. Loss in productivity is mainly due to leaf loss. Chemical control of the pest is effective but increases cost of production and has associated environmental risks. Through classic breeding and molecular selection techniques, coffee cultivars with resistance to the pest have been and continue to be developed. In Brazil, using genes from *C. racemosa*, varieties resistant to *L. coffeella* have been developed (Filho et al. 1999; Filho 2006).

## *Coffee White Stem Borer:* **Xylotrechus quadripes** *Chevrolat*

The coffee white stem borer, *Xylotrechus quadripes* (Coleoptera: Cerambycidae) is a severe pest of Arabica coffee reported in China, India, Sri Lanka, Nepal, Vietnam, and Thailand. The adult is a slender, elongated beetle with a grayish pubescence on the head, thorax, and abdomen. Characteristic of this species is the inverted "V" markings on the forewing (Lan and Wintgens 2009). The larvae of this beetle bore into the stem causing yellowing, defoliation, and wilting. Beetles are active only during the day when the weather is bright and hot, and they are inactive in cloudy and moist conditions. Older plants can withstand for a few seasons, whereas it is detrimental to young plants. Yield is reduced and there is higher occurrence of "floats." Plants are easily broken by wind and die.

Control of this pest requires an integrated approach utilizing cultural, physical, chemical, and biological methods. Field sanitation by uprooting and burning infested plants will reduce spread of the pest. Since the beetles are attracted to sunny, bright areas, maintaining optimum shade is recommended. Scrubbing the bark with a brush has been effective in India. The smooth stem surfaces discourage the beetles from laying eggs, and any eggs or grubs present are dislodged. Pheromone traps and swabbing chemicals on stem are also effective (Lan and Wintgens 2009).

## *Coffee Leaf Rust:* Hemileia vastatrix *Berkeley and Broome*

One of the most devastating coffee diseases, the coffee leaf rust, is caused by the obligate parasitic fungus *Hemileia vastatrix*. It causes considerable economic losses to coffee producers (Diola et al. 2011), especially in *C. arabica* and is currently found in all coffee growing regions of the world. This disease was first observed near Lake Victoria in 1861and has now spread throughout coffee-growing countries. In 1868, infection by this disease led to significant economic impact in Sri Lanka (Silva et al. 2006). In India, crop losses due to coffee rust accounts for about 70% in susceptible *C. arabica* cultivars (Prakash et al. 2004). The 2012–2013 outbreak of coffee rust in Central America resulted in crop devastation in Nicaragua, El Salvador, Guatemala, Dominican Republic, and Honduras, impacting over 1.08 million hectares and in Mexico, more than 60% of the trees exhibited 80% defoliation (Cressey 2013; ICO 2013). According to the International Coffee Organization, the 2012–2013 outbreak of coffee rust in Central America caused crop losses of $500 million with 374,000 jobs lost (ICO 2013).

The first observable symptoms occur on the upper surface of the leaves as small, pale yellow spots. The spots gradually increase in diameter and masses of yellowish-orange to bright orange uredospores are seen on the undersurfaces of the leaves. The centers of the spots eventually turn brown and dry, while the margins continue producing uredospores and expand. They eventually spread to limbs covering significant areas. Due to loss of physiological activity, the leaves fall. Severe infection can cause branches to wither completely. Spread of the uredospores is by both wind and rain, with splashing rain serving as an important means of local dispersal. Long-range dispersal is primarily by wind. Good cultural management is key in achieving good control of this disease though many factors dictate cultural methods such as varieties grown, soil characteristics, and amount and distribution of rainfall. People working in the farms can also serve as vectors for spread of the disease as well as any infected leaf litter from the previous infection. Chemical control using copper-based products is effective if applied at regular intervals as preventative measures. When using copper-based fungicides, care should be taken to ensure that it does not accumulate in the soil and reach levels toxic to plants and other organisms (Amerson 2000; Muller et al. 2009).

Development and cultivation of disease resistant or tolerant varieties is the most effective option for disease management in order to be economically viable and for minimization of chemical input for environmental sustainability. Hence, breeding for varieties resistant to coffee leaf rust has been one of the highest priorities in many countries (Prakash et al. 2004).

### *Coffee Berry Disease:* **Colletotrichum kahawae** *Bridge and Waller*

Currently restricted to East, Central, and South African coffee-growing countries (as cited in Hindorf and Omondi 2011), the coffee berry disease (CBD) is caused by the fungus *Colletotrichum kahawae.* It was first detected in Kenya in 1922 around Mt. Elgon west of the Rift Valley. From Kenya, the disease spread to Angola in 1930, Zaire in 1937, Cameroon between 1955 and 1957, Uganda in 1959, Tanzania in 1964, Ethiopia in 1971, and Malawi in 1985. Causing up to 70–80% crop losses if no control measures are adopted, it infects all stages of the crop from flowers to ripe fruits and occasionally leaves. Maximum crop losses occur following infection of green berries, leading to formation of dark sunken lesions and premature dropping and mummification of the fruits (as cited in Silva et al. 2006). Crop losses by CBD and cost of chemical control in Africa results in an estimated annual economic impact of US$ 300–500 million (van der Vossen and Walyaro 2009). Although CBD is currently restricted to Africa, in order to prevent introduction of this disease to other coffee-producing countries, precautions should be taken (Silva et al. 2006).

Control of the disease can be achieved through an integrated cultivation approach utilizing chemical control linked to improved cultivation practices and genetic control (Muller et al. 2009). It is reported that CBD resistance appears to be complete in *C. canephora* and partial in *C. arabica* (Silva et al. 2006). In response to severe disease epidemics about 35–40 years ago in Kenya, Ethiopia, and Tanzania, breeding for CDB resistance in *C. arabica* was initiated, with release of resistant cultivars to coffee growers since 1985 (van der Vossen and Walyaro 2009). Under field and laboratory conditions, differences in resistance of coffee trees to CBD have been observed with higher resistance in Geisha 10, Blue Mountain, K7, Rume Sudan, and progenies of Hibrido de Timor than in Harar and Bourbon in Kenya (Silva et al. 2006).

## 6 Genetics of Coffee Quality

Quality is a trait that varies along the producer to consumer value chain characterized by something specific for each segment of the value chain. Leroy et al. (2006) classified this into four levels:

- Farmer level—quality is a combination of level of production, price, and ease of culture.
- Exporter or importer level—quality is linked to bean size, lack of defects, availability, and price.
- Roaster level—quality is tied to moisture content, stability of characteristics, biochemical compounds, organoleptic quality, origin, and price.

- Consumer level—quality deals with taste, flavor, price, geographic origin, environmental and social sustainability aspects, and effects on health.

A long chain of chemical transformations occurs from seed to the cup. The final expression and distinguishable result of this transformation is the flavor of a cup of coffee (Joet and Dussert 2018). Hence, coffee quality will be discussed in the context of flavor here. Of the two cultivated coffees, *C. arabica* is characterized as a fruit-flavored, acidulous, milder beverage whereas, *C. canephora* is described as a weak-flavored, neutral coffee with occasional pronounced, and strong bitterness (Bertrand et al. 2003). More than 900 volatile compounds have been identified that contribute to the flavor of roasted coffee beans. The green coffee beans themselves do not have a distinguishable smell, but contain the precursors that generate the coffee aroma during roasting at which time the beans undergo chemical modifications generating these volatile compounds (Joet and Dussert 2018). Many parameters along the whole value chain of coffee impact the chemical composition and hence the flavor and quality of seeds: coffee species and cultivars, soil, climate, agronomic practices, ripening stage at harvest, postharvest processing, storage, transportation, roasting, and storage of roasted beans.

The four main biochemical compounds that contribute to coffee flavor are chlorogenic acids (CGA), caffeine, sucrose, and trigonelline. Chlorogenic acids (CGA) are phenolic compounds found in green coffee beans and their presence modifies cup taste. The higher the CGA content, the lower the quality of the beverage (Ky et al. 1999). Caffeine is an alkaloid that impacts beverage quality; higher caffeine content increases the bitterness of the drink. The caffeine content in green coffee beans varies between and within-species with *C. pseudozanguebariae* having 0% dry matter basis (dmb) of caffeine to 3.3% dmb in *C. canephora* (Campa et al. 2005). Sucrose in raw coffee bean is an important precursor of flavor and aroma (Geromel et al. 2006). Although the free sugars glucose, fructose, and sucrose are present in young coffee beans, only sucrose is present in mature beans. Mature Arabica endosperm has higher sucrose levels than robusta endosperm (Rogers et al. 1999), giving the Arabica coffee a better cup quality (Geromel et al. 2006). Sucrose content in *C. arabica* varies from 7% to 11% dmb compared to lower values in *C. canephora* ranging from 4% to 7% dmb (Ky et al. 2001). Trigonelline, an alkaloid, is an aroma precursor in coffee with higher levels in *C. arabica* ranging from 0.88 to 1.77% dmb and 0.75 to 1.24% dmb in *C. canephora* (Ky et al. 2001).

Significant genetic variability for chemical composition and organoleptic characteristics of beans exists between and within-species levels in coffee. Through interspecific or within-species hybridization strategies, genetic gain for quality can be achieved (Leroy et al. 2006). Increasing robusta cup quality has been one of the targets of several breeding programs. Achieving this would involve increasing sucrose and trigonelline contents while decreasing CGA and caffeine contents. Evaluation of biochemical polymorphism and inheritance indicates that plants can be bred to decrease CGA and caffeine contents and increase sucrose contents (Ky et al. 2001). In a recent study comparing three *C. canephora* genotypes with *C. arabica* in Brazil, Lemos et al. (2020) report one of the robusta genotypes, Diamante

108 having aroma and flavor characteristics which scored above the *C. arabica* used for comparison in the study, sparking interest in national and international commercialization.

## 7 Breeding Approaches

The objectives of breeding programs for Arabica coffee vary for each region/country based on the challenges faced by that specific region/country. Some of the objectives of breeding programs are improving yields and plant vigor, pest and disease resistance, nematode resistance, improving quality, drought tolerance, uniform ripening to enable mechanical harvesting, and compact growth habit. In the early 1920s, breeding for disease resistance in combination with productivity, vigor, and quality started in India (Van der Vossen 2018). Most breeding programs between 1950 and 1978 involved intraspecific and interspecific hybridization between *C. arabica* and *C. canephora*, with the use of some wild species (Pinto-Maglio 2006). Spontaneous interspecific hybrids that occur occasionally such as the Timor Hybrid, a spontaneous cross between *C. arabica and C. canephora* that was discovered in the island of Timor, have also been widely used for improving pest and disease resistance in Arabica coffee (Pearl et al. 2004). In India, introgressed lines of *C. liberica* from a spontaneous hybrid between *C. arabica* and *C. liberica* have been used in breeding programs as a main source of rust resistance (Prakash et al. 2002).

Breeding methods in self-pollinating *C. arabica* include (Van der Vossen 2018):

- Line selection utilizing available varieties—e.g., Caturra (Brazil), Kents (India), SL28 (Kenya), Java (Cameroon).
- Pedigree selection post intervarietal crossing and backcrossing—e.g., Mundo Novo, Catuai, Tupi (Brazil), Catimor, Sarchimor (Costa Rica), Castillo (Colombia).
- Multiple crosses (three- and four-way, etc.) and backcrosses, repeated selfing, or cloning—e.g., Batian (Kenya), N39–3, TaCRI3F (Tanzania).
- F1 hybrids between genetically divergent genotypes—e.g., Ruiru 11 (Kenya), Ababuna, Melko-CH2 (Ethiopia), H1–H4 (Central America).
- Interspecific hybridization followed by backcrossing and pedigree selection—e.g., Icatu (Brazil), S2828 (India).

Van der Vossen (2018) identifies the following priorities for next-generation cultivar development of Arabica coffee: maintaining productivity under a changing climate, high beverage quality, efficient and cost-effective multiplication, improving the adoption rate of disease-resistant cultivars, and application of genomics-assisted breeding. Due to the nature of the coffee crop, centralized development and global dissemination of varieties is not an option and hence breeding and multiplication of new varieties adapted to local conditions should be undertaken at national and regional levels.

Utilizing F1 hybrids is a new development is coffee breeding. F1 hybrids are created by crossing genetically distant Arabica genotypes with traits for high cup quality, high yield, and plant disease resistance. Farmers typically tend to propagate coffee plants using seeds from the plants on their farms. The challenge with using F1 hybrids and distributing to farmers is the need to educate them that F1 hybrids need to be clonally propagated instead of propagating coffee by seeds which will lead to segregation and loss of hybrid characteristics. The variety Starmaya is an F1 hybrid that was produced using a male sterile parent making it amenable for mass propagation by seeds (Georget et al. 2019). In a study comparing F1 hybrids with conventional coffee cultivars to assess the genotype by environment interactions on yield and quality, the F1 hybrids Centroamericano, Mundo Maya® and Starmaya outperformed conventional cultivars such as Caturra and Marsellesa® (Marie et al. 2020).

The sequencing of the genomes of both *C. arabica* and *C. canephora* offers access to previously inaccessible genetic diversity information that can be utilized in breeding programs. Molecular genomics can be used in coffee for assessing introgression, determining mode of inheritance of pest and disease resistance, analyzing quantitative trait loci (QTL), characterizing and managing germplasm resources, enhanced understanding of the genetic control of traits of interest, and identifying candidate genes or tightly linked genomic regions underlying important traits (Lashermes et al. 2009; Prakash et al. 2004).

## 8 Impacts of Climate Change

The effects of climate change induced stresses are already being experienced by coffee growers worldwide, especially in *C. arabica* production. These impacts are only going to be exacerbated in the coming decades. Some of the effects include higher temperatures, lower and more erratic precipitation, more frequent and prolonged droughts, and severe hurricanes. The direct impacts from climate change leads to limited flowering and berry development, stressed growth of trees, poor yield and poor quality of beans, and increased incidence and severity of pests and diseases and damages caused by them (Bramel et al. 2017; Schroth et al. 2009).

Combined with economic impacts of coffee production, climate change impacts can lead to devastating outcomes in farmer incomes, livelihoods, and household food security. As an example, the coffee leaf rust epidemic in Central and South America from 2008 to 2013 had devastating economic and livelihood impacts on farmers in those regions. The main drives of these epidemics were both economic and meteorological; economic due to low coffee prices resulting in non-affordability of key farm inputs leading to suboptimal farm management making them susceptible to pests and diseases and meteorological due to reduction in diurnal thermal amplitude with increased minimum temperatures during winter leading to earlier incidence of rust than has been reported before (Avelino et al. 2015).

From a database of georeferenced production locations, Bunn et al. (2015a) projected current and future climate suitability for both *C. arabica* and *C. canephora* production. Their multi-model ensemble suggests reduced yields of *C. arabica* at higher temperatures and negative impacts of increasing variability of intra-seasonal temperatures on *C. canephora*. Global area suitable for coffee will be reduced by 50% with highest impacts at lower latitudes and lower altitudes. Impacts at higher latitudes and higher altitudes will also be negative, but less pronounced. Significantly impacted will be Brazil and Vietnam, currently the two largest producers, experiencing considerable reductions in area suitable for coffee. Even though some regions in East Africa and Asia may become suitable, these are in forested areas posing challenges of deforestation and loss of biodiversity.

Adaptation strategies recommended for coping with climate change include use of drought and heat-resistant varieties, improved agronomic practices, sustainable management of resources, improving shade canopy for maintaining soil moisture and protecting against landslides and soil erosion, implementing agroforestry practices, and irrigation. More severe measures will entail switching to other crops or migrating to higher altitudes (Laderach et al. 2010), though at higher altitudes, land conducive for agriculture is lower and face challenges such as difficulty of access, too steep to cultivate, and shallow soils (Bunn et al. 2015b). Hence research priority for climate change adaptation of coffee will have to be on making Arabica coffee batter adapted to heat and drought stress (Bunn et al. 2015b). De Sousa et al. (2019) stress the urgency in developing lines of research on land use planning and the assessment of new combinations of agroforestry species adapted to coffee systems in Mesoamerica.

Not only will climate change affect coffee in cultivation, but also the native forests of Arabica coffee and other wild *Coffea* species. Davis et al. (2012) predicted a 65% to almost 100% reduction in bioclimatically suitable localities of indigenous Arabica coffee in their natural habitats in Ethiopia by 2080 using locality analysis and bioclimatic modeling. This bioclimatic unsuitability would cause severe stress to populations leading to high extinction risk in the short term. The Extend of Occurrence (EOO) and Area of Occupancy (AOO) of wild Arabica coffee are projected to decline by around 30–50%, with a few models showing over 80% by 2088. When the International Union for the Conservation of Nature's (IUCN) Red List of Threatened Species criteria were applied, under a broad range of climate change projections, wild Arabica coffee shifts from Least Concerned to Endangered for a generation length of 21 years with population reductions within a 63-year window (2025–2088) projected to be greater than 70% (Moat et al. 2019). In addition to Arabica coffee, many of the other wild *Coffea* species are also vulnerable to climate change. A recent study by Davis et al. (2019) found that of the 124 described *Coffea* spp., 75 coffee species (60%) are threatened with extinction, 45% of which are not held in any ex situ genebank collections, and 28% are not known to occur in any in situ protected areas.

## 9 Future Prospects and Conclusions

Coffee is a multibillion-dollar industry that provides livelihoods for millions of people around the world. Despite the economic importance of coffee, research and development has lagged behind other crops. There is concern that within the next decade demand will exceed production. In order to cope with the threats facing coffee, concerted effort should be made to ensure the sustainability of the crop and the farmers who depend on the crop for their livelihoods. The Global Conservation Strategy for Coffee Genetic Resources developed in 2017 (Bramel et al. 2017) provides a framework to engage stakeholders at all levels in building awareness, capacity and engagement in conserving the genetic diversity, and use of coffee genetic resources for the long term. Genetic resources are under threat, both in situ and ex situ. Global and national investments in securing collections already existing in ex situ collections as well as facilitation of in situ preservation through designation of protected areas in countries of origin need to be prioritized. The responsibility lays not just with the producing countries, but also the consuming countries. It is these resources which will ensure the future of the coffee crop.

## References

Alkimim ER, Caixeta ET, Sousa TV, Pereira AA, de Oliberia ACB, Zambolim L, Sakiyama NS (2017) Marker-assisted selection provides arabica coffee with genes from other *Coffea* species targeting on multiple resistance to rust and coffee berry disease. Mol Breed 37:6. https://doi.org/10.1007/s11032-016-0609-1

Amerson PA (2000) Coffee rust. Plant Health Instr. http://download.ceris.purdue.edu/file/2759. Accessed 29 Feb 2020

Anagbogu C, Bhattacharjee R, Iloro C, Tongyoo P, Dada KE, Muyiwa AA, Gepts P, Beckles DM (2019) Genetic diversity and re-classification of coffee (*Coffea canephora* Pierre ex A. Froehner) from South Western Nigeria through genotyping-by-sequencing-single nucleotide polymorphism analysis. Genet Resour Crop Evol 66:685–696

Andrianasolo DN, Davis AP, Razafanarivo NJ, Hamon S, Rakotomalala J-J, Sabatier S-A, Hamon P (2013) High genetic diversity of in situ and ex situ populations of Madagascan coffee species: further implications for the management of coffee genetic resources. Tree Genet Genomes 9:1295–1312. https://doi.org/10.1007/s11295-013-0638-4

Anthony F, Bertrand B, Quiros O, Wilches A, Lashermes P, Berthaud J, Charrier A (2001) Genetic diversity of wild coffee (*Coffea arabica*) using molecular markers. Euphytica 118:53–65

Anthony F, Combes MC, Astorga C, Bertrand B, Graziosi G, Lashermes P (2002) The origin of cultivated *Coffea arabica* L. varieties revealed by AFLP and SSR markers. Theor Appl Genet 104:894–900

Anthony F, Dussert S, Dulloo E (2007a) Coffee genetic resources. In: Engelmann F, Dulloo ME, Astorga C, Dussert S, Anthony F (eds) Conserving coffee genetic resources: complementary strategies for *ex situ* conservation of coffee (*Coffea arabica* L.) genetic resources. A case study in CATIE, Costa Rica. Bioversity International, Rome, pp 12–22

Anthony F, Astorga C, Avendaño J, Dulloo E (2007b) Conservation of coffee genetic resources in the CATIE field genebank. In: Engelmann F, Dulloo ME, Astorga C, Dussert S, Anthony F (eds) Conserving coffee genetic resources: complementary strategies for *ex situ* conservation of coffee (*Coffea arabica* L.) genetic resources. A case study in CATIE, Costa Rica. Bioversity International, Rome, pp 23–34

Avelino J, Cristancho M, Georgiou S, Impach P, Aguilar L, Bornemann G, Laderach P, Anzueto F, Hruska AJ, Morales C (2015) The coffee rust crises in Colombia and Central America (2008-2013): impacts, plausible causes and proposed solutions. Food Secur **7**(2):303–321. https://doi.org/10.1007/s12571-015-0446-9

Baruah A, Naik V, Hendre PS, Rajkumar R, Rajendrakumar P, Aggarwal RK (2003) Isolation and characterization of nine microsatellite markers from *Coffea arabica* L., showing wide cross-species amplifications. Mol Ecol Notes 3:647–650

Bertrand B, Guyot B, Anthony F, Lashermes P (2003) Impact of the *Coffea canephora* gene introgression on beverage quality of *C. arabica*. Theor Appl Genet 107:387–394

Bramel P, Krishnan S (2019) Systematic assessment for conservation and utilisation of crop genetic resources. In: Zeigler RS (ed) Sustaining global food security, the nexus of science and policy. CSIRO Publishing, Clayton, pp 1–14

Bramel P, Krishnan S, Horna D, Lainoff B, Montagnon C (2017) Global conservation strategy for coffee genetic resources. The Crop Trust and World Coffee Research, Bonn

Breitler J-C, Dechamp E, Campa C, Rodrigues LAZ, Guyot R, Marraccini P, Etiemme H (2018) CRISPR/Cas9-mediated efficient targeted mutagenesis has the potential to accelerate the domestication of *Coffea canephora*. Plant Cell Tissue Organ Cult 134(3):383–394

Bunn C, Laderach P, Rivera OO, Kirschke D (2015a) A bitter cup: climate change profile of global production of Arabica and Robusta coffee. Clim Change 129:89–101

Bunn C, Laderach P, Jimenez JGP, Montagnon C, Schillling T (2015b) Multiclass classification of agro-ecological zones for Arabica coffee: an improved understanding of the impacts of climate change. PLoS One 10:e0140490. https://doi.org/10.1371/journal.pone.0140490

CABI (2020) Coffee berry borer datasheet. Invasive species compendium. https://www.cabi.org/isc/datasheet/51521. Accessed 29 Feb 2020

Campa C, Doulbeau S, Dussert S, Hamon S, Noirot M (2005) Diversity in bean caffeine content among wild *Coffea* species: evidence of a discontinuous distribution. Food Chem 91:633–637

Charrier A, Berthaud J (1985) Botanical classification of coffee. In: Clifford MN, Willson KC (eds) Coffee: botany, biochemistry and production of beans and beverage. The Avi Publishing Company, Westport, pp 13–47

Charrier A, Lashermes P, Eskes A B (2009) Botany, genetics and genomes of coffee. In: Wintgens JN (ed) Coffee: botany, growing, processing, sustainable production—a guidebook for growers, traders, and researchers, second revised edition. Wiley-VCH, Weinheim, pp 25–60

Christensen A (2019) The forgotten species: specialty Liberica takes root in Borneo. Daily Coffee News by Roast Magazine June 3, 2019. https://dailycoffeenews.com/2019/06/03/the-forgotten-species-specialty-liberica-takes-root-in-borneo/. Accessed 4 Feb 2020

Combes MC, Andrzejewski S, Anthony F, Bertrand B, Rovelli P, Graziosi G, Lashermes P (2000) Characterization of microsatellite loci in *Coffea arabica* and related coffee species. Mol Ecol 9:1178–1180

Cressey D (2013) Coffee rust gains foothold. Nature 494:587

Cros J, Combes MC, Trouslot P, Anthony F, Hamon S, Charrier A, Lashermes P (1998) Phylogenetic analysis of chloroplast DNA variation in *Coffea* L. Mol Phylogenet Evol 9(1):109–117

Crowe TJ (2009) Coffee pests in Africa. In: Wintgens JN (ed) Coffee: growing, processing, sustainable production—a guidebook for growers, processors, traders, and researchers, 2nd edn. Wiley-VCH Verlag GmbH & Co. KGaA, Weinheim, pp 425–462

Cubry P, Musoli P, Legnate H, Pot D, de Bellis F, Poncet V, Anthony F, Dufour M, Leroy T (2008) Diversity in coffee assessed with SSR markers: structure of the genus *Coffea* and perspectives for breeding. Genome 51:50–63

da Silva BSR, Sant'Ana GC, Chaves CL, Androcioli LG, Ferreira RV, Sera GH, Charmetant P, Leroy T, Pot D, Domingues DS, Pereira LFP (2019) Population structure and genetic relationships between Ethiopian and Brazilian Coffea arabica genotypes revealed by SSR markers. Genetica 147(2):205–216

Damania AB (2003) The early history and spread of coffee. Asian Agric Hist 7(1):67–74
Davis AP, Govaerts R, Bridson DM, Stoffelen P (2006) An annotated taxonomic conspectus of the genus *Coffea* (Rubiaceae). Bot J Linn Soc 152:465–512
Davis AP, Chester M, Maurin O, Fay MF (2007) Searching for the relatives of *Coffea* (Rubiaceae, Ixoroideae): the circumscription and phylogeny of Coffeeae based on plastid sequence data and morphology. Am J Bot 94(3):313–329
Davis A, Tosh J, Ruch N, Fay MF (2011) Growing coffee: *Psilanthus* (Rubiaceae) subsumed on the basis of molecular and morphological data: implications for the size, morphology, distribution and evolutionary history of *Coffea*. Bot J Linn Soc 167:357–377
Davis AP, Gole TW, Baena S, Moat J (2012) The impact of climate change on indigenous arabica coffee (*Coffea arabica*): predicting future trends and identifying priorities. PLoS One 7(11):e47981. https://doi.org/10.1371/journal.pone.0047981
Davis AP, Chadburn H, Moat J, O'Sullivan R, Hargreaves S, Lughadha EN (2019) High extinction risk for wild coffee species and implications for coffee sector sustainability. Sci Adv 5:eaav3473
De Sousa K, van Zonneveld M, Holmgren M, Kindt R, Ordonez JC (2019) The future of coffee and cocoa agroforestry in a warmer Mesoamerica. Sci Rep 9:8828. https://doi.org/10.1038/s41598-019-45491-7
Diola V, de Brito GG, Caixeta ET, Maciel-Zambolim E, Sakiyama NS, Loureiro ME (2011) High-density genetic mapping for coffee leaf rust resistance. Tree Genet Genomes 7:1199–1208. https://doi.org/10.1007/s11295-011-0406-2
Dulloo ME, Guarino L, Engelmann F, Maxted N, Newbury JH, Attere F, Ford-Lloyd BV (1998) Complementary conservation strategies for the genus *Coffea*: a case study of Mascarene *Coffea* species. Genet Resour Crop Evol 45:565–579
Dulloo ME, Ebert AW, Dussert S, Gotor E, Astorga C, Vasquez N et al (2009) Cost efficiency of cryopreservation as a long-term conservation method for coffee genetic resources. Crop Sci 49:2123–2138
Dussert S, Vasquez N, Salazar K, Anthony F, Engelmann F (2007) Cryopreservation of coffee genetic resources. In: Engelmann F, Dulloo ME, Astorga C, Dussert S, Anthony F (eds) Complementary strategies for *ex situ* conservation of coffee (*Coffea arabica* L.) genetic resources. A case study in CATIE, Costa Rica. Tropical reviews in agricultural biodiversity. Bioversity International, Rome, pp 49–58
Eira MTS, da Silva EAA, de Castro RD, Dussert S, Walters C, Bewley JD, Hilhorst HWM (2006) Coffee seed phusiology. Braz J Plant Physiol 18(1):149–163
Fazuoli LC, Maluf MP, Filho OG, Filho HM, Silvarolla MB (2000) Breeding and biotechnology of coffee. In: Sera T, Soccol CR, Pandey A, Roussos S (eds) Coffee biotechnology and quality. Kluwer Academic Publishers, Dordrecht, pp 27–45
Filho OG (2006) Coffee leaf miner resistance. Braz J Plant Physiol 18(1):109–117
Filho OG, Silvarolla MB, Eskes AB (1999) Expression and mode of inheritance of resistance in coffee to leaf miner *Perileucoptera coffeella*. Euphytica 105:7–15
Georget F, Marie L, Alpizar E, Courtel P, Bordeaux M, Hidalgo JM, Marraccini P, J-C B, Dechamp E, Poncon C, Etienne H, Bertrand B (2019) Starmaya: the first Arabica F1 coffee hybrid produced using genetic male sterility. Front Plant Sci 10:1344. https://doi.org/10.3389/fpls.2019.01344
Geromel C, Ferreira LP, Guerreiro SMC, Cavalari AA, Pot D, Pereira LFP, Leroy T, Vieira LGE, Mazzafera P, Marraccini P (2006) Biochemical and genomic analysis of sucrose metabolism during coffee (*Coffea arabica*) fruit development. J Exp Bot 57:3243–3258
Herrera JC, Combes MC, Anthony F, Charrier A, Lashermes P (2002) Introgression into the allotetraploid coffee (*Coffea arabica* L.): segregation and recombination of the *C. canephora* genome in the tetraploid interspecific hybrid (*C. arabica* x *C. canephora*). Theor Appl Genet 104:661–668
Herrera JC, Combes MC, Cortina H, Lashermes P (2004) Factors influencing gene introgression into the allotetraploid *Coffea arabica* L. from its diploid relatives. Genome 47:1053–1060

Herrera PJC, Alvarado AG, Cortina GHA, Combes MC, Romero GG, Lashermes P (2009) Genetic analysis of partial resistance to coffee leaf rust (*Hemileia vastatrix* Berk & Br.) introgressed into the cultivated *Coffea arabica* L. from the diploid *C. canephora* species. Euphytica 167:57–67

Hindorf H, Omondi CO (2011) A review of three major fungal diseases of *Coffea arabica* L. in the rainforests of Ethiopia and progress in breeding for resistance in Kenya. J Adv Res 2:109–120

ICO (2013) Report on the outbreak of coffee leaf rust in Central American and Action Plan to combat the pest. International Coffee Organization, London. 6 pages

ICO (International Coffee Organization) (2020) International Coffee Organization Trade Statistics. http://www.ico.org/trade_statistics.asp. Accessed 4 Feb 2020

Infante F, Perez J, Vega FE (2012) Redirect research to control coffee pest. Nature 489:502

Joet T, Dussert S (2018) Environmental and genetic effects on coffee seed biochemical composition and quality. In: Lashermes P (ed) Achieving sustainable cultivation of coffee. Volume 1: Breeding and quality traits. Burleigh Dodds Science Publishing, Cambridge. https://doi.org/10.19103/AS.2017.0022.03

Krishnan S (2013) Current status of coffee genetic resources and implications for conservation. CAB Rev 8(16):1–9. https://doi.org/10.1079/PAVSNNR20128016_2013

Krishnan S (2014) Marker assisted selection in coffee. In: Benkeblia N (ed) OMICS technologies and crop improvement. CRC Press, Boca Raton, pp 209–218

Krishnan S (2017) Sustainable coffee production. Oxford Research Encyclopedia of Agriculture and the Environment. https://doi.org/10.1093/acrefore/9780199389414.013.224. http://environmentalscience.oxfordre.com/view/10.1093/acrefore/9780199389414.001.0001/acrefore-9780199389414-e-224?rskey=F5I3bP&result=24

Krishnan S (2018) Ensuring the genetic diversity of coffee. In: Lashermes P (ed) Achieving sustainable cultivation of coffee. Volume 1: Breeding and quality traits. Burleigh Dodds Science Publishing, Cambridge, pp 69–82. https://doi.org/10.19103/AS.2017.0022.04

Krishnan S, Ranker T, Davis AP, Rakotomalala JJ (2013a) The study of genetic diversity patterns of *Coffea commersoniana*, an endangered coffee species from Madagascar: a model for conservation of other littoral forest species. Tree Genet Genomes 9:179–187. https://doi.org/10.1007/s11295-012-0545-0

Krishnan S, Ranker T, Davis AP, Rakotomalala JJ (2013b) An assessment of the genetic integrity of *ex situ* germplasm collections of three endangered species of *Coffea* from Madagascar: implications for the management of field germplasm collections. Genet Resour Crop Evol 60:1021–1036. https://doi.org/10.1007/s10722-012-9898-3

Krishnan S, Bramel P, Horna D, Lainoff B, Montagnon C, Schilling T (2018) Development of a global conservation strategy for coffee genetic resources. Acta Horticult 1205:505–508. https://doi.org/10.17660/ActaHortic.2018.1205.62

Ky CL, Louarn J, Guyot B, Charrier A, Hamon S, Noirot M (1999) Relations between and inheritance of chlorogenic acid contents in an interspecific cross between *Coffea pseudozanguebariae* and *Coffea liberica* var '*dewevrei*'. Theor Appl Genet 98:628–637

Ky CL, Louarn J, Dussert S, Guyot B, Hamon S, Noirot M (2001) Caffeine, trigonelline, chlorogenic acids and sucrose diversity in wild *Coffea arabica* L. and *C. canephora* P. accessions. Food Chem 75:223–230

Labouisse J-P, Bellachew B, Kotecha S, Bertrand B (2008) Current status of coffee (*Coffea arabica* L.) genetic resources in Ethiopia: implications for conservation. Genet Resour Crop Evol 55:1079–1093

Laderach P, Haggar J, Lau C, Eitzinger A, Ovalle O, Baca M, Jarvis A, Lundy M (2010) Mesoamerican coffee: building a climate change adaptation strategy. CIAT policy brief no. 2. Centro Internacional de Agricultura Tropical (CIAT), Cali, Colombia. 4 p.

Lan CC, Wintgens JN (2009) Major pests of coffee in the Asia-Pacific region. In: Wintgens JN (ed) Coffee: growing, processing, sustainable production—a guidebook for growers, processors, traders, and researchers, 2nd edn. Wiley-VCH Verlag GmbH & Co. KGaA, Weinheim, pp 463–477

Lashermes P, Trouslot P, Anthony F, Combes MC, Charrier A (1996) Genetic diversity for RAPD markers between cultivated and wild accessions of *Coffea arabica*. Euphytica 87:59–64

Lashermes P, Combes MC, Trouslot P, Charrier A (1997) Phylogenetic relationships of coffee-tree species (*Coffea* L.) as inferred from ITS sequences of nuclear ribosomal DNA. Theor Appl Genet 94:947–955

Lashermes P, Combes M-C, Robert J, Trouslot P, D'Hont A, Anthony F, Charrier A (1999) Molecular characterization and origin of the *Coffea arabica* L. genome. Mol Gen Genet 261:259–266

Lashermes P, Andrzejewski S, Bertrand B, Combes MC, Dussert S, Graziosi G, Trouslot P, Anthony F (2000a) Molecular analysis of introgressive breeding in coffee (*Coffea arabica* L.). Theor Appl Genet 100:139–146

Lashermes P, Combes MC, Topart P, Graziosi G, Bertrand B, Anthony F (2000b) Molecular breeding in coffee (*Coffea arabica* L.). In: Sera T, Soccol CR, Pandey A, Roussos S (eds) Coffee biotechnology and quality. Kluwer Academic Publishers, Amsterdam, pp 101–112

Lashermes P, Bertrand B, Etienne H (2009) Breeding coffee (*Coffea arabica*) for sustainable production. In: Jain SM, Priyadarshan PM (eds) Breeding plantation tree crops: tropical species. Springer Science+Business Media, LLC, Berlin, pp 525–543

Lemos MF, Perez C, Pereira de Cunha PH, Filgueiras PR, Pereira LL, Almeida da Fonseca AF, Ifa DR, Scherer R (2020) Chemical and sensory profile of new genotypes of Brazilian *Coffea canephora*. Food Chem 310:125850. https://doi.org/10.1016/j.foodchem.2019.125850

Leroy T, Ribeyre F, Bertrand B, Charmetant P, Dufour M, Montagnon C, Marraccini P, Pot D (2006) Genetics of coffee quality. Braz J Plant Physiol 18(1):229–242

Leroy T, De Bellis F, Legnate H et al (2011) Improving the quality of African robustas: QTLs for yield- and quality-related traits in Coffea canephora. Tree Genet Genomes 7:781–798

Maluf MP, Silvestrini M, Ruggiero LMC, Filho OG, Colombo CA (2005) Genetic diversity of cultivated *Coffea arabica* inbred lines assessed by RAPD, AFLP, and SSR marker systems. Sci Agric 62(4):366–373

Marie L, Abdallah C, Campa C, Courtel P, Bordeaux M, Navarini L, Lonzarich V, Bosselmann AS, Turreira-Garcia N, Alpizar E, Georget F, Breitler J-C, Etienne H, Bertrand B (2020) G × E interactions on yield and quality in *Coffea arabica*: new F1 hybrids outperform American cultivars. Euphytica 216:78. https://doi.org/10.1007/s10681-020-02608-8.

Maurin O, Davis AP, Chester M, Mvungi EF, Jaufeerally-Fakim Y, Fay MF (2007) Towards a phylogeny for *Coffea* (Rubiaceae): identifying well-supported lineages based on nuclear and plastid DNA sequences. Ann Bot 100:1565–1583

Meyer FG (1965) Notes on wild *Coffea arabica* from southwestern Ethiopia, with some historical considerations. Econ Bot 19:136–151

Mishra MK, Nishani S, Gowda M, Padmajyothi D, Suresh N, Sreenath H, Raghuramulu Y (2014) Genetic diversity among Ethiopian coffee (*Coffea arabica* L.) collections available in Indian gene bank using sequence related amplified polymorphism markers. Plant Breed Seed Sci 70:29–40

Missio RF, Caixeta ET, Zambolim EM, Zambolim L, Sakiyama NS (2009) Development and validation of SSR markers for *Coffea arabica* L. Crop Breed Appl Biotechnol 9:361–371

Missio RF, Caixeta ET, Zambolim EM, Zambolim L, Cruz CD, Sakiyama NS (2010) Polymorphic information content of SSR markers for *Coffea* spp. Crop Breed Appl Biotechnol 10:89–94

Missio RF, Caixeta ET, Zambolim EM, Pena GF, Zambolim L, Dias LAS, Sakiyama NS (2011) Genetic characterization of an elite coffee germplasm assessed by gSSR and EST-SSR markers. Genet Mol Res 10(4):2366–2381

Moat J, Gole TW, Davis AP (2019) Least concern to endangered: applying climate change projections profoundly influences the extinction risk assessment for wild Arabica coffee. Glob Chang Biol 25:390–403. https://doi.org/10.1111/gcb.14341

Moncada P, McCouch S (2004) Simple sequence repeat diversity in diploid and tetraploid *Coffea* species. Genome 47:501–509

Muller RA, Berry D, Avelina J, Bieysse D (2009) Coffee diseases. In: Wintgens JN (ed) Coffee: growing, processing, sustainable production—a guidebook for growers, processors, traders, and researchers, 2nd edn. Wiley-VCH Verlag GmbH & Co. KGaA, Weinheim, pp 495–549

Musloi P, Cubry P, Aluka P, Billot C, Dufour M, De Bellis T, Pot D, Bieysse D, Charrier A, Leroy T (2009) Genetic differentiation of wild and cultivated populations: diversity of *Coffea canephora* Pierre in Uganda. Genome 52:634–646

OEFCCA (Oromia Environment, forest and Climate Cgnage Authority) and OFWA (Oromia Forest and Wildlife Enterprise) (2018) Yayu coffee forest biosphere reserve management plan. 80 pages. https://www.phe-ethiopia.org/pdf/yayu-coffee-forest.pdf

Pearl HM, Nagai C, Moore PH, Steiger DL, Osgood RV, Ming R (2004) Construction of a genetic map for arabica coffee. Theor Appl Genet 108:829–835

Pinto-Maglio CAF (2006) Cytogenetics of coffee. Braz J Plant Physiol 18(1):37–44

Poncet V, Hamon P, Minier J, Carasco C, Hamon S, Noirot M (2004) SSR cross-amplification and variation within coffee trees (*Coffea* spp). Genome 47:1071–1081

Poncey V, Dufour M, Hamon P, Hamon S, de Kochko A, Leroy T (2007) Development of genomic microsatellite markers in *Coffea canephora* and their transferability to other coffee species. Genome 50:1156–1161

Prakash NS, Combes MC, Somanna N, Lashermes P (2002) AFLP analysis of introgression in coffee cultivars (*Coffea arabica* L.) derived from a natural interspecific hybrid. Euphytica 124:265–271

Prakash NS, Marques DV, Varzea VMP, Silva MC, Combes MC, Lashermes P (2004) Introgression molecular analysis of a leaf rust resistance gene from *Coffea liberica* into *C. arabica* L. Theor Appl Genet 109:1311–1317

Prakash NS, Muniswamy B, Hanumantha BT, Sreenath HL, Sundaresha KD, Suresh N, Santosh P, Soumya PR, Bhat AM, Bhat SS, Jayarama (2011) Marker assisted selection and breeding for leaf rust resistance in coffee (*Coffea arabica* L.)—some recent leads. Indian J Genet 71:1–6

Priyono, Sumirat U (2012) Mapping of quantitative trait loci (QTLs) controlling cherry and green bean characters in the robusta coffee (*Coffea canephora* Pierre). J Agric Sci Technol A2:1029–1039

Pruvot-Woehl S, Krishnan S, Solano W, Schilling T, Toniutti L, Bertrand B, Montagnon C (2020) Authentication of *Coffea arabica* varieties through DNA fingerprinting and its significance for the coffee sector. J AOAC Int 103(2):325–334. https://doi.org/10.1093/jaocint/qsz003

Razafinarivo NJ, Rakotomalala J-J, Brown SC, Bourge M, Hamon S, De Kochko A, Poncet V, Dubreuil-Tranchant C, Couturon E, Guyot R, Hamon P (2012) Geographical gradients in the genome size variation of wild coffee trees (*Coffea*) native to Africa and Indian Ocean islands. Tree Genet Genomes 8(6):1345–1358

Razafinarivo NJ, Guyot R, Davis AP, Couturon E, Hamon S, Crouzillat D, Rigoreau M, Dubreuil-Tranchant C, Poncet V, De Kochko A, Rakotomalala J-J, Hamon P (2013) Genetic structure and diversity of coffee (*Coffea*) across Africa and the Indian ocean islands revealed using microsatellites. Ann Bot 111:229–248

Rogers WJ, Michaux S, Bastin M, Bucheli P (1999) Changes to the content of sugars, sugar alcohols, myo-inositol, carboxylic acids and inorganic anions in developing grains from different varieties of Robusta (*Coffea canephora*) and Arabica (*C. arabica*) coffees. Plant Sci 149:115–123

Rovelli P, Mettulio R, Anthony F, Anzueto F, Lashermes P, Graziosi G (2000) Microsatellites in *Coffea arabica* L. In: Sera T, Soccol CR, Pandey A, Roussos S (eds) Coffee biotechnology and quality. Kluwer Academic Publishers, Amsterdam, pp 123–133

Santana-Buzzy N, Rojas-Herrara R, Galaz-Avalos RM, Ku-Cauich JR, Mijangos-Cortes J, Gutierrez-Pancheco LC, Canto A, Quiroz-Figueroa F, Loyola-Vargas VM (2007) Advances in coffee tissue culture and its practical applications. In Vitro Cell Dev Biol Plant 43:507–520

Scalabrin S, Toniutti L, Di Gaspero G, Scaglione D, Magris G, Vidotto M, Pinosio S, Cattonaro F, Magni F, Jurman I, Cerutti M, Liverani FS, Navarini L, Del Terra L, Pellegrino G, Ruosi MR, Vitulo N, Valle G, Pallavicini A, Graziosi G, Klein PE, Bentley N, Murray S, Solano W,

Al Hakimi A, Schilling T, Montagnon C, Morgante M, Bertrand B (2020) A single polyploidization event at the origin of the tetraploid genome of *Coffea arabica* is responsible for the extremely low genetic variation in wild and cultivated germplasm. Sci Rep 10:4642. https://doi.org/10.1038/s41598-020-61216-7

Schroth G, Laderach P, Dempewolf J, Philpott S, Haggar J, Eakin H, Castillejos T, Morena JG, Pinto LS, Hernandez R, Eitzinger A, Ramirez-Villegas J (2009) Towards a climate change adaptation strategy for coffee communities and ecosystems in the Sierra Madre de Chiapas, Mexico. Mitig Adapt Strateg Glob Chang 14:605–625

Silva MC, Varzea V, Guerra-Guimaraes L, Azinheira HG, Fernandez D, Petitot A-S, Bertrand B, Lashermes P, Nicole M (2006) Coffee resistance to the main diseases: leaf rust and coffee berry disease. Braz J Plant Physiol 18:119–147

Silvestrini M, Junqueira MG, Favarin AC, Guerreiro-Filho O, Maluf MP, Silvarolla MB, Colombo CA (2007) Genetic diversity and structure of Ethiopian, Yemen and Brazilian *Coffea arabica* L. accessions using microsatellite markers. Genet Resour Crop Evol 54:1367–1379

Silvestrini M, Maluf MP, Silvarolla MB, Guerreiro-Filho O, Medina-Filho HP, Vanini MMT, Oliveira AS, de Gaspari-Pezzopane C, Fazuoli LC (2008) Genetic diversity of a *Coffea* germplasm collection assessed by RAPD markers. Genet Resour Crop Evol 55:901–910

Teressa A, Crouzillat D, Petiard V, Brouhan P (2010) Genetic diversity of coffee (*Coffea arabica* L.) collections. Ethiop J Sci Technol 1(1):63–79

Tesfaye K, Borsch T, Govers K, Bekele E (2007) Characterization of *Coffea* chloroplast microsatellites and evidence for the recent divergence of *C. arabica* and *C. eugenoides* chloroplast genomes. Genome 50:1112–1129

Tesfaye K, Govers K, Bekele E, Borsch T (2014) ISSR fingerprinting of *Coffea arabica* throughout Ethiopia reveals high variability in wild populations and distinguishes them from landraces. Plant Syst Evol 300:881–897

Tessema A, Alamerew S, Kufa T, Geredew W (2011) Genetic diversity analysis for quality attributes of some promising *Coffea arabica* germplasm collections in Southwestern Ethiopia. J Biol Sci 11(3):236–244

Thomas AS (1942) The wild Arabica coffee on the Boma Plateau, Angle-Egyptian Sudan. Empire J Exp Agric 10:207–212

USDA FAS (United States Department of Agriculture Foreign Agricultural Service) (2020) Coffee: world markets and trade. https://apps.fas.usda.gov/psdonline/circulars/coffee.pdf

Van der Vossen HAM (2018) Developing varieties of Arabica coffee. In: Lashermes P (ed) Achieving sustainable cultivation of coffee. Volume 1: Breeding and quality traits. Burleigh Dodds Science Publishing, Cambridge. https://doi.org/10.19103/AS.2017.0022.05

Van der Vossen HAM, Walyaro DJ (2009) Additional evidence for oligogenic inheritance of durable host resistance to coffee berry disease (*Colletotrichum kahawae*) in arabica coffee (*Coffea arabica* L.). Euphytica 165:105–111

Vega FE, Ebert AW, Ming R (2008) Coffee germplasm resources, genomics, and breeding. In: Janick J (ed) Plant breeding reviews, vol 30. Wiley, New York, pp 415–447

Vega FE, Infante F, Castillo A, Jaramillo J (2009) The coffee berry borer, *Hypothenemus hampei* (Ferrari) (Coleoptera: Curculionidae): a short review, with recent findings and future research direction. Terres Arthropod Rev 2:129–147

Vega FE, Davis AP, Jaramillo J (2012) From forest to plantation? Obscure articles reveal alternate host plants for the coffee berry borer, *Hypothenemus hampei* (Coleoptera: Curculionidae). Biol J Linn Soc 107:86–94

Vieira ESN, Von Pinho EVR, Carvalho MGG, Esselink DG, Vosman B (2010) Development of microsatellite markers for identifying Brazilian *Coffea arabica* varieties. Genet Mol Biol 33(3):507–514

Yan L, Ogutu C, Huang L, Wang X, Zhou H, Lv Y, Long Y, Dong Y, Han Y (2019) Genetic diversity and population structure of coffee germplasm collections in China revealed by ISSR markers. Plant Mol Biol Report 37(3):204–213

# Part II
# Industrial Crops

# Origin, Genetic Diversity, Conservation, and Traditional and Molecular Breeding Approaches in Sugarcane

**Danilo Eduardo Cursi, Raul Oswaldo Castillo, Yusuke Tarumoto, Makoto Umeda, Amarawan Tippayawat, Werapon Ponragdee, Josefina Racedo, Maria Francisca Perera, Hermann Paulo Hoffmann, and Monalisa Sampaio Carneiro**

**Abstract** Modern sugarcane cultivars are highly polyploid and have giant genomes (10 giga bases (Gb)) derived from interspecific hybridization between the cultivated species S. *officinarum* L. and the wild species S. *spontaneum* L. Genetic resources could be useful for developing new varieties, and therefore, plant breeding programs are assembling a germplasm collection to increase the number of possible novel gene combinations. The use of wild relatives in sugarcane breeding started at the beginning of introgression during the nobilization process, and it is still used in

---

Danilo Eduardo Cursi, Raul Oswaldo Castillo, Yusuke Tarumoto, Makoto Umeda, Amarawan Tippayawat, Werapon Ponragdee, Josefina Racedo, Maria Francisca Perera, Hermann Paulo Hoffmann and Monalisa Sampaio Carneiro contributed equally with all other contributors.

---

D. E. Cursi (✉)
Sugarcane Breeding Program of RIDESA/UFSCar, Araras, São Paulo, Brazil
e-mail: daniloc@ufscar.br

R. O. Castillo
Centro de Investigación de la Caña de Azúcar del Ecuador—CINCAE, Guayaquil, Ecuador

Y. Tarumoto · M. Umeda
Sugarcane Breeding Group, Kyushu Okinawa Agricultural Research Center (KARC), National Agriculture and Food Research Organization (NARO), Kumamoto, Japan

A. Tippayawat · W. Ponragdee
Khon Kaen Field Crops Research Center, Khon Kaen, Thailand

J. Racedo · M. F. Perera
Instituto de Tecnología Agroindustrial del Noroeste Argentino (ITANOA), Estación Experimental Agroindustrial Obispo Colombres (EEAOC)—Consejo Nacional de Investigaciones Científicas y Técnicas (CONICET), Tucumán, República Argentina

H. P. Hoffmann · M. S. Carneiro
Sugarcane Breeding Program of RIDESA/UFSCar, Araras, São Paulo, Brazil

Department of Biotechnology, Vegetal and Animal Production, Federal University of São Carlos, Araras, São Paulo, Brazil


P. M. Priyadarshan, S. M. Jain (eds.), *Cash Crops*,
https://doi.org/10.1007/978-3-030-74926-2_4

breeding programs primarily to search for varieties that are more tolerant to biotic and abiotic stresses. The success of a traditional sugarcane-breeding program relies on several factors, among which the appropriate parental selection must be made to maximize the chance of genetic enhancement. This choice will be determined by the short- and long-term goals, the availability of materials, flowering synchronism, breeding values, and the amount of data available from any parent or combination. In general, the process of developing a new cultivar is long and complex. Genetic resistance to diseases has been successfully achieved through traditional breeding, although this approach is challenging and takes a long time. Several studies have been conducted to unravel and understand the genetic basis of disease resistance and complex traits (e.g., sugar and fiber, among others) through QTL and association mapping. In recent decades, important advances have been made in understanding the sugarcane genome and the gene expression associated with agronomic traits. Furthermore, transgenic sugarcane has been produced in several countries, and there have been numerous initiatives to employ genome editing technology. New breeding technologies and strategies are required to boost genetic improvements significantly in future crop cultivars. Genomic selection has the potential to increase the rate of genetic gain significantly in sugarcane, primarily by (1) reducing the breeding cycle length, (2) increasing the prediction accuracy for clonal performance, and (3) increasing the accuracy of the breeding values for parent selection. This chapter describes the origin, genetic diversity, conservation, and traditional and molecular breeding approaches associated with sugarcane.

**Keywords** *Saccharum* spp. · Germplasm collection · Wild relatives · Parental selection · Biotechnology

## 1 Origin of Sugarcane

Sugarcane has an interesting origin and long history, including countless millennia of geological habitat change, natural evolution among wild grasses, long centuries of association with primitive communities, and natural and human selection towards intensive hybridization in almost all sugarcane-producing countries. Physiologically, the sugarcane plant has evolved a unique source-to-sink system that makes it the only plant with its specific pathways for sugar synthesis and translocation as well as massive sugar quantity storage (Alexander 1973). Daniels and Roach (1987) stated that sugarcane is a large grass in the *Gramineae* family of the genus *Saccharum* within the *Andropogoneae* tribe. This tribe typically comprises tropical and subtropical species including food crops (*Sorghum* and *Zea*), fodder (*Iseilema* and *Themeda*), sugar (*Sorghum*, *Saccharum*, and *Zea*), essential oils (*Cymbopogon*), and archery, thatching, or housebuilding materials (*Imperata* and *Miscanthus*). Molecular data using chloroplast DNA markers have allowed for the establishment of more definitive relationships within the *Saccharinae* group, a broad relationship among the *Andropogoneae* and the likely polyphyletic origin of *Sorghum*, *Miscanthus,* and *Saccharum* (Spangler et al. 1999).

An extensive review of the origin and distribution of the *Saccharum* genus was proposed by Mukherjee in Mukherjee 1957 within the grouping of the "*Saccharum* complex," which includes the genera *Saccharum*, *Erianthus* (=sect. *Ripidium*), *Sclerostachya*, and *Narenga* in a closely related interbreeding group. Daniels and Roach (1987) suggested that all members of the *Saccharum* complex and its species occur in the Indo-Burma-China border region, suggesting that this area is the center of diversity (origin) for the group, and phylogenetic studies indicate that the *Miscanthus* sect. *Diandra* of the *Saccharum* complex is distributed from Tahiti in the Pacific to eastern Indonesia, Indochina and the north of China, Siberia, and Japan. Boundary species within the *Saccharum* genus are generally accepted as the ancestors of sugarcane, namely *S. officinarum* L. (thick-stalk sugarcane); *S. barberi* Jesw. (sugarcane of India); *S. sinense* Roxb. (sugarcane of China); *S. spontaneum* L. (wild cane found from New Guinea to the Mediterranean and Africa); *S. robustum* Brandes and Jesw. (wild cane from Indonesia/New Guinea); and S. *edule* Hassk. (edible inflorescence sugarcane, New Guinea and Melanesia). There are several hypotheses that *S. edule* was related to the origin and use of sugarcane; Artschwager and Brandes (1958) as cited by James (2004) suggested that a common ancestor of sugarcane originated in southern Asia and spread southeastwards across an ancient land mass that extended from Asia to Australia during the early Cretaceous period (60 million years ago). There were two periods of sugarcane movement, with the earliest one dating back to the Cretaceous period when a large continent provided a continued land bridge where primitive Asiatic canes could have moved to Melanesia with a common wild ancestor. The second period is related to the flood conditions during the later Cretaceous, which imposed restrictions on the movement of flora, and many plants survived thanks to the establishment of seed repositories. Seed produced by early *Saccharum* may not have survived, and therefore, it is better concluded that distinct wild forms of sugarcane evolved in isolated spots in Asia, New Guinea, and other Melanesian islands. Considerable changes might have taken place by natural selection and hybridization. Primitive humans also took part in the process by selecting and disseminating materials during later periods as appreciated today regarding advances in chewing canes, which were planted by local communities in New Guinea (James 2004).

Although no evidence is presented, Barnes (1974) wrote a taxonomic discussion maintaining that India was the home of sugarcane based on Hindu mythology. In addition, Parthasanrathy wrote in Parthasanrathy 1946 and was cited by James (2004) suggesting two broad groups: (1) northern Indian canes with *S. barberi* and *S. sinense* parentage; these species are characterized by thin, hardy stalks and (2) noble canes of *S. officinarum* parentage, which have thick stalks. In their detailed taxonomic and evolutionary treatment, Daniels and Roach (1987) suggested that the oldest and most acceptable center of origin of *S. officinarum* was the Polynesian islands in the South Pacific, based on botanical samples from Cook's second voyage (1772–1775).

## 2 Genetic Diversity

The *Saccharum* genus (Table 1) traditionally comprises six species: *S. spontaneum*, *S. officinarum*, *S. robustum*, *S. edule*, *S. barberi*, and *S. sinense* (Daniels and Roach 1987; D'Hont et al. 1998). However, Irvine (1999) suggested that the genus should be reduced to just two species, by grouping *S. robustum*, *S. edule*, *S. barberi*, *S. sinense,* and *S. officinarum* together in the species *S. officinarum* and leaving *S. spontaneum* as a separate species.

The range of variability and its ecogeographical distribution and adaptation include all related gene pools associated with sugarcane. Functional classifications of plant genetic resources or the genetic diversity of land races, advanced cultivars, weeds, and wild relatives of domesticated plants are also applied when studying sugarcane (Frankel and Bennett 1970). All genetic resources are useful in developing new varieties, and therefore, plant breeding programs are assembling germplasm collections to increase the possibilities of novel gene combinations. In accordance with the biological species concept coined by Harlan and de Wet (1971), to understand the implications of taxonomy, evolution, and practical plant breeding, the ability of unreduced gametes to operate in *Saccharum* and its complex must be considered, and Daniels and Roach (1987) proposed the following gene pools (GP):

**GP1**. Crossing is easy among members of the pool, with *S. officinarum*, commercial hybrids, *S robustum*, *S. spontaneum*, and any genera involved in the origin of *S. officinarum* (*Erianthus* sect. *Ripidium*, *Miscanthus*, and *Sclerostachya*). This group lies within the classical biological species concept and has been used most frequently in sugarcane-breeding programs since their beginning (Fig. 1).

**GP2**. Gene transfer is possible by crossing, but there is difficulty with the barrier separating biological species, including the remaining species from the *Saccharum* complex and potentially from *Saccharastrae*.

**GP3**. Other genera that might be crossed with *Saccharum* are *Sorghum* and *Zea*. The crosses might be difficult or require special in vitro techniques to assist in plant development. However, some progenies might be sterile.

**Table 1** Chromosome number, sugar content, and species type of *Saccharum* genus members

| Species | Chromosome number | Sugar content | Type |
|---|---|---|---|
| *S. spontaneum* L. | $2n = 40$–$120$ | Very low to low | Wild species |
| *S. robustum* Brandes/ Jeswiet ex Grassl | $2n = 60$–$200$ | Very low | Wild species |
| *S officinarum* L. | $2n = 80$ | High | Noble canes |
| *S. barberi* Jeswiet | $2n = 111$–$120$ | Low | Ancient hybrid |
| *S. sinense* Roxb. | $2n = 80$–$124$ | Low | Ancient hybrid |
| *S. edule* Hassk | $2n = 60$–$80$ (included aneuploid forms) | Low<br>The inflorescence is eaten as a vegetable | Cultivated species |

Source: Daniels and Roach (1987)

**Fig. 1** *S. officinarum*, *S robustum*, *S. spontaneum,* and *Erianthus* accesses (**a**) commercial hybrids derived from different genetic backgrounds (**b**, **c**)

With current biotechnological techniques, all species of any genetic resource from the distinct gene pools may be used to transfer genes from any species. For instance, genetic transformation by inserting exogenous rearranged genes into a plant is one method; CRISPR technology is another powerful tool that can be used for editing genomes by altering DNA sequences and modifying gene function. Gene editing using CRISPR-Cas9 can silence or activate certain genes of interest (Bolotin et al. 2005; Martin et al. 2012; Mohan 2016). Therefore, in terms of diversity, all breeders must recognize that every genus or species is a useful plant genetic resource for conservation and use.

Modern cultivars ($2n = 100$–120) are highly polyploid and aneuploid, with meiotic abnormalities and a giant genome estimated to be approximately >10 giga bases (Gb) in size. The *Saccharum* complex (*Saccharum* spp.) was derived from interspecific hybridization performed a century ago between the sugar-producing autopolyploid species *S. officinarum* L. ($2n = 80$) and the wild species *S. spontaneum* L ($2n = 40$–128) (Piperidis et al. 2010a; Vieira et al. 2018). In addition, all the hom(oe)ologous genes were predicted to be functional (except for one gene fragment) and showed signs of evolving under purifying selection, except for genes within segmental duplications. By contrast, compared with *Sorghum*, its close relative in the *Andropogonea* tribe with transposable elements, displayed a general absence of collinearity among hom(oe)ologous haplotypes and appeared to have undergone dynamic expansion in *Saccharum* (Piperidis et al. 2010b). Among the chromosomes of cultivar R570 ($2n = 107$–115), approximately 10% were identified as originating from *S. spontaneum* and approximately 10% were identified as recombinant chromosomes between the two species *S. officinarum* and

*S. spontaneum*. In another previous report, D'Hont et al. (1996) were the first to demonstrate the occurrence of recombination between the chromosomes of these two species. To decrypt the genome structure, Piperidis and D'Hont (2020) used specific chromosome oligo probes to validate a basic chromosome number of $x = 10$ for *S. officinarum* and $x = 8$ for *S. spontaneum* with several rearrangements, and they showed that *S. spontaneum* contribution to modern cultivars varied between 13 and 20%. Studies by Garsmeur et al. (2018) involving sequence analysis using bacterial artificial chromosome (BAC) systems for the highly complex genome of sugarcane confirmed that the two species involved in modern cultivars differ with respect to their LTR retrotransposons (*S. officinarum* and *S. spontaneum*). The most abundant group of transposable elements (TEs) in plants is found inside or close to genes and therefore influence their expression and evolution. Modern cultivars differ in terms of their transposable elements and a few large chromosomal rearrangements, explaining their distinct genome size and distinct basic chromosome numbers while also suggesting that polyploidization arose in both lineages after their divergence in the *Saccharum* genus.

Hybridization and the development of new varieties in sugarcane is a recent event compared to other crops. The process of introgression in sugarcane breeding involves performing crosses between *S. officinarum* and *S. spontaneum* and a series of backcrossing; the resulting hybrids with *S. officinarum* were developed in Java at the start of the twentieth century by a group of Dutch breeders. Interspecific crosses were initially performed by Soltwedel and Jeswiet in 1885 in Java (James 2004), which is known as "nobilization," followed by a series of hybridizations primarily using *S. spontaneum* in introgression breeding. Since 1912, the Sugarcane Breeding Institute at Coimbatore has been engaged in improving varieties for the subtropical regions of India. In 1926, it expanded its work into tropical regions. The Barbados Cane-Breeding Program in the Caribbean has flourished since 1887, which was when the Caribbean Sugar Industry was revived (Galloway 1996); it was later named The West Indies Central Sugar Cane Breeding Station Incorporated (WICSCBS Inc). It is an internationally recognized sugarcane-breeding operation that provides fuzz (true seed) or varieties to regional and international breeding partners.

Mechanisms of polyploidization and mechanisms of $2n$ transmission through egg cells have been studied in sugarcane and were reported by Price (1963), suggesting few hypotheses to explain the mechanism of $2n + n$ transmission in sugarcane. They are: (1) the formation of unreduced egg cells, chromosome doubling through endo-duplication either at the dyad or at the tetrad stage, the postmeiotic fusion of the two innermost megaspores, and postmeiotic endomitosis in the egg cell; (2) the incompatibility of $n + n$ gametes due to either selective fertilization or a combination of selective fertilization and parthenogenesis, i.e., the differential survival of $2n + n$ and $n + n$ zygotes; and (3) the failure of certain zygotic combinations due to faulty endosperm development (selective survival). However, while these approaches account for $2n + n$ zygotes, they do not explain the transmission of the n chromosome number in *S. officinarum* on selfing or intraspecific crossing; $2n + n$ transmission occurs only after pollination and fertilization following interspecific crossing. Therefore, it could be inferred that *S. officinarum* produces both reduced (haploid) and unreduced egg cells. These mechanisms are important properties to note in breeding programs for genetic diversification, the introgression of new varieties, and the use of wild germplasm.

## 3 Ex Situ and In Situ Germplasm Collections of the *Saccharum* Group

Germplasm collections can range from collections of wild species to elite domesticated breeding lines that have undergone extensive human selection. Germplasm collections are important for the maintenance of biological diversity and food security. In 1974, the foundations working on genetic resources realized that 10 years after the Green Revolution, modern, high-yielding crop varieties would sweep through the world's fields, wiping out traditional varieties of the major crops. The success of modern breeding was undermining its own future by eliminating the genetic diversity on which breeders depend. An organization called the International Board for Plant Genetic Resources, or IBPGR (later named IPGRI and Bioversity International), set out on a race against time to collect as much genetic diversity as possible for priority crops (primarily major food crops) and to ensure that it was conserved for posterity. In 1987, the Food and Agriculture Organization (FAO) of the United Nations (UN) and national governments responded to this shift by working on an International Treaty for Plant Genetic Resources for Food and Agriculture and a Multilateral System for sharing germplasm (Bioversity International 2019). Sugarcane is an important industrial cash crop and was therefore also included in the International Treaty, and today, germplasm access through many of the established collections can be cumbersome. However, no country is self-sufficient, and each country depends on others for approximately 70% of the genetic resources for their primary crops.

Germplasm collections from the *Saccharum* complex were sponsored by IBPGR, while others were performed based on local plant breeding needs and research projects. A comprehensive review of those collections is described by Berding and Roach (1987), including a position regarding the conservation of the *Saccharum* complex of the two major global collections. They stated that the Miami, Florida, the USA collection is not optimally located due to its environmental stress effects on plant growth in comparison with the Kannur, Coimbatore, India collection, with its much better location. However, the Kannur collection has not been as available as expected. In addition, the Miami collection is not well maintained due to a lack of resources, resulting in a high risk of genetic erosion and a direct loss of accessions.

There has been concern regarding the conservation and distribution of the global collections held in India and the USA. During the International Society of Sugarcane Technologists (ISSCT) Workshop held in Brazil (Castillo et al. 2013), a panel discussion on this topic was arranged with the two curators, Dr. Vijayan Nair, Director of Sugarcane Breeding Institute, Kannur, India and Dr. Tomas Ayala, the curator at ARS, USDA. Both germplasm collections are available for the ISSCT members. The report from India shows that the collection consisting of 1819 clones is well maintained and is clonally propagated every year. The clones are characterized, and the catalogs are available from the Sugarcane Breeding Institute upon request. The report shows that the germplasm has been sent to more than 256 countries. The varieties can be obtained following an official request before June (one shipment of not more than 25 clones per year per country). For the Miami collection, Dr. Ayala, curator at ARS, USDA, informed at the meeting that a database on the 887 sugarcane clones is available and can be consulted at the following website address: www.ars-grin.gov. Clones can also be obtained from this collection, but freight charges must be paid by the importing country. Following recommendations from the ISSCT workshop with the support of the ISSCT Breeding Committee, the group has launched web access to review the current status of the germplasm and the commercial varieties in each country (www.issct.org/ or www.sugarcanevariety.org/default.aspx). National breeding programs in each country also maintain collections of different numbers, and some of them exchange materials after specific memorandum of understanding (MOU) arrangements. An active exchange source of sugarcane genetic clones/varieties comes from CIRAD through the Visacane system, including the quarantine services (Guinet-Brial et al. 2016).

Ex situ conservation using in vitro techniques is an alternative to guarantee the safe maintenance of germplasm collections since in field conditions there are risks of genetic erosion due to biotic and abiotic factors. In vitro preservation has several advantages, such as: small space required to maintain access, insects and disease-free plants, simplicity, and efficiency in germplasm exchange. However, this alternative can be expensive and cumbersome, as it requires the construction of specific facilities and the need for specialized manpower to maintain the accesses. In vitro collection options have been developed for sugarcane using storage methodologies through slow growth and cryopreservation, including axillary buds, meristem tip micropropagation, and developing adventitious shoot formation and embryogenesis callus (Maretski 1987; Engelmann et al. 2006; Volk et al. 2019). Several techniques have been developed for a massive plant propagation operation that can be used for conservation based on media and temperature parameters (Sarwar and Siddiqui 2004; Temam and Alemayehu 2017). However, some limitations such as somaclonal variation, in vitro contamination, and low efficiency rates for tissue regeneration maintained for a long term still remain a challenge for a greater adoption of this strategy in the conservation of sugarcane germplasm. Research on in vitro conservation techniques using minimal culture media and with osmo-protectant are underway at ICAR—Sugarcane Breeding Institute, Coimbatore, India by Dr. Chandran K. (personal communication).

The use of wild sugarcane relatives was employed from the beginning of introgression during the nobilization process, and it is still used in breeding programs. Therefore, the in situ conservation of wild sugarcane relatives becomes very important. However, no specific reports on in situ conservation have been found for sugarcane so far, and several examples for home gardens and in situ conservation of plant genetic resources in farming systems, including cultivated sugarcane clones, are reported in Watson and Eyzaguirre (2002) and Maxted et al. (2011). The FAO and the International Conservation Union (IUCN) suggested preserving wild relatives in protected areas by management categories that would be compatible with genetic reserve nominations, with Category I—strictly protected reserves (often small) set aside and left untouched to protect particular species under threat; Category II—large ecosystem-scale protected areas maintained to allow wild sugarcane to continue to flourish and evolve under natural conditions; and Category III—small reserves managed to maintain specific species; for example, through controlled use or cutting to retain important grassland habitat, coppicing to maintain woodland ground flora, or sometimes even intervening to restore the habitat of threatened *Saccharum* complex species (Maxted and Kell 2009; Stolton et al. 2006).

## 4 Wild Cane *Saccharum spontaneum* for Biotic and Abiotic Resilience

The sugarcane industry is strongly influenced by the impacts of biotic and abiotic stresses, and an understanding of these potential impacts is essential for plant breeding. Some adversities are commonly faced by different sugarcane-producing countries, such as drought, low temperatures, and low soil fertility. In addition, some countries also suffer from natural disasters, e.g., typhoons in Japan. There, this natural phenomenon frequently causes serious damage to sugarcane yields and sugar contents through losses of green leaves, lodging, and broken stalks.

To maintain stable cane production under those environment conditions, since 1980, the Kyushu Okinawa Agricultural Research Center, National Agriculture and Food Research Organization (KARC/NARO) has been working in collaboration with the Tokunoshima branch, Kagoshima Prefectural Institute for Agricultural Development (TB/KIAD), Okinawa Prefectural Agricultural Research Center (OPARC) and Tropical Agriculture Research Front, Japan International Research Center for Agricultural Sciences (TARF/JIRCAS) on several crosses between commercial cane varieties and wild cane (*S. spontaneum*) (Nagatomi 1982). *S. spontaneum* is an important breeding material because of its high tolerance to biotic and abiotic stresses as well as its high ratooning ability. However, it has unfavorable traits such as a thin stalk diameter and low sugar content. This wild cane grows naturally in Japan (primarily in coastal regions), and the exploration of Japanese *S. spontaneum* started in 1977 (Nagatomi and Degi 2007); an additional advantage of this wild Japanese cane is its cold tolerance. This effort has resulted in the collection of more than 500 accessions, which are currently preserved by the Ministry of Agriculture, Forestry and Fisheries (MAFF) of Japan gene bank project. These accessions have been distributed throughout Japan from 24° to 36° north latitude, and this region is the northernmost area of the world where *S. spontaneum* grows naturally (Nagatomi and Degi 2007). More recently,

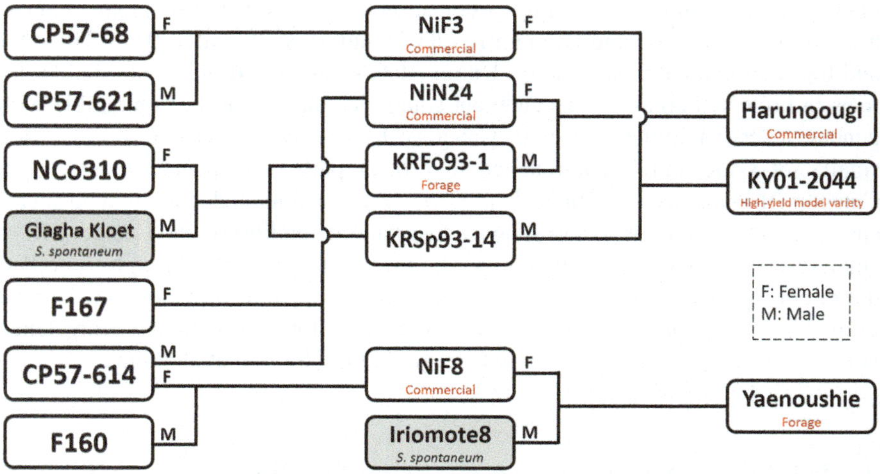


**Fig. 2** Pedigree record of varieties derived from S. *spontaneum*

the discovery of smut resistance in the Japanese *S. spontaneum* collection has led to the more frequent use of these wild canes in breeding (Sakaigaichi et al. 2019). A new forage sugarcane variety has been developed, "Yaenoushie," using one of the high smut-resistant accessions, "Iriomote8," in 2018 (Fig. 2). The establishment of pollen preservation technique contributed to this effort. The use of these smut-resistant accessions will improve the resulting smut-resistance level dramatically because most of the breeding clones derived from "Glagah Kloet" (an *S. spontaneum* originated from Indonesia) are susceptible to this disease. Nevertheless, to use these accessions more efficiently, molecular techniques are used in joint research with the Toyota Mortar Corporation. Recently, a QTL for smut resistance inherited from "Iriomote8" with a markedly high LOD score (Umeda et al. 2019) was detected. This result implies the existence of a major gene that controls smut resistance. A marker-assisted selection system (MAS) using this knowledge is under development.

In addition, *Erianthus arundinaceus* (Retz.) Jeswiet, a related genus of sugarcane, has received attention as a breeding material because of its high biomass productivity in ratoon crops and its excellent adaptability to adverse environmental factors such as drought (Jackson and Henry 2011). Therefore, many sugarcane-breeding countries have been crossing it with sugarcane (D'Hont et al. 1995; Besse et al. 1997; Piperidis et al. 2000, 2010b; Cai et al. 2005; Nair et al. 2006; Krishnamurthi et al. 2007; Fukuhara et al. 2013). However, a low production efficiency has been reported among intergeneric hybrids (Piperidis et al. 2000), and few examples of E. *arundinaceus* germplasms have been considered. Hence, effective crossing techniques to use diverse E. *arundinaceus* germplasm for sugarcane breeding have been important (Terajima et al. 2019).

## 5 Traditional Breeding Approach

The success of a traditional sugarcane breeding program relies on several factors, among which is the appropriate parental selection for maximizing the chance of genetic progress (Mbuma et al. 2020). An important feature of most effective breeding programs is the rapid renew of the parental population through the introduction of elite clones that have been identified by their good performance for key traits during advanced stages of selection, the introduction of elite foreign clones as well as the discarding of unproductive parents (Jackson 2018). The parental selection is one of the most challenging and important decisions for plant breeders. This choice will be determined by the short- and long-term goals (commonly, the main objectives of breeding programs are high cane and sugar yield, disease resistance, sprouting, and ratoonability, among others), the availability of materials, flowering synchronism, breeding values, and the amount of data available from any parent or combination (Heinz and Tew 1987).

In sugarcane, additive genetic variance is more important than nonadditive variance for most traits of economic importance, including sucrose content, fiber, and disease resistance. The primary exception is for cane yield, for which additive and nonadditive genetic variances appear to be of equal importance (Hogarth 1987; Jackson 2005). According to Bastos et al. (2003), when there is a predominance of additive genetic effects, genotypes can be selected based on its per se performance or their general combining ability (GCA). Conversely, when the nonadditive genetic effects are important for a trait, there is a possibility of increasing the genetic gain by exploiting heterosis through the selection of crosses with high specific combining abilities (SCA). The disconnected factorial scheme is an alternative crossing design that allows for the possibility of identifying superior crosses and the ability to evaluate a large number of parents due to better flexibility regarding the number of crosses per parent (Resende and Barbosa 2005; Mendes et al. 2020). Burdon and van Buijtenen (1990) suggest the formation of small groups of crossings, usually 3 × 3 or 4 × 4, to provide a reliable estimate of both sources of variation. Another appropriate strategy that should be considered by sugarcane breeding programs to exploit both additive and nonadditive effects is the reciprocal recurrent selection (RRS) (Barbosa et al. 2005; Jackson 2018). According to Barbosa (2000), RRS can be applied at an individual level (IRRS), where the additive genetic effects would be better estimated by topcross experiments, i.e., a group of clones of a certain population receives pollen from a clone of another population and vice versa, resulting in predictions of the interpopulation additive effects, that is, information on the interpopulation GCA. The clones with the best combining ability would be used in crossings aimed at SCA exploitation. It is suggested that the individuals originating from superior crossing should be self-fertilized to produce inbred populations (S1), from which superior individuals would be selected to be part of an individual RRS program. Moreover, RRS can be applied at a population level, involving the crossing of several individuals from one population with individuals from the reciprocal population (Resende and Barbosa 2005).

These procedures commonly address unbalanced data. The Restricted Maximum Likelihood (REML) and Best Linear Unbiased Prediction (BLUP) have been widely adopted by animal and plant breeding programs (Bernardo 2002; Piepho et al. 2008) since they allow for the simultaneous correction of environmental effects, estimations of variance components, and predictions of genetic values; they address complex data structures (repeated measurements, different years, locations, and designs); and they can be applied to unbalanced data and nonorthogonal designs (Resende 2004).

These statistical methods are commonly used in family experiments, which in turn focus on the appropriate parental selection as well as a strategy to select or reject an entire group of individuals with respect to family structure according to the average phenotypic value of the family (Cursi et al. 2020). Among the widely practiced different family selection methods, the Australian proposal (McRae et al. 1998; Stringer et al. 2011), the Modified Sequential Selection (Bressiani 2001), and the Selection by Simulated Individual BLUP—BLUPis (Resende and Barbosa 2006) stand out.

The Australian proposal of family selection is a combination of selections between and within families, in which the selection of the best families ("between") is performed at the plant cane to assess the performance of each family at producing tons of cane, sugar, and fiber. The selection is then concentrated on the families that performed the best ("within"). Only the top 40% families are considered for selection from the first ratoon crop, while the families with the worst performances (below 40%) are discarded as a whole. In addition, the number of individuals to be selected above 40% (best families) will depend on their performance in terms of cane plants. For example, 32 out of a total of 80 individuals can be selected from the top 10% families (i.e., 40% of the individuals per family), 24 between 10 and 20% (30% of selection intensity), 16 between 20 and 30% (20% of selection intensity), and 8 between 30 and 40% (10% of selection intensity). This approach indicates that the best families are selected with the highest selection intensities (Stringer et al. 2011).

The Modified Sequential Selection uses the idea of the Australian model and simultaneously seeks to remove the existing subjectivity from the definition of the total classes and from the combination of the selection rates between and within each class. In this approach, each family becomes a class, and the selection rate among individuals, within each family, is defined in such a way that the average of the selected individuals in each family is the same for all. In this way, the expected progress through the selection is kept constant, in absolute values, for all families (Bressiani 2001).

According to Resende and Barbosa (2006), the optimal strategy for selecting sugarcane individuals during the early breeding stages would be through the genotypic values predicted by the individual BLUP (BLUPi), which simultaneously includes family and individual information for selection. However, this method has not been used in sugarcane due to its practical difficulties in obtaining data from individual plants (Cursi et al. 2020). In seeking to overcome this adversity, Resende and Barbosa (2006) proposed a new selection strategy via simulated individual

BLUP (BLUPis), which promotes the dynamic allocation of individuals to be selected within each full-sib family by using BLUP as a basis for the genotypic effects and plot effects for that family. In other words, the BLUPis methodology determines the number of individuals to be selected per family, the total number of clones to be advanced to the next generation, and the number of families to contribute to the selected individuals. In addition, BLUPis allows for the identification of which repetition the best individuals are present in. For this purpose, the data are obtained at the family level, through the mechanical harvesting and total weighing of the plots. Basically, the form of data collection in the BLUPis procedure is similar to that performed in the Australian sequential selection (BLUPseq); according to Cursi et al. (2020), both methods are highly correlated.

Despite the proven efficiency of family selection methods, mass (or individual) selection is still widely used by sugarcane breeding programs during the early breeding stages (Cox 1996; Bressiani 2001; Almeida 2010). Since the phenotype is the only selection criterion, its effectiveness is obviously a function of the heritability of the selected trait. According to Borojevic (1990), the success of mass selection depends on the following factors:

1. Genetic action for a desired trait: In this case, if the effect is additive, the success of the selection will be greater than if it is dominant, since it is a heritable effect that passes from generation to generation. This factor does not apply to sugarcane, since cloning is involved.
2. Heritability: When high, an individual's phenotype proves to be a predictor of its breeding value.
3. Genotype x environment interaction: When this type of interaction is high, which is the case for low heritability traits, there are greater challenges in identifying the superior genotypes.
4. Sample size: This is another factor of great importance in mass selection, primarily for opening pollinated plants that must be sampled on a large scale to avoid endemic depression.

According to Stringer (2006), early-stage experiments are subject to variations due to spatial variability and due to competition between plots, which can seriously compromise genotype estimates and therefore reduce genetic progress. An alternative is to use analytical methods in which a plot is adjusted for spatial variability using information from the closer neighbors. This method of analysis, which is commonly known as spatial analysis, has been successfully and routinely used in grain breeding programs in Australia (Gilmour et al. 1997).

In the following stages, the experimental precision must be increased due to the use of a greater number of repetitions and evaluation sites. Typically, a randomized block design is used, and it is arranged in plots that are subdivided over time (different cuts). Stability and adaptability studies should also be performed, facilitating the cultivar recommendations.

Finally, when the experimental precision allows for reasonably secure recommendations after evaluations of at least four cuts from different years and cultivation environments, the cultivars can be released commercially (Matsuoka et al. 1999).

## 6 Biotechnology Tools

### *Omic Approaches and Databases*

Biotechnology research offers excellent opportunities for crop improvement, especially for sugarcane, if the complex genetics of the commercial varieties are considered. In recent years, numerous initiatives to understand the complexity of the sugarcane genome have produced several sequencing projects around the world. A reference for the euchromatin gene-rich part of the genome in cultivar R570 was assembled, providing an essential genome template for aligning sequencing data. In this project, a minimum tiling path of 4660 sugarcane BACs that best covers the gene-rich part of the sorghum genome was sequenced and then assembled in a 382 Mb single tiling path. More than 25,000 protein-coding gene models were predicted, 17% of which display no collinearity with their sorghum orthologs (Garsmeur et al. 2018). Additionally, the United States completed the assembly of one haploid *S. spontaneum* genome derived from SES208 (Zhang et al. 2018). Furthermore, a gene space assembly of the Brazilian variety SP80-3280, including 373,869 putative genes (>99% of the total genes) and their potential regulatory regions, showed that this genotype has specific regulatory elements involved in sucrose synthesis that are not found in the ancestor *S. spontaneum* (Souza et al. 2019). In addition, genomes of varieties from Thailand (Khon Kaen KK3) and Colombia (CC01-1940) were also sequenced (Shearman Jeremy et al. 2018; www.cenicaña.org, 2020).

Over the past decade, the use of several high-throughput methods has exponentially increased the amount of data, providing useful information for expression studies in sugarcane. The SUCEST-FUN Project (http://sucest-fun.org) database is an important resource for managing sugarcane genome data with information about transcriptome and phenotype analyses of the primary Brazilian cultivars and the identification of genes associated with the sucrose content, yield, lignin, and drought (Dal-Bianco et al. 2012; Waclawovsky et al. 2010; Papini-Terzi et al. 2009; Ferreira et al. 2016). Another important database is the Sugarcane Genome Hub (https://sugarcane-genome.cirad.fr), which is an integrated web-based database providing centralized access to the monoploid reference genome sequence of the R570 cultivar.

Several transcriptome analyses have been conducted on different varieties for different purposes. For example, in the transcriptome profile of lignin content from varieties IACSP04-065 and IACSP04-627, more than 2000 transcripts that showed differential expression were found, including several genes involved in the lignin biosynthetic pathway (Vicentini et al. 2015). Cardoso-Silva et al. (2014) studied the leaf transcriptome for six sugarcane genotypes, which produced 72,269 unigenes that showed significant similarity to more than 28,788 sorghum proteins, including 5272 unigenes in which a large number of molecular markers were identified. Notably, Dantas et al. (2020) studied the differences between the rhythmic transcriptomes under circadian and field conditions, and they highlighted the importance of field experiments in understanding how the plant circadian clock works in natura. Cunha et al. (2017) studied ethylene-induced responses, and they discovered a correlation between hormone levels and transcriptional changes that suggests ethylene as a trigger of multiple hormone signal cascades. In addition, posttranscriptional gene regulation during water stress has been described in sugarcane (Ferreira et al. 2017). The analyses that focused on the expression of sugarcane miRNAs (Ferreira et al. 2012; Gentile et al. 2013) demonstrated that the microtranscriptome is modulated in different genotypes and growth phases to cope with different stress intensities (Gentile et al. 2015).

Sugarcane proteome research is quickly evolving from merely cataloguing proteins under static conditions to comparing dynamic changes in the proteome in response to abiotic and biotic stress factors (Thelen and Peck 2007). Differential proteomics of sugarcane varieties under different physiological conditions would indicate the alterations in the proteome and would allow researchers to unravel the cellular and molecular mechanisms governing its ability to adapt to the stress situation, e.g., drought-tolerant mechanisms (Sugiharto et al. 2002) and salt-tolerance mechanisms (Murad et al. 2014). Even more notably, in physiological and proteomic analyses by Pacheco et al. (2013) on drought tolerant and susceptible sugarcane varieties under salt stress (200 mM), the proteins involved in growth, development, carbohydrate metabolism, ROS scavenging, and membrane stabilization were identified, indicating that their expression would be crucial in salinity tolerance. Although sugarcane proteomes of stalk tissue have not been explored due to the rigidity, fibrous nature, the presence of oxidative enzymes, phenolic compounds, and extreme levels of carbohydrates of stalk tissue, Amalraj et al. (2010) established a reference map for the sugarcane stalk tissue proteome after optimizing the extraction method.

The metabolic composition exerts effects on several important agronomic traits, and metabolomics, which represents the chemical composition in a cell, is a powerful tool for bridging phenotype–genotype interactions (Ferreira et al. 2018). For example, the metabolic networks in buds and culms, the tissues involved in the vegetative propagation of sugarcane, were studied. The authors noted that primary metabolite profiling provided information on the variability of metabolic features, even under a narrow genetic background (Ferreira et al. 2018). In another study, proteomics and metabolomics data were combined to understand how two contrasting commercial varieties, CTC15 (tolerant) and SP90-3414 (susceptible), adjust their leaf metabolism in response to drought. A network analysis revealed that different biological processes had a stronger impact on each variety (Budzinski et al. 2019). In addition, physiological and metabolic alterations in two contrasting genotypes grown under high- and low-nitrogen conditions were assessed. Bassi et al. (2018) found that the N supply and the responsiveness of the genotype determined the degree of senescence, the carboxylation process mediated by phosphoenolpyruvate carboxylase, and the differential accumulation of soluble sugars.

## *Molecular Markers for Disease Resistance*

Different bacterial, fungal, and viral diseases occur annually in sugarcane production areas worldwide, negatively affecting yields and causing severe phytosanitary problems. Although sugarcane breeders try to select resistant cultivars, it is very difficult to introduce resistance against all pathogens at the same time through conventional breeding approaches. In that context, several studies have been conducted to unravel and understand the genetic basis of complex traits and detect the molecular markers responsible for disease resistance in sugarcane (Aitken et al. 2008; Nibouche et al. 2012; Aljanabi et al. 2007). For example, Borrás-Hidalgo et al. (2005) and You-Xiong et al. (2011) identified genes that could contribute to disease resistance to sugarcane smut, which is caused by *Ustilago scitaminea*, through cDNA-AFLP analysis. A more recently developed approach, RNA-Seq, was used to investigate the changes in temporal transcript expression in the immune system when sugarcane was challenged with *Puccinia kuehnii* (orange rust). The authors revealed that *P. kuehnii* initially suppressed sugarcane genes involved in plant defense systems, and the late overexpression of specific regulatory pathways also revealed the existence of an inefficient recognition system by a susceptible sugarcane genotype (Correr et al. 2020). Other pathosystems such as sugarcane-ratoon stunting (*Leifsonia xyli* subs. *xyli*) (Cia et al. 2018) were also studied.

Taking into account the multilocus architecture of sugarcane, the development of diagnostic markers that can be applied routinely within a breeding scheme as an MAS approach is a challenge. However, some studies provide useful knowledge for the development of future applications, such as the identification of molecular markers tightly linked to the first well-characterized major Mendelian dominant gene, *Bru1* (Daugrois et al. 1996), which contributed greatly to both the scientific community and sugarcane breeding programs focused on brown rust resistance caused by *P. melanocephala*. The frequency and effectiveness of *Bru1* was studied extensively in diverse germplasm around the world (Costet et al. 2012; Glynn et al. 2012; Molina et al. 2013; Racedo et al. 2013; Parco et al. 2014; Li et al. 2017; Barreto et al. 2017; Neuber et al. 2017). Diagnostic markers tagging *Bru1* locus allowed researchers to identify additional sources of brown rust resistance, alerting them about threatening situations that can lead to the breakdown of *Bru1* resistance (Glynn et al. 2012; Neuber et al. 2017; Li et al. 2018). Furthermore, two different studies that shed light on the genetic resistance of sugarcane to orange rust identified molecular markers linked to resistant genes. By analyzing a biparental mapping population, Yang et al. (2018) revealed three QTLs that control orange rust resistance, and together, they can reduce the disease severity by 56.4%. These QTLs were then confirmed in the diverse sugarcane panel of an associated study (Yang et al. 2019). According to these studies, gene-derived marker G1 was significantly associated with orange rust resistance, being recently evaluated as a successful molecular tool by Brazilian breeding programs (Fier et al. 2020). Notably, McCord et al. (2019) identified SNPs that control a portion of the variability in orange rust resistance in the sugarcane breeding germplasm at Canal Point, Florida. They concluded that the markers they found may not be sufficient for selecting individual clones that are resistant to orange rust, since the single most important marker explained approximately only 11% of the trait variation. However, they proposed these markers as a useful tool for choosing parents and parental combinations. Another diagnostic molecular marker for sugarcane resistance screening is available for smut. It is a 300-bp RAPD-derived Sequence Characterized Amplified Region (SCAR) marker developed by Khan et al. (2017). Interestingly, Wei et al. (2006) detected associations between markers and several sugarcane diseases (23 markers explained 62.9% of the phenotypic variation in smut resistance; five markers explained 26% of variation in leaf scald resistance; four markers explained 32% of the phenotypic variability in *Pachymetra* root rot; and five markers accounted for 26% of variations in resistance to Fiji leaf gall). Costet et al. (2012) used biparental mapping to reveal the first major QTA known to control a viral disease in sugarcane, *Ryl*1, accounting for 32% of the phenotypic variance in *sugarcane yellow leaf virus* (SCYLV) resistance. Although no diagnostic molecular markers linked to this QTA were developed, MQ76-53 can be useful for introgressing the SCYLV resistance derived from *Ryl*1 into elite germplasm. After that, in an association mapping study involving a cultivar panel ($n$ = 189), which is representative of the global sugar-producing germplasm, six independent markers significantly associated with the SCYLV-resistance phenotype were detected (Débibakas et al. 2014). In addition, Gutierrez et al. (2018) detected a marker from a biparental population that could be used, in combination with other linked SNPs, as leaf scald (*Xanthomonas albilineans* (Ashby) Dowson) resistance diagnostic markers in MAS.

## *Genetic Diversity Studies*

Knowledge of the genetic diversity in sugarcane provides useful information concerning its genotype value and contributes to the improved use and conservation of genetic resources (Perera et al. 2012). In addition, it is very helpful to breeders not only to plan the crossings but also to broaden the genetic base. In that context, molecular markers are powerful tools for estimating genetic diversity and for understanding the complex genetics of sugarcane (D'Hont et al. 1997). This technology has been used in several genetic diversity studies, varietal identification, estimations of the hybrid character of progeny obtained through crossings, and genetic mapping (de Morais et al. 2015). For diversity studies, different types of molecular markers such as RAPD, AFLP, SSR, ETS-SSR, TRAP, ISSR, SRAP, and SNP have been used around the world (Besse et al. 1998; Cordeiro et al. 2003; Fontana et al. 2003; Alwala et al. 2006; Suman et al. 2008; Creste et al. 2010; Devarumath et al. 2012; Perera et al. 2012, 2016; Singh et al. 2013; Manechini et al. 2018). Recently, Medeiros et al. (2020) evaluated 254 accessions of the *Saccharum* complex with TRAP markers and showed that even when considering genes under the selection process, it is still possible to introgress new favorable alleles. In addition, Junior et al. (2020) used TRAP markers to reveal that both parental species contributed significantly to the variability in genes involved in sugar metabolism, and considerable variability in the lignin metabolism genes remains underexplored by local Brazilian breeding programs. TRAP markers were also useful for elucidating the male parent that intervened in polycrossings made by the EEAOC breeding program of Argentina (Perera et al. 2020). In addition, to identify new sugarcane varieties in Argentina, TRAP markers are employed together with the morphological traits proposed by UPOV (2005) (Perera et al. 2020).

## *Genetic Maps, QTL Mapping, and GWAS Analysis*

Increasing our understanding of the genetic architecture underlying complex traits by analyzing genotype–phenotype associations and genes/polymorphisms accounting for trait variation is crucial to improving the integration of molecular markers into breeding efforts (Bartholomé et al. 2016). Molecular markers associated with relevant agronomic traits could significantly reduce the time and cost involved in developing new varieties since they could aid in selecting the best parents as well as in accelerating the rate of genetic gain by breeding programs (Racedo et al. 2016). High-density genetic maps are valuable tools for understanding the genetic and genomic organization of this complex polyploid crop (Aitken et al. 2014); although substantial efforts have been made, most genetic maps are still incomplete and unsaturated. Garcia et al. (2006) developed the first integrated Brazilian genetic map from a cross between SP80-180 and SP80-4966, in which the 357 SSR, RFLP, and AFLP markers were mapped. In 2007, Oliveira et al. added to the preexisting genetic map, using the 664 EST-SSR and EST-RFLP markers, increasing the length and the average marker density. Another genetic map of the Australian sugarcane variety Q165 incorporated 2267 DArT, AFLP, SSR, SNP, RFLP, and RAPD markers, which were placed into eight homology groups (Aitken et al. 2014). Other molecular markers such as TRAP, DArT, and SNP *scIvana_1*-based markers have also been used to construct maps (Da Silva et al. 1995; Aitken et al. 2005; Balsalobre et al. 2017; Garcia et al. 2013; Palhares et al. 2012).

Quantitative trait locus (QTL) mapping and genome-wide association studies (GWAS) are strategies for understanding the genetic architecture of complex traits affecting MAS (Cao et al. 2016). Mapping QTL based on biparental populations has been widely used in sugarcane and constitutes an effective method for detecting markers linked to a trait. However, compared with traditional QTL mapping, GWAS has higher resolution and allows the identification of genes for multiple traits in a single population, but population substructure and genotyping errors can result in high false-positive rates. Therefore, the traditional QTL mapping powerfully compares pairs of alleles with low resolution, whereas association analysis provides a high-resolution evaluation of numerous alleles with uneven statistical power (Li et al. 2016). Both strategies have been successfully developed for several trait in sugarcane (Table 2).

**Table 2** QTL studies in sugarcane for yield traits

| References | Strategy | Population derived from | Type of marker | N° of markers | Yield-related traits | N° of QTLs detected | Phenotypic variation % max explained by a model |
|---|---|---|---|---|---|---|---|
| Sills et al. (1995) | L | LA Purple × Mol 5829 | apPCR | 83 | SN | 18 total | 58 |
| | | | | | TS | | |
| | | | | | SD | | |
| Ming et al. (2001) | L | Green German × IND 81–146 and PIN 84–1 × Muntok Java | RFLP | 1255 | CCS | 36 | 68.3 |
| Hoarau et al. (2002) | L | R570 (selfing) | AFLP | 939 | SD | 9 | 35 |
| | | | | | SL | 12 | 35 |
| | | | | | SN | 10 | 51 |
| | | | | | BR | 10 | 55 |
| Ming et al. (2002) | L | Green German × IND 81–146 and PIN 84–1 × Muntok Java | RFLP | 735 | SY | 102 total | 32 |
| | | | | | POL | | 39.9 |
| | | | | | SW | | 71.6 |
| | | | | | SN | | 13.9 |
| | | | | | FIB% | | 60.6 |
| | | | | | Ash | | 41.4 |
| Aitken et al. (2006) | L | IJ76-514 × Q165 | AFLP/SSR | 1000 | BR | 37 total | 65 |
| | | | | | POL | | 66 |
| Aitken et al. (2008) | L | IJ76-514 × Q165 | AFLP/SSR | 2238 | SN | 38 total | 54 |
| | | | | | SW | | 57 |
| | | | | | SD | | 60 |
| | | | | | SL | | 29 |
| | | | | | TCH | | 59 |
| Pinto et al. (2010) | L | SP80-180 × SP80-4966 | RFLP | 222 | FIB | 20 | 10.9 |
| | | | | | POL | 19 | 12.6 |
| | | | | | TCH | 22 | 18.12 |
| | | | | | TSH | 25 | 8.75 |
| Wei et al. (2010) | LD | 480 sugarcane accessions | DART | 1531 | CCS | 136 | NA |
| | | | | | TCH | 144 | NA |
| Pastina et al. (2012) | L | SP80-180 × SP80-4966 | RFLP/EST-RFLP/EST-SSR | 741 | TCH | 13 | NA |
| | | | | | TSH | 14 | NA |
| | | | | | FIB% | 11 | NA |
| | | | | | POL | 8 | NA |

(continued)

**Table 2** (continued)

| References | Strategy | Population derived from | Type of marker | N° of markers | Yield-related traits | N° of QTLs detected | Phenotypic variation % max explained by a model |
|---|---|---|---|---|---|---|---|
| Banerjee et al. (2015) | LD | 108 accessions | SSR/EST-SSR | 989 | NN | 11 | 19 |
| | | | | | NMC | 7 | 57 |
| | | | | | POL | 6 | 20 |
| | | | | | SD | 4 | 34 |
| | | | | | SL | 7 | 27 |
| | | | | | SW | 5 | NA |
| Gouy et al. (2015) | LD | 183 accessions | AFLP/DART/SSR | 3327 | SuY | 1 | NA |
| Margarido et al. (2015) | L | SP80-180 × SP80-4966 | RFLP/SSR/AFLP | 1680 | FC | | NA |
| | | | | | POL | 13 | |
| | | | | | TCH | | |
| Liu et al. (2016) | L | LCP 85-384 (selfing) | SSR/TRAP/AFLP | 967 | SuC | 24 | 25.68 |
| Siraree et al. (2017) | LD | 92 accessions | SSR | 1546 | BR | 21 total | 38.8 |
| | | | | | CCS | | 37.3 |
| | | | | | FIB% | | 22.1 |
| | | | | | NMC | | 33.1 |
| | | | | | POL | | 36.2 |
| | | | | | PU | | 33.5 |
| | | | | | SD | | 35 |
| | | | | | SL | | 37.5 |
| Costa et al. (2016) | L | IACSP95-3018 × IACSP93-3046 | AFLP/SSR/SNP | 668 | SD | 6 | |
| | | | | | SW | 5 | |
| | | | | | SL | 4 | 3.87 |
| | | | | | FIB% | 5 | |
| | | | | | POL | 2 | |
| | | | | | BR | 3 | |
| Racedo et al. (2016) | LD | 88 accessions | DART/TRAP | 1745 | CCS | 20 | NA |
| | | | | | CY | 23 | NA |
| Balsalobre et al. (2017) | L | SP80-3280 × RB835486 | SNP/SSR/TRAP | 7678 | POL% | 2 | 8.10 |
| | | | | | BR | 3 | 9.19 |
| | | | | | FIB% | 1 | 2.71 |
| | | | | | SD | 1 | 5.38 |

(continued)

**Table 2** (continued)

| References | Strategy | Population derived from | Type of marker | N° of markers | Yield-related traits | N° of QTLs detected | Phenotypic variation % max explained by a model |
|---|---|---|---|---|---|---|---|
| Barreto et al. (2019) | LD | 134 accessions | SSR | 100 | BR | 6 | |
| | | | | | SH | 3 | |
| | | | | | SN | 7 | |
| | | | | | SW | 4 | |
| | | | | | TCH | 3 | |

*L* linkage, *LD* linkage disequilibrium, *BR* Brix, *CCS* commercial cane sugar, *FC* fiber content, *NMC* number of millable canes per row, *NM* number of nodes, *TS* number of tasseled stalks, *POL* Pol, *PUR* purity, *SD* stalk diameter, *SL* stalk length, *SN* stalk number, *SW* stalk weight, *SD* stalk weight, *SuC* sucrose content, *SuY* sucrose yield, *SY* sugar yield, *TCH* tons of cane per ha, *TSH* tons of sugar per ha, *NA* not available

## *New Breeding Techniques*

New breeding technologies and strategies are required to significantly boost genetic improvements in future crop cultivars (Voss-Fels et al. 2019). Genomic selection (GS) has arisen as a modern breeding tool that has been very successfully applied in animal and plant breeding. GS approaches that can accurately capture nonadditive genetic effects and potentially improve the accuracy of genomic estimated breeding values are particularly promising for the adoption of GS in sugarcane breeding. This approach has the potential to increase the rate of genetic gain significantly in sugarcane, primarily by (1) reducing the breeding cycle length, (2) increasing the prediction accuracy for clonal performance, and (3) increasing the accuracy of breeding values for parent selection (Yadav et al. 2020). Genetic values were predicted in two independent panels, in which 167 accessions were genotyped with 1499 DArT markers and phenotyped at two sites. Depending on the given trait, within-panel cross validation yielded median correlations ranging from 0.29 to 0.62 in the Reunion Island panel and from 0.11 to 0.5 in the Guadeloupe panel. Cross validation between panels yielded correlations ranging from 0.13 for smut resistance to 0.55 for Brix. This level of correlations is promising for future implementations (Gouy et al. 2013). In addition, Deomano et al. (2020) implemented GS for cane yield and commercially extractable sucrose (CCS), obtaining prediction accuracy levels ranging 0.25–0.45. Although the first steps are already being taken, more studies are necessary in order to implement GS effectively in sugarcane breeding programs.

## *Transgenic Sugarcane*

In the past few years, numerous initiatives to produce transgenic sugarcane have been implemented in several sugarcane-producing countries, and transformations mediated by *Agrobacterium* and biobalistics have been the most commonly adopted method for generating transgenic events (Arencibia et al. 1997; Islam et al. 2016). For transgenic sugarcane development, numerous traits are being introduced, such as those for herbicide, pest and disease resistance, tolerance to abiotic stresses such as drought, higher accumulations of sugar and biomass, and flowering suppression, among others. Regarding the commercial release of transgenic sugarcane around the world, the first was a drought-tolerant variety released in Indonesia in 2013, which contains a bacterial gene responsible for producing betaine, a compound that stabilizes plant cells under water stress (Parisi et al. 2016). In 2015, the official approval for commercially releasing the first transgenic sugarcane variety that was resistant to glyphosate was issued by the corresponding regulatory bodies in Argentina (Noguera et al. 2015a). However, the commercial release was put on hold by the industrial sugar sector of the country due to economic reasons (Nikpay et al. 2020). In late 2017, the Comissão Técnica Nacional de Biossegurança (CTNBio, or National Commission of Biosafety 2017), the Brazilian government regulatory authority responsible for evaluating biotechnology-derived products, approved the commercialization of CTC20 developed by Centro de Tecnologia Canavieira (CTC, or Sugarcane Technology Center). This transgenic commercial event was obtained by bombarding the embryogenic callus of the immature inflorescences of the CTC20 sugarcane cultivar with microprojectiles containing a purified DNA fragment harboring the cry1Ab and nptII genes. Currently, there are at least three more commercially approved insect-resistant GM sugarcane varieties in Brazil harboring Cry1Ab or Cry1Ac genes. The insecticidal proteins produced from these genes are widely used to obtain transgenic plants to control insects because they are not toxic to vertebrates and nontarget insects (Baranek et al. 2017).

## *Genome Editing*

Regarding genome editing (GE) technology in sugarcane, only a few studies have been performed around the world. GE is a type of genetic modification in which DNA is inserted, deleted, or replaced within the genome of an organism using engineered nucleases. Currently, four families of engineered nucleases are available (meganucleases, ZFNs, TALENs, and the CRISPR/Cas9 technology) (Mohan 2016). Jung and Altpeter (2016) reported a TALEN-mediated approach for the first time in sugarcane to modify the cell wall characteristics to increase lignocellulosic ethanol for biofuel production. Additionally, Kannan et al. (2018) showed that field-grown TALEN-mediated mutants showed up to 19.7% lignin reduction and significantly decreased syringyl to guaiacyl (S/G) ratios resulting in up to 43.8% improved saccharification efficiency. However, among the genome editing tools, CRISPR (clustered regularly interspaced short palindromic repeat)-Cas9 (CRISPR-associated nuclease 9) has surpassed the existing GE tools through its simplicity, cost-effectiveness, and efficient ability (Augustine 2017) to provide precise genome modification in many different tissues and organisms.

### *In Vitro Multiplication*

Micropropagation has emerged as one of the most important global agricultural biotechnology techniques that is routinely used for the rapid generation of high-quality, uniform, and disease-free plant material (Thorpe 2007). It is used to supplement commercial sugarcane propagation in many countries including Brazil, the USA, India, Cuba (Pandey et al. 2011), and Argentina (Noguera et al. 2015b). In addition to the conventional micropropagation static system, the temporary immersion bioreactor has been found to be an important tool for sugarcane micropropagation, allowing higher shoot formation rates and cost reduction (Lorenzo et al. 2001).

In previous studies by using in vitro culture with modified MS medium (Murashige and Skoog, 1962) containing 2,4-dichlorophenoxyacetic acid (2,4-D), calli were produced after multiple steps following osmotic or temperature stress applied to panicles prior to culturing of anthers (Fitch and Moore 1996). In that sense, Jalaja and Sreenivasan (1992) obtained several doubled haploids plants from anther culture of *S. spontaneum*, *Erianthus*, and Indian commercial sugarcane cultivars.

Although micropropagation is an in vitro plant cloning method, genetic and epigenetic changes (somaclonal variation) can occur in regenerated plants (Larkin and Scowcroft 1981). Somaclonal variation is a problem when it is necessary to maintain maximum genetic fidelity (Martínez-Estrada et al. 2017); however, it has useful applications in sugarcane genetic improvement when variability should be generated. Another well-known way to induce variability is by mutagenesis. For example, two chemical mutagens were employed to induce variability in some of the sugarcane production traits, and the mutagenic effects were positive for the internode number and Brix content (Hrishi et al. 1968). In addition, Gamma ray-induced in vitro mutagenesis and selection for salt (NaCl) tolerance produced mutants that exhibited improved sugar yields with increased Brix% and numbers of millable canes (Nikam Ashok et al. 2015). The results suggested that mutagenesis offers an effective way to increase genetic variation in sugarcane.

## 7 Conclusions and Prospects

The high genetic complexity of modern sugarcane cultivars creates enormous challenges for both traditional breeding and biotechnology approaches. In traditional breeding, several efforts have been made jointly between most breeding programs around the world, especially with regard to the incorporation of new sources of genetic variability through germplasm exchange processes. Initiatives have also been taken towards the conservation of the global collections, to diminish the risk of genetic erosion and the direct loss of accessions. However, the so-called "world collections" under the ISSCT informal agreements needs more attention for conservation and variety exchange from the sugarcane scientific community. In the search for varieties that are more tolerant to biotic and abiotic stresses, several initiatives based on wild relative introgression are still used in breeding programs. In addition, parental selection strategies as well as more accurate hybridization and clone selection methodologies have been possible with the use of more robust statistical tools, enabling more efficient explorations of complex and unbalanced data. In biotechnology, several advances have been made in recent years, such as the development of high-density genetic mapping and diagnostic markers for diseases of economic importance. Furthermore, studies of genetic resources together with QTL analysis and associative mapping have generated relevant information about the genetic architecture of agronomic traits in sugarcane. Genomic selection is still beginning in sugarcane, but it can be an important instrument for revolution in the genetic improvement programs on this crop. Additionally, numerous efforts in different countries are being employed towards the sequencing of structural genomes and transcriptome and metabolomics studies. Currently, transgenic sugarcane is already a reality for simple inheritance traits (such as herbicide tolerance and insect resistance). In the future, in addition to the incorporation of genome editing technology, sugarcane may be genetically modified for complex traits (such as sugar, drought). These biotechnological tools have great potential for integration with the traditional breeding framework, and the complete combination of both areas is expected so that the levels of productivity can be increased as well as the acquisition of more in-depth knowledge about the crop genetics.Conflicts of InterestThe authors have no conflicts of interest to declare.

## References

Aitken KS, Jackson PA, McIntyre CL (2005) A combination of AFLP and SSR markers provides extensive map coverage and identification of homo(eo)logous linkage groups in a sugarcane cultivar. Theor Appl Genet 110:789–801. https://doi.org/10.1007/s00122-004-1813-7

Aitken KS, Jackson PA, McIntyre CL (2006) Quantitative trait loci identified for sugar related traits in a sugarcane (Saccharum spp.) cultivar× Saccharum officinarum population. Theor Appl Genet 112(7):1306–1317

Aitken KS, Hermann S, Karno K, Bonnett GD, McIntyre LC, Jackson PA (2008) Genetic control of yield related stalk traits in sugarcane. Theor Appl Genet 117(7):1191–1203

Aitken KS, McNeil MD, Hermann S et al (2014) A comprehensive genetic map of sugarcane that provides enhanced map coverage and integrates high-throughput Diversity Array Technology (DArT) markers. BMC Genomics 15:152. https://doi.org/10.1186/1471-2164-15-152

Alexander AG (1973) Sugarcane physiology: a comprehensive study of the Saccharum source -to-sink system. Elsevier Scientific, The Netherlands, pp 4–66

Aljanabi SM, Parmessur Y, Kross H, Dhayan S, Saumtally S, Ramdoyal K et al (2007) Identification of a major quantitative trait locus (QTL) for yellow spot (Mycovellosiella koepkei) disease resistance in sugarcane. Mol Breed 19(1):1–14

Almeida LM (2010) Seleção de famílias de irmãos-completos de cana-de-acúcar e estimativa da diversidade genética via marcador de DNA (ISSR). Universidade Estadual do Norte Fluminense – UENF, Campos dos Goytacazes

Alwala S, Suman A, Arro JA, Veremis JC, Kimbeng CA (2006) Target region amplification polymorphism (TRAP) for assessing genetic diversity in sugarcane germplasm collections. Crop Sci 46(1):448–455

Amalraj RS, Selvaraj N, Veluswamy GK et al (2010) Sugarcane proteomics: establishment of a protein extraction method for 2-DE in stalk tissues and initiation of sugarcane proteome reference map. Electrophoresis 31(12):1959–1974. https://doi.org/10.1002/elps.200900779

Arencibia A, Vázquez RI, Prieto D, Téllez P, Carmona ER, Coego A, Hernández L, Gustavo A, Selman-Housein G (1997) Transgenic sugarcane plants resistant to stem borer attack. Mol Breed 3:247–255

Augustine SM (2017) CRISPR-Cas9 system as a genome editing tool in sugarcane. In: Mohan C (ed) Sugarcane biotechnology: challenges and prospects. Springer, Cham. https://doi.org/10.1007/978-3-319-58946-6_11

Balsalobre TWA, da Silva Pereira G, Margarido GRA, Gazaffi R, Barreto FZ, Anoni CO et al (2017) GBS-based single dosage markers for linkage and QTL mapping allow gene mining for yield-related traits in sugarcane. BMC Genomics 18(1):72

Banerjee N, Siraree A, Yadav S, Kumar S, Singh J, Pandey DK, Singh RK (2015) Marker-trait association study for sucrose and yield contributing traits in sugarcane (Saccharum spp. hybrid). Euphytica 205(1):185–201

Baranek J, Konecka E, Kaznowski A (2017) Interaction between toxin crystals and vegetative insecticidal proteins of bacillus thuringiensis in lepidopteran larvae. BioControl 62:649–658

Barbosa MHP (2000) Perspectivas para o melhoramento da cana-de-açúcar. In: Raposo FV, Lambert ES, Alves GF, Mendonça HA, Faria MV, Gomes MS (eds) Anais do IV Simpósio de Atualização e Melhoramento de Plantas. Editora UFLA, Lavras, pp 1–17

Barbosa MH, Resende MD, Bressiani JA, Silveira LC, Peternelli LA (2005) Selection of sugarcane families and parents by Reml/Blup. Crop Breed Appl Biotechnol 5:443–450

Barnes AC (1974) The sugarcane. Leonard Hills Books, London, 572 p

Barreto FZ, Balsalobre TWA, Chapola RG, Hoffmann HP, Carneiro MS (2017) Validation of molecular markers associated with brown rust resistance in sugarcane. Summa Phytopathol 43(1):36–40. https://doi.org/10.1590/0100-5405/168917

Barreto FZ, Rosa JRBF, Balsalobre TWA, Pastina MM, Silva RR, Hoffmann HP, Pereira de Souza A, Garcia AAF, Carneiro MS (2019) A genome-wide association study identified loci for yield component traits in sugarcane (Saccharum spp.). PLoS One 14(7):e0219843

Bartholomé J, Bink MC, van Heerwaarden J et al (2016) Linkage and association mapping for two major traits used in the maritime pine breeding program: height growth and stem straightness [published correction appears in PLoS One. 2017 Jan 30;12 (1): e0171439]. PLoS One 11(11):e0165323. https://doi.org/10.1371/journal.pone.0165323

Bassi D, Menossi M, Mattiello L (2018) Nitrogen supply influences photosynthesis establishment along the sugarcane leaf. Sci Rep 8:2327. https://doi.org/10.1038/s41598-018-20653-1

Bastos IT, Barbosa MHP, Cruz CD, Burnquist WL, Bressiani JA, Silva FD (2003) Análise dialélica em clones de cana-de-açúcar. Bragantia 62:199–206

Berding N, Roach BT (1987) Germplasm collection, maintenance and use. In: Heinz D (ed) Sugarcane improvement through breeding. Elsevier, New York, pp 143–211

Bernardo R (2002) Breeding for quantitative traits in plants. Stemma Press, Woodbury

Besse P, McIntyre CL, Burner DM, de Almeida CG (1997) Using genomic slot blot hybridization to assess intergeneric Saccharum×Erianthus hybrids (Andropogoneae -Saccharinae). Genome 40:428–432

Besse P, Taylor G, Carroll B, Berding N, Burner D, McIntyre CL (1998) Assessing genetic diversity in a sugarcane germplasm collection using an automated AFLP analysis. Genetica 104(2):143–153

Bioversity International (2019) Celebrating 45 years of agrobiodiversity research. Rome, Bioversity International, 200p

Bolotin A, Quinquis B, Sorokin A, Ehrlich SD (2005) Clustered regularly interspaced short palindrome repeats (CRISPRs) have spacers of extrachromosomal origin. Microbiology 151(pt 8):2551–2561

Borojevic S (1990) Principles and methods os plant breeding. Elsevier B.V, Amsterdam

Borrás-Hidalgo O, Thomma BP, Carmona E, Borroto CJ, Pujol M, Arencibia A, Lopez J (2005) Identification of sugarcane genes induced in disease-resistant somaclones upon inoculation with Ustilago scitaminea or Bipolaris sacchari. Plant Physiol Biochem 43(12):1115–1121

Bressiani JA (2001) Seleção Sequencial em Cana-de-Açúcar. Ph.D. thesis, Universidade de São Paulo - USP/ESALQ

Budzinski IGF, de Moraes FE, Regiani CT, Maria FLÃ, Alberto LC (2019) Network analyses and data integration of proteomics and metabolomics from leaves of two contrasting varieties of sugarcane in response to drought. Front Plant Sci 10:1574. https://doi.org/10.3389/fpls.2019.01524

Burdon RD, van Buijtenen JP (1990) Expected efficiencies of mating designs for reselection of parents. Can J For Res 20:1664–1671

Cai Q, Aitken K, Deng HH, Chen XW, Fu C, Jackson PA, McIntyre CL (2005) Verification of the introgression of Erianthus arundinaceus germplasm into sugarcane using molecular markers. Plant Breed 124:322–328

Cao K, Zhou Z, Wang Q, Guo J, Zhao P, Zhu G et al (2016) Genome-wide association study of 12 agronomic traits in peach. Nat Commun 7:13246. https://doi.org/10.1038/ncomms13246. PMID: 27824331

Cardoso-Silva CB, Costa EA, Mancini MC et al (2014) De novo assembly and transcriptome analysis of contrasting sugarcane varieties. PLoS One 9(2):e88462. Published 2014 Feb 11. https://doi.org/10.1371/journal.pone.0088462

Castillo RO, Dookun-Saumtally A D'Hont A, Jackson P, Rossi-Machado G, Hale A Sampaio M, Hermann PH, Haihua D, Mirkov E, Mota C, Souza G (2013) Breaking breeding and biotechnology paradigms-towards a complementary approach. In: Sugar cane research: a summary of the tenth ISSCT breeding and germplasm and seventh molecular biology workshops. Proceedings of International Society of Sugar Cane Technology, vol 28, pp 1–9

Cia MC, de Carvalho G, Azevedo RA, Monteiro-Vitorello CB, Souza GM, Nishiyama-Junior MY, Lembke CG, Raphael Severo da Cunha R, de Faria A, Rodrigues Marques JP, Melotto M, Camargo LEA (2018) Novel insights into the early stages of ratoon stunting disease of sugarcane inferred from transcript and protein analysis. Phytopathology 108(12):1455–1466

Cordeiro GM, Pan YB, Henry RJ (2003) Sugarcane microsatellites for the assessment of genetic diversity in sugarcane germplasm. Plant Sci 165(1):181–189

Correr FH, Hosaka GK, Gómez SGP et al (2020) Time-series expression profiling of sugarcane leaves infected with Puccinia kuehnii reveals an ineffective defense system leading to susceptibility. Plant Cell Rep. https://doi.org/10.1007/s00299-020-02536-w. [published online ahead of print, 2020 Apr 20]

Costa EA, Anoni CO, Mancini MC, Santos FRC, Marconi TG, Gazaffi R et al (2016) QTL mapping including codominant SNP markers with ploidy level information in a sugarcane progeny. Euphytica 211(1):1–16

Costet L, Raboin LM, Payet M, D'Hont A, Nibouche S (2012) A major quantitative trait allele for resistance to the sugarcane yellow leaf virus (Luteoviridae). Plant Breed 131(5):637–640

Cox M (1996) Optimum family selection for Net Merit Grade in stage 2 trials. Technical report. SRA - Sugar Research Australia, Bundaberg

Creste S, Sansoli DM, Tardiani ACS, Silva DN, Gonçalves FK, Fávero TM, Medeiros CNF, Festucci CS, Carlini-Garcia LA, Pinto LR (2010) Comparison of AFLP, TRAP and SSRs in the estimation of genetic relationships in sugarcane. Sugar Tech 12(2):150–154

CTNBio (2017) Comissão Técnica Nacional de Biossegurança, Extrato De Parecer Técnico N°5.483/2017. https://www.in.gov.br/materia/-/asset_publisher/Kujrw0TZC2Mb/content/id/19141678/do1-2017-06-27-extrato-de-parecer-tecnico-n-5-483-2017-19141574. Accessed 20 Sept 2020

Cunha C, Roberto G, Vicentini R et al (2017) Ethylene-induced transcriptional and hormonal responses at the onset of sugarcane ripening. Sci Rep 7:43364. https://doi.org/10.1038/srep43364

Cursi DE, Cox MC, Anoni CDO, Hoffmann HP, Gazaffi R, Garcia AAF (2020) Comparison of different selection methods in the seedling stage of sugarcane breeding. Agron J. https://doi.org/10.1002/agj2.20431. Accepted Author Manuscript

D'Hont A, Rao PS, Feldmann P, Grivet L, Islam-Faridi N, Taylor P, Glaszmann JC (1995) Identification and characterization of sugarcane intergeneric hybrids, Saccharum officinarum×Erianthus arundinaceus, with molecular markers and DNA in situ hybridization. Theor Appl Genet 91:320–326

D'Hont A, Grivet L, Feldmann P, Rao S, Berding N, Glaszmann JC (1996) Characterization of the double genome structure of modern sugarcane cultivars (Saccharum spp.) by molecular cytogenetics. Mol Gen Genet 250(4):405–413

D'Hont A, Seshagiri Rao P, Alleyne S, Glaszmann JC, Feldmann P (1997) Application of molecular markers in sugarcane breeding. In: Proceedings West Indies sugar technologists, pp 77–79

D'Hont A, Ison D, Alix K, Roux C, Glaszmann JC (1998) Determination of basic chromosome numbers in the genus Saccharum by physical mapping of ribosomal RNA genes. Genome 41:221–225.8

Da Silva J, Honeycutt RJ, Burnquist W, Al-Janabi SM, Sorrells ME, Tanksley SD et al (1995) Saccharum spontaneum L. 'SES 208' genetic linkage map combining RFLP-and PCR-based markers. Mol Breed 1(2):165–179

Dal-Bianco M, Carneiro MS, Hotta CT, Chapola RG, Hoffmann HP, Garcia AA et al (2012) Sugarcane improvement: how far can we go? Curr Opin Biotechnol 23:265–270

Daniels J, Roach BT (1987) Taxonomy and evolution. In: Heinz D (ed) Sugarcane improvement through breeding. Elsevier, New York, pp 7–84

Dantas LLDB, Almeida-Jesus FM, de Lima NO et al (2020) Rhythms of transcription in field-grown sugarcane are highly organ specific. Sci Rep 10:6565. https://doi.org/10.1038/s41598-020-63440-7

Daugrois JH, Grivet L, Roques D, Hoarau JY, Lombard H, Glaszmann JC, D'Hont A (1996) A putative major gene forrust resistance linked with a RFLP marker in sugarcane cultivar "R570". Theor Appl Genet 92:1059–1064

Débibakas S, Rocher S, Garsmeur O, Toubi L, Roques D, D'Hont A, Hoarau JY, Daugrois JH (2014) Prospecting sugarcane resistance to sugarcane yellow leaf virus by genome-wide association. Theor Appl Genet 127(8):1719–1732

Deomano E, Jakson P, Wei X, Aitken K, Kota R, Pérez-Rodríguez P (2020) Genomic prediction of sugar content and cane yield in sugar cane clones in different stages of selection in a breeding program, with and without pedigree information. Mol Breed 40

Devarumath RM, Kalwade SB, Kawar PG, Sushir KV (2012) Assessment of genetic diversity in sugarcane germplasm using ISSR and SSR markers. Sugar Tech 14(4):334–344

Engelmann F, Arnao MTG, Paulet F, Roques D (2006) Cryopreservation as a tool for long-term conservation of sugarcane germplasm. In: Eighth ISSCT breeding and germplasm workshop. Proceedings of the International Society of Sugar Cane Technology, pp 37–38

Ferreira TH, Gentile A, Vilela RD, Costa GG, Dias LI, Endres L et al (2012) microRNAs associated with drought response in the bioenergy crop sugarcane (Saccharum spp.). PLoS One 7:e46703. https://doi.org/10.1371/journal.pone.0046703

Ferreira SS, Hotta CT, Poelking VGD et al (2016) Co-expression network analysis reveals transcription factors associated to cell wall biosynthesis in sugarcane. Plant Mol Biol 91:15–35. https://doi.org/10.1007/s11103-016-0434-2

Ferreira et al (2017) Sugarcane water stress tolerance mechanisms and its implications on developing biotechnology solutions. Front Plant Sci. https://doi.org/10.3389/fpls.2017.01077

Ferreira DA, Martins MCM, Cheavegatti-Gianotto A et al (2018) Metabolite profiles of sugarcane culm reveal the relationship among metabolism and axillary bud outgrowth in genetically related sugarcane commercial cultivars. Front Plant Sci 9:857. https://doi.org/10.3389/fpls.2018.00857

Fier I, Balsalobre TWA, Chapola RG, Hoffmann HP, Carneiro MS (2020) Field resistance and molecular detection of the orange rust resistance gene linked to G1 marker in Brazilian cultivars of sugarcane. Summa Phytopathol 46(2):92–97

Fitch MMM, Moore PH (1996) Haploids of sugarcane. In: Fain SM, Sopory SK, Veilleux RE (eds) In vitro haploid production in higher plants, vol 3. Kluwer Academic, Dordrecht, pp 1–16

Fontana PD, García MG, Cuenya MI, Chavanne E, Ontivero M, Díaz Ricci JC, Castagnaro AP (2003) Utilidad de la técnica de RAPD de alta resolución para la caracterización molecular de dos genotipos emparentados de caña de azúcar (Saccharum spp). Avance Agroindustrial 24(3):25–26

Frankel OH, Bennett E (1970) Geneetic resources in plants – their exploration and conservation. In: International biological program handbook no. 11. Blackwell Scientific, Oxford, 554p

Fukuhara S, Terajima Y, Irei S, Sakaigaichi T, Ujihara K, Sugimoto A, Matsuoka M (2013) Identification and characterization of intergeneric hybrid of commercial sugarcane (Saccharum spp. hybrid) and Erianthus arundinaceus (Retz.) Jeswiet. Euphytica 189:321–327

Galloway JH (1996) Botany in the service of empire: the Barbados cane-breeding program and the revival of the Caribbean sugar industry, 1880s-1930s. Ann Assoc Am Geogr 86(4):682–706

Garcia AAF, Kido EA, Meza AN, Souza HMB, Pinto LR, Pastina MM et al (2006) Development of an integrated genetic map of a sugarcane (Saccharum spp.) commercial cross, based on a maximum-likelihood approach for estimation of linkage and linkage phases. Theor Appl Genet 112(2):298–314

Garcia AA, Mollinari M, Marconi TG, Serang OR, Silva RR, Vieira ML et al (2013) SNP genotyping allows an in-depth characterisation of the genome of sugarcane and other complex autopolyploids. Sci Rep 3:3399. https://doi.org/10.1038/srep03399

Garsmeur O, Droc G, Antonise R et al (2018) A mosaic monoploid reference sequence for the highly complex genome of sugarcane. Nat Commun 9:2638. https://doi.org/10.1038/s41467-018-05051-5

Gentile A, Ferreira TH, Mattos RS, Dias LI, Hoshino AA, Carneiro MS et al (2013) Effects of drought on the microtranscriptome of field-grown sugarcane plants. Planta 237:783–798. https://doi.org/10.1007/s00425-012-1795-7

Gentile A, Dias LI, Mattos RS, Ferreira TH, Menossi M (2015) MicroRNAs and drought responses in sugarcane. Front Plant Sci 6:58. https://doi.org/10.3389/fpls.2015.00058

Gilmour A, Cullins R, Verbyla P (1997) Accounting for natural and extraneous variation in the analysis of field experiments. J Agric Biol Environ Stat 2:269–293

Glynn NC, Laborde C, Davidson RW, Irey MS, Glaz B, D'Hont A, Comstock JC (2012) Utilization of a major brown rust resistance gene in sugarcane breeding. Mol Breed. https://doi.org/10.1007/s11032-012-9792-x

Gouy M, Rousselle Y, Bastianelli D et al (2013) Experimental assessment of the accuracy of genomic selection in sugarcane. Theor Appl Genet 126:2575–2586. https://doi.org/10.1007/s00122-013-2156-z

Gouy M, Rousselle Y, Chane AT, Anglade A, Royaert S, Nibouche S, Costet L (2015) Genome wide association mapping of agro-morphological and disease resistance traits in sugarcane. Euphytica 202(2):269–284

Guinet-Brial I, Bousquet JF, Darroussat MJ, Fernandez E, Daugrois JH (2016) Visacane, an overview of the distribution of disease-free sugarcane varieties for the last 10 years. Proc Int Soc Sugar Cane Technol 29:58–62

Gutierrez AF, Hoy JW, Kimbeng CA, Baisakh N (2018) Identification of genomic regions controlling leaf scald resistance in sugarcane using a bi-parental mapping population and selective genotyping by sequencing. Front Plant Sci 9:877

Harlan JR, de Wet JMJ (1971) Towards a rational classification of cultivated plants. Taxon 20(4):509–517

Heinz DJ, Tew TL (1987) Hybridization procedures. In: Heinz DJ (ed) Sugarcane improvement through breeding, chap 8, 11th edn. Elsevier B.V., Amsterdam, pp 313–342

Hoarau JY, Grivet L, Offmann B, Raboin LM, Diorflar JP, Payet J et al (2002) Genetic dissection of a modern sugarcane cultivar (Saccharum spp.). II. Detection of QTLs for yield components. Theor Appl Genet 105(6–7):1027–1037

Hogarth DM (1987) Genetics of sugarcane. In: Heinz DJ (ed) Sugarcane improvement through breeding. Elsevier B.V., Amsterdam, pp 255–271

Hrishi N, Krishnamurthy TN, Marimuthammal S (1968) Studies in mutagenesis in sugarcane. Proc Indian Acad Sci Sect B 68:181–189. https://doi.org/10.1007/BF03051766

Irvine JE (1999) Saccharum species as horticultural classes. TAG Theor Appl Genet 98:186–194

Islam N, Laksana C, Chanprame S (2016) Agrobacterium-mediated transformation and expression of bt gene in transgenic sugarcane. J ISSAAS 22:84–95

Jackson PA (2005) Breeding for improved sugar content in sugarcane. Field Crop Res 92:277–290

Jackson P (2018) Advances in conventional sugarcane breeding programmes. In: Rott P (ed) Achieving sustainable cultivation of sugarcane: breeding, pests and diseases, 1st edn. Burleigh Dodds Science Publishing, Cambridge, pp 75–98

Jackson P, Henry RJ (2011) Erianthus. In: Kole C (ed) Wild crop relatives: genomic and breeding resources, industrial crops. Springer, Berlin, pp 97–107

Jalaja NC, Sreenivasan TV (1992) Geneties and cytogeneties: anther culture. In: Mohan Naidu K, Arulraj S (eds) Annuual report of Sugarcane Breeding Institute, Coimbatore. Sugarcane Breeding Institute, Coimbatore, pp 34–35

James G (2004) Sugarcane, 2nd edn. Blackwell Science, Oxford, 211 p

Jung JH, Altpeter F (2016) TALEN mediated targeted mutagenesis of the caffeic acid O-methyltransferase in highly polyploid sugarcane improves cell wall composition for production of bioethanol. Plant Mol Biol 92:131–142

Junior CADK, Manechini JRV, Corrêa RX, Pinto ACR, da Costa JB, Favero TM, Pinto LR (2020) Genetic structure analysis in sugarcane (Saccharum spp.) using target region amplification polymorphism (TRAP) markers based on sugar- and lignin-related genes and potential application in core collection development. Sugar Tech 22, 641–654. https://doi.org/10.1007/s12355-019-00791-0

Kannan B, Jung JH, Moxley GW, Lee SM, Altpeter F (2018) TALEN-mediated targeted mutagenesis of more than 100 COMT copies/alleles in highly polyploid sugarcane improves saccharification efficiency without compromising biomass yield. Plant Biotechnol J 16(4):856–866

Khan M, Pan YB, Iqbal J (2017) Development of an RAPD-based SCAR marker for smut disease resistance in commercial sugarcane cultivars of Pakistan. Crop Prot 94:166–172

Krishnamurthi M, Sundaram KS, Sekar S, Rajeswari S, Kawar PG (2007) Introgression of Erianthus for the development of commercial sugarcane cultivars. Proc Int Soc Sugar Cane Technol 26:678–680

Larkin PJ, Scowcroft WR (1981) Somaclonal variation: a novel source of variability from cell cultures for plant improvement. Theor Appl Genet 60:197–214

Li X, Zhou Z, Ding J, Wu Y, Zhou B, Wang R et al (2016) Combined linkage and association mapping reveals QTL and candidate genes for plant and ear height in maize. Front Plant Sci 7:833

Li WF, Wang XY, Huang YK, Zhang RY, Shan HL, Yin J, Luo ZM (2017) Molecular detection of *Bru*1 gene and identification of brown rust resistance in Chinese sugarcane germplasm. Sugar Tech 19(2):183–190

Li WF, Shan HL, Zhang RY, Pu HC, Wang XY, Cang XY, Yin J, Luo ZM, Huang YK (2018) Identification of field resistance and molecular detection of the brown rust resistance gene *Bru*1 in new elite sugarcane varieties in China. Crop Prot 103:46–50

Liu P, Chandra A, Que Y, Chen PH, Grisham MP, White WH et al (2016) Identification of quantitative trait loci controlling sucrose content based on an enriched genetic linkage map of sugarcane (Saccharum spp. hybrids) cultivar 'LCP 85-384'. Euphytica 207(3):527–549

Lorenzo JC, Ojeda E, Espinosa A et al (2001) Field performance of temporary immersion bioreactor-derived sugarcane plants. In Vitro Cell Dev Biol Plant 37:803–806. https://doi.org/10.1007/s11627-001-0133-8

Manechini JRV, Costa JB, Pereira BT, Carlini-Garcia L, Xavier MA, Landell MGA et al (2018) Unraveling the genetic structure of Brazilian commercial sugarcane cultivars through microsatellite markers. PLoS One 13(4):e0195623. pmid:29684082

Maretski A (1987) Tissue culture: its prospects and problems. In: Heinz D (ed) Sugarcane improvement through breeding. Elsevier, Amsterdam, p 343

Margarido GRA, Pastina MM, Souza AP, Garcia AAF (2015) Multi-trait multi-environment quantitative trait loci mapping for a sugarcane commercial cross provides insights on the inheritance of important traits. Mol Breed 35(8):175

Martin J, Chylinski K, Fonfara I, Hauer M, Doudna J, Charpentier E (2012) A programmable dual-RNA–guided DNA endonuclease in adaptive bacterial immunity. Science 337(6096):816–821. https://doi.org/10.1126/science.1225829. Originally published online June 28, 2012

Martínez-Estrada E, Caamal-Velázquez JH, Salinas-Ruíz J et al (2017) Assessment of somaclonal variation during sugarcane micropropagation in temporary immersion bioreactors by inter-simple sequence repeat (ISSR) markers. In Vitro Cell Dev Biol Plant 53:553–560. https://doi.org/10.1007/s11627-017-9852-3

Matsuoka S, Garcia AAF, Arizono H (1999) Melhoramento de Cana-de-açúcar. In: Borém A (ed) Melhoramento de Espécies Cultivadas. Universidade Federal de Viçosa, Viçosa, pp 205–252

Maxted N, Kell S (2009) Establishment of a global network for the in-situ conservation of crop wild relatives: status and needs. Background study paper no. 39. FAO, Rome, 203p

Maxted N, Kell S, Brehm JM (2011) Options to promote food security: on-farm management and in-situ conservation of plant genetic resources for food and agriculture. Commission on genetic resources for food and agriculture. Background study paper no. 51. FAO, Rome

Mbuma NW, Zhou MM, van der Merwe R (2020) Estimating breeding values of genotypes for sugarcane yield using data from unselected progeny populations. Euphytica 216:2

McCord P, Glynn N, Comstock J (2019) Identifying markers for resistance to sugarcane orange rust (Puccinia kuehnii) via selective genotyping and capture sequencing. Euphytica 215(9):150

McRae TA, Erquiaga DL, Jensen LF, Rattey AR, Stringer JK (1998) BSES sugarcane breeding program in the Burdekin. In: Australian Society of Sugar Cane Technologists conference, Ballina, 1998. Proceedings. Watson Ferguson, Brisbane, pp 196–203

Medeiros C, Balsalobre TWA, Carneiro MS (2020) Molecular diversity and genetic structure of Saccharum complex accessions. PLoS One 15(5):e0233211. https://doi.org/10.1371/journal.pone.0233211

Mendes de Paula TO, Brasileiro BP, Cursi DE et al (2020) Establishment of gene pools for systematic heterosis exploitation in sugarcane breeding. Agron J 2020:1–12

Ming R, Liu SC, Moore PH, Irvine JE, Paterson AH (2001) QTL analysis in a complex autopolyploid: genetic control of sugar content in sugarcane. Genome Res 11(12):2075–2084

Ming R, Wang Y, Draye X, Moore P, Irvine J, Paterson A (2002) Molecular dissection of complex traits in autopolyploids: mapping QTLs affecting sugar yield and related traits in sugarcane. Theor Appl Genet 105(2–3):332–345

Mohan C (2016) Genome editing in sugarcane: challenges ahead. Front Plant Sci 7:1542. https://doi.org/10.3389/fpls.2016.01542

Molina L, Queme JL, Rosales F (2013) Comparative analysis between phenotype and *Bru*1 marker for incidence to brown rust in sugarcane. Proc Int Soc Sugar Cane Technol 28:738–743 de Morais LK, de Aguiar MS, de Albuquerque e Silva P, Câmara TMM, Cursi DE, Júnior ARF, Chapola RG, Carneiro MS, Filho JCB (2015) Breeding of sugarcane. In: Cruz VMV, Dierig DA (eds) Industrial crops: breeding for bioenergy and bioproducts. Springer, New York, pp 29–42

Mukherjee SK (1957) Origin and distribution of Saccharum. Bot Gaz 119:55–61

Murad AM, Molinari HBC, Magalhàes BS, Franco AC et al (2014) Physiological and proteomic analyses of Saccharum spp. grown under salt stress. PLoS One 9:e98463

Nagatomi S (1982) Methods and problem of basic sugarcane breeding in Okinawa. Bull Okinawa Agric Exp Station 7:1–13. [In Japanese]

Nagatomi S, Degi K (2007) Collection and description of wild sugarcane species indigenous to Japan. Proc Int Soc Sugar Cane Technol 26:745–748

Nair NV, Selvi A, Sreenivasan TV, Pushpalatha KN, Sheji M (2006) Characterization of intergeneric hybrids of Saccharum using molecular markers. Genet Resour Crop Evol 53:163–169

Neuber AC, Santos FRC, Costa JB, Volpin M, Xavier MA, Perecin D, Burbano RCV, Landell MGA, Pinto LR (2017) Survey of the Bru1 gene for brown rust resistance in Brazilian local and basic sugarcane germplasm. Plant Breed 136:182–187. https://doi.org/10.1111/pbr.12463

Nibouche S, Raboin LM, Hoarau JY, D'Hont A, Costet L (2012) Quantitative trait loci for sugarcane resistance to the spotted stem borer Chilo sacchariphagus. Mol Breed 29(1):129–135

Nikam Ashok A, Devarumath RM, Ahuja A, Babub H, Shitole MG, Suprasanna P (2015) Radiation-induced in vitro mutagenesis system for salt tolerance and other agronomic characters in sugarcane (Saccharum officinarum L.). Crop J 3(1):46–56. https://doi.org/10.1016/j.cj.2014.09.002

Nikpay A, Vejar-Cota G, Budeguer F, Goebel FR, Perera MF, Qin ZQ (2020) Agroecological management of stem borers for healthy seed production in sugarcane. In: Tiwari AK (ed)Advances in seed production and management. En prensa https://doi.org/10.1007/978-981-15-4198-8_3

Noguera AS, Enrique R, Perera MF, Ostengo S, Racedo J, Costilla D, Zossi S, Cuenya MI, Filippone MP, Welin B, Castagnaro AP (2015a) Genetic characterization and field evaluation to recover parental phenotype in transgenic sugarcane: a step towards commercial release. Mol Breed 35:115. https://doi.org/10.1007/s11032-015-0300-y

Noguera AS, Paz NV, Díaz ME, Perera MF, Díaz Romero C, García MB, Filippone MP, Welin B, Cuenya MI, Digonzelli PA, Castagnaro AP (2015b) Production of healthy seed cane in Tucumán, Argentina. Int Sugar J (6)112–116

Oliveira KM, Pinto LR, Marconi TG, Margarido GRA, Pastina MM, Teixeira LHM et al (2007) Functional integrated genetic linkage map based on ESTmarkers for a sugarcane (Saccharum spp.) commercial cross. 2007. Mol Breed 20:189–208

Pacheco CM, Pestana-Calsa MC, Gozzo FC, Nogueira RJMC, Menossi MTC (2013) Differentially delayed root proteome responses to salt stress in sugar cane varieties. J Proteome Res 12:5681–5695

Palhares AC, Rodrigues-Morais TB, Van Sluys MA, Domingues DS, Maccheroni W, Jordão H et al (2012) A novel linkage map of sugarcane with evidence for clustering of retrotransposon-based markers. BMC Genet 13:51

Pandey RN, Rastogi J, Sharma ML, Singh RK (2011) Technologies for cost reduction in sugarcane micropropagation. Afr J Biotechnol 10:7814–7819

Papini-Terzi FS, Rocha FR, Vencio RZ, Felix JM, Branco DS, Waclawovsky AJ, Del Bem LE, Lembke CG, Costa MD, Nishiyama MY Jr et al (2009) Sugarcane genes associated with sucrose content. BMC Genomics 10:120

Parco AS, Avellaneda MC, Hale AH, Hoy JW, Kimbeng CA, Pontif MJ, Gravois KA, Baisakh N (2014) Frequency and distribution of the brown rust resistance gene *Bru*1 and implications for the Louisiana sugarcane breeding programme. Plant Breed 133:654–659

Parisi C, Tillie P, Rodríguez-Cerezo E (2016) The global pipeline of GM crops out to 2020. Nat Biotechnol 34:32–36

Parthasanrathy N (1946) The probable origin of north Indian sugarcanes. J Indian Bot Soc 1946:133–150

Pastina MM, Malosetti M, Gazaffi R, Mollinari M, Margarido GRA, Oliveira KM et al (2012) A mixed model QTL analysis for sugarcane multiple-harvest-location trial data. Theor Appl Genet 124(5):835–849

Perera MF, Arias ME, Costilla D, Luque AC, García MB, Cuenya MI, Racedo J, Ostengo S, Filippone MP, Castagnaro AP (2012) Genetic diversity assessment and genotype identification in sugarcane based on DNA markers and morphological traits. Euphytica 185(3):491–510. https://doi.org/10.1007/s10681-012-0661-9

Perera MF, Racedo J, García MG, Pardo EM, Rocha CML, Orce IG, Chiesa MA, Filippone MP, Welin B, Castagnaro AP (2016) Use of molecular markers to improve the agro-industrial productivity in the north west of Argentina. Mol Biol 5:153. https://doi.org/10.4172/2168-9547.1000153

Perera MF, Costilla D, Ovejero NS, Aybar Guchea M, Figueroa MF, Noguera AS, Cuenya MI, Castagnaro AP (2020) Caracterización botánica mediante marcadores fenotípicos y moleculares de variedades de caña de azúcar obtenidas por el Programa de Mejoramiento Genético de la EEAOC. Revista Industrial y Agrícola de Tucumán. Tomo 97(2):5–9

Piepho HP, Mohring J, Mechinger AE, Buchse A (2008) BLUP for phenotypic selection in plant breeding and variety testing. Euphytica 161:209–228

Pinto LR, Garcia AAF, Pastina MM, Teixeira LHM, Bressiani JA, Ulian EC et al (2010) Analysis of genomic and functional RFLP derived markers associated with sucrose content, fiber and yield QTLs in a sugarcane (Saccharum spp.) commercial cross. Euphytica 172(3):313–327

Piperidis N, D'Hont A (2020) Sugarcane genome architecture decrypted with chromosome-specific oligo probes. Plant J. https://doi.org/10.1111/tpj.14881

Piperidis G, Christopher MJ, Carroll BJ, Berding N, D'Hont A (2000) Molecular contribution to selection of intergenic hybrids between sugarcane and the wild species Erianthis arundinaceus. Genome 43:1033–1037

Piperidis G, Piperidis N, D'Hont A (2010a) Molecular cytogenetic investigation of chromosome composition and transmission in sugarcane. Mol Gen Genomics 284(1):65–73. https://doi.org/10.1007/s00438-010-0546-3

Piperidis N, Chen JW, Deng HH, Wang LP, Jackson P, Piperidis G (2010b) GISH characterization of Erianthus arundinaceus chromosomes in three generations of sugarcane intergeneric hybrids. Genome 53:331–336

Price S (1963) Cytogenetics of modern sugar canes. Econ Bot 17(2):97–106

Racedo J, Perera MF, Bertani R, Funes C, González V, Cuenya MI, D'Hont A.; Welin B. and Castagnaro, A.P. (2013) Alternative sources of resistance to sugarcane brown rust disease. Euphytica 191:429–436. https://doi.org/10.1007/s10681-013-0905-3

Racedo J, Gutierrez L, Perera MF, Ostengo S, Pardo EM, Cuenya MI, Welin B, Castagnaro AP (2016) Genome-wide association mapping of quantitative traits in a breeding population of sugarcane. BMC Plant Biol 16:142. https://doi.org/10.1186/s12870-016-0829

Resende MDV (2004) BLUP for phenotypic selection in plant breeding and variety testing. Documentos Embrapa Floresta, Colombo

Resende MDV, Barbosa (2005) Melhoramento genético de plantas de propagação assexuada. Embrapa Florestas, Colombo, 130p

Resende MDV, Barbosa MHP (2006) Selection via simulated individual BLUP based on family genotypic effects in sugarcane. Pesq Agrop Brasileira 41:421–429

Sakaigaichi T et al (2019) Evaluation of sugarcane smut resistance in wild sugarcane (Saccharum spontaneum L.) accessions collected in Japan. Plant Prod Sci 22(2):327–332

Sarwar M, Siddiqui SU (2004) In vitro conservation of sugarcane (Saccharum officinarum L.) germplasm. Pak J Bot 36(3):549–556

Shearman Jeremy R, Chutima Sonthirod, Chaiwat Naktang, et al (2018) Assembling a hybrid sugarcane genome. In: PAG conference

Sills GR, Bridges W, Al-Janabi SM, Sobral BW (1995) Genetic analysis of agronomic traits in a cross between sugarcane (Saccharum officinarum L.) and its presumed progenitor (S. robustum Brandes & Jesw. Ex Grassl). Mol Breed 1(4):355–363

Singh RK, Jena SN, Khan S, Yadav S, Banarjee N, Raghuvanshi S et al (2013) Development, cross-species/genera transferability of novel EST-SSR markers and their utility in revealing population structure and genetic diversity in sugarcane. Gene 524(2):309–329
Siraree A, Banerjee N, Kumar S, Khan MS, Singh PK, Kumar S, Sharma S, Singh RK, Singh J (2017) Identification of marker-trait associations for morphological descriptors and yield component traits in sugarcane. Physiol Mol Biol Plants 23(1):185–196. https://doi.org/10.1007/s12298-016-0403-x
Souza GM, Van Sluys MA, Lembke CG, Lee H, Margarido GRA, Hotta CT et al (2019) Assembly of the 373k gene space of the polyploid sugarcane genome reveals reservoirs of functional diversity in the world's leading biomass crop. GigaScience 8(12):giz129. https://doi.org/10.1093/gigascience/giz129
Spangler R, Zaitchik B, Russo E, Kellogg E (1999) Andropogoneae evolution and generic limits in Sorghum (Poaceae) using ndhF sequences. Syst Bot 24(2):267–281
Stolton S, Maxted N, Ford-Lloyd B, Kell SP, Dudley N (2006) Food stores: using protected areas to secure crop genetic diversity. WWF arguments for protection series. WWF, Gland
Stringer JK (2006) Joint modelling of spatial variability and interplot competition to improve the efficiency of plant improvement. PhD thesis, University of Queensland, Brisbane
Stringer JK, Cox MC, Atkin FC, Wei X, Hogarth DM (2011) Family selection improves the efficiency and effectiveness of selecting original seedlings and parents. Sugar Tech 13:36–41
Sugiharto B, Ermawati N, Mori H, Aoki K et al (2002) Identification and characterization of a gene encoding drought-inducible protein localizing in the bundle sheath cell of sugarcane. Plant Cell Physiol 43:350–354
Suman A, Kimbeng CA, Edmé SJ, Veremis J (2008) Sequence-related amplified polymorphism (SRAP) markers for assessing genetic relationships and diversity in sugarcane germplasm collections. Plant Genet Resour 6(3):222
Temam AH, Alemayehu LA (2017) Sugarcane (Saccharum Officinarum L) tissue culture in Ethiopia: opportunities for Ethiopia's sugar industries. Int J Sci Technol Res 6(8):398–406
Terajima Y, Tippayawat A, Ponragdee W, Sansayawichai T, Irei S, Sugimoto A, Takagi H, Hayashi H (2019) Application of a delayed heading technique to early-heading Erianthus arundinaceus native to Thailand for intergeneric crossing with sugarcane. Trop Agric Dev 63(1):1–11
Thelen JJ, Peck SC (2007) Quantitative proteomics in plants: choices in abundance. Plant Cell 19:3339–3346
Thorpe TA (2007) History of plant tissue culture. Mol Biotechnol 37:169–180
Umeda M et al (2019) QTL detection for smut resistance inherited from a Japanese wild sugarcane in the progeny of an interspecific cross. Proc Int Soc Sugar Cane Technol 30:1348–1353
Vicentini, Renato et al (2015) Large-scale transcriptome analysis of two sugarcane genotypes contrasting for lignin content. PLoS One 10(8):e0134909. https://doi.org/10.1371/journal.pone.0134909
Vieira MLC, Almeida CB, Oliveira CA, Tacuatiá LO, Munhoz CF, Cauz-Santos LA, Pinto LR, Monteiro-Vitorello CB, Xavier MA, Forni-Martins ER (2018) Revisiting meiosis in sugarcane: chromosomal irregularities and the prevalence of bivalent configurations. Front Genet 9:213. https://doi.org/10.3389/fgene.2018.00213
Volk GM, Jenderek MM, Staats E, Shepherd A, Bonnart R, Ledo A, Ayala-Silva T (2019) Challenges in the development of a widely applicable method for sugarcane (Saccharum spp.) shoot tip cryopreservation. Acta Hortic 1234:335–342
Voss-Fels KP, Cooper M, Hayes BJ (2019) Accelerating crop genetic gains with genomic selection. Theor Appl Genet 132:669–686. https://doi.org/10.1007/s00122-018-3270-8
Waclawovsky AJ, Sato PM, Lembke CG, Moore PH, Souza GM (2010) Sugarcane for bioenergy production: an assessment of yield and regulation of sucrose content. Plant Biotechnol J 8:263–276
Watson JW, Eyzaguirre PB (eds) (2002) Proceedings of the second international home gardens workshop: contribution of home gardens to in situ conservation of plant genetic resources in

farming systems, Witzenhausen, Federal Republic of Germany, 17–19 July 2001. International Plant Genetic Resources Institute, Rome, 117p

Wei X, Jackson PA, McIntyre CL, Aitken KS, Croft B (2006) Associations between DNA markers and resistance to diseases in sugarcane and effects of population substructure. Theor Appl Genet 114(1):155–164

Wei X, Jackson PA, Hermann S, Kilian A, Heller-Uszynska K, Deomano E (2010) Simultaneously accounting for population structure, genotype by environment interaction, and spatial variation in marker–trait associations in sugarcane. Genome 53(11):973–981

Yadav S, Jackson P, Wei X, Ross EM, Aitken K, Deomano E, Atkin F, Hayes BJ, Voss-Fels KP (2020) Accelerating genetic gain in sugarcane breeding using genomic selection. Agronomy 10:585

Yang X, Islam MS, Sood S, Maya S, Hanson EA, Comstock J, Wang J (2018) Identifying quantitative trait loci (QTLs) and developing diagnostic markers linked to orange rust resistance in sugarcane (SaccharumA spp.). Front Plant Sci 9(March):1–10. https://doi.org/10.3389/fpls.2018.00350

Yang X, Sood S, Luo Z, Todd J, Wang J (2019) Genome-wide association studies identified resistance loci to orange rust and yellow leaf virus diseases in sugarcane (Saccharum spp.). Phytopathology 109(4):623–631

You-Xiong Q, Jian-Wei L, Xian-Xian S, Li-Ping X, Ru-Kai C (2011) Differential gene expression in sugarcane in response to challenge by fungal pathogen Ustilago scitaminea revealed by cDNA-AFLP. J Biomed Biotechnol 2011:160934

Zhang J, Zhang X, Tang H, Zhang Q, Hua X, Ma X et al (2018) Allele-defined genome of the autopolyploid sugarcane Saccharum spontaneum L. Nat Genet 50(11):1565–1573

# Genic Conservation and Genetic Improvement of *Hevea brasiliensis*

**M. B. Mohamed Sathik and T. Gireesh**

**Abstract** Domestication of *Hevea* which descended from seeds transported from Brazil to South-East Asia started since 1860 with minuscule of genetic diversities within the Wickham germplasm, and the eventual progress that happened during the last 70 years resulted in primary and secondary clones with exceptional yielding abilities. South East Asian countries succeeded in developing productive cultivars, but challenges are many including climatic vagaries that are likely to challenge its cultivation in the future. This review deals with the genic conservation of *Hevea* and its vitality in sustainable breeding approaches aimed to improve its productivity and its adaptability to varying agroclimatic conditions. The molecular breeding approaches that started with genome analysis and exploration for genomic markers such as RAPD, RFLP, AFLP, EST-SSR or microsatellite markers and SNPs with the advent of new generation sequencing technologies, resulted in sudden spike in the rate of information generated on such markers. These abundant genomic data also contributed for the phylogenetic analysis apart from designing breeding strategies. Studies on G × E interaction by simultaneous incorporation of phenotypic as well as genotypic information contributed to shorten the length of the selection cycle. The phenotypic data collected over decades can also contribute for the genomic selection of superior varieties. This responsibility entirely lies with the breeder who can only glean the apt information from the vast available data and translate into superior varieties of *Hevea*. This review presents the overall latest development in genomic information and its implications in crop improvement aspects of *Hevea*.

**Keywords** *Hevea brasiliensis* · Wickham gene pool · Nuclear and cytoplasmic genetic diversity · Marker-assisted selection · Conventional and molecular breeding

---

M. B. M. Sathik (✉) · T. Gireesh
Rubber Research Institute of India, Kottayam, India
e-mail: sathik@rubberboard.org.in



P. M. Priyadarshan, S. M. Jain (eds.), *Cash Crops*,
https://doi.org/10.1007/978-3-030-74926-2_5

# 1 Introduction

Rubber is a polymer material obtained either from the exudates of certain tropical plants or derived from petroleum and natural gas (synthetic rubber). From mere rubber bands to catheters to condom to latex threads, rubber makes more than 50,000 products. Natural rubber (NR) is produced from over 7500 plant species (Compagnon 1986), confined to 300 genera of seven families *viz.*, *Euphorbiaceae*, *Apocynaceae*, *Asclepiadaceae*, *Asteraceae*, *Moraceae*, *Papaveraceae* and *Sapotaceae* (Cornish et al. 1993). At least two fungal species are also known to make NR (Stewart et al. 1955). *Hevea brasiliensis* (Willd. Ex. A. de. Juss. Müll-Arg.) is the major contributor of NR produced worldwide (Greek 1991). *Hevea* trees descended from seedlings that were transplanted from Brazil to South and South East Asia that have undergone several cycles of breeding are now the prime source of the modern world's NR.

The latex found in the latex vessels of inner bark of *H. brasiliensis* is obtained by tapping (shaving the bark with a sharp knife) and collecting the latex in cups (Fig. 1). Addition of acid, such as formic acid, solidifies rubber through manipulating the ionic balance. The solidified rubber is then pressed between twin rollers to remove excess water to form sheets, the primary produce. The sheets are commonly visually graded and packed in bales for shipping. NR is also commonly transported in the form of concentrated latex for specific purposes. History of the *Hevea* rubber is documented in ancient religious documents from Mexico dating back to 600 AD

**Fig. 1** General appearance of a *Hevea* plantation of a latex and timber clone (inset: a tapped panel of *Hevea* tree)

(Serier 1993) while Columbus wrote the first mention of rubber in 1496. de la Condamine, an astronomer, was the first to scientifically describe and to send samples of the elastic substance called 'caoutchouc' (the French word meaning "weeping wood"), from Peru to France in 1736 with full details about the habit and habitat and procedures for processing (Dijkman 1951; Baker 1996). The English chemist Joseph Priestley coined the name 'rubber' in 1770 when he found its wonderful property of rubbing out the pencil marks. As a botanist, Fusée Aublet described the genus *Hevea* in 1775. Rubber derived from *Hevea brasiliensis* is predominantly constituted of *cis*-1, 4 polyisoprene $(C_5 H_8)_n$ where $n$ may range from 150 to 2,000,000. Carbonyl groups were also detected which significantly contribute to the degree of cross linking and storage hardening (Pushparajah 2001). Many reviews have appeared in the past highlighting the importance of genic diversity and utility in breeding programmes (Priyadarshan and Clément-Demange 2004; Priyadarshan et al. 2008; Priyadarshan 2015, 2016; Gireesh et al. 2017). This review deals with genic conservation in *Hevea* and its vitality in breeding in tune with recent developments in the era of genomics.

## 2 Introduction of Rubber to South East Asia

The trios, Clement Markham of British India Office, Joseph Hooker, Director of Kew Botanic Gardens and Henry Wickham, Naturalist, played a key role in early exploration of the species. Also, Henry Ridley (Director of Singapore Botanic Gardens) and R. M. Cross (Kew Gardner), with Kew Botanic Gardens, played pivotal role in rubber procurements and distribution. As per directions of Markham, Wickham collected 70,000 seeds from Rio Tapajos region of Upper Amazon (Boim District) and transported them to Kew Botanical Gardens during June 1876 (Wycherley 1968; Schultes 1977; Baulkwill 1989). Of the 2700 seeds germinated, 1911 were sent to Botanical Gardens, Ceylon (now Sri Lanka), during 1876, and 90% of them survived. During September 1877, 100 *Hevea* plants specified as 'Cross material' were also sent to Ceylon. Earlier, in June 1877, 22 seedlings not specified either as Wickham or Cross were sent from Kew to Singapore, which were distributed in Malaya and formed the prime source of 1000 tappable trees found by Ridley during 1888. An admixture of Cross and Wickham materials might have occurred, as the 22 seedlings were unspecified (Baulkwill 1989). One such parent tree planted during 1877 was available in Malaysia even after 100 years (Schultes 1987). Seedlings from Wickham collection of Ceylon were also distributed worldwide. Rubber trees covering millions of hectares in South East Asia are believed to be derived from very few plants of Wickham's original stock from the banks of the Tapajos (Imle 1978). After reviewing the history of rubber tree domestication into East Asia, Thomas (2001) reported the origin of modern clones invariably from the 1911 seedlings that were sent to Ceylon (now Sri Lanka) during 1876. Also, Charles Farris could transport some seedlings to Kolkata in India (erstwhile Calcutta) during 1873. Hence, the contention that the modern clones were derived from "22 seedlings" is debatable. Moreover, if the modern clones are derived from 1911 seedlings, then the argument that they originated from a 'narrow genetic base', as believed even now, needs to be reviewed (Thomas 2002).

First introduction of rubber to India was during 1879 from Ceylon when 28 *Hevea* plants were planted in Nilambur Valley of Kerala State in South India (Haridasan and Nair 1980). During 1880–1882, plantations on experimental scale were raised in different parts of South India and the Andaman islands. *Hevea* was first introduced to Vietnam in 1897 by French, but got rejuvenated only after 1975 because of long-lasting war (Priyadarshan et al. 2005). Developments in domestication of rubber after 1880 commenced in Singapore Botanical garden, one of the world's finest in terms of both its aesthetic appeal and the quality of its botanical collection. Approximately 3000 species of tropical and subtropical plants and a herbarium of about 500,000 preserved specimens are the hallmark of this garden. Under the direction of Henry N. Ridley, who took over as superintendent in 1888, the garden became a centre for research on *Hevea brasiliensis*. Ridley developed an improved method of tapping rubber trees that resulted in better latex yield that revolutionized the region's economy. His persistence resulted in establishment of the first rubber estate in 1896, and thereon the rubber industry grew into one of the economic mainstays of the Malay states.

Synthetically speaking, yield improvement through breeding was initiated with a very strict mass selection among the trees at the beginning of the twentieth century. With the introduction of bud-grafting, 'generative' and 'vegetative' selection methodologies simultaneously resulted in development of seedlings and grafted clones (Dijkman 1951). Around 1950, the advantages of grafted clones proved to be overwhelming for yield potential compared to genetically improved seedlings and the focus shifted to derivation of clones for latex productivity. With all these cultural developments, *Hevea brasiliensis* soon ousted many other rubber producing species: *Castilla, Manihot glaziovii,* (ceara or manicoba rubber tree), *Ficus elastica, Landolphia* and *Clitandra* vines (African rubber).

Once the *Hevea* had been successfully transplanted to South East Asia, the development of rubber plantation industry was rapid, and considerable quantity of NR-based commodities were available in the market by 1910. Factors like availability of labour, favourable soil and climate contributed to this development. With the steady increase in global demand, the total area of plantation in the East amounted to 5000 acres by the turn of 1900. In 1910, it was 1 million acres, and in 1920, four million acres. After the end of World War II in 1945, the total acreage exceeded nine million, and by the mid-1960s, it was 11.5 million. According to the Food and Agriculture Organization of the United Nations, in 1996 the total land area of harvested NR in Asia amounted to about 15.6 million acres. Rubber produced from *Hevea* in Asian countries, ranging from Philippines to Sri Lanka, accounted for almost 95% of the world's NR supply (9.2 million tons from 9.5 million ha, IRSG 2015). Worldwide, there was a 37% increase in yield from 1995 to 2007 (IRSG 2015). There was always a constant correlation in the prices of oil and NR. World economic recessions also contributed for the downfall in the prices of NR. An extensive survey of the history and development of NR is out of scope of this review. Readers interested in such details may refer Baulkwill (1989) for an extensive account.

## 3 Narrow Genetic Base

Extent of variability and access to varied gene pool is basic requirement prime for the improvement of any crop species. Utilization of Amazonian germplasm and in-situ conservation strategies of wild germplasm are the added advantages of *Hevea* rubber. Conservation and utilization of allied gene resources are vital for the improvement of crop species, and NR has been noted as an undeniably beneficial commodity for the past 100 years (Priyadarshan and Goncalves 2003). Progress in yield improvement over past 70 years resulted in primary and hybrid clones with exceptional yielding abilities. Basic philosophy of genetic improvement programme of any crop in the world lies in the enhancement of variability in its gene pool. NR improvement programme is not an exception that needs new variability and diversity to circumvent new challenges in future, especially in the changing climatic era. Adaptability is the capacity for genetic response to selection that results in adaptation (Simmonds 1962). Genetic diversity remains as fuel for all crop improvement activities of plants to tide over new challenges. Importance of NR was well determined and even attained attention over 450 years since its historically known domestication. Later, accidental discovery of vulcanization by Goodyear in 1839 followed by value addition and utilization fuelled by the industries in the developed world grew its importance into many folds.

While primary introduction of *H. brasiliensis* is believed be originated from the Para region of Amazon rain forest of Brazil, it does not represent the entire genetic resources available in the tropical rain forests south of the Amazon River (Schultes 1977). World War II led second Amazon rubber boom followed by invasion of Japan during 1942 in Malaysia and Indonesia brought tremendous changes in world's NR supply. The wide species diversity available in the region was not covered fully during the initial exploration and collection. About 70,000 seeds collected by Sir Henry Wickham ended up with 22 seedlings at Singapore Botanic Gardens through a series of shipments and transfers right from the Amazon to Kew Botanic Garden and then to Ceylon (now Sri Lanka). Most of the initial collection was restricted to *H. brasiliensis* because of predominance of the species in the Para region. Thus, the 'Wickham' gene pool (Simmonds 1989) by origin itself is represented as 'very narrow'. Genetic improvement of the species rewarded tremendous progress in many important agronomic traits, particularly dry rubber yield from a meagre 250 to 3000 kg $ha^{-1}$. But the theoretical yield remained to a tune of 7000–12,000 kg $ha^{-1}$ (Pardekooper 1989). This remains as a distant goal mainly due to the constraints like long-breeding cycle, strong G × E interactions for useful traits (Meenakumari et al. 2018) and availability of land resources for laying out wide space multilocation trials and absence of variability towards major diseases and abiotic stresses. But fortunately, additional variations were inducted through the conserved wild germplasm which facilitated the introgression of genes into the Wickham gene pool in a slow pace.

The genetic variability among the present cultivated clones and other wild species could be improved considerably through limited interspecific hybridization (Figs. 2 and 3). Canopy and branching pattern of rubber tree are important traits that often determine the tree-level and stand-level productivity (McCrady and Jokela 1996; Cilas et al. 2004) and its vulnerability towards wind fastness (CleÂment-Demange et al. 1995). Retention of tree stands for a long period is affected by many factors like wind and planting density. It is presumed that compact canopy genotypes can escape wind damage to a certain level and can contribute for the optimal utilization of land through deploying suitable intercropping systems. Canopy variability which is obviously available in the present gene pool (Fig. 4) exists rarely in mutants spotted from the seedling and segregating populations (Gireesh and Mydin 2014).

Molecular markers played a crucial role in delineating the genetics of germplasm lines. DNA fingerprinting studies using RFLP and ribosomal DNA variations were used to assess the genetic variability among *H. brasiliensis* genetic resources. A genetic analysis using random nuclear probes and isozymes of the 168 individuals (including 73 cultivated Wickham clones and 95 wild clones) (Besse et al. 1993a, b, 1994) revealed a relatively high level of polymorphism in the cultivated clones, despite their narrow genetic base due to high level of polymorphism in the cultivated clones and high level of inbreeding and same level of polymorphism. Molecular interventions revealed higher level of conservation mtDNA in modern

**Fig. 2** The picture depicts the set fruit after hybridization by hand pollination

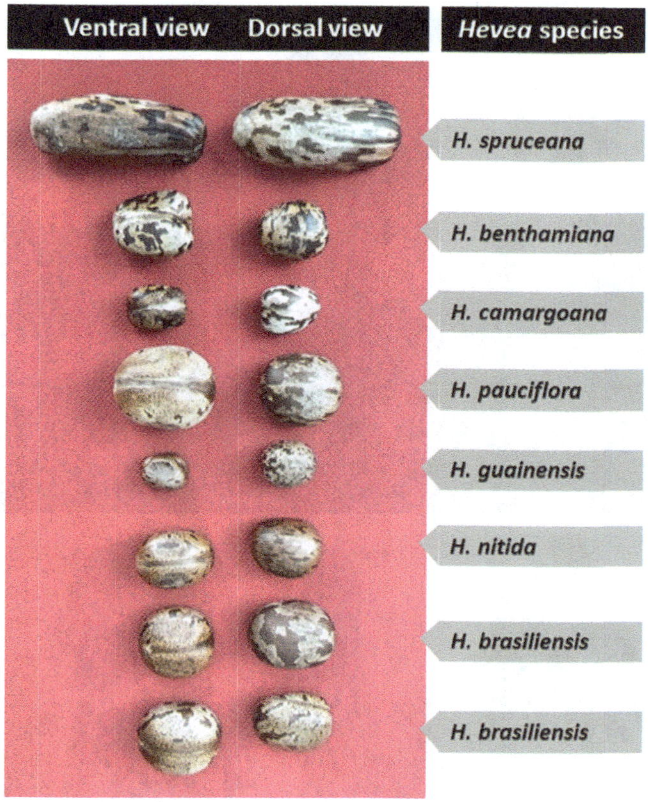


**Fig. 3** Seeds of different species of *Hevea* genera and their comparative seed morphology, characterized with unique mottling patterns

clones was contributed by only two clones (PB 56 and Tjir 1), and the nuclear DNA was due to breeding and selection under varies geo-climates (Priyadarshan and Goncalves 2003).

Gouvêa et al. (2010) observed the heterozygosity to range between 0.05 and 0.96 based on multivariate techniques and microsatellite marker studies in a group of clones. Both high total diversity ($H_{T'} = 0.58$) and high gene differentiation ($G_{st'} = 0.61$) were observed which indicated high genetic variation among the 60 genotypes that may be useful in breeding programs. Results found agreeable with that obtained from crosses involved within Wickham gene pool as selections imparted promising yield improvement when compared to their parents (Priyadarshan and Clément-Demange 2004). But long breeding cycle and genotype by environmental interaction of multigene traits often fool the selection of superior clones which needs to be confirmed through multilocation trial for introgressed traits from the wild sources to express. Genotype by environment interactions in perennial species has received considerable attention recently through the use of progeny tests set up in different locations. Depending on whether they are controlled or not, such

**Fig. 4** This picture depicts different and distinct canopy patterns observed in *H. brasiliensis*

interactions can lead to gains or losses in tree breeding programs (Zobel and Talbert 1984). But studies indicated that use of two sites is more profitable when the gains in efficiency of selection are greater than 10% (Costa et al. 2000).

## 4 Addition of New Amazonian Germplasm

Collection of new genotypes from the centre of origin had been happening since 1890. Through many planned expeditions, Malaysia received 1614 seedlings belonging to five species during 1952 collection (Brookson 1956; Tan 1987). Tan (1987) reported combined efforts by USDA, IAN (Brazil) and Liberia between 1957 and 1959 which led to more accessions reaching Malaysia.

The 1981 international collection of wild genetic resources from *H. brasiliensis* species, organized by IRRDB in the three Amazonian states *viz.*, Acre, Rondonia and Mato Grosso with the essential partnership of Brazil, was the predominant operation aimed at enlarging the genetic base of the cultivated *H. brasiliensis* species since the beginning of rubber cultivation one century before (Clément-Demange et al. 2000). The approach of directional selection for commercial yield followed by vegetative propagation devastated the then existed genetic diversity in the gene pool. Polyclonal seedling populations are also continuously diminishing which needs *in-situ* conservation at least in the secondary centre of origin. International Board of Plant Genetic Resources (IBPGR 1984) identified rubber as priority crop among the others for conservation of entire gene pool. Amazonian populations introduced to the main rubber-growing countries are still under experimental stages. Performance of hybrid generated from artificial hybridisations between Wickham and Amazonian accessions (belonging to Rondonia, Mato Grosso and Acre) is being evaluated under various trials (Chandrasekhar et al. 2004) and has superior siring ability as a male parent. Pollination success rate in W × A crosses ranged from nil to 14.2% while Wickham × Wickham crosses 0.8–19%. The wild accession MT 2226 as a male parent success rate of 12.1% was highest and was in concurrent with earlier reports (Morris 1929; Ross 1960; Rao 1961; Attanayake and Dharmaratnae 1984; Olapade and Omokhafe 1990; Gireesh and Pravitha 2009). Current level of rubber yield from undomesticated Amazonian population is around 12% of that of domesticated Wickham clones in addition to the fairly high resistance to *Microcyclus* or *Corynespora* (Clément-Demange et al. 2001).

During the last two decades, India made many attempts to characterize the IRRDB 1981 germplasm at the field level. Breeders used wild germplasm accessions for integrating through many crossing programmes. Mercy et al. (2014) documented most of the hybridization campaigns involving two Rondonian accessions (RO 87 and RO 1420) with RRIM 600 as female parent during 1990 and could identify 26 F1 hybrids superior to RRII 105 in terms of growth parameters. Highest yield was observed from the family of RRII 105 × MT 196. From another breeding programme, using Rondonian clones (RO 24, RO 26, RO 34, RO 87 and RO 132) and MT 196 as male parent with RRII 105, seven hybrid clones reached on farm trials recently. In 1997, RRIM 600 and RRII 105 as female parent and five Mato Grosso clones (MT 1021, MT 999, MT1027, MT 1014 and MT 1005); three Acre clones (AC 495, AC 498 and AC 817) and one Rondonian accession clone (RO 380) as male parent. Five promising preliminary selections were identified from the hybridization programmes happened during 2007, 2009, 2011, 2012 and 2013 utilized high yielding Wickham clones RRII 105, RRII 429 and RRII 414 as female parent and drought-tolerant wild accessions as male parent in four cross combinations. The resultant hybrids are under different stages of evaluation.

Seguin et al. (2003) proposed a general organization of *Hevea brasiliensis* germplasm with six genetic groups: group 1 made up with the two districts AC/T

(Tarauaca) and AC/F (Feijo) in the western part of Acre, and with the Calima component of the Schultes collection; group 2 made up with the three districts AC/B (Brasileia), AC/S (Sena Madureira) and AC/X (Xapuri) in the eastern part of Acre; group 3 made up with the six following districts of Rondonia: RO/A (Ariquemenes), RO/C (Calama), RO/CM (Costa Marques), RO/J (Jaru), RO/JP (Jiparana) and RO/OP (Ouro Preto), the district MT/VB (Vila Bella) of Mato Grosso and accessions MDF (Madre de Dios Firestone) from the Firestone collection in Peru; group 4 made up with three districts MT/A (Aracatuba), MT/C (Juruena) and MT/IT (Itauba) of Mato Grosso and the district RO/PB (Pimenta Bueno) of Rondonia; group 5 made up with the Palmira component of the Schultes collection and group 6 made up with the domesticated Wickham population. Even if no prediction can be made about the progenies of crosses between these groups, they can be used as a base for managing the genetic variability in the long term and organizing the recombination process.

## 5 Nuclear vs. Cytoplasmic Genetic Diversity

Most of the breeding and selection activities being carried out in progenies are based on phenotypic values expressed in consequence to the nuclear genome. However, no much efforts have been made to explore the cytoplasmic genome influence on phenotype particularly on sterility, incompatibility and reproductive biology of the species. Previous studies in *Hevea* had made use of molecular markers such as RAPD, RFLP, microsatellite or SSR and SSCP for clone identification and genetic variability assessment (Besse et al. 1994; Varghese et al. 1997; Venkatachalam et al. 2002; Mathew et al. 2005; Luo et al. 1995; Roy et al. 2004; Saha et al. 2005; Lekawipat et al. 2003; Priyadarshan 2003). All these efforts mainly focused on nuclear genome. Besse et al. (1994) categorized accessions of Brazil Amazonia according to their geographic origin (Acre, Rondonia and Mato-Grosso) based on nuclear DNA polymorphism. In contrast, the cultivated clones exhibited relatively higher level of polymorphism, despite its narrow genetic base and continuous assortative mating and selection. As expected, polymorphism is very prudent among allied species of *Hevea*. A comparison of isozyme analysis (Lebrun and Chevallier 1990) with demonstrated DNA markers revealed much similarity (Besse et al. 1994). Identification of all Wickham clones with 13 probes associated with restriction enzyme *EcoRI* (Besse et al. 1993b). The cultivated clones are genetically close to the Mato-Grosso genotypes. Rondonia and Mato-Grosso clones are more polymorphic as per RFLP data (Besse et al. 1994). A Rondonia clone (RO/C/8/9) showed eight specific restriction fragments and a unique malate dehydrogenase (MDH) allele, indicating this clone's interspecific origin. Such molecular markers are useful in rubber tree improvement since no distinct morphological traits exist. In a major study on mitochondrial DNA (mtDNA) polymorphism in 345 Amazonian accessions, 50 Wickham clones and two allied species (*H. benthamiana, H. pauciflora*) by Luo et al. (1995), variation in wild accessions was evident while the cultivated clones formed only two clusters.

## *Potentiality of Organelle Genome*

***mtDNA:*** In the initial attempts, selections were based on nuclear DNA polymorphism, and expressed phenotypic traits to evolve modern clones. The geographic specificity towards nuclear and mtDNA (Fig. 5) polymorphisms are due to great level of genetic structuring among natural populations in the Amazon forests in relation to hydrographic network (Luo et al. 1995). In wild accessions, seed dispersal and selection are as per the environmental conditions. Most of the phenotypic variations occurred in the natural habitat were being lost due to natural selection pressure prevailed during the evolution. It was observed that the wild accessions have not been introgressed for developing high yielding clones, after they had been introduced to South East Asian countries. In contrast, the Wickham clones exhibited high nuclear DNA polymorphism, perhaps due to directional breeding under different agro climates. Thanks to the nuclear genome that played main role in enhancing the variation in clones according to the diverse climatic conditions prevailed in the newly introduced areas. Certain primary clones alone provided common lineage for development of new genetic stocks which resulted in lesser genetic variation in improved Wickham clones.

In general, the primary clones like PB 56 or Tjir1 were the cytoplasmic donors for most of the improved clones. For example, PB 56 is the donor of PB 5/51 while Tjir 1 is the donor of RRII 105, RRIM 600 and RRIM 605. In most of the conventional breeding systems of *Hevea*, best parents were selected for next cycle of breeding (Simmonds 1989), which led to the development of two clusters of mtDNA profile in the clones developed, while the wild accessions displayed greater polymorphism since they evolved through possible interspecific hybridization. Obviously, this is the reason for the mtDNA profile exhibiting only two clusters (Priyadarshan and Goncalves 2003). A possible explanation for greater polymorphism in mtDNA of wild accessions is that many might have been evolved through interspecific hybridization. More investigations are needed to study the mtDNA polymorphism across wild accessions to ascertain the relatedness between wild and domesticated clones through employing molecular marker approaches.

Mitochondrial genome expansion in land plants is primarily due to large intergenic regions, repeated segments, intron expansion and incorporation of foreign DNA such as plastid and nuclear DNA (Turmel et al. 2003; Bullerwell and Gray 2004). Accumulation of repetitive sequences in plant mitochondrial genomes causes frequent recombination events and dynamic genome rearrangements within a species (Chang et al. 2011; Allen et al. 2007). Several mutations in the mitochondrial genes were found associated with cytoplasmic male sterility (CMS). Mutation in T-urf13 gene in maize (Dewey et al. 1986), pcf gene (a fusion of atp9 and cox2 portions) in petunia (Young and Hanson 1987), cox1 in rice (Wang 2006) and

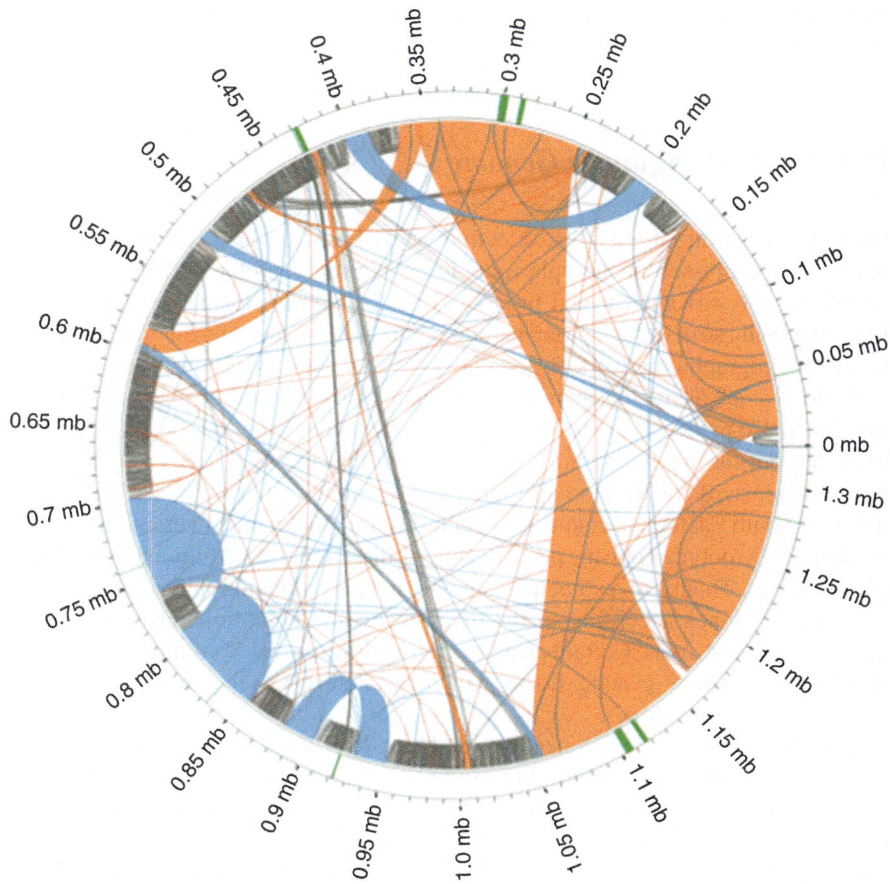


**Fig. 5** Annotated map of mitochondrial genome (outer circle) of *Hevea brasiliensis*. Grey arches indicate the mapping of each pair of the Illumina paired-end sequence data (inner circle). Direct repeats are shown as blue arches and inverted repeats as orange arches. (Adopted with permission from Shearman et al. 2014)

mutations in ATPase subunits in sunflower (Laver et al. 1991) and Brassica (Landgren et al. 1996) are few examples. RNA processing also plays an important role in controlling CMS as evidenced in *orf355/orf77* (*atp9*) and *T-urf13* in maize (Gallagher et al. 2002; Dill et al. 1997). Clone BPM 24 exhibits cytoplasmic male sterility that was inherited from GT 1. Shearman et al. (2014) identified a unique rearrangement in mitochondrial genome of BPM 24 that codes for a novel transcript containing a portion of atp9. This unique rearrangement led to slight reduction in ATP production efficiency and thus causing cytoplasmic male sterility. With the development of next-generation sequencing (NGS) technologies, new strategies were used to obtain plant mitochondrial genomes. A combination approach of shot-gun and paired-end NGS from non-enriched whole genome DNA libraries could be successfully used to obtain the mitochondrial genomes (Rahman et al. 2013; Shearman et al. 2014).

***cpDNA:*** Chloroplast genomes are sufficiently larger and complex to include structural and point mutations that are useful for evolutionary studies from intraspecific to interspecific levels (Neale et al. 1988; McCauley 1992; Graham and Olmstead 2000; Provan et al. 2001). Since the first report on complete chloroplast (cp) genome sequence of liverwort (*Marchantia polymorpha*) in 1986 (Ohyama et al. 1986), more than 150 chloroplast genomes have been sequenced and characterized thus disclosing an enormous amount of evolutionary and functional information of chloroplasts (Fig. 6). Tangphatsornruang et al. (2010, 2011) reported the complete chloroplast genome sequence of rubber tree with a 161,191 bp length including a pair of inverted repeats of 26,810 bp separated by a small single copy region of 18,362 bp and a large single copy region of 89, 209 bp. It contains 112 unique genes, 16 of which are duplicated in the inverted repeat. Of the 112 unique genes, 78 are predicted protein-coding genes, four are ribosomal RNA genes and 30 are tRNA genes. Relative to other plant chloroplast genomes, Tangphatsornruang et al. (2011) observed a unique rearrangement of a 30-kb inversion between the *trnE(UUC)-trnS(GCU)* and the *trnT(GGU)-trnR(UCU)*. A comparison between the rubber tree chloroplast genes and cDNA sequences revealed 51 RNA editing sites in which most (48 sites) were located in 26 protein-coding genes while the rest of the three sites were in introns.

This information on plastome can be better utilized to determine phylogenetic relationships among angiosperms. Phylogenetic analysis of chloroplast genes using 144 orthologus clusters from 17 sequenced plant genomes showed *Hevea*'s closest ancestry with *Manihot esculenta* within the family and with *Populus trichocarpa* outside *Euphorbiaceae* (Rahman et al. 2013). The latest report on chloroplast genome of *H. benthamiana*, a SALB resistant and close relative of *H. brasiliensis* species also illustrates its close relationship thus indicating the possibility of utilizing the same in breeding programmes to obtain SALB-resistant *H. brasiliensis* clones (Niu et al. 2020). As a synthesis of these diversity studies, closer relationships were found from the studies using different genetic markers. Even if the contribution of isozymes is important by itself, molecular markers provided important clarifications for the distinction between different groups. There would be no barrier to migration of *Hevea* genes within the Amazonian basin. However, the vastness of the area and the limited dispersion of *Hevea* seeds enabled the preservation of the current structure, which is assumed to have resulted from the fragmentation of the Amazonian forest during the Pleistocene period, according to the refuge theory presented by Haffer (1982). Moreover, the *Hevea* germplasm genetic structure clearly appears as geographically structured in relationship with the hydrographic network of the Amazonian forest, which confirms the role of rivers and inundated zones in the transport of seeds and dissemination of the species (Besse et al. 1993a; Luo et al. 1995; Seguin et al. 1996).

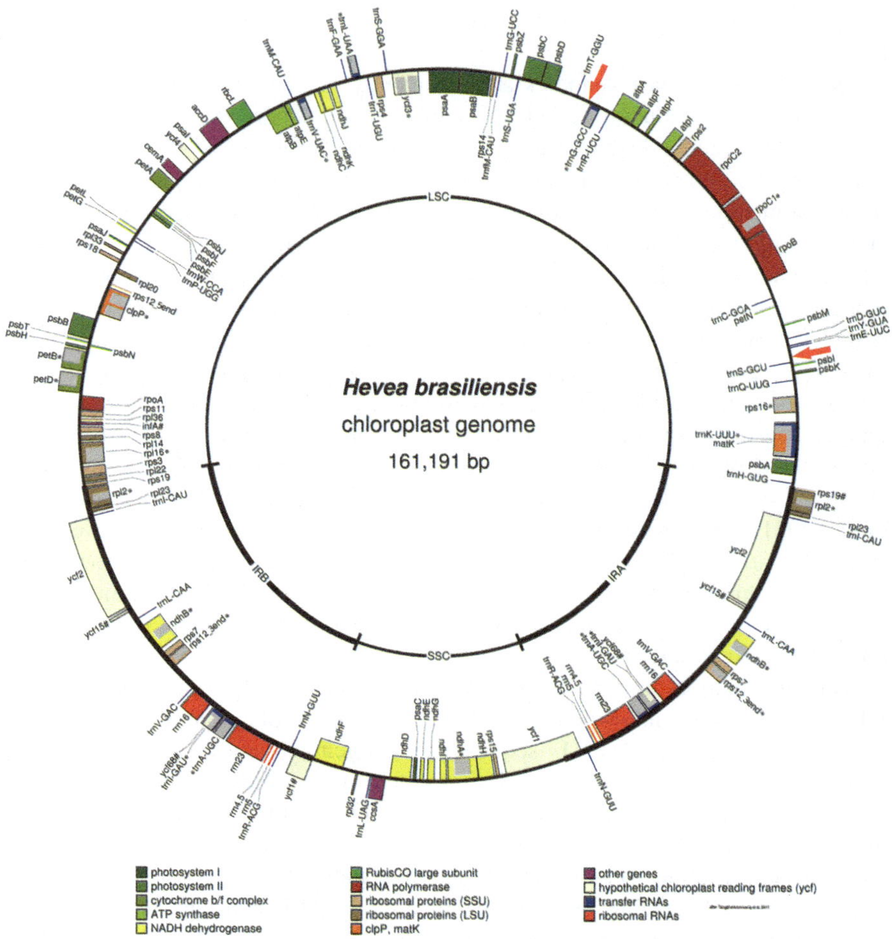


**Fig. 6** Annotated map of the chloroplast genome of *Hevea brasiliensis*. Thick lines indicate the extent of the inverted repeats (IRa and IRb) which separate the genome into small and large single copy regions. Genes on the outside of the map are transcribed clockwise and those on the inside of the map are transcribed counter clockwise. Genes containing introns and pseudogenes are marked with * and #, respectively. Arrows indicate the positions of a 30-kb unique rearrangement in relative to the Cassava chloroplast genome. (Adopted with permission from Tangphatsornruang et al. 2011)

## 6 Perspectives for New Germplasm

The objective of utilization of germplasm is the introgression of genes to present domesticated resources of *Hevea*. Wild accessions can be included in the research programmes to improve or develop superior clones with high latex yield potential, biotic, abiotic stress tolerance, etc. Recently, IRRDB initiated sharing of frontline improved genetic resources among member countries. In situ conservation of species in their natural habitats and creation of field gene bank is considered the most appropriate way of conserving biodiversity in easily manageable way. Protecting the areas where populations of species exist naturally is difficult for an introduced crop like *Hevea brasiliensis.* In situ conservation of species is necessary for preserving genetic diversity in ecosystem level. Many-fold increase in the global demand for latex and dry rubber products is expected in the coming years (Smith and Burger 1992), mainly by the automobile industry in emerging economies. In order to bridge the gap between production and consumption, cultivation has to be extended to marginal areas with limiting climatic conditions forced breeders to develop new clones with tolerance to environmental constraints like drought, cold, wind, higher altitude, moisture stress deficit and diseases. Vast areas have been identified in major rubber-producing countries like India, Thailand, Vietnam, China, Brazil, Cote d'Ivoire and other African countries. IPR issues and related restrictions may restrict accessing the germplasm transfer, in the coming years, especially for collection of new entries from the centre of origin. It is highly essential to safeguard the precious germplasm accessions available in each country for further utilization in crop breeding programmes. Mydin and Gireesh (2016) made an attempt to improve the diversity and heterosis among the present germplasm by creating new combinations through cross breeding which yielded hybrids with improved yield and other secondary traits. Success of heterosis breeding looks promising when compared to other breeding methods in *Hevea.* Such new combinations have great potential to contribute towards improving future breeding tools that may ultimately end up with pyramiding of economic traits in at least few genotypes. Studies at ploidy levels of germplasm accessions of *Hevea* were very meagre with few attempts (Chen et al. 1979). But unfortunately, there has not been any follow-up studies exploiting this system like production of dihaploids and so on for crop improvement in *Hevea.*

## 7 Genic Diversity and Clone Development

The early genetic resources are high yielding and highly adapted to current cropping systems in Asia, Africa and non-SALB areas of Latin America. Breeders believe that the reduced genetic variability of Wickham base is weak and insufficient to meet the demand for the improvement of growth and latex yield traits. *Hevea brasiliensis* originated from miniscule of germplasm, about 22 seedlings recovered from consignment originated from Singapore to Malaysia during 1876, popularly known as l Wickham germplasm. Extensive use of these seedlings resulted in the origin of present-day cultivars. Breeders used to select elite progenies based mainly on latex yield, eventually sidelined other traits of importance, resulted in directional selection, appears to be the major reason for the genetic erosion. The fact that populations were subjected to several rounds of controlled crossing further narrowed the diversity. But the strategy followed by the breeders to select only the desirable genotypes and to reject the unwanted ones (without assessing the utility other than yield) is the main reason that reduced diversity. The strategy for utilization of wild gene pool involves gene introgression from germplasm, pyramiding favourable gene and gene combinations and improvement of the cultivated clones using IRRDB wild collection. The recently conducted plant breeders meeting by IRRDB at Malaysia (2019) reviewed the progress of bilateral clone exchange programme between the member countries and suggested to intensify exchange of germplasm between all the stake holders to maximize the genetic diversity and to improve the gene pool.

Possibility of genetic diversification using other species as sources of poly isoprene also needs to be exploited. Apart from the latex, availability and inheritance pattern of wood production has not been attempted seriously either in cultivated or wild Amazonian gene pool. Related species, genera and wild accessions from the original habitat available in all rubber-growing nations can also be integrated and networked into breeding programmes in a global scale to overcome the narrow genetic base. Utilizing allied species as excellent resources of timber and possible pre-breeding activity are the other options for widening the genetic base of rubber.

Classic plant breeding programs under changing climatic conditions mainly depend upon evaluation of phenotypes in varying environmental conditions. Selection of genotypes that are resistant to environmental variations is basically done by monitoring their physiological responses under adverse environmental conditions. In breeding programmes, the common physiological parameters employed for selecting the tolerant varieties include photosynthesis, content of photosynthetic pigments (chlorophyll *a* and *b* and carotenoids) and chlorophyll fluorescence (Holá et al. 2010; Sterling et al. 2019). Since photosynthesis is the key process that determines the plant productivity, it is important to assess its photosynthetic capacity under limiting environmental conditions as marker for tolerance. The recent developments in genome-based data and molecular markers have enabled the selection of such climate resilient genotypes much easier which can significantly reduce the time and cost required to identify or develop such genotypes. Several genomic prediction models combining genotype × environment (G × E) interactions have been developed for application in genomic selection in tree breeding programmes (Lopez-Cruz et al. 2015; Cuevas et al. 2016; Souza et al. 2019; Cros et al. 2019; Meuwissen et al. 2001). It is highly essential to employ physiological markers like gas exchange parameters for validating the genetic markers to identify potential markers associated with abiotic stress tolerance in *Hevea*.

## 8 Molecular Advances in *Hevea* Research

Conventional genetic analysis in *Hevea* is cumbersome and time consuming because of its perennial nature, long breeding cycles and difficulties in raising $F_2$ progenies. The recent development of molecular markers like isozymes, restriction fragment length polymorphisms (RFLPs), amplified fragment length polymorphisms (AFLPs), SSR markers, microsatellites, and ESTs has helped to understand the genetic diversity in *Hevea*. The recent advances in gene engineering and in vitro culture methods have increased the possibility of developing new genotypes with improved latex yield, tolerance to various diseases and tapping panel dryness (TPD) syndrome, growth rate and wood quality or reduction in undesirable traits (Supriya and Priyadarshan 2019). Gene editing is a viable and potential option to develop genotypes with altered gene expression that facilitates expression of the desired traits. In *Hevea,* CRISPR/Cas9-based targeted mutagenesis was first reported by Fan et al. (2020) who could successfully target *FLOWERING LOCUS T* (*FT*) and *TERMINAL FLOWER1* (*TFL1*) genes with five sgRNAs through one transformation step in protoplasts. This recently developed system of genome editing with Cas9 nuclease directed by target-specifying single-guide RNA (sgRNA) has proven efficient in functional analysis of endogenous genes and breeding of new varieties (Cong et al. 2013; Francis et al. 2017). This approach minimizes off-target effects and prevents genomic integration of foreign DNA without the need for codon optimization and species-specific promoters. This study opens up the possibility of large-scale production of DNA-free genome-edited plants of *Hevea* with specific traits from protoplasts. Conventional breeding in combination with the recent molecular techniques can be best utilized to develop *Hevea* clones for new environments (Saha and Priyadarshan 2012).

Many studies in *Hevea* have made use of RAPD, RFLP, microsatellite or SSR and SSCP markers for clone identification and genetic variability studies (Besse et al. 1994; Varghese et al. 1997; Venkatachalam et al. 2002; Mathew et al. 2005; Luo et al. 1995; Roy et al. 2004; Saha et al. 2005; Lekawipat et al. 2003). RAPD markers were also used to study the genetic diversity of a vast collection of germplasm collections from Amazonian region under the aegis of International Rubber Research Development Board (IRRDB) during 1981 and to classify them into five groups that would further facilitate crosses to utilize heterosis and abundant genetic diversity (Lam et al. 2009). Though RAPD markers were used for genotyping, they are less preferred when compared to microsatellite markers due to their low reproducibility and low discrimination power (García et al. 2011). In contrast, the RFLP markers are considered efficient in terms of genetic polymorphism and are evenly distributed across the *Hevea* genome, and their applicability to genotyping of large progenies is expensive in terms of time and cost (Seguin et al. 2008). While the AFLP markers allow rapid identification and mapping of hundreds of markers, they are not as informative as RFLP. The microsatellite markers or simple sequence repeats (SSRs) are considered most efficient because of their properties like co-dominant multi-allele nature and random distribution throughout the genome with high polymorphism (Saha et al. 2005). They combine the advantage of RFLP (in terms of polymorphism, its abundance in genome and its locus specificity) and AFLP (PCR-based high output genotyping) and are dependable apart from their high efficiency (Seguin et al. 2008).

Many attempts have been made to create dense SSR map of *Hevea* for genetic mapping and to locate QTLs (obtained from different mapping projects) in chromosomes. The pioneering studies of genetic mapping for agronomic traits include SALB resistance (Seguin et al. 1996; Lespinasse et al. 2000b; Le Guen et al. 2007) and latex yield and growth (Clément-Demange et al. 2006, 2008). These studies indeed explained the diploid nature of *H. brasiliensis* and *H. benthamiana* genomes with a limited number of duplicated chromosome segments. Initial studies with 108 *Hevea* accessions identified SSR M574 as highly polymorphic (Lekawipat et al. 2003). Four markers (HMAC4, HMAC5, HMCT1 and HMCT5) could discriminate 27 *Hevea* clones (Saha et al. 2005) while HMGR marker was found suitable for studying genetic variability in wild *Hevea* (Saha et al. 2007). In a similar attempt, microsatellite markers were used to analyze genetic diversity of 307 clonally propagated individuals from 19 different collection areas to understand the existence of a moderate differentiation among a population of 220 individuals and larger genetic distance in Mato Grosso populations (Le Guen et al. 2009).

The focus of subsequent research shifted from anonymous markers (AFLPs and microsatellite markers) to single-nucleotide polymorphisms (SNPs) which are abundant in plant genomes. Initial attempts had contributed to the identification of SNP loci that could be utilized in the construction of high-density (10 SNP loci belonging to 5 loci) genetic maps (Pootakham et al. 2011; Mantello et al. 2014). The subsequent high-density genetic map developed by Pootakham et al. (2015a) had one SNP marker per 0.89 cM. Genetic variation plays a significant role in maintaining heterosis in *Hevea*. Saturated genetic maps are essential to identify the genomic regions harbouring major genes and QTLs that control important agronomic traits for employing in the breeding programs. The development of next-generation sequencing (NGS) technologies provided ample opportunities to extract large number of SNP markers without the need of a reference genome sequence. The advantages of NGS are: higher sensitivity to detect low-frequency variants, faster turnaround time for sample volumes, comprehensive genomic coverage, higher throughput with sample multiplexing and the ability to sequence hundreds to thousands of genes or gene regions simultaneously (Shendure and Ji 2008; Schuster 2008).

Salgado et al. (2014) reported *H. brasiliensis* transcriptome, covering a wide range of tissues and organs, leading to the production of the first developed SNP markers. Silva et al. (2014), identified 172 SNPs that are suitable for breeding studies from a set of 17,166 contigs developed from cDNA of cold stressed and different tissues of *Hevea*. This study identified two highly upregulated cold stress-responsive ESTs from 5025 unigenes and 912 novel ESTs along with 169 novel EST-SSR markers and also 43 SNP markers in 13 ESTs associated with stress response, latex biosynthesis and developmental processes. During 2015, two reports were published based on advanced technology like genotyping-by-sequencing (GBS) to develop high-density genetic map. Shearman et al. (2015) reported the first high-density genetic map of rubber with an average marker density of 1.90 cM. They employed the identified SNPs and Indels to genotype 149 progenies from a cross between RRIM 600 and RRII 105. They included 20,143 genes to construct the linkage map. In a study of gene expression analysis and SNP Indel discovery for yield heterosis in a $F_1$ hybrid population (RRIM 600 vs. PR 107) by Li et al. (2016), higher yield in F1 population was found positively associated with their higher genome heterozygosity. This study indicated the role played by variation of SNPs in the genic region in manifestation of heterosis that contribute for better yield in F1 populations. The haplotype analysis of about 25 SNPs done for three genes such as farnesyl diphosophate synthase (FPPS), hydroxymethylglutaryl-CoA synthase (HMGS) and *cis*-prenyl transferase (CPT) in various *Hevea* clones (of diverse parentage) identified seven FPPS, eight HMGS and eight CPT haplotypes (Uthup et al. 2013, 2016, 2018). The segregation and linkage disequilibrium analysis revealed recombination events as the reason for the generation of allelic diversity rather than point mutations. These studies emphasized the major role of SNPs in evolution of candidate genes coding for functional traits in plants. This information can even be best utilized to trace the effect of phenotypic selection on changes at the genome level and further at the level of successive generations of crosses (Chow et al. 2020).

A first study on genome-wide association mapping of latex yield and girth of 170 Amazonian accessions grown under suboptimal environments (limited rainfall and lengthy dry season) using 14,155 high-quality filtered SNPs (from transcripts) was reported by Chanroj et al. (2017). With a mixed linear model, they could detect three significant SNPs in three candidate genes associated with plant adaptation to drought stress, individually explaining 12.7–15.7% of phenotypic variance. This study identified one SNP marker (SNP 53285) associated with stem growth and two SNP (6672 and 14857) markers associated with yield. De Souza et al. (2018) obtained a total of 77,660 and 21,283 SNPs from 626 genotypes (including 368 germplasm accessions and 254 individuals from a mapping population, respectively) through GBS approach. The previously mapped mapping population constructed with 1062 SNPs had only 576 SNPs from GBS approach thus reducing the average interval between markers to 4.4 cM (de Souza et al. 2018).

## *Saturated Genetic Linkage Mapping and QTLs*

Genetic linkage map describes the linear order of markers whether genes or any other small DNA sequences in their respective linkage groups depicting their relative chromosomal locations by their pattern of inheritance (Priyadarshan 2017). It enhances our understanding on specific segments of the genome associated with a trait. Based on segregation analysis, Lespinasse et al. (2000b) assembled the first ever saturated genetic map of *Hevea,* encompassing 717 loci (covered 18 chromosomes) which eventually served as a reference genetic map. Initially, SSR loci/markers were identified both from published ESTs and through transcriptome sequencing. Rattanawong et al. (2008) constructed a genetic linkage map with 229 SSR and 198 AFLP markers. They could also identify a QTL directly associated with rubber yield (Hbg16a131) and another (Hbg3a312) associated with girth of the trunk. Triwitayakorn et al. (2011) developed EST-SSR markers through transcriptome sequencing of shoot apical tissue to construct linkage map (97 loci with a mean interval of 11.9 cM) and to identify traits of commercial interest. The subsequent molecular genetic map obtained by Souza et al. 2013, revealed 284 markers distributed among 23 linkage groups with a total of 2688.8 cM. The length of each group ranged from 2.7 to 228.7 cM. They also detected a total of 18 QTLs for growth traits during summer and winter seasons. Nine QTLs (7 for height and 2 for girth) for summer, five for winter (2 for height and 3 for girth) and four for both height (2) and girth (2) were detected in 11 linkage groups. The average distance between adjacent markers on these maps was between 8 and 11.9 cM, which was larger for downstream marker-trait association analyses. This led to research on achieving high-density linkage maps to directly study the analysis of sequence variations based on single-nucleotide polymorphisms (SNPs). Identification and genotyping of common SNPs became possible by the invention of genotyping-by -sequencing (GBS), a technique which combined the next-generation sequencing, genome complexity reduction techniques and barcoding. Using this technique, Pootakham et al. (2015b) constructed genetic linkage maps for two populations comprising 1704 and 1719 markers with a coverage of 2041 and 1874 cM, respectively. The average marker densities ranged between 1.23 and 1.25 cM. This saturated map showed an improved SNP marker density of 0.89 cM. However, the average distances between these adjacent markers were still large, and demanded a genetic linkage map with still higher density.

In a similar study by Shearman et al. (2015), a high-density genetic map constructed from 149 progenies of RRIM 600 and RRII 105 generated 12,326 SNPs from 4244 contigs with an average marker density of 1.90 cM. In a study by Chanroj et al. (2017), two SNP markers (SNP7772 and SNP14857) associated with yield and one SNP marker (SNP53285) associated with stem growth were identified. Hou et al. 2017, identified 180 primer pairs of EST-SSR markers (based on 38815 EST sequences) to exhibit polymorphism. In 2018, the first ultra-high-density genetic linkage map constructed by Xia et al. (2018) in *Hevea* identified 571,267 SNPs and 134,184 indels and could elucidate 6940 SNP markers with a density of 0.30 cM and a recombination rate of 0.97 cM/Mb. They could obtain 17 QTLs for dry latex yield (DLY) among which QTLs qDFY-10 and qDFY-18-4 could explain up to 38.3% and 33.3%, respectively. This study also identified highly promising QTL candidate genes such as thioredoxin h, plastin-like protein, calmodulin-binding protein, cytochrome c oxidase and methylglutaconyl-CoA hydratase that have significant association with yield. QTL mapping for stem diameter, tree height and number of whorls was reported from a sibling population of GT1 × RRIM 701 cross (Conson et al. 2018). The base map with 18 linkage groups was constructed using 225 SSR and 186 SNP markers and the dense linkage map with 671 SNPs generated through genotyping by sequencing. This study investigated on key dominant genes associated with growth of *Hevea* under water stress conditions and could identify 24 QTLs for stem diameter, seven for tree height and seven for number of whorls. Rosa et al. (2018) constructed a saturated multipoint integrated genetic map containing 354 microsatellite and 151 SNP markers to identify QTL for growth and latex production in a full-sib population (progenies of RRIM 600 × PB 217) cultivated under suboptimal conditions. In a recent study, An et al. (2019) genotyped 206 F1 progenies of CATAS 8-79 and MT/C/119/67 through Locus Amplified Sequencing Technology (SLAF-Seq) using 268,592 SNPs to construct a high-density genetic map with 4543 SNPs covering 2670.27 cM of the whole rubber genome with an average marker density of 0.59 cM. They could detect 11 QTLs for stem growth and

12 QTLs for latex yield distributed in 15 linkage groups. The QTLs for stem growth and latex yield (qzSG-8-2 and qLD-8-3, respectively) were closely linked with a genetic distance of 3.0 cM.

The additional information generated through the recently developed ultra-high-density genetic linkage maps have immensely contributed to enrich the database on genome-wide EST-SSR markers, SNP markers and genomic regions containing major genes and QTLs containing agronomically important traits thus having direct application in molecular breeding through marker-assisted selection. Since the beginning of *Hevea* cultivation, attaining higher yield potential had been the major breeding objective through improving the yield traits like trunk biomass, bark thickness and number of latex vessel rows (Priyadarshan 2003). The recent reports have identified few QTLs for stem growth and height (Rattanawong et al. 2008; Souza et al. 2013; Chanroj et al. 2017; Rosa et al. 2018; An et al. 2019) and yield (Rattanawong et al. 2008; Rosa et al. 2018; Xia et al. 2018; An et al. 2019) which can now be used for MAS in *Hevea*. The winter stress responsive QTLs that influence phenotypic variation in *Hevea* can be best employed for MAS to develop cold stress-tolerant clones of *Hevea*. The map constructed by Conson et al. 2018, provides information on genomic regions related to growth under water stress which can be best used for MAS for drought tolerance.

The first ultra-high-density genetic linkage map constructed by Xia et al. 2018, identified 17 most reliable QTLs for dry yield among which two QTLs had phenotypic variation of about 33 and 38%. The additionally found QTLs associated with defence mechanisms, energy metabolism and rubber biosynthesis pathways in this study also revealed significant association with dry yield (Fig. 7). The traits like latex yield and stem growth are complex and are polygenically or quantitatively controlled (Simmonds 1989). This was addressed by An et al. 2019, who did conditional QTL mapping to identify six QTLs for stem growth, seven QTLs for latex yield and five and six QTLS for stem growth and latex yield, respectively by conventional mapping. More studies are being carried out in this aspect as well as on QTLs for biotic and abiotic tolerance. All these information generated (Table 1) could be used as precious resources by the breeders to explore the possibility of utilizing them in the MAS programmes in *Hevea*.

Apart from employing markers for growth, yield and tolerance to the biotic and abiotic stresses, these microsatellite markers are also used to study the genetic variance in *Hevea* (Hou et al. 2017; Antwi-Wiredu et al. 2019). Mantello et al. (2011) identified 69 microsatellite loci from 384 clones and reported the existence of highly conserved flanking regions of microsatellites between species of *Hevea*. In a similar study, Souza et al. (2009) characterized 27 polymorphic microsatellite loci in a GA-CA enriched genomic library of *Hevea* and obtained cross-species amplification in wild *Hevea* species (*H. guianensis, H. rigidifolia, H. nitida, H. pauciflora, H. benthamiana* and *H. camargoana*) thus revealing the highly conserved nature of loci/markers and the complete colinearity of *Hevea* genome among its species (Souza et al. 2009; Prapan et al. 2006). There were many attempts to study the genetic diversity using EST-SSR markers. Cubry et al. (2014) characterized 164 polymorphic EST-SSRs for diversity and SALB resistance. A genetic diversity

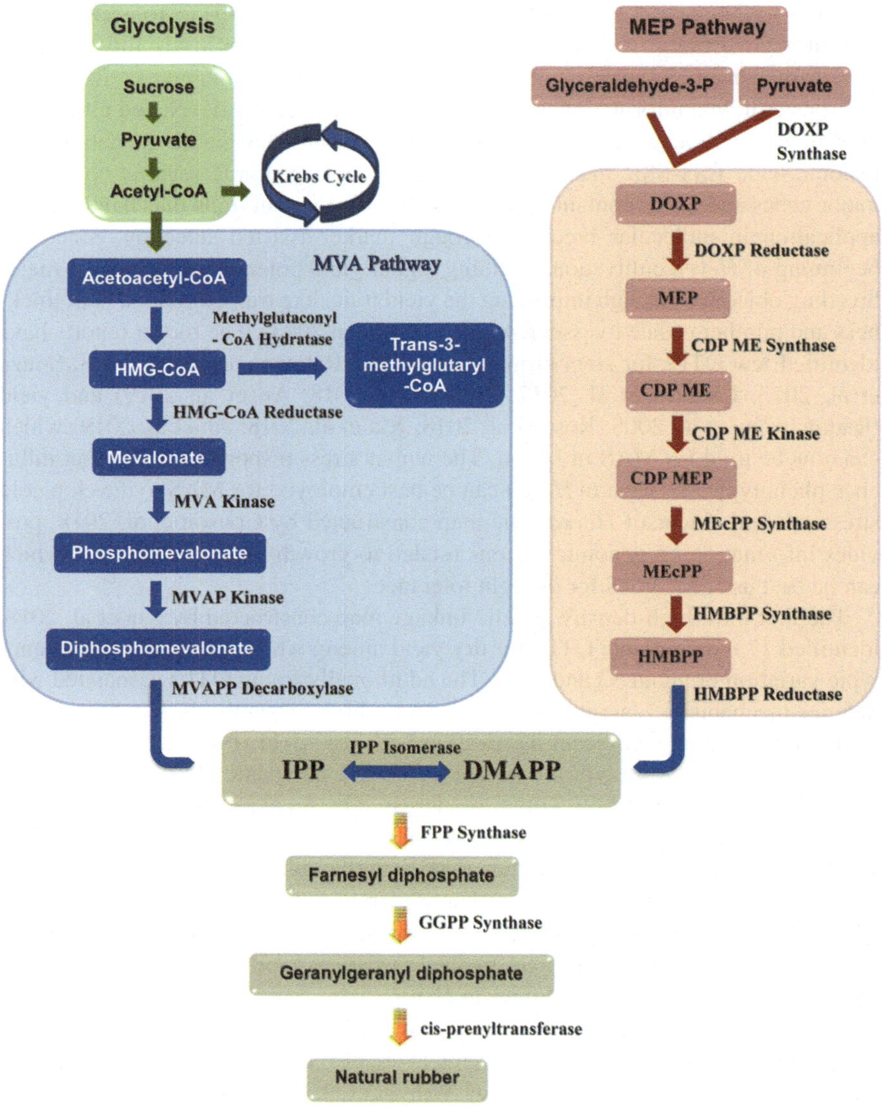


**Fig. 7** The candidate QTL genes in rubber biosynthesis pathway as described by Xia et al. (2018)

study by genotyping 1117 accessions with 13 microsatellite markers identified a total of 408 alleles among which 319 were shared between groups while 89 were confined to different groups of accessions (de Souza et al. 2015). A high frequency of gene flow between groups with high genetic distance was observed, and a total of 99 accessions with maximum genetic diversity were further identified for genomic breeding. In a similar study, Feng et al. (2008) identified 799 SSR loci from *Hevea* ESTs and studied genetic diversity in 12 clones and four related species. Analysis on 74 alleles using 30 randomly selected primer pairs indicated medium

**Table 1** Details of published genetic linkage maps of *Hevea brasiliensis*

| | | Markers | | | | | | | | |
|---|---|---|---|---|---|---|---|---|---|---|
| Map length (cM) | LG number | Isozyme | RFLP | AFLP | RAPD | SSR | SNP | Total | Average marker distance (cM) | References |
| 2144 | 18 | 10 | 301 | 388 | – | 18 | – | 717 | 3 | Lespinasse et al. (2000a, b) |
| 2075 | 18 | | | 198 | | 229 | | 427 | 4.86 | Rattanawong et al. (2008) |
| 1937.06 | 18 | – | – | – | – | 91 | – | 91 | 21.29 | Feng et al. (2010) |
| 2471.20 | 23 | – | – | – | – | 225 | – | 225 | 11 | Souza et al. (2011) |
| 2441 | 18 | – | – | 187 | – | 195 | – | 383 | 6.37 | Le Guen et al. (2011) |
| 842.9 | 23 | – | – | – | – | 97 | – | 97 | 11.9 | Triwitayakorn et al. (2011) |
| 2688.8 | 23 | – | – | – | – | 284 | – | 284 | 10 | Souza et al. (2013) |
| 2052 | 18 | – | – | – | – | – | 2321 | 2,321 | 0.89 | Pootakham et al. (2015a, b) |
| 4160 | 18 | – | – | – | – | – | 2186 | 2186 | 1.9 | Shearman et al. (2015) |
| 3779.70 | 18 | – | – | – | – | 224 | 855 | 1079 | 3.5 | Conson et al. (2018) |
| 2094.10 | 18 | – | – | – | – | – | 6940 | 6940 | 0.3 | Xia et al. (2018) |
| 3525 | 23 | | | | | 354 | 151 | 505 | 7.4 | Rosa et al. (2018) |
| 2670.27 | 18 | – | – | – | – | – | 4543 | 4543 | 0.59 | An et al. (2019) |

polymorphism of the EST-SSRs. This study also provided evidence for cross-species/genera transferability of the EST-SSR markers developed.

## *Transcriptome Studies by High-Throughput Sequencing*

Recent high-throughput sequencing technology and various genomic tools have hugely facilitated the data acquisitions in several crops in general and rubber in particular over the last two decades (Table 2). High-throughput genomic techniques are associated with innovative bioinformatics tools that assume much importance in rubber tree breeding programmes. Recent technological developments made RNA sequencing a cost-effective tool in gene expression profiling while providing qualitative and quantitative information at whole-genome level compared to other conventional methods (Zhang et al. 2010; Xia et al. 2011). Some of the earlier studies in *Hevea* dealt with ESTs from latex transcriptome, e.g. Ko et al. (2003) identified 1176 ESTs from latex transcriptome. Chow et al. (2007) sequenced 10,040 ESTs and obtained 3441 unique transcripts. With the advent of transcriptome sequencing technologies, large-scale EST studies followed which led to the development of plenty of EST-derived SSR markers for rubber (Xia et al. 2011; Triwitayakorn et al. 2011; Pootakham et al. 2011; Li et al. 2012). Investigations on ESTs of various stress and latex biosynthesis-related cDNA libraries have revealed a set of genes that can be employed as marker for important traits. A systematic approach on the transcriptome of bark was reported by Li et al. (2012) using RNA from the high yielding genotype RY7-33-97 (CATAS, China) through Illumina paired-end sequencing technology study that identified a total of 22,756 unigenes and predicted 39,257 EST-SSRs.

Liang et al. (2009) reported the constitutive expression of a translationally controlled tumour protein, *HbTCTP* in latex, leaves and barks upon ethylene (ET) treatment. Deng et al. (2012) further analyzed the gene structure and developed molecular markers of *HbTCTP*. Li et al. (2013) cloned *HbTCTP1*, another *TCTP* gene in *Hevea*. *HbTCTP1* is expressed in different tissues throughout the developmental stages of leaves and is regulated by drought, low temperature, high salt, ethylene stimulation, wounding, $H_2O_2$ and methyl-jasmonate (Deng et al. 2012). Isolation and molecular characterization of 1-aminocyclopropane-1-carboxylic acid synthase (*ACS*) genes in *Hevea brasiliensis* revealed the presence of nine *ACS*-like genes (Zhu et al. 2015). Structure and expression profile of sucrose synthase gene family in *Hevea* and their role in stress response and sucrose utilization in the laticifers was discussed by Xiao et al. (2013).

*Latex and Rubber Biosynthetic Pathway*: The first report on the ethylene-stimulated induction of laticifer-specific expression of genes was by Kush et al. (1990). Later, Miao and Gaynor (1993) isolated *MnSOD* gene, which was found to express in leaf, petiole, root, latex and callus and at higher levels in younger leaves. Thanseem et al. (2003, 2005) cloned and characterized β-1, 3-glucanase gene from various clones of *Hevea*. Priya et al. (2006) characterized promoter sequence of *REF* (rubber elongation factor) gene from *Hevea*. This was followed by reports

**Table 2** List of genome and transcriptome studies using new generation sequencing technologies

| Sequencing method | Transcriptome type | References |
|---|---|---|
| PE-RNA-Seq (Illumina) | Latex and leaf combined; clone RY7-33-97 | Xia et al. (2011) |
| 454 Pyrosequencing (Roche) | Leaf tissue; clone RRIM 600 RRII 105 | Pootakham et al. (2011) |
| 454 Pyrosequencing (Roche) | Shoot apical meristem; clone RRIM 600 | Triwitayakorn et al. (2011) |
| RNA-Seq (Illumina) | Latex; clone RRIM 600 | Chow et al. (2012) |
| PE-RNA-Seq (Illumina) | Bark; clone RY7-33-97 | Li et al.(2012) |
| 454 Pyrosequencing (Roche) | Leaf, bark, latex, root, embryogenic tissues; clone PB 260 | Duan et al. (2013) |
| PE-RNA-Seq (Illumina)<br>454 Pyrosequencing (Roche) | Leaf; clone RRIM 600<br>Leaf; clone RRIM 600 | Rahman et al. (2013) |
| 454 Pyrosequencing (Roche) | Pooled RNA extracts of different tissues and open pollinated seedlings; clone RRIM 600 | Salgado et al. (2014) |
| RNA-Seq (Illumina) | Bark samples; clones GT1 and PR255 | Mantello et al. (2014) |
| RNA-Seq (Illumina) | Latex; clone Reyan 7–33–97 | Zhiyi et al. (2015) |
| RNA-Seq (Illumina) | Bark; clone PR107 | Liu et al. (2015) |
| Illumina paired-end sequencing | Latex; clone CATAS8-79 and PR107 | Chao et al. (2015) |
| Illumina deep-sequencing | Bark; clone PR107 | Liu et al. (2016) |
| RNA-Seq (Illumina) | Latex; clones CATAS7-33-97 and HAIKEN 2 | Li et al. (2016) |
| Illumina HiSeq 2000 | Bark; clone RY 7–33–97 | Li et al. (2016) |
| Illumina HiSeq 2500 | Leaf, bark, latex; clone RRIM 600 | Lau et al. (2016) |
| IlluminaGA2 and HiSeq2000 | Different tissues and organs; clone Reyan7-33-97 | Tang et al. (2016) |
| 454 GS FLX + and Illumina HiSeq 2000 and PacBio RSII | Genome and transcriptome analysis | Pootakham et al. (2017) |
| RNA-Seq (Illumina) | Drought and cold-responsive transcripts from leaves of clone RRIM 600 and RRII 105 | Sathik et al. (2018) |
| RNA-Seq (Illumina) | Ethylene and TPD-responsive transcriptome from clone PB 260 | Montoro et al. (2018) |
| RNA-Seq (Illumina) | Cold stress-tolerant transcriptome from Reyan 7-33-97 | Gong et al. (2018) |
| RNA-Seq (Illumina) and degradome analysis | laticifers of TPD trees | Zhang et al. (2019) |
| RNA-Seq (Illumina) | Defence/Corynespora responsive genes | Roy et al. (2019) |

on a full-length cDNA encoding cysteine protease, *HbCP1* (Peng et al. 2008), full-length aquaporin cDNAs HbPIP2*; 1* and *HbTIP1; 1* (Tungngoen et al. 2009) and coronatine-insensitive 1 protein (*HbCOI1*) (Peng et al. 2009). Chow et al. (2007) obtained more than 10,000 ESTs from latex (clone RRIM 600) from which 3441 unique transcripts were identified. Through quantitative PCR (qPCR), they reported REF (rubber elongation factor) and SRPP (small rubber particle protein) as the most abundant transcripts in latex. In a similar study, Han et al. (2000) identified 245 ESTs in which REF and SRPP were the most abundant followed by defence and protein metabolism-related genes. Sando et al. (2008) reported full-length cDNA of genes encoding all the enzymes of MVA pathway which were abundant in latex. Yield stimulation through ethrel application diluted the latex, and the study revealed a positive correlation between expression of HbPIP2.1, (an aquaporin) and latex yield. This study also observed a normally downregulated HbTIP1.1 an aquaporin gene in laticifers getting up-regulated under ethrel treatment (Tungngoen et al. 2009). Promoter sequences of these genes were found to harbour ethylene, auxin, copper and sulphur responsive elements that are known to be associated with yield.

The transcriptome profiling by Wei et al. (2015) observed downregulation of almost all genes associated with energy metabolism indicating the long-time latex flowing trees as more energy intensive than normal flowing types. To support the long-term flow of latex (more intensive tapping), it is necessary to improve energy supplies (oxidative phosphorylation and photosynthesis process to generate more ATP). They also observed the possibility of using copper ion of latex as a standard to estimate tapping intensity as copper (Cu) is an essential micronutrient that is involved in essential functions, metalloenzymes, non-electron transfer reactions and ethylene-responsive receptor. Availability of copper ions has been observed as one of the control mechanisms of rubber tree latex isoprene biosynthesis metabolism (Fan and Yang 1991). Mantello et al. (2014) identified rubber biosynthesis-related ESTs from bark samples of GT1 and PR255 clones that are known as high yielders and tolerant to cold and wind. They validated a total of 78 SNPs of MVA and MEP pathways of rubber biosynthesis in 36 genotypes. Comparative analysis of latex transcriptome between clone PR107 (low yielder) and CATAS8-79 (high yielder) was performed to uncover the molecular mechanism behind the regulation of latex regeneration and duration of latex flow and to identify differential expression of several genes related to rubber biosynthesis, cellulose and lignin biosynthesis (Chao et al. 2015). They observed the influence of higher level expression of unigenes such as HbLOX and HbOPDR in the JA biosynthesis pathway, HbSUT, HbBAM and HbPK mediating carbohydrate metabolism, HbHMGR1 in MVA pathway and HbHRT2 in rubber biosynthesis pathway with enhanced latex regeneration. In turn, prolonged latex flow upregulated HbACO-mediated ethylene biosynthesis (Chao et al. 2015).

The first ever genome sequencing in *Hevea* reported by Rahman et al. (2013) identified 12 distinct sub-metabolic pathways represented by 417 genes involved in carbon assimilating steps in rubber biosynthesis (Fig. 8). This study reported representation of 417 genes in connection with the 12 distinct sub-metabolic pathways of carbon assimilatory mechanisms. The study enlists the genes involved in metabolic pathways that contribute for the production of isoprenoid precursors required for the rubber biosynthesis pathway (fructan metabolism—54 genes; starch metabolism—38 genes; glycolysis—52 genes; alternate pentose phosphate pathway—14 genes; acetyl CoA biosynthetic pathway—120 genes). Genome sequencing report by Lau et al. (2016) explained a comprehensive genome-wide analysis of clone RRIM600 and attributed to *H. brasiliensis*'s capacity to produce higher levels of

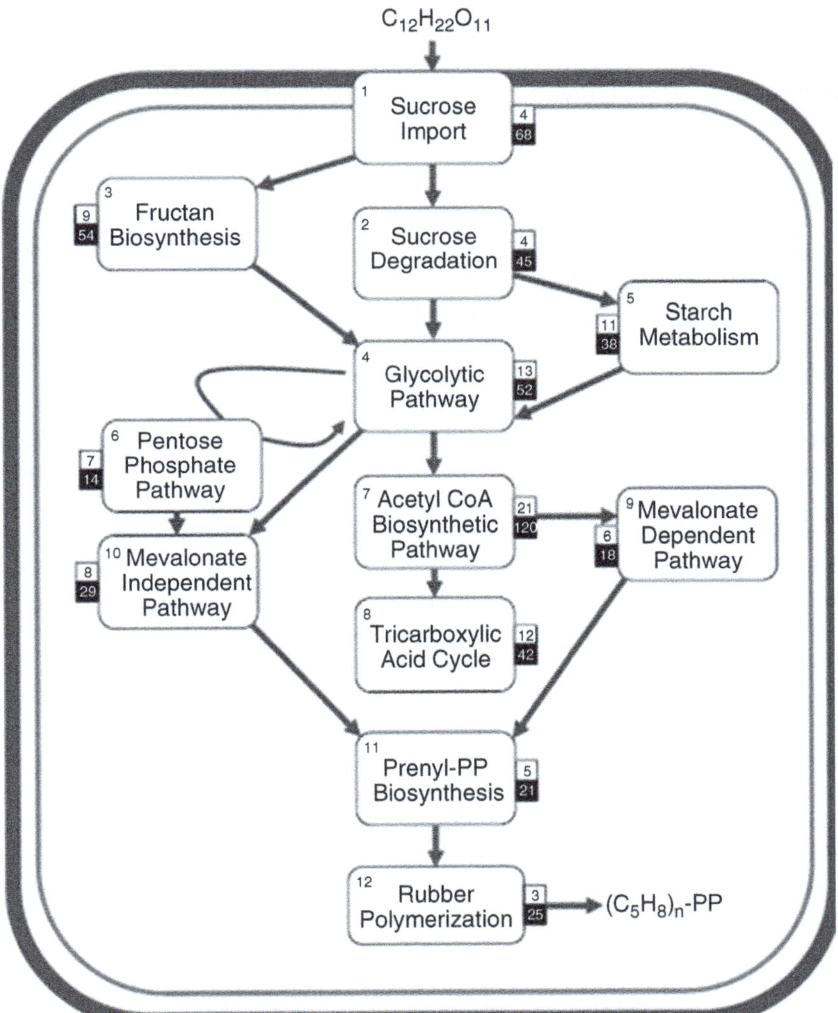


**Fig. 8** Schematic representation of metabolic pathway leading to natural rubber biosynthesis. Import of sucrose until biosynthesis of rubber involves 12 sub-metabolic pathways represented in the large boxes. The number of enzymes and associated proteins in each individual pathway is shown in small white boxes and the number of orthologs in *Hevea* in the grey boxes. (Adopted from Rahman et al. 2013)

latex to the expansion of rubber biosynthesis-related genes in its genome and their enhanced expression. They observed increased levels of MVA pathway genes in latex (acetyl CoA acetyltransferase, HMG-CoA synthase and mevalonate diphosphate decarboxylase gene families) than leaf and bark. They also observed that MEP pathway that contributes for the intermediates required for rubber biosynthesis is not preferentially expressed in latex. A high-quality genome assembly of *H. brasiliensis* by Tang et al. (2016) reported 84 rubber biosynthesis genes from 20 families (in which 18 and 22 genes belonging to cytosolic mevalonate pathway and plastidic MEP pathway were identified, respectively), to contribute for IPP synthesis. They also identified 15 initiator synthesis genes in cytosol and 39 genes involved in rubber elongation on rubber particles. REF/SRPP family of *Hevea* is the largest one with 18 members and is the most abundant proteins in latex. Of the 24 MEP genes, only two 1-deoxy-D-xylulose 5-phosphate synthase genes (DXS7 and DXS10) were found preferentially expressed in latex.

Wu et al. (2018) analysed the relationship between rubber yield and expression of nine major latex metabolism-related genes, i.e. HMG-CoA synthase (HMGS), HMG-CoA reductase (HMGR), diphosphomevalonate decarboxylase (PMD), farnesyl diphosphate synthase (FPS), *cis*-prenyltransferase (CPT), rubber elongation factor (REF), small rubber particle protein (SRPP), dihydroxy acid dehydratase (DHAD) and actin depolymerizing factor (ADF), and found most of the genes to express predominantly except *HbHMGR1, HbPMD* and *HbDHAD*. Most of these genes (except *HbDHAD*) were greatly influenced by both ethephon (ETH) and methyl-jasmonate (MeJA). This study also found positive correlation between the expression level of *HbCPT, HbFPS, HbHMGS, HbHMGR1* and *HbDHAD* and rubber yield and/or yield characteristics while *HbREF* exhibited a negative correlation. HbSUT3, a sucrose transporter gene, was observed to express predominantly in latex (Tang et al. 2010, 2013). Expression of HbSUT3 got induced by tapping, ethrel stimulation and jasmonic acid and was positively correlated with latex yield. Efficient sucrose loading in laticifer cells is important for improved rubber production in *Hevea,* and this function is carried out mainly by sucrose transporters encoded by SUT (sucrose transporter) genes. A study by Sui et al. (2017) suggested the role of SWEET proteins in sugar efflux transport as response to plant development and stress response based on a genome-wide analysis for the first time. From a total of 127 SWEET genes studied, HbSWEET1a, 2e, 2f and 3b proteins were involved in phloem loading while HbSWEET10a and 16b in laticifer sugar transport and HbSWEET9a in nectary-specific sugar transport. SWEET proteins also have versatile role in plant development, stress response and plant–microbe interactions and transport sugar substrates like sucrose, fructose and glucose selectively.

Jasmonate (JA) signalling regulates secondary laticifer differentiation (Hao and Wu 2000) and NR biosynthesis in *Hevea* through jasmonate ZIM-domain (JAZ) proteins which are the master regulators of jasmonate signalling. In a recent study by Loh et al. (2019), influence of JA on secondary laticifer differentiation through Bel5-GA2 oxidase 1-KNOTTED-like homeobox complex was observed based on 450 genes unique to jasmonic acid (JA) and linolenic acid (LA). Chao et al. (2019)

identified 18 JAZ proteins in *Hevea* among which the *HbJAZ5.0* and *HbJAZ10.0b* were presumed to be associated with differentiation of laticifer while the *HbJAZ8.0b* negatively regulates rubber biosynthesis. Latex yield depends upon the duration of latex flow after tapping and the capability of latex regeneration between two consecutive tappings (d'Auzac et al. 1997). Latex regeneration involves multiple numerous biological processes including transcription and translation, intracellular trafficking and signalling pathways and movement of water and biochemical intermediates across the laticifer cells (Tang et al. 2013). The elaborate studies being made through NGS technology have generated plenty of information on rubber biosynthesis genes and their regulatory mechanisms in *Hevea*. Almost all the genes involved in general metabolic pathways leading to rubber biosynthesis have been identified in latex and bark, some have been even cloned and further characterized. More studies are coming up on what exactly decides a high yielder. For example, a historical and a detailed study by Tang et al. (2013) in a super high yielding rubber tree attributed improved sugar loading capability, rubber biosynthesis-preferred sugar utilization, enhanced general metabolism and timely stress alleviation as the reason for high yield. More investigations are needed in this line to illustrate the yield-enhancing factors and in increasing the productivity.

*Ethylene Stimulation and Tapping Panel Dryness:* Ethephon, an ethylene releaser applied to tapping panel, stimulates latex production by inducing several biochemical changes in laticifers like sucrose loading, water uptake and nitrogen assimilation or expression of ethylene-responsive genes/defence proteins (Putranto et al. 2015). Initially, Kush et al. (1990) reported laticifer-specific genes induced by ethylene in *H. brasiliensis*. The tapping induced endogenous ethylene production and ethephon induced exogenous ethylene are the major sources of oxidative stress in *Hevea* that results in the development of the tapping panel dryness (TPD) syndrome. Though ethylene is important for the optimum latex yield, it becomes deleterious in excess levels. Ethylene signal is perceived through ethylene response factors (ERFs) which belong to AP2/ERF superfamily. ERFs are trans-acting factors that bind to GCC or DRT/CRT cis-acting elements in the promoter region of target genes (Shoji et al. 2013) and have been reported as biotic and abiotic stress responsive (Mizoi et al. 2012). In latex of high yielding clone PB 217, ethylene stimulation induces expression of sugar transporter HbSUT1B than inner bark tissues while it was vice versa in clone PB260, a low yielder. Tang et al. (2010) identified predominant expression of HbSUT3, a functional sugar transporter in latex of *Hevea*, was found to be ethylene responsive and have positive correlation with yield.

Putranto et al. (2015) studied the expression of abiotic stress and hormone responsive ERF genes in latex and identified 21 *HbERF* genes regulated by ethylene, methyl-jasmonate and dehydration. This study made an important observation that tapping induces expression of mostly dehydration-responsive genes except for few wounding-related genes. In addition, they also found that most of the ethephon-induced ERFs were responsive to osmotic stress influenced dehydration indicating the occurrence of tapping-induced osmotic stress. Twenty differentially expressed genes belonging to ERF groups I, II, IV, VI, VIII, IX and X were identified as stress responsive and thus can be considered as expression marker genes. Study by

Putranto et al. (2015) found various cis-acting regulatory elements in promoters of HbERF-IX that could be induced by ethylene, JA, auxin, cytokinin, gibberellins, abscisic acid and oxidative stress. Moreover, these genes were intron free (Nakano et al. 2006) and had better chance for their rapid and constitutive expression required under stress (Jeffares et al. 2008).

Transcriptome analysis in ethylene-induced cDNA library (Liu et al. 2016) indicated the upregulation of certain regulatory enzymes in the glycolytic pathway as well as genes in the carbon fixation pathway (Calvin cycle). It was presumed that rapid acceleration of glycolytic pathway required to supply precursors for the biosynthesis of IPP, and subsequently NR instead of rubber biosynthesis per se may be responsible for ethylene stimulation of latex in *Hevea*. Tang et al. (2016) reported 509 ethylene-responsive genes in latex which included 53 transcription factors and 31 protein kinase both of which are essential for ethylene-responsive signal transduction in higher plants. A total of 225 ethylene-responsive element (AP2/ERF) binding factors were also observed. The transcription factors involved in ethylene signalling pathways like AP2/ERF domain-containing transcription factors and ethylene-responsive transcription factor RAP2 had direct association with latex yield (Xia et al. 2018). Ethylene biosynthesis and signalling genes were also identified in *Hevea* (Kuswanhadi et al. 2010; Piyatrakul et al. 2012; Duan et al. 2013). Three MADS-box genes from *Hevea, HbMADS1, HbMADS2 and HbMADS3,* were cloned and characterized by Li et al. (2011).

Molecular studies in bark and latex of tapping panel dryness (TPD)-affected rubber trees indicated downregulation of *HbMyb1* transcription factor (Chen et al. 2002). Initially, TPD-associated genes were identified by suppression subtractive hybridization (Sathik et al. 2006; Venkatachalam et al. 2007; Li et al. 2010). Subsequent mRNA differential display study revealed *HbTOM20*'s association with TPD (Venkatachalam et al. 2009). A comparative transcriptome analysis between healthy and TPD-affected rubber trees indicated differential expression of genes related to rubber biosynthesis and jasmonate synthesis (Liu et al. 2015). In another study, transcriptome analysis between healthy and TPD trees revealed differential expression of genes associated with ROS metabolism, jasmonate and ethylene biosyntheses, ubiquitin proteosomal pathway (UPP), programmed cell death (PCD) and rubber biosynthesis (Li et al. 2016).

*Abiotic Stress*: Abiotic stresses like drought, low temperature, wind, hailstorm are the major factors that affect the yield of *Hevea* in major rubber-producing countries. Drought and low temperature stresses hinder plant growth and development by hindering several metabolic processes including stomatal conductance, nutrient uptake and photosynthetic assimilation that eventually results in significant yield loss (Buttery and Boatman 1976; Sethuraj et al. 1984; Sreelatha et al. 2007, 2011; Shinozaki et al. 2003). Plants have also evolved multilevel stress perception and signalling and acclimation mechanisms to cope up with such limiting conditions. To mitigate the adverse effect of stress, plants often respond by altering at physiological, biochemical and gene expression levels. Drought tolerance is a complicated phenomenon that involves multiple genes and their interaction with the environmental factors (Zandalinas et al. 2020). Developing resilience to multiple stresses in crops is the real challenge which can be achieved by identification of key genes associated with such traits and signal transduction pathways and by incorporating in the breeding programmes to develop varieties with those specific traits without compromising on yield (Bailey-Serres et al. 2019).

Initial studies on cold stress indicated significant increase in cold-responsive genes like carbonic anhydrase, glutathione peroxidase, metallothionein, chloroplastic Cu/Zn SOD, serine/threonine protein kinase, transcription factor, and DNA-binding protein. (Saha et al. 2010). In a similar study, gene expression analysis in cold-stressed leaf samples revealed higher levels of expression of LEA 5 protein, NAC tf and peroxidase and suggested association of genes *viz LEA 5* protein, peroxidase, *ETR1, ETR2* and *NAC* transcription factor with cold tolerance (Sathik et al. 2012). In a transcriptome study by Sathik et al. (2018), ethylene-responsive transcription factor (ERF) was found strongly associated with cold tolerance. Cheng et al. (2015) functionally characterized CRT/DRE binding factor 1 (*HbCBF1*) gene in *H. brasiliensis* when transgenic *Arabidopsis* overexpressing *HbCBF1* was found to have enhanced cold tolerance. Expression analysis of stress-responsive transcripts and their association with drought tolerance in different *Hevea* clones was studied (Thomas et al. 2012) and found upregulation of NAC tf in a drought-tolerant genotype (RRIM 600) than a relatively drought-susceptible one (RRII 105) indicating its strong association with drought tolerance (Luke et al. 2017). Differential gene expression analysis in several *Hevea* genotypes with varying levels of drought tolerance indicated the association of MAP Kinase, Myb tf, CRT/DRE binding factor and NFYA with drought tolerance (Luke et al. 2015). In another study, there was strong association between expression of ferritin, DNA-binding protein, NAC tf and aquaporin with drought tolerance (Sathik et al. 2018). Significant upregulation of genes related to energy biosynthesis and ROS scavenging systems including *HbCuZnSOD, HbMnSOD, HbAPX, HbCAT, HbCOA, HbATP* and *HbACAT* was also observed under drought stress (Wang et al. 2014).

Cheng et al. (2016) found *ErbB-3 binding protein 1* (*EBP1*) *to* undergo changes in its expression in response to cold, drought stress and ABA treatment in *Hevea*. Overexpression of *EBP1* in *Arabidopsis* enhanced resistance to freezing and drought stress. Long et al. (2015) identified four glucose-6-phosphate dehydrogenase (*HbG6PDH*s) gene family in *Hevea* that respond to temperature and drought stresses in root, bark and leaves were found to be involved in redox balance maintenance and defence against oxidative stress. Gong et al. (2018) identified 1457 and 2328 DEGs from leaves of cold-tolerant clone Reyan 7-33-97 in response to 2 and 24 h of cold stress treatment. Flavonoid biosynthesis, phenylpropanoid biosynthesis, plant hormone signal transduction, cutin, suberin and wax biosynthesis, pentose and glucuronate interconversions, phenylalanine metabolism and starch and sucrose metabolism were the significantly expressed main KEGG pathway genes. This study also observed differential expression of 239 cold-responsive transcription factors (TFs). This included TFs *viz.*, ARR-B, B3, BES1, bHLH, C2H, CO-like, Dof, ERF, AR1, G2-like, GRAS, GRF, HD-ZIP, HSF, LBD, MIKC-MADS, M-type MADS, MYB, MYB-related, NAC, RAV, SRS, TALE, TCP, Trihelix, WOX, WRKY, YABBY and ZF-HD. ATP-binding cassette proteins (ABC proteins) in the laticifers of *H. brasiliensis* was first characterized by Zhiyi et al. (2015). All eight plant ABC protein paralog subfamilies were identified in latex among which ABCB, ABCG and ABC1 were the most abundant. Expression of HbABCB15, HbABCB19, HbABCD1 and HbABCG21 was significantly regulated by ethylene, JA and wound stresses.

In plants, gene expression is regulated by non-coding small RNAs (Storz 2002) like microRNA (miRNA), small interfering RNA (siRNA), small nucleolar RNA (snoRNA), small rDNA-derived RNA (srRNA), small nuclear RNA (U-RNA), tRNA-derived small RNA (tsRNA) in response to development, biotic and abiotic stresses (Meister et al. 2013). miRNAs play main role in negatively regulating gene expression by binding with the complementary mRNAs to either cleave or repress translation (Chinnusamy et al. 2007). Role of miRNAs in *Hevea* was first reported by Zeng et al. (2010) followed by Gebelin et al. (Gébelin et al. 2012; Gebelin et al. 2013a, b) and Lertpanyasampatha et al. (2012). Gébelin et al. (2012) reported drought-responsive miRNAs by deep sequencing (48 conserved and 10 novel miRNA families) while Lertpanyasampatha et al. (2012) identified 115 miRNAs belonging to 56 families and 20 novel miRNAs through high-throughput sequencing in high and low yielding clones (PB 260 and PB 217, respectively). miRNAs that regulate MYB transcription factor, auxin-responsive factor (ARF) and type III HD-Zip transcription factors were abundantly expressed in leaves. In response to cold stress and JA, MIR159b was found enhanced in leaves and bark (Gebelin et al. 2013a). Enhanced expression of eMIR159b gene was observed in TPD-affected trees (Gebelin et al. 2013b). Four miRNAs (miR482, miR164, miR167 and HbmiRn_42) were found drought responsive while miR482 had strong association with drought tolerance (Kuruvilla et al. 2016). A differential expression analysis of cold-responsive miRNAs indicated miR169, miR159 and miR482 to have strong association with cold tolerance (Kuruvilla et al. 2017). Under cold stress, miR169 got downregulated while its corresponding target mRNA (NF-YA) associated with stress tolerance got upregulated (Luke et al. 2015). A novel miRNA (HbmiRn_42) that targeted HMG-CoA reductase was found highly upregulated under drought conditions in tolerant clones probably indicates the suppressed rubber biosynthesis under drought (Kuruvilla et al. 2016). These studies also indicate the possibility of using miRNA as markers for stress tolerance and in crop improvement programmes (Sathik and Kuruvilla 2017) while more investigations are required to study the relationship between miRNAs and their corresponding mRNAs in the context of biotic and abiotic stress tolerance. A recent report by Leclercq et al. (2020) indicates maximum level of post-transcriptional regulation in latex biosynthetic pathway and hormonal signalling in latex by mRNA cleavage through miRNAs and ta-siRNAs (trans-activated small interfering RNA).

*Biotic Stress: Hevea* is vulnerable to many fungal pathogens. *Oeidium heveae, Colletotrichum* spp., *Phytophthora* spp., *Corynespora cassicola* and *Microcyclus ulei* are some of the fungal pathogens that cause leaf diseases. *Corticium salmonicolor* causes stem infection called pink disease. South American leaf blight (SALB) caused by *M. ulei* is the most devastating in Brazil. It has not yet come to Asia region, but there is always a possibility that it may reach this region and breeders should be ready with SALB-resistant/tolerant clones to tackle the situation. Corynespora leaf disease caused by *Corynespora cassicola* is one of the major diseases that leads to significant yield loss in *Hevea*. Through transcriptomic sequencing approach, Roy et al. (2019) could identify transcripts related to defence response, response to stimulus and stresses from RRII 105 (susceptible) and GT1 (moderately resistant). They could identify enhanced expression of defence-related genes, receptor-like kinases and transcription factors (TFs). TFs such as NAC, ERF, MYB, GATA, WRKY, LEA and bZIP are important upstream regulatory proteins which play crucial role in regulating plant responses to stresses and enhanced disease resistance against pathogens. NAC TFs play critical role in plant immune responses. This study also identified a total number of 32,323 SSRs with a relative abundance of 503.59 SSRs/Mb and SSR polymorphism between clones RRII 105 and GT 1 in control and challenged conditions. Information on these SSRs and the SNPs would be a valuable source for the construction of linkage maps, QTL mapping, genetic diversity and MAS breeding. RAPD markers were used to identify *Oidium*-resistance gene (Chen et al. 1994, 2003). In a study on powdery mildew, Li et al. (2016) reported 78 differentially expressed genes from RRIC 52 (an *Oidium*-tolerant clone). The ESTs expressed at least 78 h post infection were found *Oidium* responsive among which ESTs related to membrane pathway, transcription factors, signal transduction, phytoalexin biosynthesis and other metabolic pathways were observed. Increased chitinase and B-1,3 glucanase activity could be observed in both tolerant and partially susceptible clones.

Páez et al. (2015) identified 86 differentially expressed SALB-responsive genes in clone FX3864 (a cross between *H. brasiliensis* PB 86 and *H. brasiliensis* B 38) infected with *M. ulei* for 48 h and observed induction of SA-dependent defence genes that mediates resistance to SALB. The results indicated downregulation of seven putative gene members of ethylene-dependent AP2/ERF family and upregulation of three SA-associated genes involved in cell wall synthesis. Fang et al. (2016) identified 3905 differentially expressed genes from leaves of clone Reyan 7-37-97 at four developmental stages and revealed the distinct expression pattern of genes associated with cyanogenic metabolism, lignin and anthocyanin biosynthesis and its correlation with change in SALB resistance between the stages of leaf development. Three defence-responsive genes such ascorbate oxidase, NADPH oxidase and lipoxygenase and six genes coding for ROS scavenging such as thioredoxins, glutaredoxin, caleosin-related peroxygenase, polyamine oxidase and thioredoxin were found differentially expressed in latex (Tang et al. 2016). More such disease resistance genes have to be identified from *Hevea* using high-throughput sequencing to employ in breeding for disease-tolerance programme.

*DNA Methylation*: Under biotic and abiotic stress conditions, plants undergo DNA methylation to alter gene expression in numerous biochemical pathways associated with stress acclimatization and molecular adaptation (Finnegan et al. 1998). Hypomethylation in genome induced by cold stress (Chinnusamy et al. 2008) and drought stress (Labra et al. 2002) have proven that it is a well-synchronized strategy of the plants to alter gene expression to cope up with the changing environment (Uthup et al. 2011). In *Hevea*, abiotic stress influenced methylation was observed in the promoter region of rubber biosynthesis genes and disease-resistance gene (Uthup et al. 2011) while the non-stressed plants (clone RRII 105) remained unmethylated. The clone RRIM 600 and PB 260 were found epigenetically more stable. Methylation in coronatine-insensitive 1 (COI1) gene, a defence-related gene (Xie et al. 1998) in clone RRII 105 and RRIM 600, indicates the suppression of its expression thus making it susceptible to pathogen attack. This study indicates that methylation can be used as a parameter to identify environmentally stable clones for molecular breeding (Uthup et al. 2011).

## 9 Genome Sequencing

Construction of *Hevea* genome and generating multi-transcriptome database can accelerate the process of genome analysis and facilitate genome data-based marker identification for application in crop improvement programs. Till date, four draft *Hevea* genome have been published between 2013 and 2017. A comparative table on these sequencing is furnished in Table 3. First draft genome sequencing data of *Hevea* was made available by Malaysian team (Rahman et al. 2013) in RRIM 600, a progeny of genotypes Tjir1 and PB 86. The whole-genome shotgun (WGS) approach generated ~43× coverage from Roche/454, Illumina and SOLiD platforms. The assembly covered about 1.1 Gb of the estimated 2.15 Gb haploid genome, and they could predict 68,955 genes of which 12.7% were unique. Based on 154 microsatellite markers, they could anchor 143 scaffolds and associated 1325 genes onto the 18 *H. brasiliensis* linkage groups. They could identify 9516 cluster of genes common to *Euphorbiaceae* among which 2708 clusters comprising 8748 genes are specific to *Hevea*. Expression analysis indicated plastidic mevalonate independent (MVA) pathway (29 genes) contributing as an alternate source of IPP precursor to rubber biosynthesis.

The second genome sequencing was reported by Rubber Research Institute, CATAS, China, (Tang et al. 2016) in genotype Reyan7-33-97 based on WGS and pooled BAC clones. This assembly covered a length of 1.37 Gb covering about 94% of the predicted genome size (1.46 Gb). This study observed *Hevea* to have largest genome size and greatest repeat content (71%) among the five genera of *Malpighiales* (*Hevea, Manihot, Ricinus, Populus* and *Linum*). This study identified 84,241 unique transcripts and 43,792 protein-coding genes (39,919 were found in NCBI) from which 84 rubber biosynthetic genes from 20 families were found. Eighteen genes belonging to cytosolic mevalonate pathway and 22 genes of plastidic MEP pathway were found to contribute for IPP synthesis. This study reported the latex biased expression of at least one gene each of the MVA pathway towards rubber biosynthesis. Eighteen-member REF/SRPP family of *Hevea* was found as the largest with the other 17 sequenced plants thus indicating the larger representation of this family of genes in *Hevea*. Ethylene-responsive DGE analysis in latex indicated expression of 342 annotated DEGs, 53 transcription factors and 31 protein kinases (signal transduction and response). Among the eight ETRs expressed in response to

**Table 3** A comparison of four published draft genome data of *Hevea brasiliensis*

| Genome details | Rahman et al. (2013) | Tang et al. (2016) | Lau et al. (2016) | Pootakham et al. (2017) |
|---|---|---|---|---|
| Genotype | RRIM 600 | Reyan 7-37-97 | RRIM 600 | BPM 24 |
| Genome size (Gb) | 1.1 | 1.46 | 1.55 | 1.26 |
| Scaffold number | 6,08,017 | 7453 | 1,89,316 | 5,92,580 |
| Scaffold length | 1119 Mb | 1.37 Gb | – | – |
| Scaffold N50 | 2972 bp | 1.28 Mb | 67.24 kb | 96.8 kb |
| Longest scaffold | 531.5 kb | 6.41 Mb | 871.19 kb | 2020 kb |
| Contigs number | 12, 23, 364 | 84,285 | 2,62,709 | – |
| Total size of contigs (Gb) | – | 1.29 | 1.55 | 1.25 |
| Contigs N50 (kb) | – | 30.6 | 20.75 | 96.8 |
| Longest contigs | – | 312.7 kb | 325.50 kb | 2,026,921 |
| No. of protein coding genes | 68,955 | 43,792 | 84,443 | 43,868 |
| Mean gene length (bp) | 1332 | 3913 | | 2747 |
| Mean transcript length | – | 1483 | 3069 | – |
| Average coding sequence length | 696 | 1123 bp | 971 bp | |
| Average number of exons per gene | – | – | 4.97 | 4 |
| Average exon length (bp) | 238 | 308 | 196 | 223 |
| Exon GC content | | 41.54 | | 43.1 |
| Average intron length | 332 | 677 bp | 478 bp | |
| intron GC content | | 32.61 | | 32.3 |
| Repeat sequence (%) | 71 | 72 | 72.5 | 69.2 |

ethylene, four got highly upregulated (ETR1, CTR1, EIN2 and EIN3/EIL1). Higher levels of EIN2 and EIL1 indicates the occurrence of active ethylene signalling in laticifers. A total of 225 AP2/ERF members of ethylene-responsive element binding factors were also identified. Ethylene triggered downstream biochemical cascade of events related to latex production such as sugar loading and its catabolism, water uptake, energy availability, cytosolic alkalinization, nitrogen assimilation and defence responses were also identified. Five out of six DEGs related to ROS production and scavenging were also reported to be highly upregulated. While expression of ascorbate oxidase and NADPH oxidase were at higher levels, lipoxygenase got downregulated.

The third report on *Hevea* genome sequencing was by Lau et al. (2016), a joint venture by Malaysia (Centre for Chemical Biology, University of Malaysia) and Japan (RIKEN Centre for sustainable Resource Science) in clone RRIM 600. This assembly has a total length of 1.55 Gb with 72.5% repetitive DNA sequences and represented more than 93.7% of the transcriptome sequences. They could predict a total of 84,450 high-confidence protein-coding genes. This study also revealed the presence of highest number (483) of disease-resistance–related genes. Proteome comparison indicated sharing of 12,406 gene families among *Euphorbiaceae* species such as *Ricinus communis, Manihot esculenta* and *Jatropha curcas*. Among the 3142 putative transcription factors (tfs) identified in this study, MYB tfs were the most abundant group which is mainly involved in hormonal signal transduction, disease resistance and secondary metabolism. The CAGE analysis (cap analysis gene expression) indicated preferential expression of MVA pathway genes in latex at higher levels than the MEP pathway.

Consequently in the following year, the fourth report on *Hevea* genome assembly was published by Pootakham et al. (2017) from National Centre for Genetic Engineering and Biotechnology and Rubber Authority of Thailand. The genotype BPM 24 which is resistant to major fungal pathogens like *Phytophthora* and *Corynespora* was sequenced using three platforms (Roche 454 GS FLX + and Illumina HiSeq 2000 and PacBio RSII). The final assembly had a length of 1.26 Gb (N50 = 96.8 Kb) and contained 69.2% repetitive sequences with a GC content of 34.31%. Apart from predicting a total of 43,868 genes coding for proteins, 623 tRNA, 274 rRNA, 282 snoRNA, 164 snRNA and 193 miRNA were also reported. This study found 2308 markers (from 2321 markers reported by Pootakham et al. 2015a, b, by GBS-SNP–based linkage map) unambiguously located on the genome assembly and could anchor on the 862 scaffolds spanning approximately 310 Mb on 18 linkage groups of *Hevea*. With an additional 708 scaffolds from literature, a total of 1568 anchored scaffolds were found associated with about two-thirds of the predicted coding genes.

The assembly by Pootakham et al. (2017) is slightly larger than 1.1 Gb reported by Rahman et al. (2013) in RRIM 600 but smaller than 1.37 Gb assembly of Reyan7-37-97 by Tang et al. (2016) and 1.55 Gb of RRIM 600 reported by Lau et al. (2016). The percentage of repeat sequences (69.2%) reported by Pootakham et al. (2017) is also somewhat similar to the previous reports (71, 72 and 72.15%) by Rahman et al. (2013), Tang et al. (2016) and Lau et al. (2016), respectively. This study also revealed the common ancestry of *Hevea* and *Manihot* genome (Pootakham et al. 2017). *Hevea* also was found to have a large number of nucleotide-binding site (NBS) and leucine-rich repeat (LRR) containing R genes similar to its close relatives such as *Manihot* and *Ricinus*. An important finding of this study is the identification of 164 syntenic blocks containing 2951 paralogous gene pairs in *Hevea* that led to a paleotetraploidization event. Distribution of these paralogous gene pairs across the 18 linkage groups evidently revealed one-to-one synteny between five pairs of homeologous linkage groups that might have evolved from a recent whole-genome duplication event. The presence of conserved syntenic genes between the *Hevea* and *Manihot* indicates the possibility of occurrence of paleotetraploidy event prior to the speciation of these two species.

Makita et al. (2018) created a *Hevea* genome and multi-transcriptome–based database with an aim to help *Hevea* researchers and breeders make use of vast information on genome and transcriptome, gene expression and precise gene structure such as transcription start sites (TSSs) and isoforms of *Hevea*. They employed cap analysis gene expression (CAGE) method that captures 5′ end of the transcribed and capped mRNAs to study the major expressed TSS in different tissues. Gene annotation becomes very difficult in non-model plants with genes of unknown functions and to overcome this problem (especially for the 22,991 *Hevea* specific genes), Makita et al. (2018) carried out co-expressed gene analysis to predict functionally related gene groups that are regulated by the same transcription factor (TF) based on the fact that genes regulated by the same TF often display similar pattern of expression. They prepared an NR biosynthetic pathway by this method to clearly illustrate information on the genes involved (Fig. 9) by integrating three types of transcriptome data such as RNA-Seq, Full-length cDNA (FL-cDNA) and CAGE.

Currently, four draft genome data are available for three clones *viz.* RRIM 600 (Rahman et al. 2013; Lau et al. 2016), Reyan 7-33-97 (Tang et al. 2016) and BPM 24 (Pootakham et al. 2017) of *Hevea brasiliensis*. Though the most advanced sequencing platforms like PacBio which can generate reads of about 10–15 kb length has made high-quality genome-sequencing assembly possible, the raw reads may suffer from high error rate. This was addressed by Pootakham et al. (2017) by employing long-range Chicago data along with PacBio reads to scaffold contigs which improved sequence contiguity considerably. Further, a recently developed Hi-C–based sequencing approach (Lieberman-Aiden et al. 2009; van Berkum et al. 2010) has made possible to decipher the relationship between chromosome organization and genome activity to understand genomic processes like transcription and replication. This does not require population data for map construction and can be done in relatively shorter time and with ease (Cheng et al. 2020). It is also possible to detect finer interactions between enhancers, silencers and insulators as the cost of sequencing is getting cheaper (Lieberman-Aiden et al. 2009). Similar approach is required in *Hevea* to generate higher quality genome data in chromosome level with a contig N50 size more than 5 Mb, about 20 Mb of scaffold N50 and less than 100 scaffolds of which 95% should be mounted and ordered onto the whole set of 18 chromosomes (Cheng et al. 2020). There is no doubt that data from these efforts should make whole-genome association studies, identification of potential markers, studying marker–trait relationship at whole-genome level, genetic modification of *Hevea* with traits of interest along with high yield and improvement of germplasm possible.

## 10 Conclusions and Future Perspectives

Unlike other crops, *Hevea* got domesticated around 150 years ago, and the progress attained in *Hevea* breeding has been meticulous, especially after 1950s. Through conventional breeding itself, the crop yield could be remarkably and tremendously improved from a mere 456 kg $ha^{-1}$ $year^{-1}$ to more than 2500 kg $ha^{-1}$ $year^{-1}$, and in exceptional cases even more than 3500 kg $ha^{-1}$ $year^{-1}$ could be attained which is still far below the theoretically possible yield potential of 7000–12,000 kg $ha^{-1}$ $year^{-1}$. It is worth appreciating the enormous efforts put in by the breeders to raise the yield

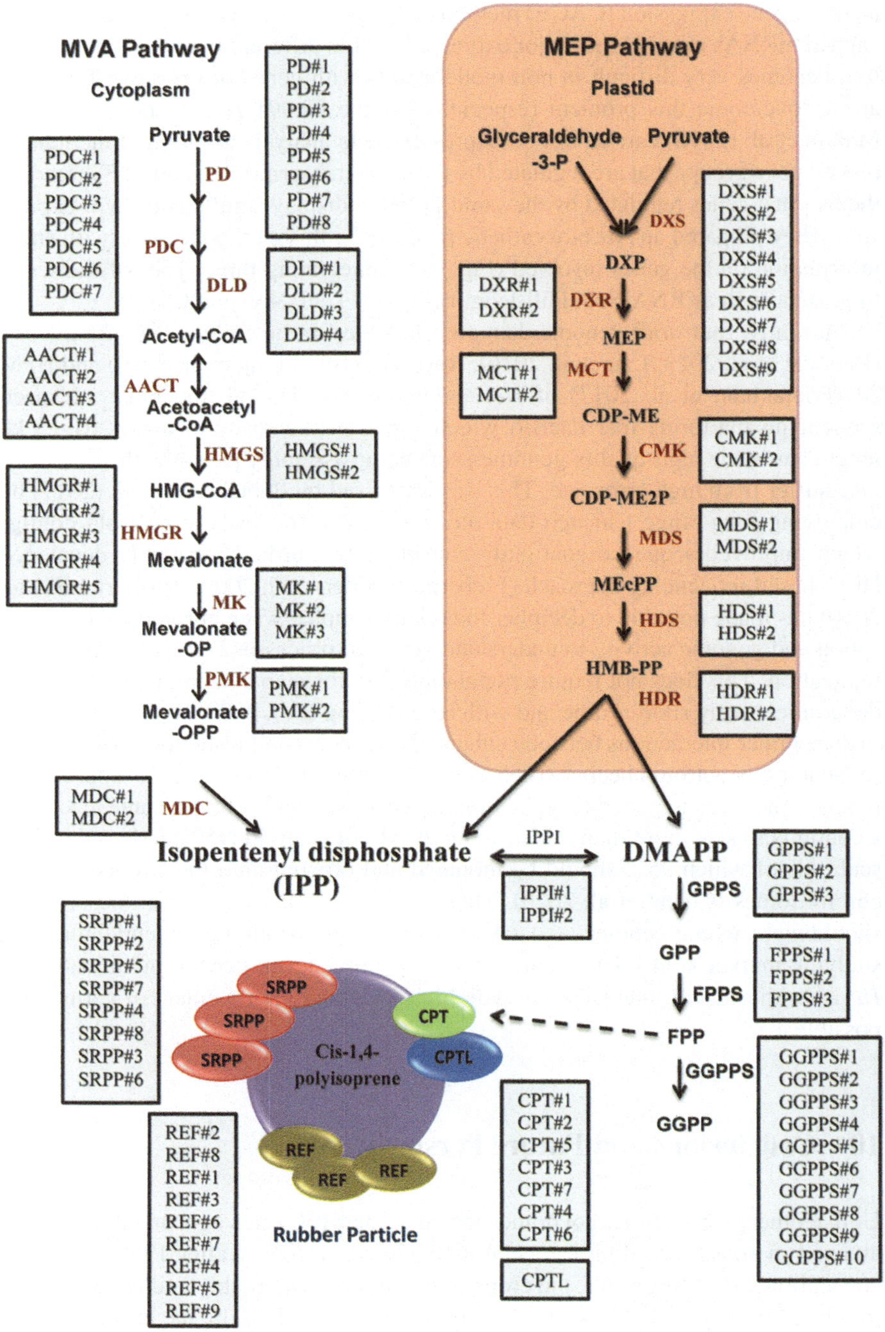


**Fig. 9** Database for rubber biosynthesis pathway search. Genes involved in each step is furnished. For easy access of each gene, the reader can visit the link provided in the original paper. (Adopted from Makita et al. 2018)

potential to such higher levels through recombination from a mere stock of few dozen seedlings from the Wickham base that survived the travel to Asian countries then. The subsequent IRRDB initiated germplasm collection conducted during 1980s from the primary centre of origin also injected a vast collection of new Amazonian germplasm into the breeding programmes which is still in infancy only. Much of the variation could not be utilized for want of techniques that can transfer traits into the existing superior clones. The absence of feasible juvenile selection process makes breeding more cumbersome and time consuming. The spurge in molecular techniques and information generated from molecular investigations in *Hevea* has opened the avenue of rapid progress in crop improvement aspects. Genome data along with transcriptome data enables us to investigate on the gene expression at the levels of rubber biosynthesis, disease and abiotic stress tolerance, *etc*. Further, incorporating the proteomic and metabolomics data can provide comprehensive picture in every respect. Markers are supposed to play the main role in early selection, and this property can be best utilized to develop early selection methods for accurate prediction of mature phenotype at juvenile stage itself which is the main objective of *Hevea* breeding. Incorporating genomics into breeding programmes to identify high yielding genotypes would minimize the requirement of both space and time.

Genomic selection (GS), a new approach that uses whole-genome molecular markers, has the potential to improve the quantitative traits in shorter time in a large population. By the availability of genomic data, G × E can be modelled by means of M × E when effect of marker varies among environments or groups of environments, and a correlation is possible between these effects. This is fast emerging as an alternate genome-wide marker-based method to predict future genetic responses that allows the estimation of genetic value of large number of selection of candidates at an early stage. It requires genotype × environment (G × E) interactions for developing prediction models. The phenotypic data collected over decades can be used for genomic selection to develop superior clones of *Hevea*. The whole responsibility of crop improvement entirely lies with the breeder who can only glean the apt information from the vast data available to develop superior *Hevea* clones with most of the required traits incorporated. Probably, the decade to come will witness such multidimensional approach in crop breeding programmes to develop new varieties of *Hevea* augmented with all the economically as well as ecologically important traits.

## References

Allen JO, Fauron CM, Minx P, Roark L, Oddiraju S, Lin GN, Meyer L, Hui Sun H, Kim K, Wang C, Du F, Xu D, Gibson M, Cifrese J, Clifton SW, Newton KJ (2007) Comparisons among two fertile and three male-sterile mitochondrial genomes of maize. Genetics 177(2):1173–1192. https://doi.org/10.1534/genetics.107.073312

An Z, Zhao Y, Zhang X, Huang X, Hu Y, Cheng H, Li X, Huang H (2019) A high-density genetic map and QTL mapping on growth and latex yield-related traits in *Hevea brasiliensis* Müll. Argic Ind Crops Prod 132:440–448. https://doi.org/10.1016/j.indcrop.2019.03.002

Antwi-Wiredu A, Amiteye S, Asare AT, Kusi-Adjei R, Diawuoh RG, Aryee CO, Klu GYP (2019) The use of microsatellite markers in genetic variation analysis of some introduced rubber tree (*Hevea brasiliensis*) Clones Cultivated in Ghana. J Agric Sci Sri Lanka 14(1):8. https://doi.org/10.4038/jas.v14i1.8452

Attanayake DPSTG, Dharmaratnae RS (1984) Preliminary observations on flowering, pollen germination, fruit set in *Hevea* Species. J Rubb Res Inst Sri Lanka 62:41–46

Bailey-Serres J, Parker JE, Ainsworth EA et al (2019) Genetic strategies for improving crop yields. Nature 575:109–118. https://doi.org/10.1038/s41586-019-1679-0

Baker CSL (1996) NR—ab iniio ad futurum. Rubb Dev 49:40–44

Baulkwill WJ (1989) The history of NR production. In: Webster CC, Baulkwill WJ (eds) Rubber. Longman Scientific and Technical, London, pp 1–56

van Berkum NL, Lieberman-Aiden E, Williams L et al (2010) Hi-C: a method to study the three-dimensional architecture of genomes. J Vis Exp 39:1869. https://doi.org/10.3791/1869

Besse P, Seguin M, Lebrun P, Lanaud C (1993a) Ribosomal DNA variations in wild and cultivated rubber tree (*Hevea brasiliensis*). Genome 36:1049–1057

Besse P, Lebrun P, Seguin M, Lanaud C (1993b) DNA fingerprints in *Hevea brasiliensis* (rubber tree) using human minisatellite probes. Heredity 70:237–244

Besse P, Seguin M, Lebrun P, Chevallier MH, Nicolas D, Lanaud C (1994) Genetic diversity among wild and cultivated populations of *Hevea brasiliensis* assessed by nuclear RFLP analysis. Theor Appl Genet 88:199–207

Brookson EV (1956) Importation and development of new strains of *Hevea brasiliensis* by the Rubber Research Institute of Malaya. J Rubb Res Inst Malaya 14:423–448

Bullerwell CE, Gray MW (2004) Evolution of the mitochondrial genome: protist connections to animals, fungi and plants. Curr Opin Microbiol 7:528–534

Buttery BR, Boatman SG (1976) Water deficit and flow of latex. In: Kozlowski (ed) Water deficit and plant growth. Academic, New York, pp 233–289

Chandrasekhar TR, Alice J, Gireesh T, Rao GP, Saraswathiamma CK (2004) Observations on pollination fecundity/siring ability and seed germination in *Hevea brasiliensis*. J Rubb Res 7(4):265–280

Chang S, Yang T, Du T et al (2011) Mitochondrial genome sequencing helps show the evolutionary mechanism of mitochondrial genome formation in *Brassica*. BMC Genomics 12:497. https://doi.org/10.1186/1471-2164-12-497

Chanroj V, Rattanawong R, Phumichai T, Tangphatsornruang S, Ukoskit K (2017) Genome-wide association mapping of latex yield and girth in Amazonian accessions of *Hevea brasiliensis* grown in a suboptimal climate zone. Genomics 109:475–484. https://doi.org/10.1016/j.ygeno.2017.1007.1005

Chao J, Chen Y, Wu S, Tian W-M (2015) Comparative transcriptome analysis of latex from rubber tree clone CATAS8-79 and PR107. Genomics Data 5:120–121. https://doi.org/10.1016/j.gdata.2015.05.034

Chao J, Zhao Y, Jin J, Wu S, Deng X, Chen Y, Tian W-M (2019) Genome-wide identification and characterization of the JAZ gene family in rubber tree (*Hevea brasiliensis*). Front Genet 10:372. https://doi.org/10.3389/fgene.2019.00372

Chen CH, F-t C, C-f C, Wang C-h, S-c C, H-e H, H-h O, Y-t H, Lu T-m (1979) A process of obtaining pollen plants of *Hevea brasiliensis* Muell.-Arg. Sci Sin 22:81–90

Chen S, Shao H, Hu D, Lin S, Zheng X (1994) Identification of mildew resistance from *Hevea* tree by RAPD technique. Chin J Trop Crops 15(2):26

Chen S-C, Peng S-Q, Huang G-X, Wu K, Fu X, Chen Z (2002) Association of decreased expression of a *Myb* transcription factor with the TPD (tapping panel dryness) syndrome in *Hevea brasiliensis*. Plant Mol Biol 51:51–58. https://doi.org/10.1023/A:1020719420867

Chen S, Shao H, Hu D, Wu K, Zheng XF (2003) Use of RAPD marker to screen mildew-resistance germplasm from *Hevea brasiliensis*. In: Chen Q, Zhou J (eds) Proceedings of the IRRDB rubber conference, 1999, pp 345–350

Cheng H, Cai H, Fu H, An Z, Fang J, Hu Y, Huang H (2015) Functional characterization of *Hevea brasiliensis* CRT/DRE binding factor 1 gene revealed regulation potential in the CBF pathway of tropical perennial tree. PLoS One 10(9):e0137634. https://doi.org/10.1371/journal.pone.0137634

Cheng H, Chen X, Zhu J, Huang H (2016) Overexpression of a *Hevea brasiliensis* ErbB-3 binding protein 1 gene increases drought tolerance and organ size in Arabidopsis. Front Plant Sci 7. https://doi.org/10.3389/fpls.2016.01703

Cheng H, Tang C, Huang H (2020) The Reyan 7-33-97 rubber tree genome: insight into its structure, composition and application. In: Matsui M, Chow K-S (eds) The rubber tree genome, compendium of plant genomes. Springer, New York. https://doi.org/10.1007/978-3-030-42258-5_2

Chinnusamy V, Zhu J, Zhou T, Zhu JK (2007) Small Rnas: big role in abiotic stress tolerance of plants. In: Jenks MA, Hasegawa PM, Jain SM (eds) Advances in molecular breeding toward drought and salt tolerant crops. Springer, Dordrecht. https://doi.org/10.1007/978-1-4020-5578-2_10

Chinnusamy V, Gong Z, Zhu JK (2008) Abscisic acid-mediated epigenetic processes in plant development and stress responses. J Integr Plant Biol 50(10):1187–1195. https://doi.org/10.1111/j.1744-7909.2008.00727.x

Chow K-S, Wan K, Isa MNM, Bahari A, Tan S, Harikrishna K, Yeang H (2007) Insights into rubber biosynthesis from transcriptome analysis of *Hevea brasiliensis* latex. J Exp Bot 58(10):2429–2440. https://doi.org/10.1093/jxb/erm093

Chow, K-S., Bahari, A., Taylor, M.A., Marshall, D.F. (2020). Genomics of rubber biosynthesis in *Hevea brasiliensis*. In: The rubber tree genome (Compendium of plant genomes), Eds. Matsui and Chow. Springer New York. ISBN 978-3-030-42257-8

Cilas C, Costes E, Milet J, Legnate H, Gnagne M, CleÂment-Demange A (2004) Characterization of branching in two *Hevea brasiliensis* clones. J Exp Bot 55(399):1045–1051

CleÂment-Demange A, Chapuset T, Legnate H, Costes E, Doumbia A, Obouayeba S, Nicolas D (1995) Wind damage: the possibilities of an integrated research for improving the prevention of risks and the resistance of clones in Rubber tree. In: Proceedings of the IRRDB symposium on physiological and molecular aspects of the breeding of *Hevea brasiliensis*, Penang, Malaysia, 6–7 Nov 1995, pp 182–199

Clément-Demange A et al (2000) Rubber tree. In: Charrier A, Jacquot M, Hamon S, Nicolas D (eds) Tropical plant breeding. Collection Reperes. CIRAD-ORSTOM, Montpellier, pp 455–480

Clément-Demange A, Legnate H, Seguin M, Carron MP, Le Guen V, Chapuset T, Nicolas D (2001) Rubber tree. In: Charrier A, Jacquot M, Hamon S, Nicolas D (eds) Tropical plant breeding. Collection Repères. CIRAD-ORSTOM, Montpellier, pp 455–480

Clément-Demange A, Prapan K, Rattanawong R, Teerawatanasuk K (2006) Molecular genetic markers and rubber breeding in Thailand: 2 – Field study of the family RRIM600 x PB217 for QTL identification. In: Franco-Thai Project 2005-2008: towards the improvement of rubber tree productivity, Bangkok, Thailand, 1–2nd June 2006, 21p

Clément-Demange A, Rattanawong R, Prapan K, Lekawipat N, Teerawatanasuk K, Kasemsap P, Seguin M (2008) Identification de 2 QTLs explicatifs de la croissance et de la production de latex chez *Hevea brasiliensis*. In: Biologie Moléculaire des Ligneux, BML10, Nancy, 5–7 May 2008, Poster

Compagnon P (1986) Le Caoutchouc Naturel: Biologie - Culture - Production. In: Maisonneuve et Larose (ed) Techniques agricoles et productions tropicales, 595p

Cong L, Ran FA, Cox D, Lin S, Barretto R, Habib N, Hsu PD, Wu X, Jiang W, Marraffini LA, Zhang F (2013) Multiplex genome engineering using CRISPR/Cas systems. Science (New York, N.Y.) 339(6121):819–823. https://doi.org/10.1126/science.1231143

Conson ARO, Taniguti CH, Amadeu RR, Andreotti IAA, de Souza LM, dos Santos LHB, Rosa JRBF, Mantello CC, da Silva CC, José Scaloppi Junior E, Ribeiro RV, Le Guen V, Garcia AAF, de Souza Gonçalves P, de Souza AP (2018) High-resolution genetic map and QTL analysis of growth-related traits of *Hevea brasiliensis* cultivated under suboptimal temperature and humidity conditions. Front Plant Sci 9(1255):1–16. https://doi.org/10.3389/fpls.2018.01255

Cornish K, Siler DJ, Grosjean O, Goodman N (1993) Fundamental similarities in rubber particle architecture and function in three evolutionarily divergent plant species. J NR Res 8:275–285

Costa R, De R, De Araujo A, Goncalves P, Higa A (2000) Selection and genetic gain in rubber tree (Hevea) populations using a mixed mating system. Genet Mol Biol 23. https://doi.org/10.1590/S1415-47572000000300028

Cros D, Mbo-Nkoulou L, Bell JM, Oum J, Masson A, Soumahoro M, Tran DM, Achour A, Le Guen V, Clement-Demenge A (2019) Within-family genomic selection in rubber tree (*Hevea brasiliensis*) increases genetic gain for rubber production. Ind Crops Prod 138(2019):111464. https://doi.org/10.1016/j.indcrop.2019.111464

Cubry P, Pujade-Renaud V, Garcia D, Espeout S, Le Guen V, Granet F, Seguin M (2014) Development and characterization of a new set of 164 polymorphic EST-SSR markers for diversity and breeding studies in rubber tree (*Hevea brasiliensis*Müll. Arg.). Plant Breed 133(3):419–426. https://doi.org/10.1111/pbr.12158

Cuevas J, Crossa J, Soberanis V, Pérez-Elizalde S, Pérez-Rodríguez P, de los Campos, G., et al. (2016) Genomic prediction of genotype × environment interaction kernel regression models. Plant Genome 9:3. https://doi.org/10.3835/plantgenome2016.03.0024

d'Auzac J, Jacob J-L, Prévôt J-C (1997) The regulation of cis-polyisoprene production (natural rubber) from *Hevea brasiliensis*. Research Singpost, Trivandrum

Deng Z, Qin B, Feng F, Li D (2012) Gene structure analysis and molecular marker development of *HbTCTP*. Chin J Trop Crops 33:1450–1455. https://doi.org/10.3969/j.issn.1000-2561.2012.08.023

Dewey RE, Levings CS, Timothy DH (1986) Novel recombinations in the maize mitochondrial genome produce a unique transcriptional unit in the Texas male-sterile cytoplasm. Cell 44:439–449

Dijkman MJ (1951) Hevea: thirty years of research in the Far East. University of Miami Press, Coral Gables, p 329

Dill CL, Wise RP, Schnable PS (1997) Rf8 and Rf* mediate unique T-urf13-transcript accumulation, revealing a conserved motif associated with RNA processing and restoration of pollen fertility in T-cytoplasm maize. Genetics 147:1367–1379

Duan C, Argout X, Gébelin V, Summo M, Dufayard J-F, Leclercq J, Hadi K, Piyatrakul P, Pirrello J, Rio M, Champion A, Montoro P (2013) Identification of the *Hevea brasiliensis* AP2/ERF superfamily by RNA sequencing. BMC Genomics 14:30

Fan S, Yang S (1991) Ethrel physiology of rubber tree. Chin J Trop Agric 3:71–79

Fan Y, Xin S, Dai X, Yang X, Huang H, and Hua Y (2020). Efficient genome editing of rubber tree (Hevea brasiliensis) protoplasts using CRISPR/Cas9 ribonucleoproteins. Ind Crops Prod 146. https://doi.org/10.1016/j.indcrop.2020.112146

Fang Y, Mei H, Zhou B, Xiao X, Yang M, Huang Y, Long X, Hu S, Tang C (2016) De novo transcriptome analysis reveals distinct defense mechanisms by young and mature leaves of *Hevea brasiliensis* (para rubber tree). Sci Rep 6(1). https://doi.org/10.1038/srep33151

Feng SP, Li WG, Huang HS, Wang JY, Wu YT (2008) Development, characterization and cross-species/genera transferability of EST-SSR markers for rubber tree (*Hevea brasiliensis*). Mol Breed 23(1):85–97. https://doi.org/10.1007/s11032-008-9216-0

Feng S et al (2010) Conservation and divergence of methylation patterning in plants and animals. Proc Natl Acad Sci USA 107:8689–8694. https://doi.org/10.1073/pnas.1002720107

Finnegan EJ, Genger RK, Peacock WJ, Dennis ES (1998) DNA methylation in plants. Annu Rev Plant Physiol Plant Mol Biol 49:223–247. https://doi.org/10.1146/annurev.arplant.49.1.223

Francis D, Finer JJ, Grotewold E (2017) Challenges and opportunities for improving food quality and nutrition through plant biotechnology. Curr Opin Biotechnol 44:124–129. https://doi.org/10.1016/j.copbio.2016.11.009

Gallagher LJ, Betz SK, Chase CD (2002) Mitochondrial RNA editing truncates a chimeric open reading frame associated with S male-sterility in maize. Curr Genet 42:179–184
García R, González S, Montoya C, Aristizabal (2011) Identification *in silico* of SSR markers for genotyping *Hevea* sp. clone gardens in Colombia. Agron Colomb 29(3):359–366
Gébelin V, Argout X, Engchuan W, Pitollat B, Duan C, Montoro P, Leclercq J (2012) Identification of novel microRNAs in *Hevea brasiliensis* and computational prediction of their targets. BMC Plant Biol 12:18. https://doi.org/10.1186/1471-2229-12-18
Gebelin V, Leclercq J, Hu S, Tang C, Montoro P (2013a) Regulation of *MIR* Genes in Response to Abiotic Stress in *Hevea brasiliensis*. Int J Mol Sci 14(10):19587–19604
Gebelin V, Leclercq J, Kuswanhadi, Argout X, Chaidamsari T, Hu S, Tang C, Sarah G, Yang M, Montoro P (2013b) The small RNA profile in latex from *Hevea brasiliensis* trees is affected by tapping panel dryness. Tree Physiol 33(10):1084–1098. https://doi.org/10.1093/treephys/tpt076
Gireesh T, Pravitha M (2009) Juvenile selection in progenies from controlled pollinations of new generation clones of *Hevea brasiliensis*. Nat Rubb Res 22(1&2):81–92
Gireesh T, Mydin KK (2014) Field evaluation of progenies of a canopy mutant of Hevea brasiliensis. Rubb Sci 27(1):61–68
Gireesh T, Meenakumari T, Mydin KK (2017) Fast track evaluation and selection of *Hevea brasiliensis* clones from a clonal nursery. Ind Crop Prod 103:195–201. https://doi.org/10.1016/j.indcrop.2017.04.001
Gong X-X, Yan B-Y, Hu J, Yang C-P, Li Y-J, Liu J-P, Liao W-B (2018) Transcriptome profiling of rubber tree (*Hevea brasiliensis*) discovers candidate regulators of the cold stress response. Genes Genomics 40(11):1181–1197. https://doi.org/10.1007/s13258-018-0681-5
Gouvêa LRL, Rubiano LB, Chioratto AF, Zucchi MI, de Souza Gonçalves P (2010) Genetic divergence of rubber tree estimated by multivariate techniques and microsatellite markers. Genet Mol Biol 33(2):1415–4757
Graham SW, Olmstead RG (2000) Utility of 17 chloroplast genes for inferring the phylogeny of the basal angiosperms. Am J Bot 87:1712–1730
Greek BF (1991) Rubber demand is expected to grow after 1991. C &E News 69:37–54
Haffer J (1982) General aspects of the refuge theory. In: Prance GT (ed) Biological diversification in the tropics. Columbia University Press, New York, pp 6–26
Han K-H, Shin DH, Yang J, Kim IJ, Oh SK, Chow KS (2000) Genes expressed in the latex of *Hevea brasiliensis*. Tree Physiol 20(8):503–510. https://doi.org/10.1093/treephys/20.8.503
Hao B-Z, Wu J-L (2000) Laticifer differentiation in *Hevea brasiliensis*: induction by exogenous jasmonic acid and linolenic acid. Ann Bot 85(1):37–43. https://doi.org/10.1006/anbo.1999.0995
Haridasan V, Nair VKG (1980) History of rubber plantation industry in India. In: Pillay PNR (ed) Handbook of NR production in India. Rubber Research Institute of India, Kottayam, pp 3–12
Holá D, Benešová M, Honnerová J, Hnilička F, Rothová O, Kočová M et al (2010) The evaluation of photosynthetic parameters in maize inbred lines subjected to water deficiency: Can these parameters be used for the prediction of performance of hybrid progeny? Photosynthetica 48:545–558
Hou B, Feng S, Wu Y (2017) Systemic identification of *Hevea brasiliensis* EST-SSR markers and primer screening. J Nucl Acids 2017:1–9. https://doi.org/10.1155/2017/6590902
IBPGR (1984) Annual report. International Board of Plant Genetic Resources, Rome, p 73
Imle EP (1978) *Hevea* rubber: past and future. Econ Bot 32:264–277
International Rubber Study Group (IRSG) (2015) Statistical summary of world rubber situation. Quarterly statistics
Jeffares DC, Penkett CJ, Bahler J (2008) Rapidly regulated genes are intron poor. Trends Genet 24:375–378. https://doi.org/10.1016/j.tig.2008.05.006. PMID: 18586348
Ko J-H, Chow K-S, Han K-H (2003) Transcriptome analysis reveals novel features of the molecular events occurring in the laticifers of *Hevea brasiliensis* (para rubber tree). Plant Mol Biol 53(4):479–492. https://doi.org/10.1023/b:plan.0000019119.66643.5d

Kuruvilla L, Sathik MM, Thomas M et al (2016) Expression of miRNAs of *Hevea brasiliensis* under drought stress is altered in clones with varying levels of drought tolerance. Indian J Biotechnol 15:153–160

Kuruvilla L, Sathik MM, Thomas M et al (2017) Identification and validation of cold responsive microRNAs of *Hevea brasiliensis* using high throughput sequencing. J Crop Sci Biotechnol 20:369–377. https://doi.org/10.1007/s12892-017-0062-0

Kush A, Goyvaerts E, Chye ML, Chua NH (1990) Laticifer specific gene expression of *Hevea brasiliensis* (rubber tree). Proc Natl Acad Sci 87:1787–1790

Kuswanhadi JL, Rio M, Tregear J, Ducamp-Collin MN, Montoro P (2010) Isolation of three members of the multigene family encoding ACC oxidases in *Hevea brasiliensis* and investigation of their responses to ethylene stimulation and wounding. J Rubb Res 13:185–205

Labra M, Ghiani A, Citterio S, Sgorbati S, Sala F, Vannini C, Ruffini-Castiglione M, Bracale M (2002) Analysis of cytosine methylation pattern in response to water deficit in pea root tips. Plant Biol 4(6). https://doi.org/10.1055/s-2002-37398

Lam LV, Thanh T, Chi VTQ, Tuy LM (2009) Genetic diversity of *Hevea* IRRDB'81 collection assessed by RAPD markers. Mol Biotechnol 42(3):292–298. https://doi.org/10.1007/s12033-009-9159-7

Landgren M, Zetterstrand M, Sundberg E (1996) Alloplasmic male-sterile Brassica lines containing *B. tournefortii* mitochondria express an ORF 3' of the atp6 gene and a 32 kDa protein. off. Plant Mol Biol 32:879–890

Lau N-S, Makita Y, Kawashima M, Taylor TD, Kondo S, Othman AS, Shu-Chien AC, Matsui M (2016) The rubber tree genome shows expansion of gene family associated with rubber biosynthesis. Sci Rep 6(1):1–14. https://doi.org/10.1038/srep28594

Laver HK, Reynolds SJ, Moneger F, Leaver CJ (1991) Mitochondrial genome organization and expression associated with cytoplasmic male sterility in sunflower (*Helianthus annuus*). Plant J Cell Mol Biol 1:185–193

Le Guen V, Garcia D, Mattos CRR, Doaré F, Lespinasse D, Seguin M (2007) Bypassing of a polygenic *Microcyclus ulei* resistance in rubber tree, analyzed by QTL detection. New Phytol 173:335–345

Le Guen V, Doaré F, Weber C, Seguin M (2009) Genetic structure of Amazonian populations of Hevea brasiliensis is shaped by hydrographical network and isolation by distance. Tree Genet Genomes 5(4):673–683. http://dx.doi.org/10.1016/j.indcrop.2020.112146

Le Guen V, Gay C, Xiong TC, Souza LM, Rodier-Goud M, Seguin M (2011) Development and characterization of 296 new polymorphic microsatellite markers for rubber tree (*Hevea brasiliensis*). Plant Breed 130(2):294–296. https://doi.org/10.1111/j.1439-0523.2010.01774.x

Lebrun P, Chevallier MH (1990) Starch and polyacrylamide gel electrophoresis of *Hevea brasiliensis*: a laboratory manual. IRCA/CIRAD, Montpellier

Leclercq J, Wu S, Farinas B, Pointet S, Favreau B, Vignes H, Kuswanhadi K, Ortega-Abboud E, Dufayard J-F, Gao S, Droc G, Hu S, Tang C, Montoro P (2020) Post-transcriptional regulation of several biological processes involved in latex production in *Hevea brasiliensis*. PeerJ 8:e8932. https://doi.org/10.7717/peerj.8932

Lekawipat N, Teerawatanasuk K, Vanavichit A, Toojinda, T. And Tragoonrung, S. (2003) Evaluating the genetic relatedness of wild and cultivated *Hevea brasiliensis* accessions with sscp markers. SABRAO J Breed Genet 35(2):123–134

Lertpanyasampatha M, Gao L, Kongsawadworakul P, Viboonjun U, Chrestin H, Liu R, Chen X, Narangajavana J (2012) Genome-wide analysis of microRNAs in rubber tree (*Hevea brasiliensis* L.) using high-throughput sequencing. Planta 236(2):437–445. https://doi.org/10.1007/s00425-012-1622-1

Lespinasse D, Grivet L, Troispoux V et al (2000a) Identification of QTLs involved in the resistance to South American leaf blight (Microcyclus ulei) in the rubber tree. Theor Appl Genet 100:975–984. https://doi.org/10.1007/s001220051379

Lespinasse D, Rodier-Goud M, Grivet L, Leconte A, Legnate H, Seguin M (2000b) A saturated genetic linkage map of rubber tree (*Hevea* spp.) based on RFLP, AFLP, microsatellite and isozyme markers. Theor Appl Genet 100:975–984. https://doi.org/10.1007/s001220051379

Li D-J, Deng Z, Chen C-L, Xia Z-H, Wu M, He P et al (2010) Identification and characterization of genes associated with tapping panel dryness from *Hevea brasiliensis* latex using suppression subtractive hybridization. BMC Plant Biol 10:140. https://doi.org/10.1186/1471-2229-10-140

Li HH, Wang Y, Guo D, Tian WM, Peng SQ (2011) Three MADS-box genes of *Hevea brasiliensis* expressed during somatic embryogenesis and in the laticifer cells. Mol Biol Rep 38:4045–4052. https://doi.org/10.1007/s11033-010-0523-2

Li D, Deng Z, Qin B, Liu X, Men Z (2012) De novo assembly and characterization of bark transcriptome using Illumina sequencing and development of EST-SSR markers in rubber tree (*Hevea brasiliensis* Muell. Arg.). BMC Genomics 13(1):192. https://doi.org/10.1186/1471-2164-13-192

Li D, Deng Z, Liu X, Qin B (2013) Molecular cloning, expression profiles and characterization of a novel translationally controlled tumor protein in rubber tree (*Hevea brasiliensis*). J Plant Physiol 170(5):497–504. https://doi.org/10.1016/j.jplph.2012.11.014

Li D, Zeng R, Li Y, Zhao M, Chao J, Li Y, Wang K, Zhu L, Tian W-M, Liang C (2016) Gene expression analysis and SNP/InDel discovery to investigate yield heterosis of two rubber tree F1 hybrids. Sci Rep 6(1):24984. https://doi.org/10.1038/srep24984

Liang XL, Li HL, Peng SQ (2009) Cloning and expression of *HbTCTP* from *Hevea brasiliensis*. Mol Plant Breed 7:188–193. https://doi.org/10.3969/j.issn.1672-416X.2009.01.031

Lieberman-Aiden E, van Berkum NL, Williams L, Imakaev M, Ragoczy T, Telling A, Dekker J (2009) Comprehensive mapping of long-range interactions reveals folding principles of the human genome. Science 326(5950):289

Liu J-P, Xia Z-Q, Tian X-Y, Li Y-J (2015) Transcriptome sequencing and analysis of rubber tree (*Hevea brasiliensis* Muell.) to discover putative genes associated with tapping panel dryness (TPD). BMC Genomics 16(1):398. https://doi.org/10.1186/s12864-015-1562-9

Liu J-P, Zhuang Y-F, Guo X-L, Li Y-J (2016) Molecular mechanism of ethylene stimulation of latex yield in rubber tree (*Hevea brasiliensis*) revealed by de novo sequencing and transcriptome analysis. BMC Genomics 17(1). https://doi.org/10.1186/s12864-016-2587-4

Loh SC, Othman AS, Veera Singham G (2019) Identification and characterization of jasmonic acid- and linolenic acid-mediated transcriptional regulation of secondary laticifer differentiation in *Hevea brasiliensis*. Sci Rep 9(1):14296. https://doi.org/10.1038/s41598-019-50800-1

Long X, He B, Wang C, Fang Y, Qi J, Tang C (2015) Molecular identification and characterization of the pyruvate decarboxylase gene family associated with latex regeneration and stress response in rubber tree. Plant Physiol Biochem 87:35–44. https://doi.org/10.1016/j.plaphy.2014.12.005

Lopez-Cruz M, Crossa J, Bonnett D, Dreisigacker S, Poland J, Jannink JL et al (2015) Increased prediction accuracy in wheat breeding trials using a marker x environment interaction genomic selection model. G3 (Bethesda) 5:569–582. https://doi.org/10.1534/g3.114.016097

Luke LP, Sathik MBM, Thomas M, Kuruvilla L, Sumesh KV, Annamalainathan K (2015) Quantitative expression analysis of drought responsive genes in clones of *Hevea* with varying levels of drought tolerance. Physiol Mol Biol Plants 21(2):179–186. https://doi.org/10.1007/s12298-015-0288-0

Luke LP, Sathik MBM, Thomas M, Kuruvilla L, Sumesh KV (2017) Expression of NAC transcription factor is altered under intermittent drought stress and re-watered conditions in *Hevea brasiliensis*. J Plant Biotechnol 44:142–148. https://doi.org/10.5010/JPB.2017.44.2.142

Luo H, Van Coppenolle B, Seguin M, Boutry M (1995) Mitochondrial DNA polymorphism and phylogenetic relationships in *Hevea brasiliensis*. Mol Breed 1:51–63

Makita Y, Kawashima M, Lau NS, Othman AS, Matsui M (2018) Construction of Pará rubber tree genome and multi-transcriptome database accelerates rubber researches. BMC Genomics 19(S1):922. https://doi.org/10.1186/s12864-017-4333-y

Mantello CC, Cardoso-Silva CB, da Silva CC, de Souza LM, Scaloppi Junior EJ, de Gonçalves PS, Vicentini R, de Souza AP (2014) De novo assembly and transcriptome analysis of the rubber tree (*Hevea brasiliensis*) and SNP markers development for rubber biosynthesis pathways. PLoS One 9(7):e102665. https://doi.org/10.1371/journal.pone.0102665

Mantello CC, Suzuki FI, Souza L M, Gonçalves P de S, Pereira de Souza A (2011) Development of microsatellite molecular markers and genetic diversity in Hevea braziliensis. BMC Proceedings 5(S7):11. https://doi.org/10.1186/1753-6561-5-s7-p11

Mathew R, Roy BC, Ravindran M, Nazeer MA, Saha T (2005) Phylogenetic relationships of *Hevea* species based on molecular markers. Indian J Nat Rubb Res 18:14–25

McCauley DE (1992) The use of chloroplast DNA polymorphism in studies of gene flow in plants. Trends Ecol Evol 10:198–202

McCrady RL, Jokela EJ (1996) Growth phenology and crown structure of selected loblolly pine families planted at two spacings. For Sci 42:46–57

Meenakumari T, Meenattoor JR, Thirunavoukkarasu M, Vinod KK, Krishan B, Gireesh T, Thomas V, Mydin KK, Jacob J (2018) Dynamics of long-term adaptive responses in growth and rubber yield among *Hevea brasiliensis* genotypes introduced to a dry sub-humid climate of Eastern India. Ind Crops Prod 119:294–303. https://doi.org/10.1016/j.indcrop.2018.02.066

Meister B, Herzer S, Silahtaroglu A (2013) MicroRNAs in the hypothalamus. Neuroendocrinology 98:243–253. https://doi.org/10.1159/000355619

Mercy MA, Narayanan C, Reju MJ, Rao GP, Abraham T, Gireesh T, Madhavan J, Reghu CP (2014) Utilization of wild *Hevea* germplasm for hybridization programmes in India. In: IRRDB-RRII workshop on multilateral clone exchange, extended abstract, Rubber Research Institute of India, Kottayam, 4–7 June 2014, pp 17–48

Meuwissen TH, Hayes BJ, Goddard ME (2001) Prediction of total genetic value using genome-wide dense marker maps. Genetics 157:1819–1829

Miao Z, Gaynor J (1993) Molecular cloning, characterization and expression of Mn-superoxide dismutase from the rubber tree (*Hevea brasiliensis*). Plant Mol Biol 23:267–277

Mizoi J, Shinozaki K, Yamaguchi-Shinozaki K (2012) AP2/ERF family transcription factors in plant abiotic stress responses. Biochim Biophys Acta 1819(2):86–96. https://doi.org/10.1016/j.bbagrm.2011.08.004

Montoro P, Wu S, Favreau B, Herlinawati E, Labrune C, Martin-Magniette M-L, Pointet S, Rio M, Leclercq J, Ismawanto S, and Kuswanhadi (2018) Transcriptome analysis in Hevea brasiliensis latex revealed changes in hormone signalling pathways during ethephon stimulation and consequent Tapping Panel Dryness. Sci Rep 8(1). https://doi.org/10.1038/s41598-018-26854-y

Morris LE (1929) Field observations and experiments on the pollination of *Hevea brasiliensis*. J Rubb Res Inst Malaya 1:41–49

Mydin KK, Gireesh T (2016) Diversity and heterosis by recombination breeding of *Hevea brasiliensis* in India. Rubb Sci 29(1):20–35

Nakano T, Suzuki K, Fujimura T, Shinshi H (2006) Genome-wide analysis of the ERF gene family in Arabidopsis and rice. Plant Physiol 140:411–432. PMID: 16407444

Neale DB, Saghai-Maroof MA, Allard RW, Zhang Q, Jorgensen RA (1988) Chloroplast DNA diversity in populations of wild and cultivated barley. Genetics 120:1105–1110

Niu Y-F, Hu Y-S, Liu Z-Y, Zheng C, Liu J (2020) Complete chloroplast genome of *Hevea benthamiana*, a SALB-resistant relative wild species of rubber tree. Mitochondrial DNA B 5(3):2062–2064. https://doi.org/10.1080/23802359.2020.1763868

Ohyama K, Fukuzawa H, Kohchi T, Shirai H, Sano T (1986) Chloroplast gene organization deduced from complete sequence of liverwort *Marchantia polymorpha* chloroplast DNA. Nature 322:572–574

Olapade EO, Omokhafe KO (1990) Hand pollination of Hevea brasiliensis in Nigeria (1984-1990): Comparison with open pollinated field data. J Rubb Res Inst Sri Lanka, pp.27–34

Pardekooper EC (1989) Exploitation of the rubber tree. In: Webster CC, Baulkwill WL (eds) Rubber. Longman Scientific & Technical, Singapore, pp 349–414

Páez UAH, García Romero IA, Restrepo Restrepo S, Aristizábal Gutiérrez FA, Montoya Castaño D (2015) Assembly and analysis of differential transcriptome responses of *Hevea brasiliensis* on interaction with *Microcyclus ulei*. PLoS One 10(8):e0134837. https://doi.org/10.1371/journal.pone.0134837

Peng S, Zhu J, Li H et al (2008) Cloning and characterization of a novel cysteine protease gene (*HbCP1*) from *Hevea brasiliensis*. J Biosci 33:681–690. https://doi.org/10.1007/s12038-008-0088-6

Peng SQ, Xu J, Li HL, TianWM (2009) Cloning and molecular characterization of *HbCOI1* from *Hevea brasiliensis*. Biosci Biotechnol Biochem 73:665–670

Identification of a mitochondrial protein associated with cytoplasmic male sterility in Petunia. https://www.researchgate.net/publication/20592928_Identification_of_a_Mitochondrial_Protein_Associated_with_Cytoplasmic_Male_Sterility_in_Petunia. Accessed 8 Jan 2021

Piyatrakul P, Putranto RA, Martin F, Rio M, Dessailly F, Leclercq J, Dufayard F, Lardet L, Montoro P (2012) Some ethylene biosynthesis and AP2/ERF genes reveal a specific pattern of expression during somatic embryogenesis in *Hevea brasiliensis*. BMC Plant Biol 12:244

Pootakham W, Chanprasert J, Jomchai N, Sangsrakru D, Yoocha T, Therawattanasuk K, Tangphatsornruang S (2011) Single nucleotide polymorphism marker development in the rubber tree, *Hevea brasiliensis* (Euphorbiaceae). Am J Bot 98(11):e337–e338. https://doi.org/10.3732/ajb.1100228

Pootakham W, Jomchai N, Ruang-Areerate P, Shearman JR, Sonthirod C, Sangsrakru D, Tragoonrung S, Tangphatsornruang S (2015a) Genome wide SNP discovery and identification of QTL associated with agronomic traits in oil palm using genotyping-by-sequencing (GBS). Genomics 105:288–295. https://doi.org/10.1016/j.ygeno.2015.02.002

Pootakham W, Ruang-Areerate P, Jomchai N, Sonthirod C, Sangsrakru D, Yoocha T, Theerawattanasuk K, Nirapathpongporn K, Romruensukharom P, Tragoonrung S, Tangphatsornruang S (2015b) Construction of a high-density integrated genetic linkage map of rubber tree (*Hevea brasiliensis*) using genotyping-by-sequencing (GBS). Front Plant Sci 6:367. https://doi.org/10.3389/fpls.2015.00367

Pootakham W, Sonthirod C, Naktang C, Ruang-Areerate P, Yoocha T, Sangsrakru D, Theerawattanasuk K, Rattanawong R, Lekawipat N, Tangphatsornruang S (2017) De novo hybrid assembly of the rubber tree genome reveals evidence of paleotetraploidy in *Hevea* species. Sci Rep 7(1). https://doi.org/10.1038/srep41457

Prapan K, Lekawipat N, Weber C, Rodier-Goud M, Clément-Demange A, Seguin M (2006) Molecular genetic markers and rubber breeding in Thailand. 1 – genetic mapping of the family RRIM600 x PB217 by use of microsatellite markers. In: The Second Thai-French seminar "rubber: from tree to end-products", 1–2 June 2006. Century Park Hotel, Bangkok, 10p

Priya P, Venkatachalam P, Thulaseedharan A (2006) Molecular cloning and characterization of the rubber elongation factor gene and its promoter sequence from rubber tree (*Hevea brasiliensis*): a gene involved in rubber biosynthesis. Plant Sci 171:470–480. https://doi.org/10.1016/j.plantsci.2006.05.009

Priyadarshan PM (2003) Breeding *Hevea brasiliensis* for environmental constraints. Adv Agron 79:351–400

Priyadarshan PM (2015) Breeding vis-à-vis genomics of tropical tree crops. In: Al-Khayri JM et al (eds) Advances in plant breeding strategies: agronomic, abiotic and biotic stress traits. Springer, Switzerland. https://doi.org/10.1007/978-3-319-22518-0_6

Priyadarshan PM (2016) Genetic diversity and erosion in *Hevea* rubber. In: Ahuja M, Jain SM (eds) Genetic diversity and erosion in plants, sustainable development and biodiversity. Springer, Switzerland. https://doi.org/10.1007/978-3-319-25954-3_6

Priyadarshan PM (2017) Refinements to *Hevea* rubber breeding. Tree Genet Genomes 13:20. https://doi.org/10.1007/s11295-017-1101-8

Priyadarshan PM, Clément-Demange A (2004) Breeding *Hevea* rubber: formal and molecular genetics. Adv Genet 52:51–115

Priyadarshan PM, Goncalves PDS (2003) *Hevea* gene pool for breeding. Genet Resour Crop Evol 50(1):101–114

Priyadarshan PM, Hoa TTT, Huasun H, de Gonçalves PS (2005) Yielding potential of rubber (Hevea brasiliensis) in sub-optimal environments. J Crop Improve 14(1–2):221–247. https://doi.org/10.1300/j411v14n01_10
Priyadarshan PM, Dey SK, Nazeer MA, Varghese YA, Kang MS (2008) Adaptability analysis of rubber Hevea brasiliensis Muell.-Arg.) Clones via GGE Biplot. J Rubb Res 11(4):237–244
Provan J, Powell W, Hollingsworth PM (2001) Chloroplast microsatellites: new tools for studies in plant ecology and evolution. Trends Ecol Evol 16(3):142–147. https://doi.org/10.1016/s0169-5347(00)02097-8
Pushparajah E (2001) NR. In: Last FT (ed) Tree crop ecosystems, vol 19: Ecosytems of the world. Elsevier Science, Amsterdam, pp 379–407
Putranto R-A, Duan C, Kuswanhadi, Chaidamsari T, Rio M, Piyatrakul P, Herlinawati E, Pirrello J, Dessailly F, Leclercq J, Bonnot F, Tang C, Hu S, Montoro P (2015) Ethylene response factors are controlled by multiple harvesting stresses in *Hevea brasiliensis*. PLoS One 10(4):e0123618. https://doi.org/10.1371/journal.pone.0123618
Rahman AYA, Usharraj AO, Misra BB, Thottathil GP, Jayasekaran K, Feng Y, Hou S, Ong SY, Ng FL, Lee LS, Tan HS, Sakaff MKLM, Teh BS, Khoo BF, Badai SS, Aziz NA, Yuryev A, Knudsen B, Dionne-Laporte A, Mchunu NP, Yu Q, Langston BJ, Freitas TAK, Young AG, Chen R, Wang L, Najimudin N, Saito JA, Alam M (2013) Draft genome sequence of the rubber tree *Hevea brasiliensis*. BMC Genomics 14(1):75. https://doi.org/10.1186/1471-2164-14-75
Rao BS (1961) Pollination of *Hevea* in Malaya. J Rubb Res Inst Malaya 17:14–18
Rattanawong R, Prapan K, Lekawipat N, Teerawatanasuk K, Kasemsap P, Seguin M, Clément-Demange A (2008) QTLs detection for growth and initial latex production in rubber (*Hevea brasiliensis*). IRRDB, Kuala Lumpur
Rosa JRBF, Mantello CC, Garcia D, de Souza LM, da Silva CC, Gazaffi R, da Silva CC, Toledo-Silva G, Cubry P, Garcia AAF, de Souza AP, Le Guen V (2018) QTL detection for growth and latex production in a full-sib rubber tree population cultivated under suboptimal climate conditions. BMC Plant Biol 18(1). https://doi.org/10.1186/s12870-018-1450-y
Ross JM (1960) Observations on the 1959 Hand pollinations at Rubber Research Institute of Malaya. In: Proceedings of NR research conference, Kuala Lumpur, pp 392–408
Roy CB, Nazeer MA, Saha T (2004) Identification of simple sequence repeats in rubber (*Hevea brasiliensis*). Curr Sci 87(6):807
Roy CB, Liu H, Rajamani A, Saha T (2019) Transcriptome profiling reveals genetic basis of disease resistance against *Corynespora cassiicola* in rubber tree (*Hevea brasiliensis*). Curr Plant Biol. https://doi.org/10.1016/j.cpb.2019.02.002
Saha T, Priyadarshan PM (2012) Genomics of *Hevea* Rubber. In: Schnell RJ, Priyadarshan PM (eds) Genomics of tree crops. New York, Springer, pp 261–298. https://doi.org/10.1007/978-1-4614-0920-5
Saha T, Bindu CR, Nazeer MA (2005) Microsatellite variability and its use in the characterization of cultivated clones of *Hevea brasiliensis*. Plant Breed 124:86–92
Saha T, Roy B, Ravindran M, Bini K, Nacer MA (2007) Alltheic diversity revealed through SSR polymorphisms at the locus encoding HMG-coa reductase in rubber (*Hevea brasiliensis*). Silvae Genet 56:58–65
Saha T, Alam B, Ravindran M, Thomas KU (2010) Cold-induced gene expression pro fi ling in rubber (*Hevea brasiliensis*). In: Proceedings of the international workshop on climate change and rubber cultivation. RRII, Kottayam
Salgado LR, Koop DM, Pinheiro DG, Rivallan R, Le Guen V, Nicolás MF, de Almeida LGP, Rocha VR, Magalhães M, Gerber AL, Figueira A, de Mattos Cascardo JC, de Vasconcelos AR, Silva WA, Coutinho LL, Garcia D (2014) De novo transcriptome analysis of *Hevea brasiliensistissues* by RNA-seq and screening for molecular markers. BMC Genomics 15(1):236. https://doi.org/10.1186/1471-2164-15-236
Sando T, Takaoka C, Mukai Y, Yamashita A, Hattori M, Ogasawara N, Fukusaki E, Kobayashi A (2008) Cloning and characterization of mevalonate pathway genes in a NR producing plant,

*Hevea brasiliensis*. Biosci Biotechnol Biochem 72(8):2049–2060. https://doi.org/10.1271/bbb.80165

Sathik MBM, Kuruvilla L (2017) Developments in research on abiotic stress responsive microRNAs of *Hevea brasiliensis*. Int J Plant Physiol 22(4):470–483. https://doi.org/10.1007/s40502-017-0342-8

Sathik MBM, Thomas M, Jose RM, Thomas RA, Jacob J (2006) Differentially expressed transcripts of *Hevea brasiliensis* experiencing TPD. Presented in 'International natural rubber conference', Ho Chi Minh City, Vietnam, 13–14 Nov 2006

Sathik MBM, Kuruvilla L, Thomas M, Luke LP, Satheesh PR, Annamalainathan K, Jacob J (2012) Quantitative expression analysis of stress responsive genes under cold stress in *Hevea brasiliensis*. Rubb Sci 25:199–213

Sathik MBM, Luke LP, Rajamani A, Kuruvilla L, Sumesh KV, Thomas M (2018) *De novo* transcriptome analysis of abiotic stress responsive genes of *Hevea brasiliensis*. Mol Breed 38:32. https://doi.org/10.1007/s11032-018-0782-5

Schultes RE (1977) Wild *Hevea* : an untapped source of germplasm. J Rubb Res Inst Sri Lanka 54:227–257

Schultes RE (1987) Studies in the genus *Hevea* VIII. Notes on intrageneric variants of *Hevea brasiliensis* (Euphorbiaceae). Econ Bot 41:125–147

Seguin M, Lespinasse D, Rodier-Goud M, Legnate H, Troispoux V, Pinard F, Clément-Demange A (1996) Genome mapping in connection with resistance to the South American Leaf Blight in rubber tree (*Hevea brasiliensis*). In: Third ASAP conference on agricultural biotechnology, Hua-Hin, Thailand, vol 1, 10–15 Nov 1996, pp 1–8

Schuster SC (2008) Next-generation sequencing transforms today's biology. Nat Methods 5:16–18, 77–88

Seguin M, Flori A, Legnaté H, Clément-Demange A (2003) Rubber tree. In: Hamon, Seguin, Perrier et Glaszmann (eds) Genetic diversity of cultivated tropical plants. Cirad, Ird, Collection "Repères". ISBN 2-87614-541-3. (English translation of above reference), pp 277–306

Seguin M, Weber C, Fouet O, Rodier-Goud M (2008) Integration of microsatellite (SSR) markers in Hevea spp. reference genetic map: a powerful tool for genetic analysis of qualitative and quantitative agronomic traits. IRRDB Conference, Malaysia

Serier JB (1993) Histoire du caoutchouc. Paris, Éditions Desjonquères

Sethuraj MR, Rao CG, Raghavendra AS (1984) The pattern of latex flow from rubber tree (*Hevea brasiliensis*) in relation to water stress. J Cell Biochem Suppl 8B:236

Shearman JR, Sangsrakru D, Ruang-areerate P, Sonthirod C, Uthaipaisanwong P, Yoocha T, Poopear S, Theerawattanasuk K, Tragoonrung S, Tangphatsornruang S (2014) Assembly and analysis of a male sterile rubber tree mitochondrial genome reveals DNA rearrangement events and a novel transcript. BMC Plant Biol 14(1):45. https://doi.org/10.1186/1471-2229-14-45

Shearman JR, Sangsrakru D, Jomchai N, Ruang-areerate P, Sonthirod C, Naktang C, Theerawattanasuk K, Tragoonrung S, Tangphatsornruang S (2015) SNP identification from RNA sequencing and linkage map construction of rubber tree for anchoring the draft genome. PLoS One 10(4):e0121961. https://doi.org/10.1371/journal.pone.0121961

Shendure J, Ji H (2008) Next-generation DNA sequencing. Nat Biotechnol 26:1135–1145

Shinozaki K, Yamaguchi-Shinozaki K, Seki M (2003) Regulatory network of gene expression in the drought and cold stress responses. Curr Opin Plant Biol 6:410–417. https://doi.org/10.1016/S1369-5266(03)00092-X

Shoji T, Mishima M, Hashimoto T (2013) Divergent DNA-binding specificities of a group of ethylene response factor transcription factors involved in plant defense. Plant Physiol 162:977–990. https://doi.org/10.1104/pp.113.217455

Silva CC, Mantello CC, Campos T, Souza LM, Gonçalves PS, Souza AP (2014) Leaf-, panel- and latex-expressed sequenced tags from the rubber tree (*Hevea brasiliensis*) under cold-stressed and suboptimal growing conditions: the development of gene-targeted functional markers for stress response. Mol Breed 34(3):1035–1053. https://doi.org/10.1007/s11032-014-0095-2

Simmonds NW (1962) Variability in crop plants, its use and conservation. Biol Rev 37:442–465

Simmonds NW (1989) Rubber breeding. In: Webster CC, Baulkwill WJ (eds) Rubber. Longmann Scientific and Technical, Essex, pp 85–124
Smith HP, Burger K (1992) The outlook for NR production and consumption. In: Sethuraj MR, Mathew NM (eds) NR. Elsevier Science, Amsterdam, pp 26–49
Souza LM, Mantello CC, Santos MO, de Souza Gonçalves P, Souza AP (2009) Microsatellites from rubber tree (*Hevea brasiliensis*) for genetic diversity analysis and cross-amplification in six *Hevea* wild species. Conserv Genet Resour 1(1):75–79. https://doi.org/10.1007/s12686-009-9018-7
Souza LM, Gazaffi R, Mantello CC, Silva CC, Garcia D, Le Guen V et al (2013) QTL mapping of growth-related traits in a full-sib family of rubber treetrees (*Hevea brasiliensis*) evaluated in a sub-tropical climate. PLoS One 8:e61238. https://doi.org/10.1371/journal.pone.0061238
de Souza LM, Le Guen V, Cerqueira-Silva CBM, Silva CC, Mantello CC, Conson ARO, Vianna JPG, Zucchi MI, José Scaloppi Junior E, de Freitas Fialho J, de Moraes MLT, de Gonçalves PS, de Souza AP (2015) Genetic diversity strategy for the management and use of rubber genetic resources: more than 1,000 wild and cultivated accessions in a 100-genotype core collection. PLoS One 10(7):e0134607. https://doi.org/10.1371/journal.pone.0134607
de Souza LM, dos Santos LHB, Rosa JRBF, da Silva CC, Mantello CC, Conson ARO, Scaloppi Junior EJ, Fialho JDF, de Moraes MLT, de Souza Gonçalves P, Margarido GRA, Garcia AAF, Le Guen V, de Souza AP (2018) Linkage disequilibrium and population structure in wild and cultivated populations of rubber tree (*Hevea brasiliensis*). Front Plant Sci 9:815. https://doi.org/10.3389/fpls.2018.00815
Souza LM, Francisco FR, Gonçalves PS, Scaloppi Junior EJ, Le Guen V, Fritsche-Neto R, Souza AP (2019) Genomic selection in rubber tree breeding: a comparison of models and methods for managing G×E interactions. Front Plant Sci 10(1353):1–14. https://doi.org/10.3389/fpls.2019.01353
Souza LM, Mantello CC, Suzuki F et al (2011) Development of a genetic linkage map of rubber tree (Hevea braziliensis) based on microsatellite markers. BMC Proc 5, P39. https://doi.org/10.1186/1753-6561-5-S7-P39
Sreelatha S, Simon SP, Kurup GM, Vijayakumar KR (2007) Biochemical mechanisms associated with low yield during stress in *Hevea* clone RRII 105. J Rubb Res 10(2):107–115
Sreelatha S, Mydin KK, Simon SP, Krishnakumar R, Jacob J, Annamalainathan K (2011). Seasonal variations in yield and associated biochemical changes in RRII 400 series clones of Hevea brasiliensis. Nat Rub Res 24:117–123.
Sterling A, Rodríguez N, Quiceno E, Trujillo F, Clavijo A, Suárez-Salazar JC (2019) Dynamics of photosynthetic responses in 10 rubber tree (*Hevea brasiliensis*) clones in Colombian Amazon: Implications for breeding strategies. PLoS One 14(12):e0226254. https://doi.org/10.1371/journal.pone.0226254
Stewart WD, Wachtel WL, Shipman JJ, Yanko JA (1955) Synthesis of rubber by fungi. Science (New York, N.Y.) 122(3183):1271–1272. https://doi.org/10.1126/science.122.3183.1271
Storz G (2002) An expanding universe of noncoding RNAs. Science 296(5571):1260–1263. https://doi.org/10.1126/science.1072249
Sui J-L, Xiao X-H, Qi J-Y, Fang Y-J, Tang C-R (2017) The SWEET gene family in *Hevea brasiliensis* - its evolution and expression compared with four other plant species. FEBS Open Biol 7(12):1943–1959. https://doi.org/10.1002/2211-5463.12332
Supriya R, Priyadarshan PM (2019) Genomic technologies for *Hevea* breeding. Adv Genet 104:1–73. ISSN 0065-2660. https://doi.org/10.1016/bs.adgen.2019.04.001
Tan H (1987) Strategies in rubber tree breeding. In: Abbott AJ, Atkin RK (eds) Improving vegetatively propagated crops. Academic, London, pp 28–54
Tang C, Huang D, Yang J, Liu S, Sakr S, Li H, Zhou Y, Qin Y (2010) The sucrose transporter HbSUT3 plays an active role in sucrose loading to laticifer and rubber productivity in exploited trees of *Hevea brasiliensis* (para rubber tree). Plant Cell Environ 33(10):1708–1720. https://doi.org/10.1111/j.1365-3040.2010.02175.x

Tang C, Xiao X, Li H, Fan Y, Yang J, Qi J, Li H (2013) Comparative analysis of latex transcriptome reveals putative molecular mechanisms underlying super productivity of *Hevea brasiliensis*. PLoS One 8(9):e75307. https://doi.org/10.1371/journal.pone.0075307

Tang C, Yang M, Fang Y, Luo Y, Gao S, Xiao X, An Z, Zhou B, Zhang B, Tan X, Yeang H-Y, Qin Y, Yang J, Lin Q, Mei H, Montoro P, Long X, Qi J, Hua Y, He Z, Sun M, Li W, Zeng X, Cheng H, Liu Y, Yang J, Tian W, Zhuang N, Zeng R, Li D, He P, Li Z, Zou Z, Li S, Li C, Wang J, Wei D, Lai C-Q, Luo W, Yu J, Hu S, Huang H (2016) The rubber tree genome reveals new insights into rubber production and species adaptation. Nat Plants 2(6). https://doi.org/10.1038/nplants.2016.73

Tangphatsornruang S, Birch-Machin I, Newell CA, Gray JC (2010) The effect of different 3′ untranslated regions on the accumulation and stability of transcripts of a gfp transgene in chloroplasts of transplastomic tobacco. Plant Mol Biol 76(3–5):385–396. https://doi.org/10.1007/s11103-010-9689-1

Tangphatsornruang S, Uthaipaisanwong P, Sangsrakru D, Chanprasert J, Yoocha T, Jomchai N, Tragoonrung S (2011) Characterization of the complete chloroplast genome of *Hevea brasiliensis* reveals genome rearrangement, RNA editing sites and phylogenetic relationships. Gene 475(2):104–112. https://doi.org/10.1016/j.gene.2011.01.002

Thanseem I, Venkatachalam P, Thulaseedharan A (2003) Sequence characterization of β-1,3-glucanase gene from *Hevea brasiliensis* through genomic and cDNA cloning. Indian J Nat Rubb Res 16:106–114

Thanseem I, Joseph A, Thulaseedharan A (2005) Induction and differential expression of β-1,3-glucanase mRNAs in tolerant and susceptible *Hevea* clones in response to infection by *Phytophthora meadii*. Tree Physiol 25:1361–1368

Thomas KK (2001) Role of Clement Robert Markham in the introduction of *Hevea* rubber into the British India. Planter 77:287–292

Thomas KK (2002) Role of James Collins in the development of rubber plantation industry in British India: a historical reappraisal. Planter 78:33–41

Thomas M, Sathik MBM, Luke LP, Sumesh, KV, Satheesh PR, Annamalainathan K, Jacob J (2012) Stress responsive transcripts and their association with drought tolerance in Hevea brasiliensis. J Plant Crop 40:180–187

Triwitayakorn K, Chatkulkawin P, Kanjanawattanawong S, Sraphet S, Yoocha T, Sangsrakru D, Chanprasert J, Ngamphiw C, Jomchai N, Therawattanasuk K, Tangphatsornruang S (2011) Transcriptome Sequencing of *Hevea brasiliensis* for Development of Microsatellite Markers and Construction of a Genetic Linkage Map. DNA Res 18(6):471–482. https://doi.org/10.1093/dnares/dsr034

Tungngoen K, Kongsawadworakul P, Viboonjun U, Katsuhara M, Brunel N, Sakr S, Narangajavana J, Chrestin H (2009) Involvement of HbPIP2;1 and HbTIP1;1 aquaporins in ethylene stimulation of latex yield through regulation of water exchanges between inner liber and latex cells in *Hevea brasiliensis*. Plant Physiol 151(2):843–856. https://doi.org/10.1104/pp.109.140228

Turmel M, Otis C, Lemieux C (2003) The mitochondrial genome of *Chara vulgaris*: insights into the mitochondrial DNA architecture of the last common ancestor of green algae and land plants. Plant Cell 15(8):1888–1903. https://doi.org/10.1105/tpc.013169

Uthup TK, Ravindran M, Bini K, Saha T (2011) Divergent DNA methylation patterns associated with abiotic stress in *Hevea brasiliensis*. Mol Plant 4:996–1013. https://doi.org/10.1093/mp/ssr039

Uthup TK, Saha T, Ravindran M, Bini K (2013) Impact of an intragenic retrotransposon on the structural integrity and evolution of a major isoprenoid biosynthesis pathway gene in *Hevea brasiliensis*. Plant Physiol Biochem 73:176–188

Uthup TK, Rajamani A, Ravindran M, Saha T (2016) Molecular evolution and functional characterisation of haplotypes of an important rubber biosynthesis gene in *Hevea brasiliensis*. Plant Biol 18(4):720–728. https://doi.org/10.1111/plb.12433

Uthup TK, Rajamani A, Ravindran M, Saha T (2018) Distinguishing CPT gene family members and vetting the sequence structure of a putative rubber synthesizing variant in *Hevea brasiliensis*. Gene. https://doi.org/10.1016/j.gene.2018.12.001

Varghese YA, Knaak C, Sethuraj MR, Ecke W (1997) Evaluation of random amplified polymorphic DNA (RAPD) markers in *Hevea brasiliensis*. Plant Breed 116(1):47–52. https://doi.org/10.1111/j.1439-0523.1997.tb00973.x

Venkatachalam P, Thomas S, Priya P, Thanseem I, Gireesh T, Saraswathyamma CK, Thulaseedharan A (2002) Identification of DNA polymorphism within the clones of *Hevea brasiliensis* (Muell.) Arg. using RAPD analysis. Indian J Nat Rubb Res 15:172–181

Venkatachalam P, Thulaseedharan A, Raghothama K (2007) Identification of expression profiles of tapping panel dryness (TPD) associated genes from the latex of rubber tree (*Hevea brasiliensis* Muell. Arg.). Planta 226(2):499–515. https://doi.org/10.1007/s00425-007-0500-8

Venkatachalam P, Thulaseedharan A, Raghothama K (2009) Molecular identification and characterization of a gene associated with the onset of tapping panel dryness (TPD) syndrome in rubber tree (*Hevea brasiliensis* Muell.) by mRNA differential display. Mol Biotechnol 41(1):42–52. https://doi.org/10.1007/s12033-008-9095-y

Wang Z (2006) Cytoplasmic male sterility of rice with Boro II cytoplasm is caused by a cytotoxic peptide and is restored by two related ppr motif genes via distinct modes of mRNA silencing. Plant Cell Online 18(3):676–687

Wang J, Yang Y, Liu X, Huang J, Wang Q, Gu J et al (2014) Transcriptome profiling of the cold response and signaling pathways in Lilium lancifolium. BMC Genomics 15:203. pmid:24636716

Wei F, Luo S, Zheng Q, Qiu J, Yang W, Wu M, Xiao X (2015) Transcriptome sequencing and comparative analysis reveal long-term flowing mechanisms in *Hevea brasiliensis* latex. Gene 556(2):153–162. https://doi.org/10.1016/j.gene.2014.11.048

Wu C, Lan L, Li Y, Nie Z, Zeng R (2018) The relationship between latex metabolism gene expression with rubber yield and related traits in *Hevea brasiliensis*. BMC Genomics 19(1):897. https://doi.org/10.1186/s12864-018-5242-4

Wycherley PR (1968) Introduction of *Hevea* to the Orient. Planter 4:1–11

Xia Z, Xu H, Zhai J, Li D, Luo H, He C, Huang X (2011) RNA-Seq analysis and de novo transcriptome assembly of *Hevea brasiliensis*. Plant Mol Biol 77(3):299–308. https://doi.org/10.1007/s11103-011-9811-z

Xia Z, Liu K, Zhang S, Yu W, Zou M, He L, Wang W (2018) An ultra-high density map allowed for mapping QTL and candidate genes controlling dry latex yield in rubber tree. Ind Crop Prod 120:351–356. https://doi.org/10.1016/j.indcrop.2018.04.057

Xiao X, Tang C, Fang Y, Yang M, Zhou B, Qi J, Zhang Y (2013) Structure and expression profile of the sucrose synthase gene family in the rubber tree: indicative of roles in stress response and sucrose utilization in the laticifers. FEBS J 281:291–305

Xie DX, Feys BF, James S, Nieto-Rostro M, Turner JG (1998) COI1: an Arabidopsis gene required for jasmonate-regulated defense and fertility. Science 280(5366):1091–1094. https://doi.org/10.1126/science.280.5366.1091

Young EG, Hanson MR (1987) A fused mitochondrial gene associated with cytoplasmic male sterility is developmentally regulated. Cell 50:41–49

Zandalinas SI, Fichma Y, Devireddy AR, Sengupta S, Azad RK, Mittler R (2020) Systemic signalling during abiotic stress combination in plants. PNAS. https://doi.org/10.1073/pnas.2005077117/-/DCSupplemental

Zeng C, Wang W, Zheng Y et al (2010) Conservation and divergence of microRNAs and their functions in Euphorbiaceous plants. Nucleic Acids Res 38(3):981–995. https://doi.org/10.1093/nar/gkp1035

Zhang G, Guo G, Hu X, Zhanga Y, Li Q, Li R, Zhuang R, Lu Z, He Z, Fang X, Chen L, Tian W, Tao Y, Kristiansen K, Zhanga X, Li S, Yanga H, Wang J, Wang J (2010) Deep RNA sequencing at single base-pair resolution reveals high complexity of the rice transcriptome. Genome Res 20(5):646–654

Zhang Y, Leclercq J, Wu S, Ortega-Abboud E, Pointet S, Tang C, Hu S, Montoro P (2019) Genome-wide analysis in Hevea brasiliensis laticifers revealed species-specific post-transcriptional regulations of several redox-related genes. Sci Rep 9(1):5701. https://doi.org/10.1038/s41598-019-42197-8

Zhiyi N, Guijuan K, Yu L, Longjun D, Rizhong Z (2015) Whole-transcriptome survey of the putative ATP-binding cassette (ABC) transporter family genes in the latex-producing laticifers of *Hevea brasiliensis*. PLoS One 10(1):e0116857. https://doi.org/10.1371/journal.pone.0116857

Zhu JH, Xu J, Chang WJ, Zhang ZL (2015) Isolation and molecular characterization of 1-aminocyclopropane-1-carboxylic acid synthase genes in Hevea brasiliensis. Int J Mol Sci 16(2):4136–49. https://doi.org/10.3390/ijms16024136. PMID: 25690030; PMCID: PMC4346948

Zobel BJ, Talbert J (1984) Applied forest tree improvement. Wiley, New York

# Breeding and Biotechnology of Jute

**Kanti Meena, A. Anil Kumar, and Maruti R. T.**

**Abstract** Jute (*Corchorus* spp.) is a natural fibre crop with highly versatile nature and has been utilized as an ingredient in small handicraft industries to big geotextiles for commercial purposes. Lack of male sterile lines, strong sexual incompatibility between two cultivated species, compatible genomic resources for breeding purposes and narrow genetic diversity are the major drawbacks with this crop. This chapter describes jute and its improvement through germplasm collection, characterization, exploitation and utilization in breeding programme and biotechnological tools and value addition of the fibre for commercial purposes. Efforts were made to screen germplasm lines including wild relatives for yield and major biotic stress. Germplasm lines of both cultivated and wild species of jute were screened for stem rot, Bihar hairy caterpillar and nematodes and identified different resistant sources. Interspecific crosses between wild and *C. olitorius* generated stable resistant lines for stem rot disease. Identification of multiple disease and pest resistant lines are more valuable, as many useful traits can be introgressed easily. Non-availability of sufficient markers in jute provided limited success to understand the crop diversity, evolutionary pattern, etc. In this endeavour, efforts have been made for the development of advanced molecular marker systems such as EST-SSRs, SSRs, SNPs, RAD and InDels in past many years. Utilization of these markers for different application in jute has provided useful information related to evolutionary studies, genetic relationship, gene or QTL mapping, genetic diversity, etc. Transcriptome analysis in *C. capsularis* identified genes related to cellulose and lignin biosynthesis, defence-related genes, genes related to salt and drought tolerance. Due to high tensile strength and flexibility, jute fibre is being blended with other textile fibres and is utilized commercially in different textile industries. Apart from fibre, jute leaves are enriched with high nutritional value.

**Keywords** Abiotic and biotic stress · *Corchorous* · Fibre and nutritional quality · Germplasm · Jute · Molecular markers · Strongest vegetable fibre · Value addition

K. Meena (✉) · A. Anil Kumar · Maruti R. T.
ICAR-Central Research Institute for Jute and Allied Fibres (CRIJAF), Kolkata, India



P. M. Priyadarshan, S. M. Jain (eds.), *Cash Crops*,
https://doi.org/10.1007/978-3-030-74926-2_6

# 1 Introduction

Jute (*Corchorus* spp. family: *Malvaceae*) is a versatile natural bast fibre crop and considered as one of the natures strongest vegetable fibres (Mahapatra et al. 2012). It is an annual and herbaceous crop used for enormous end use products from simple rope to textile purposes. Jute ranks second, next to cotton as a natural fibre and occupies an important place in the Indian economy. In the genus *Corchorus,* about 30–40 species have been reported; however, it is represented by only two cultivated species, which are used for commercial fibre production: *C. capsularis* L. (commonly known as white jute with $2n = 2x = 14$) and *C. olitorius* L. (commonly known as dark jute, $2n = 2x = 14$). Both the cultivated jute species are mainly self-fertilized; however, 15% natural outcrossing is observed in *C. olitorius* (Mir et al. 2008b; Sinha et al. 2011). The fruit and seed colour of both the cultivated species varies (Fig. 1). Jute grows in tropical low-land areas with about 60–90% relative humidity (RH) and 24–37 °C temperature (Rowell and Stout 2007). In addition, it also plays an important role in reducing carbon footprints from the environment by consuming about 15 t of $CO_2$ from the atmosphere and release up to 11 t of $O_2$ $ha^{-1}$ (Palit and Meshram 2008).

This natural fibre is endowed with high tensile strength, high insulating property with low extensibility, moderate moisture regime and low thermal conductivity (Rowell and Stout 2007); completely biodegradable and recyclable (Palit and Meshram 2010). Biodegradable and flexible property of jute makes it moisture absorbent and drains well, thus used in geotextiles. Due to unique versatility and durability of jute fibre, demand for jute-blended yarns is increasing (Mahapatra et al. 2009). Fibre from dark jute (golden colour) is more lustrous, soft and stronger than white jute (Kundu et al. 1959), thus demand for dark jute fibre for the production of jute-based quality products is increasing. Increasing concern for climate change has increased use of jute in composites, which has opened up new possibilities for value-added diversified products developed from jute. Jute-growing countries produced 3098 thousand tons of raw jute in 2017–2018 (Table 1) and Bangladesh and India produced 1655 and 1430 thousand tons of raw jute, respectively (FAO Statistical Bulletin, 2018). India earned around 151 millions of US dollars by exporting value-added jute products.

# 2 Origin of Cultivated *Corchorus* Species

Among the two cultivated species: the primary centre of origin for *C. olitorius* L. is believed to be African countries, whereas India or Indo-Burma and south China is considered as secondary centre of diversity. *C. capsularis* L. is believed to be originated from Indo-Burma. Genus *Corchorus* belongs to *Malvaceae* family with haploid chromosome number of $n = 7$ except a few *Corchorus* species (*Corchorus junodii* $2n = 48$). The origin of cultivated jute species is still an enigma except few

**Fig. 1** Variation in two cultivated species

reports indicating *C. aestuans* L. as common progenitors of *C. olitorius* and *C. capsularis* and *C. urticifolius* as nuclear progenitor of *C. olitorius* (Kundu et al. 2013). Presence of setae is the main identification feature of genus *Corchorus* except *C. urticifolius* and *C. depressus*, and similar five sepals and petals will be observed in most of the *Corchorus* species except *C. urticifolius* and *C. depressus* (Fig. 2).

**Table 1** Jute area, production, yield and export in major jute-growing countries for the year 2017–2018

| Country | Area (thousand ha.) | Jute fibre production (thousand tons) | Yield (kg/ha) | Export of raw jute (thousand tons) |
|---|---|---|---|---|
| Bangladesh | 715.3 | 1655.8 | 2.3 | 231.8 |
| India | 746.7 | 1430 | 2.0 | 22.4 |
| Myanmar | 1.3 | 0.3 | 0.3 | – |
| Nepal | 8 | 12 | 1.5 | – |
| Total | *1471.3* | *3098.1* | *1.5* | *254.2* |

**Fig. 2** (**a**) Variation in the presence of setae, flower traits and (**b**) pod shape among corchorus species

**Table 2** *Corchorus* germplasm status at ICAR-CRIJAF, Kolkata (2018–2019)

| Species | Indigenous | Exotic | Mutants | Total |
|---|---|---|---|---|
| *C.olitorius* | 1383 | 346 | 61 | 1790 |
| *C. capsularis* | 660 | 244 | 61 | 965 |
| Wild *Corchorus* | 404 | 157 | – | 561 |
| *Total germplasm lines* | | | | *3316* |

**Fig. 3** Cultivated and wild corchorus species maintained at ICAR-CRIJAF

Major jute-growing countries are India and Bangladesh accounting for 93% of the global production, and other countries of considerable importance are Nepal and Myanmar.

## 3 Germplasm

For any crop improvement programme, availability of diverse germplasm plays a vital role and can be explored and collected nationally and internationally from different regions. Germplasm collections from the centre of diversity play a major role as they conserve maximum genetic diversity. In jute, Africa being the centre of diversity and home for nearly 30 *Corchorus* species, and in term of number of collections Bangladesh has approximately 4111 *Corchorus* collections followed by India (Table 2). In India, till today, only nine *Corchorus* species were reported and in the national gene bank total ten species were conserved (Fig. 3). Figure 3 shows ten different *Corchorus* species conserved at ICAR-CRIJAF, Kolkata, India.

All *Corchorus* germplasm lines were stored in medium- and long-term gene banks as ex situ collections; among the *Corchorus* collections, exotic collections play vital role as major jute-growing countries (India, Bangladesh, and China) have limited species diversity (Table 3). In 1984, International Jute organization (IJO)

**Table 3** Links to access jute germplasm collections

| Country | Institute name | Type of collection | Web link |
|---|---|---|---|
| Bangladesh | BJRI | Global | http://www.bjri.gov.bd |
| India | ICAR-NBPGR | Global | http://www.nbpgr.ernet.in |
| Taiwan | AVRDC | Global | http://seed.worldveg.org |

was formed in Dhaka, Bangladesh. Its first major objective was "collection of *Corchorus* germplasm" from member countries, especially from African countries. A total of 790 germplasm collections were made through five explorations, and these seeds were distributed to member countries for conservation, evaluation and utilization in breeding programmes.

## *Diversity in Cultivated and Wild Relatives*

Genetic diversity of any crop species is broadly measured both at morphological and molecular levels. Both strategies are complementary to each other, which provide holistic information about the species diversity. Morphological or phenotypic markers are of great value and easy to use in studies of germplasm characterization and evaluation. Palit et al. (1996) used four morpho-physiological attributes viz., plant height, harvest index, cambial activity and fibre strength for the estimation of genetic diversity of 192 *C. olitorius* and 216 *C. capsularis* accessions. In another study, 60 mutants of *C. olitorius* were derived from different doses of X-ray and gamma radiation in different germplasm lines of *C. olitorius* were grouped into different clusters using morphological traits related to fibre and biomass productivity, anatomical traits related to fibre distribution and development in the bark tissues of jute. Results of these study showed that morpho-anatomical characterization revealed high variability in the population and facilitated in identifying promising mutant genotypes with superior fibre recovery and fibre anatomy traits (Satya et al. 2014a). However, the phenotypic markers have a limitation as their expression is influenced by both genotype and environmental factors. The use of morphological characters alone is not sufficient to portray the existing genetic diversity.

In jute, cytological techniques were also employed to understand the genetic relatedness between cultivated and wild species. Based on karyotype analysis, Datta et al. (1975); Maity and Datta (2009) and Saha et al. (2014) reported different morphological variations within cultivated and wild species, and variation in chromosome banding pattern of cultivated species was reported by Alam and Rahman (2000).

## 4 Application of Molecular Markers

Genetic diversity in the crop in terms of yield, resistance to biotic or abiotic stresses is the major criteria for the selection of parents and use them in any crop improvement programme. Jute has narrow genetic base, thus breeders' main focus has been to broaden the genetic base without much achievements. Molecular markers have played a pivotal role in studying genetic diversity, population structure, linkage map constructions, genetic resource conservation etc. in many crops (Akkaya et al. 1995; Castro et al. 2016; Nadeem et al. 2018; Ali et al. 2019). The use of genetic markers begins with the extraction of proteins or chemicals (for biochemical markers) or DNA (for molecular markers) from plant tissues (leaf, root, seeds). Their stability, ease of use and cost-effectiveness have made them a very popular tool for a variety of applications including genome mapping, gene tagging, genetic diversity, phylogenetic analysis in many crops including cash crops. In the last few decades, many molecular marker systems have been developed and exploited throughout the world in different crops. However, only few molecular markers are global acceptance (AFLPs, RFLPs, RAPDs, ISSRs, SSRs and SNPs) may be due to their availability, reproducibility, polymorphism and other reasons.

In jute, a variety of DNA markers have been developed in past decades; however, later major classes of DNA markers developed and validated in jute are isoenzymes, amplified fragment length polymorphism (AFLP), random amplified polymorphic DNA (RAPD), inter simple sequence repeat (ISSR), simple sequence repeat (SSR), expressed sequence tag-derived simple sequence repeat (EST-SSR), sequence characterized amplified region (SCAR), sequence-related amplified polymorphism (SRAP), restriction-site-associated DNA (RAD), sequence-tagged microsatellite site (STMS), single-nucleotide polymorphism (SNP), start codon-targeted (SCoT) polymorphism, chloroplast-derived simple sequence repeat (cpSSR) and InDel markers.

Several efforts were made in the past for the assessment of genetic diversity in both wild and cultivated jute species using molecular markers. Isozymes or isoenzymes (multiple forms of enzymes) are enzymes that differ in amino acid sequence but catalyse the same chemical reaction. Study of genetic diversity at enzyme level using electrophoresis was most widely used to study population genetics. As a supplement to morphological characters, isozyme markers are also important due to its co-dominant nature, environmental independent system, easy to use, low cost and easy applicability. However, limited number of markers has forced for the development of advance and more informative DNA-based marker. Still isozyme remains quickest and cheapest marker systems to identify low levels of genetic variation. Isozyme banding pattern has been used to study genetic diversity in many crops. Very few reports on utilization of isozymes as marker system in jute have been obtained, and the details are listed (Table 4).

**Table 4** Application of isozyme markers in jute

| S. No. | Isozyme marker | Species studied | Results | References |
|---|---|---|---|---|
| 1 | Esterase, acid phosphatase and peroxidase | Wild species: *C. pseudo-olitorius* and *C. trilocularis* | Showed both are completely different species | Khatun and Alam (2010) |
| 2 | Esterase, acid phosphatase and peroxidase | Wild species: *C. aestuans* | Generated very stable and species-specific five esterase, one peroxidase and two acid phosphatase bands | Alam et al. (2011) |
| 3 | Esterase, acid phosphatase and peroxidase | Wild species: *C. tridens*, *C. fascicularis* and *C. pseudo-capsularis* | Acid phosphatase showed two common bands in *C. fascicularis* and *C. pseudo-capsularis* Esterase band makes *C. tridens* unique | Khatun et al. (2011) |
| 4 | Peroxidase and esterase | Mutants of *C. olitorius* (6) with their parent JRO 632 | Classified into two groups of peroxidases and five groups of esterases | Ali et al. (2012) |

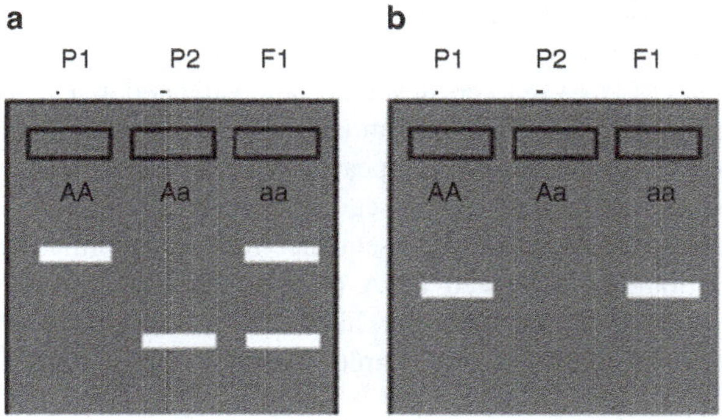


**Fig. 4** (**a**) Codominant and (**b**) dominant markers. Codominant markers clearly differentiate homozygotes and heterozygotes, however dominant markers are inefficient. Source: Collard et al. (2005)

DNA-based markers can be dominant, which allows analysing many loci at one time and does not distinguish homo and heterozygous individuals. RFLP and RAPD markers have been utilized to distinguish cold sensitive and heat tolerant jute. Co-dominant markers express both alleles when co-occurring in an individual, thus they can be used to distinguish heterozygous from homozygous (Fig. 4).

Basu et al. (2016) developed eight markers cpSSRs from *Nicotiana tabacum* and used them for genetic diversity analyses in jute which was the first report on codominant cpSSR marker development in jute, which revealed that the two cultivated species have different maternal origin. Topdar et al. (2013) detected 26 definitive QTLs for bast fibre quality, yield and yield-related traits in jute. A high-density genetic map was constructed using SNPs, and 13 minor QTLs involved in salt

**Table 5** Utilization of molecular markers in improving jute

| S. No. | DNA markers | Species studied | Application | References |
|---|---|---|---|---|
| 1. | Jute cpRFLP | *C. capsularis* and *C. olitorius* | Characterization of interspecific somatic hybrid cell lines in cultivated jute | Saha et al. (2008) |
| 2. | AFLP | *C. capsularis* and *C. olitorius* | Genetic diversity, DNA polymorphism, species relationship and population structure | Hossain et al. (2003), Basu et al. (2004), Benor (2011), Das et al. (2011), Benor et al. (2012), Ghosh et al. (2017) |
| 3. | RAPD | *C.capsularis, C. olitorius, C. incisifolus, C. aestuans* and *C. trilocularis* | Genetic diversity, species relationship, DNA fingerprinting, phylogenetic relationship and linkage mapping | Hossain et al. (2002, 2003), Qi et al. (2003a, 2004), Roy et al. (2006), Haque et al. (2007, 2008), Ogunkami et al. (2010), Mir et al. (2011), Chen et al. (2015), Adeyemo and Abati (2015), Heikal et al. 2015 Youssef et al. (2019) |
| 4. | ISSR | *C. capsularis, C. olitorius, C. aestuans, C. fascicularis, C. depressus, C. pseudocapsularis, C. pseudo-olitoriustridens, C. trilocularis* and *C. urticifoliuss* | Genetic diversity, species relationship and linkage mapping | Qi et al. (2003b, 2004), Roy et al. (2006), Sultana et al. (2006), Tao et al. (2012), Chen et al. (2015), Saha et al. (2014), Nag et al. (2018) |
| 5. | SCoTs | *C. capsularis* and *C. olitorius* | Genetic diversity | Rana et al. (2012) |
| 6. | SSR and EST-SSR | *C. capsularis* and *C. olitorius* | Genetic diversity, DNA fingerprinting, and QTL mapping | Das et al. (2012), Topdar et al. (2013), Satya et al. (2014b), Zhang et al. (2014, 2015b, c) |
| 7. | SRAP | *C. capsularis* and *C. olitorius* | Genetic diversity, DNA fingerprinting and linkage mapping | Tao et al. (2012), Rana et al. (2013, Chen et al. (2014), Nag et al. (2014), Youssef et al. (2019) |
| 8. | RAD | *C. capsularis* and *C. olitorius* | QTL identification, comparative genomics and linkage mapping | Kundu et al. (2015), Islam et al. (2017) |

(continued)

**Table 5** (continued)

| S. No. | DNA markers | Species studied | Application | References |
|---|---|---|---|---|
| 9. | SCAR | *C. capsularis* | Screening for *Macrophomina phaseolina*-resistance and linkage mapping | Mir et al. (2011) |
| 10. | STMS | *C. capsularis, C. olitorius* and *C. aestuans* | Genetic diversity | Roy et al. (2006) |
| 11. | SNP | *C. olitorius* | QTL identification, comparative genomics, linkage mapping, population structure and genetic differentiation | Kundu et al. (2015), Sarkar et al. (2019), Yang et al. (2019) |
| 12. | CBDP | *C. capsularis* and *C. olitorius* | Genetic diversity and relationship between species | Singh et al. (2014) |
| 13. | cpSSR | *C. capsularis, C. olitorius C. aestuans, C. fascicularis, C. pseudoolitorius, C. tridens, C. trilocularis, C. pseudocapsularis* and *C. urticifolius* | Relationship between species and chloroplast haplotyping | Basu et al. (2004), Kundu et al. (2013) |
| 14. | GBD | *C. capsularis, C. olitorius C. aestuans, C. fascicularis, C. pseudoolitorius, C. tridens, C. trilocularis* and *C. siliquosus* | Genetic analysis, phylogenetic analysis and population structure | Satya et al. (2014a), Tanmoy et al. (2015) |
| 15. | InDel | *C. capsularis* and *C. olitorius* | Development and identification of markers related to cellulose content of bast fibre | Zhang et al. (2017) |
| 16. | Organelle-specific markers | *C. capsularis* and *C. olitorius* | Genetic diversity | Basu et al. (2016) |
| 17. | SLAF | *C. capsularis* | QTL identification and genetic mapping | Tao et al. (2017) |

tolerance were identified (Yang et al. 2019). Comprehensive linkage maps for *Corchorus capsularis* based on SNP was reported by Chinmay et al. (2015).

Application of DNA-based markers in jute and its wild species till date are summarized in Table 5 and represented using few markers (Fig. 5).

In an attempt to develop molecular resources in jute, 4509 eSSR loci were identified from unigene sequences of *C. capsularis* and all the information-related eSSRs were compiled in a user-friendly web database, Jute Marker db (http://jutemarkerdb.icar.gov.in/), for easy access to this resource and for use in jute genetic

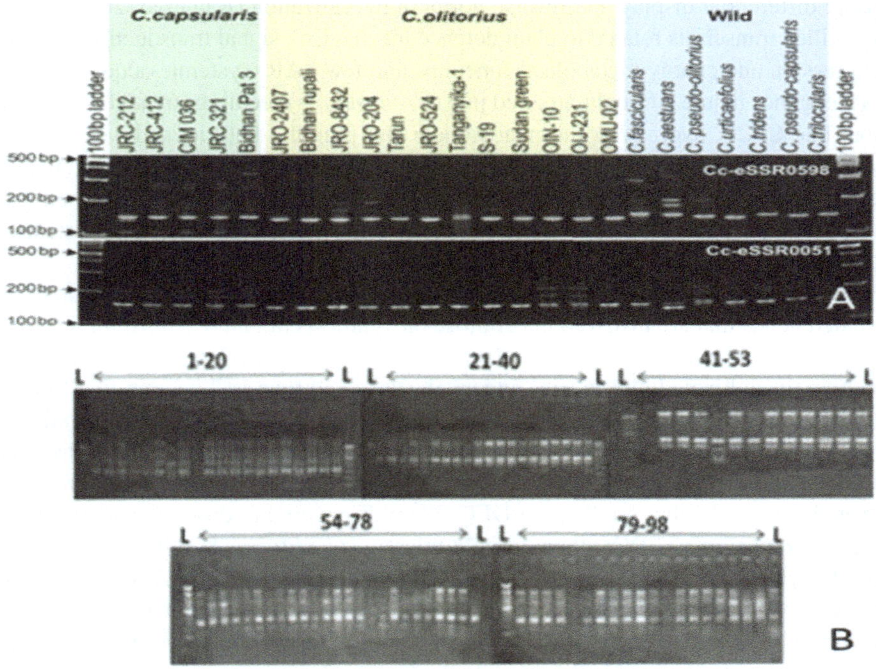


**Fig. 5** Showing genetic diversity in jute using (**a**) eSSRs. Source: Saha et al. (2017), ICAR-CRIJAF and (**b**) ISSR. Source: Nag et al. (2018), ICAR-CRIJAF

improvement. (Saha et al. 2017). Further, a random set of 74 primer pairs from these collections produced 58.1% polymorphism in a panel of 24 jute accessions comprising of *C. capsularis, C. olitorius* and wild species (Saha et al. 2017). This suggests that these markers can be effectively utilized for estimating genetic diversity and mapping of quantitative trait loci (QTLs) in jute.

## 5 Transcriptomics of Jute

Genomic research in jute is very recent and has been delayed due to many limitations including narrow genetic diversity, genomic resources very high sexual incompatibility (Sarkar et al. 2016). However, recent development of advance tools and technologies related to genomic studies has fasten the research in this endovour (Sarkar et al. 2016). The first reference bast transcriptomes for jute (*C. capsularis*) by Illumina (HiSeq 2000 platform) paired-end sequencing and de novo assembly discovered 81 isoforms of 26 genes involved in lignin biosynthesis (Chakraborty et al. 2015). *Macrophomina phaseolina* (causes stem rot disease) infected jute (*C. capsularis* cv. JRC-412) leaves, with or without treating with β-amino butyric acid (BABA) (an elicitor activating defence response in plants), were used for transcriptomic analysis using differential display and reverse northern hybridization (Ray et al. 2011). They identified transcripts related to plant defence mechanism, signal transduction or gene expression and energy metabolism functions and few SAR (systemic-acquired resistance) genes in jute. Drought-stressed jute (*C. capsularis*) seedlings treated with elicitors like $CaCl_2$, hydroxyapatite nanoparticle and β-amino butyric acid was used to study for transcripts and enzyme profiles in jute. Upregulation of genes related to late embryogenesis abundant protein, dehydration responsive element and biochemical markers like proline and peroxidase associated with the priming and systemic-acquired resistance of the jute seedlings was reported (Das et al. 2016).

De novo transcriptome analysis was reported for identification of major bast-related genes that are involved in cellulose biosynthesis in jute (*C. capsularis* L.) by Zhang et al. (2015a)). Yang et al. (2017a) identified transcripts related to salt stress in both the cultivated jute species under salt stress and found 127 common differentially expressed genes (DEGs) and 4489 and 492 common DEGs in the root and leaf tissues, respectively, of both *Corchorus* species. Transcriptome analysis for drought tolerant in two species of jute in response to polyethylene glycol (PEG)-induced drought stress. Higher numbers of DEGs were discovered in drought-sensitive species (*C. capsularis* L., YY) than drought tolerant (*C. olitorius* L., GF). 14 unigenes in phenylpropanoid biosynthesis and peroxisome pathway, involved in scavenging of reactive oxygen species (ROS), showed differential expression in response to increase of superoxide (Yang et al. 2017b). The first comprehensive transcriptome analysis was studied in *C. olitorius* for the leaves, flowers, stem bast and fruit and identified 4772 novel genes in these tissues. In the metabolic pathway for biosynthesis of secondary metabolites, 26 genes in stem bast were up and downregulated. Upregulation of 53 genes in the phenylpropanoid biosynthesis pathway in flower was reported (Yang et al. 2020).

## 6 Retention of Diversity in Land Races

In jute, information regarding landraces was sparse, as seed is not the economical part and whole crop is harvested for fibre purpose. Chaudhury (2000) reported about 70 land races in present-day India and Bangladesh in both the cultivated species and their descriptions. In India, only 44 land races were preserved, and there are no reports of land races in other countries, except in Egypt by Youssef et al. 2019. In the present-day intensive agriculture and shifting to remunerative crop cultivation, special encouragement is needed for the farmers to conserve valuable land races.

## 7 Disease and Insect Pest Resistance

Jute is grown in hot and humid conditions which are congenial for diseases and insect pest infestation. The major disease that causes tremendous loss for jute production is stem rot caused by *macrophomina phaseolina* (Tassi) Goid. Similarly,

**Fig. 6** Major biotic stresses in jute (**a**) stem rot disease (**b**) Bihar hairy caterpillar (**c**) Yellow mite

among insect pests Bihar hairy caterpillar (BHC) and yellow mite (Fig. 6) are the major, other pests include stem weevil and semi looper. Very few resistance sources were identified for the above biotic stresses. For stem rot disease, germplasm material was screened and one-line OIN-154 was identified as moderate resistance in *C. olitorius* (De et al. 2019); however, two germplasm lines CIM-36 and CIN-109 were found to be resistant in *C. capsularis* (Mir et al. 2011).

Screening under sick plot conditions for multiple years resulted in identification of stable resistance sources in wild *Corchorus* germplasm lines. Among *C. aestuans* (WCIN-136-1 and WCIN-183A) and *C. fascicularis* (WCIJ-28 and WCIJ-150-1) (ICAR-CRIJAF Annual report 2018–2019) were identified as resistant. An interspecific population was developed to introgress stem rot resistant sources from WCIN-136-1 into *C. olitorius,* which resulted in the development of 35 stable-resistant lines. Those lines were screened under sick plot conditions for 2 consecutive years (unpublished data). Among the wild species, *C. aestuans* germplasm line WCIN-179 was identified as the resistant source for BHC (Gotyal et al. 2015), and the resistance was attributed to non-preference and antibiosis mechanism. Identification and introgression of multiple biotic stresses are useful in reducing the cost of cultivation of resource poor jute farmers; in this context, resistant to multiple biotic stresses was identified in wild *Corchorus* germplasm lines, mostly in *C. aestuans.* WCIN-136-1 was identified as a resistance source for stem rot and BHC, and WCIN-183A aestuans germplasm line showed multiple resistance for stem rot, BHC and nematode (Fig. 7).

## 8 End-Use Quality and Nutritional Quality

Jute is a lingo-cellulosic fibre mainly used for coarse textile purpose, and most of the breeding programmes were focused on improving the fibre yield neglecting the quality parameters which are useful for jute diversified products. Few capsularis

**Fig. 7** Stem rot and nematode resistance in *Corchorus aestuans* L

**Table 6** Chemical composition of different bast fibre crops

| Different bast fibre crops | | | | | | | |
|---|---|---|---|---|---|---|---|
| Chemical components | Jute | Flax | Ramie | Mesta | Roselle | Sisal | Sunnhemp |
| α-Cellulose | 60.7–61.0 | 70.3 | 96–97 | 60.6 | 59.7 | 63.9 | 78.3 |
| β-Cellulose | – | – | 4–5 | – | – | – | – |
| Hemicellulose | 15.6–15.9 | 4.4 | 3.9 | 15.0 | 15.0 | 17.9 | 3.6 |
| Lignin | 12.5–13.5 | 5.5 | 0.5 | 10.1 | 9.9 | 8.6 | 4.0 |

jute varieties possess fine fibre quality properties suitable for diversified products. In tossa jute, Chaudhury et al. (2014) developed few olitorius lines with moderate improvement in tensile strength and fineness. Few wild *Corchorus* germplasm lines were identified with high fibre cell length and length–breadth ratio (*C. trilocularis* L/B ratio 216.4 and *C. aestuans* L/B ratio 211.3) and tried several interspecific crosses to improve jute fibre quality amenable for textile purpose. As fibre fineness is governed by multiple traits, i.e. fibre strength, ultimate fibre cell length, length–breadth ratio and low lignin and high cellulose fibres, breeding programmes should focus on these traits (Jain 2004). Comparison of lignin, hemicelluloses and cellulose content of one of the most versatile crops with other fibre crop is represented (Table 6). Presence of lignin in the jute fibres imparts brittleness and changes the colour over the time reducing the suitability for textile purpose, in this direction few mutant lines with low lignin were developed in capsularis (*dlpf*) by Sengupta and Palit (2004), and in olitorius (*llpf*) (Choudhary et al. 2017), but intensive breeding programmes should be aimed to improve fibre quality parameters (low lignin, high cellulose content, high L/B ratio of fibre cell and reduced meshyness) that suit the textile purpose. Fibre of both the cultivated species of *Corchorus* and variation in the colour of fibre can be seen (Fig. 8).

**Fig. 8** Fibre of cultivated jute and variation in fibre colour of Corchorus species

## *Nutritional Quality*

Jute is mainly grown as a fibre crop; however, leaves of these two cultivated jute species are consumed as leafy vegetable (Furumuto et al. 2002). There is limited available information with respect to nutritional aspects of jute leaves due to lack of attention given by the researchers. Jute plant parts are palatable and are endowed with high nutritional value, containing crude fibre, carbohydrate, lipids, protein, vitamins (like A, C, E) and the most of the minerals like sodium, calcium, potassium, phosphorus and iron (Odhav et al. 2007; Dansi et al. 2008; Choudhary et al. 2013). Study on nutritional profile of cultivated and wild *Corchorus* species revealed that cultivated germplasm lines were rich in most of the nutrients (Choudhary et al. 2013) except in iron content. The plant is also enriched with some flavonoids, saponins, triterpenes, glucoside, β-sitosterol, scopoletin, fusidic acid and steroids (Joy 2014; Ramadevi 2014). Isolation of some unknown triterpene called glycoside and capsugenin 30-*O*-glucopyranoside from leaves of *C. capsularis* was reported (Zakaria 2007; Joy 2014). Due to highly nutritional value, jute leaves are being used for preparation of soup, curry and are blended with other vegetables too (Fig. 9). Consuming 87 g of jute leaves provide sufficient number of vitamins, mineral and amino acids mentioned (Table 7). Nutritional composition of jute leaf in both cultivated and wild species was studied by Zablul et al. 2019, and reported presence of more vitamins and minerals in wild species.

# 9 Conventional Breeding Approaches

Initial jute breeding programme was started with release of Chinsura green in *C. olitorius* and D-154 in *C. capsularis* through pure line selection, and later varieties were developed by selections from mutation breeding, single cross hybrids and multiple cross hybrids through pedigree method. Any breeding programme aimed to develop a high yielding variety will result in narrow genetic base among the cultivars and jute is not an exception, to develop premature flowering tolerant varieties in tossa jute which fits into jute rice-based cropping system resulted in repeated use

**Fig. 9** Jute leaves as a part of diet due to high nutritional value

**Table 7** Nutrition value for jute leaves

| Vitamins | | Amino acids | | Minerals | |
|---|---|---|---|---|---|
| Vitamin K | 94 μg | Tryptophan | 0.021 g | Iron | 2.73 mg |
| Vitamin B6 | 0.496 mg | Threonine | 0.113 g | Copper | 0.222 mg |
| Vitamin A, | 225 μg | Isoleucine | 0.152 g | | |
| Vitamin C | 28.7 mg | Leucine | 0.266 g | | |
| | | Lysine | 0.151 g | | |
| | | Methionine | 0.044 g | | |

of Sudan green and Tanganyika-1 (pre-mature flowering resistant lines) and narrowing the genetic base of jute cultivars (Kar et al. 2009). In conventional jute breeding programmes, most of the varieties were developed by intraspecific crossing, as no successful interspecific crossing was reported. Initial interspecific crossing was carried out between capsularis and olitorius lines resulted in unsuccessful crossing (Datta et al. 1960). Successful interspecific crossing was reported by Maity and Datta (2008) and Mandal et al. (2013) in capsularis and trilocularis lines by cytomorphological studies. Recently successful interspecific crossing was reported in *C. olitorius* with *C. aestuans* resulting wide variation in morphological traits, yield attributing traits and flowering tolerance (Anil Kumar et al. 2018).

## 10 Jute Fibre at Commercial Level

Jute is used mainly for weaving; gunny bags are designed from thick cloth made from jute fibre. Fine type of cloth prepared from jute fibre is mainly used for sleep on purposes, whereas coarse cloth is maximum used for making the sails of country boats, bags to hold heavy material. Jute yarn and twines are used for manufacturing curtains, decorative, carpets, rugs, chair coverings, shopping and handbags, linoleum backing, household fabrics (Hazra et al. 2008) shirtings and are also mixed with silk or used for imitating silk fabrics. In addition to diversified uses in

**Fig. 10** Value addition of jute fibre for the production of diversified products

agriculture and other sectors, jute-based geosynthetics are being promoted as they prevent soil erosion and landslides. Slowly jute is gaining importance in paper and pulp industries due to its nature of hardwood and lignin contents and is being used to some extent in paper making (Fig. 10).

## 11 Conclusions and Future Outlook

Limited gene pool are one of the major constraints for the breeders that had limited genetic improvement and enhance production through conventional breeding approaches and utilization of molecular marker-assisted selection. Despite many efforts not much success has been made in broadening the genetic diversity of jute, in improving the physical fibre quality traits (fibre fineness and tensile strength), in development of low lignin jute. Lack of male sterile lines in jute is a major roadblock for the development of good hybrids. Development of male sterile using transgenic approaches have been reported in any crops, however due to its recalcitrant nature and non-availability of a reproducible protocol for in vitro plantlet regeneration of jute. Genomics and proteomics information related to fibre quality, low lignin content and expression of genes during the development of fibre can provide basic knowledge. TILLING (targeting induced local lesions in genomes) is tissue culture independent technology can be utilized for developing new breeding lines or varieties in jute. It can also be used for generating mutations to generate genetic diversity in jute. Sexual incompatibility is between the two cultivated species is another major issue needs to be addressed for broadening the genetic base of jute.

## References


Adeyemo O, Abati A (2015) Genetic diversity in Corchorus olitorius L. grown in south-West Nigeria inferred from RAPD data. Jordan J Agric Sci 11:329–337

Akkaya MS, Shoemaker RC, Specht JE, Bhagwat AA, Cregan PB (1995) Integration of simple sequence repeat DNA markers into a soybean linkage map. Crop Sci 35:1439–1445

Alam S, Khatun M, Sultana S (2011) Differential chromosome banding and isozyme assay in *Corchorus aestuans* L. Bang J Bot 40(1):47–52. https://doi.org/10.3329/bjb.v40i1.7997

Alam S, Rahman A (2000) Karyotype Analysis of Three Corchorus Species. Cytologia 65:443–446

Ali MN, Ghosh A, Sasmal BG, Sarkar HK, Das PK (2012) Isozyme diversity in selected leaf mutants of 'tossa' jute (Corchorus olitorius L.). Indian J Biotechnol 11:333–336

Ali A, Pan YB, Wang QN, Wang JD, Chen JL, Gao SJ (2019) Genetic diversity and population structure analysis of *Saccharum* and *Erianthus* genera using microsatellite (SSR) markers. Sci Rep 9:395. https://doi.org/10.1038/s41598-018-36630-7

Anil Kumar A, Choudhary SB, Sharma HK, Maruthi, RT, Karmakar PG, Mitra J (2018). Development of inter-specific *Corchorus olitorius* L. x *C. aestuans* L. population for important tossa jute breeding traits. In: International conference on agriculture and allied sciences: the productivity, food security and ecology

Basu A, Ghosh M, Meyer R, Powell W, Basak SL, Sen SK (2004) Analysis of genetic diversity in cultivated jute determined by means of SSR markers and AFLP profiling. Crop Sci 44:678–685. https://doi.org/10.2135/cropsci2004.6780

Basu T, Satya P, Sarkar D, Kar CS, Mitra J, Karmakar PG (2016) Organelle genetic diversity in a global collection of Jute (*Corchorus capsularis* and *C. olitorius*, Malvaceae). South Afr J Bot 103:54–60

Benor S (2011) Phylogeney of the genus Corchorus (Malvaceae s. l.) and diversity analyses in selected species: evidence from morphology, flow cytometry, and molecular data. Dissertation, Universität Kassel, Kassel

Benor S, Demissew S, Hammer K, Blattner FR (2012) Genetic diversity and relationships in Corchorus olitorius (Malvaceae s. l.) inferred from molecular and morphological data. Genet Resour Crop Evol 59:1125–1146. https://doi.org/10.1007/s10722-011-9748-8

Castro JA, Oliveira EJ, Jesus ON, Soares TL, Margarido GRA (2016) Molecular markers for conservation genetic resources of four Passiflora species. Sci Hortic 212:251–261. https://doi.org/10.1016/j.scienta.2016.10.005

Chakraborty A, Sarkar D, Satya P, Karmakar PG, Singh NK (2015) Pathways associated with lignin biosynthesis in lignomaniac jute fibres. Mol Genet Genomics 290(4):1523–1542. https://doi.org/10.1007/s00438-015-1013-y

Chaudhury NC (2000) Jute and substitutes, 4th edn. Biotech Books, New Delhi

Chaudhury SK, Sinha MK, Chattopadhyay S, Karmakar PG (2014) Genetic manipulations for fibre quality improvement in jute. In: Hazra SK, Karmakar PG (eds) 2004 in Proceedings of national seminar on "Diversified uses of jute and allied fibre crops". CRIJAF, NIRJAFT and IFS(ER), Barrackpore, p 178

Chen M, Wei C, Qi J, Chen X, Su J, Li A, Tao A, Wu W (2014) Genetic linkage map construction for kenaf using SRAP, ISSR and RAPD markers. Plant Breed 130:679–687

Chen Y, Zhang L, Qi J, Chen H, Tao A, Xu J, Lin L, Fan P (2015) Genetic linkage map construction for white jute (Corchorus capsularis L.) using SRAP, ISSR and RAPD markers. Plant Breed 133:777–781. https://doi.org/10.1111/pbr.12205

Chinmay B, Dey P, Karmakar P, Satpathy S (2015) Discovery of large- scale SNP markers and construction of linkage map in a RIL population of jute (Corchorus capsularis). Mol Breed 35:1–10. https://doi.org/10.1007/s11032-015-0311-8

Choudhary SB, Sharma HK, Karmakar PG, Anil Kumar A, Saha AR, Hazra P, Mahapatra BS (2013) Nutritional profile of cultivated and wild jute (*Corchorus*) species. Aust J crop Sci 7(13):1973–1982

Choudhary SB, Chowdhury I, Singh RK, Pandey SP, Sharma HK, Anil Kumar A, Karmakar PG, Kumari N, Souframanien J, Jambhulkar SJ (2017) Morphological, histobiochemical and molecular characterisation of low lignin phloem fibre (llpf) mutant of dark jute (Corchorus olitorius L.). Appl Biochem Biotechnol 183(3):980–992. https://doi.org/10.1007/s12010-017-2477-5

Collard BCY, Jahufer MZZ, Brouwer JB, Pang ECK (2005) An introduction to markers, quantitative trait loci (QTL) mapping and marker-assisted selection for crop improvement: The basic concepts. Euphytica 142:169–196. https://doi.org/10.1007/s10681-005-1681-5

Dansi A, Adjatin A, Adoukonou-Sagbadja H, Faladé V, Yedomonhan H, Odou D, Dossou B (2008) Traditional leafy vegetables and their use in the Benin Republic. GenetResour Crop Evol 55:1239–1256. https://doi.org/10.1007/s10722-008-9324-z

Das M, Banerjee S, Topdar N, Kundu A, Sarkar D, Sinha MK, Balyan HS, Gupta PK (2011) Development of large-scale AFLP markers in jute. J Plant Biochem Biotechnol 20:270–275. https://doi.org/10.1007/s13562-011-0058-1

Das M, Banerjee S, Topdar N, Kundu A, Mir RR, Sarkar D, Sinha MK, Balyan HS, Gupta PK (2012) QTL identification for molecular breeding of fibre yield and fibre quality traits in jute. Euphytica 187(2):175–189

Das A, Ray R, Mandal N, Chakrabarti K (2016) An analysis of transcripts and enzyme profiles in drought stressed jute (*Corchorus capsularis*) and rice (*Oryzasativa*) seedlings treated with $CaCl_2$, hydroxyapatite nano-particle and β-amino butyric acid. Plant Growth Regul 79:401–412. https://doi.org/10.1007/s10725-015-0144-9

Datta RM, Dana SK, Banerjee SN (1960) Investigation on the interspecific hybridization between autotetraploids of the cultivated jute species (*Corchorus olitorius* L. and *C. capsularis* L.) and the failure of viable seed formation in them. Genet Iber 12:1–32

Datta RM, Mukhopadhaya D, Panda BS, Sasmal PK (1975) Cytotaxonomic studies of different Corchorus (jute) species II. Cytologia 40(3–4):685–692

De Rajib Kumar (2019) Jute Diseases: Diagnosis and Management,Tech. Bull. No. 04/2019. ICAR-Central Research Institute for Jute and Allied Fibres, (Indian Council of Agricultural Research), Barrackpore, Kolkata 700120, India. pp 38.

Furumuto T, Wang R, Okazaki K, Hasan FA, Ali IM (2002) Anti-tumour promoters in leaves of jute (*Corchorus capsularis* and *Corchorus olitorius*). Food Sci Technol Res 8:239–243

Ghosh S, Meena K, Sinha MK, Karmakar PG (2017) Genetic diversity in Corchorus olitorius genotypes using jute SSRs. Proc Natl Acad Sci India Sect B Biol Sci 87:917–926. https://doi.org/10.1007/s40011-015-0652-4

Gotyal BS, Selvaraj K, Meena PN, Satpathy S (2015) Host plant resistance in cultivated jute and its wild relatives towards jute hairy caterpillar *Spilosomaobliqua* (Lepidoptera: Arctiidae). Fla Entomol 98:721–727

Haque S, Begum S, Sarker RH, Khan H (2007) Determining genetic diversity of some jute varieties and accessions using RAPD markers. Plant Tissue Cult Biotechnol 17:183–191. https://doi.org/10.3329/ptcb.v17i2.3239

Haque S, Ashraf N, Begum S, Sarkar RH, Khan H (2008) Construction of genetic map of jute (Corchorus olitorius L.) based on RAPD markers. Plant Tissue Cult Biotechnol 18:165–172. https://doi.org/10.3329/ptcb.v18i2.3647

Hazra SK, Saha A, Karmakar PG (2008) Unexploited potentials of jute and allied fibre crops: whole plant, plant parts, stick and waste fibre. In: Karmakar PG, Hazra SK, Ramasubramanian T, Mandal RK, Sinha MK, Sen HS (eds) Jute and allied fibre updates. Central Research Institute for Jute and Allied Fibres, Kolkata, pp 297–319

Heikal HA, Abd El-Azeem RH, El-Wakil HE (2015) Molecular fingerprinting and phylogenetic relationships among three Egyptian Moulokhyia genotypes (*Corchorus olitorius*L.) using RAPD, ISSR and SRAP markers. Indian Streams Res J 5(1):1–12

Hossain MB, Haque S, Khan H (2002) DNA fingerprinting of jute germplasm by RAPD. J Biochem Mol Biol 35:414–419

Hossain MB, Awal A, Rahman MA, Haque S, Khan H (2003) Distinction between cold-sensitive and -tolerant jute by DNA polymorphism. J Biochem Mol Biol. 36:427–432

Islam MS, Saito JA, Emdad EM, Ahmed B, Islam MM et al (2017) Comparative genomics of two jute species and insight into fibre biogenesis. Nat Plants 3:16223

Jain KC (2004) Fibre quality requirement for diversified uses of jute. In: Hazra SK, Karmakar PG (eds) 2004 in Proceedings of national seminar on "Diversified uses of jute and allied fibre crops". CRIJAF, NIRJAFT and IFS(ER), Barrackpore, p 178

Joy A (2014) Encyclopedia medicinal plants, herbs and medicinal plant usage and identification database. DK Publishing, New York. http://herbs.indianmedicinalplants.info/index.php/sanskrit-names-of-plants/44-2012-02-24-07-34-36/427–corchorus-capsularis

Kar CS, Kundu A, Sarkar D, Sinha MK, Mahapatra BS (2009) Genetic diversity in jute (Corchorus spp.) and its utilization: a review. Indian J Agric Sci 79:575–586

Khatun M, Alam SS (2010) Confirmation of species status of *Corchorus trilocularis* and *C. pseudo-olitorius* by differential chromosome banding and isozyme assay. Cytologia 75(1):83–88

Khatun M, Sultana SS, Ara H, Islam MN, Alam SS (2011) Differential chromosome banding and Isozyme assay of three *Corchorus* spp. Cytologia 76(1):27–32

Kundu BC, Basak KC, Sarkar PB (1959) Jute in India. The Indian Central Jute Committee, Calcutta

Kundu A, Topdar N, Sarkar D, Sinha MK, Ghosh A, Banerjee S, Das M, Balyan HS, Mahapatra BS, Gupta PK (2013) Origins of white (Corchorus capsularis L.) and dark (*C. olitorius* L.) jute: a reevaluation based on nuclear and chloroplast microsatellites. J Plant Biochem Biotechnol 22:372–381. https://doi.org/10.1007/s13562-012-0165-7

Kundu A, Chakraborty A, Mandal NA, Das D, Karmakar PG, Singh NK, Sarkar D (2015) A restriction-site-associated DNA (RAD) linkage map, comparative genomics and identification of QTL for histological fibre content coincident with those for retted bast fibre yield and its major components in jute (*Corchorus olitorius* L., Malvaceae s. l.). Mol Breed 35:19. https://doi.org/10.1007/s11032-015-0249-x

Mahapatra BS, Mitra S, Ramasubramanian T, Sinha MK (2009) Research on jute (Corchorus olitorius and C. capsularis) and kenaf (Hibiscus cannabinus and H. sabdariffa): present status and future prospects. Indian J Agric Sci 79:951–967

Mahapatra BS, Mitra S, Kumar M, Ghorai AK, Sarkar SK, Kar CS, Kundu DK, Karmakar PG (2012) An overview of research and development in jute and allied fibre crops in India. Indian J Agron 57:72–82

Maity S, Datta AK (2008) Cytomorphological studies in F1 hybrids (*Corchorus capsularis* L. X *Corchorus trilocularis* L.) of jute (Tiliaceae). Comp Cytogenet 2(2):143–149

Maity S, Datta AK (2009) Karyomorphology in nine species of jute (Corchorus L.,Tiliaceae). Cytologia 74(3):273–279

Mandal A, Datta AK, Datta S, Gupta S, Sengupta S, Ghosh BK (2013) Assessment of stability and viability of a jute hybrid (*Corchorus capsularis* L. × *C. trilocularis* L.) and its trueness based on RAPD markers. Nucleus 56:163–170

Mir R, Rustgi S, Sharma S, Singh R, Goyal A, Kumar I, Gaur A, Tyagi AK, Khan H, Sinha MK, Balyan HS, Gupta PK (2008b) A preliminary genetic analysis of fibre traits and the use of new genomic SSRs for genetic diversity in jute. Euphytica 161:413–427

Mir JI, Roy A, Ghosh SK, Karmakar PG (2011) Development of linkage map in $F_2$ population of selected parents with respect to *Macrophominaphaseolina* resistance trait using screened polymorphic RAPD and developed SCAR markers of jute. Arch Phytopathol Plant Protect 44(7):671–683. https://doi.org/10.1080/03235400903308883

Nadeem MA, Nawaz MA, Shahid MQ, Dogan Y, Comertpay G, Yıldız M, Hatipoglu R, Ahmad F, Alsaleh A, Labhane N, Ozkan H (2018) DNA molecular markers in plant breeding: current status and recent advancements in genomic selection and genome editing. Biotechnol Biotechnol Equip 32(2):261–285

Nag S, Mitra J, Satya P, Kar CS, Karan M, Ali N (2014) A comparison of efficiency parameters of SSR primers and genetic diversity analysis in jute (Corchorus spp.). Trends Biosci 7:2882–2885

Nag S, Mitra J, Satya P, Kar CS, Karmakar PG (2018) Genetic diversity assessment in jute (*Corchorus* species) utilizing inter simple sequence repeat and simple sequence repeat markers. IJBT 17(2):316–326

Odhav B, Beekrum S, Akula US, Baijnath H (2007) Preliminary assessment of nutritional value of traditional leafy vegetables in KwaZulu-Natal. South Afr J Food Compos Anal 20(5):430–435. https://doi.org/10.1016/j.jfca.2006.04.015

Ogunkami LA, Okunowo WO, Oyelakin OO, Oboh BO, Adesina OO, Adekoya KO, Ogundipe OT (2010) Assessment of genetic relationships between two species of jute plants using phenotypic and RAPD markers. Int J Bot 6:7–11

Palit P, Meshram JH (2008) Physiology of jute yield and quality. In: Karmakar PG, Hazra SK, Ramasubramanian T, Mandal RK, Sinha MK, Sen HS (eds) Jute and allied fibre updates. Central Research Institute for Jute and Allied Fibres, Kolkata, pp 112–124

Palit P, Meshram JH (2010) Production and utilization of jute and allied fibre: potentialities and problems. In: Palit P, Sinha MK, Meshram JH, Mitra S, Laha S, Saha AR, Mahapatra BS (eds) Jute and allied fibre production, utilization and marketing. Indian Fibre Society (Eastern Region), Kolkata, pp 22–26

Palit P, Sasmal BC, Bhattacharryya AC (1996) Germplasm diversity and estimate of genetic advance of four morpho-physiological traits in a world collection of jute. Euphytica 90:89–110

Qi JM, Zhou D, Wu W, Lin L, Fang P, Wu J (2003a) The application of RAPD technology in genetic diversity detection of jute. Yi Chuan Xue Bao 30:926–932

Qi JM, Zhou D, Wu W, Lin L, Wu J, Fang P (2003b) Application of ISSR technology in genetic diversity detection of jute. Ying Yong Sheng Tai Xue Bao 14:1473–1477

Qi JM, Zhou D, Wu WR, Wu W, Lin L, Fang P, Wu JM, Wu J (2004) A comparison between RAPD and ISSR technology in detection of genetic diversity of jute. Sci Agric Sin 37:2006–2011

Ramadevi D (2014) Phytochemical examination of Corchorus capsularis leaves. Int J Curr Pharm Rev Res 4(2):42–46

Rana MK, Singh S, Singh AK, Kak A (2012) Genetic diversity assessment in Indian jute (Corchorus spp) cultivars using gene-targeted molecular markers. India J Agric Sci 82:660–666

Rana MK, Arora K, Singh S, Singh AK (2013) Multi-locus DNA fingerprinting and genetic diversity in jute (Corchorus spp.) based on sequence-related amplified polymorphism. J Plant Biochem Biotechnol 22:1–8. https://doi.org/10.1007/s13562-012-0104-7

Ray R, Ghosh A, Bera A, Dutta N, Chattopadhyay D, Chakrabarti K (2011) Analysis of differentially expressed transcripts in jute upon fungal infection and beta-amino butyric acid treatment. Physiolo Mol Plant Pathol 76:59. https://doi.org/10.1016/j.pmpp.2011.05.001

Rowell RM, Stout HP (2007) Jute and kenaf. In: Lewin M (ed) Handbook of fibre chemistry, 3rd edn. CRC Press, Boca Raton, pp 405–452

Roy A, Bandyopadhyay A, Mahapatra AK, Ghosh SK, Singh NK, Bansal KC, Koundal KR, Mohapatra T (2006) Evaluation of genetic diversity in jute (Corchorus species) using STMS, ISSR and RAPD markers. Plant Breed. 125:292–297. https://doi.org/10.1111/j.1439-0523.2006.01208.X

Saha T, Majumdar S, Banerjee NS, Sen SK (2008) Development of interspecific somatic hybrid cell lines in cultivated jute and their early characterization using jute chloroplast. RFLP Marker 120(5):439–444. https://doi.org/10.1046/j.1439-0523.2001.00634.x

Saha P, Sarkar D, Kundu A, Majumder S, Datta SK, Datta K (2014) Karyotype analysis and chromosomal evolution in Asian species of Corchorus (Malvaceaesl). Genet Resour Crop Evol 61(6):1173–1188

Saha D, Rana RS, Chakraborty S, Datta S, Anil Kumar A, Chakraborty AK, Karmakar PG (2017) Development of a set of SSR markers for genetic polymorphism detection and interspecific hybrid jute breeding. Crop J. https://doi.org/10.1016/j.cj.2017.02.006

Sarkar D, Satya P, Mandal NA, Das D, Karmakar PG, Singh NK (2016) Jute genomics: emerging resources and tools for molecular breeding. In: Ramawat KG, Ahuja MR (eds) Fiber plants, sustainable development and biodiversity, vol 13. Springer, Switzerland. https://doi.org/10.1007/978-3-319-44570-0_9

Sarkar D, Kundu A, Das D, Chakraborty A, Mandal NA, Satya P, Karmakar PG, Kar CS, Mitra J, Singh NK (2019) Resolving population structure and genetic differentiation associated with RAD-SNP loci under selection in tossa jute (*Corchorus olitorius* L.). Mol Genet Genomics 294:479–492

Satya P, Banerjee R, Biswas C, Karan M, Ghosh S, Ali N (2014a) Genetic analysis of population structure using peroxidase gene and phenylalanine ammonia-lyase gene-based DNA markers: a case study in jute (spp.). Crop Sci 54:1609–1620. https://doi.org/10.2135/cropsci2013.08.0518

Satya P, Banerjee R, Ghosh S, Karmakar PG (2014b) Morpho-anatomical and SSR diversity in mutant gene pool of jute (Corchorus olitorius L.). Indian J Genet 74(4):478–486. https://doi.org/10.5958/0975-6906.2014.00873.6

Sengupta G, Palit P (2004) Characterization of a lignified secondary phloem fibre-deficient mutant of jute (Corchorus capsularis). Ann Bot 93:211–220. https://doi.org/10.1093/aob/mch029

Singh A, Rana MK, Singh S, Kumar S, Kumar R, Singh R (2014) CAAT box-derived polymorphism (CBDP): a novel promoter-targeted molecular marker for plants. J Plant Biochem Biotechnol 23:175–183. https://doi.org/10.1007/s13562-013-0199-5

Sinha MK, Kar CS, Ramasibramanian T, Kundu A, Mahapatra BS (2011) Corchorus. In: Kole C (ed) Wild crop relatives: genomic and breeding resources, industrial crops. Springer, Berlin, pp 29–61. https://doi.org/10.1007/978-3-642-21102-7_2

Sultana N, Khan H, Ashraf N, Sharkar MTK (2006) Construction of an interspecific linkage map of jute. Asian J Plant Sci 5:758–762

Tanmoy AM, Alam MM, Moosa MM, Ghosh A, Quarni W, Ahmed F, Zaman NR, Sharmin S, Islam MT, Islam MS, Hossain K, Ahmed R, Khan H (2015) Corchorus L. and Hibiscus L.: molecular phylogeny helps to understand their relative evolution and dispersal routes. Bioresour Commun 1:1–10

Tao A, Qi J, Li M, Fang P, Lin L, Xu J (2012) Origin and evolution of jute analysed by SRAP and ISSR methods. Sci Agric Sin 45:16–35. https://doi.org/10.3864/j.issn.0578-1752.2012.01.003

Tao A, Huang L, Wu G, Afshar RK, Qi J, Xu J, Fang P, Lin L, Zhang L, Lin P (2017) High-density genetic map construction and QTLs identification for plant height in white jute (Corchorus capsularis L.) using specific locus amplified fragment (SLAF) sequencing. BMC Genomics 18:355. https://doi.org/10.1186/s12864-017-3712-8

Topdar N, Kundu A, Sinha MK, Sarkar D, Das M, Banerjee S, Kar CS, Satya P, Balyan HS, Mahapatra BS (2013) A complete genetic linkage map and QTL analyses for bast fibre quality traits, yield and yield components in jute (*Corchorus olitorius L.*). Tsitol Genet 47(3):129–137

Yang Z, Yan A, Lu R, Dai Z, Tang Q, Cheng C et al (2017a) De novo transcriptome sequencing of two cultivated jute species under salinity stress. PLoS One 12(10):e0185863. https://doi.org/10.1371/journal.pone.0185863

Yang Z, Dai Z, Lu R, Wu B, Tang Q, Xu Y, Cheng C, Su J (2017b) Transcriptome analysis of two species of jute in response to polyethylene glycol (PEG)-induced drought stress. Sci Rep 7:16565. https://doi.org/10.1038/s41598-017-16812-5

Yang Z, Yang Y, Dai Z, Xie D, Tang Q, Xu Y, Liu C, Deng C, Chen C, Su J (2019) Construction of a high-resolution genetic map and identification of quantitative trait loci for salt tolerance in jute (*Corchous* spp.). BMC Plant Biol 19:391

Yang Z, Wu Y, Dai Z, Chen X, Wang H, Yang S, Xie D, Tang Q (2020) Comprehensive transcriptome analysis and tissue-specific profiling of gene expression in jute (*Corchorus olitorius* L.). Ind Crops Prod 146:112101

Youssef AF, Younes NA, Youssef M (2019) Genetic diversity in *Corchorus olitorius* L. revealed by morpho-physiological and molecular analyses. Mol Biol Rep 46:2933–2940

Zablu T, Kazi B, Amin M, Muhammad S, Moniruzzaman M, Mohammad SA, Islam M (2019) Nutritional composition of some jute genotypes as vegetables. Int J Veg Sci 26:1–10. https://doi.org/10.1080/19315260.2019.1658686

Zakaria ZA (2007) Free radical scavenging activity of some plants available in Malaysia. Iran J Pharm Ther 6(1):87–91

Zhang L, Yuan M, He X, Liu X, Fang P, Lin L, Tao A, Xu J, Qi J (2014) Development and universality evaluation of EST-SSR markers in jute (Corchorus spp.) from GenBank database. Acta Agron Sin 40:1213–1219. https://doi.org/10.3724/SP.J.1006.2014.01213

Zhang L, Ming R, Zhang J, Tao A, Fang P, Qi J (2015a) De novo transcriptome sequence and identification of major bast-related genes involved in cellulose biosynthesis in jute (Corchorus capsularis L.). BMC Genomics 16:1062. https://doi.org/10.1186/s12864-015-2256-z

Zhang L, Cai R, Yuan M, Tao A, Xu J, Lin L, Fang P, Qi J (2015b) Genetic diversity and DNA fingerprinting in jute (Corchorus spp.) based on SSR markers. Crop J 3:416–422. https://doi.org/10.1016/j.cj.2015.05.005
Zhang L, Li Y, Tao A, Fang P, Qi J (2015c) Development and characterization of 1,906 EST-SSR markers from unigenes in jute (Corchorus spp.). PLoS One 10:e0140861. https://doi.org/10.1371/journal.pone.0140861
Zhang L, Gao Z, Wan X, Xu Y, Zhang L, Tao A, Fang P, Qia J, Zhang L (2017) Development of novel small InDel markers in jute (Corchorus spp.). Trop Plant Biol 10:169–176. https://doi.org/10.1007/s12042-017-9193-8

# Part III
# Oil Seeds

# Conservation and Utilization of Genetic Diversity in Coconut (*Cocos nucifera* L.)

**Anitha Karun, S. V. Ramesh, M. K. Rajesh, V. Niral, R. Sudha, and K. S. Muralikrishna**

**Abstract** Genetic diversity is an important determinant of future yield security and hence is a significant factor for crop varietal improvement. Genetic diversity in coconuts is evidenced by differences in morphological characters such as plant height (tall and dwarf), nut size, nut shape and colour, leaf and leaflet length, width and colour, husk and endosperm/kernel thickness. The coconut diversity is threatened because of several reasons such as the cultivation of improved varieties and hybrids possessing a narrow genetic base, natural calamities, high incidence of pests and diseases and urbanization. Hence, there is an urgent need for the conservation of coconut diversity to avoid genetic erosion. Existing approaches to conserve coconut germplasm such as in situ and ex situ conservation, encompassing in vitro conservation techniques such as cryopreservation of zygotic embryos, pollen and embryogenic calli, in conjunction with the advancements in coconut biotechnology for sustainably conserving coconut genetic resources, will facilitate addressing of the limitations faced in the conservation of coconut genetic resources. These multi-functional approaches could enable the strengthening of strategies which would guarantee efficient conservation.

**Keywords** Arecaceae · Coconut · Biotechnology · Cryo-preservation · Genomics-assisted breeding · In vitro recalcitrance · Genomics

## 1 Introduction

Coconut is one of the principal plantation crops cultivated in humid tropical regions. It is commonly referred to as *Kalpavriksha* (translating to the 'tree of life') as all palm parts find myriad of uses. The palm supplies nutrition-rich food (the endosperm), refreshing drink (tender nut water), inflorescence sap and oil suitable for edible and industrial purposes. The fibre obtained from the palm is of immense value commercially. The shell finds use as a fuel and in various industries, and the

A. Karun (✉) · S. V. Ramesh · M. K. Rajesh · V. Niral · R. Sudha · K. S. Muralikrishna
ICAR-Central Plantation Crops Research Institute, Kasaragod, Kerala, India
e-mail: anitha.karun@icar.gov.in



P. M. Priyadarshan, S. M. Jain (eds.), *Cash Crops*,
https://doi.org/10.1007/978-3-030-74926-2_7

timber provides diverse commodities for domestic and industrial uses. Of late, coconut and its products are being touted as health enhancers with virgin coconut oil (VCO), tender coconut water and inflorescence sap gaining popularity as functional foods. Coconut is currently grown in about 12.08 million hectares, of which the chief growing regions are in the tropical Asia-Pacific, with annual production in the range of 69 million nuts (ICC 2019). Although coconut is monotypic, substantial diversity is encountered in the existing ecotypes/landraces. The various diverse forms of the coconut can be described in terms of either (1) qualitative differences: tall or dwarf, sizes, colour and shapes of the nuts and resistance to pests and disease or (2) quantitative differences: flowering precocity, number of bunches and nuts, the weight of the nut, composition of nuts, etc (Menon and Pandalai 1960; Bourdeix et al. 2010).

Genetic variability in crop species is an indispensable component for exploitation by researchers in crop improvement programmes in the development of new varieties and hybrids. Generally, it is accomplished by rigorous molecular and phenotypic characterization of existing genetic resources (Arunachalam and Rajesh 2017). However, when the germplasm size is large, a subset of germplasm called core or mini core collection, which is a representation of the genetic diversity of the species, could suffice. Coconut diversity is encountering challenges due to changes in climatic conditions, rise in sea levels, soil salinization, and the outbreak of pests/diseases. Conservation of coconut genetic diversity is imperative to food and nutritional security of masses in the wake of climate change. The urbanization and industrialization of traditional coconut areas have seriously caused tremendous genetic erosion. In this context, conservation of indigenous ecotypes/germplasm before replanting is considered of utmost importance. Therefore, systematic conservation of the native coconuts is a timely need to protect the existing populations.

This chapter provides the present status, strategem and advances in the conservation of coconut genetic diversity, in situ and ex situ collections, traditional and modern breeding techniques being followed and molecular and in vitro approaches that are being applied in the coconut-producing countries, in general, and India, in particular.

## 2 Origin

The coconuts, a member of order Arecales; family *Arecaceae* under the sub-family *Cocoidae*, which encompasses 600 species under 27 genera. The crop is a diploid having 32 chromosomes ($2n = 32$) and is the only species under the genus *Cocos*. The traditional means of propagation is via seed-derived seedlings, with the palm attaining its reproductive maturity at 3–10 years, depending on the genotype and the environment grown.

The palm is found in most tropical and subtropical areas (Chan and Elevitch 2006; Maloney 1993). Owing to its natural ability to circulate for a long distance, the dispersal of coconut palm could not be identified with certainty. The nuts or fruits generally float in oceans for months, and on reaching dry land, it can germinate. Regions around the western Pacific or eastern Indian oceans are considered to be their origin. Based on Western historical records, the first report of coconut was made in 545 AD by an Egyptian, Cosmos, who observed them in Sri Lanka and India. Similar observations were recorded by Marco Polo when he visited Indonesia in 1280. During his travel to India via the southern tip of Africa in 1498, Vasco de Gamma reported observing 'coquos' at Malindi coast of Kenya. Coconut was introduced to Puerto Rico in 1549 by Diego Corenco (McCormack 2005).

Nevertheless, the research evidences based on the early Spanish records indicate that the palm was carried to America either by explorers or through oceanic currents (McCormack 2005). Palynological studies suggest that at least the cultivated *Cocos* might have originated in India and reached the Pacific region due to natural dispersion (Maloney 1993). Another study suggests that Malay-Indonesia may be the centre of origin from where it spread to the Indian and Pacific islands. Fossil studies in Queensland, described in 1995, suggest that the coconut palm might be in existence for around 2 million years. Thus, the palm had enough time to disperse worldwide, as suggested by H.C. Harries (McCormack 2005). Identification of coconut pollen from the Cook Islands, sediments from Lake Mangaia that dated 5300 BC, Lake Atiu dated 6600 BC, Polynesian settlement dated back to 900 AD, in Raratonga, of the Southern Cook Islands suggest that coconut palm predates human settlement.

Harries (1978) described two prominent types of coconut palms, viz. *Niu Vai* and *Niu Kafa*. Nuts of Niu Vai types are spherical, possess thick flesh (endosperm) and thin husk and germinate early with origins in Indonesia and South East Asia. Such types are widely distributed due to human activities. On the other hand, Niu Kafa is characterized by an angular nut, predominant husk with thinner flesh and germinate later than Niu Vai. Though these types originated in regions around Indian and Pacific Oceans, the nuts were dispersed by floatation via oceanic currents owing to their typical nut characteristics.

Preliminary molecular analysis using markers have unearthed the genetic diversity and population structure of coconut suggesting two genetically different groups representing Pacific Ocean basin and the Indian and Atlantic Oceans (Meerow et al. 2009; Lebrun et al. 1998; Perera et al. 1999, 2003; Teulat et al. 2000).

Gunn et al. (2011), based on a comprehensive molecular analysis, have reported that *Cocos nucifera* had its origins in regions around Indian and Pacific Oceans since the palm has a recognized evolutionary occurrence in these two oceanic basins. This observation is supported by fossil data from both these two regions (Rigby 1995; Tripathi et al. 1999). Besides, Gunn et al. (2011) stated that the genetic differentiation amidst the Indian Ocean and Pacific Ocean coconut lineages supports the observation that the palm was brought into independent domestication in these two regions. In the Pacific region, the diversity in phenotypic traits lies from the Malay Peninsula to New Guinea ('Malesia'), consistent with the assertion that this region is the centre of domestication (Harries 1990). In Indian Ocean region, archaeological and archaeobotanical investigations (Wheeler et al. 1946), Proto-South Dravidian lingual features and antique Ayurvedic manuscripts (Menon and Pandalai 1958) imply that domestication of coconuts in the southern regions of the Indian subcontinent occurred around 3000 years ago. It can be concluded that this region, covering the fringes of Southern India and Lakshadweep Islands, Sri Lanka and the Maldives, is a possible centre where coconut was domesticated.

## 3 Diversity

Coconut is grown across the tropical belts, with its enormous genetic diversity explained first, at worldwide and regional levels (Ashburner et al. 1997). A compilation of 18 coconut varieties by Miquel (1855), including names in Latin and detailed descriptions, was utilized to describe coconut cultivars in Dutch settlements. FAO published a detailed list of coconut descriptors way back in 1970 based on materials gathered from various sources (Harries 1970). Satyabalan (1997) made a compilation of available details of coconut populations from various regions where coconut is grown.

The first attempt to classify coconut in a systematic manner was made by Narayana and John (1949), who classified coconut varieties based on their stature as 'talls' and 'dwarfs'. Tall palms were further classified into three: var. typica and two mutants, viz. var. androgena and var. Spicata. Typica was further classified into nine forms. Dwarfs, on the other hand, were classified as var. nana and var. javanica. This classification was later used by Liyanage (1958) to categorize accessions from Sri Lanka, based on fruit sizes and appearance, as 'typica', 'nana' and 'aurantiaca'.

The type of pollination system encountered in coconut palms was used by Fremond et al. (1966) to classify coconuts as (i) cross-pollinating or allogamous (the majority of tall accessions) and (ii) self-pollinating or autogamous (the majority of dwarf accessions).

Harries (1978) reported a practical technique to identify Niu Kafa and Niu Vai coconut varieties, based on the traits of fruits and germination characteristics of seed nuts, which permits an assessment of coconut varieties regardless of country of origin and growth conditions. Dwarf types did not find a place as they are thought to have evolved much later. Nevertheless, the involvement of the human factor, via migration and cultivation, has brought the tall and dwarf types together, and cross-pollination between these two types has resulted in intermediate forms. Fruit component traits were used by Satyabalan (1997) to categorize coconut varieties into five tall groups and three dwarf groups, the latter based on the three prominent colour types. But Niu Leka Dwarf, possessing a large size of fruits and high copra content similar to tall types, did not find a place. Multiple variants for phenotypic traits are provided in Table 1, while distinctive characters of coconut morphotypes from different countries are provided in Table 2.

**Table 1** Coconut: multiple variants for phenotypic traits

| Trait | Variants |
|---|---|
| Habit | Dwarf; intermediate forms; talls |
| Stem | Branching; polyembryony; suckering |
| Vegetative portions | Albinism; chimaera; rosette-type seedlings; the fusion of leaflets ('plicata'); forking of leaves |
| Inflorescence | Twin spadix; multiple spathes; incomplete spike suppression; secondary splitting of spikes; spikes unbranched ('spicata'); secondary spikelets; viviparous germination; bulbils; midgets; inflorescences found persistent on the stem |
| Sex expression | Either male or female ('dioecy'); sterility of male flowers; hermaphrodites; variation in the number of pistillodes, stamens, perianth |
| Fruit | Endosperm variants such as jelly-like; exhibiting fragrance; sweet forms; edible husk; variants in overall fruit size, shape and colour; variations in the colour of the husk, fibre quality and the relative proportion of husk; horned structures; double ovary; mono, bi and tetracarpic fruits |

**Table 2** Characters of coconut morphotypes from different countries (Niral and Jerard, 2017, 2018)

| Sl. No. | Variety name | Characteristics | Country of report | References |
|---|---|---|---|---|
| 1. | Makapuno | The endosperm is light curd-like and largely filled within the cavity. Makapuno is a delicacy in the Philippines. Variants have been reported in India ('Thairu Thengai') and Sri Lanka ('Dikiripol') | The Philippines, Malaysia, Sri Lanka and India | Omar (1919); Copeland (1931); Jerard et al. (2013) |
| 2. | Klapa Wangi | Endosperm emits fragrance. Similar forms: Aromatic Dwarf, Nam-hom/Aromatic Green Dwarf | Thailand, the Philippines | Omar (1919); Arunachalam and Rajesh (2008) |
| 3. | Coconino | Very dwarf comparable to micro- and nana-forms. Similar forms: mangipod in the Philippines, Bodiri in Sri Lanka, Mapraw-Puong in Thailand | The Philippines | Copeland (1931) |
| 4. | Spicata | Absence of spikelets in inflorescence keeps it unbranched; though an inflorescence bears 100 and 50 female and male flowers yields around 15 fruits per inflorescence | India | Jacob (1941); Sankaran et al. (2015) |
| 5. | San Ramon | High yielder and possessing large nuts<br>Similar forms: 'Kamandala' in Sri Lanka, 'Lupisan' in the Philippines, 'Kappadam' in India and 'Markan' in Papua New Guinea | The Philippines | Copeland (1931) |

(continued)

**Table 2** (continued)

| Sl. No. | Variety name | Characteristics | Country of report | References |
|---|---|---|---|---|
| 6. | Pugai | Dwarf in stature, having a short juvenile period | The Philippines | Copeland (1931) |
| 7. | Ran Tembili | A dwarf form with pink endosperm; commonly referred to as king coconut<br>Guelle Rose of Mauritius, pinkish mesocarp types observed in India, China, Fiji, Indonesia | South Asia | Watt (1889); Trimen (1898) |
| 8. | Male coconut | Inflorescence characterized with male flowers only (around 5000) | – | John and Narayana (1949) |
| 9. | Freaks or abnormalities | Vivipary and suckering | – | John and Narayana (1949) |
| 10. | Mangipod | Intermediate of Coconino and Pugai; possesses a brief juvenile phase | – | Copeland (1931) |
| 11. | Niu Leka | Dwarf forms; stout stem; close leaf scars; prolific bearer | Fiji | Jerard et al. (2014a) |
| 12. | King Coconut | Sweet nut water | Sri Lanka | – |
| 13. | Nudam | Thin husk and hardy | Fiji | – |
| 14. | Kaithathali | Tall palm characterized with sweet/edible immature mesocarp<br>Similar forms: 'Navasi' in Sri Lanka; 'Tamisan', 'Taban', 'Cayamis' in the Philippines; 'Kalapa Tebu' in Indonesia; 'Cay Dua Sap' in Viet Nam, 'Klapa Logo' in Malaysia | India, Sri Lanka, the Philippines, Indonesia, Vietnam, Malaysia | Patel (1938) |
| 15. | Coco Raisin | Tiny grape-like and spherical nuts<br>Similar forms: Laccadive Micro, Ayiramkachi in India | Seychelles, India | – |
| 16. | Mini Micro | Very small nuts | India | Jerard et al. (2014b) |
| 17. | Mohachao Narel | Sweet kernel | India | Samsudeen et al. (2013); Karun et al. (2017) |

## 4 Germplasm Collection and Conservation

The International Plant Genetic Resources Institute (IPGRI) has set up the International Coconut Genetic Resources Network (COGENT) in 1991 to optimize the conservation and utilization of genetic resources of coconut and encourage the research activities aimed at achieving sustainable productivity goals. COGENT was set up in the background of diminishing productivity levels of coconut plantations spread across the world. Furthermore, 96% of coconut farmers worldwide belong to smallholders class. Hence, COGENT has an overall objective of sustaining the livelihood security of the marginalized and resource-poor coconut farmers. Considering the importance of coconut genetic resources for livelihood security of the coconut communities and to advance the accessibility and to ensure the secure movement of the coconut germplasm, COGENT established five multi-site International Coconut Gene banks (ICGs), namely (1) Indonesia—Southeast and East Asia; (2) India—South Asia and the Middle East; (3) Papua New Guinea—South Pacific; (4) Côte d'Ivoire—Africa and the Indian Ocean; (5) Brazil—Latin America and the Caribbean. The COGENT has also suggested the requisite population size of tall (90) and dwarf (45) palms for effective conservation efforts in the ICGs. These ICGs constitute an International Network of ex situ collections of plant species as agreed under Article 7 of the International Undertaking on Plant Genetic Resources with the Food and Agriculture Organization (FAO) of the United Nations working as the trustee. Such ex situ collections are brought under the FAO Commission on Plant Genetic Resources for Food and Agriculture following the International Treaty for Food and Agriculture. Thus, ICG has the major mandates of conserving the coconut diversity at both national and regional levels; and to devise efforts in conserving the identified diversity at the international arena, assessing the existing diversity, performing germplasm evaluation and timely distribution of diversity information and to supply the requisite germplasm to member nations according to established protocols and also to organize research and training in these areas.

## *Complementary Conservation Strategies*

Complementary conservation approaches in coconut comprise long-term conservation of zygotic embryos and pollen, thus providing an adjunct to field gene banks.

### Embryo Culture

Despite a number of scientific efforts in standardizing the zygotic embryo culture, the successful protocol developed by ICAR-CPCRI is widely used in the culture of zygotic embryos derived from 8- to 11-month-old coconuts (Karun et al. 2004). An artificial medium was standardized, which sustains the zygotic embryo's growth into a complete plantlet under in vitro conditions. Endosperm plugs are stored either in KCl for a period of up to 2 weeks (Assy-Bah et al. 1987) or in sterile water for 2 months (Karun and Sajini 1994). This latter approach is employed in the germplasm collection expeditions of ICAR-CPCRI. Also, it was deduced that germination of the embryo was arrested in a media which is free of sucrose and activated charcoal (Assy Bah et al. 1987).

The protocol developed by ICAR-CPCRI (Karun et al. 1993) was effectively deployed in transferring exotic coconut germplasm (six accessions from Pacific Ocean regions) from the World Coconut Germplasm Centre (WCGC), the Andaman

Islands, India, to the mainland. Later, many international expeditions were conducted by ICAR-CPCRI; 4182 embryos, comprising 45 accessions from eight countries, viz. Bangladesh, Comoros, Madagascar, Maldives, Mauritius, Seychelles, Reunion and Sri Lanka, were brought in as embryos, retrieved and planted in the International Coconut Gene Bank (ICG-SAME). Genotypic and geographic origin variations were observed. The percentage of germination varied from 54 (Sri Lanka) to 82.2 (Bangladesh), and 3–4 folds of more embryos are required for success in in vitro retrieval and ex vitro establishment.

Embryo rescue is an in vitro technique that enables getting plantlets from the embryos with difficulty in germinating under natural conditions such as Makapuno (De Guzman 1970) and 'Dikiri Pol' possess jelly-like endosperm (Vidhanaarachch et al. 1998). In India, successful application of embryo rescue technique has been conducted for the regeneration of coconut accessions from the Konkan region of Maharashtra possessing sweet kernel (CPCRI 2011; Karun et al. 2017) and accession Horned Coconut from the Andaman Islands, which have poor germination owing to horn-like structures (CPCRI 2012; Jerard et al. 2014b).

## Cryopreservation

In coconut, traditional conservation approach involves growing of palms under ex situ conditions in field gene banks. However, these necessitate large-scale financial, infrastructural and human resources and vast land. Hence, the technique of cryopreservation of coconut zygotic embryos and pollen is suggested as a long-term conservation strategy either alone or concomitantly with the establishment of gene banks. Being recalcitrant, the zygotic embryos succumb to prolonged desiccation and can be conserved only for a shorter period under the optimum moisture. Pre-treatments including desiccating utilizing either silica gel or air current of the laminar flow chamber, high sucrose, use cryoprotectants such as DMSO, glycerol, sorbitol, propylene glycol or formamide, followed by encapsulation with 3% sodium alginate, are mandatory before conserving the coconut tissues in liquid nitrogen at −196 °C.

Since coconut pollen's viability does not last for many days, other for short-term conservation techniques explored, including freeze-drying. Karun et al. (2014) have developed protocols for long-term pollen cryopreservation of two cultivars, viz. West Coast Tall and Chowghat Orange Dwarf, and the possibility of setting up a pollen cryobank in coconut. Hence, along with the field gene banks, embryo and pollen cryobanks serve as an alternative or complementary genetic resources conservation measures. The salient achievements in coconut cryopreservation are presented in Table 3.

**Table 3** Salient achievements in coconut cryopreservation

| Tissue | Methodology | Features | References |
|---|---|---|---|
| Embryo | | | |
| Immature embryos (West Coast Tall) | Partially dehydrated embryos treated with cryoprotectant solution comprising of 7% DMSO and 7% sucrose in MS liquid medium, covered in sterile aluminium foil, frozen in liquid nitrogen (5 min), warm water wash (35–40 °C) followed by culturing on MS medium incorporated with growth regulators, viz. 2,4-D (0.2 mg $L^{-1}$), NAA (0.5 mg $L^{-1}$) and kinetin (0.1 mg $L^{-1}$) | Retrieved embryos exhibited a lag period of 4 months | Bajaj (1984) |
| Single embryos | Cryoprotection using DMSO and slow freezing | Embryos survived 15 months after cryo treatment | Chin et al. (1989) |
| Plumular tissue | Encapsulated tissues pre-cultured for 72–96 h in nutrient medium with 0.75 M sucrose followed by desiccation using silica gel | Callus growth observed on retrieval | Hornung et al. (2001) |
| Immature embryos from hybrid PB 121 (7–8 months after pollination) | Standard basal medium with 600 g $L^{-1}$ glucose<br>Medium with the cryoprotectants, viz. glycerol, sorbitol or polyethylene glycol (PEG) 6000 at different concentrations such as 5, 10 or 15% | Varied survival rates. Survival rate of 25 and 10% in the pre-growth treatment of 10 and 15% glycerol, and 43% survival with 10% sorbitol<br>15% glycerol pre-treatment yielded a rooted plantlet | Assy-Bah and Engelmann (1992a) |
| Mature embryos (Hybrid PB 121, Cameroon Red Dwarf, Indian Tall, Rennell Island Tall) | Simple desiccation in the laminar air current (4 h) and incubation in vitrification medium comprising of 600 g $L^{-1}$ glucose and 15% glycerol for 11–20 h | Recovery rate of 33 and 93% | Assy-Bah and Engelmann (1992b) |
| Mature embryos of West Coast Tall variety | Desiccation pre-treatments using silica gel or laminar airflow | Healthy plantlets from embryos treated for 18 h of silica gel or 24 h of laminar airflow desiccation | Karun et al. (2005) |
| Plumular tissues from mature zygotic embryos of Malayan Yellow Dwarf | Pre-culturing in standard medium (of sucrose for 3 days) and encapsulation (3% Na–alginate and 0.15 M sucrose or 0.1 M calcium chloride containing 0.15 M sucrose) followed by drying on sterile filter paper (40 g Silica gel) | Dehydration for 14 and 16 h, yielded 21%, and 20% tissue regrowth<br>Pregrowth of encapsulated plumule beads in 1 M sucrose for 16 h yielded 20% leafy shoot production | Nan et al. (2008) |

(continued)

**Table 3** (continued)

| Tissue | Methodology | Features | References |
|---|---|---|---|
| Indonesian cultivars | Rapid dehydration (autoclavable fan and activated silica gel-680 g), quick cooling, speedy rewarming and recovery in vitro and acclimatization | 20–40% seedlings resurrection | Sisunandar et al. (2010a) |
| | Rapid dehydration of embryos (activated silica gel for 8 h), reducing the moisture content to 20% | Recovered plantlets showed no morphological, karyotypic and genetic variations | Sisunandar et al. (2010b) |
| Coconut plumules | ABA supplementation in the encapsulation dehydration process | Incorporation of 40 μM of ABA in the sucrose pre-treatment yielded 84% survival and 39% recovery | Bandupriya et al. (2007) |
| Mature zygotic embryos for plumule excision | Endosperm plug containing embryos kept at the refrigerator for 10 days, embryos in agar medium or KCl solution (stored at 27 °C in the dark for 10 days) | Storage of mature embryo in KCl and solidified agar medium resulted in a significantly higher recovery rate<br>Pre-treatment of frozen plumules with 1.0 M sucrose enhanced the recovery rate (40%) | Bandupriya et al. (2010) |
| Effect of pre-culture and vitrification | Pre-culture of mature zygotic embryos in Y3 medium supplemented with 0.6 M sucrose, followed by vitrification in PVS3 solution for 16 h in an orbital shaker and cooling rapidly in liquid nitrogen and faster rewarming in 40 °C water for 2 min and unloading in 1.2 M sucrose liquid medium for 1.5 h and inoculating in recovery medium (Fig. 1) | Survival rates of 70–80% and recovery of 20–25% with a normal shoot and root growth | Sajini et al. (2011) |
| Pollen | | | |
| Coconut pollen | Storage in liquid $N_2$ (24 h) | No alterations under in vitro germination | Karun and Sajini (2010) |
| West Coast Tall and Chowghat Orange Dwarf cultivars | Long-term cryopreservation of coconut pollen (4 years) (Fig. 2) | Pollen retained their viability and fertility and normal seed set<br>One hundred per cent germination of embryos | Karun et al. (2014) |

**Fig. 1** Cryopreservation, retrieval, regeneration and field establishment of plantlets from mature zygotic embryos of coconut following PVS3 vitrification method. (**a**) Scooping of zygotic embryo along with endosperm plug, (**b**) mature zygotic embryo, (**c**) pregrowing of embryo, (**d**) vitrification in PVS3, (**e**) vitrified embryo in cryovial, (**f**) immersion of cryovial in liquid nitrogen cryotank, (**g**) unloading of cryopreserved embryo post thawing, (**h**, **i**) regeneration, (**j**, **k**) hardening, (**l**) field establishment

**Fig. 2** Collection, cryopreservation, in vitro viability assay and fertility studies of pollen in coconut. (**a**) spikelets from WCT and COD coconut cultivars, (**b**) sieve used for the extraction of pollen, (**c**) desiccation of male flowers at 40 °C in an incubator, (**d**) collection of pollen from desiccated male flowers, (**e**) immersion of cryovials filled with desiccated pollen in to liquid nitrogen cryotank, (**f**) in vitro germination assay of desiccated pollen, (**g**) in vitro germination assay of cryopreserved pollen, (**h**) field fertility study of cryopreserved COD pollen in WCT palm, (**i**) field fertility study of cryopreserved WCT pollen in COD palm

## *Tissue Culture*

Rapid multiplication of elite coconut genotypes characterized with yield attributes, tolerance to various biotic/abiotic stresses and speciality characters is pertinent. Application of various in vitro multiplication techniques to achieve this has met with varying degrees of success in coconut. After the success of tissue culture of oil palm, coconut tissue culture also received a greater fillip. It was also presumed that the accomplishments in oil palm could be recreated in coconut. Nevertheless, it was observed that coconut was highly recalcitrant to tissue culture-based manipulations. The culture media designed and developed for oil palm could not replicate similar success in coconut. The in vitro recalcitrance in coconut irrespective of explants, genotypes or culture conditions could be attributed to various factors such as maturity stage of explants, culture conditions-adsorption of nutrients and phytohormones or growth regulators supplied, generation of compact calli, meagre regeneration potential, the poor performance of the regenerated plantlets and slow growth response (Fernando et al. 2010). Hence, optimization and standardization of tissue culture media, stage or age and kind of explants, type and concentrations of plant growth regulators, sub-culturing intervals, other additives in the media are very pertinent to achieve a reproducible tissue culture protocol. Coconut explants contain secondary metabolites, especially plant phenolics, which cause browning. To circumvent this problem, the use of activated charcoal in the culture medium was suggested. Addition of activated charcoal, however, is fraught with other concerns. It hampers the potential of the supplemented plant growth regulators and other media supplements, which leads to uncertainties in the composition of growth media used for culture.

Selection and utilization of appropriate explants determine the success of tissue culture. Initially, Eeuwens (1976) had standardized Y3 media and successfully induced callus using various explants such as stem, leaf and inflorescence. It has spurred the coconut in vitro culturing techniques in laboratories around the world. However, the lack of friable calli formation is a serious impediment coupled with the undesirable formation of abnormal tissues. Immature inflorescence harbours numerous meristematic points and hence could yield a potential source explant. Branton and Blake (1984) successfully obtained calli (termed as 'colloids') in culture media using inflorescence rachillae as explants. On the other hand, Verdeil et al. (1994), working with immature inflorescence explants, successfully obtained somatic embryos, which possessed both root and shoot poles ('bipolar'). It was observed that the basal media composition, besides the size of the immature inflorescence explants, is crucial for success. Furthermore, altering the media composition and prolonged exposure of explants to auxin–cytokinin combination could generate vegetative primordial from coconut's floral tissues. However, the frequency of conversion is too little (Raju 2006). Refining this protocol, Shareefa et al. (2019) could successfully obtain plantlets from immature explants of 2–12 cm sizes of WCT cultivar (Fig. 3).

**Fig. 3** Immature inflorescence culture in coconut. (**a**) inflorescence explants of different length, (**b**) rachillae used for the inoculation, (**c**) inoculation of rachilla bits in medium, (**d–g**) incubation of cultures in dark condition and initiation of shoot-like structures, (**h–l**) growth and development of plantlet in light, (**m–p**) hardening. (Photo credit: Dr. Shareefa M)

Anther culture in coconut has been reported by Iyer (1981) and several other workers. Iyer (1981) demonstrated the formation of embryoids from the coconut anthers under in vitro conditions and defined when anthers and microspores are suitable for in vitro manipulations. Somatic embryo-like organs having typical root and shoot apices were obtained via anther culture (Perera et al. 2008). Media composition for successful anther culture was further refined by supplementation with sucrose (9%) (Perera et al. 2009), and the indispensable role of activated charcoal (0.1%) in media aimed at lessening the callus necrosis was reported. Successful induction of calli from microspores could be attained using a culture media comprising 100 μM of 2,4-D, 9 μM of TDZ and 100 μM NAA and incubating under dark conditions. Similarly, the induction of the somatic embryo maturation occurs in a culture media containing ABA (5 μM) and $AgNO_3$ (10 μM). Enhanced germination of somatic embryos was achieved when callus was transferred to a hormone-free medium followed by growth in a media supplemented with 5 μM BAP. Further enhancement of somatic embryo germination frequency could be achieved by supplementing the media with 0.35 μM gibberellic acid and 5 μM BAP (Perera et al. 2009). One half of the anther-culture-derived plantlets were haploids, suggesting to obtain double haploids for further genetic and molecular analysis.

Ovary culture using unfertilized ovary as an explant with relatively good regeneration potential has been suggested. Further, callogenesis efficiency could be enhanced in CRI72 medium infused with 2,4-D (Perera et al. 2008). Perera et al. (2009) reported high callusing and regeneration frequency in indigenous tall accessions from Sri Lanka. Embryos of age 6–7 months post-anthesis yielded callus in a culture medium supplemented with 2,4-D. Over one-half of them have grown into globular embryos with 22% of germination in these cultures.

Considering the consistent results and promising outcomes it renders, plumular explants focus on coconut tissue culture experiments (Perez-Nunez et al. 2006; Rajesh et al. 2005, 2014a). It is crucial to devise suitable in vitro protocols that achieve rapid multiplication of the embryogenic calli from plumular explants to propagate elite coconut genotypes. Rapid multiplication of embryogenic calli was attained by repeated sub-culturing technique (Perez-Nunez et al. 2006). Rajesh et al. (2005) devised an in vitro protocol using the West Coast Tall variety's plumular explants. Callus induction was attained using Y3 media supplemented with either an auxin alone (2, 4-D; 74.6 μM) or in combination with a cytokinin (TDZ; 4.54 μM). This study has proven that greater frequencies of embryogenic calli, somatic embryoids and meristemoids could be achieved in the Y3 medium by manipulating spermine or BA concentrations. Though histological evidence showed promising and regenerated plantlets were established in the field conditions, multiplication to achieve commercial-scale remains elusive since a large proportion of somatic embryos could not be converted into plantlets (Fig 4).

**Fig. 4** Callus initiation, somatic embryogenesis and plantlet regeneration from plumular explants of coconut. (**a**, **b**) zygotic embryo extracted from coconut, (**c**) preconditioning, (**d**) slicing of the preconditioned embryo, (**e–g**) callus initiation, (**h–m**) plantlet regeneration

# 5 Crop Improvement

## *Traditional Breeding Approaches*

### Selection

Evaluating coconut accessions for their performance under diverse ecological situations is a potential approach to obtain ecotypes suited to different areas. In India, evaluation and selection of elite coconut accessions conserved at ICAR-CPCRI, coordinating centres of All India Coordinated Research Project on Palms (AICRPP) and State Agricultural Universities (SAU), have resulted in the development and release of 30 improved varieties (Fig. 5, Table 4).

**Fig. 5** Coconut varieties and hybrids developed by ICAR-CPCRI. (**a**) Kalpasree, an early flowering dwarf accession, (**b**) Kalpa Jyothi a dwarf variety, (**c**) COD, a dwarf tender nut variety, (**d**) Kalpa Haritha, (**e**) Kalpa Ratna, a multipurpose variety, (**f**) Kera Sankara, a T × D hybrid variety, (**g**) Kalpa Sreshta, a D × T hybrid variety

**Table 4** Development of improved coconut varieties for cultivation following selection

| Variety | Traits | Released by |
|---|---|---|
| Chandra Kalpa | Besides drought-tolerance trait, high oil—72% | ICAR-CPCRI |
| Kalpa Mitra | Nut and oil yield along with drought-tolerance trait | ICAR-CPCRI |
| Kalpa Dhenu | High oil and nut yield and tolerant to drought | ICAR-CPCRI |
| Kalpatharu | Besides high yield, tolerant to drought and amenable for ball copra production | AICRPP, UHS Bagalkot, Karnataka |
| Pratap | High yield | Dr. BSKKV, Maharashtra |
| Kamarupa | High yield | AAU, Assam |
| Aliyarnagar Tall 1—ALR (CN) 1 | High yield | TNAU |
| Kera Bastar | High yield | AICRPP, IGAU, Chhattisgarh |
| Kera Keralam | High yield | AICRPP |
| Aliyarnagar Tall 2—ALR (CN) 2 | High yield | TNAU |
| VPM-3 | High yield, drought tolerant | TNAU |
| Kera Sagara | High yield | KAU |
| Double century | High yield | APAU |
| Kera Chandra | High yield | ICAR-CPCRI |
| Kalpa Pratibha | High nut, oil yield, tender nut, drought tolerant | ICAR-CPCRI |
| Kalpa Haritha | Dual-purpose variety amenable for both the copra and tender nut | ICAR-CPCRI |
| Kalyani Coconut | High yield | BCKV, West Bengal |
| Gautami Ganga | Dwarf, green fruits | ANGRAU, Andhra Pradesh |
| Kera Madhura | Semi-tall, dual purpose for copra and tender nut | KAU, Kerala |
| CARI-C1 (Annapurna) | High copra content, tender nut purpose, green colour fruit | CARI, Andamans |
| Chowghat Orange Dwarf | Dwarf, orange colour fruit; tender nut purpose | ICAR-CPCRI |
| Kalpa Jyothi | Dwarf, yellow colour fruit; tender nut purpose | ICAR-CPCRI |
| Kalpa Surya | Dwarf, orange colour fruit; tender nut purpose | ICAR-CPCRI |
| Kalparaksha | Suitable for root (wilt) disease-affected areas, green nuts with high oil yield and suited for tender nut purpose | ICAR-CPCRI |
| Kalpasree | High yield in the root (wilt) disease-affected areas | ICAR-CPCRI |
| CARI-C2 (Surya) | Ornamental palm, orange colour fruit | ICAR-CIARI |
| CARI-C3 (Omkar) | Ornamental purpose, yellow colour fruit | ICAR-CIARI |
| CARI-C4 (Chandan) | Ornamental purpose, orange colour fruit | ICAR-CIARI |
| Kalpa Shatabdi | Large nuts suitable for oil and tender nut purposes | ICAR-CPCRI |
| Kalpa Ratna | Multipurpose variety suitable for oil, tender nut purposes and inflorescence sap | ICAR-CPCRI |

## Hybridization

Though hybridization in coconut was first explored in Fiji (Marechal 1928), Patel (1937) in India was the first to exploit hybrid vigour in coconut in a hybridization programme involving three intra-varietal and one inter-varietal cross at the Coconut Research Station, Nileshwar, in the year 1932. This technological breakthrough has led to evaluating over 150 hybrid combinations involving indigenous and exotic parents over the years at the ICAR-CPCRI and under the All India Coordinated Research Project on Palms. These research efforts have yielded 21 coconut hybrids for cultivation, and their yield potential varies from 95 to 167 nuts/palm/year and 13.20–35.90 kg copra palm$^{-1}$ year$^{-1}$ (Niral et al. 2014) (Fig. 5 and Table 5).

## Breeding for Specific Traits

In coconut, the main focus of breeding programmes has been developing speciality varieties that can withstand drought, pest and disease attacks, besides yield improvement (Jerard et al. 2016). Farias Neto et al. (2003) utilized statistical methods and described high variability in yield and economical traits. Further, it was also shown that these characters exhibit a high heritability and genetic advance. Accordingly, these traits were considered important while exercising breeding selection processes to improve copra yield (Ganesamurthy et al. 2002a). Later, fruit component trait studies, by Niral et al. (2009), revealed a higher variability for fruit weight and endosperm content; in contrast, lesser variability was observed for the characters such as the thickness of the endosperm and oil content. These studies thus unravelled that yield of oil (ha$^{-1}$). The yield of copra (ha$^{-1}$) is under the influence of nut size and endosperm content and not the percentage of oil content or thickness of endosperm as such. The accessions based on copra content were classified into high (>300 g; San Ramon Tall, Markham Valley Tall and Malayan Tall) and low [>100 g; Surinam Brown Dwarf, Chowghat Green Dwarf, Pattukottai Green Dwarf, Laccadive Micro Tall and Nu Fella Tall] copra content categories. Overall dwarf cultivars showed low copra and oil content. Furthermore, correlation studies of the fruit component traits revealed a need to select accessions showing lesser husk percentage when the breeding objective is to develop high endosperm/oil yielder. Moreover, the yield of copra (ha$^{-1}$) and yield of oil (ha$^{-1}$) exhibited positive correlation, suggesting that these two characters are influenced by the fruit component traits and the nut yield. Hence, it was suggested that fruit weight, percentage of the husk and the nut yield are the major determinants of coconut oil productivity. The selection of accessions having these traits is a research priority.

Copra is major economic produce meant to extract oil or to be consumed directly as in ball copra. The latter is considered the best quality copra and fetches a premium trade price. Niral et al. (2010) assessed 17 coconut accessions for ball copra production showed that the recovery was highest in tall accessions.

**Table 5** List of coconut hybrids for commercial cultivation in India

| Hybrid variety | The source population of parents | Important traits | Agency responsible for the release |
|---|---|---|---|
| Chandra Sankara | COD × WCT | High yield | ICAR-CPCRI |
| Kera Sankara | WCT × COD | High yield, drought tolerant | ICAR-CPCRI |
| Chandra Laksha | LCT × COD | High yield, drought tolerant | ICAR-CPCRI |
| Kalpa Samrudhi | MYD × WCT | Drought-tolerance trait, dual-purpose genotype, with high nutrient use efficiency | ICAR-CPCRI |
| Kalpa Sankara | CGD × WCT | Exhibits tolerance to root (wilt) disease, high yield | ICAR-CPCRI |
| Kalpa Sreshta | MYD × TPT | Dual-purpose variety, high yield | ICAR-CPCRI |
| Laksha Ganga | LCT × GBGD | High yield | Kerala Agricultural University (KAU) |
| Ananda Ganga | ADOT × GBGD | High yield | KAU |
| Kera Ganga | WCT × GBGD | High yield | KAU |
| Kera Sree | WCT × MYD | High yield | KAU |
| Kera Sowbhagya | WCT × SSAT | High yield | KAU |
| VHC-1 | ECT × MGD | High yield | Tamil Nadu Agricultural University (TNAU) |
| VHC-2 | ECT × MYD | High yield | TNAU |
| VHC-3 | ECT × MOD | High yield | TNAU |
| Godavari Ganga | ECT × GBGD | High yield | Acharya N. G. Ranga Agricultural University (ANGRAU) |
| Konkan Bhatye coconut hybrid 1 | GBGD × ECT | High yield | Dr. Balasaheb Sawant Konkan Krishi Vidyapeeth (Dr. BSKKV) |
| Kalpa Ganga | GBGD × FJT | High yield, suitable for ball copra production | University of Horticultural Sciences (UHS), Bagalkot |
| Vasista Ganga | GBGD × PHOT | High yield | Dr. Y.S.R. Horticultural University (Dr. YSRHU) |
| Ananta Ganga | GBGD × LCT | High yield | Dr. YSRHU |
| VHC-4 | LCT × CCNT | High yield | TNAU |

[a]Yield estimated at 7.5 m × 7.5 m spacing with a population of 177 palms $ha^{-1}$

## *Drought Tolerance and Climate Resilience Studies*

An average of 15 cm monthly rainfall is indispensable for the optimal growth of coconut palm and reasonable nut yield. Drought spells affect the current nut productivity; they also exert its physiological and biochemical effects until subsequent 3 years to adversely affect the economic yield. Hence, the development of drought-tolerant coconut genotypes or varieties that can withstand moisture stress is imperative.

Earlier reports have suggested the plausibility of screening the genotypes and identifying anatomical, physiological and biochemical traits that ultimately confer drought tolerance (Chempakam et al. 1993; Kasturi Bai et al. 1996; Naresh Kumar et al. 2000; Rajagopal et al. 1991, 2007). The parameters correlated with the drought-tolerance trait are the considerable accumulation of epicuticular leaf wax, reduced frequency of stomata and water potential besides the upregulated activities of enzymes like acid phosphatase and glutamate oxaloacetate transaminase. Based on these criteria, genotypes such as West Coast Tall, Java Tall, Federated Malay States Tall, Andaman Giant Tall, Fiji Tall, Chandra Laksha (LCT × COD) and Laksha Ganga (LCT × GBGD) were found to be tolerant to drought (Rajagopal et al. 1988, 1990). Genetic dissection of drought tolerance traits using line × tester analysis [two dwarf lines (CGD, MYD) and four testers (ECT, PHOT, LCT, FMST)] revealed that the genotypes exhibited differential drought sensitivity for indicative traits, namely rate of transpiration, photosynthetic rates (Pn), water potential and lipid peroxidation status. Further gene action studies also deduced the interplay of both the additive and non-additive actions (Rajagopal et al. 2007). Specific combining ability (SCA) was high for the trait transpiration rate, suggesting this character's heterotic nature. It was also suggested to utilize the trait Pn in heterosis breeding since it exhibited non-additive gene action. Thus, this study formed a basis for identifying the type of gene action that underlies the drought-sensitive traits that could be utilized selectively for breeding programmes for drought tolerance. Furthermore, the genotypes identified as a result of these studies are currently being exploited in crop improvement programmes of ICAR-CPCRI to develop drought-tolerant varieties.

In the background of climate change, the predictions in the rise of GHG (380 to 500–970 ppm) would cause a rise in global temperature of 1–5.5 °C for the year 2100 (IPCC 2007). Coconut is a C3 crop; hence, it is expected to be benefited due to the increase in atmospheric $CO_2$ concentrations. The Infocrop coconut model predicts that coconut yield levels would increase by 10, 16 and 36% over present-day yields during 2020, 2050 and 2080, respectively (Naresh Kumar et al. 2008). Most Southern states in India would witness a yield increase due to climate change.

In contrast, Gujarat, Odisha, Telangana and Andhra Pradesh would experience a significant decline in coconut yields. Owing to the projected increase in atmospheric temperature due to climate change, it is pertinent to investigate the effect of elevated temperature and its interaction effect with drought stress in coconut. On the other hand, the enhanced biomass production due to elevated $CO_2$ concentration is expected to cause a huge demand for plant nutrients (Cavagnaro et al. 2011). The effects of the interaction of elevated $CO_2$ concentrations and increased temperature with drought and nutrient utilization in coconut uncovered that coconut showed high biomass production under elevated $CO_2$ (Hebbar et al. 2013). The physiological parameters such as transpiration and stomatal conductance in seedlings under elevated $CO_2$ concentrations were low with no compromise in the photosynthetic processes, leading to a significant increase in WUE of coconut seedlings. In contrast to annual crops (Hebbar et al. 1994), coconut expressed low WUE under drought and elevated temperature conditions. Even under elevated $CO_2$ conditions, yield levels of coconut subjected to drought was low since a considerable quantum of energy is being exhausted in non-photochemical quenching (qN).

## *Cold Tolerance*

Coconut being a tropical crop is highly susceptible to low-temperature stresses; hence, there is a need to develop cold-tolerant lines. It was observed that temperatures below 13 °C affect flowering and fruit set. Also, spear leaf damage, leaf drying, wrinkled endosperms/kernels are commonly observed due to cold injury (Mao 1986). Accessions such as the Chinese Hainan Tall and Indian Kamrup Tall have been recorded to possess tolerance to cold conditions (Chowdhury et al. 2001; Mao and Lai 1993).

## *Insect Resistance*

Among the insect pests that inflict severe damage in coconut palms, red palm weevil and rhinoceros beetle are considered major threats. Currently, there are no specific or designated breeding programmes aimed at developing insect-resistant cultivars. However, initial results of screening of accessions against *Nephantis serinopa* Meyr. (leaf-eating caterpillar) (Kapadia 1981) and *Oryctes rhinoceros* Linn. (rhinoceros beetle) (Sumangala Nambiar 1991) suggested there exist variations among cultivars concerning susceptibility.

### Eriophyid Mite (*Aceria guerreronis* Keifer)

Eriophyid mite (*Aceria guerreronis* Keifer) has a great bearing on the nut yield and quality of the nuts (Nair 2000). Various fruit characteristic features, viz. colour and shape of the fruit, tightness of tepals around the nut, space between the rim of the fruit and aestivation of tepal have been associated with the resistance to coconut mites (Howard and Rodriguez 1991; Julia and Mariau 1979; Moore and Alexander 1990, 1990). Application of digital phenotyping of tender fruits revealed that colour of the inner tepal and the tightness of the younger tepal (3 months old) are crucial morphological traits linked with the mite infestation (Arunachalam et al. 2013). Accordingly, tolerance to mites was recorded in an indigenous accession from Cambodia possessing tight tepals. Studies at ICAR-CPCRI revealed that nuts of Kulasekharam Green Dwarf (KGD) and Chowghat Orange Dwarf (COD) attracted a little mite infestation as compared to those of Laccadive Micro Tall as well as Straits Settlement Apricot (Nair et al. 2009). Accordingly, a tall selection of KGD named as Kalpa Haritha showed decreased infestation of mites compared to the serious infestation in other genotypes in the vicinity (Niral et al. 2014). While Kalpa Haritha recorded the least eriophyid mite incidence, relatively lesser mite incidence was seen in a few other accessions, viz. Navasi Tall, Gangapani Tall, Jamaica Tall and East Coast Tall (ICAR-CPCRI 2019).

### Rhinoceros Beetle *(Oryctes rhinoceros)*

Earlier studies indicated that coconut hybrids, with COD as a pollen parent, were found to have a high susceptibility to rhinoceros beetle infestation (Nambiar 1988). In India, screening trials indicated that West Coast Tall (WCT), East Coast Tall (ECT) and hybrid combinations of Laccadive Ordinary Tall × Cochin China Tall and Gangabondam Green Dwarf × East Coast Tall showed significantly lesser damage. Insects' damage was serious whenever a hybrid involved dwarf genotypes as parents (Muthiah and Bhaskaran 2000).

### Red Palm Weevil *(Rhynchophorus ferrugineus)*

An initial field survey of coconut accessions against the red palm weevil infestation divulged the susceptibility of most of the genotypes studied (Sadakathulla and Ramachandran 1993). Similarly, ovipositional preference of RPW revealed that CGD, COD and Benaulium invited the highest cumulative egg laying (Faleiro and Rangnekar 2001), suggesting their susceptible nature.

### Coconut Scale Insect

Galvez et al. (2018) screened Tambulilid, a Javanica coconut variety and natural outcross Laguna Tall × Coco Nino for host resistance. They reported plant resistance against coconut scale insect (CSI) in a limited host–insect interaction assay. To validate and further characterize the resistance mechanism, the associated leaf glandular trichome was evaluated among the Tambulilid, Coco Nino and Laguna Tall coconut populations conserved at the Philippine Coconut Authority, Albay Research Center. Among these coconut genotypes, Coco Nino was observed to have the highest average number of trichomes per leaflet area (194.4 trichomes per sq. cm) and significantly different from the susceptible control, Laguna Tall by 81.6

trichomes per sq. cm. On the other hand, Tambulilid shows no significant difference in trichome density (104.9 trichomes sq. $cm^{-1}$) vs. Laguna Tall (113.2 trichomes sq. $cm^{-1}$).

## *Disease Resistance*

Coconut plantations are affected by the various diseases such as *Phytophthora* bud rot, lethal yellowing and root (wilt) disease, stem bleeding and Ganoderma wilt, which seriously undermine the production potential of the crop.

### Root (Wilt) Disease

Absence of efficient plant protection measures against root (wilt) disease warrants the development of resistant cultivars. Varghese (1934) initiated initial studies to screen or identify coconut genotypes that show resistance/tolerance to root (wilt) disease. Even though most of the cultivars or hybrids recorded susceptibility to the disease (Menon et al. 1981), CGD exhibits over 90% field tolerance. Also, WCT and CGD palms free from the disease were located in the disease 'hot spots' (Iyer et al. 1979). Later studies on the root (wilt) endemic regions of Kerala, India identified disease-free COD palms. Screening of exotic genotypes for root (wilt) disease yielded only susceptible reaction even though genotypic variations in disease intensity was documented (Mathai et al. 1991). As a follow-up, inter se and selfed nuts of exotic germplasm from the South Pacific Ocean region were screened for disease resistance/tolerance in the 'hot spot' areas (Jacob and Rawther 1991).

Various genotypic cross combinations involving WCT × WCT, tall × dwarfs (CGD × WCT, WCT × CGD, COD × WCT and WCT × COD), and the selfed individuals of WCT, CGD and COD were performed in a quest to breed for tolerant varieties. To generate a collection of palms tolerant to root (wilt) disease, pollination was effected using the mixed pollen from the healthy palms. Around 2455 seedlings have been screened at the root (wilt) disease hot spots to identify any genetic resources exhibiting considerable resistance. A major breakthrough in the disease breeding of coconut was achieved when the variety Kalparaksha was recommended for release for cultivation in root (wilt) disease tracts by ICAR-CPCRI (Nair et al. 2009). The variety is a selection of a cultivar Malayan Green Dwarf that exhibits characteristics of the root (wilt) disease resistance, along with comparably high nut yield and sweet tender nut water (Nair et al. 2009).

Furthermore, screening trials involving local and exotic accessions divulged Chowghat Green Dwarf (CGD) to be the promising donor of the resistance-conferring gene. It was further corroborated by the natural survey (Nair et al. 2006). It has led to the development of a variety 'Kalpasree' for root (wilt) disease affected tracts in 2009. Furthermore, hybrid vigour of D × T cross (CGD × WCT) was exploited to develop a hybrid variety Kalpa Sankara, with not only disease tolerance but also high yield potential and early bearing (Nair et al. 2006).

Towards the application of molecular breeding techniques to root (wilt) disease resistance breeding, Devakumar et al. (2011) investigated the apparently disease-free and susceptible palms of WCT selected from various Southern Kerala regions using SSR-based molecular markers to decipher their population structure. This study unravelled the two-population structure of resistant WCT palms and suggested to utilize these two diverse populations as parents for the development of disease-resistant/tolerant progenies.

## Lethal Yellowing

Lethal Yellowing disease (LYD), a phytoplasmal disease, has rapidly destroyed huge tracts of coconut plantations in many countries in West and East Africa. The disease was first observed in the late nineteenth century in Cuba, Jamaica and Cayman Islands (Arellano and Oropeza 1995). Since the disease is known as many species of *Arecaceae*, eradication of the disease is a research priority. Appropriate disease management strategies are not available; hence, the development of cultivars resistant to the disease or planting of genotypes that show field resistance is the suitable disease management intervention strategies available (Dery and Philippe 1995). Hence, Maypan hybrids [Malayan Yellow Dwarf (MYD) × Panama Tall (PNT)] were planted in the disease afflicted areas in the 1980s (Been 1981). Quite interestingly, the genotypes from South East Asia showed some level of tolerance (<35% infection), suggesting the exploration of these cultivars for *R* genes.

Among the genotypes screened for LYD resistance or tolerance in Ghana, it was found that only two cultivars, viz. Vanuatu Tall and the Sri Lanka Green Dwarf, were relatively disease-free (Mariau et al. 1996). However, Vanuatu Tall was highly susceptible to the disease in Jamaica, but MYD showed promise due to the very small proportion of diseased palms. In Tanzania too, dwarf cultivars, viz. Cameroon Red Dwarf, Equatorial Guinean Dwarf and Brazilian Green Dwarf, were relatively free from the disease compared to talls. Studies in Mexico by Zizumbo Villarreal et al. (2002) revealed the high susceptibility of Atlantic Tall compared to the low susceptibility of Pacific Talls. In this context, it was suggested to utilize a broad range of partially resistant cultivars, and the genotypes of Southeast Asia and Mexico are promising to date (Zizumbo et al. 1999).

Niu Leka Dwarf exhibited variability in resistance to LYD in Florida (Flores 2008). Generally, LYD recovery occurs in genotypes involving Malayan Dwarfs or their hybrids with tall cultivars (Sherman and Maramorosch 1977). Some of the resistance sources are Maypan hybrid (90%), Malayan Dwarf (96%), (Eden-Green 1997), Panama Tall, offspring of Malayan Dwarf × Fiji Dwarf and other Indian and Sri Lankan dwarf genotypes (Been 1981).

Initial investigations on LYD incidence in Africa carried out by Sangare et al. (1992) divulged that both the morphoforms (dwarfs and tall varieties) were found to be resistant. Nevertheless, widely grown West African Tall and its hybrids were found to be highly susceptible. Around 81% of yield losses were reported, and Talls were severely inflicted by the disease (Mpunami et al. 1992). Excluding East African Tall from Tanga all the local genotypes were found to be susceptible to the disease. Hence, the hybrids between Pemba Red Dwarf and East African Tall showed some degree of resistance. Also, the resistance/susceptibility reactions of the genotypes varied with the geographic conditions (Schuiling et al. 1992), temperature, presence of insect vectors or unfavourable growth conditions (Harries 1998), or variations in the phytoplasmal strains (Eden-Green 1997).

Considering the resistant sources of coconut accessions from Malaysia and South East Asia, hybrids of parents originating from these regions hold promise for developing hybrids possessing LYD resistance (Harries 2001). Hence, it was planned to combine the resistant dwarf forms with the tall landraces in Mexico and the Caribbean regions. For instance, the Donají hybrid (resistant to LYD), was derived from a cross involving Malayo Enano Amarillo cv. Acapulco and a landrace Alto Pacífico cv. Escondido (Serrano et al. 2011).

Analysis of coconut varieties under LYD endemic areas of Nigeria revealed that dwarf varieties (MGD, MYD and MOD) were found to be resistant in comparison to WAT (West African Tall) (Odewale et al. 2013).

## Phytophthoral Diseases

*Phytophthora* spp. are the causative agents of major diseases of palms cultivated worldwide. Among the disease's various symptoms, bud rot or heart rot is one of the most widespread infection features. Coconut plantations in South Asia are infected by *P. palmivora,* whereas palms in Marc Delorme Research Station, Côte d'Ivoire, and Hawaii exhibit the infection due to *P. katsurae.* The disease caused by the latter is expressed as premature nut fall and severe reduction in yields up to 80% although it does not lead to palm mortality. The biochemical or molecular reasons underlying the differential behaviour of palms regarding the disease-induced mortality remain unexplored.

## Stem Bleeding

Caused by the fungi *Ceratocystis paradoxa*, characteristic symptoms of stem bleeding disease include bleeding from the stem during initial stages, leading to wilting of palms and ultimately the drying of leaves. Disease severity index was negatively correlated with economic yield levels. Six hybrids derived from Gangabondam (GB) were evaluated for resistance against stem bleeding disease (Radhakrishnan and Balakrishnan 1991). The least infected genotype was found in a cross involving Cochin China Tall × Gangabondam Green Dwarf (Radhakrishnan and Balakrishnan 1991). Similarly, Ramanujam et al. (1998) used the detached petiole inoculation method to screen 26 coconut genotypes against the disease and found that none of the cultivars was resistant. However, the degree of lesion size was varying in the genotypes with the least lesion size in Banawali Green Round and the maximum in Malayan Green Dwarf.

### Leaf Spot and Leaf Blight

Among the minor diseases of coconut, leaf spot (causal agent *Bipolaris incurvata*) and leaf blight (caused by *Pestalotia palmarum*) have attained greater importance. During 2000–2002, 14,000 palms were infected in Brazil due to *B. incurvata* caused leaf spot. Three juvenile hybrids, namely PB 121, PB 132 and PB 123, were evaluated for resistance against leaf spot disease. They revealed that PB 121 was more promising than the other two, which exhibited more susceptibility (Gomez-Navarro et al. 2009). Similarly, the search for cultivars resistant to leaf blight by Ray and Kaiser (1990) resulted in identifying 11 genotypes. The search for resistant sources in Kerala State, India, revealed that none of the varieties was resistant; however, MYD and CGD showed relatively low disease incidences of 4.44 and 6.66, respectively (Govindan et al. 1991). Evaluation of eight coconut genotypes in coastal Odisha, India against grey leaf spot disease identified Tiptur tall and Chowghat Orange Dwarf as resistant and susceptible sources, respectively (Ghose et al. 2006). Further, the disease occurrence rate was affected by weather factors such as high rainfall and drop in temperature (Ghose et al. 2006).

Screening of 28 coconut accessions and hybrids, for resistance against grey leaf spot or leaf blight (*Pestalotiopsis palmarum*), reported that none of them possessed resistance to the disease (Suriachandraselvam et al. 2000), even though the disease severity, measured in terms of leaf spot intensity, varied from 23.7 to 57% among the accessions/hybrids, with hybrids being more susceptible. Results of screening and involving coconut hybrids, introduced into Brazil and screened for resistance against *Botryodiplodia theobromae*, revealed that the hybrid PB 141 exhibited better disease tolerance (Warwick et al. 1991). Similarly, inoculation techniques were utilized to study Côte d'Ivoire varieties against *Helminthosporium halodes* infection (Quillec and Renard 1975). Interestingly, Malayan Green Dwarf was susceptible to *Helminthosporium incurvatum* compared to the cultivar's red and yellow variants (Anonymous 1976).

Though *Ganoderma* wilt or Thanjavur wilt is a major threat to coconut cultivation, the disease is managed following appropriate phytosanitary measures and hence a specific breeding programme to develop resistant cultivars have not been initiated.

## *Nut Water Quality*

Nut water quality is an important trait that warrants considerable attention due to its increasing popularity as a nourishing drink. In India, the cultivar COD, with estimated total sugar (7 g 100 $mL^{-1}$) and reducing sugars (4.7 g 100 $mL^{-1}$), optimal quantities of mineral nutrients (sodium and potassium) and acceptable organoleptic features, was recommended for release of a variety suitable for tender nut purposes (Dhamodaran et al. 1993; Anonymous 1991). Ganesamurthy et al. (2002b) evaluated the changes in the biochemical features such as total sugars, reducing sugars, nutrients mean potassium and sodium content of 40 coconut accessions and concluded that Zanzibar Tall, WCT, COD and MOD were found to be superior.

## *Novel Traits*

Various novel traits resulting from spontaneous mutations, with desirable traits, have been documented in coconut with immense potential for future breeding interventions (Arunachalam et al. 2001). Spicata, a mutant characterized with multiple female flowers on an unbranched inflorescence; Thairu Thengai, having soft endosperm and edible-husked variant, and Kaithathali, with edible mesocarp, have great utility in breeding programmes. Among these, the soft endosperm variant is known by various names in different countries: Makapuno (from the Philippines), Thairu Thengai (from India), Coco Gra (from Seychelles), Mapharao Khati (from Thailand), Kopyor (from Indonesia), Dikiri Pol (from Sri Lanka), Dua Sap (from Vietnam) and Niu Garuk (from Papua New Guinea). Cultivation and conservation of these soft endosperm variants utilizing embryo culture and the exploitation of these lines in breeding programmes offer novel ways of product diversification efforts. Coconut characterized with sweet kernel or endosperm ('Mohacho Narel') has immense potential for using the kernel for table purpose. In the absence of a biochemical basis, genetics and inheritance patterns of this trait development of a considerable population bearing this trait are challenging (Niral et al. 2013). Extremely dwarf mutants identified from Lakshadweep, India (CPCRI 2010) would provide parent genotypes to develop compact coconuts with potential use in high-density planting. The aroma of tender nut water and the accompanying flavour is a trait of economic importance. Genotypes such as Aromatic Green Dwarf and Klapawangi Tall possess this character; however, the development of huge populations with this trait is difficult to achieve because of heterozygosity. Production of bulbils in coconut is a rare occurrence with a great potential for clonal propagation (Mohandas et al. 1976). Bulbils producing palms serve as an invaluable resource for developing mapping population and marker identification which is difficult with the heterogenous population currently available (Jerard et al. 2013). Similarly, the formation of suckers is another way of obtaining clonal multiplication in coconut. Presence of naturally occurring haploids (Whitehead and Chapman 1962) opens the gateway for developing dihaploids in coconut with a potential breeding application.

# 6 Genomic Resources

Perennial nature of the crop, long juvenile phase, lengthy breeding cycle, difficulties raising F2 progenies and heterozygosity lead conventional breeding to be difficult in coconut. Therefore, the genetics of the desirable traits have been inadequately investigated in coconut. Application of biotechnological tools will speed up the laborious and time-consuming process of traditional coconut breeding.

## *Molecular Markers*

A key component of any successful breeding programme is comprehensive information about the genetic diversity of the crop. Though morphological parameters were utilized earlier to characterize coconut genetic diversity, they had serious limitations such as non-availability of sufficient markers, extended juvenile phase, and need for large-scale, expensive and resource-consuming field trials. Hence, the search for molecular markers has gained momentum as these markers are very abundant, exhibit polymorphism, function independent of the environmental factors and are amenable to be undertaken in a rapid manner (Perera et al. 1998). In coconut, the use of molecular markers has aided characterization of germplasm and their management, assessing the extent of genetic diversity and population structure and identifying QTLs for marker-assisted selection (MAS), linkage mapping, estimation of linkage disequilibrium and association analysis. Molecular markers in coconut include biochemical markers (proteins, polyphenols and isozymes) and DNA-based markers (Rajesh et al. 2019). The utility of these molecular techniques has unravelled novel avenues for molecular evolutionary genomics analysis besides providing an appropriate toolbox for effective conservation and utilization of coconut germplasm.

## *Biochemical Markers*

Analysis of coconut germplasm, for assessing the extent of genetic diversity, was performed during the initial years, utilizing biochemical markers, viz. proteins, isozymes, polyphenols or carotenoids. Isozyme analysis of nine plant enzymes was initiated by characterizing the coconut pollen at IRHO in 1978 (Benoit and Ghesquiere 1984). Analysis of four-enzyme systems in eight coconut ecotypes revealed a weak polymorphic pattern and less polymorphic loci per system of two alleles per locus. Similarly, the studies on six tall genotypes from Pacific Ocean region, utilizing 15 enzymatic systems, exhibited polymorphism in four enzymatic systems only with low variability (Moran 1991). The 17 isozyme systems used by Meunier et al. (1992) differentiated the diverse genotypes of coconut, especially the landraces from South East Asia, Africa and regions along the Pacific Ocean (Bourdeix et al. 1993). Sri Lankan coconut germplasm was analysed based on six-enzyme systems in horizontal starch gel electrophoresis (Fernando and Gajanayake 1997) which yielded a couple of polymorphic loci, viz. peroxidases and esterases. Cardeña et al. (1998) devised an electrophoretic pattern of proteins, leaf peroxidases and endopeptidases of six cultivars/hybrids to discriminate them, testing progeny authenticity, contamination of pollen and in evaluating the breeding systems. Similarly, molecular diversity analysis of 22 indigenous and exotic populations from Mexico, performed using various enzyme systems, showed very low polymorphism (Zizumbo Villarreal et al. 2002). Fixation indices revealed lower overall heterozygosity within the populations, indicating endogamy and genetic drift are prevailing in the population; however, a high diversity between the populations of Pacific and Gulf of Mexico coast indicated their genetic distinctness. Forty coconut accessions, derived from various regions worldwide, and six hybrids were analysed using eleven isozymes (Parthasarathy et al. 2004). Such studies, unfortunately, encounter the problems of mono-morphism and are found to be very less active. Researchers working with leaf proteins suggested that the leaf extracts did not provide good-quality enzyme activities since the extracts were easily oxidized. Moreover, the reproducibility of the results was a major concern. Despite these drawbacks, studies on leaf polyphenol-based polymorphism revealed variability in geographical origins and between the tall and dwarf palms (Jay et al. 1989). Later, Chempakam and Ratnambal (1993) demonstrated significant variations in the leaf polyphenol contents of 36 coconut accessions originating from eight geographical areas. Nevertheless, the banding pattern of secondary metabolites such as polyphenol has serious limitation since the environmental factors deeply govern them. Therefore, these biochemical markers are currently only of academic interest as the DNA-based molecular markers replace them.

## *DNA-Based Markers*

Various DNA-based molecular markers such as AFLP, RAPD, ISSR, SSRs and SCoT have been utilized in the genetic diversity analysis of coconut accessions and trait-market linkage studies (Table 6).

## *Linkage Mapping and Identification of QTLs*

Linkage maps are an indispensable component for efficient marker-assisted breeding as linkage maps provide the location of specific genes on a particular chromosome based on their recombination frequencies. The effective use of molecular markers and sophisticated statistical tools has led to the uncovering of these markers' inheritance pattern and helped in precisely locating the chromosomal loci of a

**Table 6** Application of various DNA-based molecular markers in coconut genomics and instances of marker-trait association studies

| Sl. No. | DNA-based markers | Genotypes | Probe or primers | Results | Inference(s) | References |
|---|---|---|---|---|---|---|
| 1. | Restriction fragment length polymorphism (RFLP) | Diverse genotypes | cDNA probes of oil palm, rice, maize and coconut and wheat cytoplasmic probe | Differentiated two genetically distinct population, viz. coconut ecotypes of the Far East and the Southern Pacific and those from Indian sub-continent and West Africa | Pacific and Far East region identified as the origin of coconut | Lebrun et al. (1998) |
| 2. | Randomly amplified polymorphic DNA (RAPD) | Seventeen South Pacific populations | 14 RAPD primers | – | Genetic drift and possible bottleneck in the evolutionary history were attributed for low inter-population diversity | Ashburner et al. (1997) |
| | | Genetic diversity analysis of coconut accessions from Sri Lanka | – | – | Diversity analysis | Everard et al. (1996) |
| | | Genetic diversity analysis of coconut accessions from the Philippines | – | – | Diversity analysis | Rodriguez et al. (1997) |
| | | Dwarf populations from East Java | – | – | High genetic variability of the population attributed to them being raised from open-pollinated seed nuts | Hayati et al. (2000) |
| | | 15 Indian and five exotic accessions | – | – | High heterozygosity, polymorphism and high genetic diversity in talls compared to dwarfs | Upadhyay et al. (2004) |
| | | Dwarf populations characterized by yellow fruits | 16 polymorphic RAPD primers | MYD, Andaman Yellow Dwarf (AYD) and Kulashekaram Yellow Dwarf (KYD) | Genetic diversity | Ritto et al. (2008) |

(continued)

**Table 6** (Continued)

| Sl. No. | DNA-based markers | Genotypes | Probe or primers | Results | Inference(s) | References |
|---|---|---|---|---|---|---|
| 3. | Amplified fragment length polymorphism (AFLP) | 42 Sri Lankan genotypes | Eight primer pairs | – | Variation in tall accessions (*typica*), rather than intermediate (*aurantica*) and dwarf (*nana*) accessions | Perera et al. (1998) |
| | | AFLP and SSRs | 31 palms of 14 coconut populations from various ecological zones | – | – | Teulat et al. (2000) |
| 4. | Inverse sequence-tagged repeat (ISTR) | | Coconut genotypes of world | | Large genetic loci with abundant polymorphism | Rohde et al. (1992) |
| | | | East African Tall coconut populations | | Evidence for truncated, *copia*-like repetitive elements in coconut genome | Duran et al. (1997) |
| 5 | Inter simple sequence repeats (ISSR) | 33 accessions, representing worldwide collections, maintained at ICG-SAME (India) | 19 ISSR primers | Out of 199 ISSR markers, 154 were polymorphic | Accessions from South and South-East Asia and South Pacific clustered distinctly | Manimekalai and Nagarajan (2006) |
| 6. | Simple sequence repeats (SSR ) | Samples (20) from the Philippine Coconut Authority (PCA) | 38 informative SSR markers derived from genomic libraries | – | SSRs detected substantial polymorphism (5.2 alleles per SSR) | Rivera et al. (1999) |
| | | 40 coconut samples | Eight SSRs | – | Eight alleles per SSR | Rivera et al. (1999) |
| | | Sri Lankan coconut populations | Eight SSRs from pre-cloning enrichment procedure | High diversity of *Typica* than the *nana* and the *aurantiaca* | – | Perera et al. (1999) |
| | | 130 samples (75 tall and 55 dwarfs) of coconut | Eight pairs of SSR primers | – | A phenetic tree constructed clustered individuals into five groups | Perera et al. (2000) |

| | | | | | | |
|---|---|---|---|---|---|---|
| | | 33 tall population from Sri Lanka | – | – | Genetic diversity study for conservation strategies | Perera et al. (2001) |
| | | | Fourteen microsatellite markers | Discrimination of varieties | CIRAD developed a microsatellite-based kit in collaboration with COGENT | Baudouin and Lebrun (2002) |
| | | Eight cultivars maintained at two areas in Florida, USA | Fifteen SSRs | – | Genetic distinctiveness of Red Malayan Dwarfs from the Green and Yellow Dwarfs. 'Red Spicata' clustered with 'Fiji Dwarf' | Merrow et al. (2003) |
| | | 21 Indian and 24 exotic coconut accessions | Eight SSRs | A total of 48 alleles, with a mean of six alleles per locus | Genetic diversity values were low for most of the dwarfs in comparison to tall accessions Kulasekharam Orange Dwarf exhibited high genetic diversity among talls | Devakumar et al. (2006) |
| | | Lethal yellowing disease tolerant (Vanuatu Tall *and Sri Lankan Green Dwarf) and susceptible* genotypes (West African Tall) | 12 SSRs | 58 alleles | Microsatellites helped in *identification of suitable* parents for developing a mapping population and in MAS of lethal yellowing resistant genes | Konan et al. (2007) |
| | | MYD from Jamaica and a reference DNA of MYD from five different countries, viz. Ghana, Malaysia, the Philippines, Mexico and India | 34 SSR markers | Explored if the Lethal Yellowing disease affected Maypan population in Jamaica is true to type | Variation at 34 SSR loci in MYD samples from Jamaica compared to the rest of the world | Lebrun et al. (2008) |

(continued)

**Table 6** (Continued)

| Sl. No. | DNA-based markers | Genotypes | Probe or primers | Results | Inference(s) | References |
|---|---|---|---|---|---|---|
| | | 26 coconut accessions from the Andaman and Nicobar Islands, India | 14 SSR markers | 103 alleles were obtained with an average of 7.35 alleles per locus | Observed and expected heterozygosity levels were 0.29 and 0.66, respectively. High levels of differentiation of populations were inferred from the mean fixation index (0.49) and predominance of rare alleles in Tall accessions of the Nicobar Islands | Rajesh et al. (2008a) |
| | | Ten landraces representing three Indian coconut-growing communities | 14 SSRs | Ninety alleles; an average of 6.42 alleles $locus^{-1}$ | Highest heterozygosity recorded in tall landraces | Rajesh et al. (2008b) |
| | | Forty-three conserved ex situ in the gene banks of CRI (Sri Lanka) | – | – | The narrow genetic base of Sri Lankan coconut germplasm with diversity was mostly limited to tall accessions | Dasanayaka et al. (2009) |
| | | Characterization of two ecotypes of coconut from Kerala, India, namely Annur and Bedakam | – | – | Distinct nature of the ecotypes in comparison to local WCT populations | Rajesh et al. (2014b) |
| | | Coconut populations selecting Laccadive Ordinary Tall (LCT) and Laccadive Micro Tall (LMT) from Amini and Kadamat Islands from Lakshadweep | 20 SSRs | – | Pear-shaped nuts from both islands showed affinity and presumed to have formed by introgression between the elliptical and the round types | Rajesh et al. (2014a) |

| | | | | | | |
|---|---|---|---|---|---|---|
| | | Five indigenous accessions ('Banawali', 'Gangabondam Green Dwarf', 'Pratap', 'Konkan', 'East Coast Tall' and Bhatye Coconut Hybrid-I') from the Konkan area, Maharastra, India | 14 SSRs | Levels of polymorphism among the five accessions ranged from 85.7 to 100% | | Rasam et al. (2016) |
| | | Brazilian coconut accessions ('Praia do Forte' and 'Merpe') and seven accessions introduced from different geographic areas around the globe, maintained at ICG-LAC, Brazil | 19 SSRs | Primers could detect four and ten alleles per locus, with an average of 6.57 | Genetic relationships between Brazilian and African accessions and among the South East Asian and the South Pacific accessions, confirming the common origin of the accessions | Loiola et al. (2016) |
| | | Genetic diversity and the population structure estimation of four distinct coconut accessions (Laccadive Giant Tall, Laccadive Ordinary Tall, Laccadive Micro Tall and Laccadive Mini Micro Tall types) of Minicoy Island, Lakshadweep, India | – | Seventy alleles; an average of 3.68 alleles per locus | Genetic distinctiveness of Laccadive Mini Micro Tall | Jerard et al. (2017) |
| | | | Diagnostic SSR marker | – | Hybridity testing in coconut | Perera (2010); Rajesh et al. (2012) |
| | | | EST-SSRs | – | Genetic diversity assessment and hybrid authentication in coconut | Preethi et al. (2014, 2016, 2020) |
| 7. | Start codon targeted polymorphism (SCoT) | Twenty three coconut accessions representing both talls and dwarfs | – | – | Potential of these markers for genetic diversity assessment in coconut | Rajesh et al. (2015b) |
| 8. | *WRKY* loci | Fifteen genotypes of six accessions | SNPs in WRKY transcription factors | Ten informative markers generated | Potential of these markers for genetic diversity assessment in coconut | Mauro-Herrera et al. (2006) |
| | | Sequence variations of *WRKY* of Kopyor (a mutant) | Eight SNP-based primers | – | Differentiated five Kopyor variants | Pesik et al. (2017) |

(continued)

**Table 6** (Continued)

| Sl. No. | DNA-based markers | Genotypes | Probe or primers | Results | Inference(s) | References |
|---|---|---|---|---|---|---|
| 9. | Marker-trait association studies | | | | | |
| | *Lethal* yellowing disease | Resistant and susceptible coconut populations (susceptible West African Tall (WAT), resistant Malayan Yellow Dwarf (MYD) and a resistant Atlantic Tall (AT)) | RAPD | – | A total of 12 RAPD markers were identified and for use in MAS | Cardena et al. (2003) |
| | Resistance to coconut mite (*Aceria guerreronis* Keifer) | Mite-resistant and mite-susceptible accessions from India | Thirty two SSRs and seven RAPD markers | Three SSRs and two RAPD markers explained 100% association with mite resistance | $OPE18_{600}$, $CnCirE2_{151}$, $OPP16_{1200}$ markers linked to mite resistance | Shalini et al. (2007) |
| | Palm habit | Tall, dwarf and dwarf × tall hybrids | | RAPD converted into a SCAR marker | OPA09 linked to tall morphoform | Rajesh et al. (2013) |
| | Palm habit | Tall, dwarf and dwarf × tall hybrids | | RAPD primer OPBA3 differentiates both the tall and dwarf bulks | The purity of the Dwarf × Tall hybrids was tested using RAPD-derived sequence characterized amplified region (SCAR) marker | Rajesh et al. (2014a) |

gene or gene cluster to identify quantitative trait loci (QTLs). Saturated linkage maps enable mapping of QTLs of economic importance, their exact phenotypic effects and gene action mode. Furthermore, characterization of economically important trait and its linkage with an identified marker and co-segregation of that marker with a QTL would greatly assist in coconut genomics-aided breeding efforts. In coconut, AFLP, ISSR, ISTR, RAPD and SSR-based molecular markers were utilized to dissect QTLs governing various traits such as germination, yield attributes and epicuticular wax content for improved climate-smart coconut, etc. and to develop linkage maps (Table 7). Furthermore, association mapping analysis was performed by Geethanjali et al. (2017) in 79 diverse coconut genotypes using 48 SSR primer combinations. The allelic analysis revealed a mean of 4.1 alleles per locus and overall two to seven alleles among the genotypes. Also, hierarchical clustering classified genotypes from Indo-Atlantic and SA regions under a cluster followed by both the talls and dwarfs of Indo-Pacific and SEA in another cluster. Besides, association analysis identified an SSR locus linked with fruit yield component traits corroborating Baudouin et al. (2006) findings.

**Table 7** Salient achievements in the field of QTL mapping in coconut

| Sl. No. | Mapping population | Markers and map length | QTLs mapped | References |
|---|---|---|---|---|
| 1. | Malayan Yellow Dwarf (MYD) × Laguna Tall (LAG) | ISTR | – | Rhode et al. (1995) |
| 2. | Laguna Tall × Malayan Yellow Dwarf | AFLP, ISSR, ISTR & RAPD<br>Laguna tall: 2226 cM<br>MYD 1266 cM | Six QTLs linked to precocious germination and yield | Herran et al. (2000) |
| 3. | Cameroon Red Dwarf (CRD) × Rennell Island Tall (RIT) | AFLP & SSR<br>1971 cM | Nine QTLs for yield and yield attributing characters such as bunch number and nut number | Lebrun et al. (2001) |
| 4. | Cameroon Red Dwarf (CRD) × Rennell Island Tall (RIT) | Oil palm derived AFLP, SSRs and some new coconut SSRs<br>1849.8 cM | 48 QTLs for fruit component traits | Baudouin et al. (2006) |
| 5. | African Tall (EAT) × Rennell Island Tall (RIT) | 241 AFLP, 64 SSR and 22 SSRs from oil palm<br>2739 cM | 46 QTLs associated with wax content and components traits | Riedel et al. (2009) |
| 6. | 79 genotypes | SSRs | Linkage disequilibrium-based mapping identified SSR locus CnCir73 is linked to fruit component traits | Geethanjali et al. (2017) |

## *Multiomics Approaches (Genomics and Transcriptomics)*

### Discovery of Gene(s)

To comprehend the molecular basis of recalcitrance in coconut, various investigations were conducted. Pérez-Núñez et al. (2009) characterized the nucleotide sequence features of plumule-derived *SOMATIC EMBRYOGENESIS RECEPTOR-like KINASE* (*SERK*) gene. Gene expression analysis revealed the spatial expressional differences in *CnSERK* gene in meristematic tissues. Further, the expression of *CnSERK* precedes the somatic embryogenesis process in the embryogenic calli, whereas little expression was detected in non-embryogenic calli. Thus, this study has delineated that *CnSERK* is an important biomarker that confers potency to the somatic cells to transform into embryos under in vitro conditions.

A protein involved in cell cycle control [cyclin-dependent kinase (*CDKA*)] in plumule was characterized, and its expression pattern was deciphered during the process of somatic embryogenesis (SE) (Montero-Cortés et al. 2010). It was found that *CDKA* showed enhanced expression during callus formation in SE and gradually transcript abundance declined when somatic embryos developed. Hence, *CDKA* could be a potential biomarker linked to in vitro embryogenic competence under in vitro conditions. Analysing the transcriptome sequence data obtained from embryogenic calli, Rajesh et al. (2016) performed expression profiling of 14 putative genes, including *CLAVATA1* (CLV), *WUSCHEL* (WUS) and *WRKY*, among others, of plumular explants involved in various stages of regeneration during the process of somatic embryogenesis. Initial analysis by Bhavyashree et al. (2015) recorded an abundance of transcriptional factors (TFs) such as ECP, *GST*, *LEAFY* and *WUS* in a relatively old (21 weeks) embryogenic calli. In contrast, other TFs (namely *SERK*, *WRKY*, among others) showed high expression in 21 days old embryogenic calli, implying the long-term effect of embryogenesis in calli derived from plumule tissues. As a follow-up, differences in the in vitro regeneration potential of two popular coconut genotypes, namely WCT and COD and the underlying differential gene expression pattern were documented (Bhavyashree et al. 2016).

Further search for novel genes in coconut identified a gene (*CnANT*-AINTEGUMENTA-like), encoding AP2 domains. Comparative analysis of *CnANT* gene of two morphoforms of coconut resulted in the designing of an SSR marker. Further analysis revealed that the gene's enhanced expression in tissues like matured zygotic embryos and in callus with embryogenic potential compared to cells in various SE pathway suggests its role in cellular proliferation (Bandupriya et al. 2014). It was further validated in the ectopic expression of the gene in *Arabidopsis* (Bandupriya et al. 2014). Molecular characterization of AP2/EREBP domain-containing transcription factor, *Wrinkled* (CoWRI1), and validation in *Arabidopsis* and rice divulged its regulatory role in oil accumulation (Sun et al. 2017).

Similarly, another important gene of oil biosynthesis encoding for enzyme lysophosphatidyl acyltransferase (LPAAT) was cloned, and promoter region characterized (Xu et al. 2010). Besides individual genes, the expression profile of small non-coding RNAs such as miRNAs and their role in endosperm development were also investigated (Li et al. 2009). To complement these studies, Liang et al. (2014)

used the SSH technique to unravel 737 unigenes and their differential expression pattern in coconut endosperm. The whole-transcriptome data of Chowghat Green Dwarf, a genotype which is known to possess resistance against coconut root (wilt) disease, was analysed for resistant gene analogues (RGAs) which yielded six distinct classes of 243 RGA sequences, and the expression of NBS-LRR kind of RGAs was validated (Rajesh et al. 2015a; Rajesh et al. 2018). Earlier, NBS-LRR category of RGAs was identified from the genotypes resistant and susceptible to lethal yellowing disease and their expressional differences in response to salicylic acid was documented (Puch-Hau et al. 2015). Analysing the conserved regions of *NBS-LRR* genes, Rachana et al. (2016) designed oligonucleotides and amplified putative RGA in CGD cultivar followed by its structural and functional characterization. Another interesting gene discovery pertains to identifying and characterizing an ortholog of rice aromatic gene, which confers pandan-like aroma in coconut (*CnAMADH2*) (Saensuk et al. 2016). Comparative analysis of coding regions of *CnAMADH2* gene of genotype Aromatic Green Dwarf from Thailand and its non-aromatic variant revealed guanine (G) to cytosine (C) conversion causing alanine to proline amino acid substitution resulting in the characteristic aroma. This study had led to the development of a PCR marker to distinguish alleles which confers and nullifies aroma in coconut.

## Coconut Genome

Flow cytometer analysis revealed that coconut has a mean chromosomal DNA content of $5.6 \pm 0.2$ pg, roughly $5.4 \times 10^9$ bp. Analysis by Gunn et al. (2015) revealed that the genome of a typical coconut is 5.96 pg. Later, experiments of Neto et al. (2016) divulged that mean genome content of talls and dwarfs accessions are 5.59 and 5.55 pg, respectively. The first complete nuclear genome of coconut (genotype-Hainan Tall) was characterized by utilizing NGS technology (Xiao et al. 2017). It also revealed that the genome contains 28,039 protein-coding genes which are less compared to related palm genera *Elaeis guineensis* (34,802) and *Phoenix dactylifera* (28,889–41,660).

Interestingly, over 70% of the coconut genome is characterized as transposable elements predominated by long-terminal repeats (LTRs), suggesting the mobile nature of a major portion of coconut genome (Xiao et al. 2017). Furthermore, Bayesian molecular clock-based evolutionary analysis has uncovered that the palm species had diverged from its relative genera oil palm approximately 46 million years ago. The comparative genomic analysis further uncovered that coconut had expanded gene families belonging to plasma membrane transporters such as carnitine/acylcarnitine translocases, exchangers involved in $K^+$ and $Ca^{2+}$, and $Na^+/H^+$ antiporters. Prevalence of these genes in coconut suggests their importance in adaptation to a saline environment.

The complete nuclear genome of the cultivar 'Catigan Green Dwarf' (CATD) (2.15 Gb) was obtained utilizing various sequencing technologies (Lantican et al. 2019). The predicted genome size of the crop is relatively large than date palm (660 Mb), *Elaeis guineensis* (1.8 Gb) and *Elaeis oleifera* (1.8 Gb). However, the previously sequenced tall cultivar 'Hainan Tall' (HAT) had the estimated genome size (2.42 Gb) (Xiao et al. 2017). Quite interestingly, this study has compared the dwarf genome sequence with its tall counterpart and identified novel 7139 SSR markers, 58,503 SNP variants and 13 gene-linked SSRs (Lantican et al. 2019). It predicted a total of 34,958 gene models associated with various traits with high confidence. Other genes involved in drought response, biotic stress resistance and oil biosynthesis were tagged with the SSRs, which has a great potential for molecular assisted breeding. Molecular evolutionary genomic analysis with genera of *Arecaceae* suggests that at least three rounds of genomic duplications might have occurred in coconut (Lantican et al. 2019).

Recently, Rajesh et al. (2020) have characterized the complete nuclear and organellar genome of a dwarf cultivar (Chowghat Green Dwarf) by combining the short and long reads Illumina HiSeq 4000 and Pacific Biosciences RSII platforms, respectively. Stringent filtering condition applied predicted that the genome harbours 13,707 genes which encode 11,181 proteins. The evidence for the predicted genes was validated through transcriptome sequences obtained from various coconut tissues. Further, the chloroplast and mitochondrial genomes of CGD variety were also characterized. Availability of CGD genome, having enhanced disease resistance trait, is a valuable genetic resource for accelerating genomics-assisted breeding in coconut (Rajesh et al. 2020).

The chloroplast genome of dwarf coconut describing gene organization, repeat structures and RNA editing role was presented (Huang et al. 2013). The chloroplastic genome of coconut is the smallest (among the palms) with 154,731 bp in length. Nonetheless, genome organization, gene content and repeat features are conserved among the palms. The chloroplast genome encodes 130 genes, among others. Similarly, the mitochondrial genome of coconut (genotype: Oman Local Tall) was deduced by Aljohi et al. (2016). The mitochondrial genome (678.65 kb in length) is characterized with 45.5% GC content and codes for 72 proteins, nine pseudogenes, 23 tRNAs and three rRNAs. Only a little over 5% of the mitochondrial genome of coconut is derived from the chloroplastic regions.

In contrast, in the related genera like the date palm, more than 93% of the genome sequence had its origin from the chloroplast. Among the various mitochondrial genes, *sdh* is unique to the coconut among the monocots. Further, the monocot specific *rps3* gene is conspicuously absent in coconut. Moreover, it is also found that those of cp-derived tRNA are replacing mt tRNA genes.

## Transcriptome Analysis

Whole-transcriptome analysis utilizes NGS techniques to dissect the molecular features, especially the gene expression pattern, underlying any particular trait of importance. In coconut, genome-wide transcriptome analysis has been carried to analyse the various traits summarized in Table 8.

**Table 8** Application of transcriptomic approaches to dissect the molecular responses of coconut and in generating novel genomic resources

| Sl. No. | Trait | Features | Inferences | References |
|---|---|---|---|---|
| 1. | RNA-seq analysis of leaf and fruit tissue | Hainan Tall cultivar | Accumulation of medium-chain fatty acids (MCFAs) such as lauric acid is correlated with the expression pattern of enzyme fatty acyl-ACP thioesterase | Fan et al. (2013) |
| 2. | Maturing gelatinous endosperm, embryo and young leaf | Aromatic dwarf coconut | The role of small non-coding RNAs mediated epigenetic regulation during seed development | Huang et al. (2014) |
| 3. | Coconut yellow decline disease | Diseased and healthy Malayan Red Dwarf palms | Reprogramming of defence-related gene(s), upregulation of GA-*2ox* (gibberellin-2-oxidase) caused reduced gibberellins, leading to stunted growth, necrosis inflorescence and premature nut fall. Upregulation of ABC transporter genes was linked to sugar import to maintain the energy source of Phytoplasma | Nejat et al. (2015) |
| 4. | Embryogenic calli | West Coast tall cultivar | Identified transcripts coding for receptor-like kinases, protein kinases, TFs and extracellular proteins involved in somatic embryogenesis | Rajesh et al. (2016) |
| 5. | Immature embryo, mature embryo, microspore derived embryo and mature leaves | Comparative transcriptomic analysis | Transcripts encoding for proteins, viz. chitinase, β-1,3-glucanase, ATP synthase CF0 subunit, a thaumatin-like protein involved in embryogenesis, were identified | Bandupriya and Dunwell (2016) |
| 6. | 'Pandan-like' aroma | Aromatic Green Dwarf coconut of Thailand | Allelic differences in the transcripts encoding 2AP in aromatic and non-aromatic variants of coconut | Saensuk et al. (2016) |
| 7. | Host–pathogen interaction during root (wilt) disease in the variety Chowghat Green Dwarf (CGD) | Healthy and diseased CGD palms sequenced, utilizing Illumina HiSeq 2000 platform | Significant differentially expressed genes (2718) between the diseased and healthy CGD palms | Rajesh et al. (2018) |

(continued)

**Table 8** (continued)

| Sl. No. | Trait | Features | Inferences | References |
|---|---|---|---|---|
| 8. | Water-deficit stress | Chowghat green dwarf and Kalpatharu (Tall) having contrasting water use efficiency trait | Genotypic differences in molecular response to water-deficit stress; genic SSRs and the role of long non-coding RNAs (lnc RNAs) deciphered | Ramesh et al. (2020) |

Whole-transcriptome analysis of the leaves of coconut genotype characterized 243 resistance gene analogue (RGA) sequences categorized into six classes (Rajesh et al. 2015a). Phylogenetic analysis and qRT-PCR-based expression studies of NBS-LRR-type RGAs were performed. This investigation provided useful resources for identifying and utilizing the candidate genes that could confer disease resistance in coconut. Furthermore, the sequence features of NBS-LRR genes of Chowghat Green Dwarf (CGD) was characterized along with their expression profile using comparative genomic approaches utilizing the sequences of date palm (Rachana et al. 2016).

### Transformation Studies

Though genetic transformation studies in coconut is a rarity successful introgression of foreign genes in coconut genome is possible by combining the techniques of biolistics and *Agrobacterium*-mediated transformation. The initial attempt of genetic transformation was made by Samosir (1999) to incorporate the *GUS* gene in embryogenic calli and young leaf tissues. However, results revealed that the constitutive promoters such as actin (*Act1*) and ubiquitin (*Ubi*) exhibited a relatively strong transient expression. Further, various coconut tissues, including anthers, zygotic embryos, embryogenic calli and SE-derived tissues such as roots and leaves, were subjected to *Agrobacterium*-mediated transformation (Andrade-Torres et al. 2011). However, in the absence of reliable and scalable in vitro regeneration protocol, the application of genetic transformation studies in coconut is a distant dream.

## 7 Conclusion and Perspectives

Despite the considerable improvements made in the collection and conservation of coconut genetic diversity since the inception of COGENT, data availability and inclusive representation are far from complete. There is a need for conservation of indigenous germplasm since many genotypes with unique features are under the threat of genetic erosion. The focus has to shift on rapid use of this germplasm in breeding programmes to justify the investment made in collection and conservation.

Use of modern breeding techniques and molecular tools are warranted to develop core or mini core collections of germplasm so that financial and human resources are optimally utilized. Farmer participatory approaches require to be followed wherever it is necessary to document the unique features of palms and indigenous traditional knowledge being followed to improve the knowledge base of the crop.

## References


Aljohi HA, Liu W, Lin Q, Zhao Y, Zeng J, Alamer A, Alanazi IO, Alawad AO, Al-Sadi AM, Hu S, Yu J (2016) Complete sequence and analysis of coconut palm (*Cocos nucifera*) mitochondrial genome. PLoS One 11(10):e0163990. https://doi.org/10.1371/journal.pone.016399

Andrade-Torres A, Oropeza C, Sáenz L, González-Estrada T, Ramírez-Benítez JE, Becerril K, Chan JL, Rodríguez-Zapata LC (2011) Transient genetic transformation of embryogenic callus of *Cocos nucifera* L. Biologia 66(5):790–800

Anonymous (1976) Annual Report 1975-76. Central Plantation Crops Research Institute, Kasaragod

Anonymous (1991) Proceedings of tenth All India Coordinated Project on Palms, Kasaragod

Arellano J, Oropeza C (1995) Lethal yellowing. In: Oropeza C, Howard FW, Ashburner GR (eds) Lethal yellowing: research and practical aspects, vol 5: Development in Plant Pathology. Springer, Netherlands, pp 1–15

Arunachalam V, Rajesh MK (2008) Breeding of coconut palm (*Cocos nucifera* L.). In: CAB reviews: perspectives in agriculture, veterinary science, nutrition and natural resources no. 053. https://doi.org/10.1079/PAVSNNR20083053

Arunachalam V, Rajesh MK (2017) Coconut genetic diversity, conservation and utilization. In: Ahuja MR, Jain M (eds) Biodiversity and conservation of woody plants. Springer, Cham, pp 3–36. https://doi.org/10.1007/978-3-319-66426-2_1

Arunachalam V, Jerard BA, Elangovan M, Ratnambal MJ, Dhanapal R, Rizal SK, Damodaran V (2001) Unexploited diversity in coconut palm (*Cocos nucifera* L.). Plant Genet Resour Newsl 127:39–43

Arunachalam V, Augustine Jerard B, Elain Apshara S, Jayabose C, Subaharan K, Ravikumar N, Palaniswami C (2013) Digital phenotyping of coconut and morphological traits associated with eriophyid mite (*Aceria guerreronis* Keifer) infestation. J Plant Crops 41(3):417–424

Ashburner GR, Thompson WK, Halloran GM, Foale MA (1997) Fruit component analysis of south Pacific coconut palm populations. Genet Resour Crop Evol 44(4):327–335

Assy Bah B, Durand-Gasselin T, Pannetier C (1987) Use of zygotic embryo culture to collect germplasm of coconut (*Cocos nucifera* L.). Plant Genet Resour Newsl 71:4–10

Assy-Bah B, Engelmann F (1992a) Cryopreservation of immature embryos of coconut (*Cocos nucifera* L.). Cryo-Letters 13:67–74

Assy-Bah B, Engelmann F (1992b) Cryopreservation of mature embryos of coconut (*Cocos nucifera* L.) and subsequent regeneration of plantlets. Cryo-Letters 13:117–126

Assy-Bah B, Durand-Gasselin T, Pannetier C (1987) Use of zygotic embryo culture to collect germplasm of coconut (*Cocos micifera* L.). FAOIIBPGR Plant Genet Resour Newsl 71:4–10

Bajaj YPS (1984) Induction of growth in frozen embryos of coconut and ovules of citrus. Curr Sci 53:1215–1216

Bandupriya HD, Dunwell JM (2016) Transcriptome analysis for discovering candidate genes involve in embryogenesis in coconut (*Cocos nucifera* L.) through 454 pyrosequencing. J Natl Sci Found Sri Lanka 43:4

Bandupriya HDD, Fernando SC, Verdeil JL, Malaurie B (2007) Effect of abscisic acid on survival and recovery of cryopreserved plumule explants of coconut (*Cocos nucifera* L.). Cocos 18:58–68

Bandupriya HDD, Fernando SC, Verdeil JL, Malaurie B (2010) Cryopreservation of encapsulated plumules of coconut: effect of transport/store conditions. Asia-Pac J Mol Biol Biotechnol 18:135–137

Bandupriya HDD, Gibbings JG, Dunwell JM (2014) Overexpression of coconut AINTEGUMENTA-like gene, CnANT, promotes in vitro regeneration in transgenic *Arabidopsis*. Plant Cell Tissue Org 116(I):67–79

Baudouin L, Lebrun P (2002) The development of microsatellite kit and dedicated software for use with coconuts. Burotrop Bull 17:16–20

Baudouin L, Lebrun P, Konan JL, Ritter E, Berger A, Billotte N (2006) QTL analysis of fruit components in the progeny of a Rennell Island Tall coconut (*Cocos nucifera* L.) individual. Theor Appl Genet 112(2):258–268

Been BO (1981) Observations on field resistance to lethal yellowing in Jamaica. Oleagineux 36(1):9–12

Benoit H, Ghesquiere M (1984) Electrophorèse, compte rendu cocotier. IV. Déterminisme génétique, Rapport interne, IRHO-CIRAD, France

Bhavyashree U, Lakshmi Jayaraj K, Rachana KE, Muralikrishna KS, Sajini KK, Rajesh MK, Karun A (2015) Maintenance of embryogenic potential of calli derived from shoot meristem of West Coast Tall cv. of coconut (*Cocos nucifera* L.). J Plant Crops 43(2):105–116

Bhavyashree U, Fayas T, Muralikrishna K, Sajini K, Rajesh MK, Karun A (2016) A comparative study of three different methods of shoot meristem excision for induction of embryogenic calli in coconut. Res Biotechnol 7:50–57

Bourdeix R, N'Cho YP, Sangare A, Baudouin L (1993) Coconut genetic improvement: Results and prospects. In: Proceedings of the Eurococo Seminar, Montpellier, CIRAD, France, pp 29–31

Bourdeix R, Batugal P, Oliver JT, MLC G (eds) (2010) Catalogue of conserved coconut germplasm. International Coconut Genetic Resources Network (COGENT), Bioversity International, Regional Office for Asia, the Pacific and Oceania, Serdang

Branton RL, Blake J (1984) Development of organized structures in callus derived from explants of *Cocos nucifera* L. Ann Bot 52:673–678

Cardeña R, Oropeza C, Zizumbo-Villarreal D (1998) Leaf proteins as markers useful in the genetic improvement of coconut palms. Euphytica 102:81–86

Cardena R, Ashburner GR, Oropeza C (2003) Identification of RAPDs associated with resistance to lethal yellowing of the coconut (*Cocos nucifera* L.) palm. Sci Hortic 98(3):257–263

Cavagnaro T, Gleadow R, Miller RE (2011) Plant nutrient acquisition and utilisation in a high carbon dioxide world. Funct Plant Biol 38(2):87–96

Chan E, Elevitch CR (2006) Species profiles for pacific island agroforestry: *Cocos nucifera*, traditional tree. http://agroforestry.net/tti/Cocos-coconut.pdf

Chempakam B, Ratnambal MJ (1993) Variation for leaf polyphenols in coconut cultivars. In: Nair MK, Khan HH, Gopalasundaram P, Bhaskara Rao EVV (eds) Advances in coconut research and development. Oxford and IBH Publishing, New Delhi, pp 51–53

Chempakam B, Kasturi Bai KV, Rajagopal V (1993) Lipid peroxidation in relation to drought tolerance in Coconut. Plant Physiol Biochem 20:5–10

Chin HF, Krishnapillay B, Hor YL (1989) A note on the cryopreservation of embryos from young coconuts (*Cocos nucifera* var. Mawa). Pertanika 12(2):183–186

Chowdhury D, Nath JC, Mohan NK (2001) 'Kamrupa' – a newly released coconut variety by Assam Agricultural University. Indian Coco J 31:12–13

Copeland EB (1931) The coconut. Macmillan, London

CPCRI (2010) Annual report 2009–2010. Central Plantation Crops Research Institute, Kasaragod, p 150

CPCRI (2011) Annual report 2010–2011. Central Plantation Crops Research Institute, Kasaragod, p 146

CPCRI (2012) Annual report 2011–2012. Central Plantation Crops Research Institute, Kasaragod, p 128

Dasanayaka PN, Nandadasa HG, Everard JMDT, Karunanayaka EH (2009) Analysis of coconut (*Cocos nucifera* L.) diversity using microsatellite markers with emphasis on management and utilisation of genetic resources. J Natl Sci Found Sri Lanka 37(2):99. https://doi.org/10.4038/jnsfsr.v37i2.1065

De Guzman EV (1970) The growth and development of coconut "Makapuno" embryos *in vitro*. I. Induction of rooting. Philippine Agric 53:65–78

Dery SK, Philippe R (1995) Results of preliminary study of the epidemiology of the Cape St Paul Wilt disease of coconut in Ghana. In: International workshop on lethal yellowing-like diseases on coconut, Elmina, Ghana, p 14

Devakumar K, Jayadev K, Rajesh MK, Chandrasekhar A, Manimekalai R, Kumaran PM, Parthasarathy VA (2006) Assessment of genetic diversity of Indian coconut accessions and their relationship to other cultivars using microsatellite markers. Plant Genet Resour Newsl 145:38–45

Devakumar K, Thomas RJ, Nair RV, Jerard BA, Rajesh MK, Jacob PM, Jayadev K, Parthasarathy VA (2011) Analysis of population structure and genetic relatedness among root (wilt) disease resistant and susceptible coconut palms (*Cocos nucifera* L.) cv. West Coast Tall using micro-satellite markers. Indian J Agric Sci 81(5):487–493

Dhamodaran S, Ratnambal MJ, Chempakam B, Pillai RV, Viraktamath BC (1993) In: Nair MK, Khan HH, Gopalasundaram P, EVV BR (eds) Advances in coconut research and development. Oxford & IBH Publishing, New Delhi, pp 123–128

Duran Y, Rohde W, Kullaya A, Goikoetxea P, Ritter E (1997) Molecular analysis of East African Tall coconut genotypes by DNA marker technology. J Genet Breed 51:249–288

Eden-Green SJ (1997) History, world distribution and present status of lethal yellowing disease of palms. In: Eden-Green SJ, Offori F Elmina (eds) Proceedings of international workshop on lethal yellowing like disease of coconut, Ghana, pp 9–26

Eeuwens CJ (1976) Mineral requirements for growth and callus initiation of tissue explants excised from mature coconut palms (*Cocos nucifera*) and cultured *in vitro*. Physiol Plantarum 36(1):23–28

Everard JMDT, Katz M, Gregg K (1996) Inheritance of RAPD markers in the coconut palm, *Cocos nucifera* L. Trop Agric Res 8:124–135

Faleiro JR, Rangnekar PA (2001) Ovipositional preference of red palm weevil *Rhynchophorus ferrugineus* Olive to coconut cultivars. Indian Coconut J 32(6):22–23

Fan H, Xiao Y, Yang Y, Xia W, Mason AS, Xia Z, Qiao F, Zhao S, Tang H (2013) RNA-Seq analysis of *Cocos nucifera*: transcriptome sequencing and de novo assembly for subsequent functional genomics approaches. PLoS One 8(3):e59997. https://doi.org/10.1371/journal.pone.0059997

Farias Neto JTD, Lins PMP, Müller AA (2003) Estimation of repeatability coeficients for production of fruits and solid albumen in hybrids of coconut palm. Pesquisa Agropecuária Brasileira 38(10):1237–1241

Fernando WMU, Gajanayake G (1997) Patterns of isozyme variations in coconut (*Cocos nucifera* L.) populations used for breeding improved varieties. Plantations Recherche Développement 4:256–263

Fernando SC, Vidhanaarachchi VRM, Weerakoon LK, Santha ES (2010) What makes clonal propagation of coconut difficult? Asia-Pac J Mol Biol Biotechnol 18:163–165

Flores A (2008) Fiji Dwarf sets new durability standards in coconuts. Agric Res 56(5):17

Fremond Y, Ziller R, de Nuce de Lamothe M (1966) The coconut palm. International Potash Institute, Berne, 227p

Galvez HF, Lantican DV, Sison MLJ, Gardoce RR, Caoili BL, Canama AO, Dancel MP, Manohar ANC, Latina RA, Cortaga CQ, Reynoso DSR, Guerrero MS, Rivera SM, Emmanuel EE, Cueto CA, Reaño CE, Rivera RL, Vasquez MSC, Cabangbang RPM, Fernandez ECJ, Strickler SR, Mueller LA (2018) Coconut genetics and genomics for host insect resistance. In: The abstract list of the 26th international plant and animal genome conference (PAG XXVI), held at Town and Country Hotel, San Diego, 12–18 Jan 2018. https://pag.confex.com/pag/xxvi/meetingapp.cgi/Paper/31854

Ganesamurthy K, Natarajan C, Rajarathinam S, Vincent S, Khan HH (2002a). Genetic variability and correlation of yield and nut characters in coconut. J Plant Crops 30(2):23–25
Ganesamurthy K, Natarajan C, Vaithilingam R, Giridharan S, Khan HH (2002b). Screening of coconut germplasm for tender coconut water. Ind Coconut J 33(8):31–34
Geethanjali S, Anitha Rukmani J, Rajakumar D, Kadirvel P, Viswanathan P (2017) Genetic diversity, population structure and association analysis in coconut (*Cocos nucifera* L.) germplasm using SSR markers. Plant Genet Resour 16:1–13. https://doi.org/10.1017/S1479262117000119
Ghose S, Mishra BD, Rout MK (2006) Varietal evolution and occurrence of coconut grey leaf spot disease in coastal Orissa. J Mycopathol Res 44(1):105–107
Gomez-Navarro C, Jaramillo C, Herrera F, Wing SL, Callejas R (2009) Palms (Arecaceae) from a Paleocene rainforest of northern Colombia. Am J Bot 96:1300–1312
Govindan M, Radhakrishnan TC, Sathiarajan PK (1991) Reaction of certain varieties and hybrids of coconut to natural infection of leaf blight caused by *Pestalosphaeria elaeidis*. In: Silas EG, Aravindakshan M, Jose AI (eds) Proceedings of national symposium on coconut breeding and management. Kerala Agricultural University, Thrissur, pp 156–157
Gunn BF, Baudouin I, Olsen KM (2011) Independent origins of cultivated coconut in the old world tropics. PLoS One 6(6):e21143. https://doi.org/10.1371/journal.pone.0021143
Gunn BF, Baudouin L, Beulé T, Ilbert P, Duperray C, Crisp M, Issali A, Konan JL, Rival A (2015) Ploidy and domestication are associated with genome size variation in Palms. Am J Bot 102(10):1625–1633
Harries HC (1970) The 'Malayan dwarf' supersedes the 'Jamaica tall' coconut. I. Reputation and performance. Oléagineux, 25:527–531
Harries HC (1978) The evolution, dissemination and classification of *Cocos nucifera*. Bot Rev 4:265–320
Harries HC (1990) Malesian origin for a domestic *Cocos nucifera*. In: The plant diversity of Malesia. Springer, Dordrecht, pp 351–357
Harries HC (1998) Breeding phytoplasma disease resistant coconuts: alternative strategies for field exposure trials. In: Proceedings of international cashew and coconut conference: trees for life – the key to development, Dar es Salaam, Tanzania, 17–21 Feb 1997, pp 375–379
Harries HC (2001) Coconut varieties and lethal yellowing: a regional perspective for the Americas. Palms 45:148–150
Hayati DPK, Hartana A, Suharsono, Aswidinnoor H (2000) Genetic diversity of Jombang coconut populations based on RAPD markers. In: International conference on science and technology for managing plant genetic diversity in the 21st century, 12–16 June 2000, Kuala Lumpur, Malaysia, p 11
Hebbar KB, Sashidhar VR, Udayakumar M, Devendra R, Rao CN (1994) A Comparative assessment of WUE in groundnut grown in containers and in the field under water limited conditions. J Agric Sci 122:429–434
Hebbar KB, Sheena TL, Shwetha Kumari K, Padmanabhan S, Balasimha D, Kumar M, Thomas GV (2013) Response of coconut seedlings to elevated CO2 and high temperature in drought and high nutrient conditions. J Plant Crops 41:118–122
Herran A, Estioko L, Becker D, Rodriguez MJB, Rohde W, Ritter E (2000) Linkage mapping and QTL analysis in coconut (*Cocos nucifera* L.). Theor Appl Genet 101(1):292–300
Hornung R, Domas R, Lynch PT (2001) Cryopreservation of plumular explants of coconut (*Cocos nucifera* L.) to support programmes for mass clonal propagation through somatic embryogenesis. Cryoletters 22(4):211–220
Howard FW, Rodriguez AE (1991) Tightness of the perianth of coconuts in relation to infestation by coconut mites. Florida Entomol 74:358–361
Huang YY, Matzke AJM, Matzke M (2013) Complete sequence and comparative analysis of the chloroplast genome of coconut Palm (*Cocos nucifera*). PLoS One 8(8):e74736. https://doi.org/10.1371/journal.pone.0074736

Huang YY, Lee CP, Fu JL, Chang BCH, Matzke AJ, Matzke M (2014) *De Novo* transcriptome sequence assembly from coconut leaves and seeds with a focus on factors involved in RNA-directed DNA methylation. G3: Genes Genomes Genet 4:2147–2157

ICAR-CPCRI (2019) Annual report 2011–2012. Central Plantation Crops Research Institute, Kasaragod, p 102

ICC (2019) Statistics. International Coconut Community. https://coconutcommunity.org/regular/coconut_statistical_yearbook

IPCC (2007) Climate change 2007: the physical science basis. Inter-governmental Panel for Climate Change, 324. Cambridge University Press, Cambridge

Iyer RD (1981) Embryo and tissue culture for crop improvement, especially of perennials, germplasm conservation and exchange. In: Rao AN (ed) Tissue culture of economically important plants. ANBS, Singapore, pp 229–230

Iyer RD, Rao EVVB, Kutty MPG (1979) Super yielders in coconut. Indian Fmg 28(10):3–5

Jacob KC (1941) A new variety of coconut palm (*Cocos nucifera* L.) var. Spicata. J Bombay Nat Hist Soc 41:906–907

Jacob PM, Rawther TSS (1991) Varietal resistance. In: Nair MK, KKN N, Koshy PK, Jayasankar NP (eds) Coconut root (Wilt) disease. Central Plantation Crop Research Institute, Kasaragod, pp 67–72

Jay M, Bourdeix R, Potier F, Sanslaville C (1989) Premiers resultats de l'etude du polymorphisme des polyphenols foliares du cocotier. Oleagineux 44:151–161

Jerard BA, Damodaran V, Niral V, Samsudeen K, Rajesh MK, Sankaran M (2013) Conservation and utilization of soft endosperm coconut accession from Andaman Islands. J Plant Crops 41(1):14–21

Jerard BA, Rajesh MK, Elain SE, Sajini KK, Rahman S, Fayas TP, Karun A (2014a) Scope of novel and rare bulbiferous coconut palms (*Cocos nucifera* L.). Genet Resour Crop Evol 61:1–6

Jerard BA, Niral V, Dhanapal R, Damodaran V, Arunachalam V, Rajesh MK, Devakumar K, Samsudeen K, Nair RV, Kumaran PM, Thomas GV (2014b) IND 221–Andaman horned Cocos (IC0598221; INGR13063), a coconut (*Cocos nucifera*) germplasm with distinct character of horny nuts. Indian J Plant Genet Resour 27(1):66–89

Jerard BA, Niral V, Samsudeen K, Gayathri UK (2016) Pink husked coconut selection- a trait of promise. In: Chowdappa P, Muralidharan K, Samsudeen K, Rajesh MK (eds) Abstracts – 3rd international symposium on coconut research and development, 10–12 Dec 2016

Jerard BA, Rajesh MK, Thomas RJ, Niral V, Samsudeen K (2017) Island ecosystems hosts rich diversity in coconut (*Cocos nucifera*): evidences from Minicoy islands. India Agric Res 6:214–226

John CM, Narayana G (1949) Varieties and forms of the coconut (*Cocos nucifera* Linn). Indian Coconut J 2:209–226

Julia JF, Mariau D (1979) New research on the coconut mite E. guerreronis (K), in the Ivory Coast. Oléagineux 34:181–189

Kapadia MN (1981) Further record of a host plant of *Nephantis serinopa* Meyrick in Gujarat. Indian Coconut J 12(12):12

Karun A, Sajini KK (1994) Short-term storage of coconut embryos in sterile water. Curr Sci 67(2):118–120

Karun A, Sajini KK (2010) Cryopreservation of coconut zygotic embryos and pollen. Technical Bulletin No. 63. ICAR-CPCRI, Kasaragod

Karun A, Shivasankar S, Sajini KK, Saji KV (1993) Field collection and *in vitro* germination of coconut embryos. J Plant Crops 21:291–294

Karun A, Sajini KK, Radha E, Parthasarathy VA (2004) Efficacy of CPCRI embryo culture protocol in germplasm expedition. J Plant crops 32:139–143

Karun A, Sajini KK, Parthasarathy VA (2005) Cryopreservation of mature coconut embryos by desiccation method. CORD 21:13–19

Karun A, Sajini KK, Niral V, Amarnath CH, Remya P, Rajesh MK, Samsudeen K, Jerard BA, Engelmann F (2014) Coconut (*Cocos nucifera* L.) pollen cryopreservation. Cryo-Letters 35(5):407–417

Karun A, Sajini KK, Rajesh MK, Muralikrishna KS, Samsudeen K, Kumar PA, Nagwekar DD (2017) In vitro retrieval via embryo rescue of '*Mohachao Narel*', a sweet endosperm coconut from Maharashtra. Int J Innov Hortic 6(2):147–150

Kasturi Bai KV, Rajagopal V, Chempakam B, Prabha CD (1996) Assay of enzymes in coconut cultivars and hybrids under non-stress and stress conditions. J Plantat Crops 20:548–554

Konan JNK, Koffi KE, Konan JL, Lebrun P, Dery SK, Sangare A (2007) Microsatellite gene diversity in coconut (*Cocos nucifera* L.) accessions resistant to lethal yellowing disease. Afr J Biotechnol 6:341–347

Lantican DV, Strickler SR, Canama AO, Gardoce RR, Mueller LA, Galvez HF (2019) De novo genome sequence assembly of dwarf coconut (*Cocos nucifera* L.'Catigan Green Dwarf') provides insights into genomic variation between coconut types and related palm species. G3: Genes Genomes Genet 9(8):2377–2393

Lebrun P, N'Cho YP, Seguin M, Grivet L, Baudouin L (1998) Genetic diversity in coconut (*Cocos nucifera* L.) revealed by restriction fragment length polymorphism (RFLP) markers. Euphytica 101:103–108

Lebrun P, Baudouin L, Bourdeix R, Konan JL, Barker JHA, Aldam C, Herran A, Ritter E (2001) Construction of a linkage map of the Rennell Island Tall coconut type (*Cocos nucifera L.*) and QTL analysis for yield characters. Genome 44:962–970

Lebrun P, Baudouin L, Myrie W, Berger A, Dollet M (2008) Recent lethal yellowing outbreak: why is the Malayan Yellow Dwarf Coconut no longer resistant in Jamaica? Tree Genet Genomes 4(1):125–131

Li D, Zheng Y, Wan L, Zhu X, Wang Z (2009) Differentially expressed microRNAs during solid endosperm development in coconut (*Cocos nucifera* L.). Sci Hortic 122:666–669

Liang Y, Yuan Y, Liu T, Mao W, Zheng Y, Dongdong L (2014) Identification and computational annotation of genes differentially expressed in the pulp development of *Cocos nucifera* L. by suppression subtractive hybridization. BMC Plant Biol 14:205. https://doi.org/10.1186/s12870-014-0205-7

Liyanage DV (1958) Varieties and forms of the coconut palm grown in Ceylon. Ceylon Coconut Qtly 9:1–10

Loiola CM, Azevedo AON, Diniz LREC, Aragão WM, CD de O Azevedo, Henrique P, Santos AD, Ramos HCC, Pereira MG, Ramos SRR (2016) Genetic relationships among tall coconut palm (*Cocos nucifera* L.) accessions of the International Coconut Genebank for Latin America and the Caribbean (ICG-LAC), evaluated using microsatellite markers (SSRs). PLOS Journals Published between 08–15 March 2016

Maloney BK (1993) Paleoecology and the origin of the coconut. Geo J 31(4):355–362

Manimekalai R, Nagarajan P (2006) Assessing genetic relationships among coconut (*Cocos nucifera* L.) accessions using inter simple sequence repeat markers. Sci Hortic 108:49–54

Mao Z (1986) An investigation on meteriological indices for coconut cultivation in China. Oleagineux 41:119–128

Mao ZS, Lai YT (1993) The coconut germplasm of Hainan Island, China. Plant Genet Resour Newsl 91–92:53–57

Marechal H (1928) Observations and preliminary results on the coconut palm with a view to developing improved seednuts for Fiji. Fiji Agric J 1:16–45

Mariau D, Dery SK, Sangaré A, N'Cho YP, Philippe R (1996) Le jaunissement mortel du cocotier au Ghana et tolérance du matériel végétal. Plantation, Recherche, Développement 3:105–112

Mathai G, Mathew J, Balakrishnan B (1991) Reaction of exotic cultivars of coconut (*Cocos nucifera* L.) to root (wilt) disease of Kerala. In: Silas EG, Aravindakshan M, Jose AI (eds) Proceedings of national symposium on coconut breeding and management. Kerala Agricultural University, Thrissur, pp 161–162

Mauro-Herrera M, Meerow AW, Borrone JW, Schnell RJ (2006) Ten informative markers developed from WRKY sequences in coconut (*Cocos nucifera*). Mol Ecol Notes 6(3):904–906

McCormack G (2005) The origin of the coconut palm. Cook Islands Natural Heritage Trust, Rarotonga. http://cookislands.bishopmuseum.org/species.asp?id=5960. Accessed 2008

Meerow AW, Noblick L, Borrone JW, Couvreur TLP, Mauro-Herrera M, Hahn WJ, Kuhn DN, Nakamura K, Oleas NH, Schnell RJ (2009) Phylogenetic analysis of seven WRKY genes across the Palm Subtribe Attaleinae (Arecaceae) identifies Syagrus as Sister Group of the Coconut. PLoS One 4(10):e7353. https://doi.org/10.1371/journal.pone.0007353

Menon KPV, Pandalai KM (1958) The coconut, a monograph. Indian Central Coconut Committee, Ernakulam, p 384

Menon KPV, Pandalai KM (1960) The coconut palm, a monograph. Indian Coconut Committee, Ernakulum, p 384

Menon KS, Thommen KJ, Sukumaran AS (1981) Review of research on coconut roof (Wilt) disease. ICAR-CPCRI, Kayangulam, pp 78–82

Merrow AW, Wisser RJ, Brown JS, Kuhn DN, Schnell RJ, Broschat TK (2003) Analysis of genetic diversity and population structure within Florida coconut (*Cocos nucifera* L.) using microsatellite DNA, with special emphasis on the Fiji Dwarf cultivar. Theor Appl Genet 106:715–726

Meunier J, Benoit G, Ghesquiere M, Jay M, Bourdeix R (1992) IBPGR workshop on coconut genetic resources, Cipanas, Indonesia, 8–11 Oct1992. International Crop Network Series 8, IBPGR, Rome, Italy, pp 59–62

Miquel FAW (1855) Flora van Nederlandsch Indie. III. C. G. van der Post, Amsterdam, pp. 64–72

Mohandas C, Annif PT, Pavithran K (1976) Anatomical studies on the bulbils of coconut. Curr Sci 45:310–311

Montero-Cortés M, Rodríguez-Paredes F, Burgeff C, Pérez-Nuñez T, Córdova I, Oropeza C, Verdeil JL, Sáenz L (2010) Characterisation of a cyclin-dependent kinase (CDKA) gene expressed during somatic embryogenesis of coconut palm. Plant Cell Tiss Organ 102(2):251–258

Moore D, Alexander L (1990) Resistance of coconut in St. Lucia to attack by the coconut mite Eriophyes guererronis Keifer. Trop Agric 67:33–36

Moran G (1991) Report of isozyme research of coconuts. CSIRO, Dívision of Forest Research Canberra, School of Botany, University of Melbourne, Australia (unpublished)

Mpunami A, Schuiling M, Kaiza D (1992) The lethal disease of coconut palm. Tanzania Res Train Newsl (Dar es Salaam) 7(1–3):12–15

Muthiah C, Bhaskaran R (2000) Major outbreak of eriophyid mite of coconut in India. Planter, Kuala Lumpur 76(889):243–246

Nair CRP (2000) Status of coconut eriophyid mite, *Aceria guerreronis* K. in India. In: Proceedings of International workshop on coconut Eriohyid mite. CRI, Sri Lanka, pp 9–12

Nair RV, Jacob PM, Thomas RJ, Pillai SP, Mathews C (2006) Performance of CGD × WCT hybrid in the root (wilt) disease prevalent tract. J Plant Crops 34(1):15–20

Nair RV, Thomas RJ, Jacob PM, Thomas GV (2009) Kalparaksha, a new coconut variety resistant to root (wilt) disease. Indian Coconut J LII 1:14–16

Nambiar SS (1988) Susceptibility of hybrid coconut varieties to *Oryctes rhinoceros* Linn. (Scarabaeidae) under rainfed conditions at Pilicode. In: Silas EG, Aravindakshan M, Jose AI (eds) Proceedings of national symposium on coconut breeding and management. Kerala Agricultural University, Thrissur, pp 17–18

Nan ON, Hocher V, Verdeil JL, Konan JL, Ballo K, Mondeil F, Malaurie B (2008) Cryopreservation by encapsulation dehydration of plumules of coconut. Cryo-Letters 29:339–350

Narayana GV, John CM (1949) Varieties and forms of coconut. Madras Agric J 36:349–366

Naresh Kumar S, Champakam B, Rajagopal V (2000). Fatty acid composition of coconut oil among coconut cultivars: An insight into industrial application. Indian Coconut J 31(3):25–28

Naresh Kumar S, Kasturi Bai KV, Rajagopal V, Aggarwal PK (2008) Simulating coconut growth, development and yield with the info crop-coconut model. Tree Physiol 28:1049–1058

Nejat N, Cahill DM, Vadamalai G, Ziemann M, Rookes J, Naderali N (2015) Transcriptomics-based analysis using RNA-Seq of the coconut (*Cocos nucifera*) leaf in response to yellow decline phytoplasma infection. Mol Genet Genomics 290(5):1899–1910

Neto FM, Pereira TNS, Geronimo IGC, Azevedo AON, Ramos SRR, Pereira MG (2016) Coconut genome size determined by flow cytometry: tall versus dwarf types. Genet Mol Res 15(1). https://doi.org/10.4238/gmr.15017470

Niral V, Jerard BA (2017) Botany, growth and development. In: Chowdappa P, Niral V, Jerard BA, Samsudeen K (eds) Coconut. Daya Publishing House, New Delhi, pp 73–130

Niral V, Jerard BA (2018) Botany, Origin and Genetic Resources of Coconut. In: Nampoothiri K, Krishnakumar V, Thampan P, Nair M (eds) The Coconut Palm (*Cocos nucifera* L.) - Research and Development Perspectives. Springer, Singapore. pp 57–111. https://doi.org/10.1007/978-981-13-2754-4_32

Niral V, Nair RV, Jerard BA, Samsudeen K, Ratnambal MJ (2009) Evaluation of coconut germplasm for fruit component traits and oil yield. J Oilseeds Res 26(Special issue):668–670

Niral V, Samsudeen K, Nair RV (2010) Genetic resources of coconut. In: Thomas GV, Balasimha D, Krishnakumar V et al (eds) International conference on coconut biodiversity for prosperity. CPCRI, Kasaragod, pp 25–28

Niral V, Devakumar K, Umamaheswari TS et al (2013) Morphological and molecular characterization of a large fruited unique coconut accession from Vaibhaawadi, Maharashtra, India. Indian J Genet 73(2):220–224

Niral V, Augustine Jerard B, Jayasekhar S, Chowdappa P (2014) Tender coconuts for nutritional security. Indian Hortic 59(6):25–27

Odewale JO, Ataga CD, Agho C, Odiowaya G, Okoye MN, Okolo EC (2013) Genotype evaluation of coconut (*Cocos nucifera* L.) and mega environment investigation based on additive main effects and multiplicative interaction (AMMI) analysis. Res J Agric Environ Manag 2(1):001–010

Omar ABH (1919) Races of the coconut palm. Gardens Bull. Straits Settlements 2(5):143

Parthasarathy VA, Geethalakshmi P, Niral V (2004) Analysis of coconut cultivars and hybrids using isozyme polymorphism. Acta Bot Croat 63(1):69–74

Patel JS (1937) Coconut breeding. Proc Assoc Econ Biol 5:1–16

Patel JS (1938) The coconut: a monograph. Government Press, Madras

Perera L (2010) Hybrid testing and variety identification of coconut (*Cocos nucifera* L.) in Sri Lanka using microsatellite markers. Cord 26(1):39

Perera L, Russel JR, Provan J, McNicol JW, Powel W (1998) Evaluating genetic relationships between indigenous coconut (*Cocos nucifera* L.) accessions from Sri Lanka by means of AFLP profiling. Theor Appl Genet 96:545–550

Perera L, Russell JR, Provan J, Powell W (1999) Identification and characterization of microsatellites in coconut (*Cocos nucifera* L.) and the analysis of coconut population in Sri Lanka. Mol Ecol 8:344–346

Perera L, Russell JR, Provan J, Powell W (2000) Use of microsatellite DNA markers to investigate the level of genetic diversity and population genetic structure of coconut (*Cocos nucifera* L.). Genome 43:15–21

Perera L, Russell JR, Provan J, Powell W (2001) Levels and distribution of genetic diversity of coconut (*Cocos nucifera* L, var. *Typica* form *typica*) from Sri Lanka assessed by microsatellite markers. Euphytica 122:381–389

Perera L, Russell JR, Provan J, Powell W (2003) Studying genetic relationships among coconut varieties/populations using microsatellite markers. Euphytica 132:121–128

Perera PIP, Hocher V, Verdeil JL, Bandupriya HDD, Yakandawala DMD, Weerakoon LK (2008) Androgenic potential in coconut (*Cocos nucifera* L.). Plant Cell Tissue Organ 92(3):293–302

Perera PIP, Yakandawala DMD, Hocher V, Verdeil JL, Weerakoon LK (2009) Effect of growth regulators on microspore embryogenesis in coconut anthers. Plant Cell Tissue Organ 96(2):171–180

Perez-Nunez MT, Chan JL, Saenz L, Gonzalez T, Verdeil JL, Oropeza C (2006) Improved somatic embryogenesis from Cocos nucifera (L.) plumule explants. In Vitro Cell Dev Biol-Plant 42(1):37–43

Pérez-Núñez MT, Souza R, Sáenz L, Chan JL, Zuniga-Aguilar JJ, Oropeza C (2009) Detection of a SERK-like gene in coconut and analysis of its expression during the formation of embryogenic callus and somatic embryos. Plant Cell Rep 28(1):11–19

Pesik A, Efendi D, Novarianto H, Dinarti D, Sudarsono S (2017) Development of SNAP markers based on nucleotide variability of WRKY genes in coconut and their validation using multiplex PCR. Biodiversitas J Biol Diversity 18(2):465. https://doi.org/10.13057/biodiv/d180204

Preethi P, Rajesh MK, Naganeeswaran S, Rehaman S, Karun A (2014) Identification of EST-SSRs in coconut (*Cocos nucifera* L.) by deep transcriptome sequencing. In: Book of abstracts of national seminar on 'new horizons and challenges in biotechnology and bioinformatics', 9–10 Oct 2014. ICAR-CPCRI, Kasaragod, p 101

Preethi P, Rajesh MK, Rahul CU, Jerard BA, Samsudeen K, Thomas RJ, Karun A (2016) Identification and utilization of informative EST-SSR markers for genetic purity testing of coconut hybrids. J Plant Crops 44(2):77–84

Preethi P, Rahman S, Naganeeswaran S, Sabana AA, Gangaraj KP, Jerard BA, Niral V, Rajesh MK (2020) Development of EST-SSR markers for genetic diversity analysis in coconut (*Cocos nucifera* L.). Mol Biol Rep 47(12):9385–9397

Puch-Hau C, Oropeza-Salín C, Peraza-Echeverría S, Gongora-Paredes M, Córdova-Lara I, Narvaez-Cab M, Zizumbo-Villareal D, Sáenz-Carbonell L (2015) Molecular cloning and characterization of disease-resistance gene candidates of the nucleotide binding site (NBS) type from *Cocos nucifera* L. Physiol Mol Plant Pathol 89:87–96

Quillec G, Renard JL (1975) Helminthosporiosis of the coconut palm. Preliminary studies. Oleagineux 30(5):209–213

Rachana KE, Naganeeswaran SA, Fayas TP, Thomas RJ, Rajesh MK (2016) Cloning, characterization and expression analysis of NBS-LRR-type resistance gene analogues (RGAs) in coconut. Acta Bot Croat 75(1):1–10

Radhakrishnan TC, Balakrishnan PC (1991) Evaluation of intensity of stem bleeding disease of coconut. In: Silas EG, Aravindakshan M, Jose AI (eds) Proceedings of national symposium on coconut breeding and management. Kerala Agricultural University, Thrissur, pp 163–164

Rajagopal V, Shivashankar S, Kasturi Bai KV, Voleti SR (1988) Leaf water potential as an index of drought tolerance in coconut (*Cocos nucifera* L.). Plant Physiol Biochem 15:80–86

Rajagopal V, Kasturi Bai KV, Voleti SR (1990) Screening of coconut genotypes for drought tolerance. Oleagineux 45:215–223

Rajagopal V, Shivashankar S, Kasturi BKV (1991) Physiological and biochemical basis of coconut production. Cord J VII(2):12–30

Rajagopal V, Kasturi Bai KV, Naresh Kumar S, Niral V (2007) Genetic analysis of drought responsive physiological characters in coconut (*Cocos nucifera* L.). Ind J Hort 64(2):181–189

Rajesh MK, Radha E, Sajini KK, Karun A, Parthasarathy VA (2005) Plant regeneration through organogenesis and somatic embryogenesis from plumular explant of coconut. J Plant Crops 33(1):9–17

Rajesh MK, Nagarajan P, Jerard BA, Arunachalam V, Dhanapal R (2008a) Microsatellite variability of coconut accessions (*Cocos nucifera* L.) from Andaman and Nicobar Islands. Curr Sci 94:1627–1631

Rajesh MK, Arunachalam V, Nagarajan P, Lebrun P, Samsudeen K, Thamban C (2008b) Genetic survey of ten Indian coconut landraces by simple sequence repeats (SSRs). Sci Hortic 118:282–297

Rajesh MK, Thomas RJ, Rijith J, Shareefa M, Jacob PM (2012) Genetic purity assessment of D x T hybrids in coconut with SSR markers. Indian J Genet Plant Breed 72(4):472–474

Rajesh MK, Rachana KE, Babu M, Thomas RJ, Karun A (2013) Characterization of the global transcriptome responsive to root (wilt) disease in coconut using RNA-seq. In: National sym-

posium on 'pathogenomics for diagnosis and management of plant diseases'. ICAR-CTCRI, Thiruvananthapuram
Rajesh MK, Radha E, Sajini KK, Karun A (2014a) Polyamine-induced somatic embryogenesis and plantlet regeneration *in vitro* from plumular explants of dwarf cultivars of coconut (*Cocos nucifera*). Indian J Agr Sci 84(4):527–530
Rajesh MK, Samsudeen K, Rejusha P, Manjula C, Rahma S, Karun A (2014b) Characterization of Annur and Bedakam ecotypes of coconut from Kerala State, India, using microsatellite markers. Int J Biodivers. https://doi.org/10.1155/2014/260895
Rajesh MK, Samsudeen K, Jerard BA, Rejusha P, Karun A (2014c) Genetic and phylogenetic relationships of coconut populations from Amini and Kadmat Islands, Lakshadweep (India). Emir J Food Agric 26(10):898–906
Rajesh MK, Rachana KE, Naganeeswaran SA, Shafeeq R, Thomas RJ, Shareefa M, Merin B, Karun A (2015a) Identification of expressed resistance gene analog sequences in coconut leaf transcriptome and their evolutionary analysis. Turk J Agric For. https://doi.org/10.3906/tar-1409-75
Rajesh MK, Sabana AA, Rachana KE, Rahman S, Jerard BA, Karun A (2015b) Genetic relationship and diversity among coconut (*Cocos nucifera* L.) accessions revealed through ScoT [Start Codon Targeted polymorphism] analysis. 3 Biotech. https://doi.org/10.1007/s13205-015-0304-7
Rajesh MK, Fayas TP, Naganeeswaran S, Rachana KE, Bhavyashree U, Sajini KK, Karun A (2016) *De novo* assembly and characterization of global transcriptome of coconut palm (*Cocos nucifera* L.) embryogenic calli using Illumina paired-end sequencing. Protoplasma 253(3):913–928
Rajesh MK, Rachana KE, Kulkarni K, Sahu BB, Thomas RJ, Karun A (2018) Comparative transcriptome profiling of healthy and diseased Chowghat Green Dwarf coconut palms from root (wilt) disease hot spots. Eur J Plant Pathol 151(1):173–193. https://doi.org/10.1007/s10658-017-1365-8
Rajesh MK, Karun A, Parthasarathy VA (2019) Coconut biotechnology. In: KUK N, Krishnakumar V, Thampan PK, Nair MA (eds) The coconut palm (*Cocos nucifera* L.)-research and development perspectives. Springer, Singapore, pp 191–226
Rajesh MK, Chowdappa P, Behera SK, Kasaragod S, Gangaraj KP, Kotimoole CN, Nekrakalaya B, Mohanty V, Sampgod RB, Banerjee G, Das AJ (2020) Assembly and annotation of the nuclear and organellar genomes of a dwarf coconut (Chowghat Green Dwarf) possessing enhanced disease resistance. OMICS 24(12):726–742
Raju CR (2006) Direct *in vitro* shoot development from immature rachilla explants of coconut (*Cocos nucifera* L.). CORD 22(1):51–58
Ramanujam B, Nambiar KKN, Ratnambal MJ (1998) Screening of coconut cultivars/hybrids against *Thielaviopsis paradoxa* (de Seynes) Hohnel, using petiole inoculation technique. In: Mathew NM, Kuruvilla Jacob C (eds) Proceedings of PLACROSYM XII. Allied Publishers, New Delhi, pp 284–286
Ramesh SV, Hebbar KB, Rajesh MK, Abhin Sukumar P, Gangaraj KP, Athul Bobby (2020) Transcriptome analysis of *Cocos nucifera* L. seedlings having contrasting water-use efficiency (WUE) under water-deficit stress: Molecular insights and genetic markers for drought tolerance. In: The first international electronic conference on plant science. https://doi.org/10.3390/IECPS2020-08853
Rasam DV, Gokhale NB, Sawardekar SV, Patil DM (2016) Molecular characterisation of coconut (*Cocos nucifera* L.) varieties using ISSR and SSR markers. J Hortic Sci Biotechnol 91:347–352
Ray SKD, Kaiser SAKM (1990) Observations on some elite genotypes (cultivars) of coconut palm for resistance to leaf blight (grey leaf spot) disease. J Mycopathol Res 28(1):9–11
Rhode W, Kullaya A, Rodriguez J, Ritter E (1995) Genome analysis of *Cocos nucifera* L. by PCR amplification of spacer sequences separating a subset of copia-like EcoRI repetitive elements. J Genet Breed 49:179–186

Riedel M, Riederer M, Becker D, Herran A, Kullaya A, Arana-López G, Peña-Rodríguez L, Billotte N, Sniady V, Rohde W, Ritter E (2009) Cuticular wax composition in *Cocos nucifera* L. physicochemical analysis of wax components and mapping of their QTLs onto the coconut molecular linkage map. Tree Genet Genomes 5(1):53–69

Rigby JF (1995) A fossil *Cocos nucifera* L. fruit from the latest Pliocene of Queensland, Australia. In: Pant DD (ed) Birbal Sahni centennial volume. Allahabad University, Allahabad, pp 379–381

Ritto P, Jiji G, Rajesh MK, Jerard BA, Niral V (2008) Genetic diversity in yellow dwarf populations of coconut (*Cocos nucifera* L.) assessed using RAPD markers. Indian J Hortic Res 65:60–64

Rivera R, Edwards KJ, Barker JHA, Arnold GM, Ayad G, Hodgkin T, Karp A (1999) Isolation and characterization of polymorphic microsatellites in *Cocos nucifera* L. Genome 42:668–675

Rodriguez MJB, Estioko LP, Namia TI, Soniega JA (1997) Analysis of genetic diversity in coconut by RAPD. Philipp J Coconut Stud 22:1–7

Rohde W, Salamani F, Ashburner R, Randles JW (1992) An *Eco* RI repetitive sequence family of the coconut palm (*Cocos nucifera* L.) shows sequence homology to copia-like elements. J Genet Breed 46:391–394

Sadakathulla S, Ramachandran TK (1993) Field screening of coconut cultivars to red palm weevil at Tamil Nadu. Ann Plant Res Insects Newsl 19:43

Saensuk C, Wanchana S, Choowongkomon K, Wongpornchai S, Kraithong T, Imsabai W, Chaichoompu E, Ruanjaichon V, Toojinda T, Vanavichit A, Arikit S (2016) *De novo* transcriptome assembly and identification of the gene conferring a "pandan-like" aroma in coconut (*Cocos nucifera* L.). Plant Sci 252:324–334

Sajini KK, Karun A, Amarnath CH, Engelmann F (2011) Cryopreservation of coconut (*Cocos nucifera* L.) zygotic embryos by vitrification. Cryo-Letters 32(4):317–328

Samsudeen K, Rajesh MK, Nagwaker DD, Reshmi R, Kumar PA, Devadas K, Anitha K (2013) Diversity in *Mohachao Narel*, a sweet endosperm coconut (*Cocos nucifera* L.) population from Maharashtra, India. Natl Acad Sci Lett 36(3):319–330.

Samosir YMS (1999) Optimisation of somatic embryogenesis in coconut (*Cocos nucifera* L.). PhD Thesis, The University of Queensland, Australia

Sangare A, de Taffin G, de Franqueville H, Arkhust ED, Pomier M (1992) Coconut lethal yellowing in Ghana. Initial results on planting material performance in the field. Oléagineux 47(12):699–704

Sankaran M, Damodaran V, Jerard BA, Abirami K, Roy DS (2015) Multiple *Spicata* coconut (MSC): a rare type of coconut in Andaman Islands. Transcriptomics 3:123

Satyabalan K (1997) Coconut varieties and cultivars – their classification. Asian Pacific Coconut Community, Jakarta, p 105

Schuiling M, Mpunami A, Kaiza DA, Harries HC (1992) Lethal disease of coconut palm in Tanzania. III. Low resistance of imported germplasm. Oleagineux 47(12):693–698

Serrano AV, Cortazar Ríos M, Ovando Cruz ME (2011) Donají: new coconut hybrid resistant to lethal yellowing in Mexico. Revista Mexicana de Ciencias Agrícolas 2(5):773–778

Shalini KV, Manjunatha S, Lebrun P, Berger A, Baudouin L, Pirany N, Ranganath RM, Prasad DT (2007) Identification of molecular markers associated with mite resistance in coconut (*Cocos nucifera* L.). Genome 50:35–42

Shareefa M, Thomas RJ, Sreelekshmi JS, Rajesh MK, Karun A (2019) *In vitro* regeneration of coconut plantlets from immature inflorescence. Curr Sci 117(5):813–820

Sherman KE, Maramorosch K (1977) Present status of lethal yellowing disease of the coconut palm. J Plant Crops 5(2):75–83

Sisunandar RA, Turquay P, Samosir Y, Adkins SW (2010a) Cryopreservation of coconut (*Cocos nucifera* L.) zygotic embryos does not induce morphological, cytological or molecular changes in recovered seedlings. Planta 232:435–447

Sisunandar RA, Sopade PA, Samosir YM, Rival A, Adkins SW (2010b) Dehydration improves cryopreservation of coconut (*Cocos nucifera* L.). Cryobiology 61(3):289–296

Sumangala Nambiar S (1991) Susceptibility of hybrid coconut varieties to Oryctes rhinoceros under rain fed conditions at Pilicode. In: Silas EG, Aravindakshan M, Jose AI (eds) Coconut breeding and management. Kerala Agricultural University, Thrissur, pp 158–160

Sun R, Ye R, Gao L, Zhang L, Wang R, Mao T, Zheng Y, Li D, Lin Y (2017) Characterization and ectopic expression of CoWRI1, an AP2/EREBP domain-containing transcription factor from coconut (*Cocos nucifera* L.) endosperm, changes the seeds oil content in transgenic *Arabidopsis thaliana* and rice (*Oryza sativa* L.). Front Plant Sci 8:63. https://doi.org/10.3389/fpls.2017.0006

Suriachandraselvam M, Bhaskaran R, Ramachandran TK (2000) Evaluation of coconut hybrids/varieties against grey leaf spot diseases. Madras Agric J 87(7/9):497–498

Teulat B, Aldam C, Trehin R, Lebrun P, Barker JHA, Arnold GM, Karp A, Baudouin L, Rognon F (2000) An analysis of genetic diversity in coconut (*Cocos nucifera L.*) population from across the geographical range using sequence tagged microsatellites (SSRs) and AFLPs. Theor Appl Genet 106:411–422

Trimen H (1898) Handbook of the flora of Ceylon, Part IV. Dulan & Co., London, p 337

Tripathi RP, Mishra SN, Sharma BD (1999) Cocos nucifera-like petrified fruit from the tertiary of Amarkantak MP, India. Paleobotanist 48:251–255

Upadhyay A, Jayadev K, Manimekalai R, Parthasarathy VA (2004) Genetic relationship and diversity in Indian coconut accessions based on RAPD markers. Sci Hortic 99:353–362

Varghese MK (1934) Diseases of the coconut palm. Government Press, Trivandrum, p 105

Verdeil JL, Huet C, Grosdemange F, Buffard Morel J (1994) Plant regeneration from cultured immature inflorescences coconut (*Cocos nucifera* L.): evidence for somatic embryogenesis. Plant Cell Rep 13:218–221

Vidhanaarachch VR, Weerakoon LK, Fernando SC, Gamage CKA, Santha ES (1998) Status of research on coconut embryo culture and acclimatization techniques in Sri Lanka. In: Batugal PA, Engelmann F (eds) Coconut embryo *in vitro* culture. IPGRI, APO, Serdang, pp 85–88

Warwick DRN, Bezerra APO, Renard JL (1991) Reaction of coconut hybrids to leaf blight (*Botryodiplodia theobromae* Pat.). 1. Field observations. Oléagineux 46(3):100–106

Watt G (1889) Dictionary of the economic products of India. Coconut 2:418

Wheeler REM, Ghosh A, Deva K (1946) Arikamedu: an Indo-Roman trading station on east coast of India. Ancient India 2:17–124

Whitehead RA, Chapman GP (1962) Twinning and haploidy in *Cocos nucifera*. Nature 195:1228–1229

Xiao Y, Xu P, Fan H, Baudouin L, Xia W, Bocs S, Xu J, Li Q, Guo A, Zhou L, Li J (2017) The genome draft of coconut (*Cocos nucifera*). GigaScience 6(11):1–11

Xu L, Ye R, Zheng Y, Wang Z, Zhou P, Lin Y, Li D (2010) Isolation of the endosperm-specific *LPAAT* gene promoter from coconut (*Cocos nucifera* L.) and its functional analysis in transgenic rice plants. Plant Cell Rep 29:1061–1068

Zizumbo Villarreal D, Cardena-Lopez R, Pinero D (2002) Diversity and phylogenetic analysis in *Cocos nucifera* L. in Mexico. Genet Resour Crop Evol 49(3):237–245

Zizumbo VD, Fernandez M, Torres N, Cardena R (1999) Lethal yellowing resistance in coconut germplasm from Mexico. Kluwer Academic Publishers, Dordrecht

# Oil Palm

**Alain Rival**

**Abstract** Innovative and efficient methods are required to breed the oil palm for yield maximization, in order to meet the increased global demand for vegetable oil while limiting the environmental impacts of oil palm cultivation.

Breeders have been able to develop elite oil palm planting materials from an extremely narrow genetic base. Breeding the oil palm for resilience to global change requires multidisciplinary and collaborative research involving almost all disciplines related to life sciences. Research also relies on identifying genetic variation in the plant responses to stress, and this implies the exploitation of natural variation, germplasm collections, selected genitors from breeding programmes together with material of interest collected from farmers.

Climate change can highly affect the photosynthetic efficiency of oil palm and thus it directly affect yield; in parallel, oil palm cultivation also influences climate change, as the environmental impact of the industry primarily concerns the conversion of tropical rainforests into plantations by both agro-industrial companies and smallholding farmers.

The publication of the oil palm genome sequence has been a major breakthrough in oil palm biotechnology. Even if considerable knowledge and molecular resources had already been made available before this date, the availability of the genome sequence expedited a number of research projects ranging from genomic selection to the deciphering of key components of yield such as sex ratio or oil synthesis. Metagenomics can provide a useful bridging between genomics and environmental science, while addressing key question in the sustainable management of oil palm plantations.

**Keywords** Climatic resilience · Disease resistance · Epigenetics · Genomics · Oil yield · Sex ratio

A. Rival (✉)
Focal Point for Oil Palm Research at CIRAD, Regional Direction for South East Asian Island Countries, Jakarta, Indonesia
e-mail: alain.rival@cirad.fr



P. M. Priyadarshan, S. M. Jain (eds.), *Cash Crops*,
https://doi.org/10.1007/978-3-030-74926-2_8

## 1 Origin of the Oil Palm

The Palms (*Arecaceae*) form one of the most distinctive of all plant families, and they are living emblems of the tropical regions of the world in which most of their natural biodiversity is concentrated. According to the most recent estimates, the palm family contains nearly 2500 species. A large number of them are exploited in some way for human purposes; indeed, the *Arecaceae* have been recognized as being one of the three most important plant families in terms of human usage (Tregear et al. 2011). Although a handful of species such as date palm (*Phoenix dactylifera* L.), oil palm (*Elaeis guineensis* Jacq.) and coconut (*Cocos nucifera* L.) occupy a particularly high profile due to their economic importance, many members of the *Arecaceae* have uses that extend far beyond their natural habitat. *Elaeis oleifera* is a species in the oil palm genus along with the commercial African palm *E. guineensis*, which occurs naturally in South-Central America, from Honduras to Colombia and in the Amazon region.

The large-scale cultivation of oil palm (*Elaeis guineensis* Jacq.) has been able to drastically transform the tropical landscapes and livelihoods in recent decades (Rival and Levang 2014). This outstanding crop produces very high oil yields and shows an amazing agronomic plasticity; these are valuable assets in the context of climate change and ecological intensification.

Palm oil is primarily a forest product. Indeed, before becoming the *prima donna* of agro-industry in Southeast Asia and the number one public enemy of many environmental NGOs, palm oil began its career as a simple forest product in Africa. *E. guineensis* originated in the humid tropical forests bordering the Gulf of Guinea in the both West and Central Africa. The species is endemic to forests in the tropical zones of Sierra Leone, Liberia, Côte d'Ivoire, Ghana, Togo, Benin, Nigeria and Cameroon and to the equatorial zones of Gabon, Republic of Congo (Congo-Brazzaville), Democratic Republic of Congo (Congo-Kinshasa) and Angola (Bakoumé 2016).

The natural palm groves in countries bordering the Gulf of Guinea are semi-domesticated: people just encourage the plant to develop. Because the reproductive capacity of the palm is high (several bunches each with several thousand fruit per year!), it proliferates naturally and does not need to be planted. Animals and people can disperse the seeds at harvest time. The oil palm is a pioneer species, which develops particularly well in areas cleared for slash-and-burn cultivation of crops (Rival and Levang 2014) (Fig. 1).

Oil palm has played a fundamental role in the nutrition, technological, economic and social development of many African forest peoples. For these populations, the oil palm is pivotal, as it provides wine, oil and sauce from seeds. The oil palm is also used to build houses, provide flooring and weave hunting nets and traps as well as build dams to catch fish (Cochard and Durand-Gasselin 2018). Despite its ancient relationship with the populations of African forests, the oil palm is mostly only semi-domesticated. No attempts were made to propagate it by sowing seeds and/or transplanting seedlings (Zeven 1972). Hence, the oil palm could be considered as

**Fig. 1** Wild oil palm (*Elaeis guineensis* Jacq) grove

having not been subjected to breeding historically, meaning that at the beginning of the twentieth century, it was a semi-wild crop. In the middle of the nineteenth century, oil palm was planted in different botanical gardens and breeding started at the beginning of the twentieth century in both Africa and Asia.

Despite its large canopy and characteristic shoe, the oil palm is not a tree, botanically speaking. It bears a single terminal meristem, which produces an average of two leaves per month. Vertical growth rates can vary from 30 to 75 cm per annum, depending on the genotype. The root system is composed of a large number of fasciculate adventive roots. Oil palm grows preferentially in rich and temporarily damp alluvial soils (Corley and Tinker 2016). It can reach 30 m in height, with a crown of 4–5 opened palmate leaves of 10–16 m in diameter, and therefore requires a large area for growth. The standard density in industrial plantations is generally 143 palms per hectare. The oil palm is a temporal dioecious species (Cruden 1988), which displays alternate male and female flowering cycles throughout the life of the plant (Fig. 2).

The sustainable development of oil palm cultivation is generating several questions to research, not only in the traditional domains of plant science, including breeding and biotechnologies (Goh 2016; Rival 2020), but also in the fields of

**Fig. 2** Agro-industrial oil palm plantation in Indonesia

socioeconomics and social and human sciences (Cramb and McCarthy 2016; Sodano et al. 2018).

## 2 Diversity of the Oil Palm

### *The African Oil Palm*

*Elaeis guineensis* Jacq., the African oil palm, is naturally abundant in all African rain forests. Both climate and humans shaped the modern biogeographic distribution of oil palm (Barcelos et al. 2015). Analysis of the natural genetic diversity of the species suggests that wild populations can be separated into three groups, located at the extreme west of Africa, equatorial Africa and in Madagascar Island. The highest allelic diversity was found among Nigerian palm populations, indicating the possible centre of origin of *Elaeis guineensis* (Bakoumé et al. 2015).

There are three oil palm (*E. guineensis*) varieties, namely dura, pisifera and tenera, which can be distinguished by the thickness of the fruit shell and the ratio of mesocarp to fruit (Fig. 3). Dura is the thick-shelled variety (2–8 mm) with low to medium mesocarp content (35–55%) while pisifera is the shell-less variety. Tenera fruit has a thin shell (0.5–4 mm) and a medium to high mesocarp content (60–96%). Beirnart and Vanderweyen (1941) revealed he simple Mendelian inheritance of the fruit shell thickness trait and provided the basis of modern oil palm breeding and hybrid seed production programmes: the thick shell dura being homozygous dominant, the shell-less pisifera homozygous recessive and the thin-shelled tenera heterozygous. Singh et al. (2013a) have elucidated the gene responsible for the inheritance of this character. African oil palms have different colouring of exocarp, producing either *nigrescens* or *virescens* fruit types. Type *nigrescens* accumulate

**Fig. 3** The three oil palm (*E. guineensis*) varieties, namely dura, pisifera and tenera, can be distinguished by the thickness of the fruit shell and the ratio of mesocarp to fruit

**Fig. 4** African oil palms show different colouring of exocarp, producing either nigrescens or virescens fruit types. Type nigrescens accumulate large amounts of anthocyanins, while virescens fruits are green turning orange during ripening, due to the accumulation of carotenoids

large amounts of anthocyanins, which accounts for the deep violet to black colour (Fig. 4). When unripe, *virescens* fruits are green turning orange due to the accumulation of carotenoids and chlorophyll degradation in relation to ripening (Fig. 5).

**Fig. 5** Traditional harvesting of oil palm bunches in Benin (West Africa)

## *The American Oil Palm*

*Elaeis oleifera* (Kunth, Cortés), the American oil palm, is native to and broadly dispersed in Central America and the northern regions of South America. Wild *E. oleifera* populations grow along the riverbanks, tolerating well both shade and flooding, thus indicating a broader environmental adaptability compared to the African oil palm (Corley and Tinker 2015). The Amazon River Basin is considered as a centre of secondary diversification, although there are no historical indications of artificial selection to improve oil yield, thus the oil to bunch ratio of *E. oleifera* is about 5%, as compared to 25% in teneras from *E. guineensis* (Barcelos et al. 2015).

The American oil palm shows a short, often procumbent stipe. Interestingly, parthenocarpic fruits may constitute up to 90% of *E. oleifera* fruit bunches as compared to the African species. Parthenocarpic fruits are often abortive, thus contributing to poor yield. Research is under way in order to improve fruit setting in the American oil palm, through the generation of parthenocarpic fruits using exogenous plant growth regulators (Thomas et al. 1973).

### *The Interspecific Hybrids*

African and American oil palm species are sexually compatible, and F1 hybrids show vegetative vigour and mid-parent stem growth increment. Barcelos et al. (2015) showed that *E. oleifera* leaf morphology and parthenocarpic fruit development behave as dominant traits in hybrid oil palms. Assisted pollination is required to achieve commercially acceptable yields from hybrids, and this operation hampers the large-scale cultivation of hybrid material. Moreover, depending on the genetic origin of parents, several types of developmental abnormalities could contribute to lower F1 hybrid fertility, including lower pollen yield, poor pollen germination, poor anther dehiscence (Corley and Tinker 2015). There is a growing interest in the exploitation of growth regulators for the large-scale induction of parthenocarpic fruit development in oil palm interspecific hybrids (Pandolfini et al. 2018).

Lieb et al. (2017) assessed morphological traits, total lipid contents and fatty acid profiles in fruits from several accessions of *E. oleifera*, *E. guineensis* and their interspecific hybrids. Hybrids were found to display the highest mesocarp-to-fruit ratios (77.9–78.2%), and the total lipid contents of both *E. guineensis* mesocarp and kernel were significantly higher than for *E. oleifera*. Major fatty acids comprised C16:0, C18:1n9 and C18:2n6 in mesocarp and C12:0, C14:0 and C18:1n9 in kernels. *E. oleifera* samples showed by a higher proportion of unsaturated long-chain fatty acids.

## 3 Genetic Diversity and Conservation of Genetic Resources

Breeders have developed oil palm planting materials from an extremely narrow genetic base. Breeding strategies are based on populations developed in different countries or in different estates from the 1920s, which generally originated from a limited number of founders (Cochard and Durand-Gasselin 2018). Rosenquist (1986) named such populations 'Breeding Population of Restricted Origin' (BPRO) which are still now at the basis of the distinction between the principal oil palm seed suppliers worldwide. The main BPROs originate from Africa (Cote d'Ivoire, Cameroon, Democratic Republic of Congo, Nigeria and Benin) and Southeast Asia (Indonesia and Malaysia). In 1848, the Dutch imported oil palm seeds from West Africa, and four oil palm seedlings were planted in Indonesia at the Bogor (then Buitenzorg) Botanical Gardens. From these four palms, two came directly from Bourbon Island (now La Réunion Island, a French overseas region), and two others through Amsterdam Botanic Garden. These four palms are at the origin of the famous Deli populations from Sumatra Island, which provided the genetic foundation for oil palm improvement in Malaysia and Indonesia and further generated a large proportion of the present selected planting material (Corley and Tinker 2015).

The narrowness of the gene pool has been a major obstacle towards increasing yields and resistance in genetically improved material. The need to broaden the genetic base for continuous yield improvement has been recognized by oil palm breeders worldwide. This concern provided the initial impetus for research organizations such as MPOB in Malaysia to carry out several expeditions to the various centres of diversity for oil palm in West Africa and South America (Rajanaidu et al. 2018).

## 4 Ex Situ and In Situ Collections

Major oil palm research institutions are characterizing the natural diversity of both oil palms (*E. oleifera* and *E. guineensis*) either by prospection of in situ natural populations or by the establishment of ex situ germplasm collections. Oil palm germplasm that exhibits economically important traits is a valuable resource: the first attempt to broaden the genetic base of breeding materials was accomplished in Nigeria in 1973 by the Malaysia Palm Oil Board (MPOB), and sampled material was assessed in Malaysia. Besides Nigeria, prospection was made from other countries in Africa such as Angola, Cameroon, Zaire (now known as the Democratic Republic of Congo), Ghana, Guinea, Gambia, Madagascar, Sierra Leone, Senegal and Tanzania (Rajanaidu et al. 2018). Prospections were organized to sample wild material at random, on a geographical base. The breeding populations which form the genetic base and diversity of most, if not all, of the current oil palm breeding programmes have been derived from a limited number of progenitor palms prospected by early oil palm breeders at the various centres of origin and diversity in Western Africa. In Latin and Central America, *Elaeis oleifera* was collected in Colombia, Panama, Costa Rica, Honduras, Brazil and Suriname in 1981–1982, and the palms were screened for fatty acid composition (Soh et al. 2017).

Since the 1950s, large numbers of oil palms collected from in situ natural populations have been screened to isolate just a few elite individuals that were then used in breeding programmes (Barcelos et al. 2015). This strategy was largely adopted by the French Institut de Recherches pour les Huiles et Oléagineux (IHRO), presently CIRAD. Since 1970s, to capture broader spectra of natural variation and to preserve natural biodiversity, the Palm Oil Research Institute of Malaysia (PORIM), presently MPOB, engaged in random sampling of palms from wild populations to establish ex situ germplasm collection of 1467 accessions (Rajanaidu 2018). This strategy proved more expensive than in situ prospection: as an example, MPOB Nigerian collection alone occupies 200 ha. The major limitations for the in situ conservation of oil palm are the large needed, and the risk of losses because of extreme natural events or past and diseases. However, ex situ approach allows a more thorough characterization of the traits of interest, and it provides guarantees for the safe preservation of traits that might become of interest in the future.

Phenotypic screens of the *E. guineensis* germplasm collections conducted by the major oil palm research centres MPOB (Malaysia), CIRAD (France), IOPRI (Indonesia), and the Empresa Brasileira de Pesquisa Agropecuária (EMBRAPA,

Brazil) revealed a significant phenotypic diversity for various valuable agronomical characteristics (Barcelos et al. 2015).

## 5 Improvement and Climate Resilience

Breeding the oil palm for resilience to global change requires multidisciplinary and collaborative research involving almost all disciplines related to life sciences. Research also relies on identifying genetic variation in the plant responses to stress, and this implies the exploitation of natural variation, germplasm collections, selected genitors from breeding programmes together with material of interest collected from farmers.

Climate change is undoubtedly affecting the conditions of oil palm cultivation by abiotic (i.e., rainfall, temperatures, carbon dioxide and soil salinity) and biotic (i.e., diseases, pests, pollinators and associated crops) stresses (Rival 2017). The phenotyping of selected plant material under biotic/abiotic stresses involve new methods for high-throughput phenotyping, and genomic approaches are followed for the identification of genes underlying the variation of traits that will be used as selection targets. Improvements in understanding how climate change may influence chemical and physical processes in soils, how this may affect nutrient availability, and how the plant responds to changed availability of nutrients will also influence oil palm breeding. With genomic resources becoming increasingly available for the oil palm (such as sequencing, resequencing and chips development), the exploration of the genetic basis of complex traits such as oil yield or resistance to disease is now possible (Forster et al. 2017). Consequently, the availability and sharing of such a large amount of data is currently reshaping most of oil palm breeding strategies.

Considerations of how climate change affects tropical economic plants are limited, and for oil palm, they are even fewer. Oil palm being a strictly tropical plant, the margin for adapting to higher temperatures and changing humidity is reduced (Paterson et al. 2013). It is anticipated that suitable areas for oil palm cultivation will become increasingly scarce for growing oil palm, as climate change will also affect crops growing in the same biotope. In this scenario, the plants will become stressed allowing ingress of fungal diseases and novel diseases may threaten the crop. In addition, new land will be increasingly suitable where the environmental conditions are less severe than in the tropics.

Sarkar et al. (2020) assessed the impacts of climate change on oil palm production in Malaysia by using time series data from 1980 to 2010. A negative and significant relationship was found between annual average temperature and oil palm production. If temperature rises by 1, 2, 3 and 4 °C, production of oil palm can decrease from a range of 10 to 41%. The authors also described a negative impact of sea level rise (SLR) on oil palm production. Findings reveal that if areas under oil palm production decrease by 2%, 4% and 8% due to SLR of 0.5, 1 and 2 m, oil palm production can decrease by 1.98%, 3.96% and 7.92%, respectively, indicating that climate change has a significant impact on the reduction of oil palm production in Malaysia, ultimately affecting the sustainability of the sector.

Critically, no studies have investigated the effects of climate change on pollination, despite the impacts of rain and temperature on pollination efficiency (Li et al. 2019). These authors found that coleopterans increase fruit set, with thrips and moths also contributing. Population drivers of major pollinators were found to be biotic, management and climate.

Climate change can highly affect photosynthetic efficiency in oil palm and thus directly affect yield. Neoh et al. (2017) discussed the deployment and contribution of omics platforms in photosynthesis research (genomics, transcriptomics, proteomics, metabolomics and phenomics) for yield enhancement through breeding selection and planting strategies.

Oil palm cultivation also influences climate change, as the environmental impact of the industry primarily concerns the conversion of tropical rainforests into plantations. Most of the area used for the expansion of oil palm originated from forests where the emissions from conversion exceed the potential carbon fixing of oil palm plantations (Germer and Sauerborn 2008). Oil palm is often quoted as a 'major driver' of deforestation and thus biodiversity loss, but global oil palm expansion, including non-forest sites, was equal to less than 10% of forest loss in oil palm-growing countries from 2000 to 2010 (Baron et al. 2017). Most studies do take into account delays, often of several years or decades, between deforestation and oil palm planting. However, even degraded forest contains much more carbon than an oil palm plantation, and the only way to eliminate GHG emissions from land use change, and to minimize biodiversity loss, is to avoid clearing any forest (Corley 2016).

Meijide et al. (2020) recently conducted LCA (Life Cycle Assessment) studies based on measured $CO_2$, $CH_4$ and $N_2O$ fluxes in young and mature Indonesian oil palm plantations. $CO_2$ was found to dominate the on-site GHG budgets: the young plantation is a carbon source and the mature plantation a sink. Plantation rotation-cycle extension and earlier yielding varieties can potentially decrease GHG emissions. Due to the high emissions associated with forest conversion to oil palm, results indicate that only second rotation-cycle plantations or plantations established on degraded land has the potential for pronounced GHG emission savings.

Apart from minimizing deforestation, mitigation of the damaging effects of the crop will require capturing methane from effluent digestion (Ohimain and Izah 2017), and more efficient recycling of nutrients (Corley 2016). High fertilizer use in the plantation conflicts with the need to reduce the emissions of greenhouse gases in fertilizer manufacturing, transportation and application. Therefore, improved nutrient uptake efficiency is important and can be addressed by breeding for suitable root systems. A remobilization of nitrogen and micronutrients within the plants will be necessary. Prolonged root uptake and better remobilization of nutrients are targets for breeding, provided there is sufficient plasticity of these characteristics in the oil palm for increases in efficiency to be achieved (Ollivier et al. 2016).

In order to investigate limiters of photosynthate assimilation in oil palm, Neoh et al. (2019) measured differential metabolite, gene expression and gas exchanges in leaves in an open field, for palms with distinct mesocarp oil content. The authors observed higher concentrations of glucose 1-phosphate, glucose 6-phosphate, sucrose 6-phosphate and sucrose in high-oil content palms with the greatest difference being at 11:00 am, immediately after the period of low morning light intensity. Three important photosynthetic genes were identified using differentially expressed gene analysis (DEGs) and were found to be significantly enriched through Gene Ontology (GO) and pathway enrichment: chlorophyll a-b binding protein (CAB-13), photosystem I (PSI) and ferredoxin NADP reductase (FNR). Subsequent gas exchange measurements further supported increased carbon assimilation through higher level of internal $CO_2$ concentration, stomatal conductance and transpiration rate in high-oil content palms. The selection for higher expression of key photosynthesis genes together with $CO_2$ assimilation under low light is likely to be important for oil palm improvement, in particular at full maturity and under high-density planting regimes where light competition exists between palms.

## 6 Disease Resistance

When breeding the oil palm for disease resistance, attention must be paid to the plant responses to the occurrence and spread of pest and diseases on the one hand and to changes in patho-systems on the other hand (Paterson et al. 2013). It is thus of paramount importance to explore the genetic determinisms governing tolerance of major diseases. Indeed, the oil palm is susceptible to three rather specific diseases in each of the three continents where it is cultivated.

In Africa, vascular wilt, caused by *Fusarium oxysporum elaeidis*, can cause up to 70% mortality. In Southeast Asia, *Ganoderma boninense* has been found to cause up to 80% mortality in some regions. In Latin America, a bud rot disease of yet unknown origin, although *Phytophthora* is a prime suspect, can be responsible for 100% losses in oil palm plantations (Cochard et al. 2005). For these three rather endemic diseases, genetic control strategies are being developed; indeed, the improvement of oil palm for disease tolerance is based on field observations of the susceptibility of parental material and their crosses and on the assessment of the tolerance of each cross of interest through inoculation by the pathogens at the pre-nursery stage. This process is very long and uncertain as to the genetic value of the parents and to the relationship between controlled inoculations and natural infection in the field, and this has to be assessed on a case-by-case basis. However, the data already generated and currently being processed have allowed the implementation of quantitative genetic analyses to explore the genetic basis of oil palm tolerance to these three major diseases.

QTL research is being implemented for *Fusarium* wilt tolerance through the pre-nursery inoculation of a large mapping population, while testing for *Ganoderma* tolerance is based on a multi-parental genetic trial involving 14 families evaluated in the field in Indonesia for 25 years. In Latin America, bud rot resistance is assessed on first- and second-generation interspecific backcrosses planted in zones severely affected by the disease. The identification of genetic determinisms by appropriate methods will allow (a) identification of candidate genes on the basis of variation in disease tolerance within the species; (b) elucidation of the key processes involved in tolerance through cross-checking between different diseases and between nursery tests and natural conditions; and (c) implementation of marker-assisted selection (MAS) for the rapid and relevant integration of resistance sources into elite genotypes.

In order to process field data obtained under natural conditions of infection, analyses are based on epidemiological models coupled with genetic models, such as QTL or GWAS, or bulk segregant analysis. When the disease is very aggressive, the use of early tests at the pre-nursery stage requires preliminary analyses before being implemented. In both the cases, the genetic mapping approach is coupled with transcriptomic studies on crosses identified as sensitive or resistant in order to identify genes that are differentially expressed during infection and between susceptible and resistant material. The combination of the two approaches will enable the identification of candidate genes as a basis for functional and diversity studies with the aim of involving such genes in an MAS scheme.

Multi-parental populations are promising tools for identifying quantitative disease resistance loci. Stem rot is a major threat to palm oil production, prompting premature replantation of palms. There is evidence of genetic resistance sources (Durand-Gasselin et al. 2005), but the genetic architecture of *Ganoderma* resistance has not yet been investigated. Tisné et al. (2017) identified *Ganoderma* resistance loci using an oil palm multi-parental population derived from nine major founders of ongoing breeding programmes. A total of 1200 palm trees of the multi-parental population were planted in plots naturally infected by Ganoderma, and their health status was assessed biannually over 25 years. The data were treated as survival data and modelled using the Cox regression model, including a spatial effect to take the spatial component in the spread of *Ganoderma* into account. Based on the genotypes of 757 palm trees out of the 1200 planted, and on pedigree information, resistance loci were identified using a random effect with identity-by-descent kinship matrices as covariance matrices in the Cox model. Four *Ganoderma* resistance loci were identified, two controlling the occurrence of the initial stem rot symptoms, and two controlling the death of palm trees, while favourable haplotypes were identified among a major gene pool for ongoing breeding programmes. Tisné et al. (2017) developed an efficient and flexible QTL mapping approach and generated unique valuable information for the selection of oil palm with resistance to stem rot disease.

Oil palm estates in Southeast Asia face substantial losses due to basal stem rot. Field observations have been carried out in North Sumatra, Indonesia, on planting materials of known origin (Durand-Gasselin et al. 2005). Differences in disease susceptibility have been detected within *E. guineensis* and *E. oleifera*. Within *E. guineensis*, planting material of Deli origin is highly susceptible compared to material from African origin. Differences also exist in the response to infection between parents and between crosses within a given origin. The variability of resistance to basal stem rot within the same cross is also illustrated by the diverse responses of clones derived from palms of the same origin. Such results are encouraging for the development of early selection tests for genetic resistance to basal stem rot (Mercière et al. 2015).

Early screening tests for genetic resistance to diseases at the nursery or pre-nursery stage are based on artificial inoculation and correlated with field observations. The success of nursery seedling artificial inoculation relies on parameters; defining the inoculum potential, i.e. the aggressiveness of *G. boninense* isolates, the incubation period of pre-infected rubber wood blocks (RWBs), the ratio between the size of pre-infected RWBs and the volume of soil for infection, and the quality of nursery or pre-nursery shade as pre-disposing factors. When this set of parameters was optimized, disease symptoms are apparent 3 months after the inoculation of germinated seeds (Breton et al. 2006). This avoids the transfer of seedlings from the pre-nursery to the main nursery, minimizes transplanting shock and root damage and reduces the time taken for screening and the nursery area required. This early inoculation enables the discrimination of isolates according to their aggressiveness, expressed by the quantification of external and internal disease symptoms and by using a standardized scoring scale. Development of this method will lead to the selection of isolates that can be used in screening for resistance or tolerance to basal shoot rot. Preliminary results show that it is possible to distinguish the degree of susceptibility of progenies inoculated artificially at the germinated seed stage or when transferred to polybags. A good correlation was obtained between those two stages, suggesting that the level of resistance was conserved irrespective of the physiological stage used for screening in the nursery.

The use of beneficial microbes offers an environmentally friendly alternative for the mitigation of basal stem rot (BSR) disease in oil palm. However, the biocontrol mechanisms of *Trichoderma* against the pathogenic *Ganoderma* spp. are largely unknown at the molecular level. In order to identify the transcripts involved during induced systemic resistance (ISR), Ho et al. (2018) analysed the root transcriptomes of oil palm seedlings inoculated simultaneously with both *G. boninense* and *T. harzianum*, and un-inoculated oil palm seedlings, as well as those that were inoculated with either pathogenic or beneficial fungi. Analyses revealed that the biocontrol mechanisms of *T. harzianum* against *G. boninense* involve modulation of genes related to biosynthesis of phytohormones (ethylene, MeJA and MeSA), antioxidant (l-ascorbate and myo-inositol) and unique secondary metabolites such as momilactone, cell wall metabolisms and detoxification of phytotoxic compounds; in addition to its role as a biofertilizer which improves nutritional status of host plant.

Ho et al. (2019) studied induced defense gene profiles in oil palm leaves for the identification of markers for the detection of Ganoderma-infected oil palms. The authors sequenced mRNA sampled from the leaves of *G. boninense*-infected oil palm seedlings (LG) and in control treatment (LC). Differential gene expression analysis showed 711 and 482 genes that were up- and down-regulated more than fourfold in LG, respectively, compared to the LC. Differential gene expression analyses revealed the modulation of several oil palm genes involved in the defense response. Indeed, chitinases, glucanases and thaumatin-like proteins showed up-regulation in LG. Genes for enzymes related to the biosynthesis of flavonoids, alkaloids and terpenes were found to be up-regulated, while many genes involved in photosynthesis were found to be suppressed in LG. During infection, oil palm leaves to divert their resources into defense, at the cost of other biological processes, thus contributing towards identification of candidate markers for the detection of infected oil palms.

## 7 Nutritional Quality

Palm oil is used in infant formulas in order to achieve palmitic acid levels similar to those found in human milk. Palmitic acid from palm oil is esterified predominantly at the SN-1,3 position of triacylglycerol (TAG), and infant formulas are now available in which a greater proportion of PA is in the SN-2 position (typical configuration in human milk). Bronsky et al. (2019) published a literature review based on relevant studies on possible beneficial effects or harms of either palm oil or SN-2-palmitate in infant formula on various health outcomes. Interestingly, these authors found that there is insufficient evidence to suggest that palm oil should be avoided as a source of fat in infant formulas for health reasons.

Oil palm fruits are the richest plant source of pro-vitamin A and vitamin E. Hence, the crop both alleviates poverty and could provide a simple practical solution to eliminate global pro-vitamin A deficiency (Delisle 2017). Vitamin E family members—tocotrienols and tocopherols—are widely known for their health benefits. Decades of research on tocotrienols have shown they have diverse biological activities such as antioxidant, anti-inflammatory, anticancer, neuroprotective and skin protection benefits, as well as improved cognition, bone health, longevity and reduction of cholesterol levels in plasma. Tocotrienols also modulate several intracellular molecular targets, and most importantly, they were found to improve lipid profiles and reduce total cholesterol and the volume of white matter lesions in human clinical trials. A comprehensive review from Kanchi et al. (2017) provides an update on the therapeutic potentials of tocotrienols, which tocopherols lack in a variety of inflammation-driven diseases.

Luo et al. (2019) conducted genome-wide association studies in a diversity panel of 161 *E. guineensis* accessions to identify single-nucleotide polymorphisms (SNPs) linked with vitamin E, and they validated candidate genes in these marker-associated intervals. Of these candidate genes, *EgHGGT* (homogentisate geranylgeranyl transferase) involved in the biosynthesis of tocotrienols had a higher expression level in the mesocarp compared to other tissues. Expression of the *EgHGGT* gene was positively correlated with the variation in α-tocotrienol content. Induced overexpression of the gene in Arabidopsis caused a significant increase in vitamin E content and production of α-tocotrienols compared to wild Arabidopsis.

Domonhédo et al. (2017) characterized the variability of palm oil acidity in the mesocarp of mature fruits in order to get a better understanding of the inheritance of that trait in the genus *Elaeis*. In this aim, 1183 oil palms from both species (*E. guineensis*, *E. oleifera*), interspecific hybrids and a first generation of backcross were considered in different backgrounds. Palm oil acidity (% of free fatty acids) was determined by titration using KOH. Results revealed a high variability (0–88% of acidity) with 38% of the individuals that presented low acidity and among them all the *E. oleifera* individuals (0–9% of acidity). An important variation was observed in a same class of acidity (high or low) for the different materials analysed. In various different genetic origins of groups A and B of *E. guineensis*, several families homozygous for low acidity were identified just as other families in disjunction for acidity with confirmation of 3:1 segregation type by chi square test. The acidity of hybrids (*E. guineensis* × *E. oleifera*) was the same with the *E. guineensis* parent. The supposed genetic determinism of palm oil acidity, based on a major gene with low acidity as recessive allele, would be valid for the whole genus *Elaeis*. Moreover, other minor genes could be implicated. Such findings will extend the genetic diversity used to optimize commercial seed production with low acidity.

## 8 Traditional Breeding Approaches

Understanding genetic structure of different vegetative, yield and yield component traits plays a major role in oil palm breeding programme. The oil palm (*E. guineensis* Jacq.) genome, estimated by flow cytometric analysis (Rival et al. 1997), is 3.4 × 109 bp, and Madon et al. (1998) reported the chromosome number of both *E. guineensis* and *E. oleifera* to be 32.

Interspecific *E. guineensis* × *E. oleifera* hybrids produce a specific oil with higher unsaturated fatty acids, and they display a slower vertical growth (20 cm/year rather than 60 for *E. guineensis*) and resistance to diseases (vascular wilt for Africa, and especially bud rot for Latin America). The current objective of breeders is to introgress major desirable traits of *E. oleifera* into *E. guineensis* through backcross programmes. For the last 30 years or so, the interspecific hybrid has been unavoidable when replanting palms in specific areas zones thanks to its strong resistance to bud rot in Latin America. Work is in progress to improve the interspecific hybrid not only by exploiting new *E. oleifera* accessions but also by cloning the best individuals thus taking advantage of within-progeny variability (Cochard and Durand-Gasselin 2018).

The oil palm is a versatile crop with multiple uses for its oil. The commodity chain is driven by a wide diversity of stakeholders (growers, crop researchers, oleochemists and technologists, engineers, environmentalists, anthropologists, marketers, etc.). This diverse population is able to propose many suggestions for desirable selection traits to improve the palm and its oil. It is thus important for key stakeholders to reconcile, rationalize and prioritize the selection traits for improvement as it is cumbersome and inefficient for breeders to include all the desirable traits in their breeding programme at any one time, and there may also be obvious or hidden negative trade-offs between pairs of traits (Chin et al. 2017). The recurrent cycles of selection have successfully removed many 'unwanted' alleles from the breeding populations, thereby providing a mere 1% genetic gain each year for the oil palm (Cochard et al. 2005). However, this might have affected the populations' ability to withstand environmental threats (Rival 2017) and genetic gains in the future. Realizing the important role of oil palm germplasm in harmonizing these effects, several research organizations had accumulated a large number of *E. guineensis* and *E. oleifera* accessions from their centres of origin. Such materials exhibit greater genetic diversity and can offer genes for many economically important traits for future oil palm improvement (Ithnin et al. 2017).

Although conventional breeding in combination with improved field management has increased crude palm oil yield from 2.6 to 4.0 t/ha during the past 60 years, there is still a big yield gap between that and the estimated theoretical potential of 18.2 t/ha (Barcelos et al. 2015). Breeding for oil yield still remains one of the major challenges for breeders and molecular biologists, as it is a major goal for maximizing production (Tittonell 2014; Woittiez et al. 2017). Oil yield results from a composite character made up of many traits including sex ratio, fruit to bunch ratio, weight per bunch, bunch number and oil per fruit. These characters vary continuously although some of them are highly heritable and may be amenable to quantitative trait loci (QTL) analysis using molecular markers.

The oil palm still faces formidable challenges in terms of its overall yield per hectare, its oil quality and the perceived sustainability of some aspects of plantation development and management. Thanks to recent progress in crop breeding and genomics, together with emerging recognition of the importance of improved management practices across the industry, prospects are good to confront these challenges (Murphy 2009).

Genetic improvement of the oil palm has definitely a role to play in sustainable plantation management. Indeed, the choice of the appropriate seed variety is the first step towards a sustainable cultivation system. There is an average turnover of 20 years for each planting cycle, and it is reasonable to expect at least a 20% increase in oil yield from the latest generation of tenera hybrid planting material, i.e. clones, biclonal or traditional hybrid seeds. With the rapid development of the smallholder sector, which is being facilitated by governments and donors in many countries, oil palm breeders now must meet their specific requirements (Cochard et al. 2005). Farmers are paid based on the weight of FFB (fresh fruit bunch), and not on the quantity of extracted oil, and the quality of fruits entering the mill directly impacts the purchase price through a system of bonus/penalties rewarding optimal ripening stage at the time of harvesting.

Breeding work undertaken over the last 50 years by major plantation companies and research institutions has led to ongoing genetic progress which is now estimated at 1% per annum (Cochard and Durand-Gasselin 2018). Future yields are now expected to reach and even exceed 10 tonnes of oil per hectare per annum in the near future (Corley and Tinker 2015).

# 9 Genomic Resources

Armero et al. (2017) successfully exploited the comparative genomics between a target species (coconut) and a reference species (oil palm) to improve the transcriptomic data, providing a proteome useful to answer functional or evolutionary questions. The generated workflow was found to reduce redundancy and fragmentation, two inherent problems of transcriptomic data, while preserving the functional representation of the target species. This strategy can be used as a complementary step to de novo transcriptome assembly to get a representative proteome of a target species (http://palm-comparomics.southgreen.fr/).

Oil yield is known to be controlled by variant alleles of a type II *MADS-box* gene, SHELL, that affect the presence and thickness of the endocarp. Singh et al. (2019) identified six novel SHELL alleles in non-commercial African germplasm populations. These authors investigated phenotypes in heteroallelic combinations, as well as SHELL heterodimerization and subcellular localization by yeast two-hybrid, bimolecular fluorescence complementation and gene expression analyses. Four novel SHELL alleles were found to be associated with fruit form phenotype. Candidate heterodimerization partners were identified, and interactions with EgSEP3 and subcellular localization were SHELL allele-specific. This research revealed that allele-specific mechanisms by which variant SHELL alleles impact yield, as well as speculative insights into the potential role of SHELL in single-gene oil yield heterosis.

Ooi et al. (2016) improved the sensitivity of DNA-based screening for non-tenera contamination through the addition of three novel mutant alleles of SHELL to the two previously identified alleles (Singh et al. 2013a). These five mutations, clustering within a highly conserved 19 amino acid span of the MADS-box domain, were sufficient to account for all tenera and pisifera phenotype palms within the diverse geography surveyed. Over 10,000 palms and seedlings derived from multiple sites within each of 13 geographic regions were analysed. The findings reveal that non-tenera contamination is likely a concern throughout the entire palm industry, and they demonstrate that comprehensive genetic testing will have major positive environmental and economic impacts.

The palm oil yield is mainly a function of the number of bunches of fruit harvested and their weight. Yield is thus determined by the sex ratio of flowers, and the ratio of mesocarp to the total dry mass of the bunch. In a recent study, Sonicha et al. (2018) developed 139 Intron Length Polymorphism (ILP) markers from 47 Expressed Sequence Tags (ESTs) of genes implicated in flower developmental. All ILP primers successfully amplified genomic DNA with products of the expected size and 79 were found to be polymorphic in a diversity panel of 41 palms. The efficiency of the ILP markers was demonstrated by the average polymorphism information content of 0.34 and mean observed heterozygosity of 0.42. Genotypes were found to cluster into three major groups, based on pedigree and geographical origin. Thirteen polymorphic ILP markers developed from 11 candidate genes were mapped onto 6 published oil palm linkage groups corresponding to 6 oil palm chromosomes. Quantitative trait loci mapping for traits related to oil yield identified the ILP marker developed from *AGAMOUS* of a MAS-box gene. ILP markers developed from genes involved in flower development in the oil palm provide a powerful tool for the identification of functional markers that are associated with traits determining oil yield.

The oil yield trait of oil palm is expected to involve multiple genes, environmental influences and interactions, and many of the underlying mechanisms that contribute to oil yield are still poorly understood. Wong et al. (2017) used a microarray approach to study the gene expression profiles of mesocarp tissue at different developmental stages, comparing genetically related high- and low-oil yielding palms to identify genes that contributed to the higher oil-yielding palm and might contribute to the wider genetic improvement of oil palm breeding populations. Their findings indicate that a higher carbon flux (channelled through down-regulation of the Sucrose Synthase 2 pathway) was being utilized by up-regulated genes involved in glycolysis, TCA and fatty acid biosynthesis, leading to enhanced oil production in the high-yielding oil palm. Such findings help in understanding the processes that lead to production of high-yielding oil palms.

## 10 Somaclonal Variation

Molecular approaches have been fruitful in deciphering key mechanisms in the developmental biology of the oil palms. Jaligot et al. (2011) described studies of the mantled epigenetic flowering abnormality observed in oil palms produced in vitro. The large-scale development of somatic embryogenesis-based micropropagation protocols is hampered by the occurrence, in small but significant proportions, of the *mantled* somaclonal variant, which resembles the floral B class mutants of model plants (Jaligot et al., 2011; Jaligot and Rival 2011, 2015). Abnormal palms may be sterile and therefore unproductive, thus cancelling out any potential genetic improvements made in the breeding programme from which they were obtained. Much interest therefore lies in identifying the molecular causes of the *mantled* phenotype, which has been demonstrated to be epigenetic in nature. Jaligot et al. (2011) reviewed current knowledge and priorities for future research on this complex but challenging phenomenon, which targeted molecular factors of particular interest, i.e. *MADS-box* genes and transposable elements, because of their involvement in related regulatory processes in model plants.

Beulé et al. (2011) used SSH and macroarray hybridization to compare transcriptome patterns between normal and mantled oil palm inflorescences. Two SSH libraries, enriched for complementary deoxyribonucleic acids (cDNAs) of either true-to-type or somaclonal variant material, were generated. Bioinformatics analysis of these two libraries allowed the identification of 1350 unique sequences and their annotation by a gene ontology-based approach. Macroarray hybridization was used to compare gene expression between normal and mantled inflorescences, and 32 genes were found to be differentially expressed. The temporal expression patterns of six genes were further investigated in more detail in relation to male and female inflorescence development. Full-length cDNAs were isolated and characterized for two of these genes, *EgFB1* and *EgRING1*, both of which are down-regulated in the mantled inflorescences and both of which encode proteins associated with proteolytic signalling complexes. The authors showed that EgFB1 and EgRING1 encode proteins that could potentially act in the same signalling pathway although this can only be a speculation given the lack of functional evidence. Nevertheless, it is interesting to note the similarity of the expression profiles observed for EgFB1 and EgRING1, both of which display reduced expression in the mantled inflorescence.

Shearman et al. (2013) performed RNA-Seq on developing flower and fruit samples of normal and mantled oil palm to characterize their transcriptomes. Expression data for all transcripts in normal and mantled flower and fruit samples were also studied, thus showing that many genes are differentially expressed, including several from pathways that may be the cause of the mantled phenotype if disrupted, i.e. genes involved in primary hormone responses, DNA replication and repair, chromatin remodelling and a gene involved in RNA-mediated DNA methylation. In addition, such gene expression data for developing flower and fruit can serve as a valuable resource for oil palm genetics and genomic studies. Among differentially expressed genes of normal and mantled samples, there are several good candidates for explaining the phenotype including KTF1, the chromatin remodelling genes and AT1G21780. The most interesting find is differential expression in chromatin remodelling genes and histone methylation genes and is consistent with the hypothesis that somatic embryogenesis is disrupting the methylation pathway in a non-specific manner. Furthermore, non-specific disruption of the methylation pathway provides an explanation why different studies have found different patterns of methylation and different patterns of gene expression. This finding also suggests that the mantled phenotype is not a single phenotype, but rather a collection of phenotypes that include disruptions to other organs, i.e. leaf and root, but undergo selection at the plantlet stage to exclude any disruptions resulting in an apparent phenotype.

LEAFY (LFY) is an important regulator of plant's reproductive system, regulating the transition from vegetative growth to flowering, and it acts also as an upstream regulator of floral homeotic *MADS-box* genes in Arabidopsis. Owing to its importance in flower development, oil palm LFY transcripts were isolated by Vetaryan et al. (2018) and the expression pattern was characterized to elucidate the role of LFY in oil palm. In addition, the gene expression pattern was analysed to determine a possible correlation of LFY to a type of floral abnormality in clonal palms termed mantling. In this study, three LFY transcripts were isolated from oil palm inflorescence tissue. The isolated cDNAs, OpLFY1 and OpLFY2, and its splice variant, OpLFY2v, were predicted to encode proteins each consisting of 430, 451 and 181 amino acids, respectively. OpLFY1 and OpLFY2 share 90% and 94% amino acid sequence identity to two different LFYs from *Phoenix dactylifera*, respectively. OpLFY1 was localized in all floral organ primordia. Quantitative polymerase chain reaction (qPCR) analysis revealed that *OpLFY* genes are expressed in shoot apical meristem and inflorescence tissues with the highest expression level. The expression of the *OpLFY1* gene was significantly lower in stage 1 of inflorescence development in mantled palms compared with that of normal inflorescences, suggesting possible association of the gene with this somaclonal variation.

In a recent study, Ooi et al. (2019) demonstrated that transcriptomes generated by laser capture microdissected abnormal oil palm staminodes revealed an adoption of carpel programming during organ initiation. Laser capture microdissection coupled with RNA sequencing was conducted to investigate the specific transcriptomes of male and female floral organs from normal and mantled female inflorescences. A higher number of differentially expressed genes (DEGs) were identified in abnormal versus normal male organs compared with abnormal versus normal female organs. In addition, the abnormal male organ transcriptome closely mimics the transcriptome of abnormal female organ. While the transcriptome of abnormal female organ was relatively similar to the normal female organ, a substantial amount of female DEGs encode *HEAT SHOCK PROTEIN* genes (*HSPs*). A similar high amount (20%) of male DEGs encode HSPs as well. As these genes exhibited decreased expression in abnormal floral organs, mantled floral organ development may be associated with lower stress indicators. Stamen identity genes *EgDEF1* and *EgGLO1* were the main floral regulatory genes with decreased expression in abnormal male organs or pseudocarpel initials. Expression of several floral transcription factors was elevated in pseudocarpel initials, notably LEAFY, FIL and DL orthologs, substantiating the carpel specification programming of abnormal staminodes.

With the goal of exploring the relationship between epigenetic stability and the long-term in vitro proliferation of plant tissues, Rival et al. (2013) studied changes in genomic DNA methylation levels in oil palm embryogenic suspensions. Five embryogenic callus lines were obtained from selected hybrid seeds and then proliferated as suspension cultures. Cultures proliferated for 12 months and genomic DNA was sampled at 4-month intervals for the estimation of global DNA methylation rates through high-performance liquid chromatography (HPLC) quantitation of deoxynucleosides. In vitro proliferation induces DNA hypermethylation in a time-dependent fashion. Moreover, this trend is statistically significant in several clonal lines and is shared among sub-clonal lines originating from the same genotype. Interestingly, the only clonal line undergoing loss of genomic methylation in the course of proliferation was unable to generate somatic embryos. Genome-wide DNA methylation changes in proliferating cells thus provide a powerful method for investigating mechanisms for maintaining genomic and epigenomic stability.

The mantled phenotype was associated with the DNA hypomethylation of a LINE retrotransposon (Karma) present in the intron of the B class gene *DEFICIENS*, leading to alternative splicing and premature transcript termination (Ong-Abdullah et al. 2015). This epigenome-based approach confirmed previous results on the epigenetic origin of the mantled somaclonal variation which was previously deciphered by Jaligot et al. (2014) based on the epigenetic regulation of a *MADS-box* candidate gene (Adam et al. 2006, 2007).

In order to elucidate the possible role of DNA methylation in the transcriptional regulation of EgDEF1, the APETALA3 ortholog of oil palm, Jaligot et al. (2014) studied this epigenetic marker within the gene in parallel with transcript accumulation in both normal and mantled developing inflorescences. The authors also examined the methylation and expression of two neighbouring retrotransposons that might interfere with EgDEF1 regulation. They showed that the *EgDEF1* gene was essentially unmethylated and that its methylation pattern did not change with the floral phenotype; whereas expression is dramatically different, ruling out a direct implication of DNA methylation in the regulation of this gene. Jaligot et al. (2014) observed that both the gypsy element inserted within an intron of the *EgDEF1* gene and the copia element located upstream from the promoter were heavily methylated and showed little or no expression. Interestingly a shorter, alternative transcript produced by EgDEF1 was identified, and its accumulation was characterized with respect to its full-length counterpart. Depending on the floral phenotype, the respective proportions of these two transcripts changed differently during inflorescence development. This alternative splicing raises new questions in the search for the molecular mechanisms underlying the mantled phenotype in the oil palm.

## 11 Modern Breeding Approaches

### *Micropropagation*

Mass production of selected, high-yielding palms by tissue culture could raise yields on existing plantations, reducing the need for further expansion of the cultivated area, which is often associated with negative environmental impacts (Goh 2016). The development of an efficient oil palm micropropagation protocol is needed to keep up with the increasing demand for palm oil in a sustainable way. Weckx et al. (2019) described that the limited availability of explants for oil palm micropropagation through somatic embryogenesis, combined with low somatic embryo initiation and regeneration rates, necessitate the proliferation of embryogenic structures, increasing the risk for somaclonal variants such as the mantled phenotype. Several ways to improve the efficiency of the tissue culture method and to reduce the risk of somaclonal variation were described (Weckx et al. 2019). These include the use of alternative explants and propagation techniques, the introduction of specific embryo maturation treatments and the detection of the mantled abnormality in an early stage.

Hashim et al. (2018) recently published an updated review on oil palm somatic embryogenesis. The authors noted that there is still considerable unexploited genetic variability within selected crosses, thus generating interest in vegetative propagation via tissue culture. The successful large-scale production of oil palm clones has spurred oil palm plantation companies and research institutions to set up their own tissue culture laboratories. The average estimated projection of 20% yield increase from clones compared to commercial seedlings makes clonal plantations a potentially rewarding investment.

Padua et al. (2017) analysed the somatic embryogenesis process in oil palm with amino acids and low concentrations of auxins. Histodifferentiation of somatic embryo from Tenera hybrid was analysed by cytochemical and ultrastructural methods, and somatic embryogenesis was used to regenerate oil palm plants. The authors noted that the concentration of vitamins and amino acids were crucial for oil palm somatic embryogenesis.

The somatic embryogenesis receptor kinase (*SERK*) gene has been extensively studied in many plant species due to its role in conferring embryogenic competence to somatic cells. In oil palm, full-length SERK I (EgSERK I) cDNA was isolated from cell suspension culture using RACE-PCR (Lee et al. 2019). EgSERK I was most highly expressed in leaf explants, and it was also detected in all tested tissues, including vegetative tissues, reproductive tissues, embryogenic tissues and non-embryogenic tissues, suggesting that it may have a broad role in plant growth and development. Interestingly, EgSERK I mRNA was detected in the adjacent cells of the vascular tissues in the mid-vein region of leaf explants which serves as the callus initiation point of callogenesis in oil palm. Such results suggest that *EgSERK I* gene is involved in the callus initiation stage of oil palm somatic embryogenesis by transducing the signal to switch on the dedifferentiation process, triggering cellular reprogramming to form callus.

Over the past three decades, significant research has been carried out to develop tissue culture and genetic transformation protocols for oil palm (Masani et al. 2018). Many different genetic transformation techniques, including microprojectile bombardment, Agrobacterium tumefaciens mediated, Polyethylene glycol mediated and DNA microinjection, have been developed by optimizing various parameters for the efficient genetic transformation of oil palm (Yarra et al. 2019). The development of efficient genome editing tool such as CRISPR/Cas9 looks very promising in the case of oil palm (Bao et al. 2019), with huge potential application for breeding, agronomy and climatic resilience as well.

## Marker-Assisted Selection

One way to manipulate the fatty acid composition (FAC) in palm oil is through the introgression of favourable alleles from the American oil palm, *E. oleifera*, which has a more unsaturated oil. Ting et al. (2016) studied a segregating *E. oleifera* × *E. guineensis* (O × G) hybrid population for fatty acid composition, in order to identify quantitative trait loci (QTLs) linked to iodine values and various FAs. The genomic interval contains several key structural genes in the FA and oil biosynthesis pathways such as *PATE/FATB*, *HIBCH*, *BASS2*, *LACS4* and *DGAT1* and also a relevant transcription factor (TF), *WRI1*. The candidate gene approach proved useful to identify the potential causal genes linked to fatty acid composition and can be adopted for marker-assisted selection in oil palm.

Molecular marker technologies can greatly improve the breeding efficiency of oil palm. DNA markers and genetic linkage maps are essential resources for MAS to accelerate genetic improvement. The identification of expressed sequence tags–simple sequence repeats (EST-SSRs) that detect polymorphisms in elite breeding material provides tools for molecular breeding strategies. The identification of SSRs within transcripts, in particular, those that encode proteins involved in transcriptional and post-transcriptional regulation, would allow insight into the functional roles of these proteins by revealing how phenotypic traits co-segregate with these markers. Tranbarger et al. (2012) reported 6103 non-redundant ESTs derived from cDNA libraries of developing vegetative and reproductive tissues were annotated and searched for SSRs. Primer pairs from sequences flanking 289 EST-SSRs were tested to detect polymorphisms in elite breeding parents and their crosses. Among 230 of these amplified PCR products, 88 were polymorphic within the breeding material tested. A detailed analysis and annotation of the EST-SSRs revealed the locations of the polymorphisms within the transcripts and that the main functional category was related to transcription and post-transcriptional regulation. SSR polymorphisms were found in sequences encoding AP2-like, bZIP, zinc finger, MADS-box and NAC-like transcription factors in addition to other transcriptional regulatory proteins and several RNA interacting proteins.

In order to identify QTLs for fatty acid composition of palm oil, Montoya et al. (2013) studied an Elaeis interspecific pseudo-backcross of first generation (*E. oleifera* × *E. guineensis*) × *E. guineensis*. A dense microsatellite linkage map of 362 loci spanned 1.485 cM, representing the 16 pairs of homologous chromosomes in the Elaeis genus from which the authors traced segregating alleles from both *E. oleifera* and *E. guineensis* grandparents. The relative linear orders of mapped loci suggested the probable absence of chromosome rearrangements between the *E. oleifera* and *E. guineensis* genomes. There were 19 QTLs associated with palm oil fatty acid composition. The QTL positions and the species origin as well as the estimated effects of the QTL marker alleles were in coherence with the oil biosynthesis pathway in plants and with the individual phenotypic correlations between the traits. The mapping of chosen Elaeis key genes related to oleic acid C18:1, using intragene single-nucleotide polymorphisms (SNPs), supported several QTLs underlying notably FATA and SAD enzymes. The high number of hyper-variable SSR loci of known relative linear orders and the QTL information make these resources valuable for mapping in other Elaeis breeding materials.

Several linkage maps have been constructed using dominant and co-dominant markers to facilitate mapping of QTLs in oil palm; however, dominant markers are not easily transferable among different laboratories. Lee et al. (2015) constructed a consensus linkage map for oil palm using co-dominant markers, i.e. microsatellite and SNPs, and two F1 breeding populations generated by crossing Dura and Pisifera individuals. Four hundred and forty-four microsatellites and 36 SNPs were mapped onto 16 linkage groups. The map length was 1565.6 cM, with an average marker space of 3.72 cM. A genome-wide scan of QTLs identified a major QTL for stem height on linkage group 5, which explained 51% of the phenotypic variation. Genes in the QTL were predicted using the palm genome sequence and bioinformatic tools. The linkage map supplies a base for mapping QTLs for accelerating genetic improvement and will be also useful for improving the assembly of genome sequences. Markers linked to the QTLs may be used for selecting dwarf trees, and genes within the QTL will be characterized to understand the mechanisms underlying dwarfing (Lee et al. 2015).

Suwanto et al. (2018) conducted RAD-seq on an Illumina NextSeq500 to discover genome-wide SNPs, and used the SNPs to construct a linkage map for an oil palm (Tenera) population derived from a cross between a Deli Dura and an AVROS Pisifera. The RAD-seq produced 1076 million single-end reads across the breeding population containing 155 trees. Mining this dataset detected 510,251 loci. After filtering out loci with low accuracy and more than 20% missing data, 11,394 SNPs were retained. Using these SNPs, in combination with 188 anchor SNPs and 123 microsatellites, a linkage map was constructed, containing 10,023 markers covering 16 chromosomes. The map length is 2938.2 cM with an average marker space of 0.29 cM. The large number of SNPs will supply ample choices of DNA markers in analysing the genetic diversity, population structure and evolution of oil palm.

SNPs are widespread nucleotide variations in genomes, the most abundant type of DNA marker and easily detected at high-throughput in genomes. By taking advantage of the recently developed reference genome sequence of the oil palm (Singh et al. 2013a,b), it is possible to develop genome-wide DNA markers with gene information resources (Ting 2014). Furthermore, gene-based SNP markers could themselves be causative SNPs for traits and have been used for developing transcript maps, QTL analysis, association mapping and synteny analysis (Pootakham et al. 2015).

Using both SNPs and microsatellites markers, Bai et al. (2018) constructed a linkage map containing 10,023 markers covering 16 chromosomes. The map length was 2938.2 cM with an average marker space of 0.29 cM. Such a large number of SNPs will supply ample choices of DNA markers in analysing the genetic diversity, population structure and evolution of oil palm. This high-density linkage map will contribute to mapping QTLs for important traits, thus accelerating oil palm genetic improvement. Gan et al. (2018) reported the first use of the DArTseq platform to genotype two closely related self-pollinated oil palm populations. Identification of the sh marker demonstrated the robustness of using the DArTseq platform to generate high-density genetic maps of oil palm with adequate genome coverage. Both genetic maps and integrated maps are of paramount interest for QTL analysis of important yield traits as well as potentially assisting the anchoring of genetic maps to genomic sequences.

## *Genome-Wide Association Mapping*

Zheng et al. (2019) explored the regulatory mechanisms of lipid and fatty acid (FA) metabolism in oil palm, in order to gain better understanding of small noncoding RNA (ncRNA) and mRNA. Based on the deep sequencing of small noncoding RNAs and the degradome in five developmental mesocarp stages, 452 microRNAs (miRNAs), including 170 conserved known-miRNAs (kn-miRNAs) and 282 novel-miRNA (nov-miRNAs), were identified. After predicting the targets of those miRNAs to 37 FA synthesis-related genes, the authors found that 22 kn-miRNAs and 14 nov-miRNAs might be involved in FA metabolism pathways. Several miRNAs were also predicted to indirectly regulate FA synthesis and lipid metabolism through the squamosa promoter-binding protein-like gene (SPL), NAC and MYB transcription factors. A complex and extensive miRNA–mRNA regulatory network is associated with FA metabolism in the mesocarp of the oil palm. Several potential mechanisms might be regulated by miRNAs through the modulation of the expression of FA-related target gene metabolism.

Osorio-Guarín et al. (2019) published the first association mapping study in an interspecific OxG oil palm population, which shows tolerance to pests and diseases, high oil quality, and acceptable fruit bunch production. Using genotyping-by-sequencing (GBS), these authors identified a total of 3776 single-nucleotide polymorphisms (SNPs) that were used to perform a genome-wide association analysis (GWAS) in 378 O × G hybrid population for 10 agronomic traits. Twelve genomic regions (SNPs) were located near candidate genes implicated in multiple functional categories, such as tissue growth, cellular trafficking and physiological processes. Genomic regions that mapped on candidate genes involved in plant architecture and yield were identified, and such potential candidate genes need to be confirmed for future targeted functional analyses.

Babu et al. (2020) used genotyping-based sequencing (GBS) for GWAS of four yield and seven oil yield-related traits on highly diverse African oil palm germplasm. GBS generated 325 million reads covering 50.78 Gb of sequence data, with an average of 3.4 million reads per sample. Finally, 4031 fully informative SNPs with a range between 157 on chromosome 15 and 455 on chromosome 1 were used for GWAS. Association mapping resulted in identification of 40 highly significant loci, where more genetic loci were found to be associated with oil to bunch (OB), followed by average bunch weight (ABW). The nucleotide sequences of linked genetic loci for OB were found to be similar to mitogen activated protein kinase-5 (MAPK-5) protein, which is an early flowering protein. The significant loci identified can be used to select desirable palms at early stage through marker-assisted selection.

## *Genomic Selection*

Genomic selection (GS) uses genome-wide markers as an attempt to accelerate genetic gain in breeding programmes of both animals and plants. This approach is particularly useful for perennial crops such as oil palm, which have long breeding cycles, and for which the optimal method for GS is still under debate. Kwong et al. (2017) evaluated the effect of different marker systems and modelling methods for implementing GS in an introgressed dura family derived from a Deli dura × Nigerian dura (Deli × Nigerian) with 112 individuals. This family is an important breeding source for developing new mother palms for superior oil yield and bunch characters.

The traits of interest selected for this study were fruit-to-bunch (F/B), shell-to-fruit (S/F), kernel-to-fruit (K/F), mesocarp-to-fruit (M/F), oil per palm (O/P) and oil-to-dry mesocarp (O/DM). The marker systems evaluated were simple sequence repeats (SSRs) and single-nucleotide polymorphisms (SNPs). RR-BLUP, Bayesian A, B, Cπ, LASSO, Ridge Regression and two machine learning methods (SVM and Random Forest) were used to evaluate GS accuracy of the traits.

Nyouma et al. (2019) described changes in breeding schemes for oil palm yield and its quantitative components, and the changes expected to take place with genomic selection (GS). Since the genetic improvement of oil palm yield started in the 1920s through mass selection, several disruptive improvements dramatically accelerated the fate of genetic progress. These were namely the understanding of the heredity of fruit form and the adoption of tenera, with thicker mesocarp, in plantations; the discovery of hybrid vigour and the adoption of modified reciprocal recurrent selection; and the implementation of clonal selection, exploiting intra-hybrid variability. In addition, the use of linear mixed models to estimate genetic values has made selection more efficient. Today, GS appears to be a new disruptive improvement that can speed up breeding schemes by avoiding field trials in some cycles and increase selection intensity by evaluating more candidates. The genetic potential for oil palm yield has increased considerably over one century of breeding. GS is expected to bring the rate of genetic progress to a previously unprecedented level. The future studies on oil palm GS will aim at making it efficient for all yield components. For this purpose, they should focus in particular on the optimization of training populations and on the improvement of prediction models.

Traditional progeny tests allow accurate selection but limit the number of individuals evaluated. Genomic selection (GS) could overcome this constraint. Cros et al. (2017) estimated the accuracy of GS prediction of seven oil yield components using A × B hybrid progeny tests with almost 500 crosses for training and 200 crosses for independent validation. Genotyping-by-sequencing (GBS) yielded +5000 single-nucleotide polymorphisms (SNPs) on the parents of the crosses. Preselection for yield components using GBS is the first possible application of GS in oil palm, and it will increase selection intensity, thus improving the performance of commercial hybrids.

## 12 Phenotyping Approaches Including High-Throughput Phenotyping

The effective implementation of technology that facilitates accurate and high-throughput screening of thousands of field-grown lines is critical for accelerating crop improvement and breeding strategies for higher yield and disease tolerance (Shakoor et al. 2017). Progress in the development of field-based high-throughput phenotyping methods has advanced considerably in the last 10 years through technological progress in sensor development and high-performance computing. The price of oil palm fruits after harvesting is negotiated depending some key parameters of the fresh fruit bunch (FFB). Pamornnak et al. (2017) have developed a system to grade FFB depending on its quality. Several features of palm fruits were extracted from RGB, near infrared, and depth images, captured with a Microsoft Kinect camera version 2.0 installed in a light-controlled environment on the conveyor line. Two main algorithms for classification have been developed. The first algorithm is called a volume integration scheme (SVIS), which measures the relative volume of palm bunch. The second developed algorithm classifies palm bunch into three grades (L-Grade, M-Grade and H-Grade) based on oil content from Soxhlet extraction. The system achieves 83% accuracy for grading palm bunch within 6 s per one sample, which shows the possibility of using the system in a trading market.

Ground-based LiDAR also known as Terrestrial Laser Scanning (TLS) technology is an active remote sensing imaging method for plant phenotyping. Husin et al. (2020) proposed a novel technique of Basal Stem Rot classification at the oil palm canopy analysis using the point clouds data taken from the TLS. The proposed model can—using point clouds data of TLS—classify the four severity levels of infection with an accuracy of 80%; it was found to perform better when compared to the severity classification using frond number.

In order to develop drought-tolerant cultivars through screening and selection, tolerance assessment based on growth and physiological traits provides an alternative approach for rapid screening especially in tree crops. Duangpan et al. (2018) determined the contribution of some growth and physiological parameters to drought response variation in oil palm seedlings, and they used those parameters to evaluate relative drought stress tolerance in tenera oil palm progenies. Major contributions towards the variation of control and drought-treated plants were identified and appeared to be growth traits in moderate drought stress and growth trait plus proline content in severe drought stress. Field validation of D × P progenies selected as drought tolerant is needed to establish this indirect selection approach in oil palm breeding programme.

Accelerated breeding for agriculturally relevant crop traits is key to the development of improved varieties and is critically dependent on high-resolution, high-throughput field-scale phenotyping technologies that can efficiently discriminate better performing lines within a larger population and across multiple environments (Shakoor et al. 2017).

## 13 Avenues for Future Research

Innovative and efficient methods are required to breed oil palm for yield maximization in order to meet the increased demand for palm oil while limiting environmental impacts. Metagenomics is a molecular tool used to analyse DNA acquired from environmental samples, in order to study the community of microorganisms present, without the necessity of obtaining pure cultures (Lindon et al. 2018).

Metagenomics can provide a useful bridging between genomics and environmental science, while addressing key question in the sustainable management of oil palm plantations. Understanding the intricate link between the soil microbiota and their metabolic functions is important for agricultural and ecological processes and could be used as a biomarker of soil health. Metagenomic DNA is extracted from the soil microcosm and sequenced using Miseq Illumina next generation sequencing and analysed for both its structural and functional properties. For example, soil fungi are key players in nutrient cycles as decomposers, mutualists and pathogens, but the impact of tropical rain forest transformation into oil palm plantations on fungal community structures and their ecological functions are unknown. Brinkmann et al. (2019) showed that rain forest conversion was not associated with strong diversity loss but with massive shifts in soil fungal community composition. The conversion of rain forests and current land use systems restructure soil fungal communities towards enhanced pathogen pressure and, thus, threaten ecosystem health functions.

Routine species or even strain-level identification of plant pathogens directly from symptomatic tissue could significantly improve plant disease control and prevention. In the case of complex pathogenic systems such as bud rot for the oil palm, metagenomic sequencing of plants, either artificially inoculated with a known strain of the suspected pathogen or collected in the field and showing bud rot symptoms, can provide highly valuable information. It will be then possible to determine not only the group of pathogens involved in the bud rot syndrome, but also to shed light on the sequence of events governing the decay and/or the death of the palm.

The publication of the oil palm genome sequence by Singh et al. (2013a) has been a major breakthrough in oil palm biotechnology. Even if considerable knowledge and molecular resources had already been made available before this date (Rival and Jaligot 2011), the availability of the genome sequence expedited a number of research projects ranging from genomic selection to the deciphering of key components of yield such as sex ratio or oil synthesis.

Key findings in cutting-edge areas of breeding and biotechnology are undoubtedly contributing to the sustainable cultivation of oil palm. To be successful for the whole sector and its end-users, key stakeholders (either public or private) must support research efforts and knowledge must be shared by the community of oil palm researchers and rapidly disseminated and transformed into useful innovation to stakeholders.

## References

Adam H, Jouannic S, Morcillo F, Richaud F, Duval Y, Tregear JW (2006) MADS box genes in oil palm (Elaeis guineensis): patterns in the evolution of the SQUAMOSA, DEFICIENS, GLOBOSA, AGAMOUS, and SEPALLATA subfamilies. Journal of Molecular Evolution, 62(1):15–31

Adam H, Jouannic S, Orieux Y, Morcillo F, Richaud F, Duval Y, Tregear JW (2007) Functional characterization of MADS box genes involved in the determination of oil palm flower structure. Journal of experimental Botany, 58(6):1245–1259

Armero A, Baudouin L, Bocs S, This D (2017) Improving transcriptome de novo assembly by using a reference genome of a related species: translational genomics from oil palm to coconut. PLoS One 12(3):e0173300

Babu BK, Mathur RK, Ravichandran G, Anita P, Venu MVB (2020) Genome wide association study (GWAS) and identification of candidate genes for yield and oil yield related traits in oil palm (Eleaeis guineensis) using SNPs by genotyping-based sequencing. Genomics 112(1):1011–1020

Bai B, Wang L, Zhang YJ, Lee M, Rahmadsyah R et al (2018) Developing genome-wide SNPs and constructing an ultrahigh-density linkage map in oil palm. Sci Rep 691:1–7. https://doi.org/10.1038/s41598-017-18613-2

Bakoumé C (2016) Genetic diversity, erosion, and conservation in oil palm (Elaeis guineensis Jacq.). In: Genetic diversity and erosion in plants. Springer, Cham, pp 1–33

Bakoumé C, Wickneswari R, Siju S, Rajanaidu N, Kushairi A, Billotte N (2015) Genetic diversity of the world's largest oil palm (Elaeis guineensis Jacq.) field genebank accessions using microsatellite markers. Genet Resour Crop Evol 62:349–360. https://doi.org/10.1007/s10722-014-0156-8

Bao A, Burritt DJ, Chen H, Zhou X, Cao D, Tran LSP (2019) The CRISPR/Cas9 system and its applications in crop genome editing. Crit Rev Biotechnol 39(3):321–336

Barcelos E, de Almeida Rios S, Cunha R, Lopes R, Motoike SY et al (2015) Oil palm natural diversity and the potential for yield improvement. Front Plant Sci 6:190. https://doi.org/10.3389/fpls.2015.00190

Baron V, Rival A, Marichal R (2017) No, palm oil is not responsible for 40% of global deforestation. The Conversation. https://theconversation.com/no-palm-oil-is-not-responsible-for-40-of-global-deforestation-78482

Beirnart A, Vanderweyen T (1941) Contribution à l'étude génétique et biométrique des variétés d'Elaeis guineensis Jacq. INEAC, Série Scientifique no. 27

Beulé T, Camps C, Debiesse S, Tranchant C, Dussert S, Sabau X, et al (2011) Transcriptome analysis reveals differentially expressed genes associated with the mantled homeotic flowering abnormality in oil palm (Elaeis guineensis). Tree Genet Genomes 7(1):169–182

Breton F, Hasan Y, Lubis Z, De Franqueville H (2006) Characterization of parameters for the development of an early screening test for basal stem rot tolerance in oil palm progenies. J Oil Palm Res 0406:24–36

Brinkmann N, Schneider D, Sahner J, Ballauff J, Edy N, Barus H et al (2019) Intensive tropical land use massively shifts soil fungal communities. Sci Rep 9(1):1–11

Bronsky J, Campoy C, Embleton N, Fewtrell M, Mis NF, Gerasimidis K et al (2019) Palm oil and beta-palmitate in infant formula: a position paper by the European Society for Paediatric Gastroenterology, Hepatology, and Nutrition (ESPGHAN) committee on nutrition. J Pediatr Gastroenterol Nutr 68(5):742–760

Chin S, Louise C, de Franqueville H, Cooper RM, Nouy B et al (2017) Objective traits. In: Soh AC, Mayes S, Roberts JA (eds) Oil palm breeding. CRC Press, Boca Raton, pp 85–142

Cochard B, Durand-Gasselin T (2018) Advances in conventional breeding techniques for oil palm. In: Alain R (ed) Achieving sustainable cultivation of oil palm. Volume 1: Introduction, breeding and cultivation techniques. Burleigh Dodds Science Publishing, Cambridge, pp 117–144. https://doi.org/10.4324/9781351114387. (Burleigh Dodds Series in Agricultural Science, 27) ISBN 978-1-78676-104-0.

Cochard B, Amblard P, Durand-Gasselin T (2005) Oil palm genetic improvement and sustainable development. OCL 12(2):141–147. https://doi.org/10.1051/ocl.2005.0141

Corley RHV (2016) Climate change—what does it mean for oil palm? Planter 92(1086):631–654

Corley RHV, Tinker PBH (2015) The oil palm, 5th edn. Wiley-Blackwell, Hoboken. 674 pp. ISBN: 978-1-405-18939-2

Cramb RA, McCarthy JF (eds) (2016) The oil palm complex: smallholders, agribusiness and the state in Indonesia and Malaysia. NUS Press, Singapore

Cros D, Bocs S, Riou V, Ortega-Abboud E, Tisné S, Argout X et al (2017) Genomic preselection with genotyping-by-sequencing increases performance of commercial oil palm hybrid crosses. BMC Genomics 18(1):839

Cruden RW (1988) Temporal dioecism: systematic breadth, associated traits, and temporal patterns. Bot Gaz 149:1–15

Delisle H (2017) The nutritional value of red palm oil. In: Rival A (ed) Achieving sustainable cultivation of oil palm. Burleigh-Dodds Science Publishing, Cambridge, pp 217–231

Domonhédo H, Nodichao L, Billotte N, Ahanhanzo C, Cros D (2017) Genetic variability and understanding of inheritance of palm oil acidity in mature fruits of oil palm (E. guineensis, Jacq.). J Appl Biosci 119:11871–11887

Duangpan S, Buapet P, Sujitto S, Eksomtramage T (2018) Early assessment of drought tolerance in oil palm D× P progenies using growth and physiological characters in seedling stage. Plant Genet Resour 16(6):544–554

Durand-Gasselin T, Asmady H, Flori A, Jacquemard JC, Hayun Z et al (2005) Possible sources of genetic resistance in oil palm (Elaeis guineensis Jacq.) to basal stem rot caused by Ganoderma boninense—prospects for future breeding. Mycopathologia 159:93–100

Forster BP, Sitepu B, Setiawati U, Kelanaputra ES, Nur F et al (2017) Oil palm (Elaeis guineensis). In: Campos H, Caligari PDS (eds) Genetic improvement of tropical crops. Springer, Cham. https://doi.org/10.1007/978-3-319-59819-2_8

Gan ST, Wong WC, Wong CK, Soh AK, Kilian A et al (2018) High density SNP and DArT-based genetic linkage maps of two closely related oil palm populations. J Appl Genet 59:23. https://doi.org/10.1007/s13353-017-0420-7

Germer J, Sauerborn J (2008) Estimation of the impact of oil palm plantation establishment on greenhouse gas balance. Environ Dev Sustain 10(6):697–716. https://doi.org/10.1007/s10668-006-9080-1

Goh CS (2016) Can we get rid of palm oil? Trends Biotechnol 34(12):948–950

Hashim AT, Ishak Z, Rosli SK, Ong-Abdullah M, Ooi SE, Husri MN, Bakar DA (2018) Oil palm (Elaeis guineensis Jacq.) somatic embryogenesis. In: Step wise protocols for somatic embryogenesis of important woody plants. Springer, Cham, pp 209–229

Ho CL, Tan YC, Yeoh KA, Lee WK, Ghazali AK, Yee WY, Hoh CC (2018) Transcriptional response of oil palm (Elaeis guineensis Jacq.) inoculated simultaneously with both *Ganoderma boninense* and *Trichoderma harzianum*. Plant Genet 13:56–63

Ho CL, Tan YC, Yeoh KA, Lee WK, Ghazali AK, Yee WY, Hoh CC (2019) Leaf transcriptome of oil palm (Elaeis guineensis Jacq.) infected by Ganoderma boninense. Trees 33(3):943–950

Husin NA, Khairunniza-Bejo S, Abdullah AF, Kassim MS, Ahmad D, Azmi AN (2020) Application of ground-based LiDAR for analysing oil palm canopy properties on the occurrence of basal stem rot (BSR) disease. Sci Rep 10(1):1–16

Ithnin M, Serdari NM, Abdullah N, Kushairi A, Singh R (2017) Biodiversity and conservation of Elaeis species. In: Ahuja M, Jain S (eds) Biodiversity and conservation of woody plants. Sustainable development and biodiversity, vol 17. Springer, Cham. https://doi.org/10.1007/978-3-319-66426-2_9

Jaligot E, Rival A (2015) Epigenetic variations: a trade-off between genome stability and phenotypic plasticity. In: Al-Khayri JM, Jain SM, Johnson DV (eds) Advances in plant breeding strategies: breeding, biotechnology and molecular tools. Springer, Cham. https://doi.org/10.1007/978-3-319-22521-0_6

Jaligot E, Adler S, Debladis E, Beulé T, Richaud F et al (2011) Epigenetic imbalance and the floral developmental abnormality of oil palm. Ann Bot 108:1453–1462

Jaligot E, Hooi WY, Debladis E, Richaud F, Beulé T et al (2014) DNA methylation and transcriptional activity of the EgDEF1 gene and neighboring retrotransposons in mantled somaclonal variants of oil palm. PLoS One 9:e91896

Kanchi MM, Shanmugam MK, Rane G, Sethi G, Kumar AP (2017) Tocotrienols: the unsaturated sidekick shifting new paradigms in vitamin E therapeutics. Drug Discov Today 22(12):1765–1781

Kwong QB, Teh CK, Ong AL, Chew FT, Mayes S, Kulaveerasingam H et al (2017) Evaluation of methods and marker systems in genomic selection of oil palm (Elaeis guineensis Jacq.). BMC Geneti 18(1):107

Lee M, Xia JH, Zou Z, Ye J, Alfiko Y et al (2015) A consensus linkage map of oil palm and a major QTL for stem height. Sci Rep 5:8232. https://doi.org/10.1038/srep08232

Lee FC, Ong-Abdullah M, Ooi SE, Ho CL, Namasivayam P (2019) Cloning and characterization of Somatic Embryogenesis Receptor Kinase I (EgSERK I) and its association with callus initiation in oil palm. In Vitro Cell Dev Biol Plant 55(2):153–164

Li K, Tscharntke T, Saintes B, Buchori D, Grass I (2019) Critical factors limiting pollination success in oil palm: a systematic review. Agric Ecosyst Environ 280:152–160

Lieb VM, Kerfers MR, Kronmüller A, Esquivel P, Alvarado A, Jiménez VM, et al (2017) Characterization of mesocarp and kernel lipids from Elaeis guineensis Jacq, Elaeis oleifera [Kunth] Cortés, and their interspecific hybrids. J Agric Food Chem 65(18):3617–3626

Lindon JC, Nicholson JK, Holmes E (Eds). (2018) The handbook of metabolic phenotyping. Elsevier Amsterdam

Luo T, Xia W, Gong S, Mason AS, Li Z, Liu R et al (2019) Identifying vitamin E biosynthesis genes in Elaeis guineensis by genome-wide association study. J Agric Food Chem 68(2):678–685. https://doi.org/10.1021/acs.jafc.9b03832

Madon M, Clyde MM, Cheah SC (1998) Cytological analysis of Elaeis guineensis and Elaeis oleifera chromosomes. J Oil Palm Res 10:68–91

Masani MYA, Izawati AMD, Rasid OA, Parveez GKA (2018) Biotechnology of oil palm: current status of oil palm genetic transformation. Biocatal Agric Biotechnol 15:335–347. https://doi.org/10.1016/j.bcab.2018.07.008

Meijide A, de la Rua C, Guillaume T, Röll A, Hassler E, Stiegler C et al (2020) Measured greenhouse gas budgets challenge emission savings from palm-oil biodiesel. Nat Commun 11(1):1–11

Mercière M, Laybats A, Carasco-Lacombe C, Tan JS, Klopp C, Durand-Gasselin T, Alwee SSRS, Camus-Kulandaivelu L, Breton F (2015) Identification and development of new polymorphic microsatellite markers using genome assembly for Ganoderma boninense, causal agent of oil palm basal stem rot disease. Mycol Prog 14(11):103

Montoya C, Lopes R, Flori A, Cros D, Cuellar T et al (2013) Quantitative trait loci (QTLs) analysis of palm oil fatty acid composition in an interspecific pseudo-backcross from Elaeis oleifera (HBK) Cortés and oil palm (Elaeis guineensis Jacq.). Tree Genet Genomes 9:1207–1225

Murphy DJ (2009) Oil palm: future prospects for yield and quality improvements. Lipid Technol 21:257–260

Neoh BK, Teh HF, Wong YC, Ooi TEK, Cheah SS, Appleton DR (2017) Study of oil palm photosynthesis using omics technologies. In: Crop improvement. Springer, Cham, pp 27–46

Neoh BK, Wong YC, Teh HF, Ng TLM, Tiong SH, Ooi TEK et al (2019) Diurnal biomarkers reveal key photosynthetic genes associated with increased oil palm yield. PLoS One 14(3):e0213591. https://doi.org/10.1371/journal.pone.0213591

Nyouma A, Bell JM, Jacob F, Cros D (2019) From mass selection to genomic selection: one century of breeding for quantitative yield components of oil palm (*Elaeis guineensis* Jacq.). Tree Genet Genomes 15(5):69

Ohimain EI, Izah SC (2017) A review of biogas production from palm oil mill effluents using different configurations of bioreactors. Renew Sust Energ Rev 70:242–253

Ollivier J, Flori A, Cochard B, Amblard P, Turnbull N et al (2016) Genetic variation in nutrient uptake and nutrient use efficiency of oil palm. J Plant Nutr 40:558–573. https://doi.org/10.1080/01904167.2016.1262415

Ong-Abdullah M, Ordway JM, Jiang N, Ooi SE, Kok SY, Sarpan N et al (2015) Loss of Karma transposon methylation underlies the mantled somaclonal variant of oil palm. Nature, 525(7570):533–537

Ooi LCL, Low ETL, Abdullah MO, Nookiah R, Ting NC, Nagappan J et al (2016) Non-tenera contamination and the economic impact of SHELL genetic testing in the Malaysian independent oil palm industry. Front Plant Sci 7:771

Ooi SE, Sarpan N, Aziz NA, Nuraziyan A, Ong-Abdullah M (2019) Differential expression of heat shock and floral regulatory genes in pseudocarpel initials of mantled female inflorescences from *Elaeis guineensis* Jacq. Plant Reprod 32(2):167–179

Osorio-Guarín JA, Garzón-Martínez GA, Delgadillo-Duran P, Bastidas S, Moreno LP, Enciso-Rodríguez FE et al (2019) Genome-wide association study (GWAS) for morphological and yield-related traits in an oil palm hybrid (Elaeis oleifera x Elaeis guineensis) population. BMC Plant Biol 19(1):533

Pádua MS, Santos RS, Labory CRG, Stein VC, Mendonça EG, Alves E, Paiva LV (2018) Histodifferentiation of oil palm somatic embryo development at low auxin concentration. Protoplasma 255(1):285–295

Pamornnak B, Limsiroratana S, Khaorapapong T, Chongcheawchamnan M, Ruckelshausen A (2017) An automatic and rapid system for grading palm bunch using a Kinect camera. Comput Electron Agric 143:227–237

Pandolfini T, Molesini B, Spena A (2018) Parthenocarpy in crop plants. Annu Plant Rev Online 38:326–345

Paterson RRM, Sariah M, Lima N (2013) How will climate change affect oil palm fungal diseases? Crop Prot 46:113–120

Pootakham W, Jomchai N, Ruang-areerate P, Shearman JR, Sonthirod C et al (2015) Genome-wide SNP discovery and identification of QTL associated with agronomic traits in oil palm using genotyping-by-sequencing (GBS). Genomics 105:288–295. https://doi.org/10.1016/j.ygeno.2015.02.002

Rajanaidu N, Din AM, Marjuni M, Abdullah N (2018) Diversity in the genetic resources of oil palm. In: Rival A (ed) Achieving sustainable cultivation of oil palm. Volume 1: Introduction. Breeding and cultivation techniques, 1st edn. Burleigh Dodds Science Publishing, Cambridge, pp 93–116

Rival A (2017) Breeding the oil palm (Elaeis guineensis Jacq.) for climate change. OCL 24:D107, 7 p

Rival A (2020) 4.2 Elaeis guineensis oil palm. In: Litz RE, Pliego-Alfaro F, Hormaza JI (eds) Biotechnology of fruit and nut crops, 2nd edn. CABI, Wallingford, pp 92–106

Rival A, Jaligot E (2011) Epigenetics and plant breeding. CAB Rev 6:48

Rival A, Levang P (2014) Palms of controversy: oil palm and development challenges. CIFOR, Bogor. ISBN:978-602-1504-41-3

Rival A, Beulé T, Barre P, Hamon S, Duval Y, Noirot M (1997) Comparative flow cytometric estimation of nuclear DNA content in oil palm (Elaeis guineensis Jacq.) tissue cultures and seed-derived plants. Plant Cell Rep 16:884–887

Rival A, Jaligot E, Beulé T, Finnegan J (2008) Isolation and differential expression of MET, CMT and DRM methyltransferase genes from oil palm (*Elaeis guineensis* Jacq.) in relation with the "mantled" somaclonal variation. J Exp Bot 59:3271–3281

Rival A, Ilbert P, Labeyrie A, Torres E, Doulbeau S et al (2013) Variations in genomic DNA methylation during the long-term in vitro proliferation of oil palm embryogenic suspension cultures. Plant Cell Rep 32:359–368. https://doi.org/10.1007/s00299-012-1369-y

Rosenquist EA (1986) The genetic base of oil palm breeding populations. In International Workshop on Oil Palm Germplasm and Utilisation, Bangi, Selangor (Malaysia), 26–27 Mar 1985. IPMKSM.

Sarkar MSK, Begum RA, Pereira JJ (2020) Impacts of climate change on oil palm production in Malaysia. Environ Sci Pollut Res 27:9760–9770

Shearman J R, Jantasuriyarat C, Sangsrakru D, Yoocha T, Vannavichit A, Tragoonrung S et al (2013) Transcriptome analysis of normal and mantled developing oil palm flower and fruit. Genomics 101(5):306–312

Shakoor N, Lee S, Mockler TC (2017) High throughput phenotyping to accelerate crop breeding and monitoring of diseases in the field. Curr Opin Plant Biol 38:184–192

Singh R, Low ETL, Ooi LCL, Ong-Abdullah M, Chin TN et al (2013a) The oil palm SHELL gene controls oil yield and encodes a homologue of SEEDSTICK. Nature 500:340–344

Singh R, Ong-Abdullah M, Low ET, Manaf MA, Rosli R et al (2013b) Oil palm genome sequence reveals divergence of interfertile species in old and new worlds. Nature 500:335–339

Singh R, Low ETL, Ooi LCL, Ong-Abdullah M, Ting NC, Nookiah R et al (2019) Variation for heterodimerization and nuclear localization among known and novel oil palm SHELL alleles. New Phytol 226(2):426–440. https://doi.org/10.1111/nph.16387

Sodano V, Riverso R, Scafuto F (2018) Investigating the intention to reduce palm oil consumption. CAL 19(S1):500–505

Soh AC, Mayes S, Roberts JA (2017) Oil palm breeding: genetics and genomics. CRC Press, Boca Raton

Sonicha U, Limsriwilai A, Ukoskit K (2018) Intron length polymorphism markers of flower developmental genes for genetic diversity and QTL mapping for oil yield in African oil palm. Genomics Genet 11(3):35–45

Suwanto A, Chua NH, Yue GH, Bai B, Wang L, Zhang YJ et al (2018) Developing genome-wide SNPs and constructing an ultrahigh-density linkage map in oil palm. Sci Rep 8:691

Thomas RL, Seth AK, Chan KW, Ooi SC (1973) Induced parthenocarpy in the oil-palm. Ann Bot 37(3):447–452

Ting N-C (2014) High density SNP and SSR-based genetic maps of two independent oil palm hybrids. BMC Genomics 15:309. http://www.biomedcentral.com/1471-2164/15/309

Ting NC, Yaakub Z, Kamaruddin K, Mayes S, Massawe F, Sambanthamurthi R et al (2016) Fine-mapping and cross-validation of QTLs linked to fatty acid composition in multiple independent interspecific crosses of oil palm. BMC Genomics 17(1):289

Tisné S, Pomiès V, Riou V, Syahputra I, Cochard B, Denis M (2017) Identification of Ganoderma disease resistance loci using natural field infection of an oil palm multiparental population. G3: Genes Genomes Genet 7:1683–1692

Tittonell P (2014) Ecological intensification of agriculture—sustainable by nature. Curr Opin Environ Sustain 8:53–61. https://doi.org/10.1016/j.cosust.2014.08.006

Tranbarger TJ, Kluabmongkol W, Sangsrakru D, Morcillo F, Tregear WJ et al (2012) SSR markers in transcripts of genes linked to post-transcriptional and transcriptional regulatory functions during vegetative and reproductive development of Elaeis guineensis. BMC Plant Biol 12:1. http://www.biomedcentral.com/1471-2229/12/1

Tregear JW, Rival A, Pintaud J-C (2011) A family portrait: unravelling the complexities of palms. Ann Bot 108:1387–1389

Vetaryan S, Kwan YY, Namasivayam P, Ho CL, Syed Alwee SSR (2018) Isolation and characterisation of oil palm LEAFY transcripts. Biotechnol Biotechnol Equip 32(4):888–898

Weckx S, Inzé D, Maene L (2019) Tissue culture of oil palm: finding the balance between mass propagation and somaclonal variation. Front Plant Sci 10:722

Woittiez LS, van Wijk MT, Slingerland M, van Noordwijk M, Giller KE (2017) Yield gaps in oil palm: a quantitative review of contributing factors. Eur J Agron 83:57–77

Wong YC, Teh HF, Mebus K, Ooi TEK, Kwong QB, Koo KL et al (2017) Differential gene expression at different stages of mesocarp development in high-and low-yielding oil palm. BMC Genomics 18(1):470

Yarra R, Jin L, Zhao Z, Cao H (2019) Progress in tissue culture and genetic transformation of oil palm: an overview. Int J Mol Sci 20(21):5353

Zeven AC (1972) The partial and complete domestication of the oil palm (Elaeis guineensis). Econ Bot 26:274

Zheng Y, Chen C, Liang Y, Sun R, Gao L, Liu T, Li D (2019) Genome-wide association analysis of the lipid and fatty acid metabolism regulatory network in the mesocarp of oil palm (Elaeis guineensis Jacq.) based on small noncoding RNA sequencing. Tree Physiol 39(3):356–371

# Genetic Resources of Brassicas

**Hariom Kumar Sharma, Arun Kumar, V. V. Singh, H. S. Meena, Priyamedha, B. L. Meena, Pankaj Sharma, and P. K. Rai**

**Abstract** The Family Brassicaceae has a significant diversity of plant species as it includes more than 3700 species and 338 genera distributed globally. *Brassica* species are polymorphic with diverse applications ranging from vegetable, oil, condiments, fodder, green manure, lubricants, biodiesel, and other industrial uses. Various Brassica species have been originated, evolved, and domesticated in different parts of the world as per the ecology and human needs. The genus Brassica has been extensively investigated by researchers worldwide owing to its vast diversity in form and applications. It has served as a model for scientific studies on cytogenetics, evolution and domestication, speciation, and molecular biology. Brassica and its allied genera constitute a potential germplasm pool with many desirable agronomic and horticultural traits. The genus includes many model plant species including Arabidopsis which led to accelerated research in the field of molecular cytogenetics, genomics, proteomics, comparative genomics, and bioinformatics in Brassicas. In this chapter, various aspects of Brassica are discussed including taxonomy, origin of different Brassica species, Brassica landraces, Brassica genetic resource and their ex situ and in situ conservation strategies, and oilseed Brassica improvement in India including development of varieties, hybrids, and quality mustard.

**Keywords** Brassica · Rapeseed-mustard · *Brassica coenospecies* · Origin and evolution · Brassica landrace · Brassica germplasm conservation

---

H. K. Sharma · A. Kumar · V. V. Singh · H. S. Meena · Priyamedha · B. L. Meena · P. Sharma
P. K. Rai (✉)
ICAR-Directorate of Rapeseed-Mustard Research, Bharatpur, Rajasthan, India


P. M. Priyadarshan, S. M. Jain (eds.), *Cash Crops*,
https://doi.org/10.1007/978-3-030-74926-2_9

# 1 Introduction

The Family Brassicaceae contains over 3700 species in 338 genera, having significant genetic diversity (Demeke et al. 1992; Jain et al. 1994; Das et al. 2007; Warwick et al. 2009; Hayward 2012). The genus *Brassica* includes morphologically diverse species with variable uses (Dixon 2007; Warwick et al. 2009). Brassica species were domesticated by humans long ago over large geographical regions all across the globe, especially in Europe and Asia (Dixon 2007; Snowdon et al. 2007). In India, Brassicas are thought to be cultivated as early as 4000–1500 BC (Prakash 1980; Kimber and McGregor 1995). Similarly, in China, these are being used since 1122 BC, and in Europe Brassica species were cultivated for oil during the Middle Ages (Li 1980; Kimber and McGregor 1995; Snowdon et al. 2007). The Brassica crops include oliferous species, viz. *Brassica napus* (rapeseed/swede), *B. rapa* (turnip rape; syn. *B. campestris*), *B. carinata* (Abyssinian/Ethiopian mustard) and *B. juncea* (Indian mustard), vegetable species such as *B. oleracea* (e.g., broccoli, cabbage, and cauliflower), edible root species including *Raphanus* spp., as well as condiment and herb species such as *B. nigra* and *Sinapis* spp. (mustards). Furthermore, subspecies of the oilseed Brassicas are also commonly cultivated for vegetable and/or mustard crops, namely, *B. rapa* (Chinese cabbage, pak choi and turnip) and *B. juncea* (mustard, head mustard, broccoli mustard, and leaf mustard) (Rich 1991; Dixon 2007; Snowdon et al. 2007; Warwick et al. 2009; Hayward 2012). Apart from this Brassica species are also used for animal forage and fodder, green manure and soil conditioners, as well as industrial lubricants and biodiesel. Hence, Brassicaceae is considered within the ten most economically important plant families due to its diversity of plant species and their usefulness to mankind (Dixon 2007; Snowdon et al. 2007; Warwick et al. 2009). The list of different crops of Brassica and related species is presented in Table 1.

# 2 Taxonomy

## *Tribe Brassiceae*

About 54 genera, 05 subgenera (02 in genus *Brassica*, 03 in genus *Diplotaxis*), and 14 sections [05 sections in subgenus *Brassica* (Sect. *Brassica, Rapa, Micropodium, Brassicoides, Sinapistrum*); 02 sections in subgenus Brassicaria (Sect. *Brassicaria, Nasturtiops*); 03 sections in genus *Rhynchocarpum* (section *Rhynchocarpum, Heterocarpum, Erucoides*); 04 sections in genus *Sinapis* (Sect. *Sinapis, Ceratosinapis, Hebesinapis, Chondrosinapis*)] with roughly 240 species have been reported in tribe Brassiceae of Family *Brassicaceae* (Gómez-Campo 1999) (Tables 2 and 3). Of these reported genera more number of species has been reported in *Brassica* (38 sp.), *Crambe* (35 sp.), *Diplotaxis* (31 sp.), and *Erucastrum* (21 sp.).

**Table 1** Common name and chromosome number of Brassica and related species

| Scientific name | Common name | *n* | Genome (Mb/1C) |
|---|---|---|---|
| *Arabidopsis thaliana* | Thale Cress, Mouse-ear Cress | 5 | 125 |
| *Brassica rapa* | Bird rape | 10 (A) | 529 |
| subsp. *campestris* (L.) A.R. Clapham | Summer turnip rape, wild turnip rape | 10 (A) | – |
| subsp. *oleifera* (DC.) Metzg. | Winter turnip rape | 10 (A) | – |
| var. *brown sarson* | Brown sarson | 10 (A) | – |
| var. *toria* | Toria | 10 (A) | – |
| var. *yellow sarson* | Yellow sarson | 10 (A) | – |
| subsp. *japonica* | Pot herb mustard | 10 (A) | – |
| subsp. *chinensis* (L.) Hanelt | Pak-choi or bok choy, Chinese mustard, Chinese broccoli | 10 (A) | – |
| subsp. *pekinensis* (Lour.) Hanelt | Pe-tsai, Chinese cabbage | 10 (A) | – |
| subsp. *rapifera* | Turnip | 10 (A) | – |
| subsp. *nipposinica* (L.H. Bailey) Hanelt Curled mustard | Curled mustard | 10 (A) | – |
| subsp. *parachinensis* | – | 10 (A) | – |
| subsp. *rapa* | Turnip | 10 (A) | – |
| *Brassica nigra* (L.) W.D.J. Koch | Black mustard, Banarasi rai | 8 (B) | 632 |
| *Brassica oleracea* L. | Cole vegetables | 9(C) | 696 |
| var. *viridis* L. | Kale, collard | 9 (C) | – |
| var. *botrytis* L. | Cauliflower | 9 (C) | – |
| var. *capitata* L. | Cabbage | 9 (C) | – |
| var. *gongylodes* L. | Kohlrabi | 9 (C) | – |
| var. *gemmifera* (DC.) Zenker | Brussels sprouts | 9 (C) | – |
| var. *italica* Plenck | Broccoli | 9 (C) | – |
| var. *oleracea* | Wild cabbage | 9 (C) | – |
| var. *fruticose* | Branchingbush kale | 9 (C) | – |
| var. *sabauda* | Savoy cabbage | 9 (C) | – |
| subsp. *alboglabra* L.H. Bailey | Chinese kale, kailan | 9 (C) | – |
| *Brassica juncea* (L.) Czern. | Indian mustard, brown mustard, oriental mustard | 18 (AB) | 534 |
| *B. juncea* Coss | *Hakarashina* group, oilseed type of India and China | 18 (AB) | – |
| var. *napiformis* Baile | *Nekarashina*, enlarged turnip like root, Mongolia, Manchuria, and northern China | 18 (AB) | – |
| var. *foliosa* Bailey | *Hsueh li hung*, glabrous leaves, China | 18 (AB) | – |
| var. *japonica* Bailey | *Nagan sz kaai*, glabrous leaves, China | 18 (AB) | – |
| var. *crispifolia* Bailey | *Azanina*, salads and as ornamentals, USA | 18 (AB) | – |

(continued)

**Table 1** (continued)

| Scientific name | Common name | *n* | Genome (Mb/1C) |
|---|---|---|---|
| var. *integrifolia* (Stokes) Kitam | Entire succulent leaves, China, and Asia | 18 (AB) | – |
| var. *rugosa* (Roxb) Kitam | Extremely succulent with large leaves and plants | 18 (AB) | – |
| var. *bulbifolia* Mas | *Ta hsin tsai*, succulent stems and elongated internodes | 18 (AB) | – |
| var. *multiceps* | Leaf mustard | 18 (AB) | – |
| var. *tsatsai* | Stem mustard | 18 (AB) | – |
| var. *megarrhiza* | Root mustard | 18 (AB) | – |
| var. *juncea* | Oilseed mustard | 18 (AB) | – |
| var. *utilis* | Seed stalked mustard | 18 (AB) | – |
| *Brassica napus* L. | Rapeseed, canola | 19 (AC) | 566 |
| subsp. *oleifera* (Delile) Sinskaya | Summer oilseed rape, canola | 19 (AC) | – |
| *B. napus f. biennis* (Schübl. & G. Martens) Thel | Winter oilseed rape, winter canola | 19 (AC) | – |
| var. *pabularia* (DC.) Rchb | Rape-kale | 19 (AC) | – |
| var. *napobrassica* (L.) Rchb. | Rutabaga, swede | 19 (AC) | – |
| *Brassica carinata* A. Braun | Abyssinian mustard, Ethiopian mustard, Karan rai | 17 (BC) | 642 |
| *B. tournefortii* | African mustard, Sahara mustard | 10 (T) | 791 |
| *B. elongate* | Elongated mustard, long-stalked rape | – | – |
| *B. fruiticulosa* | – | 8 | – |
| *B. villosa* | – | 9 | – |
| *B. cretica* | – | 9 | – |
| *Camelina sativa* | Gold-of-pleasure, false flax | – | |
| *Capsella bursa-pastoris* | Shepherd's purse | 8, 16 | 203 |
| *Crambe hispanica* | – | 30 | – |
| *Diplotaxis erucoides* | White wall rocket | 7 | 632 |
| *D. muralis* (L.) DC | Sand-rocket | 21 (D) | – |
| *D. tenuifolia* | Wall-rocket | 11 | – |
| *D. acris* | – | 11 | – |
| *Eruca sativa* Mill | Garden rocket, rocket salad | 11 (E) | 560 |
| *Erucastrum gallicum* (Willd.) O.E. Schulz | Dog mustard, hairy rocket | 15 | – |
| *Lepidium sativum* | Garden cress, mustard cress | – | 380 |
| *Orychophragmus violaceus* | Chinese violet cress | 12 | – |
| *Raphanus raphanistrum* L. | Wild radish, jointed charlock | 9 (R) | – |
| *Raphanus sativus* L. | Radish | 9 (R) | 402 |
| *Sinapis alba* | White mustard | 12 (Sal) | 553 |
| *Sinapis arvensis* L. | Wild mustard, Charlok | 9 (Sar) | 367 |

**Table 1** (continued)

| Scientific name | Common name | *n* | Genome (Mb/1C) |
|---|---|---|---|
| *Thlaspi arvense* | Field pennycress; fanweed, stinkweed | 7 | 539 |
| *Hirschfeldia incana* (L.) Lagr.-Foss. | Hoary mustard | 7 (H) | |

Source: Yarnell (1956), Dixon (2007), Prakash et al. ( 2009), OECD (2016), Warwick et al. (2009)

## Brassica coenospecies

*Brassica coenospecies* is defined as the part of the tribe Brassiceae of Schulz, whose members are closely related to crop Brassica species. They are potentially capable of exchanging genetic material with them. Harberd (1972) proposed a concept based on extensive cytotaxonomical investigations on Brassica crop and related wild species. According to him, nine genera (*Brassica, Coincya, Diplotaxis, Eruca, Erucastrum, Hirschfeldia, Sinapis, Sinapidendron, and Trachystoma*) from subtribe *Brassicinae* and two genera (*Enarthrocarpus* and *Raphanus*) from subtribe *Raphaninae* constitute *Brassica coenospecies* (Prakash et al. 2009). Later on, Harberd (1972) classified different coenospecies of Brassiceae into 38 cytodemes or crossing groups (Table 4). A cytodeme is a group consisting of any number of species or genera that have the same chromosome number, and crosses between them always yield fertile hybrids. Furthermore, Harberd (1976), Takahata and Hinata (1983), Warwick and Black (1991, 1997) extended the number of cytodemes up to 63 including 14 genera (Prakash et al. 1999, 2009). Based on the Cp DNA-based phylogenetic analysis and phenetic clustering, genera and species of *Brassica coenospecies* are divided into two evolutionary lineages, *viz. nigra* and *rapa/oleracea* (Warwick and Black 1991; Pradhan et al. 1992; Prakash et al. 2009) (Table 5). Some of the wild Brassica and its allied genera are presented in Fig. 1.

## 3 Origin of Crop Brassica Species

Mizushima (1980) and Prakash and Hinata (1980) were of opinion that the three basic diploid species of *Brassica* are aneuploids and evolved in ascending order from a common ancestor with a basic chromosome number of $n = 6$ (7). Song et al. (1990) proposed that Brassica diploid species originated from a common ancestor with $n = 7$ (Fig. 2). From common ancestor these species evolved through two different pathways, one is through *Hirschfeldia incana* ($n = 7$) or its close relative as the primary ancestor of species of *Nigra* lineage including *B. nigra, B. fruticulosa*, and others. While the other is through *Diplotaxis erucoides* ($n = 7$) or its close alley

**Table 2** Taxonomy of Brassica and allied genera (Source: Gómez-Campo 1999)

| | |
|---|---|
| **Tribe:** ***Brassiceae*** | |
| **Subtribe:** ***Brassicinae*** | |
| **Genus:** ***Brassica*** **L.** | |
| **Subgen.** ***Brassica*** | |
| • Sect. *Brassica* | ***B. oleracea*** L.; *B. montana* Pourret; *B. incana* Ten. subsp., *incana; B. i.* subsp., *cazzae* (Ginz. and Teyb.) Trinajstic; *B. villosa* Biv. subsp., *villosa; B. v.* subsp., *bivoniana* (Mazzola and Raimondo) R. and M.; *B. v.* subsp., *drepanensis* (Caruel) Raimondo and Mazzola; *B. v.* subsp., *tinei* (Lojac.) Raimondo and Mazzola; *B. rupestris* Rafin subsp., *rupestris; B. r.* subsp., *brevisiliqua* Raimondo and Mazzola; *B. r.* subsp., *hispida* Raimondo and Mazzola; *B. macrocarpa* Guss.; *B. insularis* Moris; *B. cretica* Lam. subsp., *cretica; B. c.* subsp., *aegaea* (Heldr. and Hal.) Snogerup et al.*; B. c.* subsp., *laconica* Gustafsson and Snogerup; *B. botteri* Vis. subsp., *botteri; B. b.* subsp., *mollis* (Vis.) Trinajstic; *B. hilarionis* Post.; ***B. carinata*** **Braun**; *B. balearica* Pets. |
| • Sect. *Rapa* (Miller) Salmeen | ***B. rapa*** **L. subsp,** ***rapa;*** *B. r.* subsp., *campestris* (L.) Clapman; *B. r.* subsp., *chinensis* (L.) Hanelt; *B. r.* subsp., *dichotoma* (Roxb.) Hanelt; *B. r.* subsp., *narinosa* (Bailey) Hanelt; *B. r.* subsp., *nipposinica* (Bailey) Hanelt; *B. r.* subsp., *pekinensis* (Lour.) Hanelt; *B. r.* subsp., *trilocularis* (Roxb.) Hanelt; ***B. napus*** **L.;** ***B. juncea*** **(L.) Czern.** |
| • Sect. *Micropodium* DC. | *B. fruticulosa* Cyr. subsp., *fruticulosa; B. f.* subsp., *djafarensis* Blanco and Matarranz; *B. f.* subsp., *dolichocarpa* Emberger and Maire; *B. f.* subsp., *glaberrima* (Pomel) Maire; *B.f.* subsp., *mauritanica* (Cosson) Maire; *B.f.* subsp., *numidica* Maire; *B. f.* subsp., *pomeliana* Maire; *B. f.* subsp., *radicata* (Desf.) Maire; ***B. nigra*** **(L.) Koch**; *B. cossoniana* Boiss. and Reuter; *B. spinescens* Pomel; *B. maurorum* Durieu; *B. procumbens* (Poiret) O.E.Schulz; *B. cadmea* O.E.Schulz; *B. desertii* Danin and Hedge |
| • Sect. *Brassicoides* Boiss. | *B. deflexa* Boiss. |
| • Sect. *Sinapistrum* Willkomm | *B. barrelieri* (L.) Janka; *B. oxyrrhina* Coss.; *B. tournefortii* Gouan |
| **Subgen.** ***Brassicaria*** **(Godr.) Gomez-Campo** | |

**Table 2** (continued)

| | |
|---|---|
| • Sect. *Brassicaria* (Godr.) Cosson | *B. repanda* (Willd.) DC. subsp., *repanda; B. r.* subsp., *africana* (Maire) Greuter and Burdet; *B. r.* subsp., *almeriensis* Gomez-Campo; *B. r.* subsp., *blancoana* (Boiss.) Heywood; *B. r.* subsp., *cadevallii* (Font Quer) Heywood; *B. r.* subsp., *cantabrica* (Font Quer) Heywood; *B. r.* subsp., *confusa* (Emb. and Maire) Heywood; *B. r.* subsp., *galissieri* (Giraud) Heywood; *B. r.* subsp., *glabrescens* (Poldini) Gomez-Campo; *B. r.* subsp., *gypsicola* Gomez-Campo; *B. r.* subsp., *latisiliqua* (Boiss. and Reut.) Heywood; *B. r.* subsp., *nudicaulis* (Lag.) Heywood; *B. r.* subsp., *saxatilis* (DC.) Heywood; *B. r.* subsp., *silenifolia* (Emberger) Greuter and Burdet; *B. desnotesii* Emb. and Maire; *B. gravinae* Ten.; *B. jordanoffii* O. E. Schulz; *B. loncholoma* Pomel; *B. nivalis* Boiss. and Heldr.; *B. elongata* Ehrh. subsp., *elongate; B. e.* subsp., *integrifolia* (Boiss.) Breistr.; *B. e.* subsp., *imdrahsiana* Quezel; *B. e.* subsp., *pinnatifida* (Smal'g) Greuter and Burdet; *B. e.* subsp., *subscaposa* Maire and Weiller; *B. setulosa* (Boiss. and Reuter) Cosson; *B. somaliensis* Hedge and Miller |
| • Sect. *Nasturtiops* (Pomel) Salmeen | *B. souliei* (Batt.) Batt. subsp., *souliei; B. s.* subsp., *amplexicaulis* (Desf.) Greuter and Burdet; *B. dimorpha* Coss. and Dur. |
| **Genus: *Erucastrum* C. Presl.** | |
| | *E. gallicum* O. E. Schulz; *E. nasturtiifolium* (Poiret) O. E. Schulz subsp., *nasturtiifolium; E. n.* subsp., *sudrei* Vivant; *E. leucanthum* Coss. and Dur.; *E. palustre* (Pir.) Vis.; *E. virgatum* C. Presl. subsp., *virgatum; E. v.* subsp., *baeticum* (Boiss.) Gomez-Campo; *E. v.* subsp., *brachycarpum* (Rouy) Gomez-Campo; *E. v.* subsp., *pseudosinapis* (Lange) Gomez-Campo; *E. varium* Durieu subsp., *varium; E. v.* subsp., *mesatlanticum* Maire and Wilczek; *E. v.* subsp., *subsiifolium* Maire; *E. littoreum* (Pau and F.Quer) Maire subsp., *littoreum; E. l.* subsp., *brachycarpum* (Maire and Weiller) Gomez-Campo; *E. l.* subsp., *glabrum* (Maire) Gomez-Campo; *E. rifanum* (Emb. and Maire) Gomez-Campo; *E. elatum* (Ball.) O. E. Sehulz; *E. brevirostre* (Maire) Gomez-Campo; *E. canariense* Webb and Berth.; *E. cardaminoides* (Webb ex Christ.) O. E. Schulz; *E. ifniense* Gomez-Campo; *E. arabicum* Fisch. and Mey.; *E. elgonense* Jonsell; *E. meruense* Jonsell; *E. abyssinicum* (A. Rich.) O. E. Schulz; *E. pachypodum* (Chiov.) Jonsell; *E. rostratum* (Balf. f.) Gomez-Campo; *E. strigosum* (Thunb.) O. E. Schulz; *E. griquense* (Brown) O. E. Schulz |
| **Genus: *Diplotaxis* DC.** | |
| **Subgen. *Diplotaxis*** | *D. tenuifolia* (L.) DC. subsp., *tenuifolia; D. t.* subsp., *cretacica* (Kotov) Sobrino-Vesperinas; *D. muralis* (L.) DC. subsp., *muralis; D. m.* subsp., *ceratophylla* (Batt.) Mart.-Lab.; *D. viminea* (L.) DC.; *D. simplex* (Viv.) Spr. |
| **Subgen. *Hesperidium* (DC.) Mart.-Lab.** | *D. harra* (Forsk.) Boiss. subsp., *harra; D. h.* subsp., *confusa* Mart.-Lab.; *D. h.* subsp., *crassifolia* (Rafin) Maire; *D. h.* subsp., *lagascana* (DC.) O.Bolos and J.Vigo; *D. antoniensis* Rustan; *D. glauca* (Schmidt) O.E.Schulz; *D. gorgadensis* Rustan subsp., *gorgadensis;* D. g. subsp., *brochmanii* Rustan; *D. gracilis* (Webb) O. E. Schulz; *D. hirta* (Chev.) Rustan and Borgen; *D. sundingii* Rustan; *D. varia* Rustan; *D. vogelii* (Webb) Cout.; *D. pitardiana* Maire; *D. acris (Forsk.)* Boiss.; *D. villosa* Boul. and Jail.; *D. griffithii* (Hook.f. and Thorns.) Boiss.; *D. nepalensis* Hara |

(continued)

**Table 2** (continued)

| | |
|---|---|
| **Subgen. *Rhynchocarpum* (Prantl) Mart.-Lab.** | |
| • Sect. *Rhynchocarpum* Prantl | *D. assurgens* (Del.) Gren.; *D. tenuisiliqua* Del. subsp., *tenuisiliqua;* D. t. subsp., *rupestris* (J. Ball.) Mart.-Lab.; *D. brachycarpa* Godr.; *D. virgata* (Cav.) DC. subsp., *virgata;* D. v. subsp., *australis* Mart.-Lab.; D. v. subsp., *rivulorum* (Br.-B1. and Maire) Mart.-Lab.; D. v. subsp., *sahariensis* (Coss.) Mart.-Lab.; *D. berthautii* Br.-B1. and Maire; *D. catholica* (L.) DC.; *D. ollivieri* Maire; *D. siifolia* G. Kunze. subsp., *siifolia;* D. s. subsp., *bipinnatifida* (Coss.) Mart.-Lab.; D. s. subsp., *vicentina* (P.Cout.) Mart.-Lab. |
| • Sect. *Heterocarpum* Mart.-Lab. | *D. ibicensis* (Pau) Gomez-Campo; *D. siettiana* Maire; *D. brevisiliqua* (Coss.) Mart.-Lab.; *D. ilorcitana* (Sennen) Aedo and Mart.-Lab. |
| • Sect. *Erucoides* Mart.-Lab. | *D. erucoides* (L.) DC. subsp., *erucoides;* D. e. subsp., *longisiliqua* (Coss.) Gomez-Campo |
| **Genus: *Sinapis* L.** | |
| • Sect. *Sinapis* | *S. alba* L. subsp., *alba; S. a.* subsp., *mairei* (H. Lindb.) Maire; *S. a.* subsp., *dissecta* (Lag.) Bonnier; *S. flexuosa* Poir. |
| • Sect. *Ceratosinapis* DC. | *S. arvensis* L. subsp., *arvensis; S. a.* subsp., *allioni* (Jacq.) Baillargeon; *S. a.* subsp., *nilotica* (O. E. Schulz) Baillargeon |
| • Sect. *Hebesinapis* DC. | *S. pubescens* L. subsp., *pubescens; S. p.* subsp., *virgata* (Batt.) Baillargeon; *S. boivinii* Baillargeon; *S. indurata* Coss.; *S. aristidis* Coss. |
| • Sect. *Chondrosinapis* O.E.Schulz | *S. aucheri* (Boiss.) O.E.Schulz |
| **Genus: *Eruca* Mill.** | *E. vesicaria* (L.) Cav. subsp., *vesicaria; E. v.* subsp., *sativa* (Miller) Thell.; *E. v.* subsp., *pinnatifida* (Desf.) Emb. and Maire |
| **Genus*:* *Coincya* Rouy** | *C. richeri* (Vill.) Greuter and Burdet; *C. wrightii* (O. E. Schulz) Stace; *C. monensis* (L.) Greuter and Burdet subsp., *monensis; C. m.* subsp., *cheiranthos* (Vill.) Aedo, Leadlay and Mufioz-Garm.; *C. m.* subsp., *nevadensis* (Willk.) Leadlay; *C. m.* subsp., *orophila* (Franco) Aedo, Leadlay and Mufioz-Garm.; *C. m.* subsp., *puberula* (Pau) Leadlay; *C. transtagana* (Cout.) Clem.-Mufioz and Hern.-Bermejo; *C. longirostra* (Boiss.) Greuter and Burdet; *C. rupestris* Porta and Rigo subsp., *rupestris; C. r.* subsp., *leptocarpa* (Gonz.-Albo) Leadlay |
| **Genus: *Hirschfeldia* Moench** | *H. incana* (L.) Lagroze-Fossat subsp., *incana; H. i.* subsp., *incrassata* (Thell.) Gomez-Campo |
| **Genus*:* *Sinapidendron* Lowe** | *S. angustifolium* (DC.) Lowe; *S. frutescens* (Sol.) Lowe *subsp.frutescens;* S. f. subsp., *succulentum* (Lowe) Rustan; *S. gymnocalyx* (Lowe) Rustan; *S. rupestre* Lowe; *S. sempervivifolium* Mnzs. |
| **Genus: *Trachystoma* O.E. Schulz** | *T. aphanoneurum* (Maire and Weiller) M. and W.; *T. balli* O.E.Schulz; *T. labasii* Maire |
| **Subtribe: *Raphaninae*** | |
| **Genus: *Raphanus* L.** | *R. raphanistrum* L. subsp., *raphanistrum; R. r.* subsp., *landra* (DC.) Bonnier and Layens; *R. r.* subsp., *maritimus* (Sm.) Thell.; *R. r.* subsp., *microcarpus* (Lange) Thell.; *R. r.* subsp., *rostratus* (DC.) Thell.; *R. sativus* L. |

**Table 3** Genera of the tribe *Brassiceae* of family Brassicaceae (Source: Gómez-Campo 1999)

| No. of species in genera | Genera |
|---|---|
| <20 | *Ammosperma* (1), *Boleum* (1), *Coincya* (6), *Cordylocarpus* (1), *Crambella* (1), *Kremeriella* (1), *Calepina* (1), *Carrichtera* (1), *Ceratocnemum* (1), *Chalcanthus* (1), *Conringia* (6), *Didesmus* (2), *Dolichorrhynchus* (1), *Douepia* (1), *Enarthrocarpus* (5), *Eremophyton* (1), *Eruca* (1), *Erucaria* (6), *Euzomodendron* (1), *Foleyola* (1), *Fortuynia* (2), *Fezia* (1), *Guiraoa* (1), *Hemicrambe* (1), *Henophyton* (2), *Hirschfeldia* (1), *Moricandia* (7), *Morisia* (1), *Muricaria* (1), *Orychophragmus* (1), *Otocarpus* (1), *Physorrhynchus* (2), *Pseuderucaria* (3), *Pseudofortuynia* (1), *Psychine* (1), *Quezeliantha* (1), *Quidproquo* (1), *Raffenaldia* (2), *Raphanus* (2), *Rapistrum* (2), *Rytidocarpus* (1), *Savignya* (2), *Schouwia* (2), *Sinapidendron* (5), *Sinapis* (8), *Succowia* (1), *Trachystoma* (3), *Vella* (5), *Zilla* (2) |
| >20 | *Brassica* (38), *Crambe* (35), *Diplotaxis* (31), *Erucastrum* (21) |

Values in parenthesis are estimated no. of species in respective genera

as the primary ancestor of Rapa/Oleracea lineage species (*B. oleracea, B. rapa*, and other species).

Theory of origin of *Rapa/Oleracea* lineage species by *D. erucoides* has been supported by many workers on the basis of studies of satellite DNA, S-locus, nuclear DNA content, and cytological studies (Harbinder and Lakshmikumaran 1990; Song et al. 1990; Warwick and Black 1991; Pradhan et al. 1992; Vyas et al. 1995; Inaba and Nishio 2002). In Rapa/Oleracea lineage *B. oleracea* is supposed to be evolved first; subsequently *B. rapa* was evolved from it or from ancestral species. It is assumed that Brassica species originated first in Mediterranean basin countries and further extended to Mediterranean, Irano-Turanian, and Saharo-Sindian phytogeographical regions (Prakash and Hinata 1980; Song et al. 1990; Prakash 2001), from where the species was disbursed to other parts of the world through various migrants and travelers. Song et al. (1988a) on the basis of RFLP data proposed that *Brassica nigra* originated from one evolutionary pathway with *Sinapis arvensis* or a close relative as the likely progenitor, whereas *B. rapa* and *B. oleracea* emanated from another pathway from a common ancestor species with $n = 9$. *B. nigra* is also an ancient species (Prakash et al. 2012). *B. rapa* (turnip rape) was the first diploid species to be domesticated several millennia ago (Gómez-Campo and Prakash 1999). *B. oleracea* was domesticated later as its center of origin (Atlantic coast) was too far from major centers of its domestication, whereas *B. carinata* and *B. napus* are recently domesticate nearly 400 years ago. Among crop brassicas, three species (*B. oleracea, B. rapa*, and *B. juncea*) are highly polymorphic and exhibit morphological variants in the form of different crops in which different plant parts (flower, stem, roots, and leaves) were modified during the course of evolution and domestication in different geographical regions of the world. Dixon (2007) proposed geographical distribution and probable places of origins of brassicas (Fig. 3). Photos of cultivated rapeseed-mustard species are presented in Fig. 4.

**Table 4** Cytodemes in *Brassica coenospecies*

| Chromosome no. (*n*) | *Coenospecies* |
|---|---|
| 7 | *Brassica deflexa, Diplotaxis erucoides, Erucastrum virgatum, E. varium, Sinapis aucheri, Hirschfeldia incana, Pseuderucaria* spp. |
| 8 | *B. nigra, B. fruiticulosa (+maurorum + spinescens), D. siettiana, E. abyssinicum, E. nasturtiifolium (+leucanthum), E. strigosum, Trachystoma* spp. |
| 9 | *B. oleracea and 8 wild Mediterranean allied species, B. oxyrrhina, D. assurgens, D. catholica, D. tenuisiliqua, D. virgata, D. berthautii, E. cardaminoides, (+canariense + ifniense), Raphanus all species and subspecies, S. arvensis (+allioni), S. pubescens* |
| 10 | *B. tournefortii, B. barrelieri, B. gravinae, B. repanda (+desnottesii), B. rapa (+ many cultivated subspecies), D. siifolia, D. viminea, Enarthrocarpus* spp., *Sinapidendron* spp. |
| 11 | *B. souliei, B. elongate, D. acris, Diplotaxis (+ pitardiana); Eruca* spp. |
| 12 | *Coincya* spp. *(syn. Hutera and Rhynchosinapis), S. alba, S. flexuosa* |
| 13 | *Diplotaxis harra (*+ several subsps.*)* |
| 14 | *E. virgatum (*subsp. pseudosinapis*)*[a], *Moricandia arvensis*[a], *M. moricandioides*[a], *Rytidocarpus moricandioides*[a] |
| 15 | *Erucastrum gallicum*[b], *E. elatum*[b] |
| 16 | *B. cossoniana* (4×)[a], *North African subspecies, B. balearica*[b], *E. nasturtiifolium*[a], *E. abyssinicum*[a] |
| 17 | *Brassica carinata*[b] |
| 18 | *Brassica juncea*[b] |
| 19 | *Brassica napus*[b] |
| 20 | *Brassica gravinae*[a] |
| 21 | *Diplotaxis muralis*[b] |
| 22 | *Brassica dimorpha*[a] |
| 23 | *Brassica dimorpha* |
| 24 | *Coincya* spp. (4×)[a] |
| 28 | *Moricandia suffruticosa*[a] |
| 42 | *Moricandia spinosa*[a] |
| 80? | *Brassica repanda*[a] |

Source: From Prakash et al. (1999)
[a]Autopolyploid cytodemes
[b]Allopolyploid cytodemes

The amphidiploid species *B. napus* and *B. juncea* have evolved through polyphyletic origins from parental diploid species with highly differentiated cytoplasms. Hypothetical scheme for the possible evolution of *Brassica* species and their domesticated variants is presented in Fig. 5 (Song et al. 1988a, 1990). *B. napus* may be considered as the most ancient amphidiploid, followed by *B. juncea* and *B. carinata* (Song et al. 1996). A considerable genetic diversity developed in three natural amphidiploids Brassica species which may be due to multiple hybridizations with different diploid parents and genome modifications following polyploidization

**Table 5** Genera and species of *Brassica coenospecies* in *Nigra* and *Rapa/Oleracea* lineage

| *Nigra* lineage | *Rapa/Oleracea* lineage |
|---|---|
| **Group I**<br>*Brassica nigra* (8), *B. fruticulosa* (8), *Sinapis arvensis* (9), *S. alba* (12), *Diplotaxis ibicensis* (8), *D. siettiana* (8), *Erucastrum littoreum* (16), *Trachystoma balii* (8)<br>**Group II**<br>*B. tournefortii* (10), *B. procumbens* (9), *D. brachycarpa* (9), *E. varium* (7), *E. virgatum* (7), *Hirschfeldia incana* (7), *S. pubescens* (9)<br>**Group III**<br>*E. canariense* (9), *E. brevirostre* (9), *D. assurgens* (9), *D. siifolia* (10), *D. berthautii* (10), *D. virgata* (9), *D. catholica* (9), *Sinapidendron* spp. (10)<br>**Group IV**<br>*Coincya* spp. (12) | **Group I**<br>*B. rapa* (10), *B. oleracea* (9), *B. deflexa* (7), *B. barrelieri* (10), *B. oxyrrhina* (9), *D. cossoneana* (9), *D. erucoides* (7), *E. abyssinicum* (16), *E. strigosum* (8), *E. nasturtifolium* (8), *S. aucheri* (7), *Enarthrocarpus lyratus* (10), *Raphanus* spp. (9)<br>**Group II**<br>*D. harra* (13), *Eruca* spp. (11), *D. tenuifolia* (11), *Rytidocarpus moricandiodes* (14)<br>**Group III**<br>*M. arvensis* (14), *M. moricandiodes* (14), *M. suffruticosa* (28)<br>**Group IV**<br>*B. gravinae* (10), *B. repanda* (10), *D. viminea* (10)<br>**Group V**<br>*B. elongata* (11) |

Values in parenthesis indicate haploid chromosome no. of the species
Source: Warwick and Black (1991, 1997), Prakash et al. ( 2009)

(Song et al. 1993, 1996). U (1935) proposed the U-triangle of origin of amphidiploid Brassica species based on cytogenetical relationships among different Brassica crop species and experimental synthesis of *B. napus*, which was further investigated and confirmed by many researchers through production of synthetic amphidiploids via artificial hybridization and protoplast fusion (Prakash et al. 2009) (Fig. 5). Morinaga (1928, 1929a, b, c, 1931, 1933, 1934a, b) proposed genome analysis of *Brassica* species based on extensive work on interspecific hybridization in different *Brassica* species. Ramanujam and Srinivasachar (1943) first successfully produced *B. juncea*. Later on, Frandsen (1943, 1947) synthesized all the three allopolyploid species.

## B. nigra

*B. nigra* is considered as semi-domesticated species since it bears both wild and domesticated type traits (tall plant stature, enormous vegetative growth, small-size pods with reduced numbers of small seeds) (Song et al. 1988a). *B. nigra* is proposed to be originated at central and south Europe and found naturally distributed to Mediterranean area extending into the Middle East and central Asia. It was valued for various medicinal properties and used for spice, condiment, seed oil, and leaf vegetables in various preparations during Mediterranean civilizations including Mesopotamia, Egyptian, Greek, Roman, Turkey, Ethiopia (Hedrick 1919; Bailey

**Fig. 1** Wild Brassica and its allied genera

1922; Zeven and Zhukovsky 1975; Tsunoda 1980; Prakash et al. 2012). Based on various agro-morphological (chemotaxonomic markers), cytological (chromosome pairing), biochemical (seed protein, fraction 1 protein), and molecular studies (nuclear DNA RFLPs, cp DNA analysis, 5S rDNA spacer, repetitive DNA, RAPD, S-locus, ITS/trnL) *Sinpais arvensis* is considered as its progenitor/close relative (Mizushima 1950; Vaughan and Denford 1968; Uchimiya and Wildman 1978; Song et al. 1988a; Yanagino et al. 1987; Warwick and Black 1991; Pradhan et al. 1992; Gupta et al. 1992; Poulsen et al. 1994; Simonsen and Heneen 1995; Cheng and Heneen 1995; Kapila et al. 1996; Wu et al. 1996; Inaba and Nishio 2002; Warwick and Sauder 2005).

## Brassica oleracea

*B. oleracea* is a highly polymorphic species evolved through many structural evolution in each plant part including leaf, stem, and inflorescence. The morphological

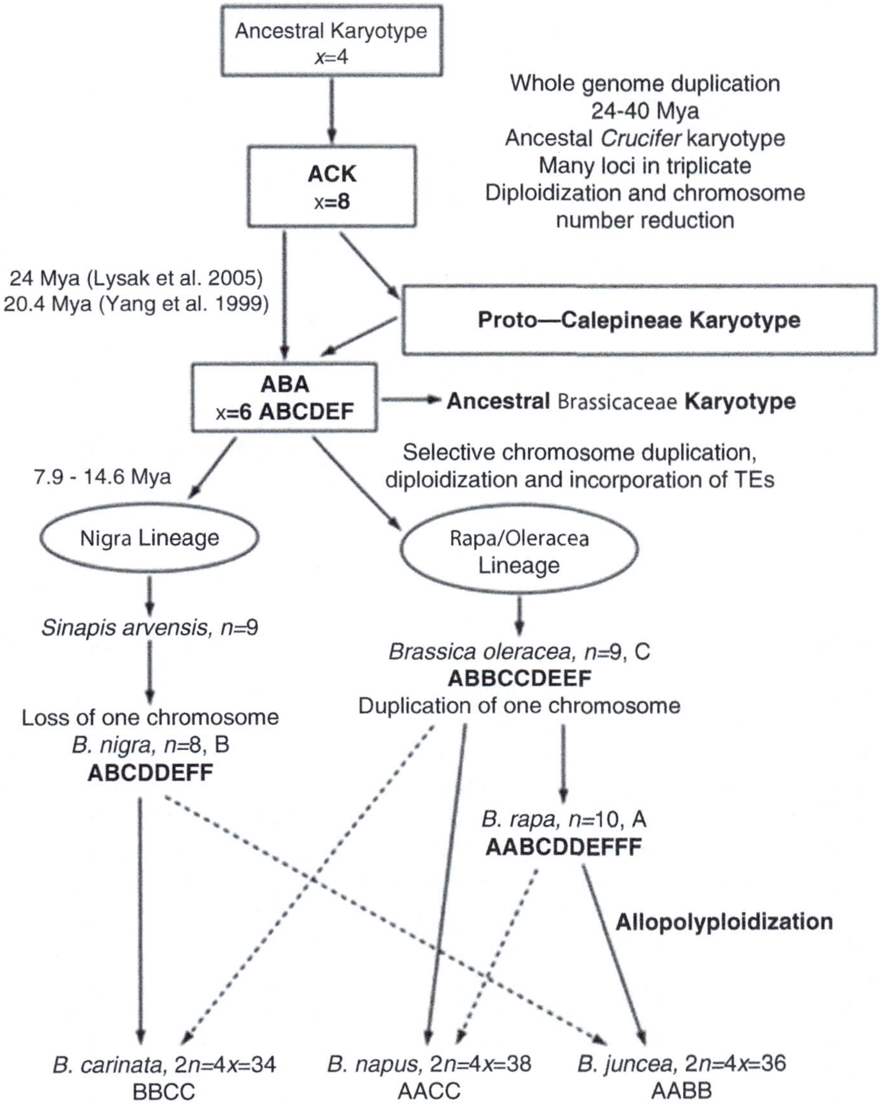


**Fig. 2** Origin of basic and alloploid Brassica species (solid line: female parents; broken lines: male parents). (Source: Prakash et al. 2012)

variants are consumed as vegetable (Lorenz and Maynard 1988; Rubatzky and Yamaguchi 1997). Bailey (1922) for the first time used the general term "cole crops" for *B. oleracea* crops in his book *Principles of Vegetable-Gardening*. Snogerup (1980) and Dixon (2007) divided *B. oleracea* complex, which includes at least six well-defined groups and designated them as varieties, namely (a) Kales (*B. oleracea*

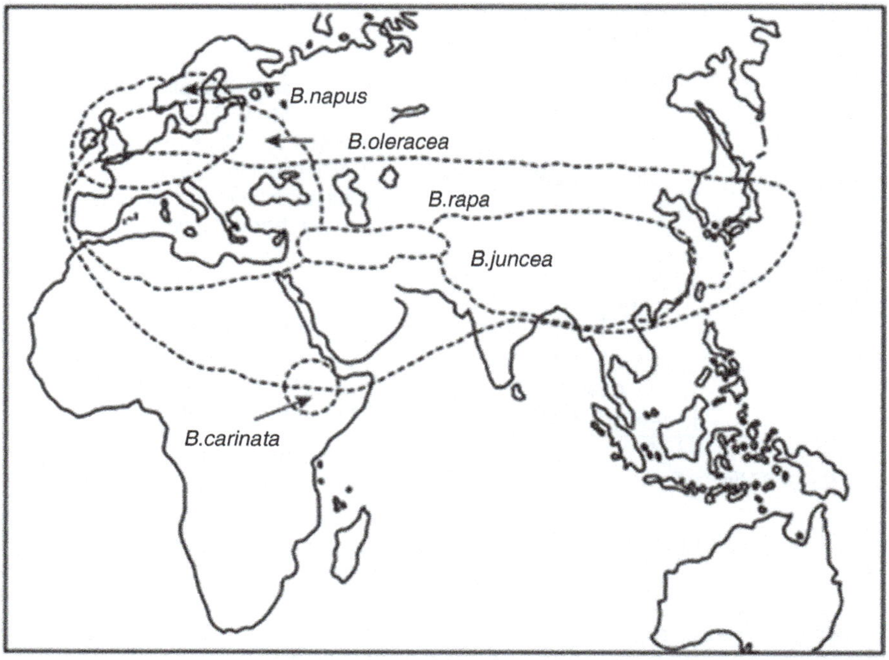


**Fig. 3** Geographical distribution and probable places of origins of brassicas. (Source: Dixon 2007)

var. *acephala* bearing edible foliage); (b) Cabbages (*B. oleracea* var. *capitata*) with formation of heads consisting of tightly packed leaves; (c) Kohlrabi (*B. oleracea* var. *gongylodes*) with thick edible stem; (d) Inflorescence kales (*B. oleracea* var. *botrytis*, var. *italica*), with thickened edible inflorescences and include cauliflower, broccoli, and sprouting broccoli; (e) Branching bush kales (*B. oleracea* var. *fruticosa*), grown for edible foliage; and (f) Chinese kale (*B. oleracea* var. *alboglabra*) with edible flower bud, flower stalk, and young leaves. *B. oleracea* var. *alboglabra* has close morphological resemblance with related wild species *B. cretica* ssp. *nivea*.

In ancient Greek, the term *Krambh* (*krambe*) was used for *B. oleracea* (before ca. 330 BCE). The first illustration of various brassicas is found in the Dioscoridean recension known as the Juliana Anicia Codex dated 512. According to earlier theories the cole crop forms evolved through mutations, human selection, and adaptation from the primitive wild kales growing along the Mediterranean coast from Greece to England, where it might have been cultivated by Celts tribe. Mediterranean kales species occur in cliffs and in rocky islets, in distinct ecogeographical centers, and mainly confined to Mediterranean region. These wild forms are closely related to cultivated forms and might have participated in their origin and domestication (Prakash et al. 2012). Snogerup (1980) has grouped these wild forms in different species with $n = 9$ (*B. cretica* ssp. *nivea; B. cretica* ssp. *cretica; B. rupestris-incana* complex; *B. drapenensis; B. insularis; B. macrocarpa; B. montana; B. oleracea;*

**Fig. 4** Cultivated rapeseed-mustard species

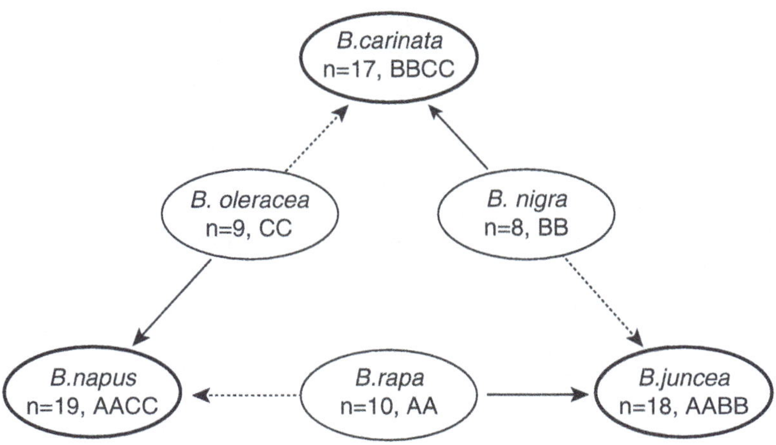


**Fig. 5** Cytogenetic relationships of crop brassicas (U, 1935). Solid and broken lines in the allopolyploids represent female and male parents, respectively. (Source: U 1935; Prakash et al. 2009)

*B. hilarionis*). These have been grouped into same cytodemes (Harberd 1972). *B. montana* is considered as the progenitor of cabbage and kales and *B. rupestris* of kohlrabi (Snogerup 1980). Bailey (1922, 1930) suggested wild *B. oleracea* as the possible progenitor of various *B. oleracea* cultivars.

Helm (1963) described the scheme of evolution of different forms of various *B. oleracea* cultivars. Gates (1953), based on studies on different wild species and cultivated forms, reported that cabbages, Brussels sprouts, and kale originated from *B. oleracea* var. *sylvestris* in Western Europe, cauliflower and broccoli in the eastern Mediterranean, and kohlrabi in the middle Mediterranean. Song et al. (1988b) believed that ancient wild progenitor was similar to wild *B. oleracea* and *B. alboglabra*, which must be the closest ancestors of cultivated forms. The earliest cultivated form was probably a leafy kale from which kales originated due to natural hybridization, gene introgression, and selection along the Mediterranean coast and north Atlantic from Greece to Wales. The relative lack of *cp* diversity in these species (Panda et al. 2003; Allender et al. 2007) indicates a monophyletic rather than polyphyletic origin and domestication of cole crops (Prakash et al. 2012).

## Brassica rapa

The cultivated *Brassica rapa* is a highly polyomorphic species, which exhibit different infra-specific types (Oleiferous, Leafy, and Turnip or root forming type). Oleiferous forms grown in Canada, Australia, and Europe are known as turnip rape while in Indian subcontinent it is called as sarson (yellow sarson, brown sarson, toria). Leafy forms are cultivated as vegetables in south-east Asian countries and as oilseeds in central and southern China. The root-forming type is grown globally for vegetable and fodder purpose and is called as Turnip (Prakash et al. 2012).

The references to oleiferous forms are found in ancient Indian Aryan literature (ca. 1500 BCE). *B. rapa* seeds were found in the stomach of Tollund man (a mummified corpse of the fourth-century BCE found in Scandinavia) (Renfrew 1973). Wild forms of *B. rapa* grow from the western Mediterranean region through Europe to central Asia and the Near East (Sinskaia 1928; Vavilov 1949; Mizushima and Tsunoda 1967; Zeven and Zhukovsky 1975; Tsunoda 1980). The Fertile Crescent region comprising present-day Iran-Iraq-Turkey is most likely the place of its origin. It probably was first domesticated at some of these places perhaps several millennia ago.

There are different opinions about the origin of *B. rapa.* Burkill (1930) suggested Europe as the place of its origin whereas Sinskaia (1928) and Vavilov (1949) strongly considered central Asia, Afghanistan, and the adjoining northwest part of the Indian subcontinent as one of the independent centers of its origin. Two races namely Western (oleiferous and turnip forms) and Eastern race (vegetable forms) have been proposed based on morphology (Sun 1946), protein, and DNA markers (Denford 1975; Song et al. 1988b; He et al. 2003; Takuno et al. 2007). However,

Europe is considered the primary center of origin of oleiferous forms. Turnip later traveled further eastward through the Middle East. China is also considered as the center of origin of a unique form of ssp. *oleifera* (oilseed form) (Li 1981). European and North American (Canadian) oleiferous forms are spring-type cool-season crops, which possess better cold tolerance. In Canada, it is often referred to as Polish rape or summer turnip rape. Chinese leafy forms of *B. rapa* are vegetable type and considered to be originated from oilseed forms after *B. rapa* was introduced into China, probably in the first century CE, through western Asia or Mongolia (Burkill 1930; Nishi 1980; Prakash et al. 2012).

The Indian oleiferous forms are divided into three ecotypes: *B. rapa* var. *brown sarson, B. rapa* var. *yellow sarson, and B. rapa* var. *toria*. Brown sarson is considered to be the oldest, and its name is derived from the Sanskrit word *sarshap*. The primitive forms of brown sarson entered India from which the present form evolved in northwestern India in the foothills of Himalayas and spread later on (Hinata and Prakash 1984). Two forms of brown sarson, i.e., *tora* (self-compatible) and *lotani* (self-incompatible) are cultivated in India. Among these *lotani* type is considered the oldest one since all the cultivated and wild forms of *B. rapa* are self-incompatible. The *tora* form is very similar to yellow sarson except seed color; hence it is considered to be originated through hybridization. Yellow sarson is supposed to be evolved from brown sarson as a mutant ca. 1200 BCE in northwestern India and mainly cultivated in Northeastern India due to short duration (Hinata and Prakash 1984). It has yellow seed color and bold seed size compared to toria and brown sarson. It was mentioned as *siddhartha* in Sanskrit literature of 1000 BCE (Prakash et al. 2012).

Turnip is of European origin, and its carbonized seeds have been excavated from Neolithic sites in Switzerland, 8000 BCE (Hyams 1971). The earliest record of its cultivation is found in a list of plants grown in the garden of Merodachbaladan (722–711 BCE) in Babylonia (Zohary and Hopf 1993). Cultivation of turnip started in Europe around 2500–2000 BCE from where it reached to Asia after 1000 BCE (De Candolle 1886). It is considered that its biennial forms with swollen hypocotyl were selected as food. The word turnip is derived from the Middle English word *nep* from *napus*, which together with *turn* (made round) became turnip (Boswell 1949), but this word appeared only after 1400. In India it is brought by invaders in eastern India and known as *salgham* (Prakash et al. 2012).

## Brassica carinata

Ethiopian mustard is cultivated in Ethiopia and parts of the east and west coasts of the African continent. It is proposed to have evolved through natural hybridization between *B. nigra* and a wild kale form (*B. oleracea*) in the highlands of Ethiopia and adjoining regions of East Africa and the Mediterranean coast where its diploid progenitors are found naturally (Mizushima and Tsunoda 1967; Prakash and Hinata

1980; Song et al. 1988a, Gómez-Campo and Prakash 1999). It has better drought tolerance and resistance to fungal pathogen.

## Brassica juncea

Carbonized mustard seeds have been excavated from Neolithic settlements of Banpo in China (ca. 4800 BCE) and from Indus Valley civilization (present-day Sindh and Punjab states of Pakistan) (ca. 3000 BCE) in the northwest of the Indian subcontinent (Institute of Archaeology Report 1963; Allchin 1969; Prakash et al. 2012), and its references are reported in ancient Indian and Chinese literature as early as 2500 BCE (Prakash and Hinata 1980; Hinata and Prakash 1984; Gómez-Campo and Prakash 1999; Manohar et al. 2009).

Brassicas were used as vegetable in China and have references in the literatures of *Xia* dynasty (about 3000 BCE) and *Zhou* dynasty (1122–247 BCE). The species was introduced into Europe during the Middle Ages as a medicinal crop and was later grown as a vegetable, condiment, and oil crop (Hemingway 1976). Different cultures have given it various names: *sinapi* (Romans), *thlaspi* (Pliny), *Sinapi alterum, S. sativum alterum* (Prakash et al. 2012). Hooker and Thomson (1861) reported polymorphism in *B. juncea* particularly the large variations in leaf shape. Many researchers divided different variants of *B. juncea* into different subspecies or varieties based on morphotypes (Bailey 1922), volatile oils and seed protein (Vaughan et al. 1963), and geographical and botanical characters (Dixon 2007). Seven groups/varieties [*B. juncea* Coss, *B. juncea* var. *napiformis* Bailey, *B. juncea* var. *foliosa* Bailey, *B. juncea* var. *crispifolia* Bailey, B. *juncea* var. *integrifolia* (Stokes) Kitam, *B. juncea* var. *rugosa* (Roxb) Kitam, *juncea* var. *bulbifolia* Mas] proposed by Dixon (2007) are presented in Table 1. Gladis and Hammer (1992) proposed five groups (leaf mustard: *B. juncea* var. *multiceps*; stem mustard: *B. juncea* var. *tsatsai*; root mustard: *B. juncea* var. *megarrhiza;* oilseed mustard: *B. juncea* var. *juncea;* seed stalk mustard: *B. juncea* var. *utilis*) (Prakash et al. 2012). At present, oil forms are simply referred to as *B. juncea* Czern. while the Chinese variants are given as varieties.

There are different opinions about the origin of *B. juncea*. Prain (1898) and Sinskaia (1928) proposed that *B. juncea* evolved in China and subsequently spread to India through a northeastern route. Many researchers strongly supported the Middle East as the place of origin since parent species naturally occurs in this region (Burkill 1930; Sun 1970; Olsson 1960a; Mizushima and Tsunoda 1967). The Russian botanist Vavilov (1949) proposed Afghanistan and adjoining regions as its primary center of origin and central and western China, eastern India, and Asia Minor through Iran as the secondary centers. Wild forms of *B. juncea* are naturally found in the plateau of Asia Minor and adjoining southern Iran (Tsunoda and Nishi 1968; Tsunoda 1980). Cytogenetical, biochemical, and molecular evidence has suggested the polyphyletic origin at many places with a sympatric distribution of parental species (Olsson 1960a; Prakash 1973a; Vaughan 1977; Prakash and Hinata 1980;

Song et al. 1988a). Oleiferous forms evolved in Middle East Indian region, whereas leafy types are developed in China (Song et al. 1988a). Chinese forms of *B. juncea* might be originated by natural hybridization of different *B. rapa* leafy morphotypes with *B. nigra*, and by mutations and/or nonhomologous recombination between A and B genome chromosomes (Prakash 1973a, b).

The earliest record of Indian forms of *B. juncea* from India is the excavated seeds from the ancient sites of the Indus Valley civilization (ca. 2300–1500 BCE) in the present-day Sindh and Punjab states of Pakistan. A description of Brassica species is well documented in Ancient Indian literatures including *Sanskrit* and *Hindi* (*Rigveda, Atharvaveda, Chandogya Upanishad, Brahmanas, Charak Samhita, Arthashastra, Sukraniti*), *Buddhist* (*Bodhyan Grhya Sutra, Sankhyan Grhya Sutra, Sankhyan Sautra Sutra, Dhammapada, Suttanipita, and Samyutta Nikaya*), and *Jainies* (*Jataka, Acharanga Sutra*). In *Sanskrit*, Brassicas were known by various names, viz. *Siddhartha, Rajika, Baja, Sarshap,* and *Svet Sharshap. Sarshap* denoted only *brown sarson* and *Siddhartha* is used for yellow sarson (Hinata and Prakash 1984) while *Svet Sarshapa* was used for *Sinapis alba* (Watt 1889). Vedic Aryans used *Rajika* for *B. juncea*. Many ancient civilizations/cultures used it for medicinal use, cooking, food seasoning, food preservation, vegetable, and body massage (Prakash et al. 2012).

Among all the Brassica species the only seed remains (carbonized seeds) of *B. juncea* were found in excavation from *Chanhu-daro*, a site of Indus Valley civilization (Allchin 1969), and in Damoder River valley site in eastern India, ca. 1000 BCE (Ghosh et al. 2006). Hinata and Prakash (1984) strongly believe that it entered into northwest India from the Middle East, through Afghanistan between 4500 and 2300 BCE. Natural hybridizations between the parental species with a sympatric distribution in Afghanistan and the adjoining region also originated new genetic stocks. However, Indian forms exclusively developed in the direction of oilseed types with lyrate pinnatisect leaves. A leafy form of *B. juncea* (vegetable type) named as *Sinapis rugosa* by Roxburgh (1832) is cultivated in the foothills of the Himalayas from Kashmir in the west to Assam and Sikkim in the east, which might have been brought by Buddhist monks in the region (Prakash et al. 2012).

## Brassica napus

*B. napus* is a recently domesticated crop just in the fifteenth century. Presently, two forms of it are being cultivated (oilseed rape, and root forming swede or rutabaga). Oilseed rape is widely grown for edible oil in Canada, Europe, China, and Australia, and rutabaga for food and fodder. The rutabagas include two forms, one with white flesh and the other with yellow. The white forms were called by various names, viz. *napobrassica, Brassica radice napiformi, navet jaune,* and *B. napus rapifera* (Bauhin 1623; Tournefort 1700; De Candolle 1821; Metzger 1833). It was an important staple food plant earlier, particularly during World War I, but now mostly grown for forage (Prakash et al. 2012). Kohlrabi might have evolved through hybridization between turnip and some forms of *B. oleracea* in farmers' fields, which has been confirmed by artificial crosses among these forms (Olsson et al. 1955; Kato et al. 1968).

Oil forms of *B. napus* are grown as winter rape and spring rape, among them the former was domesticated first in the seventeenth century and presently cultivated in European countries and China while spring rape form was developed in the late seventeenth or eighteenth century as a selection and being grown in Canada, northern Europe, and Australia. Oil forms of *B. napus* do not occur naturally in the wild. Its oil, known as raepolie, was first used for illumination and later as edible oil and for soap making. As compared to other oilseed Brassicas, distribution, cultivation, and production of oilseed rape exceeded in a brief span of 400 years since its first domestication. In India at present, its cultivation is confined to a very limited area primarily due to pod shattering and poor adaptability to dry conditions (Prakash et al. 2012).

Sinskaia (1928) and Schiemann (1932) proposed the southwest Mediterranean region as the place of origin where the constituent parents overlap in natural distribution but existence of this overlap was denied by Tsunoda (1980). Many researchers supported its multiple origins (Olsson 1960b; Prakash and Hinata 1980; Song and Osborn 1992). *B. rapa* 'spring broccoli raab' is identified as the closest extant relative of maternal genome of *B. napus* (Song et al. 1988a).

### *Origin of the Word Brassica, Mustard, and Sarson*

Hegi (1919) believed that the term Brassica originated from the Celtic word for cabbage, *Bresic* or *Bresych*. Another possible derivation is from the Greek word whose meaning is crackle, as the leaves produce a crackling noise on being broken off (Gates 1953). The word Brassica is also considered as the probable derivative of the Greek word *braske* used in southern Italy (Maggioni et al. 2010). The word Brassica first appeared in the third century BCE Latin literature by the playwright Plautus. The word mustard is believed to be a derivative of Latin *mustus ardens* because of the pungent seeds. Its Middle English/Old French word is *mustarde* and was originally an occupational name for a person dealing in hot spices. The modern Hindi word sarson appears to be a derivative of Sanskrit word *Sarshap* (Prakash et al. 2012).

## 4 Landraces of Brassica and Their Conservation

Landraces have played a fundamental role in the history of crop improvement and agricultural production, and they have been in existence since the origins of agriculture itself (Zeven 1998). During the course of evolution of the crops landraces have been subjected to genetic modification through abiotic, biotic, and human interactions (Villa et al. 2005). Since the evolution of agriculture in different geographical regions, farmers are sowing, harvesting, and conserving some seeds of crops for next season which resulted in the enrichment of genepool through evolution of more diverse landraces of crop being conserved and cultivated over the millennia (Frankel et al. 1998). This landrace evolution cycle remained until the dawn of formal plant breeding and the generation of generally higher-yielding cultivars that subsequently replaced many traditional landraces (Frankel and Bennett 1970; Frankel and Hawkes 1975; Harlan 1975).

A landrace of a seed-propagated crop is a variable population, which is identifiable and usually has a local name. It lacks "formal" crop improvement, is characterized by a specific adaptation to the environmental conditions of the area of cultivation (tolerant to the biotic and abiotic stresses of that area), and is closely associated with the traditional uses, knowledge, habits, dialects, and celebrations of the people who developed and continue to grow it (Maxted et al. 2015). Zeven (1998) after Mayr (1937) proposed two types of landraces—(1) autochthonous landraces, cultivated for more than a century in a specific region and (2) allochthonous landraces, introduced from another region and locally adapted. As a third type, the so-called Creole landraces have been defined (Bellon and Brush 1994) as material derived from a bred cultivar. Landrace has a high capacity to tolerate biotic and abiotic stress resulting in high yield stability and an intermediate yield level under a low input agricultural system (Mansholt 1909; Zeven 1998). It represents a specific evolutionary stage of crop plants.

In the early phase of germplasm conservation efforts landraces of Brassica and allied genera were collected from different geographical regions globally by researchers for their utilization in Brassicas improvement programs. Earlier developed varieties were product of selection from these landraces. In India local landraces of *B. juncea* such as '*Jatai rai*', '*Desi rai*', and '*Maghi rai*' were collected from farmer's fields in the areas bordering Bangladesh. In yellow sarson, dwarf and early types with pendulous siliqua were collected from Indo-Bangladesh border whereas tall, robust, multi-locular types were mainly collected from eastern Uttar Pradesh. Diversity of *B. tournefortii* and *B. nigra* was collected from drier parts of Haryana and Rajasthan. In explorations for wild crucifers in Pauri Garwal hills of Uttarakhand accessions of allied genera including *Capsella, Crambe, Lepidium,* and *Sisymbrium* spp. were collected (Singh 1996; Singh and Sharma 2007). Many exploration trips were conducted by the ICAR-Directorate of Rapeseed-Mustard Research, Bharatpur (Rajasthan), in which landraces and wild relatives were collected from farmers field and natural habitats of different parts of country including Assam, Haryana, Himachal Pradesh, Karnataka, Madhya Pradesh, Meghalaya, Nagaland, Punjab, Rajasthan, Uttar Pradesh, Uttranchal, and West Bengal (http://kiran.nic.in/pdf/publications/Oilseed_Brassica.pdf; accessed on May 15, 2020).

Esquinas-Alcazar (1993) wrote "The landraces or heterogeneous varieties of the past have been and still are the plant breeder's raw material". They have been a fruitful, sometimes the sole, source of genes for pest and disease resistance, adaptation to difficult environments, and other agricultural traits. The disappearance of

**Table 6** Global landrace inventories of Brassica species

| | Landrace conserved | Reference |
|---|---|---|
| IPGR Genebank, Sadovo, Plovdiv, Bulgaria | 16 accessions of *Brassica* | Krasteva et al. (2009) |
| Greek national Genebank | 150 accessions of *Brassica* and 17 accessions of *Raphanus* | Katsiotis et al. (2009) |
| Portuguese genebanks | Portuguese cabbages (***Brassica oleracea*** L.) landraces *Penca de Chaves, Penca de Mirandela, Penca da Póvoa Verde*; traditional turnip (***Brassica rapa*** L.) landraces *Greleiro Senhora da Conceição, Gandra, Greleiro Temporão, São Cosme* | Moreira and Veloso (2009) |
| | ***B. rapa var. rapa:*** *Bola de nece, das Virtudes, de Gandra, de S. Cosme, 60 dias, Saloio, Encamoesado, Ingles roxo comprido, Greleiro, Temporao de grelo, Extra temporao de grelo;* ***B. napus* var. *napus***: *Couve nabica*; ***B. oleraceae***: *Couve repolho do Algarve, Couve grande do Algarve, Couve de horto, Couve ratinha, Couve de desfolhar* | Monteiro and Dias (1996) |
| Cultivated plant inventory of Russia | *Petrovskaya* turnip (subsp. rossica Sinsk.) and *Karelian* turnips | Smekalova (2009) |
| Landrace inventory of the UK | *Shetland cabbage*, a Brassica landrace grown over centuries on the Shetland Islands | Scholten et al. (2009) |
| ICAR-NBPGR, New Delhi, India | 5537 indigenous accessions of *B. juncea*, 2575 indigenous accessions of Brassica including *B. rapa, B. nigra, B. napus, B.* sp.; 385 indigenous accessions of taramira. It includes landraces, varieties, and advanced breeding lines | http://genebank.nbpgr.ernet.in/CropgroupCrops.aspx; accessed on May 15, 2020 |
| GENESYS | 6830 landraces of genus Brassica maintained in genebanks of Portugal, Spain, France, Italy, China, Ethiopia, USA, Ukraine, UK, Korea | https://www.genesys-pgr.org/a/v2lZAXkVLlq; accessed on May 15, 2020 |
| EURISCO | 1307 landrace of *B. rapa*; 562 landrace of B. *napus*; 101 landrace of *B. nigra*; 3761 landrace of *B. oleracea*; 154 landrace of *B. juncea* | https://eurisco.ipk-gatersleben.de/apex/f?p=103:55::::::accessed on May 15, 2020 |

landraces not only means local genetic erosion but also "local cultural erosion," by which both biological and cultural evolution is hampered. Hence there is a need to conserve and protect these landraces from genetic erosion. Two strategies (ex situ and in situ) are generally used for the conservation of germplasm. In situ conservation has been suggested as a best strategy for the conservation of landraces. In situ conservation (i.e., the conservation of diversity in its natural habitat) involves the designation, management, and monitoring of the population at the location where it is currently found and within the community to which it belongs (Maxted et al. 1997). More recently, in situ conservation has been proposed as being a better conservation strategy because, in contrast to ex situ conservation, it allows a complex of populations to be preserved and evolutionary processes to be continued. In situ conservation includes conservation in Genetic Reserve and on farm conservation (Negri et al. 2009). Genetic Reserve is the location, management, and monitoring of genetic diversity in natural wild populations within defined areas designated for active, long-term conservation. Maxted et al. (1997) defined on-farm conservation as "the sustainable management of genetic diversity of locally developed crop varieties (landraces), with associated wild and weedy species or forms, by farmers within traditional agricultural, horticultural or agri-silvicultural systems." Information of ex situ conservation of Brassica germplasm landraces can be accessed from international databases and genebanks (Table 6). Collections of landraces are available from the last 100 years. Internet-accessible germplasm databases provided by Bioversity International, the Food and Agriculture Organization, the Nordic Genebank, the national genebanks of the United States (GRIN) and Canada (GRIN-CA), or GENESYS allow for searches by improvement status of the accessions, directing the seeker to landraces or other categories (Hammer and Diederichsen 2009). Genesys is global database for information of plant genetic resources. It has information of about 6830 landraces of Genus Brassica maintained in different genebanks (Portugal, Spain, France, Italy, China, Ethiopia, USA, Ukraine, UK, Korea) across the globe (https://www.genesys-pgr.org/a/v2lZAXkVLlq; accessed on May 15, 2020).

Many researchers did diversity studies in Brassica landraces. Yao et al. (2012) did diversity analysis based on SSR markers in 34 Chinese vegetable mustard landraces which were morphologically different types (root mustard, stem mustard, leaf mustard, and seed stalk mustard). Červenski et al. (2013) studied the white cabbage, landrace '*Futoški*' which was traditionally grown in Futog district in the Republic of Serbia. Ciancaleoni et al. (2014) did diversity characterization of broccoli (*Brassica oleracea* L. var. *italica* Plenck) 9 Sicilian landraces (*Sparaceddu, Friaruolo, Nero, Spigariello, Marzolo, Friaruolo, Iannarisi, Tardivo, Nostral*) based on morpho-physiological and molecular traits for their on-farm (in situ) conservation.

# 5 Ex Situ Conservation of Brassica Genetic Resources

## *Ex Situ Conservation of Brassica Globally*

Branca (2008) divided genetic resources of Brassica crops in three genepools based on their crossability. The primary gene pool includes landraces, obsolete cultivars, modern cultivars, and advanced breeding lines and genetic stocks of corresponding species of the U-triangle, which means each species has its own primary gene pool. The secondary gene pool is represented by the other species of the U-triangle, since their relationship has been confirmed by various phylogenetic and cytogenetic studies. Finally, the tertiary gene pool includes Brassica coenospecies including species and genera grouped into different cytodemes based on their crossability and cytological studies. Genera in cytodemes include *Brassica, Coincya, Diplotaxis, Eruca, Erucastrum, Hirschfeldia, Sinapis, Sinapidendron, Enarthrocarpus, and Raphanus.* A detailed list of cytodemes is presented in Table 4. *Brassica oleracea* and *B. rapa* have much more phenotypic and genetic diversity than *B. napus* and *B. juncea* (Cartea et al. 2011).

With the development of modern high yielding varieties, monocultures, changes in food habits, extensive urbanization, deforestation due to developmental activities, and natural vagaries like drought and floods led to loss of natural genetic diversity of most of the crops in the past decades. This loss of genetic diversity due to these activities and events is known as genetic drift. Genetic drift leads to loss of genetic variability of wild germplasm and landraces, which is called genetic erosion. Besides, breeding programs have led to a reduction of the allelic variation in modem elite varieties. Selection within each species for adaptation to different uses and geographical regions has resulted in different pools of genetic variation within the cultivated germplasm (Gómez-Campo and Prakash et al. 1999).

Therefore, in order to protect these valuable genetic resources, the collection and conservation programs started worldwide. During the 1970s, wild germplasm of Brassica was extensively collected and conserved (Gómez-Campo et al. 2006, 2007). With the introduction of the concept of biodiversity, ex situ and in situ conservation programs across the globe accelerated for various crop. Seed banks were created to maintain genetic diversity, to minimize genetic erosion, and to distribute these genetic resources of landraces and wild relatives for research purpose to different organizations (Branca and Cartea 2011). For ex situ conservation, very first exploration trips are made to different geographical regions, mainly to the center of diversity/center of origins of the crop from where the maximum diversity can be tapped. Hence, ex situ conservation requires expertise of the crop for proper species identification and appropriate collection strategy. Appropriate mating procedures pertaining to the respective Brassica and allied species need to be followed during regeneration to maintain the genetic integrity of the germplasm since most of the crops in Brassica are cross compatible (Boukema and van Hintum 1999).

Plant genetic resources of Brassicaceae family are conserved across the globe in seed genebanks of many countries (China, India, United Kingdom, United States,

Netherland, Spain, and Germany) and are constantly enriched by further collections through plant exploration from different geographical regions with more diversity based on identified gaps (Panjabi et al. 2019). These countries hold around 60% of total worldwide collections of crucifers (Singh and Sharma 2007). Large national collections, in USA, China, and India, are generally maintained as a network of active regional collections and a set of it is conserved in situ or ex situ in genebanks (Knee et al. 2011). Basic activity/mandate of these organization is collection, regeneration, evaluation, characterization, documentation, conservation, and sharing of these germplasm for further utilization in different breeding/research programs to different researchers. International and domestic exchange and sharing of germplasm do occur for research purposes through MTA (Material Transfer Agreement) or similar contract that governs the transfer of tangible research materials between organizations, defining the rights of the provider and the recipient with respect to the materials and any derivatives (Knee et al. 2011).

In general, most of the Brassica collections are conserved by means of seeds in medium-term and long-term storage conditions to maintain seed viability for many years (5–15 years). For active collections which are frequently used short- to medium-term storage at 4 °C is suitable while, for long-term base collections (for decades or more years), seed viability can be maintained at −20 °C (Rivero-Lepinckas et al. 2006). Seeds of accessions with moisture content 3–7% are conserved in moisture-proof sealed polyethylene aluminum pouch in medium-term and long-term genebanks (Boukema and van Hintum 1999; Rivero-Lepinckas et al. 2006; Knee et al. 2011).

One of the largest collections of wild Brassica species and allied cruciferous genera is kept by the Universidad Polite'cnica of Madrid (UPM), Spain. The Plant Germplasm Bank-UPM holds 600 crucifer accessions of rare and endangered species widespread in the western Mediterranean area (http://www.etsia.upm.es/ANTIGUA/DEPARTAMENTOS/biologia/documentos/GC-2000-Int.htm).

Since 1982, several expeditions have been carried out by Professor Gómez-Campo from the UPM to collect wild Brassica species from Mediterranean coast of Spain, Italy, Greece, Tunisia and along the Atlantic coast of northern Spain, France, and the UK (Gómez-Campo et al. 2005, 2007). These missions were supported by Bioversity International formerly known as International Board for Plant Genetic Resources (IBPGR), Rome, Italy. In 1983, that collection of Plant Germplasm Bank-UPM was designated as the global base collection for wild crucifers by the IBPGR, and in 1994, it was honored with the National Award for Environment by the Government of Spain. Recently, it has been included in the Global Biodiversity Information Facility database (http://www.gbif.es) (Branca and Cartea 2011). The Plant Germplasm Bank-UPM has one of the most complete collections of wild crucifers in the world (1027 taxa, 4863 accessions), especially the Brassiceae tribe, which has several species of economic interest including relatives of crops belonging to the genera *Brassica, Sinapis, Raphanus, Eruca,* and others. Most of the accessions were collected by Prof. Gómez-Campo and his collaborators (Cartea et al. 2011). Currently, Brassica seeds at the UPM are conserved in flame-sealed glass vials, having silica gel to ensure low moisture content, because this method is very convenient for small-sized seeds as those of Brassicas (see www.seedcontainers.net).

The Center for Genetic Resources, the Netherlands, set up European Brassica database (Bras-EDB), as per recommendation of Brassica working group which was established in 1991 under the aegis of the European Cooperative Program for Crop Genetic Resources Networks (ECP/GR) (Boukema and van Hintum 1998). Bras-EDB contains more than 19,600 accessions (http://documents.plant.wur.nl/cgn/pgr/brasedb/). The N.I. Vavilov Research Institute for Plant Industry (VIR) at St. Petersburg (Russia) comprised 6879 accessions of Brassica collection in 1996 (Shashilova 1997). In the USA, the GRIN (Germplasm Resources Information Network) database is maintained at the Beltsville Agricultural Research Center and the accessions are held by the Regional Plant Introduction Stations. This collection is duplicated and stored for a long term at the National Seed Storage Laboratory at Fort Collins (ARS-GRIN 1997).

In China, the Chinese Genetic Resources Information System (CGRIS) supports the National network of regional genebanks coordinated by the Institute of Crop Germplasm Resources, which has the responsibility for the long-term conservation of genetic resources (Branca and Cartea 2011). The Germplasm Resources Unit (GRU) based at the John Innes Centre, Norwich, UK, also hosts seed collections of Brassica crop and its wild relatives (Panjabi et al. 2019). The Multilateral System (MLS) of the International Treaty for Plant Genetic Resources for Food and Agriculture (ITPGRFA) fetches massive support from GRU with 35,000 accessions as of now (Horler et al. 2018; Panjabi et al. 2019). Furthermore, many databases have been developed globally to provide information of plant genetic resources at one place. Likewise, Genesys was developed, which allows users to explore the information about Plant Genetic Resources for Food and Agriculture (PGRFA) conserved in genebanks worldwide. It includes information on germplasm collections from three of the world's largest genebank databases and networks: EURISCO, CGIAR, and USDA NPGS. Genesys is managed by the Global Crop Diversity Trust, which manages the Crop Diversity Endowment Fund (Genesys 2020). It has information about the 40,238 accessions of Genus *Brassica* held by different genebanks across the world (Genesys 2020). The Svalbard Global Seed Vault, Spitsbergen, Norway, provides facilities for the safe deposit of seeds of germplasm accessions of importance of food security and sustainable agriculture. The Seed Vault offers storage for duplicates of all unique seed accessions conserved by national, regional, and international genebanks, research institutes, and NGOs. The seed vault is an attempt to ensure against the loss of seeds in other genebanks during large-scale regional or global crises. The seed vault is managed by the Norwegian government, the Crop Trust, and the Nordic Genetic Resource Center (NordGen) (https://www.seedvault.no/how-we-work/our-purpose/; https://en.wikipedia.org/wiki/Svalbard_Global_Seed_Vault). Information about seed holdings in the Global Seed Vault, species, depositor institutes, dates, etc. is managed through the NordGen Seed Portal database. It maintains about 17,541 accessions of Brassica and allied genera obtained from different organizations and genebanks working on crucifers (NordGen 2020). ECPGR is a collaborative program among most European countries, aiming at ensuring the long-term conservation and facilitating the utilization of plant genetic resources in Europe. It provides passport and phenotypic information of actively conserved (ex situ and in situ) European PGRFA diversity through the EURISCO catalogue. These European central crop databases are classified into six groups: Cereals; Forages; Oil and Protein Crops; Fruit; Sugar, Starch, and Fiber Crops; and Vegetables (ECPGR 2020). The European Search Catalogue for Plant Genetic Resources (EURISCO) is being maintained at the Leibniz Institute of Plant Genetics and Crop Plant Research (IPK), Gatersleben, Germany. The EURISCO has information of about 7,19,717 accessions of Brassica and allied genera including 6,85,188 accessions of *Arabidopsis* and 25,887 accessions of *Brassica* genus (EURISCO 2020). Other European countries that maintain collections of vegetable Brassica crops are Bulgaria, Croatia, the Czech Republic, Hungary, Poland, Russia, Switzerland, and Turkey. There are also collections of cole crops in Israel, Ethiopia, South Africa, India, the Philippines, and Taiwan (Swarup and Brahmi 2005). Furthermore, other genetic resources/germplasm banks which have got wild and cultivated Brassica species are Kew Gardens and the Horticulture Research Institute in Wellesbourne (United Kingdom) and the Nordic Gene Bank (Sweden) (Gupta 2012).

In the USA, Prof. P. H. Williams, at the University of Wisconsin, established the Crucifer Genetics Cooperative (CrGC) in 1982 to acquire, maintain, and distribute seed stocks and pollen of Brassicaceae. Presently, the program provides seed stocks and information of rapid-cycling Brassica. The Plant Genetic Resources Unit (PGRU) of Cornell University located at Geneva, New York, maintains some accessions from wild Brassicas and 1480 accessions of *B. oleracea* (Cartea et al. 2011). A list of important international database and genebanks is presented in Table 7 from where information regarding genetic resources of Brassica and allied genera can be accessed and accessions can be procured after placing request.

## *Collection and Conservation of Brassica Genetic Resource in India*

India has a rich diversity of oilseed Brassicas. *Brassica rapa* var. toria, *B. rapa* var. *brown sarson,* and *B. juncea* considered to be native of Indian gene center (Arora 1988) are distributed in eight agro-ecological zones of the country. Much of the diversity is concentrated in the Indogangetic plains and sub-montane Himalayas. These are widely grown in the northern plains in the states of Rajasthan, Uttar Pradesh, Madhya Pradesh, Haryana, Gujarat, West Bengal, Assam, Bihar, Jharkhand, Punjab, Himachal Pradesh, and Jammu and Kashmir (Radhamani et al. 2013). The oleiferous Brassicas are widely distributed in Indian subcontinent. The northeastern and northwestern regions in particular hold promising traits of oilseed Brassicas. Plains of Uttar Pradesh, Bihar, and West Bengal are rich in the diversity of early and dwarf types of *B. rapa* (var. *toria* and *yellow sarson)* while the hills of Jammu and

**Table 7** International database and genebanks of Brassica and allied genera

| International database for Brassicaceae germplasm information | |
|---|---|
| AEGIS—A European Genebank Integrated System | https://www.ecpgr.cgiar.org/aegis |
| AVGRIS—The AVRDC Vegetable Genetic Resources Information System | http://seed.worldveg.org/ |
| Bioversity International | https://www.bioversityinternational.org/ |
| CBD—Convention on Biological Diversity | http://www.biodiv.org |
| CGRFA—Commission on Genetic Resources for Food and Agriculture | http://www.fao.org/WAICENT/FAOINFO/AGRICULT/cgrfa/ |
| Crop Wild Relatives Global Portal | http://www.cropwildrelatives.org/ |
| CWRIS—The Crop Wild Relative Information System | http://www.pgrforum.org/index.htm |
| ECPGR—The European Cooperative Programme for Plant Genetic Resources | https://www.ecpgr.cgiar.org/ |
| ENSCOBASE—European Native Seed Conservation Network | http://enscobase.maich.gr/ |
| EU Plant variety database | https://ec.europa.eu/food/plant/plant_propagation_material/plant_variety_catalogues_databases/search/public/index.cfm |
| EURISCO—European Search Catalogue for Plant Genetic Resources | https://eurisco.ipk-gatersleben.de/apex/f?p=103:1:0::::: |
| | |
| GBIF—Global Biodiversity Information Forum | http://www.gbif.org/ |
| Genesys—PGR | http://www.genesys-pgr.org/ |
| GRAIN—Genetic Resources Action International | http://www.grain.org |
| GRIN—Global | http://www.grin-global.org/ |
| ICAR—Plant Variety Information System | http://www.nbpgr.ernet.in/pvrs/ |
| Millennium Seed Bank Seed Lists | http://apps.kew.org/seedlist/seedlist.html |
| NGRP— National Genetic Resources Program of United States Department of Agriculture | https://www.ars-grin.gov/ |
| NISM—National Information Sharing Mechanism | http://202.141.12.147/gpa/ind/main.htm |
| NPGS—National Plant Germplasm System of United States Department of Agriculture | https://www.ars-grin.gov/npgs/ |
| PGRS—Plant Germplasm Registration System, India | http://www.nbpgr.ernet.in/grpvr/login.aspx; http://www.nbpgr.ernet.in/IRCG%20Search/ircg_search.htm |

| | |
|---|---|
| SINGER—System-wide Information Network for Genetic Resources | http://singer.grinfo.net/ |
| WIEWS—World Information and Early Warning System on PGR | http://www.fao.org/wiews/en/ |
| WISM-GPA—World Information Sharing Mechanism on GPA | http://www.pgrfa.org/gpa/wismgpa.htm |
| World Information and Early Warning System on PGR (WIEWS) | http://apps3.fao.org/wiews/wiews.jsp |
| *International Genebanks for Brassicaceae germplasm* | |
| Australia—Australian Plant Genetic Resource Information Service (AusPGRIS) | http://www2.dpi.qld.gov.au/extra/asp/auspgris/ |
| Albinia—Plant Genetic Resources Center (Albania Gene Bank) | https://medomed.org/featured_item/albania-gene-bank/ |
| Australia—Australian Grains Genebank (AGG), Horsham, Victoria | http://agriculture.vic.gov.au/agriculture/innovation-and-research/research-centres/horsham |
| Austria—National Inventory of Austria | https://www.genbank.at/en.html |
| Bulgaria—National Seed Genebank, Institute for Plant Genetic Resources "K.Malkov," Bulgaria | http://ipgrbg.com/en |
| China—Oil Crops Research Institute, Chinese Academy of Agricultural Sciences, Wuhan, Hubei | http://ocri.caas.cn/en/researchareas/cropgermplasm/index.htm |
| Cyprus—National (CYPARI) Genebank, Agricultural Research Institute, Ministry of Agriculture, Rural Development and Environment | http://www.moa.gov.cy/moa/ari/ari.nsf/index_en/index_en?OpenDocument |
| Czech Republic—Gene bank, Crop Research Institute | https://grinczech.vurv.cz/gringlobal/search.aspx |
| Germany—Genebank, Leibniz Institute of Plant Genetics and Crop Plant Research | https://www.ipk-gatersleben.de/en/genebank/ |
| Germany—Genebank Information System of the IPK (GBIS/I), Gatersleben | https://gbis.ipk-gatersleben.de/gbis2i/ |
| Germany—Julius Kühn-Institut (JKI) Federal Research Centre for Cultivated Plants, Braunschweig | https://www.julius-kuehn.de/en/ |
| Germany—Informations system Genetische Ressourcen (GENRES) | https://www.genres.de/en/ |
| Hungary—The Centre for Plant Diversity (NODIK) | http://www.nodik.org/english/ |
| India—National Bureau of Plant Genetic Resource (ICAR-NBPGR), New Delhi | www.nbpgr.ernet.in/ |
| Israel—Israel Plant Gene Bank, Agricultural Research Organization (ARO), Volcanic Center | https://igb.agri.gov.il/web/?page=25&lang=en |
| Italy—CRA-Consiglio per la Ricerca e Sperimentazione in Agricoltura | http://planta-res.politicheagricole.it/pages/search_in_db.php |

(continued)

**Table 7** (continued)

| International database for Brassicaceae germplasm information | |
|---|---|
| Italy—Centro di Ricerca Cerealicoltura e Colture Industriali (CREA, sede di Bologna) | https://www.crea.gov.it/en/web/cerealicoltura-e-colture-industriali |
| Madrid—Comunidad de Madrid, Universidad Politécnica de Madrid. Escuela Técnica Superior de Ingenieros Agrónomos. Banco de Germoplasma | http://www.bancodegermoplasma.upm.es/ |
| New Zealand—Arable crop genebank and online database, New Zealand Institute for Crop & Food Research | http://www.plantandfood.co.nz/ |
| Poland—National Centre for Plant Genetic Resources: Polish Genebank, Plant Breeding and Acclimatization Institute | http://www.ihar.edu.pl/national_centre_for_plant_genetic_resources.php |
| Russian Federation—N.I. Vavilov All-Russian Scientific Research Institute of Plant Industry (VIR) | http://www.vir.nw.ru/en/ |
| Spain—INIA-Centro de Recursos Fitogenéticos-Genebank | http://webx.inia.es/web_inventario_nacional// |
| Sweden—Stored material at the Nordic Genebank, Nordic Genetic Resource Center | https://www.nordgen.org/en/ |
| Switzerland—Conservation of PGRFA – Swiss National Database | https://www.bdn.ch/ |
| The Netherlands—Centre for Genetic Resources (CGN), the Netherlands Plant Research International | https://www.wur.nl/nl/Onderzoek-Resultaten/Wettelijke-Onderzoekstaken/Centrum-voor-Genetische-Bronnen-Nederland-1.htm |
| Ukrain—Ivano-Frankivs'k Institute of Agroindustrial Production, Ukrain via EURISCO | https://eurisco.ipk-gatersleben.de/apex/f?p=103:1:0:::: |
| Ukrain—Kolomyia Experimental Station, Ukraine via EURISCO | https://eurisco.ipk-gatersleben.de/apex/f?p=103:1:0:::: |
| United Kingdom—Seed Conservation Department, Royal Botanic Gardens, Kew, Wakehurst Place | https://www.kew.org/science/collections-and-resources/collections/seed-collection |
| United Kingdom—Nottingham Arabidopsis Stock Centre (NASC), UK | http://arabidopsis.info/ |
| United States of America—National Plant Germplasm System | https://www.ars-grin.gov/npgs/ |
| United States of America—North Central Regional Plant Introduction Station (NCRPIS), Ames, Iowa, USDA-ARS | https://www.ars.usda.gov/midwest-area/ames/plant-introduction-research/# |
| United States of America—Northeast Regional Plant Introduction Station, Plant Genetic Resources Unit, USDA-ARS, Geneva, New York | https://www.ars.usda.gov/northeast-area/geneva-ny/plant-genetic-resources-unit-pgru/docs/about-pgru/ |

Source: Agrawal et al. (2012), Ougham and Thomas (2014), ECPGR (2020), Genesys (2020)

Kashmir and Himachal Pradesh are rich in local types of brown sarson. Leafy types of Indian mustard with small seed and late maturity mainly occur in northeastern parts of India, particularly Arunachal Pradesh, Mizoram, Tripura, and Sikkim. Local types of black mustard, a condiment, are available in small pockets all over the *Brassica* growing areas in India. Wild forms of *B. tournefortii* are available in drier tracts of Rajasthan, Punjab, and Haryana. Some of the wild forms of Crucifers, comprising species of *Capsella, Crambe, Lepidium,* and *Sisymbrium,* are distributed in the hills of Uttar Pradesh and the Northeastern Himalayas (Radhamani et al. 2013). Effective crop improvement program of a country is possible only if a broad genetic base of diverse origin is available for which the germplasm distributed in the center of diversity is of special interest. Hence systematic efforts made during the 1960s to collect indigenous variability of oilseed Brassicas by the erstwhile Plant Introduction Division, Indian Agricultural Research Institute, New Delhi, led to the assembly of about 2000 accessions of rapeseed-mustard and allied genera from northeastern plains and hills, northwestern plains, and central plateau (Rana and Singh 1992). The program gained momentum with the establishment of NBPGR, its regional stations, crop-based institutes, and All India Coordinated Research Projects on Rapeseed Mustard. Several explorations were undertaken leading to the collection of *B. juncea*, *B. nigra*, *B. tournefortii,* and *Eruca sativa* from western Gujarat, Chambal valley in Madhya Pradesh and Chhattisgarh, Jhalawar and adjoining parts of Rajasthan, parts of Haryana, Uttar Pradesh, Karnataka, West Bengal, Andhra Pradesh (mainly for *B. nigra*), and Arunachal Pradesh (mainly for *B. juncea*) (Chauhan et al. 2011).

India maintains plant genetic resources including mustard and rape at ICAR-National Bureau of Plant Genetic Resources (NBPGR) in New Delhi. ICAR-NBPGR ensures sustainable use of genetic resources of agri-horticultural crop plants, by coordinating with several institutions, and also conducting capacity building programs. ICAR-NBPGR has introduced over 3950 accessions of rapeseed and mustard from more than 25 countries (Singh and Sharma 2007) since last three decades. The gene bank at the NBPGR hosts a total of 7577 collections of *B. juncea*, 1111 of rapeseed (*B. napus* and *B. rapa*), 68 of *B. carinata*, 982 of mustard, 654 of *B. rapa* var. *toria*, 177 of *B. rapa* var. *yellow sarson*, and 53 of wild brassicas and allied species (NBPGR 2020). ICAR-DRMR, Bharatpur, is a national active germplasm site for rapeseed-mustard germplasm. It maintains 2452 accessions of rapeseed-mustard and allied species (Nanjundan et al. 2014). All India Coordinated Research Project on Rapeseed-Mustard maintains active collection of 6605 accessions of rapeseed-mustard (AICRPRM 2009-2019). Trait-specific germplasm of rapeseed-mustard registered and maintained at NBPGR genebank is listed in Table 8.

**Table 8** Trait-specific germplasm of rapeseed-mustard registered and maintained at National genebank at ICAR-NBPGR, New Delhi, India

| IC no. | Donor identity | Unique trait |
|---|---|---|
| *Brassica juncea* | | |
| IC296682 | CMS Trachy (Restorer) | Restorer gene for CMS Trachy with recombinant mitochondrial genome |
| IC296683 | CMS Trachy | CMS carrying recombinant mitochondrial genome |
| IC296684 | SWARNA {TERI(OE)M21} | Zero erucic acid, yellow seeded, early maturing (117 days) |
| IC296689 | NDYR 8 | Yellow seeded with high oil content (45.69%) |
| IC296690 | NDYR10 | Yellow seeded with high oil content (45.75%) |
| IC296701 | CMS | CMS carrying cytoplasm of *Brassica oxyrrhina* |
| IC296702 | CMS | CMS with *Morcandia arvensis* cytoplasm |
| IC296703 | CMS | Fertility restorer for CMS (*Morcandia arvensis*) *B. juncea* |
| IC296705 | CMS | CMS with Raphanus/ogu-*Brassica juncea* |
| IC296706 | CMS | Chlorosis corrected green, CMS mustard carrying recombinant mitochondria of Morcandia arvensis and chloroplast of B. Juncea developed through protoplast fusion |
| IC296741 | RB-1 (Raya Bawal-1) | Tetralocular silique |
| IC296746 | *Diplotaxis erucoides* (CMS) | CMS carrying cytoplasm of *Diplotaxis erucoides* |
| IC296749 | *Diplotaxic siifolia* (CMS) | CMS line carrying the cytoplasm of *Diplotaxis siifolia* |
| IC296829 | TM-1(YSM2) | Yellow seeded Indian mustard |
| IC296501 | Heera | Low glucosinolate content (16.96μmol/g of seed) and low erucic acid in oil (0.12%) |
| IC296507 | NUDH-YJ-5 | Low glucosinolate content (9.3 μmol/g of seed) and low erucic acid in oil (0.14%) |
| IC401570 | RH-8814 | Resistance to salinity |
| IC405233 | TERI-GZ-05 | Yellow seeded, double low quality and high oleic and linoleic acid |
| IC443623 | BIO-YSR | Yellow seeded, high yield and oil content and white rust resistant |
| IC546947 | PRQ-2005-1 | Low erucic acid (1.08%) in oil and yellow seed coat |
| IC546946 | PAB 9511 | Tolerant to blight (*Alternaria brassicae* and *A. brassicola*) |
| IC553910 | Divya (PPMS-1) | Dwarf, early maturity with distinct whorl around the axis |
| IC0583386 | BPR-541-4 | High water use efficiency, thermotolerance at terminal stage, salinity tolerance at juvenile stage |
| IC0583448 | BPR-543-2 | High water use efficiency, thermotolerance at juvenile stage |
| IC573439 | PRB-2006-5 | Bold seed (1000 seed weight: 6.10 g) |

(continued)

**Table 8** (continued)

| IC no. | Donor identity | Unique trait |
|---|---|---|
| IC0584669 | RH 0116 | Tolerance to salinity at 10 ds/m at seedling stage |
| IC0589778 | BPR-349-9 | Only for thermotolerance at juvenile stage |
| IC0593926 | DRMR WFM 1 | White petal color |
| IC0595268 | NDUH-YJ-6 | Low glucosinolate content in seed (<10 µmol/g of seed), high oil content (>45%), resistance to white rust, yellow seed coat color |
| IC0595525 | BPR-549-9 | Salinity tolerance at juvenile stage, high water use efficiency |
| IC0593927 | BPR-540-6 | Salinity tolerance at juvenile stage, thermo tolerance at juvenile stage |
| IC0511389 | CS 1100-1-2-2-3 | Salinity tolerance, alkalinity tolerance up to pH 9.4 |
| IC0598624 | DRMR-541-44 | Drought tolerance (high water use efficiency under rainfed conditions) |
| IC0622804 | DRMR MJA 35 | Moricandia system based CMS line. Resistant to white rust disease |
| IC0624502 | CS 15000-1-2-2-2-1 | High tolerance to salinity (ECe 12 dS/m) and alkalinity (pH 9.4) |
| IC0598622 | DRMR-2019 | White rust resistant |
| IC0598623 | DRMR-2035 | White rust resistant |
| IC0626000 | Allohexaploid (H2) | Heat tolerant, resistant to Alternaria brassiceae |
| IC0265495 | IC265495 | White rust resistant |
| IC0313380 | IC313380 | White rust resistant |
| IC0625996 | DRMR 1-5 | Double low (<2% erucic acid in oil and < 30 µmoles glucosinolate/g of defatted seed meal). White rust resistant. Yellow seed coat |
| IC0628059 | DRMRIJ 12-48 | Multiple disease resistance (white rust, powdery mildew, and Alternaria blight) |
| IC0630607 | CS 52-SPS-1-2012 | Tolerance to high salinity (ECe 14–15 dS/m) and alkalinity (pH 9.4–9.5) |
| IC0632085 | DRMR10-40 | Drought tolerance |
| IC0520764 | DRMR 2059 | High temperature tolerance at seedling and terminal stage |
| IC0632086 | DRMR 4001 | Drought tolerance |
| IC0632087 | DRMR 4005 | Thermotolerance at juvenile stage with high seed and oil yield |
| IC0628060 | Allohexaploid (H1) | Allohexaploid (*Brassica juncea* + *Sinapis alba*), resistant to Alternaria brassiceae and Sclerotinia sclerotiorum. Tolerant to temperature |
| *Brassica napus* | | |
| IC296731 | TERI-GAURAV (TERI-00-R986) | Zero erucic acid, low glucosinolate (15.3µmol/g oil), early maturing (125 days) |
| IC296732 | TERI-GARIMA (TERI-00-R985) | Zero erucic acid, low glucosinolate (12.2 m/g oil) and high oleic acid (57%) |
| IC296827 | NUDB-38 | Early maturing “00” Gobhi Sarson |

**Table 8** (continued)

| IC no. | Donor identity | Unique trait |
|---|---|---|
| IC296828 | NUDB-42 | Early maturing "00" Gobhi Sarson |
| IC296747 | *Erucastrum gallicum* | Early flowering variant (31–33 days) carrying the cytoplasm of *E. gallicum* |
| IC405232 | TERI (00) R9903 | Canola quality with early maturity (130–135 days) |
| IC296704 | CMS | CMS for *Brassica napus* |
| IC0582897 | NRC GS-1 | Early flowering (34 days) and dwarf plant type (75 cm) |
| IC296685 | Phaguni [TERI(OE) RO3] | Zero erucic acid and early maturing (136 days) |
| IC296688 | Shyamali [TERI(OE) RO9J] | Zero erucic acid, high oleic acid (70.1%), and high oil content (50.2%) |
| *Brassica carinata* | | |
| IC296748 | *Erucastrum gallicum* | Ethiopian mustard (*Brassica carinata*) with *B. juncea* type branching |
| IC0582898 | NRC KR-304 | Early maturity (135 days), long main shoot (110 cm) and bold seed |
| IC0589777; IC0590093 | MCA 1; MCB1 | For male sterility |
| *Brassica Caudatus* | | |
| IC296597 | Amphidiploid | Amphidiploid brassica |
| *Brassica rapa* var. *Brown Sarson* | | |
| IC296707 | CMS | CMS carrying cytoplasm of *Brassica oxyrrhina* |
| *Brassica rapa* var. *toria* | | |
| IC553274 | PHOT-2-2 | High oleic acid (70.1%) and low erucic acid (0.2%) and brown seed toria |
| IC553275 | PHOT-8-2-11 | Low linolenic acid (3.03%) and brown seed toria |
| IC0621691 | PT-141 | Earliness of maturity. Earliness of flowering |
| *Brassica rapa* var. *yellow sarson* | | |
| IC0393515 | WF yellow sarson | Yellow sarson genotype with white flower |
| IC0623820 | NRCPB rapa 8 | Potential parent for resynthesis of *B. juncea* |

Source: NBPGR, New Delhi; http://www.nbpgr.ernet.in:8080/ircg/index.htm accessed on February 7, 2020

## 6 In Situ Conservation of Brassica Species

The conservation of germplasm in their natural habitat and place of origin/occurrence is called in situ conservation. Such type of conservation strategy is particularly desirable for wild species, which may be difficult to regenerate ex situ (Boukema and van Hintum 1999). A strategy for in situ conservation of wild species related to *B. oleracea* has been elaborated by Maggioni et al. (1997). Plazibat (1996) has assigned the IUCN Red List Category of Rare species to *B. incana* and Endangered species to *B. botterii, B. cazzae,* and *B. mollis.* The latter three species are considered to be endemic to Croatia. Sicily is considered to be a center of diversity for ($2n$ = 18) Mediterranean Brassicas owing to the high level of inter- and intraspecies diversity (Maggioni and Eastwood 1997). Therefore, Maggioni et al. (1997) suggested Sicily, with its populations of the species *B. macrocarpa, B. villosa,* and *B. rupestris*, to be a suitable target area to gain knowledge about the population genetics and to develop in situ strategies. They further suggested following sites in Sicily for in situ conservation of Brassica species, (a) the island of Marettimo for *B. macrocarpa* populations; (b) the Zingaro reserve in the Trapani district for *B. villosa* subsp. *drepanensis;* (c) the Madonie park in the Palermo district for populations of *B. rupestris*; (d) the Pellegrino mountain in Palermo district for *B. rupestris* populations; (e) the Nebrodi natural park in the Messina district for *B. incana* populations.

In situ conservation strategy of wild species of the *B. oleracea* cytodeme has been suggested by the ECPGR Working Group on Brassica as a complementary way of preserving the diversity of these Mediterranean relatives of cultivated *Brassica* species with $n$ = 9. For the purpose the Sicilian Center of Diversity was selected as it harbors highly variable populations of *B. incana, B. macrocarpa, B. rupestris*, and *B. villosa* which are often threatened by humans. Further this group shoulders the responsibility of a thorough study (DNA analysis, morpho-molecular, quality, and nutraceutical analysis) of these species. So that these wild crop relatives could be used for breeding purposes and to select the most appropriate accessions of the future European Genebank Integrated System (Astley et al. 2007). Since 2007, AEGRO GENRES project founded by the European Union deals with a basic research for an Integrated European *in situ* management work plan to implement Genetic reserves and "on farm" concept (http://aegro.bafz.de).

## 7 Wild Species of Brassica and Allied Genera as a Source of Agronomic Traits

Wild and weedy relatives of Brassica are reservoir of various traits (Table 9). In addition to traditional breeding methods, interspecific and intergeneric transfer of genes governing qualitative and quantitative characters from wild allies to

**Table 9** List of Brassica and wild relatives as a source of different agronomic traits

| Trait | Species |
|---|---|
| *Biochemical traits* | |
| High erucic acid (>45–50%) | *B. cretica, B. incana, B. rupestris, B. villosa, Crambe abyssinica, C. hispanica, Eruca vesicaria* (*E. sativa*), *Erucastrum cardaminoides, Sinapidendron angustifolia, Sinapis alba, Sinapis arvensis* |
| High linoleic and/or linolenic acids | *Alyssum* spp., *Barbarea* spp. *B. elongate, Camelina sativa, Cardamine* spp., *Conringia* spp., *Descurainia Sophia, Diplotaxis harra, Lepidium sativum, Lepidium spinosum, Matthiola incana, Matthiola livida, Nasturtiopsis arabica, Orychophragmus violaceus, Raphanus* spp., *Sisymbrium irio* |
| High hydroxy fatty acid | *Lesquerella* and *Physaria* spp. |
| Sterols | *Camelina sativa* (high cholesterol and brassicasterol), *Crambe abyssinica, Lesquerella* spp. |
| Secondary metabolites | *Erysimum cheiri* (cardenolides), *Iberis amara* (cucurbitacins), *Lesquerella fendleri* (cinnamoyl esters-sunscreen), *Lunaria annua* (alkaloids) |
| Floral pigments | *Cheiranthus cheiri*—acylated cyanidin glucosides, *Lobularia maritima*—acylated cyanidin glucosides, *Lunaria annua*—acylated cyanidin glucosides |
| Anti-pest/allelopathy | *Alliaria petiolata, Armoracia rusticana, Brassica nigra, Crambe abyssinica, Descurainia sophia, Diplotaxis erucoides* (anti-microbial compounds), *Lepidium perfoliatum, Nasturtium officinale* (anti-nematode), *Sinapis arvensis* (anti-mosquito), *Zilla spinosa* (anti-weed and -rhizosphere fungi) |
| Flavonoids | *Brassica* spp., *Camelina sativa, Crambe* spp., *Diplotaxis acris, D. harra, Diplotaxis* 13 spp., *Erucastrum* spp., *Thlaspi arvense* |
| Tocopherols (vitamin E) | Range mg $kg^{-1}$ oil: 68 in *Diplotaxis viminea* to 2479 in *Schivereckia doerfleri* |
| Peroxidase | *Armoracia rusticana* |
| Mucilage | *Camelina sativa, Lepidium sativum, Sinapis alba* |
| *Edaphic and aquatic adaptation* | |
| Salt tolerance | *Crambe maritime, Eruca vesicaria* subsp. *sativa, Lesquerella fendleri, Lobularia maritime, Raphanus raphanistrum* ssp. *Maritimus, Thellungiella salsuginea* |
| Tolerance of calcareous (limestone) soils | *Vella bourgaeanum* and *Moricandia moricandioides* |
| Heavy metal (Ar, Cd, Ni, Pb, Se, Sr, Zn) tolerance and hyperaccumulation | *Lepidium sativum, Streptanthus* spp., *Thlaspi montanum, Thlaspi caerulescens* |
| *Cold tolerance* | |
| Mountainous habitats (elevations >2000–2500 m or above snow line) | *B. nivalis* (Mt. Olympus, Greece), *B. jordanoffii* (Mt Pirin Planina, Bulgaria), *Coincya richeri* (French and Italian Alps—1750–2500 m), *Erucastrum abyssinicum* and *E. pachypodum*, Ethiopian Highlands (>3000 m), several genera in the family, e.g., *Arabis, Crucihimalaya, Draba*, adapted to alpine areas up to 6000 m in Himalayas, Alps, Rockies, Andes, *Romanschulzia* (Mexico and Central America) and *Oreophyton* (East Africa) adapted to high mountains of the tropics |

**Table 9** (continued)

| Trait | Species |
|---|---|
| Cold tolerance studies/genes | *Barbarea vulgaris, Descurainia sophia, Thellungiella salsuginea, Thlaspi arvense* |
| *Drought tolerance* | *B. carinata, B. tournefortii, Carrichtera annua, Diplotaxis acris, D. harra, Enarthrocarpus strangulates, Eruca vesicaria, Erucaria boveana,* subsp. *sativa, Erucaria microcarpa* and *E. uncata, Lesquerella* spp., *Moricandia* spp., *Physorhynchus* spp., *Pseuderucaria clavata, Savignya parviflora, Schouwia purpurea, Thellungiella salsuginea, Zilla spinosa* |
| *Disease resistance* | |
| White rust—*Albugo candida* | *B. carinata, B. nigra, B. maurorum, Eruca vesicaria* subsp. *sativa, Raphanus sativus* |
| Black leaf spot/Alternaria spp.—*Alternaria brassicae, A. brassicicola, A. raphani* | *Alliaria petiolata, Barbarea vulgaris, Brassica nigra, Brassica* coenospecies, *B. elongate, B. fruticulosa, B. maurorum, B.nigra, B. souliei, B. spinescens, Camelina sativa, Capsella bursa-pastoris, D. catholica, D. erucoides, D. tenuifolia, Eruca vesicaria* subsp. *sativa, Hemicrambe fruticulosa, Neslia paniculata, Raphanus sativus, Sinapis alba, Sinapis avensis* |
| Blackleg—*Leptosphaeria maculans* [=*Phoma lingam*] | *Arabidopsis thaliana, B. carinata, B. nigra, B. elongata, B. fruticulosa, B. insularis, B. atlantica, B. macrocarpa, Camelina sativa, Carrichtera, Diplotaxis muralis, D. tenuifolia, Eruca vesicaria, E. pinnatifia, Hirschfeldia incana, Raphanus raphanistrum, Raphanus sativus, Sinapis alba, S. avensis, Sisymbrium loeselii, Thlaspi arvense* |
| Downy mildew (*Peronospora parasitica)* | *Eruca vesicaria* |
| Clubroot (*Plasmodiophora brassicae)* | *Arabidopsis thaliana, Armoracia rusticana, Capsella bursa-pastoris, Raphanus* spp. |
| Sclerotinia stem rot (*Sclerotinia sclerotiorum)* | *Capsella bursa-pastoris, Eruca vesicaria* subsp. sativa, *Erucastrum gallicum* |
| *Verticillium dahliae* | *Armoracia rusticana* |
| *Verticillium longisporum* | *B. napus, B.oleracea, B. rapa* |
| Black rot (*Xanthomonas campestris)* | *Alliaria petiolata, Barbarea vulgaris, B. nigra, Erysimum hieraciifolium, Matthiola incana, B. carinata, B. nigra* |
| Defensin genes | *Lepidium meyenii against Phytophora infestans* |
| *Insect resistance* | |
| Flea beetles (*Phyllotreta cruciferae* and *P. striolata)* | *Arabidopsis thaliana, B. incana, B. villosa, Camelina sativa, Capsella bursa-pastoris, Crambe abyssinica, Crambe hispanica, C. glabrata, Sinapis alba, Thlaspi arvense* |
| Diamond-back moth (*Plutella xylostella*) | *Barbarea vulgaris, Crambe abyssinica, Raphanus raphanistrum* |
| Cabbage aphid (*Brevicoryne brassicae)* | *Brassica fruticulosa, B. spinescens, Brassica* wild C genome, *B. cretica, B. incana, B. macrocarpa, B. villosa, Eruca vesicaria* subsp. *sativa, Sinapis alba* |
| Mustard aphid (*Lipaphis erysimi*) | *B. carinata, B. nigra, Brassica* spp., *Eruca vesicaria* subsp. *sativa* |

(continued)

**Table 9** (continued)

| Trait | Species |
|---|---|
| Cabbage white fly (*Aleyrodes proletella*) | *B. cretica*, *B. fruticulosa*, *B. incana*, *B. insularis*, *B. spinosa*, *B. villosa* |
| Cabbage root fly or cabbage maggot (*Delia radicum*) | *B. fruticulosa*, *B. incana*, *B. macrocarpa*, *B. spinescens*, *B. villosa*, *Camelina sativa*, *Sinapis alba* |
| Cabbage seedpod weevil (*Ceutorhynchus obstrictus*) | *B. nigra*, *B. tournefortii*, *Sinapis alba* |
| Mustard sawfly (*Athalia proxima*) | *Camelina sativa* |
| *Nematode resistance* | |
| Beet cyst nematode (*Heterodera schachtii*) | *Raphanus sativus*, *Sinapis alba* |
| Root-knot nematode (*Meloidogyne* spp.) | *Eruca sativa*, *Raphanus sativus*, *Rapistrum rugosum*, *Sinapis alba* |

Source: Warwick et al. (2009)

cultivated forms will be facilitated with various in vitro methods, such as somatic cell genetics and recombinant DNA techniques. Hence, proper collection, systematic characterization, documentation, and conservation of these species need to be done for their effective utilization in crop improvement program.

# 8 Crop Improvement in Rapeseed-Mustard: Indian Perspective

## *Varietal Development in Rapeseed-Mustard*

The research on the improvement of rapeseed-mustard started in the beginning of the twentieth century in Pusa (Bihar), the then Bengal Presidency through the collection and purification of landraces. The initial scientific work for varietal improvement of rapeseed-mustard in India started at Layallpur (now Faisalabad in Pakistan) (NRCRM 2000), which led to the development of Indian mustard variety RL 18 (Raya Layallpur 18) in 1937 and yellow sarson variety L1 through selection. Indian mustard strain RT 11 from Uttar Pradesh was released in 1936. During the course of about two decades (1947–1967), a number of high-yielding varieties of mustard (Laha 101, Varuna, Durgamani, Patan Mustard), toria (Abohar, BR 23, M 27, T 9, ITSA, T 36, DK 1), brown sarson (BSA, BSG, BSH 1, BS 2, BS 65, BS 70), yellow sarson (T 151, Patan sarson, YSPb 24, T 42), and taramira (ITSA) were developed (NRCRM 2007).

The Indian Council of Agricultural Research (ICAR) launched in April 1967 All India Coordinated Research Project on Oilseeds (AICORPO) including five major crops, namely, groundnut, rapeseed-mustard, sesame, linseed, and castor. However, Project Coordinator (Rapeseed-Mustard) was established on 28 January 1981 at

Haryana Agricultural University, Hisar, Haryana. In order to provide a strong leadership to the rapeseed-mustard research activities in the country the ICAR established National Research Centre on Rapeseed-Mustard on 20 October 1993 at Bharatpur (Rajasthan) bringing under its umbrella the All India Coordinated Research Project on Rapeseed-Mustard with 19 research centers across the country. In February 2009, the National Research Centre on Rapeseed-Mustard was redesignated as Directorate of Rapeseed-Mustard Research (DRMR) (Chauhan et al. 2006, 2011).

Initially exotic germplasm has been utilized in Indian breeding program to develop 21 varieties of rapeseed-mustard, especially for oil and seed meal quality characters. Two varieties of brown sarson and 10 varieties of gobhi sarson, 9 of Indian mustard, and 1 in karna rai were developed. Until 1970, pure line and mass selection were the only breeding methods, employed to develop 26 varieties. Later on hybridization was followed to develop varieties. Till year, 2019, 256 rapeseed-mustard varieties were developed from which 140 and 91 varieties were developed through hybridization and selection, respectively (Table 10). All the varieties developed before the inception of AICRPO/AICRP-R&M were the selections from the local germplasm (Chauhan et al. 2006, 2011).

The first notified variety was ITSA of toria (*B. rapa*) in 1973 and till 2019, among 256 varieties 200 varieties have been notified by the Central Sub-Committee on Crop Standards, Notification and Release of Varieties for Agricultural Crops. Bio 902 of Indian mustard, a somaclone, was released in 1993. The recommended varieties recorded 9–53% yield increase over national checks under varied situations (Chauhan et al. 2011). The list of trait-specific varieties is presented in Table 11. The varieties released during the last two decade (2001–2019) have been given in Table 12. Trait-specific donors for different traits (biotic stress, abiotic stress, quality, etc.) were also identified and reported in Indian mustard germplasm (Kumar and Chauhan 2005; Misra et al. 2007a, b).

**Table 10** Number of rapeseed-mustard varieties developed through different breeding methods during 1936–2019 (Source: Updated from Chauhan et al. 2011)

| | | Breeding method | | | | |
|---|---|---|---|---|---|---|
| Year | No. of varieties | Selection | hybridization | Mutation | hybrids | Tissue culture |
| 1936–1966 | 26 | 26 | – | – | – | – |
| 1967–1980 | 24 | 18 | 04 | 02 | – | – |
| 1981–2000 | 81 (78 + 3[a]) | 26 | 44 | 05 | 02 | 01 |
| 2001–2019 | 125 (124 + 1[a]) | 21 | 92 | 06 | 05 | 0 |
| **Total** | **256 (252 + 4[a])** | **91** | **140** | **13** | **7** | **01** |

[a]No information on pedigree was available for four mustard varieties (KRV74, RK 9, RK 14, Rajendra Suphalam)

**Table 11** Indian mustard varieties recommended for specific conditions/specific traits (Source: Updated from Chauhan et al. 2011)

| Trait | Variety |
|---|---|
| White rust resistant | Basanti, JM 1, JM 2 |
| Earliness | Kanti, Narendra Ageti Rai 4, Pusa Agranai, Pusa Mahak, Pusa Vijay, PM 25 (NPJ 112), PM 27 (EJ 17) |
| High oil content | Narendra Swarana Rai 8, NRCDR 02, Rohini |
| High temperature tolerant | Kanti, Pusa Agrani, RGN 13, Urvashi, NRCDR 02, PM 25, PM 27, CS 58, Pant Rai 18 (PR 2006-1), Pant Rai 19, Pusa Vijay, RH 0119, RH 406, RGN-229, RGN-236, RGN-298, PM 22 |
| Intercropping | RH 30, RH 781, Vardan |
| Late sown | Ashirwad, RLM 619, Swarnajyoti, Vardan, Navgold (YRN 6), RGN 145, NRCHB 101, CS 56, PM 26 (NPJ 113) |
| Nontraditional areas | Pusa Agrani, Pusa Jaikisan, Rajat, Gujarat Mustard 2, TKM 1, Shatabdi |
| Rainfed | Geeta, GM 1, PBR 97, Pusa Bahar, Pusa Bold, RH 781, RH 819, RGN 48, Shivani, TM 2, TM 4, Vaibhav, RB 50, RH 406, RH 725, RGN 229, RVM 2, Pant Rai-20, RGN 298, Aravali mustard |
| Salinity tolerant | CS 52, CS 54, CS 56, CS 58, CS 60 |
| Frost tolerant | RGN 13, RH 819, Swarnajyoti, RH 781, RGN 48 |
| Low erucic acid (<2%) | Pusa Karishna, PM 21, PM 22, PM 24, ELM 079 (RLC1), RLC 2, RLC 3 |
| Double low | Pusa Double Zero Mustard 1 (PDZ 1) (erucic acid <2%, low glucosinolate <30μmol/defatted oil cake) |
| Hybrid | NRCHB 506[a], DMH 1, Coral 432 (PAC 432), Coral 437 (PAC 437), Keshri (PRO 5111) |

[a]Govt. sector hybrid

**Table 12** Recently released varieties of rapeseed-mustard developed in India (2011–2019)

| Crop | Variety |
|---|---|
| *B. juncea* | Pusa Mustard 28 (NPJ 124), Coral 437 (PAC 437) (Hybrid), RLC 2 (ELM 123), RGN-229, RH 406, Divya 33, Pant Rai-19 (PR 2006-1), PBR 378, RH 749, Raj Vijay Mustard 2 (RVM-2), RRN-573, RGN-229, RGN 236, Pusa Mustard 29 (LET-36), Pusa Mustard 30 (LES-43), Giriraj (DRMRIJ 31), RGN-298, Gujarat Mustard 3 (GM-3), Albeli-1, Pant Rai-20, DRMR150-35, PBR 357, Gujarat Dantiwada Mustard 4, Gujarat Dantiwada Mustard 5, Raj Vijay Mustard-1, RLC-3, CS 1100-1-2-2-3 (CS 58), Pant Rai 21 (PRB2008-15), RH 725, CS 2800-1-2-3-5-1 (CS 60), PDZ-1, Keshri (PRO 5111) (Hybrid), DRMR 1165-40,RH 761, TBM-204 |
| *B. carinata* | PC 6 (BJC 1) |
| *B. napus* | GSC 7 (GSC 101), RSPN-25, ONK-1 |
| *B. rapa* var. *toria* | Sushree, TL17, Raj Vijay Toria 1, Pant Hill toria 1 (PT 2006-4), Pant Toria 508 (PTE 20008-2), Tapeshwari, Tripura Toria 1, RSPT-6 |
| *B. rapa* var. *yellow sarson* | Pant Sweta (PYS 2007-10) |
| *B. rapa* var. *brown sarson* | Shalimar Sarson-2 (KBS-49), Shalimar Sarson-3, HPBS-1 |
| *Eruca sativa* | Jwala Tara (RTM 1355), Jobner Tara (RTM 1351) |

### *Hybrid Development in Rapeseed-Mustard*

Reports on the availability of heterosis in rapeseed-mustard in the early 1940s (Sun 1943) raised interest among plant breeders to develop the means for harnessing hybrid vigor. Introgression of cytoplasmic male sterility developed in radish (*Raphanus sativus* L.) by Ogura in 1968 to *Brassica juncea* (Banga and Labana 1984) raised hopes to develop hybrids in rapeseed-mustard.

However, systematic and coordinated efforts for developing hybrids in rapeseed-mustard started in 1989 under the ICAR sponsored project on "Promotion of Research and Development Efforts on Hybrids in Crops" with 2 CMS systems in *B. juncea*, i.e., *ogu* and *tour* and one in *B. napus*, i.e., *polima.* The approach led to the development of seven new CMS systems, *viz. siifolia*, *oxyrrhina*, *muralis, catholica, nigra*, *moricandia,* and *trachystoma*. These systems could not be utilized for hybrid development due to several limitations, particularly non-availability of full fertility restoration (Chauhan et al. 2011).

Concerted efforts at National Research Centre for Plant Biotechnology, New Delhi, resulted in the development of fertility restorer for *mori* and *trachy* CMS systems. *Mori* CMS and fertility restorer were found to be associated with severe chlorosis and retarded growth while segregation for fertility/sterility coupled with crooked siliqua formation was observed in *trachy* system. Chlorosis associated with *mori* CMS systems was rectified and diversified into the improved background of identified heterotic combination for developing commercial hybrids (Chauhan et al. 2011).

In *gobhi sarson* (*B. napus)*, three-line technology has already been perfected with the release of first commercial hybrid PGSH 51 based on *tour* CMS system by PAU, Ludhiana, in 1994 and Hyola 401 based on *pol* CMS system was released by Advanta, India, in 1997/2000 (Chauhan et al. 2011). In Indian mustard five CMS-based hybrids, NRCHB 506 (*mori* cms based) and DMH 1 in 2009 and Coral 432 (PAC 432, *ogu* cms based) in 2010, Coral 437 (PAC 437, *ogu* cms based) in 2011, and Keshri (PRO 5111) in 2018 have been developed and tested under AICRP-RM (AICRP-RM Annual reports 2009–2019).

## 9 Quality Improvement of Rapeseed-Mustard in India

Rapeseed-mustard oil is an important dietary component in India, especially in eastern and northwestern parts. Seed meal is used as an animal feed and as organic manure and also being exported. Oil quality is determined by fatty acid composition. Linoleic and linolenic acids are essential fatty acids. The major fatty acids in rapeseed-mustard oil are oleic, linoleic, linolenic, eicosenoic, and erucic acid. Erucic acid in oil of the Indian rapeseed-mustard varieties is quite high (Chauhan et al. 2007). Consumption of higher amount (>20%) of mustard oil having high erucic acid, in the diet, caused myocardial fibrosis, impaired myocardial conductance, lipidosis, etc., and also increased blood cholesterol (Gopalan et al. 1974; Renard and Mc Gregor 1976; Ackman et al. 1977). The seed meal of rapeseed-mustard mainly comprises proteins (35–40%), carbohydrates (14–15%), fibers (10–12%), moisture (6–8%), ash (4–6%), minerals and vitamins (1.0–1.5%), glucosinolates (2–3%), tannin (1.6–3.1%), sinapin (1.0–1.5%), and phytic acid (3–6%). Among all the constituents, fibers, tannin, phytic acid, glucosinolate, and sinapin lower the feed value of seed meal (Chauhan et al. 2002). Glucosinolates are a group of plant thioglucosides found principally among members of family Cruciferae (Brassicaceae) which determine the quality of seed meal. The vegetable tissue and seed of Cruciferous plant contain one or more of over 120 known glucosinolates (Fenewick et al. 1983). Cleavage products from hydrolysis of glucosinolates are detrimental to animal health as they reduce the feed palatability and affect iodine uptake by the thyroid glands, thus reducing feed efficiency and weight gains (Bell 1984 and Fenewick et al. 1983). Glucosinolates and their derived properties have also been reported to have health beneficial effects by reducing the risk of certain cancers in humans (Zhang and Talalay 1994; Fahey et al. 2001; Shapiro et al. 2001; Mithen et al. 2003), while other glucosinolates are detrimental for human and animal consumption (Rosa et al. 1997).

Breeding efforts have been underway in India since 1970 to reduce glucosinolate content in the seed of rapeseed-mustard varieties up to 30μmol/g defatted seed meal (low or 0) and erucic acid up to 2% (low or 0) as well as combining both to develop double zero or double low varieties to meet the internationally acceptable standard of oil and seed meal. First low erucic acid variety, Pusa Karishma of Indian mustard, and first double low variety, GSC 5 of gobhi sarson, were released in 2004 and 2005, respectively. First double low variety of Indian mustard, viz. PDZ 1 was notified in 2018. Presently, seven low erucic varieties (Pusa Karishna, PM 21, PM 22, PM 24, ELM 079 (RLC1), RLC 2, RLC 3) and one double low variety (PDZ 1) have been notified in *B. juncea* (Table 13). In gobhi sarson (*B. napus*), 6 double low (GSC 5, Hyola 401, NUDB 26–11, OCN 3 (GSC 6), TERI Uttam Jawahar, GSC 7) varieties have been notified (Table 13). Apart from the varieties quality donors have been developed and registered by Plant

**Table 13** Rapeseed-mustard varieties and germplasm developed in India for oil quality

| Crop | Varieties | Germplasm |
|---|---|---|
| *B. napus* | Double low (GSC 5, Hyola 401, NUDB 26-11, GSC 6, TERI Uttam Jawahar, GSC 7) | Low erucic acid [TERI–Gaurav (TERI-00-R986); TERI-Garima (TERI-00-R985); Phaguni (TERI(OE) RO3); Shyamali (TERI(OE) RO9J)), double low (NUDB-38; NUDB-42; TERI (00) R9903) |
| *B. juncea* | Low erucic acid (Pusa Karishna, PM 21, PM 22, PM 24, ELM 079 (RLC1), RLC 2, RLC 3); Double low (PDZ 1) | Low erucic acid [Swarna {TERI(OE)M21}, PRQ-2005–1]; low glucosinolate (NDUH-YJ-6), double low (TERI- GZ-05, Heera, NUDH-YJ-5, BIO-YSR, DRMR 1–5), and high oil content (NDYR 8, NDYR10) |
| *B. rapa* var. toria | – | High oleicacid, low erucic acid (PHOT 2-2); low linolenic acid (PHOT 8-2-11) |

Germplasm Registration Committee till 2020. In Indian mustard low erucic acid (Swarna {TERI(OE)M21}, PRQ-2005–1); Low glucosinolate (NDUH-YJ-6), Double zero (TERI- GZ-05, Heera, NUDH-YJ-5, BIO-YSR, DRMR 1–5) and high oil content (NDYR 8, NDYR10) germplasm have been registered till 2019. Furthermore, in Gobhi Sarson 4 low erucic acid germplasm, *viz.* TERI–Gaurav (TERI-00-R986); TERI-Garima (TERI-00-R985); Phaguni (TERI(OE) RO3); Shyamali (TERI(OE) RO9J); and 03 double low germplasm (NUDB-38; NUDB-42; TERI (00) R9903) have been registered till 2019.

## 10 Conclusion and Prospects

The genus Brassica has been extensively investigated by researchers worldwide owing to its vast diversity in form and applications. It has served as a model for scientific studies on cytogenetics, evolution and domestication, speciation, and molecular biology. Brassica and its allied genera constitute a potential germplasm pool with many desirable agronomic and horticultural traits. The last two decades have witnessed a fastest development in molecular cytogenetics, genomics, proteomics, and bioinformatics studies in Brassicas which became possible due to the use of Arabidopsis as a model plant species. Comparative genomics based on organelle genome has helped to decipher the origin and evolution of different forms of Brassica species evolved across continents.

Many constraints are faced by Brassica researchers in improvement programs, *viz.* lack of useful variability in allopolyploid species which have been substantiated by development of many synthetic lines of *B. napus, B. juncea*, and *B. carinata* through interspecific hybridization. But, this added variability is still inadequate and more synthetic genetic variants are yet to be obtained by utilizing the enormous morphological, physiological, and geographical variability of the diploid constituent parents. Many allied genera and wild Brassica species (*Coenospecies*) have been reported as a source of various unique traits (biotic stresses, abiotic stresses, oil quality, male sterility, herbicide resistance, and photosynthetic activity). These *coenospecies* serve as donor for introgression of desirable traits through conventional hybridization and/or advanced biotechnological approaches like protoplast fusion and genetic engineering. This is evident by the development of some somatic hybrids and transgenics in Brassicaceae (Prakash et al. 2009).

At present, high-density maps are available for the Brassica species. These maps helped in tagging genes for disease resistance, abiotic stress, fertility restoration, oil quality, and morphological traits (Snowdon et al. 2007). The successful CMS and restorer developed are being used in development of commercial hybrids mainly *ogura* and *mori* cytoplasm based. Fast developing genomic tools like sequencing and bioinformatics helped to complete genome sequencing of Arabidopsis, *B. rapa*. These tools helped in the understanding of various aspects associated with the origin, evolution, and domestication of Brassica species. Development of molecular markers associated with disease and insect pest resistance, oil quality helped in faster development of varieties of rapeseed mustard. Rapid-cycling Brassica (RcBr) developed by Paul Williams of Wisconsin University (Williams and Hill 1986) have become model organisms for basic and applied research primarily because of their short life span, small size, and absence of seed dormancy (Prakash et al. 2009).

With the development of modern high yielding cultivars, monocultures, changing food habits, deforestation due to developmental activities, and natural vagaries like drought and floods led to loss of natural genetic diversity of most of the crops in the past decades. This loss of genetic diversity of wild germplasm and landraces is called genetic erosion. Therefore, in order to protect the valuable genetic resources of Brassica and allied genera, the collection and conservation programs started worldwide since 1970. Both *ex situ* and *in situ* conservation strategies are being followed for conservation of Brassica germplasm. Seed banks were created across the world to maintain genetic diversity, to minimize genetic erosion, and to distribute these genetic resources of landraces and wild relatives for research purpose to different organizations. Plant genetic resources of Brassicaceae family are conserved across the globe in seed genebanks of many countries (China, India, United Kingdom, United States, Netherland, Spain, and Germany) and are constantly enriched by further collections through plant exploration from different geographical regions with more diversity based on identified gaps. Furthermore, to get access to these genetic resources many databases including AEGIS, EURISCO, ECPGR, Genesys, GRAIN, GBIF, NPGS, CWRIS, and AVGRIS were devolved for easy access of information of Brassica species and other crops germplasm for distribution to organizations for research purposes. Sicily is considered to be a center of diversity for ($2n = 18$) Mediterranean brassicas; therefore, it is selected to develop *in situ* strategies for conservation of natural population of brassica species, *viz*. *B. macrocarpa, B. villosa,* and *B. rupestris*. According to FAO (2010), the total number of germplasm accessions maintained in *ex situ* collections of all crops worldwide is estimated as 7.4 million, but only about 30% (1.9–2.2 million) of this total are unique accessions (Bettencourt 2011). Hence there is a need to further diversify the germplasm base; hence, collecting must continue in order to fill in the diversity gap in existing germplasm collections of Brassica and allied genera. Further it is suggested to maintain ex situ and in situ conservation of landraces and wild relatives to provide safety and an equal opportunity to these resources to evolve further in natural environment according to environmental settings.

## References

Ackman RG, Eaton CA, Sipos JC, Loew FM, Hancock D (1977) Comparison of fatty acids from high levels of erucic acid of RSO and partially hydrogenated fish oil in non-human primate species in a short-term exploratory study. Nutrition Diet 25:170–85.

Agrawal RC, Sunil A, Tyagi RK (2012) An overview of biodiversity informatics with special reference to plant genetic resources. Comput Electron Agric 84(2012):92–99

AICRPRM (2009-2019) All India Coordinated Research Project on Rapeseed-Mustard. Annual Report (2009–2019). Directorate of Rapeseed-Mustard Research, Bharatpur

Allchin FR (1969) Early cultivated plants in India and Pakistan. In: Uckoand PJ, Dimbleby GW (eds) The domestication and exploitation of plants and animals. Duckworth, London, pp 323–329

Allender CJ, Allainguillaume J, Lynn J, King GJ (2007) Simple sequence repeats reveal uneven distribution of genetic diversity in chloroplast genomes of Brassica oleracea L. and (n ¼ 9) wild relatives. Theor Appl Genet 114:609–618

Arora RK (1988) The Indian Gene Centre—priorities and prospects for collection. In: Paroda RS, Arora RK, Chandel KPS (eds) Plant genetic resources: indian perspective. NBPGR, New Delhi, pp 66–75

ARS-GRIN (1997) Internet site: germplasm resources information network. National Genetic Resources Program of the USDA's Agricultural Research Service. http://www.ars-grin.gov/

Astley D, Bas N, Branca F, Daunay MC, Díez MJ, Keller J, van Dooijeweert W, van Treuren R, Maggioni L, Lipman E (2007) Report of a vegetables network. In: 2nd meeting, Olomouc, Czech Republic, 26–28 June 2007

Bailey LH (1922) The cultivated Brassicas. Gent Herb 1:53–108

Bailey LH (1930) The cultivated Brassicas. Gent Herb 2:210–267

Banga SS, Labana KS (1984) Heterosis in Indian mustard [*Brassica juncea* (L.) Coss.]. Plant Breed 92:61–70

Bauhin C (1623) Pinax Theatri Botanici. Konig, Basel

Bell JM (1984) Nutrients and toxicants in rapeseed meal: a review. J Anim Sci 58:996–1010

Bellon MR, Brush S (1994) Keepers of maize in Chiapas, Mexico. Econ Bot 48:196–209

Bettencourt E (2011) Sources of information on existing germplasm collections. Collecting plant genetic diversity: technical guidelines-2011 update

Boswell BR (1949) Our vegetable travellers. Nat Geogr Magaz 96:134–217

Boukema IW, van Hintum TJL (1998) The European Brassica database. Proceedings of an international symposium on Brassicas. Acta Horticult 459:249–254

Boukema IW, van Hintum TJL (1999) Genetic Resources. In: Gomez-Campo C (ed) Developments in plant genetics and breeding 4. Biology of *Brassica* coenospecies. Elsevier, Amsterdam

Branca F (2008) Cauliflower and broccoli. In: Prohens J, Nuez F (eds) Vegetables I: Asteraceae, Brassicaceae, Chenopodicaceae, and Cucurbitaceae. Handbook of crop breeding. Springer, Berlin, pp 151–186

Branca F, Cartea E (2011) Brassica. In: Kole C (ed) Wild crop relatives: genomic and breeding resources: oilseeds. Springer, Berlin, pp 17–36

Burkill IH (1930) The Chinese mustards in the Malay Peninsula. Gad's Bull 5

Cartea ME, Lema M, Francisco M, Velasco P (2011) Basic information on vegetable brassica crops. In: Sadowski J, Kole C (eds) Genetics, genomics and breeding of vegetable brassicas. Science Publishers, Enfield

Červenski J, Savić A, Popović A, Petrović A, Maksimović L, Takac A, Popovic C, Glogovac S (2013) Possibility of exploitation of Serbian local cultivars and landraces of cabbages (Brassica oleracea var. capitata L.): case of "Futoški Cabbage" from Futog Region. In: Branca F, Tribulato A (eds) Proceedings of the VIth IS on Brassicas and XVIIIth crucifer genetics workshop. Acta Hort. 1005, ISHS 2013

Chauhan JS, Singh KH, Kumar A (2006) Breeder seed production scenario of rapeseed-mustard varieties in India, 20 pp. NRCRM, Bharatpur

Chauhan JS, Tyagi MK, Kumar A, Nashaat NI, Singh M, Singh NB, Jakhar ML, Welham SJ (2007) Drought effects on seed yield and its components in Indian mustard (*Brassica juncea* L.). Plant Breed 126:399–402

Chauhan JS, Singh KH, Singh VV, Kumar S (2011) Hundred years of rapeseed-mustard breeding in India: accomplishments and future strategies. Indian J Agric Sci 81:1093–1109

Cheng BF, Heneen WK (1995) Satellited chromosome, nucleolus organizer regions and nucleoli of Brassica campestris L., B. nigra (L.) Koch and Sinapis arvensis L. Hereditas 122:113–118

Chauhan JS, Tyagi MK, Kumar PR, Tyagi P, Singh M, Kumar S (2002) Breeding for oil and seed meal quality in rapeseed mustard in India – A review. Agricultural Reviews 23(2):71–92.

Ciancaleoni S, Chiarenza GL, Raggi L, Branca F, Neg V (2014) Diversity characterisation of broccoli (Brassica oleracea L. var. italica Plenck) landraces for their on-farm (in situ) safeguard and use in breeding programs. Genet Resour Crop Evol 61:451–464. https://doi.org/10.1007/s10722-013-0049-2

Das S, Lagercrantz U, Lascoux M (2007) Black mustard. In: Kole C (ed) Genome mapping and molecular breeding in plants, vol 2: oilseeds. Springer, Berlin, pp 264–274

Demeke T, Adams RP, Chibbar R (1992) Potential taxonomic use of random amplified polymorphic DNA (RAPD)- a case study in *Brassica*. Theor Appl Genet 84:990–994

Denford KE (1975) Isoenzyme studies in members of the genus Brassica. Bot Notiser 128:455–462

Dixon GR (2007) Vegetable Brassicas and related crucifers. CAB International, Willingford

De Candolle AP (1821) Cruciferae. Systema Naturale. 2:139–700.

De Candolle AP (1886) Origin of cultivated plants. ( Reprint, Hafner, New York, 1964).

Dixon GR (2007) Vegetable Brassicas and Related Crucifers. Pp. 416. CAB Intl., Willingford, UK.

ECPGR (2020). https://www.ecpgr.cgiar.org/about/goals-and-objectives. Accessed 6 May 2020

Esquinas-Alcazar JT (1993) Plant genetic resources. In: Hayward MD, Bosemark NO, Romagosa I (eds) Plant breeding: principles and prospects. Chapman & Hall, London, pp 33–51

EURISCO (2020). https://eurisco.ipk-gatersleben.de/apex/f?p=103:1:0::::: Accessed 6 May 2020

Fahey JW, Zalcmann AT, Talalay P (2001) The chemical diversity and distribution of glucosinolates and isothiocyanates among plants. Phytochemistry 56:5–51

Fenewick GR, Heaney RK, Mullin J (1983) Glucosinolates and their breakdown products in food and food plants. CRC Crit Rev Food Sci Nutr 18:123–201

Frandsen KJ (1943) The experimental formation of Brassica juncea Czern et Coss. Dansk Botanisk Arkiv 11:1–17

Frandsen KJ (1947) The experimental formation of Brassica napus L. var. oleifera DC and Brassica carinata Braun. Dansk Botanisk Arkiv 12:1–16

Frankel OH, Bennett E (1970) Genetic resources—introduction. In: Frankel OH, Bennett E (eds) Genetic resources in plants—their exploration and conservation. Inter-National biological programme handbook no. 11. Blackwell, Oxford, pp 1–32

Frankel OH, Hawkes JG (eds) (1975) Crop genetic resources for today and tomorrow. Cambridge University Press, Cambridge

Frankel OH, Brown AHD, Burdon JJ (1998) The conservation of plant biodiversity, 2nd edn. Cambridge University Press, Cambridge, pp 56–78

Gates RR (1953) Wild cabbages and the effects of cultivation. J Genet 51:363–372

Genesys. 2020. http://www.genesys-pgr.org/. Accessed 6 May 2020

Ghosh R, Bera S, D'Rozario A, Banerjee M, Chakraborty S (2006) Plant remains from an archaeological site as indicators of vegetation and agricultural practice between (3,320 400) and (2,080 80) yr BP in Gangetic, West Bengal, India. J Integr Plant Biol 48:628–641

Gladis T, Hammer K (1992) The brassica collection in Gatersleben: Brassica juncea, Brassica napus, Brassica nigra, and Brassica rapa. Feddes Rep 103:469–507

Gómez-Campo C, Prakash S (1999) Origin and domestication. In: Gómez-Campo C (ed) Biology of *Brassica* coenospecies. Elsevier, Amsterdam, pp 33–58

Gómez-Campo C, Aguinagalde I, Ceresuela J, Lázaro A, Martınez-Laborde J, Parra-Quijano M, Simonetti E, Torres E, Tortosa M (2005) An exploration of wild Brassica oleracea L. germplasm in Northern Spain. Genet Resour Crop Evol 52:7–13

Gómez-Campo C, Aguinagalde I, Ceresuela J, Lázaro A, Martínez-Laborde J (2006) Erosion of genetic resources within seed genebanks: the role of seed containers. Seed Sci Res 16:291–294

Gómez-Campo C, Aguinagalde I, Arús P, Jiménez-Aguilar C, Lázaro A, Martín-Clemente JP, Parra-Quijano M, MD SY, Simonetti E, Torres E, Torcal L, Tortosa ME (2007) Geographical distribution and conservation status of Brassica montana in NE Spain. Cruciferae Newsl 27:32–34

Gómez-Campo C (1999) Biology of Brassica coenospecies. p. 489. Elsevier Science, Amsterdam
Gupta SK (2012) Brassicas. In: Gupta SK (ed) Technological innovations in major world oil crops, vol 1. Springer, New York, pp 53–83. https://doi.org/10.1007/978-1-4614-0356-2
Gupta V, Lakshmi Sita G, Shaila MS, Jagannathan V, Laksmikumaran MS (1992) Characterization of species specific repetitive DNA sequences from Brassica nigra. Theor Appl Genet 84:397–402
Gopalan CD, Krishanamurthy D, Shenolikar IS, Krishnamurthy KAVR (1974) Myocardial changes in monkey fed on mustard oil. Nutrition Metabolism 16:352–65
Hammer K, Diederichsen A (2009) Evolution, status and perspectives for Landraces in Europe. In: Veteläinen M, Negri V, Maxted N (eds) European landraces on farm conservation, management and use. Bioversity technical bulletin no. 15. Bioversity International, Rome, pp 23–44
Harberd DJ (1972) A contribution to cytotaxonomy of Brassica (Cruciferae) and its allies. Bot J Linn Soc 65:1–23
Harberd DJ (1976) Cytotaxonomic studies of Brassica and related genera. In: Vaughan JG, MacLeod AJ, Jones MG (eds) The biology and chemistry of the Cruciferae. Academic Press, London, pp 47–68
Harbinder S, Lakshmikumaran M (1990) A repetitive sequence from Diplotaxis erucoides is highly homologous to that of Brassica campestris and B. oleracea. Plant Mol Biol 15:155–156
Harlan JR (1975) Our vanishing genetic resources. Science 188:618–621
Hayward A (2012) Introduction—oilseed Brassicas. In: Edwards E, Batley J, Parkin I, Kole C (eds) Genetics, genomics and breeding of oilseed brassicas. CRC Press Taylor & Francis Group, Boca Raton. ISBN-13: 978-1-4398-8335-8
He YT, Chen BY, Fu TD, Li DR, Tu JX (2003) Origins and evolution of Brassica campestris L. in China. Acta Genet Sin 30:1003–1012
Hedrick UP (ed) (1919) Sturtevant's notes on edible plants. New York Agricultural Experiment Station J.B. Lyon, Albany
Hegi G (1919) Illustrierte Flora von Mittel-Europe. 4. Munich.
Helm J (1963) Morphologisch-taxonomische Gliederung der Kultursippen von Brassica oleracea L. Kulturpflanze 11:92–210
Hemingway JS (1976) Mustards, Brassica species and Sinapis alba (Cruciferae). In: Simmond NW (ed) Evolution of crop plants. Longman, London, pp 56–59
Hinata K, Prakash S (1984) Ethnobotany and evolutionary origin of Indian oleiferous Brassicae. Indian J Genet 44:102–112
Hooker JD, Thomson T (1861) Praecursores ad Floram Indicum—Cruciferae. J Linn Soc 5:128–181
Horler RSP, Turner AS, Fretter P, Ambrose M (2018) SeedStor: a germplasm information management system and public database. Plant Cell Physiol 59:e5. https://doi.org/10.1093/pcp/pcx195
Hyams E (1971) Cabbages and kings. In: Plants in the service of man. J.M. Dent & Sons, London, pp 33–37
Inaba R, Nishio T (2002) Phylogenetic analysis of Brassiceae based on the nucleotide sequences of the S-locus related gene SLR1. Theor Appl Genet 105:1159–1165
Institute of Archaeology of Chinese Academy of Sciences (1963) Xian Banpo Country. Special Issue. Archaeology. Archaeology Press, Beijing
Jain A, Bhatia S, Banga SS, Prakash S, Lakshmikumaran M (1994) Potential use of random amplified polymorphic DNA (RADP) technique to study the genetic diversity in Indian mustard (*Brassica juncea*) and its relationship to heterosis. Theor Appl Genet 88:116–122
Kapila R, Negi MS, This P, Delseny M, Srivastava PS, Lakshmikumaran M (1996) A new family of dispersed repeats from Brassica nigra: characterization and localization. Theor Appl Genet 93:1123–1129
Kato K, Namai H, Hosoda T (1968) Studies on the practicality of artificial rutabaga SR lines obtained from interspecific crosses between Shogoin-kabu (B. campestris ssp. rapifera) and kohlrabi (B. oleracea var. gongylodes). II. Feeding value of SR lines. J Japan Grasl Sci 14:177–181

Katsiotis A, Bebeli P, Mavromatis A, Tzivelikas A (2009) Inventory of Greek Landraces. In: Veteläinen M, Negri V, Maxted N (eds) European landraces on-farm conservation, management and use. Bioversity technical bulletin no. 15. Bioversity International, Rome, pp 97–103

Kimber DS, McGregor DI (1995) Physiology: crop development, growth and yield. In: Kimber DS, McGregor DI (eds) *Brassica* oilseeds: production and utilization. CABI, Wallingford, pp 1–10

Knee EM, Rivero L, Crist D, Grotewold E, Scholl R (2011) Germplasm and molecular resources. Genetics and genomics of the Brassicaceae. Springer, New York, pp 437–467

Krasteva L, Stoilova T, Varbanova K, Neykov S (2009) Bulgarian Landrace inventory–significance and use. In: Veteläinen M, Negri V, Maxted N (eds) European landraces on-farm conservation, management and use. Bioversity technical bulletin no. 15. Bioversity International, Rome, pp 53–68

Kumar A, Chauhan JS (2005) Status and future thrust areas of rapeseed-mustard research in India. Indian J Agric Sci 75:621–635

Li CW (1980) Classification and evolution of mustard crops (*Brassica juncea*) in China. Cruciferae Newsl 5:33–36

Li JW (1981) The origins and evolution of vegetable crops in China. Sci Agric Sin 14:90–95

Lorenz OA, Maynard DN (1988) Knott's handbook for vegetable growers, 3rd edn. Wiley, New York

Maggioni L, Eastwood A (1997) Monitoring of ($2n$=18) wild *Brassica* populations in Italy and Croatia. In: Maggioni L, Astley D, Gustafsson M, Gass T, Lipman E (eds) Report of a working group on Brassica, third meeting, 27–29 November 1996. International Plant Genetic Resources Institute, Rome

Maggioni L, Astley D, Gustafsson M, Gass T, Lipman E (1997) Report of a working group on Brassica, third meeting, 27–29 November 1996, Rome. International Plant Genetic Resources Institute, Rome

Maggioni L, von Bothmer R, Poulsen G, Branca F (2010) Origin and domestication of cole crops (Brassica oleracea L.). Linguistic and literary considerations. Ecol Bot 64:109–123

Manohar R, Pushpan R, Rohini S (2009) Mustard and its uses in Ayurveda. Indian J Trad Knowl 8:400–404

Mansholt UJ (1909) Van Pesch Plantenteelt, beknopte handleiding tot de kennis van den Neederlandschen landbouw. 3rd rev edn., pt. 2. Plantenteelt. Zwolle, 228 pp.

Maxted N, Ford-Lloyd BV, Hawkes JG (1997) Complementary conservation strategies. In: Maxted N, Ford-Lloyd BV, Hawkes JG (eds) Plant genetic conservation: the in situ approach. Chapman & Hall, London, pp 20–55

Maxted N, Avagyan A, Frese L, Iriondo JM, Magos Brehm J, Singer A, Kell SP (2015) ECPGR concept for *in situ* conservation of crop wild relatives in Europe. Wild species conservation in genetic reserves working group. European Cooperative Programme for Plant Genetic Resources, Rome

Mayr E (1937) Alpine Landsorten in ihrer Bedeutung für die praktische Züchtung. Forschungsdienst 4:162–166

Metzger J (1833) Systematische Beschreibung der Kultivirten Kohlarten. Heidelberg 2:1–68

Misra AK, Kumar A, Kumar PR, Manohar SS (2007a) Evaluation and characterization of elite germplasm of Indian mustard, *Brassica juncea* (L) Czern and Coss. Indian J Oilseeds Res 24:27–30

Misra AK, Manohar SS, Singh K, Kumar A (2007b) Characterization of germplasm for some economic traits in Indian mustard [*Brassica juncea* (L) Czern and Coss], pp 84–5. In: Extended summaries, national seminar on changing global vegetable oils scenario: issues and challenges before India. Indian Society of Oilseeds Research, DOR, Hyderabad 29–31 January 2007

Mithen R, Faulkner K, Magrath R, Rose P, Williamson G, Marquez J (2003) Development of isothiocyanate enriched broccoli and its enhanced ability to induce phase 2 detoxification enzymes in mammalian cells. Theor Appl Genet 106:727–734

Mizushima U (1950) Karyogenetic studies of species and genus hybrids in the tribe Brassiceae-Cruciferae. Tohoku J Agric Res 1:1–14

Mizushima U, Tsunoda S (1967) A plant exploration in Brassica and allied genera. Tohoku J. Agric Res 17:249–276
Monteiro AA, Dias JS (1996) Portuguese Brassica landraces: a perspective. In: Dias JS, Crute I, Monteiro AA (eds) Proceedings of the international symbosium on Brassicas, ninth crucifer genetics workshop. Acta Hort. 407. ISHS.
Moreira PMRM, Veloso MM (2009) Landrace inventory for Portugal. In: Veteläinen M, Negri V, Maxted N (eds) European landraces on-farm conservation, management and use. Bioversity technical bulletin no. 15. Bioversity International, Rome, pp 124–136
Morinaga T (1928) Preliminary note on interspecific hybridization in Brassica. Proc Imp Acad 4:620–622
Morinaga T (1929a) Interspecific hybridization in Brassica. I. The cytology of F1 hybrids of B. napella and various other species with 10 chromosomes. Cytologia 1:16–27
Morinaga T (1929b) Interspecific hybridization in Brassica. II. The cytology of F1 hybrids of B. cernua and various other species with 10 chromosomes. Jpn J Bot 4:277–289
Morinaga T (1929c) Interspecific hybridization in Brassica. III. The cytology of F1 hybrids of B. cernua and B. napella. J Dept Agric Kyushu Imp Univ 2:199–206
Morinaga T (1931) Interspecific hybridization in Brassica. IV. The cytology of F1 hybrids of B. carinata and some other species with 10 chromosomes. Cytologia 3:77–83
Morinaga T (1933) Interspecific hybridization in Brassica. V. The cytology of F1 hybrids of B. carinata and B. alboglabra. Jpn J Bot 6:467–475
Morinaga T (1934a) Interspecific hybridization in Brassica. VI. The cytology of F1 hybrids of B. juncea and B. nigra. Cytologia 6:62–67
Morinaga T (1934b) On the chromosome number of Brassica juncea and B. napus, on the +hybrid between the two, and on offspring line of the hybrid. Jpn J Genet 9:161–163
Mizushima U (1980) Genome analysis in Brassica and allied genera. pp. 89–108. In: S. Tsunoda, K. Hinata and C. Go´mez-Campo (eds.), Brassica crops and wild allies: Biology and breeding. Japan Sci. Soc. Press, Tokyo.
Nanjundan J, Singh K, Singh KH, Singh D (2014) Catalogue on rapeseed-mustard germplasm. Directorate of Rapeseed-Mustard Research, Bharatpur, Rajasthan. Pp.180
NBPGR (2020) ICAR—NBPGR Genebank. http://genebank.nbpgr.ernet.in/CropgroupCrops.aspx. Accessed July 12 2018
Negri V, Maxted N, Veteläinen M (2009) European landrace conservation: an introduction. In: Veteläinen M, Negri V, Maxted N (eds) European landraces on-farm conservation, management and use. Bioversity technical bulletin no. 15. Bioversity International, Rome, pp 1–23
Nishi S (1980) Differentiation of Brassica crops in Asia and the breeding of 'Hakuran', a newly synthesized leafy vegetable. In: Tsunoda S, Hinata K, GomezCampo C (eds) Brassica crops and wild allies: biology and breeding. Japan Scientific Society Press, Tokyo, pp 133–150
NordGen (2020). https://www.nordgen.org/en/global-seed-vault/search-seed-vault/. Accessed 4 May 2020
NRCRM (2000) Rapeseed-mustard varieties of India. NRCRM, Bharatpur, 115pp + V
NRCRM (2007) Vision 2025. Perspective Plan National Research Centre on Rapeseed-Mustard, Bharatpur, pp 1–46
OECD (2016) Brassica crops (Brassica species). In: Safety assessment of transgenic organisms in the environment, Volume 5: OECD consensus documents. OECD Publishing, Paris. https://doi.org/10.1787/9789264253018-6-en
Olsson G (1960a) Species crosses within the genus Brassica I. Artificial B. juncea Coss. Hereditas 46:171–222
Olsson G (1960b) Species crosses within the genus Brassica II. Artificial B. napus L. Hereditas 46:351–396
Olsson G, Josefsson A, Hagberg A, Ellerstrom S (1955) Synthesis of the ssp. rapifera of Brassica napus. Hereditas 41:241–249

Ougham, Thomas ID (2014) Germplasm databases and informatics Helen. In Jackson M, Ford-Lloyd B, Parry M (eds) Plant genetic resources and climate change, vol 2014. CAB International, Wallingford, pp 151–165

Panda S, Martin JP, Aguinagalde I (2003) Chloroplast and nuclear DNA studies in a few members of the Brassica oleracea L. group using PCR-RFLP and ISSR-PCR markers: a population genetic analysis. Theor Appl Genet 106:1122–1128

Panjabi P, Yadava SK, Kumar N, Bangkim R, Ramchiary N (2019) Breeding Brassica juncea and B. rapa for sustainable oilseed production in the changing climate: progress and prospects. In: Kole C (ed) Genomic designing of climate-smart oilseed crops. Springer, Berlin

Plazibat M. (1996) Brassica. In: Flora Croatica—index Florae. Croatia Natur Croatica 2.

Poulsen GB, Kahl G, Weising K (1994) Differential abundance of simple repetitive sequences in species of Brassica and related Brassicaceae. Plant Syst Evol 190:21–30

Pradhan AK, Prakash S, Mukhopadhyay A, Pental D (1992) Phylogeny of Brassica and allied genera based on variation in chloroplast and mitochondrial DNA patterns: molecular and taxonomical classifications are incongruous. Theor Appl Genet 85:331–340

Prain D (1898) The mustards cultivated in Bengal. Agric Ledger 5:1–80

Prakash S (1973a) Artificial synthesis of Brassica juncea Coss. Genetica 44:249–264

Prakash S (1973b) Non-homologous meiotic pairing in the A and B genomes of Brassica: its breeding significance in the production of variable amphidiploids. Genet Res Camb 2:133–137

Prakash S (1980) Cruciferous oilseeds in India. In: Tsunoda S, Hinata K, Gómez-Campo C (eds) *Brassica* crops and wild allies. Japan Scientific Press, Tokyo, pp 151–163

Prakash S (2001) Utilization of wild germplasm of Brassica allies in developing male sterility–fertility restoration systems in Indian mustard–Brassica juncea. In: Houli L, Fu TD (eds) Proceedings of the international symposium rapeseed. Science, Huazhong Agricultural University, China. Science Press, New York, pp 73–78

Prakash S, Hinata K (1980) Taxonomy, cytogenetics and origin of crop brassicas, a review. Opera Bot 55:1–57

Prakash S, Takahata Y, Kirti PB, Chopra VL (1999) Cytogenetics. In: Gómez-Campo C (ed) Biology of Brassica Coenospecies. Elsevier Science, Amsterdam, pp 59–106

Prakash S, Bhat SR, Quiros CF, Kirti PB, Chopra VL (2009) Brassica and its close allies: cytogenetics and evolution. Plant Breed Rev 31:21–187

Prakash S, Wu XM, Bhat SR (2012) History, evolution, and domestication of Brassica crops. Plant Breed Rev 35:19–84

Radhamani J, Singh R, Srinivasan K, Tyagi RK (2013) Conservation of trait-specific germplasm of brassicas in National Genebank. National Bureau of Plant Genetic Resources, New Delhi. 114 p

Ramanujam S, Srinivasachar D (1943) Cytogenetical investigations in the genus Brassica and the artificial synthesis of Brassica juncea. Indian J Genet 3:73–88

Rana RS, Singh R (1992) Present status of rapeseed-mustard germplasm in India. In: Kumar D, Rai M (eds) Advance in oilseed research, vol 1. Scientific Publishers, Jodhpur, pp 189–200

Renfrew M (1973) Palaeoethnobotany: the prehistoric food plants of the near East and Europe. Methuen, London, pp 186–195

Rich TCG (1991) Crucifers of Great Britain and Ireland. Botanical Society of the British Isles, London

Rivero-Lepinckas L, Christ D, Scholl R (2006) Growth of plants and preservation of seeds. In: Salinas J, Sanchez-Serrano J (eds) Methods in molecular biology. *Arabidopsis* protocols. Humana Press, Totowa

Rosa EAS, Heaney RK, Fenewick GR, Portas CAM (1997) Glucosinolates in crop plants. Hortic Rev 19:99–215

Roxburgh W (1832) Flora Indica. In: Carey W (ed.) Serampore 2nd ed, vol III, pp 117–124

Rubatzky VE, Yamaguchi M (1997) World vegetables: principles, production, and nutritive values. Chapman & Hall, New York

Renard S, Mc Gregor L (1976) Antithormbo genetic effects of erucic acid poor rapeseed oils in the rats. Rev Fr Cross 25:339–396

Schiemann E (1932) Brassica und andere Cruciferen. In: Entstehung der Kulturpflanzen Handb. Vererbwis. Lfg. 15, Berlin, pp 271–288
Scholten M, Green N, Campbell G, Maxted N, Ford-Lloyd B, Ambrose M, Spoor B (2009) Landrace inventory of the UK, pp 1–23. In: Veteläinen M, Negri V, Maxted N (eds). European landraces on-farm conservation, management and use. Bioversity technical bulletin no. 15. Bioversity International, Rome, pp 161–170
Shapiro TA, Fahey JW, Wada KL, Stephenson KK, Talay P (2001) Chemopreventive glucosinolates and isothiocyanates of broccoli sprouts: metabolism and excretion in humans. Cancer Epidemiol Biomark Prev 10:501–508
Shashilova LI (1997) The Russian *Brassica* collection. In: Report of a working group on Brassica, third meeting, 27–29 Nov 1996, Rome, Italy. International Plant Genetic Resources Institute, Rome, pp 56–57
Simonsen V, Heneen WK (1995) Inheritance of isozymes in Brassica campestris L. and genetic divergence among different species of Brassiceae. Theor Appl Genet 91:353–360
Singh R (1996) Germplasm resources. In: Chopra VL, Prakash S (eds) Oilseed and vegetable Brassica: Indian perspective. Oxford & IBH Publishing Company, New Delhi, pp 279–291
Singh R, Sharma SK (2007) Evaluation, maintenance, and conservation of germplasm. In: Gupat SK, Delseney M, Kader JC (eds) Advances in botanical research, incorporating advances in plant pathology, rapeseed breeding, vol 35. Academic Press/Elsevier, London/California, pp 466–481
Sinskaia EN (1928) The oleiferous plants and root crops of the family Cruciferae. Bull Appl Bot Genet Plant Breed 9:1–648
Smekalova T (2009) Cultivated plant inventory of Russia. In: Veteläinen M, Negri V, Maxted N (eds) European landraces on-farm conservation, management and use. Bioversity technical bulletin no. 15. Bioversity International, Rome, pp 143–154
Snogerup S (1980) The wild forms of the Brassica oleracea group (2n ¼ 18) and their possible relations to the cultivated ones. In: Tsunoda S, Hinata K, Gómez-Campo C (eds) Brassica crops and wild allies: biology and breeding. Japan Scientific Society Press, Tokyo, pp 121–132
Snowdon R, Lühs W, Friedt W (2007) Oilseed rape. In: Kole C (ed) Genome mapping and molecular breeding in plants, Oilseeds, vol 2. Springer, Berlin, pp 55–114
Song KM, Osborn TC (1992) Polyphyletic origins of Brassica napus: new evidence based on organelle and nuclear RFLP analyses. Genome 35:992–1001
Song KM, Osborn TC, Williams PH (1988a) Brassica taxonomy based on nuclear restriction fragment length polymorphisms (RFLPs). 1. Genome evolution of diploid and amphidiploid species. Theor Appl Genet 75:84–794
Song KM, Osborn TC, Williams PH (1988b) Brassica taxonomy based on nuclear restriction fragment length polymorphisms (RFLPs). 2. Preliminary analysis of subspecies within B. rapa (syn. campestris) and B. oleracea. Theor Appl Genet 76:593–600
Song KM, Osborn TC, Williams PH (1990) Brassica taxonomy based on nuclear restriction fragment length polymorphisms (RFLPs). 3. Genome relationships in Brassica and related genera and the origin of B. oleracea and B. rapa (syn. campestris). Theor Appl Genet 79:497–506
Song K, Tang K, Osbourn TC (1993) Development of synthetic *Brassica* amphidiploids by reciprocal hybridisation and comparison to natural amphidiploids. Theor Appl Genet 86:811–821
Song K, Tang K, Osbourn TC, Lu P (1996) Genome variation and evolution of *Brassica* amphidiploids. Acta Hortic 407:35–44
Sun VG (1946) The evaluation of taxonomic characters of cultivated Brassica with a key to species and varieties. I The characters Torrey. Bot Club Bull 73:244–281
Sun VG (1970) Breeding plants of Brassica. J Agric Assoc China 71:41–52
Swarup V, Brahmi P (2005) Cole crops. In: Dhillon BS, Tyagi RK, Saxena S, Randhawa GJ (eds) Plant genetic resources: horticultural crops. Narosa Publ House, New Delhi, pp 75–88
Sun FJ (1943) Hybrid vigor in Brassica. Journal of Agricultural Association, China 175:35–58
Snowdon RJ, Link K, Badani AG, Friedt W (2005) Recent advances in molecular breeding of oilseed rape (Brassica napus L.). Progress in Botany 66:144–163.

Takahata Y, Hinata K (1983) Studies on cytodemes in the subtribe Brassicineae. Tohoku J Agric Res 33:111–124

Takuno S, Kawahara T, Ohnishi O (2007) Phylogenetic relationships among cultivated types of Brassica rapa L. em. Metzg. as revealed by AFLP analysis. Genet Res Crop Evol 54:279–285

Tournefort JP (1700) Institutiones rei herbariae editio altera. I:219–227. Paris.

Tsunoda S (1980) Ecophysiology of wild and cultivated forms in Brassica and allied genera. In: Tsunoda S, Hinata K, Gómez-Campo C (eds) Brassica crops and wild allies: biology and breeding. Japan Scientific Society Press, Tokyo, pp 109–120

Tsunoda S, Nishi S (1968) Origin, differentiation and breeding of cultivated Brassica. In: Proceedings of the XII international congress genetics, Tokyo, 19–28 Aug, vol 2, pp 335–339

Uchimiya H, Wildman SG (1978) Evolution of fraction I protein in relation to origin of amphidiploid Brassica species and other member of Cruciferae. J Hered 69:299–303

U. N. (1935) Genome analysis in Brassica with special reference to the experimental formation of B. napus and peculiar mode of fertilization. Jap J Bot 7:389–452

Villa TCC, Maxted N, Scholten M, Ford-Lloyd B (2005) Defining and identifying crop landraces. Plant Genetic Resources: Characterization and Utilization 3(03): 373–384. https://doi.org/10.1079/pgr200591

Vaughan JG (1977) A multidisciplinary study of the taxonomy and origin of Brassica crops. BioScience 27:35–40

Vaughan JG, Denford KE (1968) An acrylamide gel electrophoretic study of the seed proteins of Brassica and Sinapis species with special reference to their taxonomic value. J Exp Bot 19:724–732

Vaughan JG, Hemingway JS, Schofield HJ (1963) Contributions to a study of variations in Brassica juncea Coss. et Czern. J Linn Soc Bot 58:435–447

Vavilov NI (1949) The origin, variation, immunity and breeding of cultivated plants. Chron Bot 13:1–364

Vyas P, Prakash S, Shivanna KR (1995) Production of wide hybrids and backcross progenies between Diplotaxis erucoides and crop brassicas. Theor Appl Genet 9:549–553

Warwick SI, Black LD (1991) Molecular systematics of Brassica and allied genera (subtribe Brassicinae, Brassiceae)—chloroplast genome and cytodeme congruence. Theor Appl Genet 82:81–92

Warwick SI, Black LD (1997) Phylogenetic implications of chloroplast DNA restriction site variation in subtribes Raphaninae and Cakilinae (Brassicaceae). Can J Bot 75:960–973

Warwick SI, Sauder CA (2005) Phylogeny of tribe Brassiceae (Brassicaceae) based on chloroplast restriction site polymorphisms and nuclear ribosomal internal transcribed spacer (ITS) and chloroplast trnL intron sequences. Can J Bot 83:467–483

Warwick SI, Francis A, Gugel RK (2009) Guide to wild germplasm of *Brassica* and allied crops (tribe Brassiceae, Brassicaceae), 3rd edn. Agriculture and Agri-Food Canada Technical Bulletin, Ottawa

Watt G (1889) A dictionary of the economic products of India, vol II. Government Printing, Calcutta

Wu X-M, Qian X-Z, Li RG, Zhou L, Wu N-F (1996) Evolutionary relationship between Chinese Xinjiang wild rape and Brassica as revealed by RAPDs. In: Advance of oil crop research in China. Chinese Agricultural Sciences and Technology Press, Beijing, pp 88–92

Yanagino T, Takahata Y, Hinata K (1987) Chloroplast DNA variation among diploid species in Brassica and allied genera. Japan J Genet 82:119–125

Yao QL, Chen FB, Fang P, Zhou GF, Fan YH, Zhang ZR (2012) Genetic diversity of Chinese vegetable mustard (Brassica juncea Coss) landraces based on SSR data. Biochem Syst Ecol 45(2012):41–48

Yarnell SH (1956) Cytogenetics of the vegetable crops: II. Crucifers. Bot Rev 22(2):81–166

Zeven AC (1998) Landraces: a review of definitions and classifications. Euphytica 104:127–139

Zeven AC, Zhukovsky PM (1975) Dictionary of cultivated plants and their centres of diversity. Pudoc:29–30

Zhang Y, Talalay P (1994) Anticarcinogenic activities of organic isothiocyanate: chemistry and mechanisms. Cancer Res 54(Suppl):1976s–1981s

Zohary D, Hopf M (1993) Domestication of plants in the old world: the origin and spread of cultivated plants in West Asia, Europe, and the Nile Valley, 2nd edn. Clarendon Press, Oxford

# Part IV
# Pulses

# Genetic Resources of Groundnut

**T. Radhakrishnan, A. L. Rathnakumar, M. K. Mahatma, S. Chandramohan, and S. Patel**

**Abstract** Groundnut or peanut is one of the major oilseeds and food crops cultivated globally. This oilseed can be directly consumed as a foodstuff, is a rich source of oil, protein and carbohydrates and other nutrition like tocopherol, niacin and folic acid; mineral components like Cu, Mn, K, Ca and P; dietary fibres, flavonoids, phytosterols like resveratrol, beta-sitosterol; and phenolic acids. The cultivated groundnut *Arachis hypogea* is a segmental amphidiploid ($4x = 40$), and cytogenetic and molecular evidences suggest that the origin of cultivated groundnut was from a hybridization of two diploid wild species *Arachis duranensis* (AA) and *Arachis ipaensis* (BB). Eighty-three species of Arachis have been described and most of species in the genus *Arachis* are diploid with $x = 10$ ($2n = 20$), while a few aneuploid ($2n = 2x = 18$) and tetraploid ($2n = 4x = 40$) species are also reported, and new taxa continue to be discovered. Higher yield has been the most frequently targeted trait in the breeding programmes, and the enhanced yield attained in the Indian cultivars has been attributed to the improvement in seed size, seed weight, and number of pods per plant. The other trait which have been in focus is tolerance/resistance to diseases and drought, which have been the major production constraints. Improving quality of the produce also is now being taken up in addition to enhancement of yield. Early leaf spot (ELS; caused by *Cercospora arachidicola*), late leaf spot (LLS; caused by *Phaeoisariopsis personata*) and rust (caused by *Puccinia arachidic*) are the most prevalent foliar fungal diseases in groundnut. Among the viral diseases reported in groundnut from India, peanut bud necrosis virus (PBNV), tobacco streak virus (TSV), peanut mottle virus (PeMoV), and Indian peanut clump virus (IPCV) are the economically important. Among nematodes, peanut root-knot nematode (*Meloidogyne* spp.) and the root-lesion nematode (*Pratylenchus brachyrus*) are prominent. The breeding programmes across the groundnut growing countries mainly focus on these.

Due to the tetraploid nature of the crop and the complexity of the genome, marker development in cultivated groundnut has been in a slow pace. A good amount of genomic resources for groundnut has been added during the last 15 years. Several mapping populations have been prepared and the markers were mapped into genetic maps using these populations and used to understand the genome architecture and to identify QTLs for the economically important traits. The majority of QTLs

T. Radhakrishnan (✉) · A. L. Rathnakumar · M. K. Mahatma · S. Chandramohan · S. Patel
ICAR-Directorate of Groundnut Research, Junagadh, Gujarat, India

 
P. M. Priyadarshan, S. M. Jain (eds.), *Cash Crops*,
https://doi.org/10.1007/978-3-030-74926-2_10

reported are for foliar fungal diseases followed by oil quality and stem rot resistance. Though several reports on marker associations are available, utilisation of marker-assisted breeding is very meagre.

The availability of the draft genomes have opened up the large-scale genome-wide discovery of makers and the development of an58K Axiom_Arachis SNP array By the advent of the rapid sequencing tools and assembly tools, a good number of transcripts were assembled and analysed in detail.

In India alone more than 200 varieties have been developed so far for commercial cultivation. Several programmes in genomics-assisted breeding are underway and may come out with improved cultivars which can be grown under different production constraints without compromising yield and quality.

**Keywords** Arachis · Peanut · Groundnut · Genomics-assisted breeding · Varietal improvement · Genetic resistance · QTLs · Genetic maps

## 1 Introduction

Groundnut (also known as peanut) is one of the major oilseeds and food crops cultivated globally in more than 120 countries throughout the tropics and sub-tropics in 28.5 million ha with a production and productivity of 49.9 million tons and 1611 kg per ha, respectively (FAOSTAT 2018). The semi-arid tropics accounts for about 95% of the total planting area of groundnut where day temperatures ranges between 35 and 38 °C. Production of groundnut is concentrated (90% of global groundnut area) in Asia and Africa in the semi-arid tropics (SAT) where small farmers grow the crop under rain-fed conditions with limited or no inputs. Major groundnut growing countries are India (17%), China (19%) and Nigeria (11%), and China (40%), India (15%), Nigeria (8%), USA (6%) and Indonesia (4%) are the major producers.

Groundnut seeds (kernels) contain 35.8–54.2% oil, 16.2–36.0% protein and 10.0–20.0% carbohydrate (Jambunathan et al. 1985). Groundnut kernels are used for extracting oil, consumed directly as raw, boiled or roasted and for culinary purposes. The kernels are also a good source of minerals like calcium, phosphorus and iron and vitamins like E, niacin, folacin, riboflavin and thiamine (The Peanut Institute 2006).

About 60% of the groundnut produced globally is crushed for extracting oil for edible and industrial uses, while 40% is consumed in food uses and others (such as seed for sowing the next season crop) (Birthal et al. 2010; Janila et al. 2013). Groundnut oil is a preferred cooking medium due to its high smoking point (Singh and Diwakar 1993). Groundnut oil is used as a cooking medium in India, China, Vietnam and Myanmar. The de-oiled cake from expeller cake retains about 7% oil which can be recovered by solvent extraction. The de-oiled cake is used in the animal feed industry, in preparing food for children and the aged, and as soil an amendment. In Europe and North and South America, about 75% of the production is used as food, while only 35% is used as food in Asia.

Peanut butter is the most popular among groundnut products in the USA, Canada and Australia. Groundnut kernels can be consumed raw (non-heated), boiled and roasted and also used to make confections and its flour to make baked products. The groundnut shells are used for making particle boards or used as fuel or filler in fertiliser and feed industry and as a substrate for producing pectinase (Dey et al. 2001). Haulm of groundnut is a nutritious fodder for livestock. Groundnut haulm contains carbohydrate (38–45%), minerals (9–17%), protein (8–15%) and lipids (1–3%) and has a digestibility of around 53% when fed to cattle. Being a legume crop, groundnut helps in improving soil health and fertility by leaving behind the nitrogen fixed by the root nodules and organic matter in the soil (Janila et al. 2013).

## 2 Nutritional Qualities

Groundnut being the only oilseed which can be directly consumed as a foodstuff, is a rich source of oil, protein and carbohydrates, thus has high energy value (567 kcal 100 $g^{-1}$). Direct consumption of groundnuts can provide several nutritional and functional components (Table 1) like tocopherol, niacin and folic acid; mineral components like Cu, Mn, K, Ca and P; dietary fibres, flavonoids, phytosterols like resveratrol, beta-sitosterol; and phenolic acids (Francisco and Resurreccion 2008). Approximately 80% of the total fatty acid composition of groundnut is of oleic acid, a monounsaturated fatty acid, and linoleic acid, a polyunsaturated acid (Treadwell et al. 1983). Groundnuts contain over 50% monounsaturated fats, which are easily digestible (Feldman 1999). The large seed size, high sucrose and protein contents and low glucose, RFOs and high oleic acid/linoleic acid (O/L) ratio are considered as important traits for confectionary groundnut.

### *Carbohydrates*

Carbohydrates in groundnut are present free and bound, to the extent of 16–22%, along with various other minor constituents. Presence of large amount of free sucrose mainly contributes to the sweetness. Sucrose content among 41 popular Indian groundnut cultivars was in the range of 2.6% (Kadiri-3) to 6.5% (ICGS 76) (Bishi et al. 2015). Total carbohydrates contain mono- and oligosaccharides, starch, hemicelluloses A and B, and cellulose or fibre (Tharanathan et al. 1979). Monosaccharides consist of myo-inositol, mannitol, glucose and fructose while oligosaccharides are sucrose, RFOs (raffinose, stachyose, and verbascose). Among these, sucrose (2.6–6.5%), stachyose (0.1–0.7%), myo-inositol (0.03–0.17%) and raffinose (0.01–0.12%) are major sugars in groundnut (Bishi et al. 2015). The carbohydrate composition of groundnut is variable depending on the genotype, seed maturity, climatic conditions and growth location (Chakraborty et al. 2016; Mahatma et al. 2020). The therapeutic potential of myo-inositol in

**Table 1** Nutritional and bioactive compounds in groundnut (*Arachis hypogaea*) with recommended daily allowance (RDA), all types nutritional value per 100 g

| Principle | Nutrient value | Percentage of RDA |
|---|---|---|
| Energy | 567 kcal | 29 |
| Carbohydrates | 16.13 g | 12 |
| Protein | 25.80 g | 46 |
| Total fat | 49.24 g | 165 |
| Cholesterol | 0 mg | 0 |
| Dietary fibre | 8.5 g | 22 |
| Minerals | 2.3 | – |
| Water | 6.5 | – |
| *Vitamins* | | |
| Folates | 240 μg | 60 |
| Niacin | 12.066 mg | 75 |
| Pantothenic acid | 1.767 mg | 35 |
| Pyridoxine | 0.348 mg | 27 |
| Riboflavin | 0.135 mg | 10 |
| Thiamin | 0.640 mg | 53% |
| Vitamin A | 0 IU | 0 |
| Vitamin C | 0 | 0 |
| Vitamin E | 8.33 mg | 55.5 |
| *Electrolytes* | | |
| Sodium | 18 mg | 1 |
| Potassium | 705 mg | 15 |
| *Minerals* | | |
| Calcium | 92 mg | 9 |
| Copper | 1.144 mg | 127 |
| Iron | 4.58 mg | 57 |
| Magnesium | 168 mg | 42 |
| Manganese | 1.934 mg | 84 |
| Phosphorus | 76 mg | 54 |
| Selenium | 7.2 μg | 13 |
| Zinc | 3.27 mg | 30 |

Source: USDA National Nutrient database

treating diabetic neuropathy has been highlighted by researchers (Salway et al. 1978; Clements and Darnell 1980; Croze and Soulage 2013). Addition of 50 g of groundnut in the diet can provide 65 mg of myo-inositol (Mahatma et al. 2016). The sucrose content in Virginia genotypes is 3.9–6.9% while the Spanish genotypes have 2.8–5.3%. Spanish genotypes have higher myo-inositol content (0.14%) as compared to Virginia genotypes (0.13%), and the RFOs are higher in Virginia genotypes (32.8%). The flatulence resulting on consuming groundnut has been attributed to the RFOs (Bryant et al. 2004). The crude fibre content of groundnut may range from 2 to 4%.

## *Oil and Fatty Acids*

Oil content in groundnut genotypes ranged from 44 to 54% (Table 2) in different cultivars/genotypes (Bishi et al. 2015). Spanish genotypes possess slightly higher oil content (48.6%) compared to Virginia (47.2%) genotypes (Mahatma et al. 2016). Groundnut oil is light yellow in colour with distinctive nutty taste and odour and low in viscosity. Small quantities of mucilage, albuminous materials and free fatty acids are also present in oil, but refined oil is free from these materials. The oil is non-drying in nature that solidifies from 0 to 3 °C (Padley et al. 1994; Young 1996). The unsaponifiable fraction of groundnut oil ranges from 0.7 to 1.6%. It is a mixture of group of compounds like hydrocarbons, tocopherols, sterol ester, free sterols, glycolipids, etc. Phytosterols are the main component of unsaponifiables. Hydrocarbons are responsible for characteristic colour and flavour of groundnut oil despite their very small quantities (1.8 g/ton). The percent of free fatty acids in groundnut oil varies between 0.02% and 0.6% (Guthrie et al. 1949) which plays a very important role in shelf life, nutritional value and the organoleptic properties (Lea 1962). The well-balanced fatty acid and antioxidant profile of groundnut oil protects from harmful substances especially free radicals. Inclusion of groundnut oil which has high MUFA and PUFA can reduce the risk of cardiovascular disease by lowering the low-density lipoprotein cholesterol and serum triglycerides and by maintaining high-density lipoprotein cholesterol levels (Kris-Etherton et al. 1999).

**Table 2** Variation in oil content (%) and major fatty acid composition (%) of some groundnut genotypes

| Genotype | Oil | Palmitic acid | Stearic acid | Oleic acid | Linoleic acid |
|---|---|---|---|---|---|
| AK-265 | 53.8 | 12.1 | 2.7 | 51.2 | 31.4 |
| BAU-13 | 51.6 | 11.4 | 2.4 | 50.7 | 33.9 |
| Gangapuri | 50.4 | 13.8 | 2.3 | 40.5 | 40.7 |
| GG-20 | 53.3 | 10.4 | 3.0 | 61.6 | 22.4 |
| GJG-22 | 54.0 | 9.2 | 3.3 | 65.4 | 17.4 |
| Girnar-2 | 47.7 | 13.2 | 3.4 | 44.9 | 34.6 |
| Girnar-3 | 47.5 | 13.9 | 2.1 | 51.2 | 29.3 |
| ICGV-91114 | 45.6 | 13.4 | 4.0 | 39.6 | 37.5 |
| JL-501 | 51.0 | 12.4 | 2.8 | 41.7 | 39.4 |
| Kadiri-3 | 51.4 | 14.1 | 1.9 | 45.5 | 37.3 |
| NRCGCS-587 | 55.0 | 7.4 | 2.4 | 80.0 | 4.0 |
| M-13 | 47.6 | 12.9 | 2.0 | 53.9 | 28.9 |
| Pratap mungfali-2 | 50.0 | 3.4 | 3.4 | 44.5 | 35.1 |
| SunOleic95R | 50.0 | 6.9 | 2.3 | 80.1 | 7.3 |
| TAG-24 | 50.9 | 15.9 | 3.3 | 43.0 | 35.5 |
| TG-37A | 50.4 | 15.6 | 2.5 | 37.5 | 40.2 |
| TKG-19A | 50.5 | 13.9 | 2.4 | 52.8 | 29.0 |
| TMV-10 | 51.7 | 10.7 | 3.1 | 58.2 | 26.0 |

Source: Bishi et al. (2015), Nawade et al. (2016), Bera et al. (2019)

Its major component fatty acids are oleic acid (C18:1), linoleic acid (C18:2) and palmitic acid (C16:0) in the range of 40–64%, 20–40% and 7–15%, respectively (Bishi et al. 2015; Nawade et al. 2016). The oil also contains other fatty acids (about 10%) including stearic acid (C18:0), arachidic acid (C20:0), behenic acid (C22:0), lignoceric acid (C24:0), and eicosenoic acid (C29:1) with trace amount of linolenic acid (C18:3) (Casini et al. 2003). Though the fatty acid composition in groundnut is genetically controlled (Table 2), maturity of the kernels and the growth conditions can affect the fatty acid profile (Young 1996). Generally, palmitic, oleic and linoleic acid (the three major fatty acids) are more affected than the minor fatty acids. The fatty acid composition of high-oleic varieties of groundnut have a somewhat similar to olive oil; oleic acid 80% (±2%), linoleic acid 3–4% and palmitic acid 6–9% (Bera et al. 2019).

## Protein and Amino Acids

Groundnut is a rich source of good-quality protein which is higher than that of any other nuts. Protein content varies with the genotypes ranging from 18.60 to 33.25%, while most of genotypes have 25–30% protein (Gupta et al. 1982; Misra et al. 2000; Mora-Escobedo et al. 2015; Kumar et al. 2019). Some Indian groundnut cultivars, namely JL-24, TG-5, KADIRI-3, GAG10, S-230, KADIRI-1 and J-11, have more than 30% protein content (Misra 2004). Mahatma et al. (2016) analysed 30 genotypes each from Virginia and Spanish group and reported that Virginia genotypes had 31.6% average protein content with a range of 22.6–34.8%, while Spanish genotypes possessed 26.7–34.1% with an average of 30.3% protein. After oil extraction, the protein content in the cake can reach up to 50–55% (Zhao et al. 2011). Roasting and storage of groundnut significantly decrease methionine, tryptophan and in vitro protein digestibility (IVPD) and increase the soluble proteins (Damame et al. 1990). Protein content increases in kernels under water-deficit conditions (Dwivedi et al. 1996; Chakraborty et al. 2016). Groundnut proteins are nutritionally equivalent to meat and eggs for human growth and health and rich in arginine and has the Protein Digestibility Corrected Amino Acid Score (PDCAAS) of 0.70 of 1 (Suárez López et al. 2006) while it is 0.46 for whole wheat (Schaafsma, 2000). The nutritional quality of a protein is attributed to its digestibility and amino acid composition, and hence, groundnut can be used for protein fortification (Yu et al. 2006).

## Minerals

Groundnuts are good source of different minerals that supplement human dietary requirements and the total mineral content ranges from 2 to 3.5% (Table 1). Consumption of 100 g of groundnut can provide RDA levels of 127% copper, 84% manganese, 57% iron, 54% phosphorus, 42% magnesium and 30% zinc (Arya et al. 2016).

## *Bioactive or Functional Compounds*

### Sterols

Depending on the groundnut variety, roasted groundnuts contain 61–114 mg phytosterols per 100 g. Phytosterols contain β-sitosterol (78–83%), campesterol (6–14%) and stigmasterol (3–11%). Remaining 10–25% is covered by hydrocarbons, tocopherols and others. Hundred grams of unrefined groundnut oil contains up to 434 mg phytosterols while refining and deodorisation of the oil reduce the phytosterol content (Young 1996).

### Tocopherols and Other Vitamins

Groundnut oil contains tocopherols (vitamin E) like alpha-, beta-, gamma- and delta-tocopherol (α-T, β-T, γ-T and δ-T) (Hashim et al. 1993; Carrin and Carelli 2010). Tocopherols prevent or delay oxidation of oils by interrupting the formation of free radicals, act in the same manner as synthetic antioxidants. Tocopherol content differ in Spanish and Virginia varieties of groundnut (Hashim et al. 1993). Sturm et al. (1966) reported that Virginia runner varieties are the richest source for α-T, γ-T and δ-T, while Spanish varieties had the lowest α-T. Refining deodorisation or steaming processes can remove tocopherols and tocotrienols up to 60% tocopherols (O'Brien 2004). Refined groundnut oil contains tocopherols of 53 mg/100 g and mostly as α- and γ-tocopherol. Consumption of 100 g groundnut provides about 75% RDA of niacin, 60% RDA of folate and 55% RDA of vitamin E.

### Phenolic Compounds and Antioxidant Activity

Groundnut is also a good source of flavonoids like as green and black tea, apples, red wine and soybeans. Most of the Valencia-type groundnut varieties have more phenolic content than the other types (Chukwumah et al. 2009). The total phenolic contents in blanched and unblanched kernels varied from 0.07 to 0.12 mg and 2.47 to 84.53 gallic acid equivalent (GAE)/g dry weight, respectively, in 15 Valencia-type groundnut genotypes (Senawong et al. 2012). Total phenolics ranged from 3.87 mg/g kernel (JL-24) to 1.38 mg/g (GG-7) with an average content of 2.34 mg/g kernels while antioxidant capacity ranged from 1.04 to 6.64 μmol in 41 Indian groundnut cultivars (Bishi et al. 2015). Cinnamic acid, catechol and kaempferol are predominant phenolics in groundnut kernels followed by salicylic acid, syringic acid and epicatechin. Wide variation (4–25 μg/g) was observed for cinnamic acid content in groundnut cultivars, while minimum variation detected in catechol (14–16 μg/g) and kaempferol (13–14 μg/g) content. Resveratrol content present in the range of 0.42–7.13 μg/g kernel in Indian groundnut cultivars (Mahatma et al. 2018). The resveratrol content varies widely with varieties, in products, and by processing. Resveratrol content varies from 0.01 to 1.79 mg/g (Sobolev and Cole 1999) up to 2.3–7.1 mg/g in different genotypes of groundnut (Wang et al. 2005; Mahatma et al. 2018). Higher resveratrol content in 9-day sprouted groundnut is 11.7 to 25.7 μg/g (Wang et al. 2005).

## *Anti-nutritional Compounds*

### Allergens

Groundnut seeds contain allergens which can induce the production of specific IgE antibodies in susceptible individuals. At present, 16 proteins of groundnut are officially recognised as allergens (Palladino and Breiteneder 2018). The pattern of the allergic response varies in populations from different geographical regions (Vereda et al. 2011). The allergen Ara h 1, Ara h 2 and Ara h 3 are the major elicitors of allergic reactions in USA and most of the European countries. However, Spanish patients recognised these groundnut allergens less frequently and were more often sensitised to the lipid transfer protein Ara h 9 while Swedish patients had the highest sensitisation rate to Ara h 8 (Ballmer-Weber et al. 2015). These allergen proteins perform various biological functions in plant. Ara h 1 and Ara h 3 are globulin seed storage proteins. Ara h 1 which resembles a defensin isolated from radish seeds (Terras et al. 1995) had trypsin inhibitory activity and thus might play a role in plant defense against insects (Viquez et al. 2003). The allergens Ara h 7, Ara h 6 and Ara h 2 are 2S albumin seed storage proteins, whereas Ara h 17, Ara h 16 and Ara h 9 are non-specific lipid transfer proteins. Those are involved in stabilisation of membranes, cell wall organisation, signal transduction and plant growth and development as well as in resistance to biotic and abiotic stress (Liu et al. 2015). Allergen protein Ara h 10, Ara h 11, Ara h 14 and Ara h 15 are the groundnut oleosins. Ara h 5 plays an important role in recognition processes during pollination. Ara h 8 binds the isoflavones quercetin and apigenin as well as resveratrol with high avidity (Hurlburt et al. 2013).

### Phytic Acid

Being a legume, groundnut contains phytic acid which is associated with decreasing bioavailability of other nutrients due to their binding properties. Groundnut seeds contain 0.17–4.47% phytic acid (Venkatachalam and Sathe 2006; Schlemmer et al. 2009). As compared to other legumes like soybean, the quantity of phytic acid in groundnut is smaller (Schlemmer et al. 2009). The protein digestibility will be reduced by the competition of phytate for mineral cofactors needed by peptidases to be active and in direct interaction with the protein.

### Other Antinutrients

The bitter flavour and taste of groundnut germ (embryos) are imparted by saponins (Arya et al. 2016). Saponins are a group of triterpenoid or steroid glycosides, and only some of them are suggested to reduce the digestion of proteins and bioavailability of minerals (Francis et al. 2002). Saponins are about 20 times higher in the germ of groundnut than in the cotyledons (Misra 2004). Kuang et al. (2017) reported that raw whole groundnuts of Chinese cultivars had a higher saponin content (4.7–10.7 mg/g) compared to groundnut cotyledon (2.6–4.6 mg/g) and the boiled whole groundnuts (1.8–4.2 mg/g). Groundnut seed coat contributed to the highest saponin content (108–381 mg/g). While, very high content of saponin (55.2–82.3 mg/g) in raw seeds was reported in five groundnut cultivars grown in Brazil (Lozano et al. 2019). Saponins are remarkably stable to heat processing, and their biological activity is not reduced by normal cooking (Savage 2003). The saponins contents ranged from 438 ± 2.12 to 480 ± 2.30 mg/100 g in groundnut varieties grown in arid zone of Pakistan (Shad et al. 2009). Saponins have both beneficial and adverse effects on human health.

## 3 Systematics, Centres of Origin and Diversity

Groundnut is believed to be originated in Southern Bolivia and Northwestern Argentina on the eastern slopes of the Andes, based on the occurrence of the two progenitor species *Arachis duranensis* and *Arachis ipaensis*, and the morphological variability of the landraces, as well as archaeological evidence gathered in this region (Hammons 1982; Stalker and Simpson 1995; Simpson et al. 2001). So far, 83 species have been named under the genus *Arachis* (Table 3) (Krapovickas and Gregory 1994; Valls and Simpson 2005; Valls et al. 2013; Stalker et al. 2013; Santana and Valls 2015) that include diploids and tetraploids, of which *A. hypogaea* L. is cultivated species. The cultivated groundnut *Arachis hypogea* is a segmental amphidiploid ($4x = 40$). Different species of the genus *Arachis* have originated in South America and are found in tropical and subtropical areas. Possibly the genus *Arachis* originated in the highlands in the Southwestern Mato Grosso do Sul region of Brazil near Gran Pantanal where the most ancient species like *Arachis guaranitica* Chodat. and Hassl. and *Arachis tuberosa* Bong. Ex Benth. were reported (Gregory et al. 1980; Simpson and Faries 2001). Later on the genus might have spread to the other areas of South America (Gregory and Gregory 1979; Stalker and Simpson 1995; Simpson et al. 2001). New species of Arachis are now being discovered in areas which were inaccessible previously because of poor roads and transportation.

**Table 3** List of named species of the genus *Arachis*

| Species name | 2*n* chromosome number | Genome |
|---|---|---|
| **Section Arachis** | | |
| 1. *A. decora* Krapov., W.C. Gregory & Valls | 18 | – |
| 2. *A. palustris* Krapov., W.C. Gregory & Valls | 18 | – |
| 3. *A. praecox* Krapov., W.C. Gregory & Valls | 18 | – |
| 4. *A. cardenasii* Krapov. & W.C. Gregory | 20 | A |
| 5. *A. correntina* (Burkart) Krapov. & W.C. Gregory | 20 | A |
| 6. *A. diogoi* Hoehne | 20 | A |
| 7. *A. duranensis* Krapov. & W.C. Gregory | 20 | A |
| 8. *A. helodes* Martius ex Krapov. & Rigoni | 20 | A |
| 9. *A. herzogii* Krapov., W.C. Gregory and C.E. Simpson | 20 | A |
| 10. *A. hoehnei* Krapov. & W.C. Gregory | 20 | A |
| 11. *A. kempff-mercadoi* Krapov., W.C. Gregory & C.E. Simpson | 20 | A |
| 12. *A. kuhlmannii* Krapov. & W.C. Gregory | 20 | A |
| 13. *A. linearifolia* Valls, Krapov, & C.E. Simpson | 20 | A |
| 14. *A. microsperma* Krapov., W.C. Gregory & Valls | 20 | A |
| 15. *A. schininii* Valls & C.E. Simpson | 20 | A |
| 16. *A. simpsonii* Krapov. & W.C. Gregory | 20 | A |
| 17. *A. stenosperma* Krapov. & W.C. Gregory | 20 | A |
| 18. *A. villosa* Benth | 20 | A |
| 19. *A. gregoryi* C.E. Simpson, Krapov, & Valls | 20 | B |
| 20. *A. ipaensis* Krapov., W.C. Gregory | 20 | B |
| 21. *A. magna* Krapov., W.C. Gregory & C.E. Simpson | 20 | B |
| 22. *A. valida* Krapov. & W.C. Gregory | 20 | B |
| 23. *A. williamsii* Krapov. & W.C. Gregory | 20 | B |
| 24. *A. hypogaea* L. | 40 | AB |
| 25. *A. monticola* Krapov. & Rigoni | 40 | AB |
| 26. *A. glandulifera* Stalker | 20 | D |
| 27. *A. benensis* Krapov., W.C. Gregory & C.E. Simpson | 20 | F |
| 28. *A. trinitensis* Krapov. & W.C. Gregory | 20 | F |
| 29. *A. batizocoi* Krapov. & W.C. Gregory | 20 | K |
| 30. *A. cruziana* Krapov., W.C. Gregory & C.E. Simpson | 20 | K |
| 31. *A. krapovickasii* C.E. Simpson, D.E. Williams, Valls & I.G. Vargas | 20 | K |
| **Section Caulorrhizae** | | |
| 32. *A. pintoi* Krapov. & W.C. Gregory | 20 | C |
| 33. *A. repens* Handro | 20 | C |
| **Section Erectoides** | | |
| 34. *A. archeri* Krapov. & W.C. Gregory | 20 | E |
| 35. *A. benthamii* Handro | 20 | E |
| 36. *A. brevipetiolata* Krapov. & W.C. Gregory | 20 | E |

(continued)

**Table 3** (continued)

| Species name | 2*n* chromosome number | Genome |
|---|---|---|
| 37. *A. cryptopotamica* Krapov. & W.C. Gregory | 20 | E |
| 38. *A. douradiana* Krapov. & W.C. Gregory | 20 | E |
| 39. *A. gracilis* Krapov. & W.C. Gregory | 20 | E |
| 40. *A. hatschbachii* Krapov. & W.C. Gregory | 20 | E |
| 41. *A. hermannii* Krapov. & W.C. Gregory | 20 | E |
| 42. *A. major* Krapov. & W.C. Gregory | 20 | E |
| 43. *A. martii* Handro | 20 | E |
| 44. *A. oteroi* Krapov. & W.C. Gregory | 20 | E |
| *A. paraguariensis* | | |
| 45. ssp. *paraguariensis* Chodat & Hassl. | 20 | E |
| 46. ssp. *capibarensis* Krapov. & W.C. Gregory | 20 | E |
| 47. *A. porphyrocalyx* Valls & C.E. Simpson | 18 | E |
| 48. *A. stenophylla* Krapov. & W.C. Gregory | 20 | E |
| 49. **Section Extranervosae** | | |
| 50. *A. burchellii* Krapov. & W.C. Gregory | 20 | EX |
| 51. *A. lutescens* Krapov. & Rigoni | 20 | EX |
| 52. *A. macedoi* Krapov. & W.C. Gregory | 20 | EX |
| 53. *A. marginata* Gardner | 20 | EX |
| 54. *A. pietrarellii* Krapov. & W.C. Gregory | 20 | EX |
| 55. *A. prostrata* Benth. | 20 | EX |
| 56. *A. retusa* Krapov., W.C. Gregory & Valls | 20 | EX |
| 57. *A. setinervosa* Krapov. & W.C. Gregory | 20 | EX |
| 58. *A. submarginata* Valls, Krapov. & C.E. Simpson | 20 | EX |
| 59. *A. villosulicarpa* Hoehne | 20 | EX |
| **Section Heteranthae** | | |
| 60. *A. dardani* Krapov. & W.C. Gregory | 20 | H |
| 61. *A. giacomettii* Krapov., W.C. Gregory, Valls & C.E. Simpson | 20 | H |
| 62. *A. interrupta* Valls & C.E. Simpson | 20 | H |
| 63. *A. pusilla* Benth. | 20 | H |
| 64. *A. seridoensis* Valls, C.E. Simpson, Krapov, & R. Veiga | 20 | H |
| 65. *A. sylvestris* (A. Chev.) A. Chev. | 20 | H |
| **Section Procumbentes** | | |
| 66. *A. appressipila* Krapov. & W.C. Gregory | 20 | PR |
| 67. *A. chiquitana* Krapov., W.C. Gregory & C.E. Simpson | 20 | PR |
| 68. *A. hassleri* Valls & C.E. Simpson | 20 | PR |
| 69. *A. kretschmeri* Krapov. & W.C. Gregory | 20 | PR |
| 70. *A. lignosa* (Chodat and Hassl.) Krapov. & W.C. Gregory | 20 | PR |
| 71. *A. matiensis* Krapov., W.C. Gregory & C.E. Simpson | 20 | PR |
| 72. *A. pflugeae* C.E. Simpson, Krapov, & Valls | 20 | PR |
| 73. *A. rigonii* Krapov. & W.C. Gregory | 20 | PR |

(continued)

**Table 3** (continued)

| Species name | 2*n* chromosome number | Genome |
|---|---|---|
| 74. *A. subcoriacea* Krapov. & W.C. Gregory | 20 | PR |
| 75. *A. vallsii* Krapov. & W.C. Gregory | 20 | PR |
| **Section Rhizomatosae** | | |
| Ser. Prorhizomatosae | | |
| 76. *A. burkartii* Handro | 20 | R1 |
| Ser. Rhizomatosae | | |
| 77. *A. glabrata* Benth. | 40 | R2 |
| 78. var. *hagenbeckii* Benth. (Harms ex. Kuntze) F.J. Herm. | 40 | R2 |
| 79. *A. nitida* Valls, Krapov, & C.E. Simpson | 40 | R2 |
| 80. *A. pseudovillosa* (Chodat & Hassl.) Krapov. & W.C. Gregory | 40 | R2 |
| **Section Trierectoides** | | |
| 81. *A. guaranitica* Chodat & Hassl. | 20 | TE |
| 82. *A. tuberosa* Bong. ex Benth | 20 | TE |
| **Section Triseminatae** | | |
| 83. *A. triseminata* Krapov. & W.C. Gregory | 20 | T |

Based on Stalker et al. (2013)

# 4 Species Diversity

The first species to be described in the genus *Arachis* was, *Arachis hypogaea* by Linnaeus (1753), which is the cultivated groundnut. There was a gap of nearly 100 years between Linnaeus's naming the then lone species known to Europeans, and the first taxonomic treatment of the genus by Bentham (1841). The explorations by Krapovickas and Gregory in 1959 collected 2000 plant specimens right from those distributed in South America, east of the Andes, south of Amazon, north of La Plata and from NW Argentina to NE Brazil. The collections were then assigned to the different sections on the bases of cross-compatibility and exo-morphologic character clustering (Gregory et al. 1973; Gregory and Gregory 1979). The genus *Arachis* is placed in the legume family *Fabaceae*. The genus is well defined morphologically and clearly delineated from its closest relatives by the presence of geocarpic peg developed from the hypanthium of the flower. The genus *Arachis* is related to the genera *Stylosanthes*, *Chapmannia*, *Arthrocarpum* and *Pachecoa* in the sub-tribe *Stylosanthinae* of the tribe *Aeschynomeneae* on the basis of the shared morphological characters of a staminal tube with alternately attached basal and dorsal anthers, flowers in terminal or axillary spikes or small heads (which are sometimes raceme-like), pinnate leaves and leaflets without stipules. The taxonomic treatment of the cultivated groundnut is as follows:

| | |
|---|---|
| Kingdom | *Plantae* |
| Division | *Tracheophyta* |
| Class<br>Sub-class | *Magnoliopsida*<br>*Rosidae* |
| Order | *Fabales* |
| Family | *Fabaceae* |
| Sub-family | *Faboideae* |
| Tribe | *Aeschynomeneae* |
| Genus | *Arachis* |
| Species | *hypogaea* |

Based on geographic distribution, cross compatibility, cytogenetic behaviour and morphology, *Arachis* species have been grouped to nine taxonomic sections: Trierectoides, Erectoides, Procumbentes, Rhizomatosae, Heteranthae, Caulorrhizae, Extranervosae, Triseminatae and Arachis (Fernandez and Krapovickas 1994; Krapovickas and Gregory 1994; Lavia 1999; Valls and Simpson 2005). The section Arachis includes both annual and perennial species of different chromosome numbers, ploidy levels and karyotype. The members of the sections Erectoides and Procumbentes appear to be closely related to the species in the section Arachis. The sections Extranervosae, Heteranthae and Triseminatae are adjudged as most primitive, while section Arachis as the most advanced, and that sections Caulorrhizae, Erectoides, Procumbentes, Rhizomatosae and Trierectoides are intermediate (Wang et al. 2010; Bechara et al. 2010).

Cytogenetic and molecular evidences suggest that the origin of cultivated groundnut was from a hybridisation of two diploid wild species *Arachis duranensis* (AA) and *Arachis ipaensis* (BB) followed by chromosome duplication about 3500 years. The natural introgression of wild species to *A. hypogaea* has not been reported so far (Kochert et al. 1996).

Cultivated groundnut, *Arachis hypogaea* L. is divided into two subspecies subsp. fastigiata and subsp. hypogaea based on the growth habit and flower arrangement on the main axis. The fastigiata subsp. includes the four botanical varieties, while hypogaea has two varieties. Groundnut has three market types, viz. Virginia, Spanish and Valencia as illustrated in the Table 4.

## *Centres of Diversity*

Krapovickas (1969) recognised five centres of diversity while six centres were proposed by Gregory and Gregory (1976) based on the extent of diversity in cultivated groundnut germplasm, while explorations added the seventh centre, Ecuador with the group of landraces referred as var. *aequatoriana* (Singh and Nigam 1997) (Table 5).

**Table 4** Taxonomic sub-divisions of Arachis hypogaea

| Subspecies | Variety | Market type | Distinguishing characteristics |
|---|---|---|---|
| *Hypogaea* | | | No floral axes on main stem; alternating pairs of floral and reproductive axes on branches; prostrate; branches short, less hairy; medium to late maturing |
| | Hypogaea | Virginia runner | Prostrate, profuse branching; less hairy; large seeded |
| | Hirsuta | Peruvian runner | Limited branching; more hairy, small-seeded |
| *Fastigiata* | | | Floral axes on main stem; continuous run of multi-floral axes along lateral branches; upright growth habit |
| | Fastigiata | Valencia | A few long upright branches, hairy leaf; pods with 2 or 2–4 (rarely 5) seeds, beaked; reticulation prominent; red, yellow, purple or variegated testa |
| | Peruviana | | Less hairy; pods deeply reticulated |
| | aequatoriana | | Purple stems; more branched, erect; very hairy; pods deeply reticulated |
| | Vulgaris | Spanish bunch | More branched than Valencia; pods slightly reticulated; mostly 2 seeded pods; salmon, red, purple or white testa |

**Table 5** Centre of diversity and distribution of cultivated type

| Centre of diversity | Distribution of cultivated type |
|---|---|
| 1. Guarani region (Paraguay, Parana) | Erect Valencia type, ssp. *fastigiata* and intermediate type between var. *fastigiata* and *vulgaris* |
| 2. Goias and mina Gerais region (Tocatins, San Francisco) | Erect type, ssp. *fastigiata* var. *fastigiata* and races of both var. *fastigiata* and var. *vulgaris* |
| 3. Rondonia North-West Mato Grosso (Brazil) | *Nambyquarae* type of ssp. *hypogaea* and erect type with yellow testa colour |
| 4. Eastern foothills of the Andes and Bolivia | subspecies *hypogaea* var. *hypogaea* and few races of ssp. *fastigiata* and intermediate of these two ssp. |
| 5. Peru (Upper Amazon and West Coast) | Races of ssp. *hypogaea*, with constriction, veins and beak, similar type of ssp. *fastigiata* var. *fastigiata* (var. *peruviana*) and ssp. *hypogaea* var. *hirsute* |
| 6. North East Brazil | All morphological types along with intermediate types between var. *fastigiata* and var. *hypogaea* |

*ssp* subspecies; Source: Singh and Simpson (1994)

The first five centres are considered as secondary centre of diversity, while the North–East Brazil is as a tertiary centre of diversity. Africa is also regarded as an important tertiary centre of diversity with a wide variability in the cultivated species. In Africa, the genetic diversity has been contributed by the introduction from different countries especially from Bolivia, Brazil, China and India (Smartt 1990; Singh and Simpson 1994). The collection of wild and cultivated groundnut display a very wide range of morphological variability (Fig. 1).

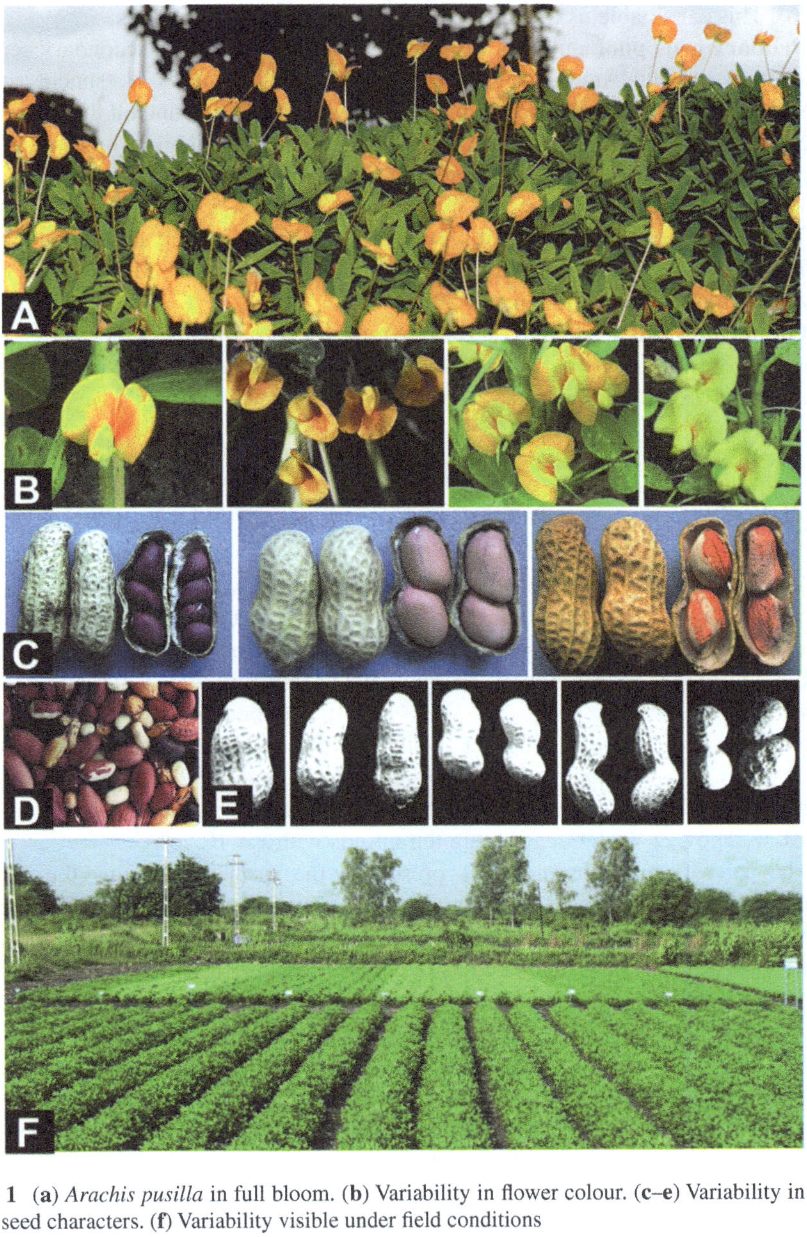


**Fig. 1** (**a**) *Arachis pusilla* in full bloom. (**b**) Variability in flower colour. (**c–e**) Variability in pod and seed characters. (**f**) Variability visible under field conditions

## *Gene Pools*

Based on cross-compatibility among the members of the genus *Arachis,* four gene pools were identified (Smartt 1990; Singh and Simpson 1994; Singh and Nigam

2016). This can enable us to develop appropriate breeding strategies.

**Primary gene pool**: cultivars of groundnut from primary and secondary centres of diversity in South America and other countries, and *A. monticola* a species which crosses freely with *A. hypogaea* found in northwest Argentina and all the landraces.

**Secondary gene pool:** diploid species from section Arachis, which are cross-compatible with *A. hypogaea*, despite ploidy differences resulting in sterile to partially fertile hybrids.

**Tertiary gene pool:** species of section *Procumbentes*, which are crossable with diploid species of section *Arachis* (Gregory and Gregory 1979; Mallikarjuna et al. 2005; Mallikarjuna and Hoisington 2009).

**Quaternary gene pool**: rest of the species of *Arachis* which are cross-incompatible or very weakly cross-compatible with the species of the section *Arachis* and are grouped into five other sections.

## *Karyology*

Though most of species in the genus *Arachis* are diploid with $x = 10$ ($2n = 20$), a few aneuploid ($2n = 2x = 18$) and tetraploid ($2n = 4x = 40$) species are also reported; four species (three belong to the section *Arachis*: *Arachis praecox* (Krapov. W.C. Greg and Valls 1994), *Arachis palustris* (Krapov. W.C. Greg. and Valls 1994) and *Arachis decora* (Krapov. W.C. Greg and Valls 1994); and one belongs to the section *Erectoides*: *Arachis porphyrocalyx* (Valls and Simpson 2005) are diploids with $x = 9$ ($2n = 18$) and five species (*A. hypogaea* L., *A. monticola* Krapov. & Rigoni; *A. glabrata*, *A. nitida* Valls, Krapov, & C.E. Simpson, *A. pseudovillosa* Chodat & Hassl., Krapov. & W.C. Gregory) are tetraploids with $x = 10$.

The pair of small chromosomes present in the species belonging the section Arachis has been designated as 'A' chromosome while the chromosome pair with the large secondary constriction present in the species *A. batizocoi* Krapov. & W. C. Gregory has been designated as 'B' chromosome (Husted 1936). The hybrids between 'A' and 'B' chromosomes containing species generally are sterile, whereas the ones between species 'A' chromosome containing species are fertile (Stalker and Simpson 1995). This has been made as a basis to the designate 'A' and 'B' genomes used in groundnut to describe the two cytological groups. Since *A. hypogea* and *A. monticola* which are considered as allotetraploids with AABB genomes have a pair of small chromosomes and another pair of chromosomes with a large secondary constriction. Stalker (1991) described *A. glandulifera* Stalker as 'D' genome species while the B genome was divided into B, F and K genomes by Seijo et al. (2004) and Robledo and Seijo (2010). As compared to A genome species the B genome species are karyotypically more diverse (Fernandez and Krapovickas 1994; Seijo et al. 2004). On the basis of cross-compatibility data, it was inferred that the evolution of the genomic groups in the genus mainly followed sectional designations (*Ambinervosae* [Am], *Triseminatae* [T], *Caulorrhizae* [C], *Extranervosae* [EX], *Erectoides* [E], *Rhizomatosae* [R], and *Arachis* [A, B and D]) (Smartt and Stalker 1982; Stalker

1991). The 30 most closely related wild species of section *Arachis* have been further sub-divided into five groups, named A, B, K, F, K (with metacentric chromosomes of similar size) and D (with an asymmetric karyotype), with the majority being A or B. All these sub-genomic classification clearly states that adequate genetic differentiation has taken place in the diploid wild species unlike the cultivated groundnut.

Based on the chromosome numbers observed in different wild species, it is apparent that the polyploidy evolved independently in sections *Arachis* and *Rhizomatosae* (Smartt and Stalker 1982). Nelson et al. (2006) concluded that ploidy evolved multiple times within section *Rhizomatosae*. Tallury et al. (2005) reported molecular evidences that the diploid species of section *Rhizomatosae* (only one known viz. *A. burkartii* Handro of *Sec. Prorhizomatosae*) did not give rise to the tetraploids. Since *A. glabrata* hybridises with the species of the sections *Erectoides* and *Arachis*, the hybridisation of the two diploids from sections *Erectoides* and *Arachis* followed by spontaneous doubling in chromosome number might have given rise to the tetraploid species of section *Rhizomatosae* (Smartt and Stalker 1982).

### *Molecular Variability*

Based on the molecular studies along with the morphological and crossability studies with species in section *Arachis,* the diversity in *A. hypogea* has been worked out. Using storage proteins (Singh et al. 1991; Bianchi-Hall et al. 1993; Liang et al. 2006), isozymes (Lu and Pickersgill 1993; Stalker et al. 1994), restriction fragment length polymorphisms (RFLPs) (Kochert et al. 1991; Paik-Ro et al. 1992), amplified fragment length polymorphisms (AFLPs) (Milla et al. 2005), simple sequence repeats (SSRs) (Hopkins et al. 1999; He et al. 2005; Hong et al. 2010; Guo et al. 2012; Nagy et al. 2012; Moretzsohn et al. 2013), randomly amplified polymorphic DNA (RAPD) (Halward et al. 1992; Lanham et al. 1992; Hilu and Stalker 1995) and in situ hybridisation (Raina and Mukai 1999; Seijo et al. 2004, 2007) have concluded that *A. hypogea* has less molecular variation than diploid species, supporting the hypothesis of the origin of *A. hypogaea* from a single hybridisation event. Apparently there has been little or no introgression from other diploid species (Kochert et al. 1996), probably leading to the narrow genetic base reported in the cultivar improvement programmes.

## 5 Ex Situ and In Situ Collections of Germplasm

The most preferred method of conservation to prevent the extension of plants in nature is the storage of seeds and plant materials collected in gene banks (Benz 2012). The ex situ conservation has got a major disadvantage that genetic variation is thought to be eroding as modern cultivars replace traditional ones and the natural habitats of wild species are being destroyed (Upadhyaya 2008). On the other hand, in situ preservation has certain advantages such as adaptation to local conditions, natural gene flow and preservation of genetic diversity though the natural habitat of many plant species is under threat or already has been demolished. Agriculture systems are also under challenge by biotic and abiotic stresses, increased demand for a particular product, culinary preferences, as the human population increases, the breeding and selection via resorting to the genes found in landraces, breeder stocks and wild relatives are also being practiced (Benz 2012). Pandey et al. (2012) has consolidated the data on the availability of groundnut accessions available in the gene banks across the world. The largest collection of groundnut germplasm is being held at ICRISAT, India (15,445 accessions from 93 countries), followed by the ICAR-National Bureau of Plant Genetic Resources (ICAR-NBPGR) (14,585 accessions), the ICAR-Directorate of Groundnut Research (ICAR-DGR) (9024 accessions) in India; the U.S. Department of Agriculture (USDA) (9917 accessions); the Oil Crops Research Institute (OCRI) of the Chinese Academy of Agricultural Sciences (CAAS) (8083 accessions) and the Crops Research Institute of the Guangdong Academy of Agricultural Sciences in China (4210 accessions). Small to medium collections are held at the Texas A&M University (TAMU) and the North Carolina State University (NCSU) in the USA; the EMBRAPA-CENARGEN and the Instituto Agronomico de Campinas in Brazil; and the Instituto Nacional de Technologia Agropecuaria (INTA) and the Instituto de Botánica del Nordeste (IBONE) in Argentina also (Table 6).

**Table 6** Major germplasm collections of groundnut

| Institution | Cultivated species | Wild species | Total accessions |
|---|---|---|---|
| International Crops Research Institute for the Semi-Arid Tropics (ICRISAT), India | 14,968 | 477 | 15,445 |
| ICAR-National Bureau of Plant Genetic Resources (ICAR-NBPGR), India | – | – | 14,585 |
| Plant Genetic Resource Conservation Unit (PGRCU), Griffin, U. S. Department of Agriculture (USDA), USA | 9310 | 607 | 9917 |
| ICAR-Directorate of Groundnut Research (ICAR-DGR), India | 8960 | 64 | 9024 |
| Oil Crops Research Institute (OCRI), Chinese Academy of Agricultural Sciences (CAAS), China | 7837 | 246 | 8083 |
| Crops Research Institute (CRI), Guangdong Academy of Agricultural Sciences (GAAS), China | 4210 | – | 4210 |
| Instituto Nacional de Technologia Agropecuaria (INTA), Argentina | 3534 | 106 | 3640 |
| EMBRAPA and CENARGEN, Brazil | 1200 | 1220 | 2420 |
| Instituto Agronomico de Campinas, Brazil | 2140 | | 2140 |
| Texas A&M University (TAMU), USA | | 1200 | 1200 |
| North Carolina State University (NCSU), USA | 740 | 406 | 1146 |
| Instituto de Botánica del Nordeste (IBONE), Argentina | – | 472 | 472 |

Source: Pandey et al. (2012)

The gene banks are also repositories of wild species which are the source of several novel alleles. Though only a limited number of wild species were available in the gene banks prior to 1980s, the expeditions made to Bolivia, Brazil, Paraguay and northern Argentina beginning in the late 1970s to 1990s initiated by W.C. Gregory and A. Krapovickas and then by C.E. Simpson and J.F.M. Valls have added a good number of accessions (Stalker 2017). Still undiscovered *Arachis* species are believed to survive in some of the more remote unexplored areas of western Brazil, south-eastern Bolivia and north-western Paraguay (Williams 2001; Singh and Simpson 1994; Singh and Nigam 2016).

Eighty-three species of Arachis have been described, and new taxa continue to be discovered. Over 3400 accessions of *Arachis* species have been documented while ~1300 are available in gene banks as plants or seeds (Stalker et al. 2002).

The largest collection of *Arachis* species (1200 accessions) are at the Brazilian Agricultural Research Corporation (EMBRAPA), and other gene banks are Brazil Texas A&M University (1200 accessions); the USDA (607 accessions); the International Crops Research Institute for the Semiarid Tropics (ICRISAT) (477 accessions); Instituto de Botánica del Nordeste (IBONE) (472 accessions) in Argentina; and at North Carolina State University (440 accessions) (Singh and Simpson 1994; Pandey et al. 2012).

Explorations have continued till 2000 AD but on a much more limited scale by scientists across the globe including in Brazil, Argentina and the United States. Since the implementation of the Convention on Biological Diversity in 1993, the exchange of groundnut germplasm has been restricted to a great extent as laws restricting access to genetic resources have been widely implemented (Holbrook 2001; Williams 2001; Williams and Williams 2001). Because most of the diversity in the cultivated groundnut is present in South America and all wild species are found there, this resulted in significant negative effects on groundnut germplasm preservation, exchange and utilisation. By the implementation of the Andean Pact in 1996 signed by the five countries (Bolivia, Colombia, Ecuador, Peru, and Venezuela) implemented strict provisions to restrict germplasm access. Although Brazil is not an Andean Pact nation, the country implemented stringent regulations for collecting and exchanging germplasm for both the international and national Brazilian scientists (Williams and Williams 2001).

Approximately 50% of wild species accessions have <50 seeds in storage, and many are propagated in greenhouses or in field as vegetative cuttings (Stalker 2017). The lifespan of groundnut seeds in storage is limited. Under ambient conditions, groundnut farmers of India store groundnut in the form of pods in traditional methods to be used as seeds till the next year (approximately 7–8 months), without loss in the viability of seeds. In general, 10–15 years is the maximum amount of time to store groundnut seeds in cold storage after which the viability dramatically declines. This timeframe means that approximately 665–1000 lines need to be regenerated annually, as being practiced by the ICAR-DGR. After drying to 5–7% moisture, seeds are stored under controlled environmental conditions following the thumb rule that the sum of temperature (F) plus relative humidity (RH) should be less than 100 to have optimal seed storage. Groundnut seeds which are frequently requested (working collection) are stored at 4 °C and 25% RH. A material which is infrequently requested (base collection) is stored at −18 °C (Holbrook 2001; Nigam 2014), and this condition slightly varies from region to region.

## 6 Core and Mini-core Collections

Availability of a large number of cultivated accessions poses a question about selection of appropriate lines for their utility in crop breeding. The 'core collection' (i.e. 10% of the entire germplasm collection) presents a manageable and cost-effective entry point into germplasm collections for identifying candidate genotypes with new sources of disease and pest resistance or abiotic stress tolerance (Brown 1989). Therefore, a core collection comprising of 1704 *A. hypogaea* accessions was developed at ICRISAT, which is similar to the U.S. core collection (831 accessions) (Holbrook et al. 1993; Upadhyaya et al. 2003). Following the same procedure, the Chinese core collection comprising 576 accessions was developed (Jiang et al. 2008). However, the size of these core collections is too large for easy exploitation by breeders. Therefore, 'mini-core collections' (i.e. 10% of the core collections and 1% of entire germplasm collection) with 184, 112 and 298 accessions have been developed at ICRISAT (Upadhyaya et al. 2002), USDA/ARS (Holbrook and Dong 2005) and China (Jiang et al. 2010), respectively. The core and mini-core collections reported in groundnut are summarised in Table 7.

Apart from the above collections, a composite collection of 1000 accessions was also developed based on available phenotypic characterisation, geographic origin and taxonomic data. A 'reference set' of 300 accessions was also developed on the basis of genetic diversity analysis and the population structure (Upadhyaya et al. 2002, 2003). These accessions were evaluated for the economically important traits like tolerance to biotic and abiotic stresses (Anderson et al. 1996; Chamberlin et al. 2010; Holbrook et al. 2009; Wang et al. 2011; Upadhyaya et al. 2014; Waliyar et al. 2016).

In addition to germplasm collections that represent naturally occurring variation, induced mutant collections for tetraploid groundnut also are being developed. Amphiploids and autotetraploids (Mallikarjuna et al. 2011), chromosome segment

**Table 7** Core and mini core collections as reported in groundnut

| Collections used | Accessions used in mini core | Traits used in forming mini core | Accessions in constituted mini core | Reference |
|---|---|---|---|---|
| US core | 7432 | 6 | 831 | Holbrook et al. (1993) |
| Asian core | 4738 | 15 | 504 | Upadhyaya et al. (2001) |
| ICRISAT mini core | 1704 | 31 | 184 | Upadhyaya et al. (2002) |
| Global core | 14,310 | 14 | 1704 | Upadhyaya et al. (2003) |
| USA mini core | 831 | 16 | 111 | Holbrook and Dong (2005) |
| Chinese core | 6390 | 15 | 576 | Jiang et al. (2008) |
| US Valencia core | 630 | 26 | 77 | Dwivedi et al. (2008) |

substitution (CSSL) lines (Foncéka et al. 2012a, b) and multiparent advanced generation intercross (MAGIC) populations (Janila et al. 2016a) have been developed which can be valuable resources for groundnut breeding.

In groundnut, the homeolog pairs Ara h 1.01/1.02, Ara h 2.01/2.02 (encoding major allergen proteins), and FAD2A/B (encoding delta-12-desaturase, the enzyme primarily controlling oleic to linoleic acid ratio in seeds) were screened in the mutagenised population to discover gene knockouts or functional mutations in at least one of the two homeologs for each gene duplicate (Knoll et al. 2011). The TILLING population will be useful for functional genomics studies as well as to recover mutations of potential value for crop improvement.

## 7 Crop Improvement

The genetic variability available in the genetic resources of *A. hypogaea* is being exploited in the cultivar development programmes. However, the variability present in the other species of *Arachis* has been sparingly utilised due to the breeding barriers like ploidy differences and cross incompatibility. The common breeding methods applied in self-pollinated crops like mass selection, single seed descent, pedigree, bulk and back-cross methods are being used in groundnut also. Most of the early generation varieties were from introduction and mass selection, and crossing and selection in segregating generations are now used extensively despite the fact that emasculation procedure for hybridisation is difficult in groundnut. Multiple crossing systems, like double or convergent cross, are also being adopted to enhance genotypic variability (Wynne and Gregory 1981). In groundnut, pedigree method to select traits with high heritability and bulk-pedigree method to select traits with low heritability are most frequently used breeding approaches. Single-seed descend method is adopted where the resources and space are limited (Isleib et al. 1994). Back cross breeding is now being frequently used in conjunction with molecular markers for varietal development (Bera et al. 2018; Nawade et al. 2019). Mutation breeding using physical and chemical mutagens also have been done to some extent in China (Qui et al. 1998) and India, leading to the release of improved varieties (Table 8). In the cultivar development process, multi-environment testing is adopted for the identification of elite genotypes having spatial adaptation (Ceccarelli et al. 2007; Vindhiyavarman et al. 2010). Recently, emphasis have also been put on farmer participatory plant breeding to pick up genotypes specifically adapted. Probably this approach may be significant for countries where the adoption rate of the new varieties is very low.

### *Targeted Traits*

Higher yield has been the most frequently targeted trait in the breeding programmes worldwide (Nigam et al. 1991), but with limited genetic gains due to the large g × e interactions observed in these traits (Zhang et al. 2011). Generally, least g × e

**Table 8** Varieties developed through mutation breeding by BARC, India

| Variety | Year of release | Salient features |
|---|---|---|
| TG-47 (RARS T1) | 2011 | Large seed, 115 days maturity, more 3 seeded pods |
| TDG-39 | 2009 | Large seed, medium maturity |
| TBG-39 | 2008 | High oleic acid, more branches |
| TG-51 | 2008 | Early maturity, high shelling % |
| TLG-45 | 2007 | Large seed, medium maturity |
| TG-38 | 2006 | High shelling %, more 3 seeded pods, stem rot tolerance |
| TG-37A | 2004 | High yield, collar rot, drought tolerance |
| TPG-41 | 2004 | Large seed, medium maturity, high oleic acid |
| TG-26 | 1996 | Earliness, high harvest index, 20 days seed dormancy |
| TKG-19A | 1996 | Large seed, 30 days seed dormancy |
| TG-22 | 1992 | Medium large seed, 50 days seed dormancy |
| TAG-24 | 1992 | Earliness, high yield, high harvest index, partitioning % |
| Somnath (TGS-1) | 1991 | Large seed, semi runner type |
| TG-3 | 1987 | High yield |
| TG-17 | 1985 | No secondary branches, 30 days seed dormancy |
| TG-1 | 1973 | High yield, large seed, more branches, 50 days seed dormancy |

interactions are preferred in selecting high-yielding genotypes despite the fact that in a location-specific cultivar, the g × e interaction is desirable for yield maximisation. Pod yield is considered as a function of crop growth rate, duration of reproductive phase and the rate of partitioning of crop growth to pod yield (Janila et al. 2013). The enhanced yield attained in the Indian cultivars has been attributed to the improvement in seed size, seed weight and the number of pods per plant (Rathnakumar et al. 2012). The other trait which has been in focus is tolerance/resistance to diseases and drought. The major criteria used for selection in these cases is also minimal reduction in pod yield under stress conditions, such as diseases and drought. However, in countries like USA where the average productivity is high, the priorities of the processing industry have also been included in the priorities. The breeding programmes on improving quality of the produce is now being taken up in addition to enhancement of yield, by researchers invariably in all groundnut growing countries. In India and China, there are breeding efforts in progress to enhance the oleic acid content in oil and to develop genetic resistance to the aflatoxin-producing fungus *Aspergillus flavus*, which is a major health hazard and trade barrier. Short duration (<100 days) is an important trait in groundnut cultivars which helps in evading the stress conditions like drought and frost and to fit in multiple cropping systems. Further, fresh seed dormancy plays a very important role in escaping in vitro germination (vivipary) when rains coincide with the crop maturity stage leading to the loss in the quality of product. In India, 205 improved groundnut varieties have been released for cultivation so far through conventional breeding approaches (Table 9).

**Table 9** Groundnut varieties released in India during last 10 years for commercial cultivation

| S. no. | Name of cultivar | Year of release | Production system | Special characters |
|---|---|---|---|---|
| 1 | VRI (Gn) 7 | 2009 | *Kharif* | Tolerant to leaf miner, LLS and rust |
| 2 | VRI (Gn) 6 (VG 9816) | 2009 | *Kharif* | Tolerant to LLS, rust, PBND |
| 3 | Jawahar Groundnut 23 (JGN 23) | 2009 | *Kharif* | Tolerant to ELS & LLS, drought tolerant |
| 4 | Kadiri 9 | 2009 | *Kharif* | Tolerant to early and late season drought |
| 5 | Greeshma | 2009 | *Kharif and rabi*-summer | Tolerant to LLS tolerant to drought, high temperature and aflatoxin, early maturing |
| 6 | Kadiri 7 | 2009 | *Kharif* | Tolerant to sucking pest and leaf spots, large seeded |
| 7 | Kadiri 8 | 2009 | *Kharif* | Tolerant to sucking pest and leaf spots, large seeded |
| 8 | Mallika (ICHG 00440) | 2009 | *Kharif* | Resistant to collar rot and PBND, large seeded |
| 9 | TGLPS 3 (TDG-39) | 2009 | *Kharif* | Tolerant to *A. flavus* and LLS, high harvest index (HI), bold kernel with less aflatoxin and confectionary type |
| 10 | JL 501 | 2010 | *Kharif* | Comparatively lowest intensity of LLS was observed in case of JL-501 as against higher in variety JL-24; suitable for early as well as late sown conditions |
| 11 | Vijetha (R 2001-2) | 2010 | *Kharif* | Resistant to PBND |
| 12 | Girnar 3 (PBS 12160) | 2010 | *Kharif* | Tolerant of leaf miner and thrips |
| 13 | Kadiri Harithandhra (K 1319) | 2010 | *Rabi*-summer | Multiple diseases (rust, ELS, LLS, stem rot, PBND) and insect pests (thrips, *Spodoptera litura,* jassids, *Helicoverpa)* resistant, Possesses fresh seed dormancy (up to 20 days) |
| 14 | GPBD-5 | 2010 | *Kharif* | Resistant to LLS and rust |
| 15 | HNG 69 | 2010 | *Kharif* | Tolerant to collar rot, stem rot and ELS |
| 16 | GJG-HPS-1 (JSP-HPS-44) | 2010 | *Kharif* | Tolerant to PBND |
| 17 | RARS-T-1 | 2011 | Kharif and rabi-summer | Tolerant to leaf spots and rust, sucking pests (thrips and jassids), bold seeded pods |
| 18 | RARS-T-2 | 2011 | Kharif and rabi-summer | Moderate tolerance to *Spodoptera litura* and leaf hoppers, early maturity |
| 19 | Pratap Raj Mungphali | 2011 | *Kharif* and summer | Tolerant to ELS, LLS, PBND, jassids, thrips and leaf miner and *Spodoptera litura*, early maturity |
| 20 | Raj Durga (RG 425) | 2011 | Kharif | Moderately resistant to tikka, rust, PBND, collar rot and stem rot |
| 21 | Divya (CSMG 2003-19) | 2011 | *Kharif* | Tolerant to PBND |

(continued)

**Table 9** (continued)

| S. no. | Name of cultivar | Year of release | Production system | Special characters |
|---|---|---|---|---|
| 22 | ICGV 00350 | 2012 | Rabi-summer | Resistant to LLS and rust, tolerant to stem rot |
| 23 | HNG 123 | 2012 | *Kharif* | Tolerant to collar rot, stem rot, LLS, *Spodoptera litura* and leaf miner |
| 24 | Raj Mungfali-1 (RG 510) | 2012 | *Kharif* | Tolerant to collar rot, stem rot, LLS, peanut stem necrosis diseases |
| 25 | CO 6 | 2012 | Kharif | Resistant to LLS and rust |
| 26 | GJG 31 (J 71) | 2012 | *Summer* | Tolerant to stem rot |
| 27 | GJG 9 (J 69) | 2012 | *Kharif* | Tolerant to stem rot |
| 28 | GJG-22 (JSSP 36) | 2013 | *Kharif* | Tolerant to collar rot |
| 29 | GJG-17 (JSP-48) | 2013 | *Kharif* | Tolerant to stem rot |
| 30 | Dharani (TCGS 1043) | 2013 | *Kharif* and summer | Tolerant to stem rot, dry root rot, drought tolerant |
| 31 | GJG 18 (JSP 49) | 2015 | *Kharif* | Moderately resistant to PBND and PSND |
| 32 | Raj Mungfali-2 (RG 578) | 2015 | *Kharif* | Resistant to LLS, dry root rot, ELS and rust, tolerant to *S. litura*, thrips, jassids and leaf miner |
| 33 | Birsa Groundnut 4 (BAU 25) | 2015 | *Kharif* | Resistant to LLS, tolerant to occasional dry spell; large seeded |
| 34 | Groundnut Co 7 | 2015 | *Kharif* and summer | Resistant to rust |
| 35 | Phule Bharti (JL 776) | 2015 | *Kharif* | Resistant to *S. litura* and rust in field condition |
| 36 | G 2-52 | 2015 | *Kharif* | Foliar disease resistant |
| 37 | GJG 19 (JSP 51) | 2016 | *Kharif* | Tolerant to stem rot, dry root rot and rust as compared to check (KDG 123) |
| 38 | Raj Mungfali 3 (RG 559-3) | 2016 | *Kharif* | Tolerant to *S. litura*, leaf miner and thrips, high yielding; large seeded |
| 39 | Phule Warna (KDG 128) | 2016 | *Kharif* | Moderately resistance to rust and leaf spot |
| 40 | Phule Morna (KDG 123) | 2016 | *Kharif* | Moderately resistance to rust and leaf spot |
| 41 | KCG 6 (CTMG 6) | 2016 | *Rabi-summer* | Moderately resistance to rust and late leaf spot |
| 42 | GKVK 5 | 2016 | *Kharif and Summer* | Moderately resistance to rust and LLS, drought tolerant |
| 43 | ALG -06-320 | 2017 | *Rabi-summer* | Tolerant to rust, LLS and Peanut Bud Necrosis disease (PBND), *S. litura*, leaf miner and thrips |
| 44 | Kadiri Amaravathi (K 1535) | 2017 | *Kharif* | Tolerant to leaf spot, sucking pests (thrips and jassids), tolerant to early and late season drought |

(continued)

**Table 9** (continued)

| S. no. | Name of cultivar | Year of release | Production system | Special characters |
|---|---|---|---|---|
| 45 | VRI 8 (VG 09220) | 2017 | *Kharif and Summer* | Moderately resistant to jassids and thrips, moderately resistant to LLS and rust |
| 46 | GJG 32 (ICGV 03043) | 2018 | *Kharif* | Tolerant to stem rot, collar rot and rust |
| 47 | GJG 33 (ICGV 07222) | 2018 | *Rabi-summer* | Tolerant to collar rot and rust |
| 48 | JL1085 (Phule Dhani) | 2019 | *Kharif* | Tolerant to LLS and rust |
| 49 | Dh256 | 2019 | *Kharif* | Moderately resistant to *S. litura*, leaf miner and leaf hopper |

**Fig. 2** (**a**) Early leaf spot. (**b**) Late leaf spot. (**c**) Rust. (**d**) Infection by *Sclerotium rolfsii* (stem rot). (**e**) Peanut bud necrosis. (**f**) Kernels colonised by *Aspergillus flavus*

## *Resistance/Tolerance to Stress*

Among the major production constraints, foliar and soil borne fungal diseases are significant. Early leaf spot (ELS; caused by *Cercospora arachidicola*), late leaf spot (LLS; caused by *Phaeoisariopsis personata*) and rust (caused by *Puccinia arachidic*) are the most prevalent foliar fungal diseases in groundnut (Fig. 2).

Very good sources of resistance to these are available in diploid wild species which remain underutilised due to breeding barriers (Table 10). Several accessions of germplasm have also been reported as disease resistant (Subrahmanyam et al. 1985; Izge et al. 2007; Kanyika et al. 2015; Janila et al. 2016b; Desmae et al. 2016).

**Fig. 3** Germplasm lines registered from ICAR-DGR, which are sources of resistance

One of the major bottlenecks in the disease resistance breeding is the linkage drag of unacceptable pod characters like deep reticulation, low yield, long duration, poor shelling outturn, etc. However, some leaf spot-resistant interspecific germplasm lines have been released in the USA (Stalker et al. 2002). The lines developed using *Arachis cardenasii* as one of the parents were found to have LLS resistance (Mallikarjuna and Varshney 2014). Singh et al. (1997) identified 69 lines developed at ICRISAT as LLS resistant. In China, 53 genotypes out of the 5700 accessions tested were identified as resistant to late leaf spot (Liao 2003). The cultivar Southern Runner with moderate resistance to LLS was released in 1984 for commercial cultivation in the United States. Moderate to high levels of resistance were reported the cultivars Florida MDR 98 (Gorbet and Shokes 2002b), C-99R (Gorbet and Shokes 2002a) and Georgia-01R (Branch 2002). The cultivar ICG 7878, released in Mali is resistant to ELS and LLS (Desmae et al. 2019). In India, the cultivar GPBD 4 and Mutant (28-2) (Motagi et al. 2014) developed by using interspecific derivative lines, released for commercial cultivation, are resistant to rust and LLS.

Very high levels of resistance to rust has been reported in the wild Arachis species in the secondary gene pool involving the sections erectoides, triseminale, extranervosae and rhizomatosae (Subrahmanyam and McDonald 1983). In cultivated groundnut also, sources of resistance to rust have been reported, and majority of those are in the subspecies fastigiata (Subrahmanyam et al. 1995; Hammons 1977; Mehan et al. 1996; Motagi et al. 1996; Singh et al. 1997; Varman 1999; Reddy et al. 2001, Varshney et al. 2014; Monyo and Varshney 2016) and some of these lines

**Table 10** Multiple disease and insect-pests resistance in wild *Arachis* sp.

| Species | EL | LL | RT | SP | HE | TH | JA | AP | PM | PSt | PS |
|---|---|---|---|---|---|---|---|---|---|---|---|
| *A*. sp. GK 12934 | – | – | – | – | HR | HR | 1 | – | – | – | – |
| *A*. sp. GK 12946 | – | – | – | – | HR | HR | HR | – | – | – | – |
| *A*. *stenosperma* HLK 410 | HR | HR | HR | – | HR | HR | HR | R | – | – | – |
| *A*. *duranensis* K 7988 | R | – | 1 | – | HR | S | 1 | R | – | – | 1 |
| *A*. *spegazzinii* GKP 10038 | HR | – | 1 | – | HR | R | HR | – | – | – | R |
| *A*. *batizocoi* K 9484 | S | – | 1 | – | HR | 1 | HR | R | – | – | – |
| *A*. *khulmanii* GK 30035 | HR | R | 1 | – | – | – | – | R | – | – | – |
| *A*. *correntina* GKP 9530 | HR | R | 1 | – | HR | HR | HR | R | R | – | R |
| *A*. *chacoense* GKP 10602 | HR | HR | 1 | R | HR | HR | HR | R | R | R | S |
| *A*. *cardenasii* GKP 10017 | HR | HR | 1 | – | HR | HR | HR | R | R | 1 | S |
| *A*. *villosa* BU 22585 | R | R | 1 | – | – | – | HR | – | – | – | 1 |
| *A*. sp. GK 30017 | HR | HR | 1 | – | – | – | HR | – | – | – | – |
| *A*. *paraguariensis* GKP 96-46 | R | MR | 1 | R | S | – | R | R | – | R | R |
| *A*. *appressipila* GKP 9990 | HR | R | 1 | R | HR | HR | R | – | – | – | R |
| *A*. sp. GKP 105773 | HR | R | 1 | R | HR | 1 | R | – | – | – | R |
| *A*. *repens* GKP 20538 | S | S | – | – | HR | 1 | HR | – | – | – | 1 |
| *A*. *kempff-mercadoi* GKP 10127 | R | – | – | R | HR | 1 | 1 | R | – | – | – |
| *A*. *villosulicarpa* | HR | HR | 1 | – | – | – | – | – | – | – | – |
| *A*. *glabrata* GKP 9830 | HR | HR | 1 | R | HR | 1 | HR | – | – | R | R |
| *A*. *hagenbeckii* HL 436 | 1 | HR | 1 | R | HR | R | 1 | – | – | R | – |
| *A*. *pusilla* GKP 12922 | HR | HR | 1 | R | HR | 1 | 1 | – | R | S | – |

*EL* early leaf spot, *LL* late leaf spot, *RT* rust, *AP* aphids, *JA* jassid, *TH* thrips, *PS* peanut stunt virus, *SP* spodoptera, *HE* heliothis, *PM* peanut mottle virus, *PSt* peanut stripe virus, *HR* highly resistant, *S* susceptible, *I* immune. (Source: Bera et al. 2004)

have resistance to both rust and leaf spots. In India, the Cv GPBD4 is being used as a major source of resistance of LLS and Rust in several breeding programmes. The cultivar Yueyou 114 developed in China has been reported to have resistance to late leaf spot and bacterial wilt (Li et al. 2007). Zhonghua 4 and Shanyou 27 having moderate resistance to rust are cultivated widely in central and southern regions in China (Liao 2003). A good number of Indian cultivars show low to moderate levels of resistance to foliar fungal diseases (Table 11).

Among the viral diseases reported in groundnut from India, peanut bud necrosis virus (PBNV), tobacco streak virus (TSV), peanut mottle virus (PeMoV) and Indian peanut clump virus (IPCV) are the economically important. Field tolerance to these viruses have been reported from extensive field screening of wild species, released varieties and other breeding lines in hotspots (Radhakrishnan et al. 2016). However, the desired level of resistance to TSV has not yet been reported in cultivated varieties of groundnut. The germplasm accessions NCAc 17,099, NCAc 17,133 (RF) and NCAc 17,536 and derivatives of the wild species *A. glabrata* have been reported resistant to IPCV. The Indian cultivars reported to have field tolerance to PBNV are Ajeya (R-2001-3), Apoorva (R-8808), Co (Gn)-5, GG-16 (JSP-39), Girnar-2

**Table 11** Released cultivars with moderate level of foliar disease resistance released for cultivation in India

| |
|---|
| *Early leaf spot* |
| K-134, Kadiri Haritandhra (K-1319), Prutha (Dh 86), HNG-69, M-522, Jawahar Groundnut-23 (JGN-23), BG-3 (Birsa Groundnut-3) |
| *Late leaf spot* |
| Kadiri-5, Kadiri-6 (K-1240), Kadiri-4 (K-150), Abhaya (TPT-25), Vikas (GPBD 4), Kadiri-9, Greeshma, Kadiri-8, Mankya (DRG-12), Apoorva (R-8808), VRI (Gn)-6 (VG-9816), ICGV-86325, AK-265, ICGV-00348, K-134, ALR-2 (ALG-56), VRI (Gn)-5, Co (Gn)-4 (TNAU-269), CO 6, Kadiri Haritandhra (K-1319), TG-26, TAG-24, Phule Unap (JL-286), Prutha (Dh 86), AK-265, GG-7 (J-38), Somnath (TGS-1), Smruti (OG-52-1), RG-141, Mukta (DRG-17), Ambar (CSMG-84-1), ICGS-1 (ICGV-87119), Girnar-2 (PBS-24030), M-548, Jawahar Groundnut-3 (JGN-3), Jawahar Groundnut-23 (JGN-23), BG-3 (Birsa Groundnut-3) |
| *Rust* |
| AK-265, ALR-2 (ALG-56), Ambar (CSMG-84-1), Co (Gn)-4 (TNAU-269), Co (Gn)-5, CO 6, Girnar-2 (PBS-24030), ICGS 5 (ICGV-87121), ICGV-00348, ICGV-86325, ICGV-00348, ICGV-86325, Jawahar Groundnut-3 (JGN-3), Kadiri Haritandhra (K-1319), Kadiri-9, Mankya (DRG-12), Mukta (DRG-17), Phule Unap (JL-286), Smruti (OG-52-1), TG 26, Utkarsh (CSMG-9510), Vikas (GPBD 4), VRI (Gn)-5, VRI (Gn)-6 (VG-9816) |
| Stem rot caused by *Sclerotium rolfsii* Sacc (*Athelia rolfsii*), which can cause yield loss ranging from 10% to 40%, especially under irrigated conditions. Efficient screening protocols were lacking, and field screening has not been effective until recently. Though sources with high level of resistance to stem rot disease is not available in cultivated groundnut, *Arachis duranensis* (NRCG 12035) and *Arachis pusilla* (NRCG 12047) are found to be resistant. A few resistant interspecific derivatives developed at the ICAR-DGR have been registered as germplasm with the National Bureau of Plant Genetic Resources (Table 12). The germplasm line CS19 which is being used in several breeding programs as a source of resistance to stem rot (Fig. 3). The interspecific pre-breeding NRCGCS-47, NRCGCS-99, NRCGCS-131, NRCGCS-319 and the three breeding lines (GG 20 × CS19)-5-8, (GG 20 × CS19)-15-1, (GG 20 × CS19)-15-3 were found to be resistant to stem rot. The Indian cultivars with moderate resistance to stem rot are: Phule Unap (JL-86), Smruti (OG-52-1), Vasundhara (Dh-101), TG 51, TG-38B (TG-38), TG-51, and HNG-69 |

(PBS-24030), ICGS 1 (ICGV-87119), ICGV-86325, JCG-88 (Jagtial-88), K-134, Kadiri Haritandhra (K-1319), Kadiri-4, (K-150), Kalahasti (TCGS-320), Mankya (DRG-12), Mukta (DRG-17), R-9251, RG-141, SG-99, TAG-24, TG 26, Vasundhara (Dh-101), Vijetha (R-2001-2), VRI (Gn)-6 (VG-9816).

Nematode infestation causes nearly 11.8% loss in groundnut yield globally. Among nematodes, peanut root-knot nematode (*Meloidogyne* spp.), and the root-lesion nematode (*Pratylenchus brachyrus*) are prominent. In United States, six interspecific germplasm lines TxAG-6 and TxAG-7 (36), GP-NC WS 5 and GP-NC WS 6 (41), and NR 0812 and NR 0817 (3) with resistance to *M. arenaria* have been registered (Dong et al. 2008). Root-knot nematode resistant cultivars COAN and NemaTAM, which are highly resistant to *M. arenaria* and *M. javanica* (Simpson and Starr 2001; Simpson et al. 2003). However, these cultivars are not been preferred for commercial cultivation due to the low yield. In India the nematode disease popularly known as *Kalahasti malady*, caused by the nematode *Tylenchorhynchus brevilineatus* has been reported in certain parts of Andhra Pradesh. Screening of nearly 1600 germplasm accessions at the hotspot area have

**Table 12** List of trait-specific pre-breeding lines registered as germplasm from ICAR-DGR

| Registered germplasm | Trait |
|---|---|
| CS-19 | Resistant to stem rot |
| NRCGCS-15, NRCGCS-74, NRCGCS-186, NRCGCS-222, NRCGCS-86, NRCGCS-77, NRCGCS-83, NRCGCS-85, NRCGCS-180, NRCGCS-124, NRCGCS-196 | Resistance to peanut bud necrosis diseases, Stem rot, late leaf spot, early leaf spot, rust and Alternaria leaf blight |
| NRCG-14326, NRCG-14336, NRCG-14350, NRCG-14409 | Spanish bunch genotypes with fresh seed dormancy |
| NRCGCS-21 | Resistant to peanut bud necrosis diseases, stem rot, resistant to late leaf spot, early leaf spot |
| NRCG-12035 | High fodder bio-mass (1.8 ton/ha/year), suitable for wasteland conditions, seed forming, semi–perennial; suitable for pasture development, higher protein content (14.8%) |
| NRCG-12990 | High fodder bio-mass (2.7 ton/ha/year), suitable for waste land conditions, perennial in nature; suitable for pasture development, higher protein (12.1%) and iron content (0.5%) |
| NRCGCS-281 | Extra-large kernel size (HPS type) |
| NRCG-11780, NRCG-14263, NRCG-17205, NRCG-17206 | High oil and sugar content |
| NRCG-12431, NRCG-14380 | Fresh seed dormancy |
| NRCG-14368 | Tolerance of seed coat colonization by *Aspergillus flavus* |
| PBS-30138 | Resistant to rust and LLS |
| PBS-29031 | High oil, large seeded |
| | High oleic acid lines |

yielded 14 accessions with resistance to this nematode (Mehan et al. 1993). However, most of these resistant accessions were with undesirable pod/seed characteristics. The only cultivar with resistance to *Kalahasti malady* released for commercial cultivation is Tirupati 3 (TCG 1518). However, the cultivars Tirupati-2, Kalahasti (TCGS-320), Kadiri-9, Prasuna (TCGS 341) also are reported to have moderate resistance to this nematode.

Breeding for resistance to insect pest received less attention but for vectors of viral diseases. This might be due to the lack of availability of good sources of the resistance in cultivated and wild *Arachis* species. The major insect pests of groundnut are Aphids (*Aphis craccivora* Koch), thrips (*Thrips palmi*, *Frankliniella schultzei*, and *F. fusca*), leaf miner (*Aproaerema modicella*), red hairy caterpillar (*Amsacta albistriga*), jassids (*Empoasca kerri* and *E. fabae*) and *Spodoptera litura.* Of these, aphids and thrips are also vectors of important virus diseases.

The wild species *Arachis kempff-mercadoi* was found to be resistant to *S. litura* (Stevenson et al. 1993). This species has been used in developing *Spodoptera*-resistant lines (Mallikarjuna et al. 2004). The Indian cultivars Abhaya (TPT-25), Co (Gn)-5, Durga (RG-382), GG-14 (JSP-28), GG-16 (JSP-39), Kadiri Haritandhra

(K-1319), Phule Unap (JL-286), Pratap Mungphali-1, Pratap Mungphali-2, Vasundhara (Dh-101) were reported to have moderate level of resistance to *S. litura.*

The cultivars Abhaya (TPT-25), Durga (RG-382), GG-14 (JSP-28), GG-16 (JSP-39), Girmar-3 (PBS-12160), Girnar-2 (PBS-24030), Jawahar Groundnut-3 (JGN-3), Kadiri Haritandhra (K-1319), Phule Unap (JL-286), Pratap Mungphali-1, Pratap Mungphali-2, Prutha (Dh 86), Vasundhara (Dh-101) released in India were found to be moderately resistant to thrips.

The Indian cultivars with moderate resistance to leaf miner are Abhaya (TPT-25), Co (Gn)-5, Durga (RG-382), GG-14 (JSP-28), GG-16 (JSP-39), Girmar-3 (PBS-12160), Girnar-2 (PBS-24030), Jawahar Groundnut-3 (JGN-3), Kadiri Haritandhra (K-1319), Kadiri-7, Kadiri-8, Phule Unap (JL-286), Pratap Mungphali-1, Pratap Mungphali-2, Ratneshwar (LGN-1).

The main storage insects in groundnut are the Bruchid beetle or Groundnut borer (*Caryedon serratus*) and rust-red flour beetle (*Tribolium castaneum*). No sources of resistance have been reported for these pest, and good storage practices have been adopted to manage their infestation.

Aflatoxin contamination is one of the major trade barriers in groundnut export in India due to stringent regimes in place on permissible levels of aflatoxins by several countries in their imports. Aflatoxins produced by *Aspergillus* spp. are potential carcinogen. In Asia and Africa, *A. flavus* is predominant while in USA *A. parasiticus* is predominant. The aflatoxin contamination results from three-point invasion, viz. pod, seed coat and kernel. There are genotypes identified for their resistance to seed coat colonisation. However, the resistance to seed coat colonisation is not able to provide fungal infection in field (Mehan et al. 1987, 1988; Utomo et al. 1990; Upadhyaya et al. 1997). So is the case with the resistance to pod level infection, and aflatoxin content was not correlated (Will et al. 1994). Genetic sources of stable resistance to neither Aspergillus colonisation nor inhibition of the toxin production has not been reported in cultivated groundnut germplasm. However, *A. duranensis* (eight accessions) and *A. cardenasii* (two accessions) from section Arachis and their derivatives were reported to have high levels of resistance to aflatoxin production (Xue 2004). So far no major developments were reported in breeding genetic resistance, which may be due to the non-availability of sources of resistance to different components of resistance and lack of robust screening methods, and limited understanding of genetics of resistance (Nigam et al. 2009).

## Drought

Maximum area under groundnut especially in the semi-arid tropics (>70%) is rainfed, and the pod yield and other growth parameters of the crop have been severely affected (Pimratch et al. 2008, Nautiyal et al. 2002, Reddy et al. 2003, Nigam et al. 2005; Pandey et al. 2014b). Inadequate or erratic distribution of rain fall affect the productivity of groundnut frequently in India. Depending upon the growth stage of the crop, severity of the drought and its duration, the yield losses can be up to

56–85% (Nageswara Rao et al. 1989). It has been also reported that the extent reductions in pod yield varies with the genotypes (Harris et al. 1988, Nageswara Rao et al. 1998). The capability of a genotype to sustain the productivity under water-deficit stress is considered as drought tolerance, while a susceptible genotype suffer reduction in productivity under the stress. Thus, yield under water-deficit stress should be considered a primary criterion for assessing the performance of a genotype. Though the genetic variability for drought tolerance has been reported by Upadhyaya (2005) and Songsri et al. (2009), the breeding programmes on drought tolerance did not make any headway as there is significant genotype and environment interactions (Wright et al. 1996). Based on the stage of occurrence, the drought pattern can be grouped into early, mid and end-of-season or terminal drought although these patterns overlap and confound the breeding strategies. End-of-season drought which directly affects the pod development is considered as more critical (Williams et al. 1985; Nigam et al. 2002).

Selection for drought tolerance has been a major challenge especially the breeding programme has been taken up in non-target areas. A good amount of research efforts has been made for associating physiological traits with drought tolerance so as to have a trait-based selection criterion for drought which are easily adoptable by the breeders. Transpiration efficiency, loss for water in transpiration, stomatal conductance, specific leaf area, stomatal conductance and osmotic adjustments have been reported to be associated with water use efficiency and drought tolerance in groundnut (Nageswara Rao et al. 2001; Clavel et al. 2005; Puangbut et al. 2009; Mafakheri et al. 2010; Devi et al. 2011; Sanogo 2016). Root traits like root dry weight, root length density, distribution and deep rooting were reported to be associated with drought tolerance (Falalou et al. 2018; Matsui and Singh 2003; Arunyanark et al. 2008; Girdthai et al. 2010; Puangbut et al. 2010). Carbon isotope discrimination ($\Delta^{13}C$), an indirect measure of transpiration efficiency, has also been reported as trait associated with water-use efficiency (Lal et al. 2006; Rowland et al. 2012). However, high SCMR and low SLA did not lead to higher pod yield always probably due to the non-additive effects and weak relationship between surrogate traits and pod yield (Krishnamurthy et al. 2007; Hamidou et al. 2012; Sanogo 2016). While SCMR and SLA were reported to have low heritability (Janila et al. 2015), high correlations of both SCMR and SLA with pod yield and other related traits were reported (Upadhyaya 2005; Songsri et al. 2009; Upadhyaya et al. 2011; Janila et al. 2015). Nigam et al. (2001) and Lal et al. (2006) reported both additive and nonadditive effects of SCMR and SLA. Empirical approach by employing selection based on pod and grain yield under imposed drought stress conditions or trait-based approach or a combination of both of these are being for phenotyping for drought tolerance (Devi et al. 2011; Janila et al. 2015; Nigam et al. 2005) as this approach can select genotypes with more water-use efficiency or partitioning of the photosynthates into economic yield.

Several germplasm accessions and breeding lines have been reported to have drought tolerance (Mayeux et al. 2003; Nigam et al. 2005; Upadhyaya 2005; Hamidou et al. 2012; Monyo and Varshney 2016). The Indian commercial cultivars

Kadiri-3, K-134, ICGS-1 (ICGV-87119), ICGS-5 (ICGV-87121), ICGS-37 (ICGV-87187) and ICGS-76 (ICGV-87141), ICGV-86031 and TAG-24 has been reported to be drought tolerant to varying extent. The commercial cultivars ICGV-91114, Prasuna, Kadiri-7, Abhaya, TMV-13, Co (Gn)-6 and VRI (Gn)-7, R-2001-3 ICGV-91114 have been released as drought-tolerant varieties released in India.

## *Quality Attributes*

The attributes like low oil, high sugar, uniform seed size and shape, pink or tan seed colour, easy blanchability, high oleic/linoleic (O/L); free from aflatoxin contamination have been considered as important for confectionary purpose (Nigam et al. 1989; Dwivedi and Nigam 1995). The stability of these characters under different growing conditions has also been a challenge for the breeders. Several high oil lines have been identified at ICRISAT. The lines ICGV 05155, ICGV 06420, ICGV 03042 and ICGV 03043 have been tested extensively for their stability in oil content. Two germplasm accessions, NRCG 7040 (55.6%) and NRCG 6999 (55.0%) had oil contents greater than or around 55%, and five accessions NRCG 1476 (54.9%), NRCG 12328 (54.9%), NRCG 3817 (54.8%) and NRCG 2390 (54.0%) had oil contents greater than or around 54% which can be used as donors in improving the oil contents of existing cultivars. The germplasm accession Ah347/7 was found to have low oil content (44.7%). The Indian commercial cultivars TMV10 and RG382 (Durga) have more than 54% oil content. Janila et al. (2012) reported a groundnut cultivar ICGV 03137 with improved blanchability. The cultivars ICGV 06099 and ICGV 06040 were reported to have high Fe and Zn content in seed (Janila et al. 2014). The germplasm accessions NRCG 7063, NRCG 7218 and NRCG 5239 were found to have 35% protein and ~6.5% sugar content. The Indian commercial cultivars Vikram (TG 1), M 335, Birsa Bold (BAU 13), B 95, M 522, TPG 41, TBG 39 (TG 39), AK 303, Mallika (ICHG 00440) and Raj Mungfali 3 (RG 559-3) are large seeded.

Only a few lines like SunOleic 95R and SunOleic 97R in the USA and PC 223 K8 and PC 223 K9 in South Africa have been reported to have high oleic acid and O/L ratio. With the discovery of the high oleic mutant F435 having more than 80% oleic acid (Norden et al. 1987) and the development of the first high oleic variety SunOleic 95R with 82% oleic acid content (Gorbet and Knauft 1997) breeding efforts have been in progress for the development of commercial cultivars with high oleic acid content. Recently, by marker-assisted back crossing two groundnut varieties Girnar 4 and Girnar 5 have been identified for commercial cultivation in India. In addition to the released commercial cultivars several trait-specific breeding also have been developed by the different groups of groundnut workers. A few of such lines developed by the plant breeders at ICAR-DGR is listed in Table 13.

**Table 13** List of trait-specific groundnut breeding lines reported from ICAR-DGR

| Breeding lines | Trait |
|---|---|
| NRCGCS-148 | Large seed |
| HOS-22,HOS-30, HOS-58, HOS-89 | High oil content (≥55%) |
| HOS-111, HOS-116, HOS-119, HOS-125, HOS-130, HOS-138, HOS-145, HOS-172, HOS-191, HOS-201, NRCGCS-587 | High oleic acid content (>78%) |
| NRCG-7040, NRCG-6999, NRCG-1476, NRCG-12328, NRCG-3817 | High oil content (>54%) |
| Ah-347/5 | Low oil content (44.4%) |
| NRCG-7063, NRCG-7128, NRCG-5239 | High protein (35%) and sugar contents (~6.5%) |
| NRCGCS-574 | Resistant to rust and LLS |
| PBS-21095, PBS-21087, PBS-16038, PBS-26019, PBS-26015, PBS-240022, PBS-11084, PBS-11058, JUN-27 | Tolerance to end-of season drought stress |
| PBS-16038 | Spanish advanced breeding line with fresh seed dormancy of 3 weeks |

# 8 Genomics

## *Genomic Resources Developed*

Due to the tetraploid nature of the crop and the complexity of the genome, marker development in cultivated groundnut has been in a slow pace. A good amount of genomic resources for groundnut has been added during the last 15 years. The early generation markers reported were isozymes mainly used for the assessment of diversity. This was followed by the other marker systems like RFLP, RAPD, AFLP. By the advent of rapid sequencing systems like NGS had paved way for the development of more robust markers like simple sequence repeats (SSRs), SNPs and DArT in groundnut (Pandey et al. 2012; Varshney 2016). The increased availability of sequences generated from complementary DNA libraries, bacterial artificial chromosome (BAC) libraries, and NGS has contributed to substantially to this surge (Yüksel and Paterson 2005; Guimarães et al. 2008, 2010; Feng et al. 2012; Guo et al. 2013). Initially, these marker systems also were predominantly used for assessment of molecular diversity (Radhakrishnan and Hariprasanna 2004). Being informative, SSR markers gained more interest and a good number of markers including the ones developed from ESTs are available (Hopkins et al. 1999; Peng et al. 2016; Bosamia et al. 2015) and used in genetic mapping and QTL analysis. Pandey et al. (2012) screened 4485 SSRs using a panel of 20 parental genotypes and found 1351 polymorphic SSRs. The highly informative SSR markers have been used in deducing genetic variability (Belamkar et al. 2011; Ren et al. 2014).

Considering the amenability, SNP markers have become more popular especially after the cost-effective high-throughput genotyping facilities become more accessible (Bertioli et al. 2014). Initially, 8486 candidate SNPs from 17 genotypes were reported (Alves et al. 2008; GCP 2011). An Illumina GoldenGate array of 1536 SNPs form 17 tetraploid genotypes in comparison with the transcriptomes of Tifrunne was developed at University of Georgia (Nagy et al. 2012) while another 768-SNP assay was reported by UC Davis (Pandey et al. 2012). From *Arachis hypogea*, the tetraploid species 53,257 SNPs were reported by Zhou et al. (2014), 263,840 SNPs and indel variants (Chopra et al. 2015), 11,902 SNPs (Peng et al. 2016) and 6965 SNPs (Peng et al. 2017a, b). Several mapping populations have been prepared and the markers were mapped into genetic maps using these populations for the genomes AA, BB and the tetraploid AABB (Guo et al. 2013; Pandey et al. 2014a). By using the markers a good number of maps have been developed to understand the genome architecture and identify QTLs for the economically important traits required for breeding. The maps developed, the populations used for the mapping, the markers used, number of the marker loci mapped and the coverage of the genome are summarised in the Table 14.

Since several maps became available, high-density consensus maps integrating the individual genetic maps from the genome AA, BB and AABB were also prepared. The resulting consensus maps can give more insight into the chromosomal rearrangements and gene duplication (Jackson et al. 2005; Wang et al. 2011). The first consensus genetic map was reported by Hong et al. (2010) integrating three genetic maps which was followed by Gautami et al. (2012), Sujay et al. (2012) and Shirasawa et al. (2013). During these developments, a good number of mapping populations were developed by hybridising genetically divergent parent(s) and with the support of the advances in bio-informatic tools and rapidly evolved marker technologies, QTLs for economically important target traits have been deduced and used to establish marker-trait association for deploying in genetic improvement of groundnut. Though Selvaraj et al. (2009) reported marker association for plant weight, the first report on QTLs were by Sarvamangala et al. (2011). Subsequently, several QTLs have been reported, but the majority of QTLs reported were minor having low levels of phenotypic variance explained (PVE). The list of QTLs reported in groundnut is summarised in the Table 15. The maximum number of QTLs are reported for foliar fungal diseases followed by oil quality and stem rot resistance. The QTLs for root knot nematode, foliar disease resistance and oil quality have been used in breeding programmes to develop commercial cultivars.

## Genomics-Assisted Breeding

Though several reports on marker associations are available, utilisation of marker-assisted breeding is very meagre. Marker associations were reported for root-knot nematode (Garcia et al. 1996; Choi et al. 1999; Simpson 2001; Chu et al. 2007; Wang et al. 2008; Nagy et al. 2010; Branch et al. 2014). Marker-assisted selection (MAS) for root-knot resistance was first used by Simpson (1991) to develop synthetic allotetraploid line (TxAG-6), and subsequently developed the first nematode-resistant cultivar COAN having the genes from the wild species (Simpson and Starr

**Table 14** A comprehensive list of genetic maps developed in groundnut

| Genome | Population used | No of loci mapped | Type of marker used | Coverage (cM) | Reference |
|---|---|---|---|---|---|
| ***AA*** | *A. stenosperma* × *A. cardenasii* | 132 | RFLP | 1063.00 | Halward et al. (1993) |
| | [*A. stenosperma* × (*A. stenosperma* × *A. cardenasii*)] | 206 | RAPD, RFLP | 800 | Garcia et al. (2005) |
| | *A. duranensis* (K7988) × *A. stenosperma* (V10309) | 204 | SSR | 1230.89 | Moretzsohn et al. (2005) |
| | *A. duranensis* (K7988) × *A. stenosperma* (V10309) | 369 | SSR, anchor, AFLP, NBS profiling, SNP | – | Leal-Bertioli et al. (2009) |
| | *A. duranensis* (PI 475887) × *A. duranensis* (Grif 15036) | 1724 | SNP, SSR, SSCP, RGC | 1081.30 | Nagy et al. (2012) |
| | *A. duranensis* (K7988) × *A. stenosperma* (V10309) | 597 | SSR, TE | 544.00 | Shirasawa et al. (2013) |
| | *A. duranensis* (K7988) × *A. stenosperma* (V10309) | 384 | SNP, SSR | 705.10 | Bertioli et al. (2014) |
| | *A. duranensis* (K7988) × *A. stenosperma* (V10309) | 502 | SNP, SSR, RGA | 1004.10 | Leal-Bertioli et al. (2016) |
| ***BB*** | *A. ipaensis* (K30076) × *A. magna* (K30097) | 149 | SSR | 1294.00 | Moretzsohn et al. (2009) |
| | *A. ipaensis* (K30076) × *A. magna* (K30097) | 798 | SSR, TE | 461.00 | Shirasawa et al. (2013) |
| | *A. ipaensis* (K30076) × *A. magna* (K30097) | 399 | SSR, TE | 678.00 | Leal-Bertioli et al. (2015) |
| | K 9484 (PI 298639) × GKBSPSc 30,081 (PI 468327) of *A. batizocoi* | 449 | SSR | 1278.60 | Guo et al. (2012) |
| ***AABB*** | Florunner × TxAG-6 {[*A. batizocoi* K9484 × (*A. cardenasii* GKP10017 × *A. diogoi* GKP10602)]4×} | 370 | RFLP | 2210.00 | Burow et al. (2001) |
| | ICG 12991 × ICGVSM 93541 | 12 | AFLP | 139.4 | Herselman et al. (2004) |
| | [Fleur 11 × (*A. ipaensis* × *A. duranensis*)4×] | 298 | SSR | 1843.70 | Foncéka et al. (2009) |
| | Yueyou 13 × Zhenzhuhei | 131 | SSR | 679.00 | Hong et al. (2008) |
| | TAG 24 × ICGV 86031 | 135 | SSR | 1270.50 | Varshney et al. (2009) |

(continued)

**Table 14** (continued)

| Genome | Population used | No of loci mapped | Type of marker used | Coverage (cM) | Reference |
|---|---|---|---|---|---|
| | TAG 24 × ICGV 86031 | 191 | SSR | 1785.40 | Ravi et al. (2011) |
| | Yueyou 13 × Zhenzhuhei | 132 | SSR | 684.90 | Hong et al. (2010) |
| | Yueyou 13 × Fu 95-5 | 109 | SSR | 540.69 | Hong et al. (2010) |
| | Yueyou 13 × J11 | 46 | SSR | 401.70 | Hong et al. (2010) |
| | TAG 24 × GPBD 4 | 56 | SSR | 462.24 | Khedikar et al. (2010) |
| | TAG 24 × GPBD 4 | 188 | SSR | 1922.40 | Sujay et al. (2012) |
| | TG 26 × GPBD 4 | 45 | SSR | 657.90 | Sarvamangala et al. (2011) |
| | TAG 24 × GPBD 4 | 181 | SSR | 1963.00 | Sujay et al. (2012) |
| | ICGS 44 × ICGS 76 | 82 | SSR | 831.40 | Gautami et al. (2012) |
| | ICGS 76 × CSMG84-1 | 119 | SSR | 2208.20 | Gautami et al. (2012) |
| | SunOleic 97R × NC94022 | 172 | SSR, CAPs | 920.70 | Qin et al. (2012) |
| | SunOleic 97R × NC94022 | 206 | SSR, CAPs | 1780.60 | Pandey et al. (2014a, b) |
| | Tifrunner × GT-C20 | 318 | SSR | 1674.40 | Wang et al. (2012) |
| | Tifrunner × GT-C20 | 239 | SSR, CAPs | 1213.40 | Qin et al. (2012) |
| | YI-0311 × Nakateyutaka | 326 | SSR, TE | 1332.90 | Shirasawa et al. (2012) |
| | Satonoka × Kintoki | 1114 | SSR, TE | 2166.40 | Shirasawa et al. (2012) |
| | VG 9514 × TAG 24 | 95 | SSR | 882.90 | Mondal et al. (2012a) |
| | *A. hypogaea* "Runner IAC 886" × (*A. ipaensis* × *A. duranensis*)4× | 1469 | SSR, TE | 1442.00 | Shirasawa et al. (2013) |
| | Tifrunner × GT-C20 | 378 | SSR, CAPs | 2487.40 | Pandey et al. (2014a) |
| | Tifrunner × GT-C20 | 418 | SSR | 1935.40 | Pandey et al. (2014a) |
| | *A. hypogaea* "Runner IAC 886" × (*A. ipaensis* × *A. duranensis*)4× | 772 | SNP, SSR | 1487.30 | Bertioli et al. (2014) |

(continued)

**Table 14** (continued)

| Genome | Population used | No of loci mapped | Type of marker used | Coverage (cM) | Reference |
|---|---|---|---|---|---|
| | Zhonghua 5 × ICGV 86699 | 1685 | SNP, SSR | 1446.70 | Zhou et al. (2014) |
| | VG 9514 × TAG 24 | 190 | SSR, ISSR, TE, RGC | 1796.70 | Mondal et al. (2014a, b) |
| | Zhonghua 10 × ICG12625 | 470 | SSR | 1877.30 | Huang et al. (2015) |
| | Zhonghua 10 × ICG12625 | 1219 | SSR, TE | 2038.75 | Huang et al. (2016) |
| | TAG 24 × GPBD 4 | 289 | SSR, TE | 1730.80 | Kolekar et al. (2016) |
| | SunOleic 97R × NC94022 | 248 | SSR | 1425.90 | Khera et al. (2016) |
| | Fuchuan Dahuasheng × ICG 6375 | 347 | SSR | 1675.60 | Chen et al. (2016a, b, c) |
| | Xuhua 13 × Zhonghua 6 | 228 | SSR | 1337.70 | Chen et al. (2016a, b, c) |
| | Florida- EP™ “113” × Georgia Valencia | 30 | SSR, SNP | 157.80 | Tseng et al. (2016) |
| | ICGV 00350 × ICGV 97045 | 1152 | DArT, DArTseq | 2423.12 | Vishwakarma et al. (2016) |
| | 79,266 × D893 | 231 | SSR | 905.18 | Li et al. (2017a) |
| | Florunner × TxAG-6 {[*A. batizocoi* K9484 × (*A. cardenasii* GKP10017 × *A. diogoi* GKP10602)]4×} | 91 | SSR | 1321.90 | Wilson et al. (2017) |
| | Yuanza 9102 × Xuzhou 68-4 | 743 | SSR | 1232.57 | Luo et al. (2017a) |
| | Yuanza 9102 × Xuzhou 68-4 | 830 | SSR | 1386.19 | Luo et al. (2017a) |
| | ICGV 07368 × ICGV 06420 | 854 | DArT, SSR | 3526.00 | Shasidhar et al. (2017) |
| | ICGV 06420 × SunOleic 95R | 1435 | DArT, DArTseq | 1869.00 | Shasidhar et al. (2017) |
| | ICGV 06420 × SunOleic 95R | 1211 | SNP | – | Liang et al. (2017) |
| | TMV 2 × TMV 2-NLM | 91 | TE | 1205.66 | Hake et al. (2017) |
| | GG20 x CS19 | 12 | SSR | – | Bera et al. (2017) |
| | ZH16 × sd-H1 | 3630 | SNP | 2098.14 | Wang et al. (2018a, b) |

(continued)

**Table 14** (continued)

| Genome | Population used | No of loci mapped | Type of marker used | Coverage (cM) | Reference |
|---|---|---|---|---|---|
| | Zhonghua 5 and ICGV86699 | 1685 | SNP, SSR | 1446.7 | Zhou et al. (2018) |
| | Xuhua 13 × Zhonghua 6 | 2595 | SNP | 2465.62 | Liu et al. (2019b) |
| | TG37A × NRCG CS85 | 266 | SNP | 1092 | Dodia et al. (2019) |
| | Tifrunner × NC 3033 | 1524 | SNP, SSR | 3382 | Chavarro et al. (2020) |
| | NC 3033 × Tifrunner | 1524 | SNP, SSR | 3381.96 | Luo et al. (2020) |
| ***Consensus*** | 3 populations | 175 | SSR | 885.40 | Hong et al. (2010) |
| | 3 populations | 293 | SSR | 2840.80 | Gautami et al. (2012) |
| | 2 populations | 225 | SSR | 1152.90 | Sujay et al. (2012) |
| | 13 maps | 3693 | SSR, TE | 2651 | Shirasawa et al. (2013) |
| | 8 populations | 5874 | SSR, TE | 2918.62 | Lu et al. (2018) |

2001). The second nematode-resistant variety was developed from COAN-derived materials, by using RFLP markers (Simpson et al. 2003) through MABC. Another nematode-resistant cultivar Tifguard was developed (Holbrook et al. 2008). The markers for rust resistance were associated by Khedikar et al. (2010), Sujay et al. (2012), Mondal et al. (2012a, b) and rust and LLS by Kolekar et al. (2016). The QTL region from cultivar 'GPBD 4' was integrated into three rust susceptible cultivars (JL24, ICGV 91114 and TAG 24) through marker-assisted backcrossing (MABC) (Varshney et al. 2014). QTLs from the chromosomes A02 and A03 governing LLS and rust resistance from GPBD 4 were introgressed into TMV 2 using five markers GM2009, GM2079, GM2301, GM1839 and IPAHM103 (Kolekar et al. 2017). One dominant (IPAHM103) and three co-dominant (GM2079, GM1536, GM2301) markers were used in the study and the field trails could select 81 introgression lines with improved rust resistance. Late leaf spot and rust markers, IPAHM103, GM2301 and pPGPseq8D09, were used to transfer resistance from to GPBD 4 to JL24 through (Yeri and Bhat 2016), and the field evaluation could identify LLS and rust-resistant lines. Markers associated with nutritional qualities especially oil quality has been reported by several authors (Chen et al. 2010; Chu et al. 2009; Sarvamangala et al. 2011; Wilson et al. 2017). Three commercial cultivars with high oleic acid have been released so far using the molecular breeding approaches (Chu et al. 2011; Mienie and Pretorius 2013; Xiuzhen et al. 2016). Bera et al. (2019) introgressed two FAD2 mutant alleles from SunOleic95R to ICGV06100, a high oil breeding line through MABC and developed introgression

**Table 15** List of QTLs identified in groundnut

| QTL for the trait | Number | Reference |
|---|---|---|
| Leaf spot resistance | 28 | Wang et al. (2013), Liang et al. (2017) |
| Late leaf spot resistance | 45 | Agarwal et al. (2018), Kolekar et al. (2016) |
| Number of sporulated lesions/ leaf area | 2 | Leal-Bertioli et al. (2015) |
| Log incubation period | 3 | Leal-Bertioli et al. (2015) |
| Susceptibility index | 4 | Leal-Bertioli et al. (2015) |
| Total number of lesions/leaf area | 3 | Leal-Bertioli et al. (2015) |
| Rust resistance | 51 | Kolekar et al. (2016), Leal-Bertioli et al. (2015), Mondal et al. (2014a, b) |
| Total number of holes on pods (bruchid) | 3 | Mondal et al. (2014a, b) |
| Percent pod weight loss (bruchid) | 1 | Mondal et al. (2014a, b) |
| Total development period (bruchid) | 4 | Mondal et al. (2014a, b) |
| Percent adult emergence (bruchid) | 4 | Mondal et al. (2014a, b) |
| Thrips resistance | 3 | Wang et al. (2013) |
| Lateral branch length | 5 | Leal-Bertioli et al. (2016), Wang et al. (2018a, b) |
| Number of lateral branches | 2 | Leal-Bertioli et al. (2016) |
| Specific leaf area_40 days | 5 | Leal-Bertioli et al. (2016) |
| Root galling index | 5 | Leal-Bertioli et al. (2016) |
| Eggs per gram of root_2011 | 3 | Leal-Bertioli et al. (2016) |
| Root-knot nematode resistance | 17 | Burow et al. (2014), Leal-Bertioli et al. (2016) |
| Linoleic acid | 52 | Sarvamangala et al. (2011), Hu et al. (2018), Pandey et al. (2014a, b) |
| Palmitic acid | 31 | Wang et al. (2015) |
| Oil content | 27 | Selvaraj et al. (2009), Sarvamangala et al. (2011), Huang et al. (2015), Pandey et al. (2014b) |
| Oleic acid | 27 | Sarvamangala et al. (2011), Hu et al. (2018) |
| SPAD chlorophyll meter reading | 12 | Faye et al. (2015) |
| SPAD chlorophyll meter reading_40 days | 7 | Leal-Bertioli et al. (2016) |
| Peg length | 11 | Foncéka et al. (2012a, b) |
| Pod beak | 10 | Foncéka et al. (2012a, b) |
| Pod constriction | 14 | Foncéka et al. (2012a, b), Leal-Bertioli et al. (2015) |
| Hundred pod weight | 6 | Foncéka et al. (2012a, b), Wang et al. (2018a, b) |
| Pod width | 21 | Foncéka et al. (2012a, b), Wang et al. (2018a, b), Chavarro et al. (2020) |
| Pod length width ratio | 4 | Wang et al. (2018a, b) |
| Protein content | 8 | Sarvamangala et al. (2011) |
| Log root dry weight | 2 | Leal-Bertioli et al. (2016) |

(continued)

**Table 15** (continued)

| QTL for the trait | Number | Reference |
|---|---|---|
| Seed length | 18 | Selvaraj et al. (2009), Foncéka et al. (2012a, b), Wang et al. (2018a, b) |
| 100-seed weight | 5 | Selvaraj et al. (2009), Foncéka et al. (2012a, b) |
| Seed size index | 3 | Chavarro et al. (2020) |
| Kernel percentage | 3 | Chavarro et al. (2020) |
| Single kernel pod | 4 | Chavarro et al. (2020) |
| Double kernel pod | 2 | Chavarro et al. (2020) |
| Pod area | 3 | Chavarro et al. (2020) |
| Pod density | 3 | Chavarro et al. (2020) |
| Weight of 10 seeds | 2 | Leal-Bertioli et al. (2015) |
| Log_weight of 10 seeds | 4 | Leal-Bertioli et al. (2016) |
| Seed width | 13 | Foncéka et al. (2012a, b), Chavarro et al. (2020) |
| Stress tolerance index, hundred pod weight | 2 | Foncéka et al. (2012a, b) |
| Stress tolerance index, seed number | 2 | Foncéka et al. (2012a, b) |
| Stress tolerance index, total biomass | 2 | Foncéka et al. (2012a, b) |
| Stress tolerance index, hundred seed weight | 2 | Foncéka et al. (2012a, b) |
| Stress tolerance index, haulm weight | 2 | Foncéka et al. (2012a, b) |
| Stress tolerance index, pod number | 3 | Foncéka et al. (2012a, b) |
| Resistance to tomato spotted wilt virus (TSWV) | 18 | Qin et al. (2012), Wang et al. (2013) |
| Tomato spotted wilt virus resistance | 12 | Agarwal et al. (2018) |
| Total biomass | 5 | Foncéka et al. (2012a, b), Leal-Bertioli et al. (2016) |
| Haulm weight | 16 | Foncéka et al. (2012a, b), Leal-Bertioli et al. (2016), Faye et al. (2015) |
| Main stem height | 11 | Foncéka et al. (2012a, b), Leal-Bertioli et al. (2015) |
| Log_main stem height | 3 | Leal-Bertioli et al. (2015) |
| Plant weight | 1 | Selvaraj et al. (2009) |
| Log weight ratio root/aerial part | 3 | Leal-Bertioli et al. (2016) |
| Growth habit | 6 | Foncéka et al. (2012a, b) |
| Harvest index | 2 | Foncéka et al. (2012a, b), Faye et al. (2015 ) |
| Shell weight | 1 | Foncéka et al. (2012a, b) |
| Pod weight | 7 | Foncéka et al. (2012a, b), Faye et al. (2015) |
| Number of pods per plant | 6 | Selvaraj et al. (2009), Chen et al. (2019a, b, c) |
| Pod number | 2 | Foncéka et al. (2012a, b) |
| Seed weight | 6 | Foncéka et al. (2012a, b), Dodia et al. (2019), Chavarro et al. (2020) |

(continued)

**Table 15** (continued)

| QTL for the trait | Number | Reference |
|---|---|---|
| Seed number | 8 | Foncéka et al. (2012a, b), Leal-Bertioli et al. (2016), Chen et al. (2019a, b, c) |
| Log_seed number | 3 | Leal-Bertioli et al. (2015) |
| Fruiting branches | 5 | Wang et al. (2018a, b) |
| Total number of branches | 4 | Wang et al. (2018a, b) |
| Internode number | 4 | Wang et al. (2018a, b) |
| Stem rot resistance | 43 | Bera et al. (2017), Dodia et al. (2019), Luo et al. (2020) |
| Root hairiness | 3 | Dodia et al. (2019) |
| Leaf shape | 2 | Dodia et al. (2019) |
| No of primary branches | 1 | Dodia et al. (2019) |
| No of secondary branches | 1 | Dodia et al. (2019) |
| Plant height | 11 | Dodia et al. (2019), Wang et al. (2018a, b), Faye et al. (2015) |
| Pod length | 4 | Dodia et al. (2019), Wang et al. (2018a, b) |
| Shelling percentage | 5 | Dodia et al. (2019), Faye et al. (2015), Huang et al. (2015) |
| Percent seed infection index (Aflatoxin) | 2 | Yu et al. (2019) |
| Aflatoxin B1 content | 7 | Yu et al. (2019) |
| Aflatoxin B2 content | 5 | Yu et al. (2019) |
| Aflatoxin resistance | 3 | Pandey et al. (2014a, b); Zuang (cited from Soni et al. 2020) |

lines with ~80% oleic acid. Huang et al. (2019) used MABC approach to introgress fatty acid desaturase gene (*FAD2*) gene into four high oil varieties, viz. 'Yuanza 9102', Yuhua 15′, 'Yuhua 9326' and 'Yuhua 9327'. Nawade et al. (2019) used marker-assisted backcross (MABC) approach to generated introgression lines using Sunoleic 95R as source and the introgression lines recorded 67.04–81.71% oleic acid content under field conditions. Janila et al. (2016a) through MABC developed 27 introgression lines with 62–83% oleic acid using SunOleic 95R as source of FAD2 allele. Mienie and Pretorius (2013) transferred *ahFAD2A* and *ahFAD2B* genes from high-oleic parents (UF-85, Guat and Atete) using MAS to develop commercial cultivars 'Akwa', 'Kwarts' and 'Harts' in South Africa. The alleles governing high-oleic-acid trait and resistance to root knot nematode were pyramided by marker-assisted breeding to develop and improve 'Tifguard' (Chu et al. 2011). They used S197 and GM565 markers for nematode resistance and CAPS assay for *FAD* gene. Using MABC, Shasidhar et al. (2020) improved the popular Indian cultivars GJG 9, GG 20 and GJGHPS 1 for foliar disease resistance combined with high oleic acid, which are now ready for multilocation testing for disease resistance and stability of oleic acid content.

Another breakthrough in the groundnut genomics was the release of the draft genomes of the two progenitors of the cultivated groundnut, namely *A. duranensis* (accession V14167; A-genome progenitor) and *A. ipaensis* (accession K30076; B-genome progenitor) (Bertioli et al. 2016). This was followed by the addition of the draft genome sequence of *A. duranensis* (accession PI475845; A-genome), which provided the better understanding of the genome architecture specifically for the traits like oil bio synthesis, allergens and geocarpy (Chen et al. 2016a, b, c). The high-quality genome assembly of the US cultivar 'Tifrunner' becomes available towards the end of 2017. Further, the availability of the draft genomes have opened up the large-scale genome-wide discovery of makers like 324,915 SSRs from A-genome (Chen et al. 2016a, b, c; Zhao et al. 2017; Luo et al. 2017b) 320,013 SSRs in the B-genome (Zhao et al. 2017), 512,900 SSRs in the AABB genome (Ma et al. 2020) and 515,223 indels were discovered (Vishwakarma et al. 2017). Pandey et al. (2017) reported the development of an Axiom_Arachis SNP array with 58 K genome-wide DNA resequencing and RNA sequencing of 41 groundnut accessions and the genomes of *A. duranensis* and *A. ipaensis*, the progenitors of the cultivated groundnut, which was used to identify tetrasomic recombinations (Clevenger et al. 2017). Zhang et al. (2017) identified 1429 candidate genes for eleven agronomic traits and 17,338 high-quality SNPs in the whole groundnut genome, adopting specific-locus amplified fragment sequencing (SLAF-seq) method. From the genome sequence of the cultivated groundnut, Chen et al. (2019a, b, c) predicted 108,604 gene models and annotated 83,087 genes by combining ab initio prediction, homologous protein data searches, and transcriptome alignments. Of these 81,414 genes were assigned to chromosomal locations. Annotation of 241 microRNAs (miRNAs), 2239 transfer RNAs (tRNAs), 25,299 small nuclear RNAs (snRNAs) and 3511 ribosomal RNAs (rRNAs) were also done. Initial reports on transcriptome analysis in groundnut were mainly based on straightforward assembly strategies centred on marker or gene discovery than homeolog resolution. These efforts has led to the identification of several genes expressed especially in response to biotic and abiotic stresses, geocarpic development, embryo development, pod development, etc. By the advent of the rapid sequencing tools and assembly tools, a good number of transcripts were assembled and analysed in detail. More than 60 reports on transcriptomics in groundnut are available which is summarised in Table 16.

As on 25th March 2020, a total of 219,242 gene sequences of groundnut are available in the NCBI database. Of these 65,888 sequences for *Arachis hypogaea*, 37,739 sequences for Arachis ipaensis and 34,576 sequences for *Arachis duranensis* are Protein-coding, annotated genes, and reference Sequences. Currently, Gene Expression Omnibus hosts about 138 gene expression datasets for groundnut including 132 for *Arachis hypogaea,* 3 for *Arachis duranensis* and 3 for *Arachis ipaensis* in NCBI database. Similarly 287,524 protein sequences for *Arachis hypogaea*, 57,734 for Arachis ipaensis and 53,015sequences for *Arachis duranensis* are present in NCBI database. Clevenger et al. (2016) reported a gene expression atlas for the subsp. *hypogaea* using 22 different tissues from all critical organs of the tetraploid cultivar Tifrunner, and identified 8816 putative homeologus genes, spotted more than 9000 alterative splicing events, and 6000 non-coding RNAs. For the subsp. Fastigiata a gene atlas was prepared using a drought-tolerant cultivar ICGV

**Table 16** List of transcriptomes reported in groundnut

| Trait | Reference |
|---|---|
| *Sclerotium rolfsii* infection/resistance | Bosamia et al. (2020) |
| Salt stress | Zhang et al. (2020) |
| Cold tolerance | Jiang et al. (2020) |
| Symbiotic pathways induced by *Metarhizium anisopliae* | Wang et al. (2020) |
| Drought responsive mechanism | Bhogireddy et al. (2020) |
| Plant height | Guo et al. (2020) |
| Anthocyanin pigment | Xia et al. (2020) |
| Foliar disease resistance | Bhat et al. (2020) |
| Meloidogyne resistance | Martins et al. (2020) |
| Seed dormancy release and germination | Xu et al. (2020) |
| Peanut allergen | Do et al. (2019) |
| Low-temperature germination | Zhang et al. (2019) |
| High oleic acid feed back | Liu et al. (2019a) |
| Embryo abortion under calcium deficiency | Chen et al. (2019d) |
| Cadmium uptake and regulation | Chen et al. (2019b) |
| Anthocyanin accumulation | Huang et al. (2019) |
| Differential calcium transport and retention | Yu et al. (2019) |
| Progress symbiosis in crack-entry (infection) | Karmakar et al. (2019) |
| Oil content | Wang et al. (2018a, b) |
| Nematode resistance | Mota et al. (2018) |
| Iron nutrition | Dai et al. (2018) |
| Post-harvest Aflatoxin contamination | Korani et al. (2018) |
| Aflatoxin production due to water stress | Fountain et al. (2018) |
| Unsaturated fatty acids under salt stress | Sui et al. (2018) |
| Drought stress | Zhao et al. (2018) |
| Allergens | Watson et al. (2017) |
| Pod development | Wan et al. (2017) |
| Stilbenoid prenyltransferases | Yang et al. (2018) |
| Calcium regulation | Yang et al. (2017) |
| Regulation of plant triacylglycerol metabolism | Zheng et al. (2017) |
| Late leaf spot | Han et al. (2017) |
| Calcium and hormone-related genes | Li et al. (2017b) |
| TSW infection | Shrestha et al. (2017) |
| Presence of beneficial root pathogens | Hao et al. (2017) |
| Plant–nematode interaction | Guimaraes et al. (2017) |
| Photoreceptor lineage gene | Aavani et al. (2017) |
| Embryogenesis and early pod development | Gao et al. (2017) |
| Nodulation | Peng et al. (2017b) |
| Seed coat cracking and pigmentation | Wang et al. (2016a) |
| Lipid accumulation and degradation | Wang et al. (2016b) |
| Developmental genes | Clevenger et al. (2016) |
| Genetic variability | Chopra et al. (2016) |
| Light regulation in embryo and pod development | Zhang et al. (2016) |

(continued)

**Table 16** (continued)

| Trait | Reference |
|---|---|
| Oil biosynthesis | Gupta et al. (2016) |
| Post-harvest aflatoxin contamination | Wang et al. (2016a) |
| Drought tolerance | Brasileiro et al. (2015) |
| Geocarpy | Chen et al. (2016a, b, c) |
| Nematode resistance | Guimaraes et al. (2015) |
| Geocarpic pod development | Zhao et al. (2015) |
| Triacylglycerol biosynthetic pathways | Yu et al. (2015) |
| Infection with the *Ralstonia solanacearum* | Chen et al. (2014) |
| Water deficit and abscisic acid | Li et al. (2014) |
| Seed abortion | Zhu et al. (2014) |
| Oil accumulation | Yin et al. (2013) |
| Gravitropic response | Li et al. (2013) |
| Early pod development in dark | Xia et al. (2013) |
| Early embryo abortion | Chen et al. (2013) |
| Drought and fungal infection | Guimarães et al. (2012) |
| Seed development | Zhang et al. (2012) |
| Oil content under stress | Deng et al. (2012) |

**Table 17** Transcriptome sequences available in groundnut

| | | Library layout | | Platform | | | | | |
|---|---|---|---|---|---|---|---|---|---|
| Species | SRA | Paired | Single | ABI Solid | BgiSeq | Illumina | Ion torrent | LS454 | PacBio SMRT |
| *A. hypogaea* | 3742 | 2895 | 847 | 1 | 43 | 3364 | 6 | 30 | 298 |
| *A. monticola* | 47 | 39 | 8 | 0 | 0 | 45 | 0 | 0 | 2 |
| *A. duranensis* | 37 | 23 | 14 | 0 | 0 | 30 | 0 | 6 | 1 |
| *A. ipaensis* | 36 | 12 | 24 | 0 | 0 | 35 | 0 | 0 | 1 |
| *A. dardani* | 16 | | 16 | 0 | 0 | 16 | 0 | 0 | 0 |
| *A. stenosperma* | 12 | 10 | 2 | 0 | 0 | 10 | 0 | 2 | 0 |
| *A. cardenasii* | 7 | 7 | 0 | 0 | 0 | 7 | 0 | 0 | 0 |
| *A. ipaensis* × *A. duranensis* | 7 | 7 | 0 | 0 | 0 | 7 | 0 | 0 | 0 |
| *A. batizocoi* | 6 | 6 | 0 | 0 | 0 | 6 | 0 | 0 | 0 |
| *A. hoehnei* | 4 | 4 | 0 | 0 | 0 | 4 | 0 | 0 | 0 |
| *A. magna* | 3 | 3 | 0 | 0 | 0 | 3 | 0 | 0 | 0 |
| *A. villosa* | 3 | 3 | 0 | 0 | 0 | 3 | 0 | 0 | 0 |
| *A. diogoi* | 2 | 2 | 0 | 0 | 0 | 2 | 0 | 0 | 0 |
| *A. paraguariensis* | 2 | 2 | 0 | 0 | 0 | 2 | 0 | 0 | 0 |
| *A. correntina* | 2 | 2 | 0 | 0 | 0 | 2 | 0 | 0 | 0 |
| *A. helodes* | 2 | 2 | 0 | 0 | 0 | 2 | 0 | 0 | 0 |
| *A. pintoi* | 2 | 2 | 0 | 0 | 0 | 2 | 0 | 0 | 0 |
| *A. batizocoi* × (*A. cardenasii* × *A. diogoi*) | 1 | 1 | 0 | 0 | 0 | 1 | 0 | 0 | 0 |
| *A. cruziana* | 1 | 1 | 0 | 0 | 0 | 1 | 0 | 0 | 0 |

*SRA* sequence read archive

91114 (Sinha et al. 2020). Using RNA-seq of 39 different tissues and imposing hormones treatments and biotic could identify 91.73% of total annotated genes (Zhuang et al. 2019).

The sequence based alignment and functional analysis reported in different species of Arachis as on March 2020 are summarised in the following Table 17.

## 9 Conclusions

Though the global groundnut area may not increase considerably, due to the market demand and increasing number value added products, the groundnut production will increase further. This increase has to come from the higher productivity to be attained rather than expansion of area of cultivation. Considerable progress has been made in groundnut genetics, genomics and breeding, thereby contributing to the enhanced productivity and production of groundnut globally. Development of improved cultivars will continue to play a pivotal role in enhancing the stability of production and productivity in the different groundnut production systems. In India alone, more than 200 varieties have been developed so far for commercial cultivation of which nearly 50 are currently in cultivation. A large number of germplasm accessions and breeding lines were identified or developed for using under the breeding programmes as sources of variability. A good amount of efforts are already on to improve the quality for diversified products for confection and to ensure food safety by minimising aflatoxin and other harmful contamination. Though globally the primary target of groundnut improvement remain higher yield, genetic resistance to biotic and abiotic stress also is a major concern. Recently much attention has also been paid to the genetic improvement of quality paraments like oil quality, protein and less pre-disposition to aflatoxin production. An invaluable treasure of genes for economically important traits especially for resistance to biotic and abiotic stress is available in the wild *Arachis* species, which remain grossly underutilised. Concerted efforts are also now on to develop pre-breeding lines using the interspecific crosses.

After an inertia, a good amount of genomic resources for groundnut have been added during the last decade, and very powerful tools have been developed for the utilisation of the resources. Marker-assisted breeding in groundnut has picked up and commercial cultivars developed through MAB and MABC started coming up already. The trait mapping approaches can bring out linked markers and diagnostic markers to be used in breeding programmes. The gene atlas of the functional genes have been prepared after the genome sequences of diploid and tetraploid groundnut become available. This will open up more opportunities to identify additional molecular markers and candidate genes. Increasing accessibility to the high-throughput sequencing techniques which is becoming cheaper and cheaper has accelerated the molecular and genomics-assisted breeding which is expected to bear fruits in the times to come.

Though several groups in India and other countries have been working on transgenics of groundnut, so far no commercial cultivars from this technology has been released (this article did exclude the developments on transgenic research in groundnut).

## References

Aavani T, Tachibana N, Wallace V, Biernaskie J, Schuurmans C (2017) Temporal profiling of photoreceptor lineage gene expression during murine retinal development. Gene Expr Patterns 23–24:32–44

Agarwal G, Clevenger J, Pandey MK, Wang H, Shasidhar Y, Chu Y, Fountain JC, Choudhary D, Culbreath AK, Liu X, Huang G, Wang X, Deshmukh R, Holbrook CC, Bertioli DJ, Ozias-Akins P, Jackson SA, Varshney RK, Guo B (2018) High-density genetic map using whole-genome resequencing for fine mapping and candidate gene discovery for disease resistance in peanut. Plant Biotechnol J 16(11):1954–1967

Alves DMT, Pereira RW, Leal-Bertioli SCM, Moretzsohn MC, Guimarães PM, Bertioli DJ (2008) Development and use of single nucleotide polymorphism markers for candidate resistance genes in wild peanuts (Arachis spp.). Genet Mol Res 7:631–642

Anderson WF, Holbrook CC, Culbreath AK (1996) Screening the peanut core collection for resistance to tomato spotted wilt virus. Peanut Sci 23:57–61

Arunyanark A, Jogloy S, Akkasaeng C, Vorasoot N, Kesmala T, Nageswara Rao RC, Wright GC, Patanothai A (2008) Chlorophyll stability is an indicator of drought tolerance in peanut. J Agron Crop Sci 194:113–125

Arya SS, Salve AR, Chauhan S (2016) Peanuts as functional food: a review. J Food Sci Technol 53:31–41

Ballmer-Weber BK, Lidholm J, Fernandez-Rivas M, Seneviratne S, Hanschmann K-M, Vogel L, Bures P, Fritsche P, Summers C, Knulst AC, Le T-M, Reig I, Papadopoulos NG, Sinaniotis A, Belohlavkova S, Popov T, Kralimarkova T, de Blay F, Purohit A, Clausen M, Jedrzejczak-Czechowcz M, Kowalski ML, Asero R, Dubakiene R, Barreales L, Clare Mills EN, van Ree R, Vieths S (2015) IgE recognition patterns in peanut allergy are age dependent: perspectives of the EuroPrevall study. Allergy 70:391–407

Bechara MD, Moretzsohn MC, Palmieri DA, Monteiro JP, Bacci Junior M, Valls JFM, Lopes CR, Gimenes MA (2010) Phylogenetic relationships in genus Arachis based on ITS and 5.8S rDNA sequences. BMC Plant Biol 10:255

Belamkar V, Selvaraj MG, Ayers JL, Payton PR, Puppala N, Burow MD (2011) A first insight into population structure and linkage disequilibrium in the US peanut minicore collection. Genetica 139:411–429

Bentham G (1841) On the structure and affinities of Arachis and Voandzeia. Trans Linn Soc London 18(2):155–162

Benz B (2012) The conservation of cultivated plants. Nat Educ Knowl 3:4

Bera SK, Rathanakumar AL, Radhakrishnan T (2004) Cytogenetics and utilization of Arachis species. In: Basu MS, Singh NB (eds) Groundnut research in India. National Research Center for Groundnut (ICAR), Junagadh

Bera SK, Kamdar JH, Kasundra SV, Ajay BC (2017) A novel QTL governing resistance to stem rot disease caused by Sclerotium rolfsii in peanut. Australas Plant Pathol 45(6):637–644

Bera SK, Kamdar JH, Kasundra SV, Dash P, Maurya AK, Jasani MD, Chandrashekar AB, Manivannan N, Vasanthi RP, Dobariya KL, Pandey MK, Janila P, Radhakrishnan T, Varshney RK (2018) Improving oil quality by altering levels of fatty acids through marker-assisted selection of ahfad2 alleles in peanut (Arachis hypogaea L.). Euphytica 214:162

Bera SK, Kamdar JH, Kasundra SV, Patel SV, Jasani MD, Maurya AK, Dash P, Ajay BC, Rani K, Manivannan N, Janila P, Pandey MK, Vasanthi RP, Dobariya KL, Radhakrishnan T, Varshney RK (2019) Steady expression of high oleic acid in peanut bred by marker-assisted backcrossing for fatty acid desaturase mutant alleles and its effect on seed germination along with other seedling traits. PLoS One 14:e0226252

Bertioli DJ, Ozias-Akins P, Chu Y, Dantas KM, Santos SP, Gouvea E, Moretzsohn MC (2014) The use of SNP markers for linkage mapping in diploid and tetraploid peanuts. G3 (Bethesda) 4:89–96

Bertioli DJ, Cannon SB, Froenicke L et al (2016) The genome sequences of Arachis duranensis and Arachis ipaensis, the diploid ancestors of cultivated peanut. Nat Genet 48:438–446

Bhogireddy S, Xavier A, Garg V, Layland N, Arias R, Payton P, Nayak SN, Pandey MK, Puppala N, Varshney RK (2020) Genome-wide transcriptome and physiological analyses provide new insights into peanut drought response mechanisms. Sci Rep 10:4071

Bhat RS, Rockey J, Shirasawa K Tilak IS, Patil MPB, Lachagari VBR (2020) DNA methylation and expression analyses reveal epialleles for the foliar disease resistance genes in peanut (Arachis hypogaea L.). BMC Res Notes 13:20. https://doi.org/10.1186/s13104-020-4883-y

Bianchi-Hall C, Keys RD, Stalker HT (1993) Diversity of seed storage proteins in wild peanuts (Arachis species). Plant Syst Evol 186:1–15

Birthal PS, Rao PP, Nigam SN, Bantilan CS, Bhagavatulu S (2010) Groundnut and soybean economies in Asia: facts, trends and outlook. International Crops Research Institute for the Semi-Arid Tropics, Patancheru

Bishi SK, Kumar L, Mahatma MK, Khatediya N, Chauhan SM, Misra JB (2015) Quality traits of Indian peanut cultivars and their utility as nutritional and functional food. Food Chem 167:107–114

Bosamia TC, Mishra GP, Radhakrishnan T, Dobaria JR (2015) Novel and stress relevant EST derived SSR markers developed and validated in peanut. PLoS One 10(6):e0129127

Bosamia T, Dodia SM, Mishra GP, Ahmad S, Joshi B, Thirumalaisamy PP, Kumar N, Rathnakumar AL, Sangh C, Kumar A, Radhakrishnan T (2020) Unraveling the mechanisms of resistance to Sclerotium rolfsii in peanut (Arachis hypogaea L.) using comparative RNA-Seq analysis of resistant and susceptible genotypes. PLoS One 15(8):e0236823

Branch WD (2002) Registration of 'Georgia-01R' peanut. Crop Sci 42:1750–1751

Branch WD, Brenneman TB, Hookstra G (2014) Field test results versus marker assisted selection for root-knot nematode resistance in peanut. Peanut Sci 41(2):85–89

Brasileiro AC, Morgante CV, Araujo AC, Leal-Bertioli SC, Silva AK, Martins AC, Vinson CC, Santos CM, Bonfim O, Togawa RC, Saraiva MA, Bertioli DJ, Guimaraes PM (2015) Transcriptome profiling of wild Arachis from water-limited environments uncovers drought tolerance candidate genes. Plant Mol Biol Rep 33:1876–1892

Brown AHD (1989) Core collections: a practical approach to genetic resource management. Genome 31:818–824

Bryant RJ, Rao DR, Ogutu S (2004) Alpha and beta-galactosidase activities and oligosaccharide content in peanuts. Plant Foods Hum Nutr 58:213–223

Burow MD, Simpson CE, Starr JL, Paterson AH (2001) Transmission genetics of chromatin from a synthetic amphiploid in cultivated peanut (A hypogaea L): broadening the gene pool of a monophyletic polyploid species. Genetics 159:823–837

Burow MD, Starr JL, Park CH et al (2014) Introgression of homeologous quantitative trait loci (QTLs) for resistance to the root-knot nematode [Meloidogyne arenaria (Neal) Chitwood] in an advanced backcross-QTL population of peanut (Arachis hypogaea L.). Mol Breed 34(2):393–406

Carrin M, Carelli A (2010) Peanut oil: compositional data. Eur J Lipid Sci Technol 112:697–707

Casini C, Dardanelli JL, Martinez MJ, Balzarini M, Borgogno CS, Nassetta M (2003) Oil quality and sugar content of peanuts (Arachis hypogaea) grown in Argentina: their relationship with climatic variables and seed yield. J Agric Food Chem 51:6309–6313

Ceccarelli S, Grando S, Baum M (2007) Participatory plant breeding in water-limited environments. Exp Agric 43:411–435

Chakraborty K, Mahatma MK, Thawait LK, Bishi SK, Kalariya KA, Singh AL (2016) Water deficit stress affects photosynthesis and the sugar profile in source and sink tissues of groundnut (Arachis hypogaea L.) and impacts kernel quality. J Appl Bot Food Qual 89:98–104

Chamberlin KDC, Melouk HA, Payton ME (2010) Evaluation of the US peanut mini core collection using a molecular marker for resistance to Sclerotinia minor Jagger. Euphytica 172:109–115

Chavarro C, Chu Y, Holbrook CC, Isleib T, Bertioli D, Hovav R, Butts C, Lamb M, Sorensen R, Jackson SA, Ozias-Akins P (2020) Pod and seed trait QTL identification to assist breeding for peanut market preferences. G3 Genes Genomes Genet 10(7):2297–2315

Chen Z, Wang ML, Barkley NA, Pittman RN (2010) A simple allele-specific PCR-assay for detecting FAD2 alleles in both A and B genomes of the cultivated peanut for high-oleate trait selection. Plant Mol Biol Rep 28:542–548

Chen X, Zhu W, Azam S, Li H, Zhu F, Li H, Hong Y, Liu H, Zhang E, Wu H, Yu S, Zhou G, Li S, Zhong N, Wen S, Li X, Knapp SJ, Ozias-Akins P, Varshney RK, Liang X (2013) Deep sequencing analysis of the transcriptomes of peanut aerial and subterranean young pods identifies candidate genes related to early embryo abortion. Plant Biotechnol J 11(1):115–127

Chen Y, Ren X, Zhou X, Huang L, Yan L, Lei Y, Liao B, Huang J, Huang S, Wei W, Jiang H (2014) Dynamics in the resistant and susceptible peanut (Arachis hypogaea L.) root transcriptome on infection with the Ralstonia solanacearum. BMC Genomics 15:1078

Chen W, Jiao Y, Cheng L, Huang L, Liao B, Tang M, Jiang H (2016a) Quantitative trait locus analysis for pod- and kernel-related traits in the cultivated peanut (Arachis hypogaea L.). BMC Genet 17:25

Chen X, Li H, Pandey MK, Yang Q, Wang X, Garg V, Yud S (2016b) Draft genome of the peanut A-genome progenitor (Arachis duranensis) provides insights into geocarpy, oil biosynthesis, and allergens. PNAS 113:6785–6790

Chen X, Yang Q, Li H, Li H, Hong Y, Pan L, Chen N, Zhu F, Chi X, Zhu W, Chen M, Liu H, Yang Z, Zhang E, Wang T, Zhong N, Wang M, Liu H, Wen S, Li X, Zhou G, Li S, Wu H, Varshney R, Liang X, Yu S (2016c) Transcriptome-wide sequencing provides insights into geocarpy in peanut (Arachis hypogaea L.). Plant Biotechnol J 14(5):1215–1224

Chen C, Cao Q, Jiang Q, Li J, Yu R, Shi G (2019a) Comparative transcriptome analysis reveals gene network regulating cadmium uptake and translocation in peanut roots under iron deficiency. BMC Plant Biol 19:35

Chen X, Lu Q, Liu H, Zhang J, Hong Y, Lan H, Li H, Wang J, Liu H, Li S, Pandey MK, Zhang Z, Zhou G, Yu J, Zhang G, Yuan J, Li X, Wen S, Meng F, Yu S, Wang X, Siddique KHM, Liu ZJ, Paterson AH, Varshney RK, Liang X (2019b) Sequencing of cultivated peanut, Arachis hypogaea, yields insights into genome evolution and oil improvement. Mol Plant 12:920–934

Chen Y, Wang Z, Ren X, Huang L, Guo J, Zhao J, Zhou X, Yan L, Luo H, Liu N, Chen W, Wan L, Lei Y, Liao B, Huai D, Jiang H (2019c) Identification of major QTL for seed number per pod on chromosome A05 of tetraploid peanut (Arachis hypogaea L.). Crop J 7(2):238–248

Chen H, Yang Q, Chen K, Zhao S, Zhang C, Pan R, Cai T, Deng Y, Wang X, Chen Y, Chu W, Xie W, Zhuang W (2019d) Integrated microRNA and transcriptome profiling reveals a miRNA-mediated regulatory network of embryo abortion under calcium deficiency in peanut (Arachis hypogaea L.). BMC Genomics 21;20(1):392

Choi K, Burow MD, Church G, Burow G, Paterson AH, Simpson CE, Starr JL (1999) Genetics and mechanism of resistance to Meloidogyne arenaria in peanut germplasm. J Nematol 31:283–290

Chopra R, Burow G, Farmer A, Mudge J, Simpson CE, Wilkins TA, Burow MD (2015) Next generation transcriptome sequencing, SNP discovery and validation in four market classes of peanut Arachis hypogaea L. Mol Gen Genomics 290:1169–1180

Chopra R, Burow G, Simpson CE, Chagoya J, Mudge J, Burow MD (2016) Transcriptome sequencing of diverse peanut (Arachis) wild species and the cultivated species reveals a wealth of untapped genetic variability. G3 (Bethesda) 6(12):3825–3836

Chu Y, Holbrook CC, Timper P, Ozias-Akins P (2007) Development of a PCR based molecular marker to select for nematode resistance in peanut. Crop Sci 47:841–845
Chu Y, Holbrook CC, Ozias-Akins P (2009) Two alleles of ahFAD2B controls the high oleic acid trait in cultivated peanut. Crop Sci 49:2029–2036
Chu Y, Wu CL, Holbrook CC, Tillman B, Person G, Ozias-Akins P (2011) Marker assisted selection to pyramid nematode resistance and high oleic trait in peanut. Plant Genome 4:110–117
Chukwumah Y, Walker LT, Verghese M (2009) Peanut skin color: a biomarker for total polyphenolic content and antioxidative capacities of peanut cultivars. Int J Mol Sci 10:4941–4952
Clavel D, Drame NK, Roy Macauley H, Braconnier S, Laffray D (2005) Analysis of early responses to drought associated with field drought adaptation in four Sahelian groundnut (Arachis hypogaea L.) cultivars. Environ Exp Bot 54:219–230
Clements RS, Darnell B (1980) Myo-inositol content of common foods: development of a high-myo-inositol diet. Am J Clin Nutr 33:1954–1967
Clevenger J, Chu Y, Scheffler B, Ozias-Akins P (2016) A developmental transcriptome map for allotetraploid Arachis hypogaea. Front Plant Sci 7:1446
Clevenger J, Chu Y, Chavarro C, Agarwal G, Bertioli DJ, Leal-Bertioli SCM, Ozias-Akins P (2017) Genome-wide SNP genotyping resolves signatures of selection and tetrasomic recombination in peanut. Mol Plant 10:309–322
Croze ML, Soulage CO (2013) Potential role and therapeutic interests of myo-inositol in metabolic diseases. Biochimie 95:1811–1827
Dai J, Qiu W, Wang N, Nakanishi H, Zuo Y (2018) Comparative transcriptomic analysis of the roots of intercropped peanut and maize reveals novel insights into peanut iron nutrition. Plant Physiol Biochem 127:516–524
Damame SV, Chavan JK, Kadam SS (1990) Effects of roasting and storage on proteins and oil in peanut kernels. Plant Foods Hum Nutr 40:143–148
Deng X, Li Z, Zhang W (2012) Transcriptome sequencing of Salmonella enterica serovar Enteritidis under desiccation and starvation stress in peanut oil. Food Microbiol 30(1):311–315
Desmae H, Moses O, Bonny N, Farid W, Jupiter J, Diarra A, Amadou A, Abdoulaye I, Kodio O, Cisse Y, Traore M, Moutari A, Zarafi AM, Mamadou CA, Ajeigbe H, Motagi B, Candidus E, Ahmed B, Alabi L, Amos (2016) Enhancing groundnut productivity and production in West and Central Africa. In: Monyo ES, Varshney RK (eds) Seven seasons of learning and engaging smallholder farmers in the drought-prone areas of sub-Saharan Africa and South Asia through Tropical Legumes, 2007–2014. ICRISAT, Patancheru
Desmae H, Janila P, Okori P, Pandey MK, Motagi BN, Monyo E, Mponda O, Okello D, Sako D, Echeckwu C, Oteng-Frimpong R, Miningou A, Ojiewo C, Varshney RK (2019) Genetics, genomics and breeding of groundnut (Arachis hypogaea L.) Plant Breed 138(4):425–444
Devi MJ, Bhatnagar-Mathur P, Sharma KK, Serraj R, Anwar SY, Vadez V (2011) Relationships between transpiration efficiency and its surrogate traits in the rd29A:DREB1A transgenic lines of groundnut. J Agron Crop Sci 197:272–283
Dey R, Pal KK, Chauhan SM, Bhatt DM, Misra JB (2001) Cellulolytic and groundnut shell decomposition potential of some microorganisms. Indian J Microbiol 42:165–167
Do AN, Watson CT, Cohain AT, Griffin RS, Grishin A, Wood RA, Wesley Burks A, Jones SM, Scurlock A, Leung DYM, Sampson HA, Sicherer SH, Sharp AJ, Schadt EE, Bunyavanich S (2019) Dual transcriptomic and epigenomic study of reaction severity in peanut-allergic children. J Allergy Clin Immunol 145(4):1219–1230
Dodia SM, Joshi B, Gangurde SS, Thirumalaisamy PP, Mishra GP, Narendrakumar D, Soni P, Rathnakumar AL, Dobaria JR, Sangh C, Chitikineni A, Chanda SV, Pandey MK, Varshney RK, Radhakrishnan T (2019) Genotyping-by-sequencing based genetic mapping reveals large number of epistatic interactions for stem rot resistance in groundnut. Theor Appl Genet 132:1001–1016
Dong WB, Holbrook CC, Timper P, Brenneman TB, Chu Y, Ozias-Akins P (2008) Resistance in peanut cultivars and breeding lines to three root-knot nematode species. Plant Dis 92:631–638

Dwivedi SL, Nigam SN (1995) Breeding for improved seed quality in groundnut (A. hypogaea L.) with special reference to export promotion: issues and opportunities. Paper presented at workshop on Status of Confectionery Groundnut Research in India and Strategies for Export Promotion 28–29 April, 1995. BARC, Trombay, India

Dwivedi SL, Nigam SN, Rao RCN, Singh U, Rao KVS (1996) Effect of drought on oil, fatty acids and protein contents of groundnut (Arachis hypogaea L.) seeds. Field Crops Res 48:125–133

Dwivedi SL, Puppala N, Upadhyaya HD, Manivannan N, Singh S (2008) Developing a Core Collection of Peanut Specific to Valencia Market Type. Crop Science 48:625–632

Falalou H, Mariama H, Achirou FB, Oumarou H, Vincent V (2018) Genotypic variation for root development, water extraction and yield components in groundnut under low phosphorus and drought stresses. Am J Agric For 6(5):122–131

FAOSTAT (2018) FAOSATAT, statistical data base. Food and Agricultural Organizations of the United Nations, Rome

Faye I, Pandey MK, Hamidou F, Rathore A, Ndoye O, Vadez V, Varshney RK (2015) Identification of quantitative trait loci for yield and yield related traits in groundnut (Arachis hypogaea L.) under different water regimes in Niger and Senegal. Euphytica 206:631–647

Feldman EB (1999) Assorted monounsaturated fatty acids promote healthy hearts. Am J Clin Nutr 70:953–954

Feng S, Wang X, Xinyou Z, Dang P, Holbrook CC, Culbreath AK, Wu Y (2012) Peanut (Arachis hypogaea) expressed sequence tag project: progress and application. Comp Funct Genomics 5013:373768

Fernandez A, Krapovickas A (1994) Cromosomas y evolucion en Arachis (Leguminosae). (In Spanish, with English abstract.). Bonplandia 8:187–220

Foncéka D, Hodo-Abalo T, Rivallan R, Faye I, Sall MN, Ndoye O, Fávero AP, Bertioli DJ, Glaszmann JC, Courtois B, Rami JF (2009) Genetic mapping of wild introgressions into cultivated peanut: a way toward enlarging the genetic basis of a recent allotetraploid. BMC Plant Biol 9:103

Foncéka D, Tossim HA, Rivallan R, Vignes H, Faye I, Ndoye O, Rami JF (2012a) Fostered and left behind alleles in peanut: inter-specific QTL mapping reveals footprints of domestication and useful natural variation for breeding. BMC Plant Biol 12:26

Foncéka D, Tossim HA, Rivallan R, Vignes H, Lacut E, de Bellis F, Rami JF (2012b) Construction of chromosome segment substitution lines in peanut (Arachis hypogaea L.) using a wild synthetic and QTL mapping for plant morphology. PLoS One 7(11):e48642

Fountain JC, Koh J, Yang L, Pandey MK, Nayak SN, Bajaj P, Zhuang WJ, Chen ZY, Kemerait RC, Lee RD, Chen S, Varshney RK, Guo B (2018) Proteome analysis of Aspergillus flavus isolate-specific responses to oxidative stress in relationship to aflatoxin production capability. Sci Rep 8(1):3430

Francis G, Kerem Z, Makkar HP, Becker K (2002) The biological action of saponins in animal systems: a review. Br J Nutr 88:587–605

Francisco ML, Resurreccion AV (2008) Functional components in peanuts. Crit Rev Food Sci Nutr 48:715–746

Gao C, Wang P, Zhao S, Zhao C, Xia H, Hou L, Ju Z, Zhang Y, Li C, Wang X (2017) Small RNA profiling and degradome analysis reveal regulation of microRNA in peanut embryogenesis and early pod development. BMC Genomics 18(1):220

Garcia GM, Stalker HT, Shroeder E, Kochert G (1996) Identification of RAPD, SCAR, and RFLP markers tightly linked to nematode resistance genes introgressed from Arachis cardenasii into Arachis hypogaea. Genome 39(5):836–845

Garcia GM, Stalker HT, Schroeder E, Lyerly JH, Kochert G (2005) RAPD-based linkage map of peanut based on a backcross population between the two diploid species Arachis stenosperma and A cardenasii. Peanut Sci 32:1–8

Gautami B, Pandey MK, Vadez V, Nigam SN, Ratnakumar P, Krishnamurthy L, Radhakrishnan T, Gowda MVC, Narasu ML, Hoisington DA, Knapp SJ, Varshney RK (2012) QTL analysis and consensus genetic map for drought tolerance traits based on three RIL populations of cultivated groundnut (Arachis hypogaea L.). Mol Breed 32:757–772

GCP (2011) CGIAR generation challenge programme: project updates (incorporating projects completed in 2010 and 2009) Texcoco, Mexico: generation challenge programme. Genome 4:110–117

Girdthai T, Jogloy S, Akkasaeng C, Vorasoot N, Wongkaew S, Holbrook CC, Patanothai A (2010) Heritability of, and genotypic correlations between, aflatoxin traits and physiological traits for drought tolerance under end of season drought in peanut (Arachis hypogaea L.). Field Crops Res 118:169–176

Gorbet DW, Knauft DA (1997) Registration of 'SunOleic 95R' peanut. Crop Sci 37:1392

Gorbet DW, Shokes FM (2002a) Registration of 'C-99R' peanut. Crop Sci 42:2207

Gorbet DW, Shokes FM (2002b) Registration of 'Florida MDR 98' peanut. Crop Sci 42:2207–2208

Gregory WC, Gregory MP (1976) Groundnut, Arachis hypogaea (Leguminosae-Papilionatae). In: Simmonds NW (ed) Evolution of crop plants. Longman, London, pp 151–154

Gregory MP, Gregory WC (1979) Exotic germplasm of Arachis L interspecific hybrids. J Hered 70:185–193

Gregory WC, Gregory MP, Krapovickas A, Smith BW, Yarbrough J (1973) Structure and genetic resources of peanuts. In: Peanuts—culture and uses. American Peanut Research Education Association, Stillwater, pp 47–133

Gregory WC, Krapovickas A, Gregory MP (1980) Structures, variation, evolution and classification in Arachis. In: Summerfield RJ, Bunting AH (eds) Advances in legume science. Royal Botanical Gardens, London, pp 469–481

Guimarães PM, Garsmeur O, Proite K, Leal-Bertioli SCM, Seijo G, Christian C, Bertioli DJ, D'Hont A (2008) BAC libraries construction from the ancestral diploid genomes of the allotetraploid cultivated peanut. BMC Plant Biol 8:14

Guimarães PM, Brasileiro ACM, Proite K, Araújo ACG, Leal-Bertioli SCM, Pic-Taylor A, Silva FR, Morgante CV, Ribeiro SG, Bertioli DJ (2010) A study of gene expression in the nematode resistant wild peanut relative, Arachis stenosperma, in response to challenge with Meloidogyne arenaria. Trop Plant Biol 3:183–192

Guimarães PM, Brasileiro AC, Morgante CV, Martins AC, Pappas G, Silva OB Jr, Togawa R, Leal-Bertioli SC, Araujo AC, Moretzsohn MC, Bertioli DJ (2012) Global transcriptome analysis of two wild relatives of peanut under drought and fungi infection. BMC Genomics 13:387

Guimaraes PM, Guimaraes LA, Morgante CV, Silva OB Jr, Araujo AC, Martins AC, Saraiva MA, Oliveira TN, Togawa RC, Leal-Bertioli SC, Bertioli DJ, Brasileiro AC (2015) Root transcriptome analysis of wild peanut reveals candidate genes for nematode resistance. PLoS One 10(10):e0140937

Guimaraes LA, Pereira BM, Araujo ACG, Guimaraes PM, Brasileiro ACM (2017) Ex vitro hairy root induction in detached peanut leaves for plant-nematode interaction studies. Plant Methods 13:25

Guo Y, Khanal S, Tang S, Bowers JE, Heesacker AF, Khalilian N, Nagy ED, Zhang D, Taylor CA, Stalker HT, Ozias-Akins P, Knapp SJ (2012) Comparative mapping in intraspesific populations uncovers high degree of macrosynteny between A- and B-genome diploid species of peanut. BMC Genomics 13:608

Guo BZ, Pandey MK, He G, Zhang XY, Liao B, Culbreath AK, Varshney RK, Nwosu V, Wilson RF, Stalker HT (2013) Recent advances in molecular genetic linkage maps of cultivated peanut (Arachis hypogaea L.). Peanut Sci 40(2):95–106

Guo F, Ma J, Hou L, Shi S, Sun J, Li G, Zhao C, Xia H, Zhao S, Wang X, Zhao Y (2020) Transcriptome profiling provides insights into molecular mechanism in Peanut semi-dwarf mutant. BMC Genomics 21:211

Gupta SK, Dhawan K, Kumar P, Yadava TP (1982) Note on chemical composition of some groundnut strains. Ind J Agric Sci 52:343–344

Gupta K, Kayam G, Faigenboim-Doron A, Clevenger J, Ozias-Akins P, Hovav R (2016) Gene expression profiling during seed-filling process in peanut with emphasis on oil biosynthesis networks. Plant Sci 248:116–127

Guthrie JD, Hoffpauir CL, Steiner ET, Stansbury MF (1949) Survey of the chemical composition of cotton fibers, cottonseed, peanuts, and sweet potatoes: a literature review. Southern Regional Research Lab, New Orleans

Hake AA, Shirasawa K, Yadawad A, Sukruth M, Patil M, Nayak SN, Bhat RS (2017) Mapping of important taxonomic and productivity traits using genic and non-genic transposable element markers in peanut (Arachis hypogaea L.). PLoS One 12(10):e0186113

Halward TM, Stalker HT, LaRue EA, Kochert G (1992) Use of single primer DNA amplification in genetic studies of peanut. Plant Mol Biol 18:315–325

Halward TM, Stalker HT, Kochert G (1993) Development of an RFLP linkage map in diploid peanut species. Theor Appl Genet 87:379–384

Hamidou F, Ratnakumar P, Halilou O, Mponda O, Kapewa T, Monyo E, Faye I, Ntare BR, Nigam SN, Upadhyaya HD, Vadez V (2012) Selection of intermittent drought tolerant lines across years and locations in the reference collection of groundnut (Arachis hypogaea L.). Field Crops Res 126:189–199

Hammons RO (1977) Groundnut rust in the United States and the Caribbean. PANS 23(3):300–304

Hammons RO (1982) Origin and early history of the peanut. In: Pattee HE, Young CT (eds) Peanuts science and technology. American Peanut Research and Education Society, Stillwater, pp 1–20

Han S, Liu H, Yan M, Qi F, Wang Y, Sun Z, Huang B, Dong W, Tang F, Zhang X, He G (2017) Differential gene expression in leaf tissues between mutant and wild-type genotypes response to late leaf spot in peanut (Arachis hypogaea L.). PLoS One 12(8):e0183428

Hao K, Wang F, Nong X, McNeill MR, Liu S, Wang G, Cao G, Zhang Z (2017) Response of peanut Arachis hypogaea roots to the presence of beneficial and pathogenic fungi by transcriptome analysis. Sci Rep 7(1):964

Harris D, Matthews RB, Nageswara Rao RC, Williams JH (1988) The physiological basis for yield differences between four genotypes of groundnut (Arachis Hypogaea) in response to drought. III. Developmental processes. Exp Agric 24:215–226

Hashim IB, Koehler PB, Eitenmiller RR (1993) Tocopheroles in runner and virginia peanut cultivars at various maturity stages. J Am Oil Chem Soc 70:633–635

He G, Meng R, Gao H, Guo B, Newman M, Pittman RN, Prakash CS (2005) Simple sequence repeat markers for botanical varieties of cultivated peanut (Arachis hypogaea L.). Euphytica 142:131–136

Herselman L, Thwaites R, Kimmins FM, Seal SE (2004) Identification and mapping of AFLP markers linked to peanut (Arachis hypogaea L.) resistance to the aphid vector of groundnut rosette disease. Theor Appl Genet 109:1426–1433

Hilu KW, Stalker HT (1995) Genetic relationships between peanut and wild species of Arachis sect Arachis (Fabaceae): evidence from RAPDs. Syst Evol 198:167–178

Holbrook CC (2001) Status of the Arachis germplasm collection in the United States. Peanut Sci 28:84–89

Holbrook CC, Dong W (2005) Development and evaluation of a mini core collection for the U S peanut germplasm collection. Crop Sci 45:1540–1544

Holbrook CC, Anderson WF, Pittman RN (1993) Selection of a core collection from the United States germplasm collection of peanut. Crop Sci 33:859–861

Holbrook CC, Timper P, Culbreath A, Craig K (2008) Registration of 'Tifguard' peanut. J Plant Reg 2(2):92–94

Holbrook CC, Guo BZ, David W, Patricia T (2009) The U.S. breeding program to develop peanut with drought tolerance and reduced aflatoxin contamination. Peanut Sci 36(1):50–53

Hong Y, Liang X, Chen X, Liu H, Zhou G, Li S, Wen S (2008) Construction of genetic linkage map based on SSR markers in peanut (Arachis hypogaea L.). Agric Sci China 7:915–921

Hong Y, Chen X, Liang X, Liu H, Zhou G, Li S, Wen S, Holbrook CC, Guo B (2010) A SSR-based composite genetic linkage map for the cultivated peanut (Arachis hypogaea L.) genome. BMC Plant Biol 10:17

Hopkins MS, Casa AM, Wang T, Mitchell SE, Dean RE, Kochert GD, Kresovich S (1999) Discovery and characterization of polymorphic simple sequence repeats (SSRs) in cultivated peanut (Arachis hypogaea L). Crop Sci 39:1243–1247

Hu XH, Zhang SZ, Miao HR, Cui FG, Shen Y, Yang WQ, Xu TT, Chen N, Chi XY, Zhang ZM, Chen J (2018) High-density genetic map construction and identification of QTLs controlling oleic and linoleic acid in peanut using SLAF-seq and SSRs. Sci Rep 8:5479

Huang L, He H, Chen W, Ren X, Chen Y, Zhou X, Xia Y, Wang X, Jiang X, Liao B, Jiang H (2015) Quantitative trait locus analysis of agronomic and quality-related traits in cultivated peanut (Arachis hypogaea L). Theor Appl Genet 128:1103–1115

Huang L, Ren X, Wu B, Li X, Chen W, Zhou X, Chen Y, Pandey MK, Jiao Y, Luo H, Lei Y, Varshney RK, Liao B, Jiang H (2016) Development and deployment of a high-density linkage map identified quantitative trait loci for plant height in peanut (Arachis hypogaea L.). Sci Rep 6:39478

Huang J, Xing M, Li Y, Cheng F, Gu H, Yue C, Zhang Y (2019) Comparative transcriptome analysis of the skin-specific accumulation of anthocyanins in black peanut (Arachis hypogaea L.). J Agric Food Chem 67(4):1312–1324

Hurlburt BK, Offermann LR, McBride JK, Majorek KA, Maleki SJ, Chruszcz K (2013) Structure and function of the peanut Panallergen Arah8. J Biol Chem 288(52):36890–33690

Husted L (1936) Cytological studies of the peanut Arachis II. Chromosome number, morphology and behaviour and their application to the origin of cultivated form. Cytologia 1:396–423

Isleib TG, Wynne JC, Nigam SN (1994) Groundnut breeding. In: Smartt J (ed) The groundnut crop: a scientific basis for improvement. Chapman & Hall, London, pp 552–623

Izge AU, Mohammed ZH, Goni A (2007) Levels of variability in groundnut (Arachis hypogaea L.) to cercospora leafspot disease-implication for selection. Afr J Agric Res 2:182–186

Jackson BN, Aluru S, Schnable PS (2005) Consensus genetic maps: a graph theoretic approach. In: Proceeding of computational systems bioinformatics conference, pp 35–43

Jambunathan R, Raju MS, Barde SP (1985) Analysis of oil content of groundnuts by nuclear magnetic resonance spectrometry. J Sci Food Agric 36:162–166

Janila P, Aruna R, Kumar JE, Nigam SN (2012) Variation in blanchability in Virginia groundnuts (Arachis hypogaea L.). Indian J Oilseeds Res 29:116–120

Janila P, Nigam SN, Pandey MK, Nagesh P, Varshney RK (2013) Groundnut improvement: use of genetic and genomic tools. Plant Sci 4:1–12

Janila P, Nigam SN, Abhishek R, Kumar VA, Manohar SS, Venuprasad R (2014) Iron and zinc concentrations in peanut (Arachis hypogaea L.) seeds and their relationship with other nutritional and yield parameters. J Agric Sci 153:975–994

Janila P, Manohar SS, Rathore A, Nigam SN (2015) Inheritance of SPAD chlorophyll meter reading and specific leaf area in four crosses of groundnut (Arachis hypogaea L). Indian J Genet 75:408–412

Janila P, Pandey MK, Shasidhar Y, Variath MT, Sriswathi M, Khera P, Manohar SS, Nagesh P, Vishwakarma MK, Mishra GP, Radhakrishnan T, Manivannan N, Dobariya KL, Vasanthi RP, Varshney RK (2016a) Molecular breeding for introgression of fatty acid desaturase mutant alleles (ahFAD2A and ahFAD2B) enhances oil quality in high and low oil containing peanut genotypes. Plant Sci 242:203–213

Janila P, Variath MT, Pandey MK, Desmae H, Motagi BN, Okori P, Manohar SS, Rathnakumar AL, Radhakrishnan T, Liao B, Varshney RK (2016b) Genomic tools in groundnut breeding program: status and perspectives. Front Plant Sci 7:289

Jiang HF, Ren XP, Liao BS, Huang JQ, Lei Y, Chen BY, Guo BZ, Holbrook CC, Upadhyaya HD (2008) Peanut core collection established in china and compared with ICRISAT mini core collection. Acta Agron Sin 34(1):25–30

Jiang HF, Ren XP, Zhang XJ, Huang JQ, Lei Y, Yan LY, Liao BS, Upadhyaya HD, Holbrook CC (2010) Comparison of genetic diversity based on SSR markers between peanut mini core collections from China and ICRISAT. Acta Agron Sin 36(7):1084–1091

Jiang C, Zhang H, Ren J, Dong J, Zhao X, Wang X, Wang J, Zhong C, Zhao S, Liu X, Gao S, Yu H (2020) Comparative transcriptome-based mining and expression profiling of transcription factors related to cold tolerance in peanut. Int J Mol Sci 21(6):1921

Kanyika BTN, Lungu D, Mweetwa A, Kaimoyo E, Njung'e VM, Monyo ES, de Villiers SM (2015) Identification of groundnut (Arachis hypogaea L.) SSR markers suitable for multiple resistance traits QTL mapping in African germplasm. Electron J Biotechnol 18(2):61–67

Karmakar K, Kundu A, Rizvi AZ, Dubois E, Severac D, Czernic P, Cartieaux F, DasGupta M (2019) Transcriptomic analysis with the progress of symbiosis in 'crack-entry' legume Arachis hypogaea highlights its contrast with 'infection thread' adapted legumes. Mol Plant-Microbe Interact 32(3):271–285

Khedikar YP, Gowda MVC, Sarvamangala C, Patgar KV, Upadhyaya HD, Varshney RK (2010) A QTL study on late leaf spot and rust revealed one major QTL for molecular breeding for rust resistance in groundnut (Arachis hypogaea L.). Theor Appl Genet 121:971–984

Khera P, Pandey MK, Wang H, Feng S, Qiao L, Culbreath AK, Kale S, Wang J, Holbrook CC, Zhuang W, Varshney RK, Guo BZ (2016) Mapping quantitative trait loci of resistance to tomato spotted wilt virus and leaf spots in a recombinant inbred line population of peanut (Arachis hypogaea L.) from SunOleic 97R and NC94022. PLoS One 11(7):e0158452

Knoll JE, Ramos ML, Zeng Y, Holbrook CC, Chow M, Chen S, Ozias-Akins P (2011) TILLING for allergen reduction and improvement of quality traits in peanut (Arachis hypogaea L.). BMC Plant Biol 11:81

Kochert G, Halward T, Branch WD, Simpson CE (1991) RFLP variability in peanut (Arachis hypogaea L.) cultivars and wild species. Theor Appl Genet 81:565–570

Kochert G, Stalker HT, Gimenes M, Galgaro L, Lopes CR, Moore K (1996) RFLP and cytogenetic evidence on the origin and evolution of allotetraploid domesticated peanut, Arachis hypogaea (Leguminosae). Am J Bot 83:1282–1291

Kolekar RM, Sujay V, Shirasawa K, Sukruth M, Khedikar YP, Gowda MVC, Pandey MK, Varshney RK, Bhat RS (2016) QTL mapping for late leaf spot and rust resistance using an improved genetic map and extensive phenotypic data on a recombinant inbred line population in peanut (Arachis hypogaea L.). Euphytica 209:147–156

Kolekar RM, Sukruth M, Shirasawa K, Nadaf HL, Motagi BN, Lingaraju S, Patil PV, Bhat RS (2017) Marker-assisted backcrossing to develop foliar disease resistant genotypes in TMV 2 variety of peanut (Arachis hypogaea L.). Plant Breed 136(6):1–6

Korani W, Chu Y, Holbrook CC, Ozias-Akins P (2018) Insight into genes regulating postharvest aflatoxin contamination of tetraploid peanut from transcriptional profiling. Genetics 209(1):143–156

Krapovickas A (1969) The origin, variability and spread of the groundnut (Arachis hypogaea). In: Ucko PJ, Falk IS (eds) The domestication and exploitation of plants and animals. Gerald Duckworth, London, pp 424–441

Krapovickas A, Gregory WC (1994) Taxonomıa del genero Arachis (Leguminosae). Bonplandia 8(1–4):1–186

Kris-Etherton PM, Pearson TA, Wan Y, Hargrove RL, Moriarty K, Fishell V, Etherton TD (1999) High-monounsaturated fatty acid diets lower both plasma cholesterol and triacylglycerol concentrations. Am J Clin Nutr. 70(6):1009–1015

Krishnamurthy L, Vadez V, Devi MJ, Serraj R, Nigam SN, Sheshshayee MS, Chandra S, Aruna R (2007) Variation in transpiration efficiency and its related traits in a groundnut. Field Crops Res 103:189–197

Kuang Q, Yu Y, Attree R, Xu B (2017) A comparative study on anthocyanin, saponin, and oil profiles of black and red seed coat peanut (Arachis hypogaea) grown in China. Int J Food Prop 20:S131–S140

Kumar N, Ajay BC, Rathanakumar AL, Radhakrishnan T, Mahatma MK, Kona P, Chikani BM (2019) Assessment of genetic variability for yield and quality traits in groundnut genotypes. Electron J Plant Breed 10:196–206

Lal C, Hariprasanna K, Rathnakumar AL, Gor HK, Chikani BM (2006) Gene action for surrogate traits of wateruse efficiency and harvest index in peanut (Arachis hypogaea). Ann Appl Biol 148:165–172

Lanham PG, Fennell S, Moss JP, Powell W (1992) Detection of polymorphic loci in Arachis germplasm using random amplified polymorphic DNAs. Genome 35:885–889

Lavia GI (1999) Caracterizacion cromosomica del germoplasma de man. PhD dissertation, Universidad Nacional de Cordoba, Argentina

Lea CH (1962) The oxidative deterioration of food lipids. In: Schultz HW, Day EA Sinnhuber RO (eds) Symposium on food: lipids and their oxidation. The AVI Publishing Company, Westport, pp 3–28

Leal-Bertioli SCM, José AC, Alves-Freitas DM, Moretzsohn MC, Guimarães PM, Nielen S, Bertioli DJ (2009) Identification of candidate genome regions controlling disease resistance in Arachis. BMC Plant Biol 9:112

Leal-Bertioli SCM, Moretzsohn MC, Roberts PA, Ballen-Taborda C, Borba TCO, Valdisser PA, Vianello RP, Araujo ACG, Guimaraes PM, Bertioli DJ (2015) Genetic mapping of resistance to Meloidogyne arenaria in Arachis stenosperma: a new source of nematode resistance for peanut. G3 (Bethesda) 6(2):377–390

Leal-Bertioli SCM, Cavalante U, Gouvea EG, Ballén Taborda C, Shirasawa K, Guimarães PM, Moretzsohn MC (2016) Identification of QTLs for rust resistance in the peanut wild species Arachis magna and the development of KASP markers for marker assisted selection. G3 (Bethesda) 5:1403–1413

Li SX, Zhou GY, Liang XQ, Li SL, Lin YK, Hong YB, Chen XP, Liu HY (2007) Breeding of a new peanut variety Yueyou 114 and its cultivation technologies. J Peanut Sci 36:34–37

Li HF, Zhu FH, Li HY, Zhu W, Chen XP, Hong YB, Liu HY, Wu H, Liang XQ (2013) Proteomic identification of gravitropic response genes in peanut gynophores. J Proteome 93:303–313

Li X, Lu J, Liu S, Liu X, Lin Y, Li L (2014) Identification of rapidly induced genes in the response of peanut (Arachis hypogaea) to water deficit and abscisic acid. BMC Biotechnol 14:58

Li Y, Li L, Zhang X, Zhang K, Ma D, Liu J, Wan Y (2017a) QTL mapping and marker analysis of main stem height and the first lateral branch length in peanut (Arachis hypogaea L.). Euphytica 213:57

Li Y, Meng J, Yang S, Guo F, Zhang J, Geng Y, Cui L, Wan S, Li X (2017b) Transcriptome analysis of calcium- and hormone-related gene expressions during different stages of peanut pod development. Front Plant Sci 8:1241

Liang XQ, Luo M, Holbrook CC, Guo BZ (2006) Storage protein profiles in Spanish and runner market type peanuts and potential markers. BMC Plant Biol 6:24

Liang Y, Baring M, Wang S, Septiningsih EM (2017) Mapping QTLs for leaf spot resistance in peanut using SNP-based next-generation sequencing markers. Plant Breed Biotech 5(2):115–122

Liao BS (ed) (2003) The groundnut. Hubei Press for Science and Technology, Wuhan, (in Chinese) ISBN 7-5352-2919-0/S.327

Linnaeus C (1753) Species Plantarum: exhibentes plantas rite cognitas, ad genera relatas, cum differentiis specificis, nominibus trivialibus, synonymis selectis, locis natalibus, secundum systema sexuale digestas

Liu F, Zhang X, Lu C, Zeng X, Li Y, Fu D, Wu G (2015) Non-specific lipid transfer proteins in plants: presenting new advances and an integrated functional analysis. J Exp Bot 66:5663–5681

Liu H, Gu J, Lu Q, Li H, Hong Y, Chen X, Ren L, Deng L, Liang X (2019a) Transcriptomic analysis reveals the high-oleic acid feedback regulating the homologous gene expression of stearoyl-ACP desaturase 2 (SAD2) in peanuts. Int J Mol Sci 20(12):3091

Liu N, Guo J, Zhou X, Zhou X, Wu B, Huang L, Luo H, Chen Y, Chen W, Lei Y, Huang Y, Boshou Liao B, Jiang H (2019b) High-resolution mapping of a major and consensus quantitative trait locus for oil content to a ~0 8-Mb region on chromosome A08 in peanut (Arachis hypogaea L.). Theor Appl Genet 133:37–49

Lozano MG, Alan Sartori GO, Bastosb DHM, Regitano-d'Arce MAB (2019) Selected nutrients and antinutrients in peanut cultivars harvested in Brazil. J Sci Food Agric 99:5334–5340

Lu J, Pickersgill B (1993) Isozyme variation and species relationships in peanut and its wild relatives (Arachis L.–Leguminosae). Theor Appl Genet 85:550–560

Lu Q, Liu H, Hong Y, Li H, Liu H, Li X, Wen S, Zhou G, Li S, Chen X, Liang X (2018) Consensus map integration and QTL meta-analysis narrowed a locus for yield traits to 0.7 cM and refined a region for late leaf spot resistance traits to 0.38 cM on linkage group A05 in peanut (Arachis hypogaea L.). BMC Genomics 19:887

Luo H, Xu Z, Li Z, Li X, Lv J, Ren X, Jiang H (2017a) Development of SSR markers and identification of major quantitative trait loci controlling shelling percentage in cultivated peanut (Arachis hypogaea L). Theor Appl Genet 130:1635–1648

Luo H, Ren X, Li Z, Xu Z, Li X, Huang L, Jiang H (2017b) Co-localization of major quantitative trait loci for pod size and weight to a 3.7 cM interval on chromosome A05 in cultivated peanut (Arachis hypogaea L.). BMC Genomics 18:58

Luo Z, Cui R, Chavarro C, Tseng Y, Zhou H, Peng Z, Chu Y, Yang X, Lopez Y, Tillman B, Dufault N, Brenneman T, Isleib TG, Holbrook C, Ozias-Akins P, Wang J (2020) Mapping quantitative trait loci (QTLs) and estimating the epistasis controlling stem rot resistance in cultivated peanut (Arachis hypogaea L). Theor Appl Genet 133:1201–1212

Ma J, Zhao Y, Chen H, Fu C, Zhu L, Zhou X, Xia H, Hou L, Li G, Zhuang W, Wang X, Zhao C (2020) Genome-wide development of polymorphic microsatellite markers and their application in peanut breeding program. Electron J Biotechnol 44:25–32

Mafakheri A, Siosemardeh A, Bahman B, Paul S, Yousef S (2010) Effect of drought stress on yield, proline and chlorophyll contents in three chickpea cultivars. Aust J Crop Sci 4(8):580–585

Mahatma MK, Thawait LK, Bishi SK, Khatediya N, Rathnakumar AL, Lalwani HB, Misra JB (2016) Nutritional composition and antioxidant activity of Spanish and Virginia groundnuts (Arachis hypogaea L.): a comparative study. J Food Sci Tech 53:2279–2286

Mahatma MK, Bishi SK, Rthanakumar AL, Singh AL (2018) In: Mahatma MK, Reddy KK, Radhakrishnana T (eds) Evaluation of nutritional and bioactive compounds of groundnut. ICAR-Directorate of Groundnut Research Annual Report 2018-19, p 66

Mahatma MK, Thawait LK, Verma A, Kumar N, Sushmita SAL (2020) Variation in quality traits of different seed sizes of groundnut. J Oilseeds Res 37:36

Mallikarjuna N, Hoisington D (2009) Peanut improvement: production of fertile hybrids and backcross progeny between Arachis hypogaea and A. kretschmeri. Food Sec 1:457–462

Mallikarjuna N, Varshney RK (2014) Genetics, genomics and breeding of peanut: an introduction. In: Mallikarjuna N, Varshney RK (eds) Genetics, genomics and breeding of peanuts. Series on genetics, genomics and breeding of crop plants. CRC Press, Boca Raton, pp 1–12. ISBN 9781482238372

Mallikarjuna N, Pande S, Jadhav DR, Sastri DC, Rao JN (2004) Introgression of disease resistance genes from Arachis kempff-mercadoi into cultivated groundnut. Plant Breed 123:573–576

Mallikarjuna N, Maheedhar G, Chandra S (2005) Genetic diversity among Arachis species based on RAPDs. Indian J Genet 65(1):5–8

Mallikarjuna N, Senthilvel S, Hoisington D (2011) Development of new sources of tetraploid Arachis to broaden the genetic base of cultivated groundnut (Arachis hypogaea L). Genet Resour Crop Eval 58:889–907

Martins ACQ, Mehta A, Murad AM, Mota APZ, Saraiva MAP, Araújo ACG, Miller RNG, Brasileiro ACM, Guimarães PM (2020) Proteomics unravels new candidate genes for Meloidogyne resistance in wild Arachis. J Proteome 217:103690

Matsui T, Singh BB (2003) Root characteristics in cowpea related to drought tolerance at the seedling stage. Exp Agric 39:29–38

Mayeux AH, Waliyar F, Ntare BR (2003) Groundnut varieties recommended by the Groundnut Germplasm Project (GGP) for West and Central Africa. ICRISAT, Patancheru, p 80

Mehan VK, McDonald D, Rajagopalan K (1987) Resistance of peanut genotypes to seed infection by Aspergillus flavus in field trials in India. Peanut Sci 14:17–21

Mehan VK, Nageswara Rao RC, McDonald D, Williams JH (1988) Management of drought stress to improve field screening of peanuts for resistance to Aspergillus flavus. Phytopathology 78:659–663

Mehan VK, Reddy DDR, McDonald D (1993) Resistance in groundnut genotypes to Kalahasti malady caused by the stunt nematode, Tylenchorhynchus brevilineatus. Int J Pest Manage 39:201–203

Mehan VK, Reddy PM, Subrahmanyam P, MacDonald D, Singh AK (1996) Identification of new sources of resistance to rust and late leaf spot in peanut. Int J Pest Manage 42:267–271

Mienie CMS, Pretorius AE (2013) Application of marker-assisted selection for ahFAD2A and ahFAD2B genes governing the high-oleic acid trait in South African groundnut cultivars (Arachis hypogaea L.). Afr J Biotechnol 12(27):4283

Milla SR, Isleib TG, Tallury SP (2005) Identification of AFLP markers linked to reduced aflatoxin accumulation in A cardenasii-derived germplasm lines of peanut. Proc Am Peanut Res Educ Soc 37:90

Misra JB (2004) Nutritive value of groundnut and composition of Indian groundnut cultivars. In: Basu MS, Singh NB (eds) Groundnut research in India. National Research Centre for Groundnut-ICAR, Junagadh, pp 273–291

Misra JB, Ghosh PK, Dayal D, Mathur RS (2000) Agronomic, nutritional and physical characteristics of some Indian groundnut cultivars. Indian J Agric Sci 70:741–746

Mondal S, Badigannavar AM, D'Souza SF (2012a) Development of genic molecular markers linked to a rust resistance gene in cultivated groundnut (Arachis hypogaea L.). Euphytica 188(2):163–173

Mondal S, Badigannavar AM, D'Souza SF (2012b) Molecular tagging of a rust resistance gene in cultivated groundnut (Arachis hypogaea L.) introgressed from Arachis cardenasii. Mol Breed 29(2):467–476

Mondal S, Hadapad AB, Hande PA, Badigannavar AM (2014a) Identification of quantitative trait loci for bruchid (Caryedon serratus Olivier) resistance components in cultivated groundnut (Arachis hypogaea L.). Mol Breed 33:961–973

Mondal S, Hande P, Badigannavar AM (2014b) Identification of transposable element markers for a rust (Puccinia arachidis Speg.) resistance gene in cultivated peanut. J Phytopathol 162:548–552

Monyo ES, Varshney RK (2016) Seven seasons of learning and engaging smallholder farmers in the drought-prone areas of sub-Saharan Africa and South Asia through tropical legumes, 2007–2014. International Crops Research Institute for the Semi-Arid Tropics, Patancheru. ISBN 978-92-9066-568-7, 236 pp

Mora-Escobedo R, Hernández-Luna P, Joaquín-Torres IC, Ortiz-Moreno A, Robles-Ramírez MC (2015) Physicochemical properties and fatty acid profile of eight peanut varieties grown in Mexico. CyTA J Food 13:300–304

Moretzsohn MC, Leoi L, Proite K, Guimarães PM, Leal-Bertioli SCM, Gimenes MA, Bertioli DJ (2005) A microsatellite-based, gene-rich linkage map for the AA genome of Arachis (Fabaceae). Theor Appl Genet 111:1060–1071

Moretzsohn MC, Barbosa AV, Alves-Freitas DM, Teixeira C, Leal-Bertioli SCM (2009) A linkage map for the B-genome of Arachis (Fabaceae) and its synteny to the A-genome. BMC Plant Biol 9:40

Moretzsohn MC, Gouvea EG, Inglis PW, Leal-Bertioli SCM, Valls JFM, Bertioli DJ (2013) A study of the relationships of cultivated peanut (Arachis hypogaea) and its most closely related wild species using intron sequences and microsatellite markers. Ann Bot 111:113–126

Mota APZ, Vidigal B, Danchin EGJ, Togawa RC, Leal-Bertioli SCM, Bertioli DJ, Araujo ACG, Brasileiro ACM, Guimaraes PM (2018) Comparative root transcriptome of wild Arachis reveals NBS-LRR genes related to nematode resistance. BMC Plant Biol 18(1):159

Motagi B, Gowda CB, Makanahally SR (1996) Mutants resistant to foliar diseases in groundnut (Arachis hypogaea L.). Curr Sci 71:582–584

Motagi BN, Naidu GK, Angadi CC, Gowda MVC (2014) Potential genotypes for resistance to foliar diseases and productivity in groundnut (Arachis hypogaea L.). Karnataka J Agric Sci 27:445–448

Nageswara Rao RC, Williams JH, Singh M (1989) Genotypic sensitivity to drought and yield potential of peanut. Agron J 81:887–893

Nageswara Rao RC, Williams JH, Sivakumar MVK, Wadia KRD (1998) Effect of water deficit at different growth phases of peanut. Response to drought during pre-flowering phase. Agron J 80:431–438

Nageswara Rao RC, Talwar HS, Wright GC (2001) Rapid assessment of specific leaf area and leaf Nitroge in peanut (Arachis hypogaea L.) using chlorophyll meter. J Agron Crop Sci 189:175–182

Nagy E, Chu Y, Guo Y, Khanal S, Tang S, Li Y, Dong WB, Timper P, Taylor C, Ozias-Akins P, Holbrook CC, Beilinson V, Nielsen NC, Stalker HT, Knapp SJ (2010) Recombination is suppressed in an alien introgression in peanut harboring Rma, a dominant root-knot nematode resistance gene. Mol Breed 26(2):357–370

Nagy ED, Guo Y, Tang S, Bowers JE, Okashah RA, Taylor CA, Knapp SJ (2012) A high-density genetic map of Arachis duranensis, a diploid ancestor of cultivated peanut. BMC Genomics 13:469

Nautiyal PC, Rachaputi NR, Joshi YC (2002) Moisture-deficit induced changes in leaf-water content, leaf carbon exchange rate and biomass production in groundnut cultivars differing in specific leaf area. Field Crops Res 74:67–79

Nawade B, Bosamia T, Radhakrishnan T, Rathnakumar AL, Kumar A, Dobaria JR, Kundu R, Mishra GP (2016) Insights into the Indian peanut genotypes for ahFAD2 gene polymorphism regulating its oleic and linoleic acid fluxes. Front Plant Sci 7:1271

Nawade B, Mishra GP, Radhakrishnan T, Sangh C, Dobariya JR, Kundu R (2019) Development of high oleic peanut lines through marker-assisted introgression of mutant ahFAD2 alleles and its fatty acid profiles under open-field and controlled conditions. 3 Biotech 9:243

Nelson A, Samuel DM, Tucker J, Jackson C, Stahlecker-Roberson A (2006) Assessment of genetic diversity and sectional boundaries in tetraploid peanuts (Arachis). Peanut Sci 33:64–67

Nigam SN (2014) Groundnut at a glance. ICRISAT, Patancheru, p 121

Nigam SN, Dwivedi SL, Reddy LJ, Rao MJV (1989) An update on groundnut breeding activities at ICRISAT Centre with particular reference to breeding and selection for improved quality. In: Proceedings of the third regional groundnut workshop (Lilongwe), pp 115–125

Nigam SN, Dwivedi SL, Gibbons RW (1991) Groundnut breeding: constraints, achievements and future possibilities. Plant Breed 61:1127–1136

Nigam SN, Upadhyaya HD, Chandra S, Rao RCN, Wright GC, Reddy AGS (2001) Gene effects for specific leaf area and harvest index in three crosses of groundnut (Arachis hypogaea). Ann Appl Biol 139:301–306

Nigam SN, Nageswara Rao RC, Wright GC (2002) Field screening for drought tolerance in groundnut. In: Field screening for drought tolerance proceedings of an international workshop on International Crops Research Institute for the Semi-Arid Tropics, Patancheru, Andhra Pradesh, India, pp 147–151. ISBN 92-9066-448-7

Nigam SN, Chandra S, Sridevi KR, Bhukta MR, Reddy AGS, Rachaputi NR, Nagda AK (2005) Efficiency of physiological trait-based and empirical selection approaches for drought tolerance in groundnut. Ann Appl Biol 146:433–439

Nigam SN, Waliyar F, Aruna FR, Reddy SV, Lava Kumar P, Craufurd PQ, Upadhyaya HD (2009) Breeding peanut for resistance to aflatoxin contamination at ICRISAT. Peanut Sci 36:42–49

Norden AJ, Gorbet DW, Knauft DA (1987) Variability in oil quality among peanut genotypes in the Florida breeding program. Peanut Sci 14:7–11

O'Brien RD (2004) Fats and oils. Formulating and processing for applications. CRC Press, Boca Raton

Padley FB, Gunstone FD, Harwood JL (1994) Occurrence and characteristics of oils and fats. In: Gunstone FD, Harwood JL, Padley FB (eds) The lipid handbook. Chapman and Hall, London, pp 47–223

Paik-Ro OG, Smith RL, Knauft DA (1992) Restriction fragment length polymorphism evaluation of six peanut species within the Arachis section. Theor Appl Genet 84:201–208

Palladino C, Breiteneder H (2018) Peanut allergens. Mol Immun 100:58–70

Pandey MK, Monyo E, Ozias-Akins P, Liang X, Guimarães P, Nigam SN, Varshney RK (2012) Advances in Arachis genomics for peanut improvement. Biotech Adv 30:639–651

Pandey MK, Guo B, Holbrook CC, Janila P, Zhang X, Bertioli DJ, Isobe S, Liang S, Varshney RK (2014a) Molecular markers, genetic maps, and QTLs for molecular breeding in peanut. In: Mallikarjuna N (ed) Genetics, genomics and breeding of peanuts. CRC Press, Boca Raton, pp 79–113. ISBN 978-1-4822-3835-8

Pandey MK, Wang ML, Qiao L, Feng S, Khera P, Wang H, Tonnis B, Barkley NA, Wang J, Holbrook CC, Culbreath AK, Varshney RK, Guo B (2014b) Identification of QTLs associated with oil content and mapping FAD2 genes and their relative contribution to oil quality in peanut (Arachis hypogaea L.). BMC Genet 15:133

Pandey MK, Agarwal G, Kale SM, Clevenger J, Nayak SN, Sriswathi M, Chitikineni A, Chavarro C, Chen X, Upadhyaya HD, Vishwakarma MK, Leal-Bertioli S, Liang X, Bertioli DJ, Guo B, Jackson SA, Ozias-Akins P, Varshney RK (2017) Development and evaluation of a high density genotyping 'Axiom_Arachis' array with 58 K SNPs for accelerating genetics and breeding in groundnut. Sci Rep 7:40577

Peng Z, Gallo M, Tillman BL, Rowland D, Wang J (2016) Molecular marker development from transcript sequences and germplasm evaluation for cultivated peanut (Arachis hypogaea L). Mol Genet Genomics 291:363–381

Peng Z, Fan W, Wang L, Paudel D, Leventini D, Tillman BL, Wang J (2017a) Target enrichment sequencing in cultivated peanut (Arachis hypogaea L.) using probes designed from transcript sequences. Mol Gen Genomics 5:1–11

Peng Z, Liu F, Wang L, Zhou H, Paudel D, Tan L, Maku J, Gallo M, Wang J (2017b) Transcriptome profiles reveal gene regulation of peanut (Arachis hypogaea L.) nodulation. Sci Rep 7:40066

Pimratch S, Jogloy S, Vorasoot N, Toomsan B, Kesmala T, Patanothai A, Holbrook CC (2008) Effect of drought stress on traits related to N2 fixation in eleven peanut (Arachis hypogaea L.) genotypes differing in degrees of resistance to drought. Asian J Plant Sci 7:334–342

Puangbut D, Sanun J, Vorasoot N, Chutipong A, Kesmalac T, Patanothai A (2009) Variability in yield responses of peanut (Arachis hypogaea L.) genotypes under early season drought. Asian J Plant Sci 8(4):254–264

Puangbut D, Jogloy S, Toomsan B, Vorasoot N, Akkasaeng C, Kesmala T, Rachaputi Rao CN, Wright GC, Patanothai A (2010) Physiological basis for genotypic variation in tolerance to and recovery from pre-flowering drought in peanut. J Agron Crop Sci 196:358–367

Qin H, Feng S, Chen C, Guo Y, Knapp S, Culbreath A, He G, Wang ML, Zhang X, Holbrook CC, Ozias-Akins P, Guo B (2012) An integrated genetic linkage map of cultivated peanut (Arachis hypogaea L.) constructed from two RIL populations. Theor Appl Genet 124:653–664

Qui QS, Li ZC, Shen YF, Wang CT, Miao HR (1998) Groundnut breeding through mutation techniques in China. Int Arachis Newsl 18:17

Radhakrishnan T, Hariprasanna K (2004) Molecular characterisation of crop genetic resources with reference to groundnut and its wild relatives. In: Thangaduarai D, Pullaiah T, Pedro K, Paltti PK (eds) Genetic resource and biotechnology, vol 2. Regency Publications, New Delhi

Radhakrishnan T, Thirumalaisamy PP, Vemana K, Kumar A (2016) Major virus diseases of groundnut in India and their management. In: Gaur R, Petrov N, Patil B, Stoyanova M (eds) Plant viruses: evolution and management. Plant viruses: evolution and management. Springer, Singapore, pp 253–271

Raina SN, Mukai Y (1999) Genomic in situ hybridization in Arachis (Fabaceae) identifies the diploid wild progenitors of cultivated (A. hypogaea) and related wild (A. monticola) peanut species. Plant Syst Evol 214:251–262

Rathnakumar AL, Hariprasanna K, Lalwani HB (2012) Genetic improvement in Spanish type groundnut, Arachis hypogaea L varieties in India over the years. J Oilseeds Res 27:1–7

Ravi K, Vadez V, Isobe S, Mir RR, Guo Y, Nigam SN, Gowda MVC, Radhakrishnan T, Bertioli DJ, Knapp SJ, Varshney RK (2011) Identification of several small main-effect QTLs and a large number of epistatic QTLs for drought tolerance related traits in groundnut (Arachis hypogaea L.). Theor Appl Genet 122:1119–1132

Reddy LJ, Nigam SN, Rao RCN, Reddy NS (2001) Registration of ICGV 87354 peanut germplasm with drought tolerance and rust resistance. Crop Sci 41:274–275

Reddy TY, Reddy VR, Anbumozhi V (2003) Physiological responses of groundnut (Arachis hypogea L.) to drought stress and its amelioration: a critical review. Plant Growth Regul 41:75–88

Ren X, Jiang H, Yan Z, Chen Y, Zhou X, Huang L, Lei Y, Huang J, Yan L, Qi Y, Wei W, Liao B (2014) Genetic diversity and population structure of the major peanut (Arachis hypogaea L.) cultivars grown in China by SSR markers. PLoS One 9(2):e88091

Robledo G, Seijo G (2010) Species relationships among the wild B genome of Arachis species (section Arachis) based on FISH mapping of rDNA loci and heterochromatin detection: a new proposal for genome arrangement. Theor Appl Genet 121:1033–1046

Rowland D, Puppala N, Beasley J Jr, Burow M, Gorbet D, Jordan D, Melouk H, Simpson C, Ferrell J (2012) Variation in carbon isotope ratio and its relation to other traits in peanut breeding lines and cultivars from U.S. trials. J Plant Breed Crop Sci 4:144–155

Salway JG, Whitehead L, Finnegan JA, Karunanayaka A, Barnett D, Payne RB (1978) Effect of myo-inositol on peripheral-nerve function in diabetes. Lancet 2:1282–1284

Sanogo O (2016) Early generation selection for drought related traits in groundnut (Arachis hypogaea L.) in Mali. PhD thesis, University of Ghana

Santana SH, Valls JFM (2015) Arachis veigae (Fabaceae), the most dispersed wild species of the genus, and yet taxonomically overlooked. Bonplandia 24:139–150

Sarvamangala C, Gowda MVC, Varshney RK (2011) Identification of quantitative trait loci for protein content, oil content and oil quality for groundnut (Arachis hypogaea L.). Field Crops Res 122:49–59

Savage GP (2003) Saponin. In: Benjamin C (ed) Encyclopedia of food sciences and nutrition. Academic Press, New York, pp 5095–5098

Schaafsma G (2000) The protein digestibility-correct amino acid score. J Nutr 130:1865S–1867S

Schlemmer U, Frølich W, Prieto RM, Grases F (2009) Phytate in foods and significance for humans: food sources, intake, processing, bioavailability, protective role and analysis. Mol Nutr Food Res 53:S330–S375

Seijo JG, Lavia GI, Fernandez A, Krapovickas A, Ducasse D, Moscone EA (2004) Physical mapping of the 5S and 18S–25S rRNA genes by FISH as evidence that Arachis duranensis and A. ipaensis are the wild diploid progenitors of A. hypogaea (Leguminosae). Am J Bot 91:1294–1303

Seijo G, Lavia GI, Fernandez A, Krapovickas A, Daniel D, Bertioli D, Eduardo M (2007) Genomic relationships between the cultivated peanut (Arachis hypogaea, Leguminosae) and its close relatives revealed by double GISH. Am J Bot 94:1963–1971

Selvaraj MG, Narayana M, Schubert AM, Ayers JL, Baring MR, Burow MD (2009) Identification of QTLs for pod and kernel traits in cultivated peanut by bulked segregant analysis. Electron J Biotechnol 12(2):E13

Senawong KST, Jogloy S, Patanothai A (2012) Comparison of total phenolic content and composition of individual phenolic acids in testae and testa-removed kernels of 15 Valencia-type peanut (Arachis hypogaea L.) genotypes. Afr J Biotechnol 11:15923–15930

Shad MA, Pervez H, Nawaz H, Khan H, Ullah MA (2009) Evaluation of biochemical and phytochemical composition of some groundnut varieties grown in arid zone of Pakistan. Pak J Bot 41:2739–2749

Shasidhar Y, Vishwakarma MK, Pandey MK, Janila P, Variath MT, Manohar SS, Varshney RK (2017) Molecular mapping of oil content and fatty acids using dense genetic maps in groundnut (Arachis hypogaea L.). Front Plant Sci 8:794

Shasidhar Y, Variath MT, Vishwakarma MK, Manohar SS, Gangurde SS, Sriswathi M, Sudini HK, Dobariya KL, Bera SK, Radhakrishnan T, Pandey MK, Janila P, Varshney RK (2020) Improvement of three popular Indian groundnut varieties for foliar disease resistance and high oleic acid using SSR markers and SNP array in marker-assisted backcrossing. Crop J 8(1):1–15

Shirasawa K, Koilkonda P, Aoki K, Hirakawa H, Tabata S, Watanabe M, Isobe S (2012) In silico polymorphism analysis for the development of simple sequence repeat and transposon markers and construction of linkage map in cultivated peanut. BMC Plant Biol 12:80

Shirasawa K, Bertioli DJ, Varshney RK, Moretzsohn MC, Leal-Bertioli SC, Thudi M, Isobe S (2013) Integrated consensus map of cultivated peanut and wild relatives reveals structures of the A and B genomes of Arachis and divergence of the legume genomes. DNA Res 20:173–184

Shrestha A, Champagne DE, Culbreath AK, Rotenberg D, Whitfield AE, Srinivasan R (2017) Transcriptome changes associated with tomato spotted wilt virus infection in various life stages of its thrips vector, Frankliniella fusca (Hinds). J Gen Virol 98(8):2156–2170

Simpson CE (1991) Pathways for introgression of pest resistance into Arachis hypogaea L. Peanut Sci 18(1):22–26

Simpson CE (2001) Use of wild Arachis species/introgression of genes into A. hypogaea L. Peanut Sci 28:114–116

Simpson CE, Faries MJ (2001) Advances in the characterization of diversity in section Arachis: archeological evidence, crossing results and their relationship in understanding the origin of Arachis hypogaea L. In: Abstracts of III SIRGEALC—Simpósio de Recursos Genéticos para a America Latina e Caribe Instituto Agronômico do Paraná, Londrina, pp 103–104

Simpson CE, Starr JL (2001) Registration of 'COAN' peanut. Crop Sci 41:918

Simpson CE, Krapovickas A, Valls JFM (2001) History of Arachis including evidence of A. hypogaea L. progenitors. Peanut Sci 28:78–80

Simpson CE, Starr JL, Church GT, Burow MD, Paterson AH (2003) Registration of NemaTAM peanut. Crop Sci 43(4):1561

Singh F, Diwakar B (1993) Nutritive value and uses of pigeonpea and groundnut. Manual. International Crops Research Institute for the Semi-Arid Tropics, Patancheru

Singh AK, Nigam SN (1997) Groundnut. In: Fuccillo D, Sears L, Stapleton P (eds) Biodiversity in trust. Cambridge University Press, Cambridge, pp 114–127

Singh AK, Nigam SN (2016) Chapter 2: Arachis gene pools and genetic improvement in groundnut. In: Rajpal VR, Rao SM, Raina SN (eds) Gene pool diversity and crop improvement. Springer International Publishing, Basel, pp 17–77

Singh AK, Simpson CE (1994) Biosystematic and genetic resources. In: Smartt J (ed) The groundnut crop, a scientific basis for improvement. Chapman and Hall, London, pp 96–138

Singh AK, Sivaramakrishnan S, Mengesha MH, Ramaiah CD (1991) Phylogenetic relations in section Arachis based on seed protein profile. Theor Appl Genet 82:593–597

Singh AK, Mehan VK, Nigam SN (1997) Sources of resistance to groundnut fungi and bacterial diseases: an update and appraisal. Information bulletin no 50. International Crops Research Institute for the Semi-Arid Tropcis, Patancheru

Sinha P, Bajaj P, Pazhamala LT, Nayak SN, Pandey MK, Chitikineni A, Huai D, Khan AW, Desai A, Jiang H, Weijian Z, Guo B, Boshou L, Varshney RK (2020) Arachis hypogea gene expression atlas (AhGEA) for fastigiata subspecies of cultivated groundnut to accelerate functional and translational genomics applications. Plant Biotechnol J 18(11):2187–2200

Smartt J (1990) Grain legumes. In: Evolution and genetic resources. Cambridge University Press, Cambridge, pp 30–84

Smartt J, Stalker HT (1982) Speciation and cytogenetics in Arachis. In: Pattee HE, Young CE (eds) Peanut science and technology. American Peanut Research Education Society, Yoakum, pp 21–49

Sobolev VS, Cole RJ (1999) Trans-resveratrol content in commercial peanuts and peanut products. J Agric Food Chem 47(4):1435–1439

Songsri P, Jogloy S, Kesmala T, Vorasoot N, Akkasaeng C, Patanothai A, Holbrook CC (2009) Heritability of drought resistance traits and correlation of drought resistance and agronomic traits in peanut. Crop Sci 48:2245–2253

Soni P, Gangurde SS, Ortega-Beltran A, Kumar R, Parmar S, Sudini HK, Lei Y, Ni X, Huai D, Fountain JC, Njoroge S, Mahuku G, Radhakrishnan T, Zhuang W, Guo B, Liao B, Singam P, Pandey MK, Bandyopadhyay R, Varshney RK (2020) Functional biology and molecular mechanisms of host-pathogen interactions for aflatoxin contamination in groundnut (Arachis hypogaea L.) and maize (Zea mays L.). Front Microbiol 11:227

Stalker HT (1991) A new species in section Arachis of peanuts with a D genome. Am J Bot 78:630–637

Stalker HT (2017) Utilizing wild species for peanut improvement. Crop Sci 57(3):1102–1120

Stalker HT, Simpson CE (1995) Germplasm resources in Arachis. In: Pattee HE, Young CT (eds) Peanuts science and technology. American Peanut Research and Education Society, Stillwater, pp 1–20

Stalker HT, Phillips TD, Murphy JP, Jones TM (1994) Variation of isozyme patterns among Arachis species. Theor Appl Genet 87:746–755

Stalker HT, Ferguson ME, Valls JFM, Pittman RN, Simpson CE, Bramel-Cox P (2002) Catalog of Arachis germplasm collection. ICRISAT, Patancheru. http://www.icrisat.org/GroundNut/Arachis/Start.htm

Stalker HT, Tallury SP, Ozias-Akins P, Bertioli D, Leal-Bertioli SC (2013) The value of diploid peanut relatives for breeding and genomics. Peanut Sci 40:70–88

Stevenson PC, Anderson JC, Blaney WM, Simmonds MJS (1993) Developmental inhibition of Spodoptera litura (Fab.) larvae by a novel caffeoylquinic acid from the wild groundnut Arachis paraguariensis (Chod et Hassl.). J Chem Ecol 19:2917–2933

Sturm PA, Parkhurst RM, Skinner WA (1966) Quantitative determination of individual tocopherols by thin layer chromatographic separation and spectrophotometry. Anal Chem 38:1244–1247

Suárez López MM, Kizlansky A, López LB (2006) Assessment of protein quality in food by calculating the amino acids score corrected by digestibility. Nutr Hosp 21:47–51

Subrahmanyam P, McDonald D (1983) Rust disease of groundnut. Inform Bull 13. International Crops Research Institute for the Semi-Arid Tropics (ICRISAT), Patancheru

Subrahmanyam P, Moss JP, McDonald D, Subba Rao PV, Rao VR (1985) Resistance to Cercosporidium personatum leafspot in wild Arachis species. Plant Dis 69:951–954

Subrahmanyam P, McDonald D, Waliyar F, Reddy LJ, Nigam SN, Gibbons RW, Subba Rao PV (1995) Screening methods and sources of resistance to rust and late leaf spot of groundnut. In: Information bulletin, no. 47. ICRISAT, Patancheru

Sui N, Wang Y, Liu S, Yang Z, Wang F, Wan S (2018) Transcriptomic and physiological evidence for the relationship between unsaturated fatty acid and salt stress in peanut. Front Plant Sci 9:7

Sujay V, Gowda MVC, Pandey MK, Bhat RS, Khedikar YP, Nadaf HL, Varshney RK (2012) Quantitative trait locus analysis and construction of consensus genetic map for foliar disease resistance based on two recombinant inbred line populations in cultivated groundnut (Arachis hypogaea L.). Mol Breed 30:773–788

Tallury SP, Hilu KW, Milla SR, Friend SA, Alsaghir M, Stalker HT, Quandt D (2005) Genomic affinities in Arachis section Arachis (Fabaceae): molecular and cytogenetic evidence. Theor Appl Genet 111:1229–1237

Terras FR, Eggermont K, Kovaleva V, Raikhel NV, Osborn RW, Kester A, Rees SB, Torrekens S, Van Leuven F, Vanderleyden J (1995) Small cysteine-rich antifungal proteins from radish: their role in host defense. Plant Cell 7:573–588

Tharanathan RN, Dharmaraj B, Wankhede DB, Rao MRR (1979) Groundnut carbohydrates—a review. J Sci Food Agric 30:1077–1084

The Peanut Institute (2006). http://www.peanut-institute.org/NutritionBasics.html

Treadwell K, Young CT, Wynne JC (1983) Evaluation of fatty acid content of forty peanut cultivars. Oleagineux 38:381–385

Tseng YC, Tillman BL, Peng Z, Wang J (2016) Identification of major QTLs underlying tomato spotted wilt virus resistance in peanut cultivar Florida-EPTM '113'. BMC Genet 17:128

Upadhyaya HD (2005) Variability for drought resistance related traits in the mini core collection of peanut. Crop Sci 45:1432–1440

Upadhyaya HD (2008) Crop germplasm and wild relatives: a source of novel variation for crop improvement. Korean J Crop Sci 53:12–17

Upadhyaya HD, Nigam SN, Mehan VK, Lenne JM (1997) Aflatoxin contamination of groundnut: prospects for a genetic solution through conventional breeding. Aflatoxin contamination problems in groundnut in Asia. In: Proceedings of the first Asia working group meet, 27-29 May, 1996. International Crop Research Institute for Semi-Arid Tropics, Patancheru, pp 81–85

Upadhyaya HD, Ferguson ME, Bramel PJ (2001) Status of the Arachis germplasm collection at ICRISAT. Peanut Sci 28:89–96

Upadhyaya HD, Bramel PJ, Ortiz R, Singh S (2002) Developing a mini core of peanut for utilization of genetic resources. Crop Sci 42:2150–2156

Upadhyaya HD, Ortiz R, Bramel PJ, Singh S (2003) Development of a groundnut core collection using taxonomical, geographical and morphological descriptors. Genet Resour Crop Evol 50:139–148

Upadhyaya HD, Sharma S, Singh S, Singh M (2011) Inheritance of drought resistance related traits in two crosses of groundnut (Arachis hypogaea L.). Euphytica 177:55–66

Upadhyaya HD, Dwivedi SL, Vadez V, Hamidou F, Singh S, Varshney RK, Liao B (2014) Multiple resistant and nutritionally dense germplasm identified from mini core collection in peanut. Crop Sci 54:679–693

Utomo SD, Anderson WF, Wynne JC, Beute MK, Hagler WM Jr, Payne GA (1990) Estimates of heritability and correlation among three mechanisms of resistance to Aspergillus parasiticus in peanut. Proc Am Peanut Res Educ Soc 22:26

Valls JFM, Simpson CE (2005) New species of Arachis (Leguminosae) from Brazil, Paraguay and Bolivia. Bonplandia 14:35–63

Valls JFM, Costa LC, Custodio AR (2013) A novel Trifoliolate species of Arachis (Fabaceae) and further comments on the taxonomic section trierectoides. Bonplandia 22:91–97

Varman PV (1999) A foliar disease resistant line developed through interspecific hybridization in groundnut (Arachis hypogaea). Indian J Agric Sci 69(1):67–68

Varshney RK (2016) Exciting journey of 10 years from genomes to fields and markets: some success stories of genomics-assisted breeding in chickpea, pigeon pea and groundnut. Plant Sci 242:98–107

Varshney RK, Bertioli DJ, Moretzsohn MC, Vadez V, Krishnamurthy L, Aruna R, Hoisington DA (2009) The first SSR-based genetic linkage map for cultivated groundnut (Arachis hypogaea L). Theor Appl Genet 118:729–739

Varshney RK, Pandey MK, Janila P, Nigam SN, Sudini H, Gowda MVC, Sriswathi M, Radhakrishnan T, Manohar SS, Nagesh P (2014) Marker-assisted introgression of a QTL region to improve rust resistance in three elite and popular varieties of peanut (Arachis hypogaea L.). Theor Appl Genet 127(8):1771–1781

Venkatachalam M, Sathe S (2006) Chemical composition of selected edible nut seeds. J Agric Food Chem 54:4705–4714

Vereda A, Van Hage M, Ahlstedt S, Ibanez MD, Cuesta-Herranz J, Van Odijk J, Wickman M, Sampson HA (2011) Peanut allergy: clinical and immunologic differences among patients from 3 different geographic regions. J Allergy Clin Immunol 127(3):603–607

Vindhiyavarman P, Manivannan N, Nigam SN, Muralidharan V (2010) Farmers' participatory varietal selection in groundnut: a case study from Tamilnadu, India. Electron J Plant Breed 1:878–881

Viquez OM, Konan KN, Dodo HW (2003) Structure and organization of the genomic clone of a major peanut allergen gene, Ara h1. Mol Immunol 40(9):565–571

Vishwakarma MK, Pandey MK, Shasidhar Y, Manohar SS, Nagesh P, Janila P, Varshney RK (2016) Identification of two major quantitative trait locus for fresh seed dormancy using the diversity arrays technology and diversity arrays technology-seq based genetic map in Spanish-type peanuts. Plant Breed 135:367–375

Vishwakarma MK, Kale SM, Sriswathi M, Naresh T, Shasidhar Y, Garg V, Varshney RK (2017) Genome-wide discovery and deployment of insertions and deletions markers provided greater insights on species, genomes, and sections relationships in the genus Arachis. Front Plant Sci 8:2064

Waliyar F, Kumar KVK, Diallo M, Traore A, Mangala UN, Upadhyaya HD, Sudini H (2016) Resistance to pre-harvest aflatoxin contamination in ICRISAT's groundnut mini core collection. Eur J Plant Pathol 145:901–913

Wan L, Li B, Lei Y, Yan L, Ren X, Chen Y, Dai X, Jiang H, Zhang J, Guo W, Chen A, Liao B (2017) Mutant transcriptome sequencing provides insights into pod development in peanut (Arachis hypogaea L.). Front Plant Sci 8:1900

Wang KH, Lai YH, Chang JC, Ko TF, Shyu SL, Chiou RYY (2005) Germination of peanut kernels to enhance resveratrol biosynthesis and prepare sprouts as a functional vegetable. J Agric Food Chem 53:242–246

Wang H, Shi Y, Ren Y, Li S, Jiao K, Yuan M, Li H (2008) Development of SSR markers for root-knot nematode resistance in peanut. J Peanut Sci 37(2):14–17

Wang CT, Wang XZ, Tang YY, Chen DX, Cui FG, Zhang JC, Yu SL (2010) Phylogeny of Arachis based on internal transcribed spacer sequences. Genet Resour Crop Evol 58(2):311–319

Wang ML, Sukumaran S, Barkley NA, Chen Z, Chen CY, Guo B, Yu J (2011) Population structure and marker-trait association analysis of the US peanut (Arachis hypogaea L.) mini core collection. Theor Appl Genet 123:1307–1317

Wang H, Penmetsa RV, Yuan M, Gong L, Zhao Y, Guo B, He G (2012) Development and characterization of BAC-end sequence derived SSRs, and their incorporation into a new higher density genetic map for cultivated peanut (Arachis hypogaea L.). BMC Plant Biol 12:10

Wang H, Pandey MK, Qiao L, Qin H, Culbreath AK, He G, Varshney RK, Guo B (2013) Genetic mapping and quantitative trait loci analysis for disease resistance using F2 and F5 generation-based genetic maps derived from 'Tifrunner' X 'GT-C20" in peanut. Plant Genome 6(3):E1–E10

Wang ML, Khera P, Pandey MK, Wang H, Qiao L, Feng S, Tonnis B, Barkley NA, Pinnow D, Holbrook CC, Culbreath AK, Varshney RK, Guo B (2015) Genetic mapping of QTLs controlling fatty acids provided insights into the genetic control of fatty acid synthesis pathway in peanut (Arachis hypogaea L.). PLoS One 10(4):e0119454

Wang H, Lei Y, Wan L, Yan L, Lv J, Dai X, Ren X, Guo W, Jiang H, Liao B (2016a) Comparative transcript profiling of resistant and susceptible peanut post-harvest seeds in response to aflatoxin production by Aspergillus flavus. BMC Plant Biol 16:54

Wang L, Li B, Pandey MK, Wu Y, Lei Y, Yan L, Dai X, Jiang H, Zhang J, Wei G, Varshney RK, Liao B (2016b) Transcriptome analysis of a new peanut seed coat mutant for the physiological regulatory mechanism involved in seed coat cracking and pigmentation. Front Plant Sci 7:1491

Wang Y, Ma X, Zhang X, He X, Li H, Cui D, Yin D (2016c) ITRAQ-based proteomic analysis of the metabolic mechanisms behind lipid accumulation and degradation during peanut seed development and post germination. J Proteome Res 15(12):4277–4289

Wang X, Xu P, Yin L, Ren Y, Li S, Shi Y, Alcock TD, Xiong Q, Qian W, Chi X, Pandey MK, Varshney RK, Yuan M (2018a) Genomic and transcriptomic analysis identified gene clusters and candidate genes for oil content in peanut (Arachis hypogaea L.). Plant Mol Biol Rep 36(3):518–529

Wang Z, Huai D, Zhang Z, Cheng K, Kang Y, Wan L, Yan L, Jiang H, Lei Y, Liao B (2018b) Development of a high-density genetic map based on specific length amplified fragment sequencing and its application in quantitative trait loci analysis for yield-related traits in cultivated peanut. Front Plant Sci 9:827

Wang F, Nong X, Hao K, Cai N, Wang G, Liu S, Ullah H, Zhang Z (2020) Identification of the key genes involved in the regulation of symbiotic pathways induced by Metarhizium anisopliae in peanut (Arachis hypogaea) roots. 3 Biotech 10, 124

Watson CT, Cohain AT, Griffin RS, Chun Y, Grishin A, Hacyznska H, Hoffman GE, Beckmann ND, Shah H, Dawson P, Henning A, Wood R, Burks AW, Jones SM, Leung DYM, Sicherer S, Sampson HA, Sharp AJ, Schadt EE, Bunyavanich S (2017) Integrative transcriptomic analysis reveals key drivers of acute peanut allergic reactions. Nat Commun 8(1):1943

Will ME, Holbrook CC, Wilson DM (1994) Evaluation of field inoculation techniques for screening peanut genotypes for reaction to preharvest A. flavus group infection and aflatoxin contamination. Peanut Sci 21:122–125

Williams D (2001) New directions for collecting and conserving peanut diversity. Peanut Sci 28:135–140

Williams KA, Williams DE (2001) Evolving political issues affecting international exchange of Arachis genetic resources. Peanut Sci 28:132–135

Williams JH, Rao RCN, Rao MHV (1985) Breeding for drought tolerance on groundnut (Arachis Hypogaea L.). In: Proceedings of the workshop on varietal improvement for rice-based farming systems, 11–15 Mar 1985, Phitsanulok, Thailand

Wilson JN, Chopra R, Baring MR, Selvaraj MG, Simpson CE, Chagoya J, Burow MD (2017) Advanced backcross quantitative trait loci (QTL) analysis of oil concentration and oil quality traits in peanut (Arachis hypogaea L). Trop Plant Biol 10:1–17

Wright GC, Nageswar Rao RC, Basu MS (1996) A physiological approach to the understanding of Genotype by environment interaction: a case study on improvement of drought adaptation in groundnut. In: Cooper M, Hammer GL (eds) Plant adaptation and crop improvement. CAB International, Wallingford, ISBN: 0-85199-108-4

Wynne JC, Gregory WC (1981) Peanut breeding. Adv Agron 34:39–72

Xia H, Zhao C, Hou L, Li A, Zhao S, Bi Y, An J, Zhao Y, Wan S, Wang X (2013) Transcriptome profiling of peanut gynophores revealed global reprogramming of gene expression during early pod development in darkness. BMC Genomics 14:517

Xia H, Zhu L, Zhao C, Li K, Shang C, Hou L, Wang M, Shi J, Fan S, Wang X (2020) Comparative transcriptome analysis of anthocyanin synthesis in black and pink peanut, Plant Signaling & Behavior, 15:2, https://doi.org/10.1080/15592324.2020.1721044

Xiuzhen W, Yueyi T, Qi W, Ligui Z, Quanxi S, Zhiwei W, Qingxuan G, Jianzhi X, Shutao Y, Chaoqi F, Chuantang W (2016) Development of a new high-oleic peanut cultivar Huayu 662 by using molecular marker aided selection and NIRS. J Nucl Agric Sci 30(6):1054–1058

Xue H (2004) Evaluation of peanut (Arachis hypogaea L.) germplasm for resistance to aflatoxin production by Aspergillus flavus Link ex Fries. PhD thesis, North Carolina State University, Raleigh, NC

Xu P, Tang G, Cui W, Chen G, Ma C-L, Zhu J Li P, Shan L, Liu Z, Wan S (2020) Transcriptional Differences in Peanut (Arachis hypogaea L.) Seeds at the Freshly Harvested, After-ripening and Newly Germinated Seed Stages: Insights into the Regulatory Networks of Seed Dormancy Release and Germination. PLoS ONE 15(1): e0219413.

Yang S, Li L, Zhang J, Geng Y, Guo F, Wang J, Meng J, Sui N, Wan S, Li X (2017) Transcriptome and differential expression profiling analysis of the mechanism of Ca2+ regulation in peanut (*Arachis hypogaea*) pod development. Front Plant Sci 8:1609

Yang T, Fang L, Sanders S, Jayanthi S, Rajan G, Podicheti R, Thallapuranam SK, Mockaitis K, Medina-Bolivar F (2018) Stilbenoid prenyltransferases define key steps in the diversification of peanut phytoalexins. J Biol Chem 293(1):28–46

Yeri SB, Bhat RS (2016) Development of late leaf spot and rust resistant backcross lines in Jl 24 variety of groundnut (Arachis hypogaea L.). Electron J Plant Breed 7:37–41

Yin D, Wang Y, Zhang X, Li H, Lu X, Zhang J, Zhang W, Chen S (2013) De novo assembly of the peanut (Arachis hypogaea L.) seed transcriptome revealed candidate unigenes for oil accumulation pathways. PLoS One 8(9):e73767

Young T (1996) Peanut oil. In: Hui YH (ed) Bailey's industrial oil and fat products. Wiley, New York, pp 377–392

Yu JM, Ahmedna M, Goktepe I, Dai J (2006) Peanut skin procyanidins: composition and antioxidant activities as affected by processing. J Food Comp Anal 19:364–371

Yu M, Liu F, Zhu W, Sun M, Liu J, Li X (2015) New features of triacylglycerol biosynthetic pathways of peanut seeds in early developmental stages. Funct Integr Genomics 15(6):707

Yu B, Huai D, Huang L, Kang Y, Ren X, Chen Y, Zhou X, Luo H, Liu N, Chen W, Lei Y, Pandey MK, Sudini H, Varshney RK, Liao B, Jiang H (2019) Identification of genomic regions and diagnostic markers for resistance to aflatoxin contamination in peanut (Arachis hypogaea L.). BMC Genet 20:32

Yüksel B, Paterson HA (2005) Construction and characterization of peanut HindIII BAC library. Theor Appl Genet 111(4):630–639

Zhang X, Han S, Tang F, Xu J, Liu H, Yan M, Dong W, Huang B, Zhu S (2011) Genetic analysis of yield in peanut (Arachis hypogaea L.) using mixed model of major gene plus polygene. Afr J Biotechnol 10(37):7126–7130

Zhang J, Liang S, Duan J, Wang J, Chen S, Cheng Z, Zhang Q, Liang X, Li Y (2012) De novo assembly and characterisation of the transcriptome during seed development, and generation of genic-SSR markers in peanut (Arachis hypogaea L.). BMC Genomics 13:90

Zhang Y, Wang P, Xia H, Zhao C, Hou L, Li C, Gao C, Wang X, Zhao S (2016) Comparative transcriptome analysis of basal and zygote-located tip regions of peanut ovaries provides insight into the mechanism of light regulation in peanut embryo and pod development. BMC Genomics 17(1):606. Erratum in: BMC Genomics 17(1):836

Zhang X, Zhang J, He X, Wang Y, Ma X, Yin D (2017) Genome-wide association study of major agronomic traits related to domestication in peanut. Front Plant Sci 8:1611

Zhang GH, Yu ST, Wang H, Wang XD (2019) Transcriptome profiling of high oleic peanut under low temperature during germination. Yi Chuan 41:1050–1059

Zhang H, Zhao X, Sun Q, Yan C, Wang J, Yuan C, Li C, Shan S, Liu F (2020) Comparative transcriptome analysis reveals molecular defensive mechanism of Arachis hypogaea in response to salt stress. Int J Genomics 2020:6524093

Zhao G, Liu Y, Zhao M, Ren J, Yang B (2011) Enzymatic hydrolysis and their effects on conformational and functional properties of peanut protein isolate. Food Chem 127:1438–1443

Zhao C, Zhao S, Hou L, Xia H, Wang J, Li C, Li A, Li T, Zhang X, Wang X (2015) Proteomics analysis reveals differentially activated pathways that operate in peanut gynophores at different developmental stages. BMC Plant Biol 15:188

Zhao C, Qiu J, Agarwal G, Wang J, Ren X, Xia H, Wang X (2017) Genome-wide discovery of microsatellite markers from diploid progenitor species, Arachis duranensis and A. ipaensis, and their application in cultivated peanut (A. hypogaea). Front Plant Sci 8:1209

Zhao X, Li C, Wan S, Zhang T, Yan C, Shan S (2018) Transcriptomic analysis and discovery of genes in the response of Arachis hypogaea to drought stress. Mol Biol Rep 45(2):119–131

Zheng L, Shockey J, Guo F, Shi L, Li X, Shan L, Wan S, Peng Z (2017) Discovery of a new mechanism for regulation of plant triacylglycerol metabolism: the peanut diacylglycerol acyltransferase-1 gene family transcriptome is highly enriched in alternative splicing variants. J Plant Physiol 219:62–70

Zhou X, Xia Y, Ren X, Chen Y, Huang L, Huang S, Liao B, Lei Y, Yan L, Jiang H (2014) Construction of a SNP-based genetic linkage map in cultivated peanut based on large scale marker development using next-generation double-digest restriction-site-associated DNA sequencing (ddRADseq). BMC Genomics 15:351

Zhou GF, Jian JB, Wang PH, Li CD, Tao Y, Li X, Renshaw D, Clements J, Sweetingham M, Yang HA (2018) Construction of an ultra-high density consensus genetic map, and enhancement of the physical map from genome sequencing in Lupinus angustifolius. Theor Appl Genet 131:209–223

Zhu W, Chen X, Li H, Zhu F, Hong Y, Varshney RK, Liang X (2014) Comparative transcriptome analysis of aerial and subterranean pods development provides insights into seed abortion in peanut. Plant Mol Biol 85(4–5):395–409

Zhuang W, Chen H, Yang M et al (2019) The genome of cultivated peanut provides insight into legume karyotypes, polyploid evolution and crop domestication. Nat Genet 51:865–876

# Lentil Gene Pool for Breeding

**Alexios N. Polidoros, Ilias D. Avdikos, Anthoula Gleridou, Stauroula D. Kostoula, Ekaterini Koura, Michalia A. Sakellariou, Evangelia Stavridou, Dimitrios Gerasopoulos, Anastasia Lagopodi, Athanasios Mavromatis, Photini V. Mylona, Irini Nianiou-Obeidat, and Dimitrios Vlachostergios**

**Abstract** Lentil is an ancient legume crop cultivated thousands of years for its nutritious seeds, its ability to improve soil colonized by nitrogen fixing symbiotic bacteria, and providing income to local farmers at semiarid areas. During the centuries, numerous landraces and traditional varieties have been developed, providing a wealth of genetic material for lentil cultivation and use by local communities worldwide. However, current improved lentil varieties suffer from many biotic and abiotic challenges, and breeding new cultivars should exploit the breadth of genetic potential reserved within the *Lens* gene pool. Landraces and wild relatives are more tolerant to adverse environmental conditions and can provide valuable genes to develop improved varieties in modern agriculture, adapted to environmental abiotic and biotic stresses, suitable as well for other industrial non-food uses, such as biomass production and use as energy crop. Molecular tools to assist breeding efforts in lentil are less well developed in comparison with other crops, although progress has been made in germplasm characterization using molecular markers. Genomic research is delayed by the large (4.3 GB) lentil genome size, and progress towards

A. N. Polidoros (✉) · I. D. Avdikos · A. Gleridou · S. D. Kostoula · E. Koura
M. A. Sakellariou · E. Stavridou · A. Mavromatis · I. Nianiou-Obeidat
Laboratory of Genetics and Plant Breeding, School of Agriculture, Aristotle University of Thessaloniki, Thessaloniki, Greece
e-mail: palexios@agro.auth.gr

D. Gerasopoulos
Laboratory of Food Processing and Engineering, School of Agriculture, Aristotle University of Thessaloniki, Thessaloniki, Greece

A. Lagopodi
Laboratory of Plant Pathology, School of Agriculture, Aristotle University of Thessaloniki, Thessaloniki, Greece

P. V. Mylona
Institute of Plant Breeding and Genetic Resources, Hellenic Agricultural Organization – DEMETER, Thermi, Greece

D. Vlachostergios
Institute of Industrial and Forage Crops, Hellenic Agricultural Organization – DEMETER, Larissa, Greece



P. M. Priyadarshan, S. M. Jain (eds.), *Cash Crops*,
https://doi.org/10.1007/978-3-030-74926-2_11

the release of the complete lentil genome sequence is expected to accelerate breeding efforts. In this chapter we review current knowledge on lentil domestication and landrace distribution, cultivar improvement and breeding, efforts to characterize abiotic and biotic stress tolerance, the research strategies and major advancements made by modern molecular technologies for identification and utilization of important markers/QTLs in lentil breeding, and future prospects for this important legume crop.

**Keywords** Crop improvement · Genetic diversity · Germplasm characterization · Landraces · Molecular markers · Phenotyping

## 1 Introduction

Lentil (*Lens culinaris* Medik.; Fabaceae) is an important food source for millions of people all over the world. It is one of the most nutritious pulses, high in protein, a good source of vitamins A and B, fiber, potassium, and iron, ideal for vegetarians, but also inexpensive food for low income people. Lentil soup was known as "the poor man's meat" and Aristophanes, the great comic playwright of ancient Athens, often used lentils in his comedies in that sense. In "Ploutos (Wealth)" he is referring to the instant lack of interest of the young lover to his older mistress after he became rich, with the comment: "When poor, he would devour anything; now he is rich, he no longer cares for lentils." The importance of lentils as staple food is also underlined in other civilizations of the old world: for example in the Bible where Esau sold his birth rights to Jacob for a dish of lentil stew and in Brihadaranyaka, one of the Hinduism sacred text, where lentils are mentioned as one of the ten grain seeds cultivated for human consumption.

Lentil seeds, with or without hulls, are cooked as dhal in India and this has been the main dish for millennia in the South Asian region. In Ayurvedic diet lentil is considered as a highly nutritious pulse, second only to the green gram or mung bean (Nene 2006). The cultivation of lentils worldwide is uninterrupted through the centuries. Due to the great landscape and climatic variation of lentil cultivation environments, many local landraces that were adapted to the microclimate of each location were empirically selected by local farmers through the centuries. As a consequence, numerous landraces were produced and cultivated in small farms at different regions for local use. Lentil is an efficient crop in rotation to cereal cropping systems and generates income to poor farmers in developing countries of the world (Kumar et al. 2015). However, in these areas, lentil suffers from many abiotic and biotic stresses and yield loss is an obstacle for sustained cultivation. Intense breeding efforts to create new cultivars aim at making progress by pyramiding traits that increase yield potential and reliability, tolerance to biotic/abiotic stresses, harvestability, improved grain quality, and general adaptation to more marginal pedoclimatic environments for lentil. However, most modern cultivars albeit high yielding are not empowered with the rest of the desirable characteristics. Breeding efforts may be currently supported by new molecular tools and technologies to enable precise mapping of valuable gene combinations and select superior genotypes for better performance under climate change conditions (Gupta et al. 2019). However, lentil breeding programs are limited in their ability to implement tools like marker-assisted selection (MAS) and genetic engineering due to a lack of genomic resources, in comparison to major legume crops such as soybean, common bean, pigeon pea, and chickpea. Next generation sequencing and genotyping by sequencing technologies are now applied to enhance our understanding of the lentil genome organization and enable the discovery of genome-wide single nucleotide polymorphism (SNP) markers. Several linkage maps have been developed in lentil, using expressed sequence tag (EST) derived simple sequence repeat (SSR) and single nucleotide polymorphic (SNP) markers. These maps have emerged as useful genomic resources to identify quantitative trait loci imparting tolerance to biotic and abiotic stresses in lentils (Kumar et al. 2015).

In this chapter, we review the current knowledge on genetic diversity, breeding, conservation, and utilization of lentils. A graphical abstract displaying basic information on lentil diversity, genetics, and breeding is shown in Fig. 1.

In this chapter, we sought to emphasize the significance of lentil genetic diversity, its characterization and utilization in crop improvement to withstand abiotic and biotic stresses through classical breeding, development of molecular tools and methodologies to identify valuable genes and traits, to characterize genetic resources and enable molecular breeding, support lentil genomic research and its potential to accelerate breeding efforts, and discuss future challenges and prospects in lentil genetic research.

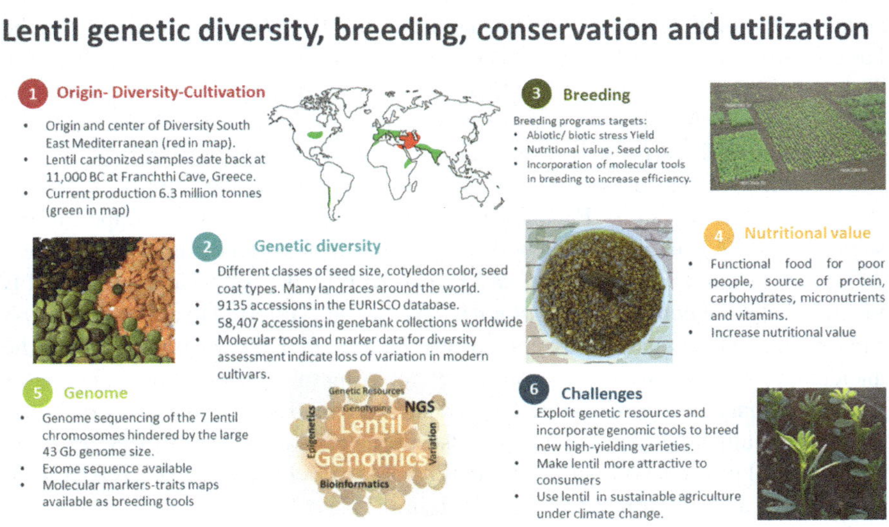


**Fig. 1** Graphical abstract presenting major points of the chapter

## 2 Origin

It is generally accepted that lentil has been domesticated in East Mediterranean and the Middle East region as its progenitor *Lens culinaris* subsp. *orientalis* (Bioss.) Ponert is present in Greece, Turkey, Syria, Lebanon, Israel, Jordan, Iraq, Iran, Afghanistan, and Uzbekistan (Sandhu and Singh 2007). The earliest archeological discovery of carbonized lentil dates back to 11,000 BC, found at the Paleolithic and Mesolithic layers of Franchthi Cave near Ermioni at Peloponnese in Southern Greece (Hansen and Renfrew 1978; Sandhu and Singh 2007; Rutter 2017). Archeological studies confirmed the presence of lentil in the Middle East region (Turkey-Syria-Iraq) as far back as 8500 BC. From Near East lentil spread west to Central Europe and east to South Asia (Nene 2006). In India lentil (mentioned as masura) has been cited in the Brahadaranyaka (c. 5500 BC), a commentary on the Rigveda (c. 8000 BC) and also in the Yajurveda (c. 7000 BC). The same word, masura, for lentil was written by Charaka (c. 700 BC), Susruta (c. 400 BC), Kautilya (c. 321–296 BC), and by later authors (Nene 2006). At the same time, *Lens* species also spread to Ethiopia, Afghanistan, Pakistan, China, and later to the New World including the Latin America (Sandhu and Singh 2007).

However, the carbonized remains found at these aceramic farming villages dated 8800–6500 BC do not allow for a clear view of the species domestication state. In these findings lentil's seed size ranged from 2.5 to 3.0 mm in diameter, similar to the wild form. Due to this morphological similarity the date of lentil domestication is hardly resolved.

Based on the seed size, lentil is classified into two main groups: (1) the small-seeded also known as *microsperma* and (2) the large-seeded, the *macrosperma*. This classification is also used as a taxonomic descriptor of the species, indicating the small-seeded plants with small pods and seeds of 3.0–6.0 mm in diameter and the large-seeded lentils with larger pods and seeds of 6.0–9.0 mm (Erskine et al. 2009). Today, *macrosperma* lentils are cultivated mainly in the Mediterranean region and the New World, whereas *microsperma* are grown in the Indian subcontinent, the Near East, and East Africa (Hawtin et al. 1980; Muehlbauer et al. 1985).

Regarding lentil taxonomy, the species belongs to the family Fabaceae (Leguminosae), subfamily Faboideae, and tribe Vicieae. Lentil is grouped in the same clade (Hologalegina) along with other cool-season legumes such as *Medicago, Trifolium, Pisum, Cicer, Lathyrus, and Vicia* (Gepts et al. 2005). Cultivated lentil is identified as *Lens culinaris* spp. *culinaris*, which is also referred to as *Lens culinaris* Medik. The species name was coined by Medikus in 1787, in translation of the ancient Greek name "*Φακός ὁ ἐσθιόμενος*" referring to the lens-shaped seed that is used in culinary. The genus name *Lens* is derived from Latin. Today *Lens culinaris* is the only cultivated species among the seven species of the genus (Van Oss et al. 1997). Lentil is a self-pollinated diploid ($2n = 14$) (Muehlbauer 1991) crop, considered to have a Mediterranean center of origin.

Based on archeological sources, domestication of lentil in the Old World is a rather simultaneous procedure closely associated with the domestication of wheat (emmer and einkorn) and barley. These cultivars were spread in the Levant region (referring to a large area in the Eastern Mediterranean) in many different sites approximately the same period. Indicatively lentil cultivation appears in about 10200–9550 cal BP (for calibrated years Before Present time, based on radiocarbon dating) of findings obtained from Neolithic settlements. Similar findings were obtained from various sites of the Fertile Crescent region such as Tell Aswad (10200–8700 cal BP), Tell Abu Hureyra (10250–9500 cal BP), Jericho (10600–9900 cal BP) Çayönü (10600–9900 cal BP) and Ali Kosh (Iran) (9600–8800 cal BP). The spread of lentil is linked with the First Agricultural Revolution, referred to also as the Neolithic agriculture featured as a wide-scale transition of many human cultures from a lifestyle of hunting and gathering to that of agriculture and settlement, making an increasingly larger population possible. Thus, the Neolithic sites of southeastern Balkans and the Aegean are key linkage sites among Europe, Africa, and the Near East in spreading lentil cultivation. Archeological excavations revealed that in the sixth millennium BC, lentil had already been distributed in many parts of Europe and Africa along with other legumes. Carbonized remnants of lentil with pea, grasspea (*Lathyrus sativus*), chickling vetch (*Lathyrus cicera* L.), and bitter vetch were found along with cereal species dated from the Mesolithic Period (5790–5630 BC) in southeast Bulgaria. Later on, in 5000–4800 BC in the area of today's Paris, France, lentil is first cultivated along with pea and other cereals. In the Bronze Age (3000–1000 BC) cultivation of lentil is widely spread along with other grain legumes such as pea and vetch. In the Far East, lentil cultivation appeared 2250–1750 BC as part of the Harappan crop assemblage in the Indian subcontinent. In the Iron Age (1600–600 BC) lentil is broadly used in Europe from the south (Mediterranean) to the north, at La Tène sites in Baden-Wurttemberg (Germany). Noteworthy that during this time the presence of lentil was obvious in everyday meals, often in combination with other legumes such as pea, vetch, or faba bean (Ljuština and Mikić 2010). In South America lentil was introduced by Spaniards in the post-Columbus era and the distribution began from Chile, Argentina, and Peru (Weiss and Zohary 2011).

## 3 Diversity of Wild Relatives and Progenitors

Lentil was initially classified in the genus *Ervum*, but later was accepted as a separated genus. It was first named as *Lens esculenta* by Moench in 1794, but soon it was abandoned (Westphal 1974). The scientific name of lentil, *Lens culinaris*, was first published by Medikus in 1787. Initially five species were recognized in the genus: *L. culinaris*, *L. ervoides*, *L. montbretii*, *L. nigricans*, and *L. orientalis*. In 1930 Barulina published the first systematic analysis of lentil species diversity using the Vavilov collection, defining the *macrosperma* and *microsperma* types as subspecies and describing regional groups according to geographic distribution of qualitative traits, mainly seed characteristics, pubescence and size of vegetative organs, phenology, height, and pod shape. Her work suggested that cultivated lentil is derived from *L. orientalis*. Later work by Williams et al. (1974) suggested *L. culinaris* and *L. orientalis* as subspecies of *L. culinaris*, and with support from Ladizinsky's work (1979) it was concluded that *L. culinaris* subsp. *orientalis* is the progenitor of the cultivated lentil, *L. culinaris* subsp. *culinaris*. In a recent re-classification, van Oss et al. (1997) suggested that the genus *Lens* consisted of seven taxa: the cultivated lentil *L. culinaris* Medikus subsp. *culinaris*, its wild progenitor *L. culinaris* subsp. *orientalis* (Boiss.) Ponert, *L. odemensis* (Ladz.), *L. ervoides* (Brign.) Grande, *L. nigricans* (M. Bieb.) Godr., and two recently recognized species, *L. tomentosus* Ladiz. and *L. lamottei* Czefr. (Sarker and Erskine 2006). Crop wild relatives (CWRs) are known to harbor a diversity of genetic traits and alleles that are indispensable to improve the genetic base of cultivated species. Thus CWRs are essential gene pools for breeding programs. A number of researchers explored the genetic diversity of wild lentil species using current molecular biology methods and tools aiming to identify genetic sites and valuable traits that could be used to improve varieties.

## 4 Genetic Diversity in Landraces and Their Conservation

Apart of CWRs, the role of landraces as key players of species' adaptation in certain regions is crucial. Landraces can be defined as "geographically or ecologically" distinctive populations, which have evolved under cultivation and become locally adapted to various environments. The term "landrace" was first used by von Rünker in 1908 (Benlioğlu and Adak 2019). Landraces, contrary to modern varieties, have not been exposed to modern plant breeding procedures and are diverse in their genetic composition both within and between populations. They provide a broad representation of the natural variation that is present in a species and constitute the primary gene pool available to breeders for improvement of crops (Bolibok-Brągoszewska and Rakoczy-Trojanowska 2015). Thus, landraces constitute a dynamic component of cultivated species genetic diversity.

Beyond their local cultural importance and heritage, landraces comprise principal characteristics: (1) high genetic diversity, (2) local genetic adaptation, (3) recognizable features (geographic identity), (4) lack of systematic plant breeding, (5) linked with traditional farming systems, and (6) history and links in a specific locality (Benlioğlu and Adak 2019). At this point it should be mentioned that the distinction among landraces and old varieties that are no longer in production can be difficult. Landraces differ from commercial varieties which are more productive and derived from the breeding process. A landrace is often characterized as "a gene pool" or "a source of genetic diversity" due to its adaptation to certain environmental conditions. The abundance of useful alleles in landraces is threatened by their replacement with commercial varieties leading to a genetic drift, known as genetic erosion. The latter denotes "the loss of genetic diversity in a plant species" and furthermore the reduction in allele number and frequency (Rogers and McGuire 2015). Therefore, conservation of plant genetic resources is indispensable in safeguarding biodiversity, and enhances food safety and plant breeding to cope with population increase and to mitigate the impact of climate change.

Nowadays lentil landraces are still an integral part of marginal lands and a stable income of remote communities. This close linkage of a landrace's adaptation and systematic cultivation to a specific geographic region and its socioeconomic importance is fundamental in preserving in situ and *on-farm* lentil genetic resources. Within the European Union (EU) regulations are implemented in provision of landraces genetic resources conservation to avert genetic erosion. Accordingly lentil products recognized as Protected Designation of Origin (PDO) and Protected Geographical Indication (PGI) aim to support the conservation through cultivation of landraces in their place of origin.

Due to the environmental fitness of a landrace to a geographic area, researchers have explored the genetic diversity and structure of lentil landraces during the last decade. The landrace Santo Stefano di Sessanio from Italy was characterized (Torricelli et al. 2012) while Ceccobelli et al. (2019) studied the genetic distinctiveness of a Protected Geographical Indication lentil landrace from Umbria, Italy, during the last 20 years. Moroccan landraces and lentil cultivars were assessed using molecular markers (Idrissi et al. 2017; Mbasani-Mansi et al. 2019b). Utilizing both molecular and morphological markers, Polidoros and coworkers (Tsanakas et al. 2018) studied the genetic diversity of the Greek lentil landrace 'Eglouvis' cultivated continuously over 400 years at an isolated plateau in the Ionian island of Lefkada. This particular landrace is an outstanding example of a diversified population recognized by its enormous morphological variety perceived in seed colors, patterns, and size. This unique diversity is also depicted by Simple Sequence Repeats (SSRs) denoting the variability of genetic alleles presence. Thus landrace 'Eglouvis' represents a significant genetic pool for breeding strategies. Nowadays, both agronomists and farmers try to enhance the preservation of local varieties. Conservation of lentil landraces could occur in two ways: by *on-farm* conservation of landraces, where farmers cultivate locally landraces, presenting in situ actions and by ex situ conservation in genebanks worldwide.

Genetic diversity of lentil is reflected to extensive phenotypic variation of leaf and seed characteristics, among others. A large collection of lentil landraces from Greece was examined for phenotypic diversity of leaf and seed characteristics. In Fig. 2 representative examples of leaf and seed variability are shown.

## 5 Ex Situ and In Situ Collections of the Species

In 1992 the United Nations conference in Rio de Janeiro established the importance of biodiversity for mankind and selected strategies to safeguard and conserve to support food safety and agriculture. Two main strategies are employed for biodiversity conservation, including in situ and ex situ. In situ refers to conservation at the natural habitats, while ex situ refers to preservation away from the natural habitat. Today a large number of lentil accessions including landraces, old and obsolete

**Fig. 2** Lentil phenotypic diversity of leaf and seed characteristics. Left: leaf shape and size of the third leaf from the top of 1-month-old growing plants for 30 landraces of Greek origin are shown; Middle: seed color range from black, yellow, red, and green in different varieties; Right: cultivar (Samos at left and Demeter at right) and landrace (Voio at left and Eglouvis at right) seeds displaying variation in size, shape, and color. Notice that cultivars are more homogeneous although variation is detected

varieties as well as species crop wild relatives are conserved ex situ. Genebanks are most widely known for ex situ preservation centers. A search in the GENESYS platform for plant genetic resources (https://www.genesys-pgr.org/c/lentil) results 31,376 lentil accessions maintained worldwide in genebanks. Of these, *Lens culinaris* includes 30,687 accessions, and *Lens ervoides*, *Lens nigricans*, *Lens esculenta*, and *Lens orientalis* 348, 164, 55, and 53 accessions, respectively (other 69). The largest genebank collection of the genus *Lens* is maintained in ICARDA (https://www.icarda.org) with 10,500 lentil accessions including crop wild relatives. The CGIAR genebank platform (https://www.genebanks.org) is an essential component of a global system on conservation and use of plant genetic resources for food and agriculture. It encircles the collaboration of 11 independent CGIAR genebanks distributed worldwide (Fig. 1). The European Cooperation Program of Plant Genetic Resources maintains the EURISCO database (https://eurisco.ipk-gatersleben.de) platform which provides information along with passport data, phenotypic data, and locality of accessions preserved in Europe, with the collaboration of more than 400 institutions. A recent search in EURISCO platform for the lentil resulted in 9135 accessions of *Lens culinaris*, 56 accessions of *Lens ervoides*, 21 accessions of *Lens lamottei*, 150 accessions of *Lens esculenta*, and 82 accessions of *Lens nigricans*. Apart from these genebanks national genebanks are maintained actively in various countries where lentil breeding programs progress. Such cases include the Canadian genebank, the Australian Grains Genebank (https://grdc.com.au/resources-and-publications/groundcover/gc110/australian-genebank), and others.

Undoubtedly genetic resources are the cornerstone to improve food safety and agriculture in response to an increasing population. Thus the value of the genetic resources remains mostly obscure until its utilization in breeding and variety development. How this is employed in lentil is presented in the following sections.

# 6 Crop Improvement and Climate Resilience and Herbicide Resistance

## *Lentil Adaptation to Abiotic Constraints*

Lentil has been traditionally cultivated in semiarid regions under rainfed conditions and is adapted to a wide range of agro-ecological environments (Kumar et al. 2016a, 2019) resulting in rich phenotypic variation. Among the main breeding objectives, with an initial selection against pod dehiscence, is the seed size, flowering response to photoperiod and temperature, as well as resistance or tolerance to abiotic and biotic stresses (Erskine et al. 2015). The lentil distribution from the center of origin has been accompanied by the selection of traits important for adaptation to environments that are defined by climate and land-use change and the imminent environmental extremes (Kumar et al. 2016a; Raza et al. 2019).

Lentil is mainly cultivated in three major climatic regions: (1) as a winter crop in the Mediterranean climatic zone, experiencing increasing temperatures and day length until harvest; (2) in south Asia, where it is sown and harvested during winter of the northern hemisphere, with less than 12 h day length and declining temperatures during the juvenile phase followed by higher temperatures during the reproductive phase; and (3) in the temperate zone, where lentil crop is sown in spring and experiences lengthening days in the first half of the growth cycle, followed by shortening days during most of the reproductive cycle (Tullu et al. 2011b). Considering the varied zones of adaptation, it is important to understand the advantages and vulnerabilities of lentil crop grown in such environments.

Lentil crop is subjected to increasing environmental constraints causing substantial reduction in crop growth, yield, and production worldwide (Kumar et al. 2015). Despite the increase in global lentil yield from 934 kg/ha in 2006 to 1038 kg/ha by 2018 and productivity of 6.3 million tonnes (FAOSTAT 2019), the current yields are still low in comparison to other crops, due to the exposure of the crop to an array of plant stressors with a major impact on plant development and productivity. Among the factors limiting yield is the low or no response to inputs, and the exposure to various biotic and abiotic stresses. However, the most important abiotic limiting factors to lentil production include the low moisture availability combined with heat stress in spring, whereas cold stress affects temperate regions, and salinity, nutrient deficiency, and toxicity occur in localized regions (Muehlbauer et al. 2006). Soil-related abiotic factors affect not only the lentil plant but also the rhizobia, and/or the rhizobia-plant symbiosis, thus having a negative impact on the symbiotic relationship formed with the rhizobia, which may lead to reduced nitrogen fixation (Sultani et al. 2007).

Lentil is mainly cultivated as a rainfed crop and often in drought-prone arid regions, in areas of the Mediterranean climate and subtropical climate, where limited rainfall and water-deficient soil conditions are encountered. Lentil shows considerable variability for drought tolerance (Idrissi et al. 2015; Singh et al. 2016c) and although it has higher water use efficiency compared to other crops, such as chickpea (Zhang et al. 2000), water availability is still critical, given that considerable losses occur due to severe water deficit (Singh et al. 2016c). Drought stress inhibits the biochemical processes of seed filling (Sehgal et al. 2017). Pod formation stage has been shown to be the most sensitive to drought stress, followed by flower initiation and mid-vegetative stages (Talukdar 2013). However, the successful crop establishment in semiarid areas is largely dependent on the ability of seeds to germinate under low water availability (Arjenaki et al. 2011; Fahad et al. 2017). The development of drought-tolerant cultivars is necessary for reducing yield losses and extending cultivation to marginal areas (Nadeem et al. 2019). Escape and avoidance strategies during the critical growth stages are crucial in overcoming the heat and drought stresses in lentil. A major drought escape strategy is the development of early maturing, short-life cycle legume crop varieties; yet, this would be an efficient strategy for regions, such as the Mediterranean, with vigorous growth in winter, when conditions are favorable, and rapid senescence induced by rising temperatures and water deficits occurring at the reproductive stage (Zhang et al. 2000; Shrestha et al. 2006). However, when the objective is to enhance yield under a range of water conditions this is a nonviable breeding strategy for lentil (Ghanem et al. 2017). Traits that determine the transpiration demand or improve the water use efficacy of the plant have been associated with drought tolerance and avoidance strategies; yet, in lentil, research is still largely unmapped on the following traits: (1) the early reduced transpiration under soil desiccation, which is advantageous for late-season water deficit, allowing the use of conserved soil water to sustain physiological activity during seed fill and (2) partial stomatal closure under high atmospheric vapor pressure deficit (VPD), which is an essential trait for environments where lentil is grown under terminal drought and high temperatures (high VPD conditions) (Ghanem et al. 2017). Nevertheless, characterization of germplasm based on drought tolerance index (DTI) revealed that the most tolerant groups outperformed the sensitive ones under both irrigated and rainfed conditions (Babayeva et al. 2014).

Major drought-associated traits include: (1) in vitro seed germination potential, water content, and root length for rapid identification and selection of drought-tolerant genotypes at early growth stages (Muscolo et al. 2014; Foti et al. 2019) and (2) early growth vigor, rapid ground cover, and leaf and root characteristics including size, morphology, depth, length, and density (Kumar et al. 2016a). Other traits associated with drought resistance in lentil are the higher chlorophyll content and increased chlorophyll stability index, lower cell membrane injury and reduced lipid peroxidation in terms of malondialdehyde (MDA) content, increased accumulation of osmotically active solutes (soluble sugars and proline), and lower hydrogen peroxide ($H_2O_2$) content (Mishra et al. 2016). It has been also observed that aboveground traits can be reliably used for indirect selection of underground traits (root) in breeding programs, and for enhanced drought tolerance (Idrissi et al. 2015), such as dry shoot biomass and chlorophyll content (Kumar et al. 2012).

A very important component of drought resistance is the plant root characteristics such as the taproot length, prolific lateral root development, and high dry root weight that enhance the capacity of a plant to extract water from deeper soil strata and absorb nutrients (Gahoonia et al. 2005, 2006; Kumar et al. 2012) and are associated with drought tolerance and seed yield in several lentil landraces (Sarker et al. 2005). High genetic variability in such traits has been observed among lentil germplasm that could be useful for the development of drought-adapted genotypes (Kumar et al. 2012, 2016a). Therefore, a well-developed root system and early biomass development are essential traits for breeding tolerant genotypes (Idrissi et al. 2015). Nevertheless, when just relying on structural aspects of root and shoot functionality might be misleading, considering that selection for deeper roots and/or rapid development of aboveground biomass might cause deprivation of moisture stored in soil and depletion of water from the upper soil strata, respectively (Ghanem et al. 2017). Research on the genetic control and linkage of SSR markers for drought tolerance in global collections of lentil accessions with different adaptation to drought conditions has led to the identification of different genes/QTLs controlling root traits; recently linked molecular markers were associated with the *Sdt* locus that controls seedling survival drought tolerance in lentil (Singh et al. 2016a, b) and a QTL-hotspot genomic region controlling a number of root and shoot characteristics linked to drought tolerance (Idrissi et al. 2015).

In Mediterranean-type climates, lentil crop matures in conditions, where low water availability occurs concurrently with increasing temperature. Among other winter pulses, lentil is highly sensitive to heat stress, especially when day/night temperatures exceed 32/20 °C (Ibrahim 2011), having a negative effect on the physiological traits and reproductive development of the plant. Heat stress also negatively affects rhizobia (Lira Junior et al. 2005), with nodulation being higher (1.8–22-fold) in heat-tolerant genotypes (Sita et al. 2017b). The crop invariably encounters drought and heat stresses at the time of podding and seed filling period leading to forced maturity and thus extensive yield losses (Bhandari et al. 2016; Kumar et al. 2016b). Exposure of lentil crop to high temperatures (30–35 °C) during the reproductive stage results not only in extensive yield losses due to pod abscission (Bhandari et al. 2016; Sita et al. 2017b) but also in reduced quality components, with severe impacts on heat-sensitive genotypes (Sita et al. 2018). Heat tolerance in lentil has been associated with pollen function and the availability of sucrose to anthers and pollen grains (Sita et al. 2017b). Lentil heat tolerance mechanisms include protection of leaf tissues through reduction of oxidative damage, optimum photosynthetic and respiratory function, and induction of nodulation (Sita et al. 2017a, b). A temperature threshold >35/25 °C is considered critical for differentiating the heat-tolerant and sensitive lentil genotypes (Gaur et al. 2015). In a combined heat and drought stress screening, seed yield was reduced to a greater extent under heat stress, whereas stress combination impaired photosynthetic ability, water relations, and carbohydrate metabolism in both leaves and seeds with a reduction in sucrose and starch-synthesizing enzymes, and thus, severely inhibited seed size and weight (Sehgal et al. 2017).

In an effort to establish an effective protocol for selecting heat-tolerant lentil genotypes under field conditions morpho-physio-biochemical traits, such as the number of filled pods and pollen viability at higher temperature, have been associated with heat tolerance (Kumar et al. 2016b) and the molecular characterization for heat stress in lentil has deciphered the genetics and pathways underlying the heat stress tolerance (Delahunty et al. 2018; Kumar et al. 2018). Simple sequence repeat (SSR) marker analysis has, thus far, been used to molecularly characterize genotypes based on the different morpho-physiological and reproductive traits under heat stress, highlighting their variability on heat adaptation and the potential to combine diverse mechanisms for the development of high-temperature tolerant genotypes (Singh et al. 2016b). In lentil under heat stress several QTL have thus far been identified for seedling survival and pod setting traits (Singh et al. 2017b), days to maturity, pods/plant, secondary branches/plant, 100 seed weight, yield/plant and reproductive duration (Kumar et al. 2018). Recently, an NGS-based transcriptome approach in lentil has revealed candidate genes expressed under heat stress conditions related to physiological and pollen phenotypes, cell wall, and secondary metabolism in lentil (Singh et al. 2019). Heat stress is considered to have more severe and adverse effects on lentil production under climate change scenarios rather than drought, if night temperature rises by at least 2 °C (Kumar et al. 2016b). However, further research on heat stress is required to critically define the effects on the growth and development of lentil and the effects on yield, especially considering that global warming has become a major challenge to rainfed agriculture.

As a winter crop, lentil encounters cold stress during the seedling and vegetative stage when temperatures are still low. The prolonged low temperatures delay growth and induce injuries on flowers leading to flower degeneration and abscission, which results in reduced seed yield (Muehlbauer et al. 2006; Kumar et al. 2016a). Cold tolerance is an essential requirement when lentils are cultivated during winter or at high altitudes where temperatures are below 4 °C. Thus, genotypes with winter hardiness and cold/frost tolerance are required and further emphasis on understanding the physiological and molecular adaptation mechanisms is a prerequisite. The effect of cold stress (4 °C) on the physiological and biochemical responses of lentil seedlings was manifested as increased malondialdehyde and proline contents, while the antioxidant enzyme activity increased in shoot tissue with superoxide dismutase being the main enzyme to mitigate the effects of ROS and ascorbate peroxidase in root tissues (Öktem et al. 2008). Sources of cold resistance have been identified among germplasm including wild species. Winter hardiness in lentil is a polygenic trait and several QTL have been identified for winter survival and winter injury (Kahraman et al. 2004a, b). Significant association has been found between leaf area and winter hardiness, with smaller leaf area being associated with improved winter survival (Kahraman et al. 2010). The differential expression of gene transcripts potentially involved in frost tolerance highlighted genes for proline-rich, glycine-rich, and transmembrane proteins as well as genes related to photosynthetic functioning as important in the response to frost temperatures (Barrios et al. 2017). However, prior to integration of genes associated with cold tolerance in breeding programs, intense field testing and marker-assisted selection are required (Kumar et al. 2016a).

Lentil is classified as one of the most salt-sensitive crop plants and legume crops when compared to moderately sensitive faba bean and soybean (Katerji et al. 2003). Salinity poses a severe threat to germination and plant growth of lentil (Bandeoğlu et al. 2004), and lentil has a salinity tolerance threshold at electrical conductivities >2.2 dS $m^{-1}$ in saturated soil (Rameshwaran et al. 2016) with approximately 3 dS $m^{-1}$ causing 88–100% yield loss (Katerji et al. 2001; Van Hoorn et al. 2001). Nonetheless, the response to salinity is dependent on the genotype, since germination in tolerant wild accessions and breeding lines was not affected at any salinity level, yet moderately tolerant and sensitive genotypes decreased germination at 40–120 mM NaCl level (Singh et al. 2017a). Salt stress also adversely affects the symbiotic relation with rhizobium and root nodulation and interferes with $N_2$ fixation and uptake (Farooq et al. 2017). Screening techniques such as germination, seedling survival, reduction in seedling growth, biomass accumulation, $Na^+$, $Cl^-$, $K^+$ contents, proline and MDA contents, $H_2O_2$ production, and antioxidant activities have been widely used for the evaluation of salt tolerance in lentil (Sidari et al. 2008; Ouji et al. 2015; Kumar et al. 2016a). However, given that salt tolerance at the germination stage is a heritable trait, screening for resistant genetic material at early growth stages could be beneficial for targeted breeding (Foti et al. 2019).

Genetic variation in salinity tolerance among cultivated lentil (El-Monem and Sharaf 2008; Ouji et al. 2015; Foti et al. 2019) and genotypes of different *Lens* species (Singh et al. 2017a) have been identified. Tolerance to increased salinity in legumes has been associated with accumulation of compatible osmolytes, increased antioxidant capacity, excretion and/or compartmentalization of toxic ions, and/or hormonal regulation (Farooq et al. 2017). In lentil, this variation may be linked to their contrasting response mechanisms, such as the ability to osmoregulate under salt stress (Sidari et al. 2008) and the activation of the antioxidant defence system (Bandeoğlu et al. 2004). Exogenous application of hormones and osmoprotectants, such as $K^+$ and $Ca^{2+}$, has mitigated the detrimental effects of high salinity on lentil growth and led to an increase in accumulation of dry matter (Kafi et al. 2012), indicating that nutrient management may improve salt tolerance in grain legumes (Farooq et al. 2017).

Lentil is also sensitive to alkaline stress and yield is drastically reduced under high alkalinity (Singh et al. 2018a, b). Adaptation of lentil to alkaline conditions is different among genotypes with diverse genetic backgrounds (Kumar et al. 2016a). Alkalinity tolerance in wild accessions and breeding lines has been associated with higher seedling survival, lesser reduction in seedling growth, biomass and seed yield, possibly due to restricted uptake of $Na^+$ by sclerenchymatous layers, intact pericycle and steler region in roots and thick epidermis in shoots, maintaining cell turgor pressure, accompanied with greater uptake of $K^+$ and lower $H_2O_2$ production (Singh et al. 2018a). Selecting for alkaline stress tolerance to high pH levels ranges between pH 9.0 and 9.1; beyond this threshold seed yield decreases with an increase in soil pH, indicating the existence of limited range of variability (Kumar et al. 2016a).

Waterlogging tolerance varies among genotypes and has been associated with the country of origin, with genotypes originating from tropical climates being more tolerant compared to those from dryer climates (Mediterranean and temperate) (Wiraguna et al. 2017). Nevertheless, it is only relevant in regions with rice-lentil relay cropping systems, despite the adverse effect on lentil germination and the consequent reduction in crop production (Wiraguna et al. 2017).

Nutrient availability and minerals, such as boron and iron imbalances, are of local importance, causing substantial yield loss. Variations in response to iron and boron deficiency and boron toxicity have been found among lentil genotypes under Fe-deficiency or excess (Gupta et al. 2017), B-deficiency, and B-toxicity (Yau and Erskine 2000; Hobson et al. 2006). This variability is necessary for developing genotypes with high yield potential in soils with mineral imbalances (Kumar et al. 2016a). Lentil reduces ferric iron at the rhizosphere which is then absorbed as ferrous iron by the root. Recently, specific DNA markers have been developed for iron metabolism related genes (Ferritin-1, BHLH-1, IRT-1) in lentil, with significant differential expression for Ferritin-1 and IRT-1 genes at the transcriptional level (Gupta et al. 2017). Transcriptome-based SSR and SNP markers have been used to develop a genetic map for elucidating the segregation of boron tolerance in lentil leading to the identification of two SNP loci associated with B-tolerance and three candidate orthologous genes with B transport-associated functions (Kaur et al. 2014).

## *Parasitic and Nonparasitic Weeds and Herbicide Tolerance*

In winter-sown lentil, parasitic and nonparasitic weeds represent a major constraint of lentil productivity worldwide, due to the short height and slow early growth (Fedoruk et al. 2011). Nonparasitic weeds compete for water, light, and nutrients, often resulting in an average yield loss up to 40% in legume crops (Kumar et al. 2016a). Besides common weeds, root parasitic weeds, such as *Orobanche*, adversely affect lentil production in some Mediterranean countries (Rubiales et al. 2006). Lentil can be parasitized mainly by *O. crenata* Forsk (crenate broomrape), and the imminent damage can reach up to 100% (Rubiales et al. 2014; Mbasani-Mansi et al. 2019a). The five general categories of weed control methods are the following: preventative, cultural, mechanical, chemical, and biological (Erskine et al. 2011). However, integrated management has been shown to be the main method to cope with the parasitic weeds (Rubiales and Mónica 2012; Fernández-Aparicio et al. 2016).

Breeding for resistance against broomrape (*Orobanche* spp.) root-holoparasitic infection in legumes is a difficult task considering the scarce and complex nature of resistance (Rubiales et al. 2014; Fernández-Aparicio et al. 2016). In the last decades, sources for resistance against *O. crenata* have been identified in Spanish lentil germplasm and in wild species including *L. ervoides*, *L. odemensis*, and *L. Orientalis* and the accessions ILWL 361 and LENS166/92 (Fernández-Aparicio et al. 2008, 2009), which could be used in lentil breeding. Recently, field trials revealed a significant effect of climatic conditions on *Orobanche* infestation levels among assessed genotypes (Mbasani-Mansi et al. 2019a), linking climate change and abiotic stresses with broomrape infection. The use of in vitro screening methods together with marker-assisted selection can be a potential tool for genetic improvement in lentil breeding for broomrape (Muehlbauer et al. 2006; Rubiales et al. 2006; Rispail et al. 2007). Opportunities for the functional characterization of genes in lentil improvement include the developing appropriate populations by the manipulation of tolerant and sensitive sources against imazethapyr (Sharma et al. 2018).

The existence of herbicide-resistant weeds imposes new challenges in cropping systems and can lead to increased weed management costs, reduced yields, grain quality, and ultimately reduced profits (Chauhan et al. 2017). Herbicide-tolerant crops have a direct impact on agriculture, controlling weed populations and reduction of crop management practices of high cost (Cokkizgin and Shtaya 2013; Singh et al. 2014a; Erskine et al. 2015). Post-emergence application of the herbicide imazethapyr appeared to be promising and effective in controlling weeds and managing seed yield of lentil (Teja et al. 2017). Genetic variability for IMI class herbicide tolerance has also been observed in lentil (Singh et al. 2014a; Sharma et al. 2018).

Although, naturally occurring herbicide tolerance in crops is the core principle of weed management, technological advancements in plant breeding through biotechnology have allowed the development of new herbicide-resistant lentil genotypes. During the last decades, various attempts were made to develop imidazolinone-tolerant crops through seed mutagenesis. Imidazoline-tolerant lentils have been developed in Canada and Australia (Slinkard et al. 2015), having great potential in expansion of lentil area under conservation agriculture (Erskine et al. 2015). Chemical mutagenesis has been used to develop herbicide-resistant lentil genotypes through seed mutagenesis (Khan and Khan 2011; Rizwan et al. 2017). The adaptation of lentil cultivars three synthetic 1,2,3 thiadiazole compounds revealed the beneficial role of GABA metabolism and ROS induction against the effect of these herbicides (Al-Quraan et al. 2015). In lentil resistance to herbicides has been achieved through: (1) the integration of the *bar* gene and the development of ammonium glufosinate herbicide-tolerant plants (Khatib et al. 2007), (2) the stable integration and expression of the mutant acetolactate synthase (ALS) gene conferring resistance to sulfonylurea herbicides (Gulati et al. 2002), and (3) the enhanced transformation via *Agrobacterium*-mediated *beta-glucuronidase* (*GUS*) gene expression (Chopra et al. 2012). Additionally, herbicide-tolerant varieties to the post-emergence herbicide, imazethapyr, have been identified, allowing the application of broad-spectrum post-emergence herbicides towards improved weed control in lentil (Sharma et al. 2018).

Lentil crop is adapted in most cases to the mixed cropping cereal-based farming systems in marginal lands (Ram et al. 2018; Liu et al. 2019) and provides additional benefits to the local communities. Therefore, developing improved lentil varieties with a high and stable yield and enhanced tolerance to abiotic stresses is a prerequisite for climate-resilient agriculture. The twofold challenge is to identify genotypes that are broadly adapted across different or specific regions and the importance for these adaptation traits. Therefore, the improved adaptation entails understanding of the underlying response mechanisms to changing environment that can be achieved through interpreting the genotype by environment interactions (G × E). This is of utter importance, especially for breeding programs, considering the diverse factors affecting a crop under widely diverse environments across locations and years. In pulses, G × E interactions contribute >70% of total phenotypic variance (Kumar et al. 2015). Recently, all available global wild lentil germplasm was characterized for various target agronomical traits, and major biotic stresses and potential accessions (gene sources) carrying important traits for enhancing genetic gains of cultivated gene pool were identified under diverse agro-ecological conditions (Singh et al. 2014a, 2020). As previously mentioned, genetic variation has been uncovered in lentil for tolerance mechanisms under abiotic stresses. Polymorphisms of gene families, such as the resistance gene analogues (RGAs), peroxidases (POX), and WRKY transcription factors, have been identified in *L. nigricans* and *L. culinaris* subsp. *orientalis* related to biotic and abiotic stress tolerance, which potentially could be used for enhancing breeding and resistance studies in lentil and developing stress-related resistance genotypes via crosses between wild types and cultivars (Sari et al. 2016). However, the mechanisms controlling abiotic stresses are dependent on the polygenic nature of the trait, the wide environmental variation associated with it, and the lack of suitable methods to efficiently select for the trait (Foti et al. 2018). Crop improvement for enhanced tolerance to abiotic stresses entails targeting germplasm with the potential for common defence mechanisms that can address multiple stress tolerance in plants (Gupta et al. 2019).

Wild species of *Lens* possess several useful genes, which can be utilized to improve agronomic traits and broadening the genetic base of cultivated lentil (Dikshit et al. 2015). Considering that the wild species have evolved under selection to survive climatic extremes, a promising strategy to increase genetic diversity and stress tolerance in new cultivars would be the introgression of favorable genes from lentil wild relatives (Kumar et al. 2019). Research has focused on the introgression of important alleles of agricultural interest from wild species to cultivars through wide crosses, towards developing tolerance to abiotic and biotic stresses. Thus far, the International Centre for Agricultural Research in the Dry Areas (ICARDA), the global mandate for lentil improvement with the largest global collection of genetic resources, is using wild species to widen the genetic base of lentil cultivars in an effort to develop drought-tolerant lentil varieties. The introgression of alien genes from wild relatives has been successfully achieved (Gupta and Sharma 2007) to broaden the genetic base of cultivated gene pool (Kumar et al. 2013). Improved stress tolerance can be achieved by selecting parental genotypes based on SSR markers before intercrossing (Kumar et al. 2015).

Environmental conditions drive natural selection towards the adaptation of lentil accessions to the adverse conditions of the growing habitat, increasing the probability of occurrence for traits and genes imparting abiotic stress tolerance, such as heat tolerance (Gaur et al. 2015; Delahunty et al. 2018; Idrissi et al. 2018). The Mediterranean Basin holds a large part of lentil biodiversity, where landraces and cultivars have been under repeated selection for adaptation to biotic and abiotic stress conditions during domestication and cultivation of lentil in a wide diversity of edaphic and climatic conditions, showing high molecular diversity and different response mechanisms to stress factors (Idrissi et al. 2015) that could help enhancing the adaptation to biotic and abiotic stresses (Khazaei et al. 2016; Idrissi et al. 2018).

The genetic structure within the Mediterranean gene pool has been defined using high-throughput genotyping by sequencing (Pavan et al. 2019). The observed high genetic heterogeneity of Mediterranean landraces from Morocco, Italy, Greece, and Turkey could be a source of adaptation to biotic and abiotic stresses (Lombardi et al. 2014) and be used for broadening the genetic base of lentil grown in South Asia (Dikshit et al. 2015).

## 7 Disease Resistance and Insect Pest Resistance

Lentil (*Lens culinaris* Medikus subsp. *culinaris*) is grown as a winter crop in all continents. Diseases are a major cause of yield losses in this crop which is affected by foliar as well as soil-borne pathogens (Chen et al. 2011). Variations in environmental conditions have a major influence on the prevalence and incidence of specific diseases in different cultivating areas. Fungal diseases are the most important biotic threat for lentil growth and productivity (Taylor et al. 2007). *Ascochyta lentis* (Ascochyta blight) and *Fusarium oxysporum* f. sp. *lentis* (Fusarium wilt) can be considered as the major fungal pathogens as they cause severe losses in most lentil growing regions globally. Other fungal pathogens such as *Uromyces viciae-fabae* (Rust), *Stemphylium botryosum* (Stemphylium blight), and *Colletotrichum truncatum* (Anthracnose) are important in specific continents or countries. In addition, Botrytis gray mold, Aphanomyces root rot, Rhizoctonia wet rot, and several viral diseases, caused by aphid transmitted viruses, are included in the most notable biotic stresses. Other pests include nematodes, insects, and parasitic plants. Management of diseases is based on integrated control, including the use of resistant cultivars, modified cultural practices, and application of fungicides (Taylor et al. 2007). In this chapter, methods and achievements in breeding for resistance to major diseases are presented.

Resistance to diseases is included in the main objectives of selection in lentil. Several breeding programs enabled selection of varieties resistant to various pests as well as other stresses (Muehlbauer et al. 2006; Sarker and Erskine 2006). Initial screening of resistance via artificial inoculations and scoring disease severity has traditionally served for the selection of genotypes which are resistant to a particular pathogen. Screening techniques can be applied under controlled conditions or in the field. Field trials allow evaluation of a large number of genotypes (Dubey and Mishra 2019; Buchwaldt et al. 2018). After phenotype screening and selection, resistant varieties are used as gene donors in conventional breeding programs. Hybridization and germplasm evaluation form the basis of conventional breeding. However, the emergence of new pathogenic races of pathogens renders previously resistant cultivars less effective or useless (Davidson et al. 2016), and loss of effective disease resistance requires development of new resistance cultivars. Marker-assisted selection (MAS) can speed up breeding programs and is deemed necessary to achieve durable resistance. Several accessions characterized as resistant are subjected to studies on the identification of molecular markers linked to resistance. Wild relatives of lentil can be used also in such programs (Singh et al. 2017c) in combination with agronomically attractive varieties. In addition, current tools can facilitate the construction of linkage maps for QTL mapping of disease resistance in lentil. Multilocation testing for the identification of durable resistance can contribute to a successful resistance breeding program.

*Fusarium oxysporum* f.sp. *lentis* (Fol) is the most important soil-borne pathogen causing wilt in lentil. Yield losses can be as high as 50% and can reach 100% in severe infections (Tiwari et al. 2018). *F. oxysporum* pathotypes persist in soil for several years due to their long-viable resting spores named chlamydospores (Chen et al. 2011). Upon appearance of a specific host, chlamydospores germinate and hyphae invade the xylem via the roots. The fungus is then established in the vascular system blocking the movement of water and nutrients via xylem vessels. Common symptoms are wilting, growth retarding, leaf discoloration, and death of the plant. The disease appears at early growth stages as well as during flowering (Chen et al. 2011).

Using resistant cultivars is the most effective means to control fusarium wilt in lentil as in most crops. Fol presents phenotypic and genotypic diversity among and within various growing areas (Tiwari et al. 2018) highlighting the need for development of race-specific wilt-resistant cultivars and for adjustment of breeding strategies to the race that prevails in each region. Earlier screenings were based on morphological/cultural characteristics of Fol isolates and virulence trials using one or various lentil varieties (Tripathi 2015). Virulence trials based on artificial inoculations are still in use. For example, using a pot screening technique and 50 local Fol isolates, Hiremani and Dubey (2018) reported evidence for the existence of eight different races. Pouralibaba et al. (2016) reported seven different patterns of virulence, named pathotypes 1–7, after testing 52 isolates from Iran, Syria, and Algeria on four resistant lentil accessions. However, molecular techniques have been implemented for the same purpose. Mohammadi et al. (2011) reported differentiation of 45 Iranian isolates and two isolates from IRCADA (*International Center for Agricultural Research in the Dry Areas*), in six groups using ISSR markers. On the contrary, after screening 101 isolates from western Iran, using SSR markers, Nourollahi and Madahjalali (2017) reported low levels of genetic variability and classified the isolates in three major groups. It is thus obvious that research results on the genetic variability of the pathogen are at present contradictory, and more accurate tools need to be developed for the elucidation of the number and virulence of races within the species.

Several resistant varieties have been made available through hybridization-based breeding programs, in ICARDA stations and in India (Choudhary and Kumar 2016). Wild relatives of lentil can also serve as gene donors in such programs (Singh et al. 2017c). Screening of resistance via artificial inoculations has been used for the selection of resistant genotypes. Hiremani and Dubey (2018) studied the susceptibility of 89 lentil genotypes to a virulent isolate of Fol, recovering seven resistant and three moderately resistant genotypes. Dubey and Mishra (2019) evaluated 130 genotypes by scoring their reaction to a pathogenic race of Fol. Fifty five genotypes were characterized as highly resistant, 25 as resistant, and 18 as moderately resistant. However, the pathogen's genetic variability described above requires more solid evidence for each cultivar's resistance.

Research on genetics of Fol wilt resistance in lentil is inadequate until present. Earlier studies reported five independent genes to govern wilt resistance in lentil (Kamboj et al. 1990), two duplicate genes in the varieties 'PL 234' and 'JL 446', two complementary genes in the variety 'PL 286', and a single dominant gene in the crosses made at ICARDA (Abbas 1995). A more recent study reported resistance conditioned by a single dominant gene and random amplified polymorphic DNA (RAPD)-marker-assisted selection for Fol resistance (Eujayl et al. 2006). Mapping resistance genes using molecular markers is preferable compared to conventional pot or plot selection; however, research towards this direction is limited and very few resistance genes (*Fw*) have been tagged (Choudhary and Kumar 2016).

Fusarium wilt is controlled by a single dominant gene, *Fw*, according to Eujayl et al. (1998). Their research revealed that *Fw* is linked to RAPD marker OPK-15900 at a distance of 10.8 cM. Hamwieh et al. (2005) reported an SSR marker and an AFLP marker at a distance of 8.0 cM and 3.5 cM, respectively, from *Fw*. More recently Choudhary and Kumar (2016) reported new markers in lentil, namely, AFLP marker p17m30710, SSR marker SSR59-2M, and RAPD marker OPK-15, at distances of 3.5 cM, 8 cM, and 10.8 cM, respectively, from the *Fw* gene. As these distances do not provide selection accuracy, it is suggested that additional more locus-specific co-dominant markers closer to the *Fw* gene are found. Choudhary and Kumar (2016) suggested that the identification of resistance-associated quantitative trait loci (QTL), especially in wild species, will facilitate breeding by targeted gene transfer.

*Ascochyta lentis* Vassilievsky is the causal agent of ascochyta blight in lentil (Chen et al. 2011). The disease is considered as one of the most important of this crop, reducing yield and seed quality and marketability. The disease is seed-borne as well as stubble-borne. Almost all aerial parts can be infected. Characteristic tan lesions with dark brown margin appear on leaves, stems, petiols, and pods. Dark pycnidia appear in the center of the lesions making easy to distinguish ascochyta blight lesions from those produced in other diseases, such as stemphylium blight. Pseudothecia (*Didymella lentis* Kaiser, Wang and Rogers) are formed on infested debris. Pod infection can result in infection of seed that shows discoloration and brown patches. Heavily infected seeds become purplish brown and shrivel. Losses due to seed staining can exceed 70% (Sudheesh et al. 2016a).

Genetic resistance to ascochyta blight was early documented in cultivars used in Australia and Canada. *Ral2*, a dominant gene, and *ral2*, a recessive gene, were reported to confer foliar resistance to cultivars Northfield (syn. ILL5588) and Indianhead, respectively (Sudheesh et al. 2016a, b and references therein). Northfield was found to bare *AbR1*, another dominant resistance gene that confers foliar resistance (Sudheesh et al. 2016a, b; Ford et al. 1999). In the same cultivar two more dominant genes responsible for major and moderate resistance, respectively, were revealed by gene mapping, whereas the same technique revealed two recessive genes in Indianhead (Ye et al. 2003). As earlier reports were inconclusive, Ye et al. (2003) re-assessed all proposed inheritance models and confirmed allelic relationships among certain resistance genes in different cultivars. Rodda et al. (2017) attributed differences in the number and nature of genes, resulting from research on ascochyta blight resistance in lentil, to the different sources of genetic resistance used, including their independent genetic control, but also to different screening assays, environmental conditions, *A. lentis* isolates, and the variable size of populations screened. Contradictory reports have provided evidence for dominant as well as recessive resistance genes in lentil seed (Rodda et al. 2017). Cultivated lentil as well as its wild relatives serve as gene pools in disease resistance breeding programs (Dadu et al. 2017; Ye et al. 2002).

Variability of virulence and pathogenicity in *A. lentis* has been documented (Davidson et al. 2016; Rodda et al. 2017; Sambasivam et al. 2017), and the emergence of new pathogenic isolates renders previously resistant cultivars less effective or useless, as in the case of Northfield and Nipper (Davidson et al. 2016). Efforts to relate resistance genes with molecular markers started early. All the abovementioned genes were connected to molecular markers such as RAPDs and sequence characterized amplified regions (SCARs) (Sudheesh et al. 2016a, b; Muehlbauer et al. 2006). Ford et al. (1999) reported two RAPD markers, RV01 and RB18, at a distance of approximately 6 cM and 14 cM, respectively, from *Ral1* (*AbR1*), in cultivar ILL5588. Chowdhury et al. (2001) identified UBC2271290 and OPD-10870, two RAPD markers, at a distance of 12 cM and 16 cM, respectively, from *ral2* in the cultivar Indianhead. Nguyen et al. (2001) used a SCAR marker (SCARW19) to select resistance to ascochyta blight in lentil accession ILL5588. Tar'an et al. (2003) used the RB18680 RAPD marker, formerly also shown to be linked to the *AbR1* gene (Ford et al. 1999), to create a SCAR marker for the same gene.

Early research on QTL analysis of the ascochyta blight resistance has been reviewed by Muehlbauer et al. (2006). More recently, SSR and SNP markers were developed and tested in linkage mapping to validate QTLs for resistance to ascochyta blight in lentil (Sharpe et al. 2013; Kaur et al. 2014; Sudheesh et al. 2016a, b). Sudheesh et al. (2016a, b) successfully used SNP markers to identify QTLs in population maps of the progenies derived from crossings between cultivars Indianahead and Northfield (IH × NF) and Indianahead and Digger (IH × DIG). Common genomic regions (AB_IH1 and AB_IH2.1/AB_IH2.2) were identified as responsible for ascochyta blight resistance in both mapping populations and were associated with the resistant parent, Indianhead, while a third genomic region was associated with Northfield parent. Such markers can be used to indicate resistance genes as candidates to be used into breeding programs and to allow pyramiding with other effective genes to confer durable resistance. Achievements in lentil genomics, related to ascochyta blight resistance such as integrated genetic linkage map, EST libraries, gene-based markers, and draft genome sequences, can be used to improve lentil cultivars (Rodda et al. 2017).

Another devastating foliar disease of lentil is stemphylim blight caused by the fungus *S. botryosum* Wallr. (Chen et al. 2011). Infected plants show tan to light brown spots on leaves which enlarge and cover the entire leaf, resulting in dryness and subsequent defoliation. At a later stage, spots develop on stems, eventually leading to the death of the plant. Severe infestation can result in flower infection and dropping but also due to infection of seeds and diminishing of seed quality (Mwakutuya and Banniza 2010). Mwakutuya and Banniza (2010) reported that seed size can be significantly reduced and seed staining similar to that caused by *A. lentis* can be induced by *S. botryosum*. However, transmission of the disease from seed to seedlings has not been reported so far on lentil. The pathogen spreads by conidia produced by conidiophores on the infected surfaces. The economic losses caused by stemphylium blight can be as high as 80% (Bakr and Ahmed 1992). However, the disease is less studied compared to those mentioned above. It has been concluded that its presence has been underestimated due to symptoms' resemblance to those of ascochyta blight on lentil (Das et al. 2019).

Till present few studies have been performed regarding pathogen's variability and the genetic base of resistance in lentil. Several massive genotype screenings have been reported (Razzak et al. 2018; Das et al. 2017; Kant et al. 2017; Mondal et al. 2017; Yadav et al. 2017) through which the degree of resistance of a number of cultivars and lines has been scored according to their reaction to pathogen inoculation. The results of such studies can be used to select genotypes for stemphylium blight resistance breeding programs. Wild relatives of lentil have also been explored as genetic sources of resistance to this disease. Podder et al. (2013) included *L. ervoides* (Brign.) Grande and other wild relatives in genotype screening under controlled conditions and proved their superiority, as compared to a resistant cultivar of the cultivated *L. culinaris*.

Genetic tools have been scarcely used for stemphylium blight resistance mapping till present. Bhadauria et al. (2017) used *Pst*I-*Msp*I-based genotyping-by-sequencing (GBS) in order to construct a genetic linkage map suitable for QTL analysis in an intraspecific biparental recombinant inbred line population ($F_9$), originating from a cross between *L. ervoides* accessions L01-827A and IG 72815. The genetic map generated expanded over 740.94 cM, and three QTLs significantly associated with resistance to *S. botryosum* were revealed by composite interval mapping. Cao et al. (2019) studied further the genetic profile of resistance of *L. ervoides* to *S. botryosum* in two recombinant inbred lines. Transcriptome sequencing revealed 8810 genes responding to the presence of *S. botryosum*, and the majority of them had a similar expression trend in both inbred lines. A great number of the remaining differentially expressed genes were connected to various metabolic processes. In a comparison of differentially expressed genes to previously observed QTLs for resistance to stemphylium blight, two genes encoding calcium transporting ATPase and glutamate receptor 3.2 were candidate resistance genes, whereas one gene with unknown function was a candidate susceptibility gene (Cao et al. 2019).

Anthracnose caused by *Colletotrichum lentis* Damm, syn. *C. truncatum* (Schwein.) Andrus and W. D. Moore (Damm et al. 2014), has occasionally been reported in Bulgaria, Bangladesh, Ethiopia, and Syria but has been characterized as the most serious disease for cultivated lentil in Canada and in the neighboring North Dakota, USA (Buchwaldt et al. 2018). Symptoms include tan lesions on leaves and stems and premature defoliation. The fungus produces microsclerotia on infected tissues which serve as overwintering propagules and source of primary inoculum. Conidia produced in acervulli during the growing season cause successive secondary infections. Management of the disease is currently based on crop rotation, reduced tilling, and fungicide application (Buchwaldt et al. 2018).

Two races of the pathogen (Ct0 and Ct1) were identified after artificial inoculations on lentil lines and scoring disease severity (Buchwaldt et al. 2004). Using quantitative RT-PCR, Bhadauria et al. (2013) suggested that temporal expression polymorphism of defense-related genes can be used to distinguish a lentil genotype as susceptible or resistant.

Tullu et al. (2003) reported *LCt-2*, a major dominant gene to be responsible for resistance in cultivar PI 320937. Buchwaldt et al. (2018) reported screening a collection of over 2300 lentil genotypes for resistance to anthracnose. Several accessions characterized as resistant served for further studies on pathogen race identification and identification of molecular markers linked to resistance. It was found that resistance to race Ct1 is based on two recessive genes, *ctr1* and *ctr2*, and three dominant genes, *CtR3*, *CtR4*, and *CtR5* which are present in different combinations among resistance lentil lines (Buchwaldt et al. 2013, 2018).

Tullu et al. (2003) identified RAPD markers OPE061250 and UBC704700, linked at 6.4 cM (in repulsion) and 10.5 cM (in coupling), respectively, to the resistance locus, *LCt-2*, after applying bulk segregant analysis of 147 $F_5$-derived $F_6$ RILs from the cross of resistant PI320937 and susceptible cultivar Eston. They also identified three AFLP markers but at longer distances form *LCt-2*. Furthermore, it has been shown that it is possible to combine resistance to *C. lentis* race Ct1 together with resistance to *A. lentis*, based on marker-assisted selection. Specifically, RAPD markers UBC2271290 and RB18680 were proved to be appropriate for selecting genotypes resistant to both anthracnose and ascochyta blight (Tar'an et al. 2003; Tullu et al. 2003).

Tullu et al. (2006b) phenotyped a biparental $F_8$ population derived from a cross between a resistant and a susceptible cultivar for ascochyta blight reaction and genotyped it with RAPD and AFLP markers. QTL analysis highlighted the linkage group 6 as a significant location for both ascochyta and anthracnose resistance (Tar'an et al. 2003; Tullu et al. 2003). Bhadauria et al. (2017) discovered five QTLs significantly associated with resistance to Ct0, and six QTLs for Ct1 resistance, in the same 740.94 cM-long map where three QTLs associated with resistance to *S. botryosum* were found.

*Uromyces viciae-fabae* (Pers.) Schroet causes rust on lentil. The disease is considered as one of the most threatening in several growing regions in Africa and Asia. Yield losses can be as high as 100% (Kumar et al. 2016a). Minute, yellowish, slightly raised pustules that enlarge and change to orange-brown in color, often surrounded by a light-colored halo, are the characteristic symptoms of the disease.

Pustules appear on leaves, stems, and pods. The fungus is autoecious overwintering with teliospores and uredospores.

In lentil, rust resistance has been shown to be of monogenic inheritance (Sinha and Yadav 1989; Singh and Singh 1992b). Kumar et al. (2001) studied resistance to rust in 23 different crosses between eight resistant and ten susceptible genotypes. They reported that resistance is governed by a single gene in each case, and there are three different resistance genes in total, two dominant (*Urf1* and *Urf2*) and one recessive (*urf3*). QTLs associated with rust resistance have been reported by Saha et al. (2010). They identified F7XEM4a, a sequence-related amplified polymorphism (SRAP), at a distance of 7.9 cM from the resistance gene. Dikshit et al. (2016) mapped and tagged the single rust resistance gene in lentil genotype L 4149, using SSR and SRAP markers. SSR marker Gllc 527 was estimated to be linked to rust-resistant locus at a distance of 5.9 cM. However, additional markers closer to rust-resistant locus are required for map-based cloning of the rust-resistance gene.

Lentil can also be infected by a range of viruses but overall the effect on yield is not as great as that caused by fungal pathogens (Taylor et al. 2007). At least 30 different viruses have been reported to infect lentil (Kumari et al. 2009). Depending on symptomatology they can be divided into two major groups: (1) viruses causing yellowing and stunting like *bean leaf roll virus* (BLRV), *beet western yellows virus* (BWYV), *faba bean necrotic yellows virus* (FBNYV), *chickpea chlorotic dwarf virus* (CpCDV), and *soybean dwarf virus* (SbDV) or *subterranean clover red leaf virus* (SCRLV). They are considered as the most important worldwide: (2) viruses causing mosaic and mottling like *pea seed-borne mosaic virus* (PSbMV), *cucumber mosaic virus* (CMV), *pea enation mosaic virus-1* (PEMV-1), *pea streak virus* (PeSV), *bean yellow mosaic* (BYMV), *alfalfa mosaic virus* (AMV), and *broad bean stain virus* (BBSV). BYMV, PeSV, and PSbMV are transmitted by aphids in a non-persistent manner whereas BLRV, FBNYV, and PEMV-1 are persistently transmitted. There are no curing means against viral infections; hence management depends on preventive measures. The most effective control of virus diseases is the cultivation of resistant varieties.

Resistant or tolerant varieties to certain viruses have been recognized by several workers through screening of genotypes (Kumari et al. 2009). Some genotypes were found with combined resistance to different viruses. In certain genotypes it was found that resistance is controlled by the single recessive gene *sbv*. No successful marker-assisted selection for resistance to virus diseases has been reported to date.

A large number of insects have been recorded to infest lentil in the field and in storage but only few of them can be considered as serious pests that warrant control (Stevenson et al. 2007). Economically important insects include aphids, *Sitona* weevils, Lygus and stinkbugs, cutworms, thrips, bud weevils, and pod borer. During storage, several species of seed beetles including *Bruchus* spp. and *Callosobruchus* spp. can cause severe damage. The importance of each pest is country and region dependent. Control of most insect pests is achieved by the use of insecticides, whereas breeding for resistance is still in its infancy. However, there are several cases where host resistance has been suggested as a potential management method. For example, several lentil genotypes are reported to be resistant to *A. craccivora*

(Kumari et al. 2009), and they could serve as a gene pool to select for resistance to the black aphid. Another example is that of *Sitona* weevils for which resistance has been documented in wild *Lens* species (Laserna-Ruiz et al. 2012; Bouhssini et al. 2008). *Sitona* adults feed on foliage while larvae feed on root nodules. It has been suggested that the production of toxic metabolites in roots could lead to reduced susceptibility to nodule feeding larvae but no such source of resistance has yet been found in lentil as in other hosts (Stevenson et al. 2007). Another case is the choice between short and long duration genotypes that seems to directly affect susceptibility to pod borer *Etiella Zinckenella* (Stevenson et al. 2007). As chemical control is not economically sustainable, efforts to select for resistance to the most important insect pests of lentil should be intensified.

# 8 End Use and Nutritional Quality

## *Nutritional Value of Lentil*

Along with common beans, lentils are considered the best source of high quality vegetable protein. In addition, lentils are characterized by low concentration of fat, calories, sodium, and the absence of cholesterol. Lentils are a great source of macro- and micronutrients and a rich source of dietary fibers (Srivastava and Vasishtha 2013). They are a rich source of folic acid, molybdenum, and a very good source of copper, iron, manganese, and phosphorus (Karagounis et al. 2011). They have a high content of thiamine (vitamin B1), pyridoxine (vitamin B6), zinc, and a significant amount of magnesium, potassium, riboflavin (vitamin B2), niacin (vitamin B3), and selenium. In addition, lentils are a very good source of isoflavonoids and phytoestrogens (De Almeida Costa et al. 2006). It becomes clear from the above that lentils play a key role in a healthy and balanced diet. Lentils are an essential ingredient in Mediterranean cuisine. In Greece they are traditionally consumed as soup with whole grains (Flint-Hamilton 1999). In other countries however, peeled grains are preferred, as they have better protein digestibility. In some areas lentils are grounded and added to cereal flour to make bread, or porridge intended for baby food (Dalgetty and Baik 2006).

Although the composition of lentil seeds has been widely studied (Dhindsa et al. 1985; Naivikul and D'Appolonia 1979; Sosulski et al. 1976), data is not always comparable due to differences in the analysis methods used and due to different environmental conditions. In many cases, the analysis is based on a single sample. Protein content appears to be very sensitive to environmental factors, such as rainfall, light intensity, the length of the cultivating period, day length, and temperature, but also to agronomic factors, such as plant density, weed management, and soil fertility (Mclean et al. 1974). Detailed chemical composition of lentil varieties has been reported (Petterson et al. 1997). Generally, lentil seed is characterized by high nutrient value. Its protein content ranges between (20% and 28%) and the average iron content is 7.0 mg/100 g of air dried seed (Papakosta-Tasopoulou 2005). Lentils are a good source of B vitamins and contain per 100 g: 0.26 mg of thiamine, 0.21 mg of riboflavin, 1.7 mg of nicotinic acid, 223 mg of choline, 107 mg of folic acid, 130 mg of inositol, 1.6 mg of pantothenic acid, 13.2 mg biotin, and 0.49 mg pyridoxine. Vitamins, in addition to folic and pantothenic acids, increase remarkably during vegetation. The dried husks contain 11.1% protein (1.3% digestible), 0.7% fat, 47.5% carbohydrates, 25.6% fiber, and 3.1% ash (Duke 1981). Lentil protein is mainly found in cotyledons (90%), with albumin and globulin being the most important. The digestibility coefficients for lentils are relatively high and range from 78% to 93%, while their biological value ranges from 32% to 58%. Oleic, palmitic, and linoleic are the predominant fatty acids (Hulse 1990). Pods, dry leaves, stems, and residues are used as animal feed. Lentil residues contain about 10.2% moisture, 1.8% fat, 4.4% protein, 50% carbohydrates, 21.4% fiber, and 12.2% ash (Muehlbauer et al. 1985). According to Muehlbauer et al. (1985), lentil residues achieved equal or better price than seed material in some countries of the Middle East, due to the reduction in production of fodder crops. Legume crops provide valuable soil fertilization due to nitrogen fixation, and their seeds are a source of starch for the textile and printing industries (Kay 1979).

Proteins are important elements of legume seeds. Their nutritional and functional properties have a significant impact on the overall yield and quality of the seed. Duranti and Gius (1997) examined pulse seed proteins, taking into account their molecular and nutritional properties, as well as other compounds of nonprotein origin that affect the use of pulses as food. These substances were tested for their involvement in the seed life cycle and human nutrition. Protein content was used as an indicator of environmental conditions for the study of genetic and environmental variability and the synthesis of inorganic elements, amino acids, and antioxidants. Four varieties of lentils were selected, with three levels of protein each. The crude protein content ranged from 24.3% to 30.2%. Analysis of variance showed that both the varieties and the environment had an effect on the starch content. Significant differences in varieties were found for ADF (acid detergent fiber), NDF (neutral detergent fiber), fat, ash, calcium (Ca), copper (Cu), potassium (K), magnesium (Mn), phosphorus (P), and zinc (Zn). Protein content showed significant differences in the amino acids arginine and tryptophan. Cultivars differed significantly in sucrose, stachyose, vegetable acid, tannins, and trypsin inhibitors (TIA) content. The important components, namely protein and starch content, were negatively correlated. Potassium, magnesium (Mn), phosphorus (P), and zinc (Zn) were negatively correlated with protein content. Tryptophan was the most deficient amino acid and the sulfur-containing amino acids were found limited in lentils. Raffinose was positively correlated with starch while negatively correlated with ADF (Wang and Daun 2006). The average crude protein in lentils produced in Canada between 1998 and 2003 ranged from 25.8% to 27.1%. The variation in protein content in individual samples was confirmed by producers and ranged from 21.4% to 30.0%. While the overall content of protein did not differ significantly by year, the large variability between individual samples within 1 year indicates the great impact that the combination of environmental conditions, agronomic practice, and genetic factors has. Significant differences in protein content were also observed between samples of the same variety. It is therefore proposed that within the same variety the content of the crude protein can be considered as an indicator of the general environmental impact. The general environmental impact includes both the uncontrolled effects of weather such as temperature and rain, as well as planting density, fertilization, and inoculation (Wang and Daun 2003).

## *Nutritional Elements of Lentil*

In order to determine the potential bioavailability of iron (Fe) and zinc (Zn) in lentil to improve human nutrition, total concentrations of Fe and Zn of 19 lentil genotypes were analyzed, which were cultivated in eight sites for 2 years in Saskatchewan, Canada. Total Fe and Zn concentrations ranged from 73 to 90 mg/kg Fe and 44–54 mg/kg Zn (Thavarajah et al. 2009a). The calculated percentages of Recommended Daily Intake (RDA) for Fe and Zn were within the RDA ranges of a 100 g portion of dry lentil. Broad-sense heritability calculations for Fe and Zn concentrations in lentil seeds were 64% to 68%, respectively. The results indicate that lentils have enormous potential as a dietary source of Fe and Zn for people affected by such nutritional deficiencies. This is the first reference to the genetic basis for the micronutrient content of Fe and Zn in lentils. These results provide some understanding of the genetic basis of Fe and Zn concentrations and will allow the development of possible strategies for genetic biodegradation (Thavarajah et al. 2009a).

Beneficial types of selenium and their impact on human health are a global issue of public health interest. Based on this, a study was carried out on the genetic potential for selenium bioassay of legume crops to improve human nutrition (Krohn et al. 2016). In a study that was carried out in Saskatchewan, Canada, the total Se concentration of 19 lentil cultivars was estimated, showing significant genotypic and environmental variation. Total Se concentration ranged from 425 to 673μg/kg, providing 77–122% of the recommended daily intake in 100 g of dry seeds. Over 70% of Se was present as selenomethionine (semet), with a lower percentage (<20%) as inorganic Se and very small amounts as selenocysteine. It was also found that the soils from the sites where the lentils were grown were rich in Se (201–313μg/kg) and that the lentils grown in Saskatchewan were able to provide an excellent natural source of this essential element. The results of the study provide a preliminary understanding of the genetic basis of Se absorption in the seeds and indicate that any possible micronutrient biodegradation strategy would require a selection of locations that minimize spatial variability in Se content (Thavarajah et al. 2008). Moreover, research results have shown greater nitrogen absorption from a high-protein diet, whether it comes from lentils or is of animal origin, which can enhance protein deposition in tissues. A cheaper, lentil-based, high-protein diet may be useful for the nutritional rehabilitation of moderately malnourished children (Hossain et al. 2009).

## *Anti-nourishing Agents of Lentil*

The acceptance of lentil as a basic food (regardless of its nutritional value) is limited because it contains anti-nourishing agents such as trypsin inhibitors, hemagglutinin, whose antitumor activity is almost nullified by boiling under steam pressure (Abu-Shakara and Tannous 1981), tannins, saponins, and oligosaccharides. These substances are responsible for causing indigestion symptoms and limiting nutrient utilization (Salunkhe and Kadam 1989). Williams et al. (1994) reported that lentils have the lowest and fava the highest concentrations of inhibitors.

## *Protein and Starch Inhibitors*

Trypsin inhibitors (TIs) are low molecular weight proteins capable of binding and inactivating a digestive enzyme, trypsin (Salunkhe and Kadam 1989). Their presence is important because they inhibit the proteolytic activity of the digestive enzyme trypsin and can lead to reduced amino acid availability and reduced growth (Liener and Kakade 1980). Legumes contain protein hydrolase inhibitors. They form stoichiometric inert complexes with enzymes and are found in microorganisms, plants, and animals. In addition to the well-studied group of proteinase inhibitors, some proteins that inhibit amylase are also known. The number of known proteinase inhibitors is large. However, only those compounds found in food are of interest to food chemistry. These inhibitors are mainly found in egg whites, vegetable seeds, and plant buds. The selectivity of proteinases varies considerably. Some inhibitors only inhibit trypsin, some act on trypsin and chymotrypsin, while others inhibit microbial or plant proteases, e.g., daptilisin or papain. Proteinase inhibitors are often found, but not exclusively, in plant seeds. Seeds of legumes contain particularly high concentrations. The concentration of the inhibitor varies significantly depending on the variety, degree of maturity, and storage time. There are often several different inhibitors found in plant matter. These inhibitors differ in their isoelectric properties, specificity of proteinases, specific activities, and thermal stability. For example, in more than 30 legumes species analyzed, nine inhibitors have been identified and five partially purified.

Foods containing inhibitors are likely to cause dietary problems. In addition, growth inhibition is observed, which can be treated by incorporation of methionine, threonine, and valine into the diet. These findings indicate that the low growth rate is probably due to a lack of amino acids, which is the result of increased nitrogen excretion. Not all the effects of proteinase inhibitors are yet understood (Belitz et al. 2005). The activity of the inhibitors is usually determined by commercial animal enzymes, e.g., bovine trypsin, or bovine chymotrypsin. To assess the potential effect of inhibitors on human health requires knowledge on inhibition of human enzymes. Evidence indicates that legume inhibitors usually inhibit human trypsin to the same degree or slightly less than bovine trypsin. On the other hand, human chymotrypsin is inhibited to a much greater extent than most legumes. Available data indicate that the average amount of trypsin and chymotrypsin produced by humans daily can be completely inhibited by 100 g of raw soybean extract or 200 g of lentil or other legume (Alonso et al. 2000).

Inactivation of proteinase inhibitors during food processing has been widely studied. In general, the inhibitors are thermally sensitive and more or less inactivated by appropriate heating treatments. In these processes, the starting material and the various parameters (time, temperature, pressure, and humidity of the sample) are of great importance. Steaming soy for 9 min at 100 °C causes 87% damage of the inhibitors.

Vegetable acids have been suggested to be sources of phosphorus, cations, phosphates, and inositol (Reddy et al. 1984). They are proven to reduce the bioavailability of metallic elements (Reddy et al. 1988) and to inhibit the activity of various enzymes (Knuckles et al. 1982). The interaction of vegetable acids with proteins, vitamins, and minerals is considered one of the limitations of the nutritional value of legume seeds (Tabekhia and Luh 1980). Vegetable acid is an antioxidant that is mainly found in seed crops, such as lentils and cereals. It has the potential to bind micronutrients to food and reduce their bioavailability (Thavarajah et al. 2009b). Concentrated tannins have been reported to be present in high concentrations in the lentil seed cover, but can be removed by appropriate treatment (Williams et al. 1994). They have the property of adhering to proteins with hydrogen bonds and with the hydrophobic interactions created, they reduce the nutritional value of lentil seeds (Hahn et al. 1984). When bound to proteins they react with lysine or methionine, making the compounds formed unavailable during digestion (Davis 1981). However, the degree of polymerization of these polyphenolic compounds plays an important role both in the effect of protein digestion and in the availability of vitamins and minerals (Suschetet 1975). Another disadvantage of the tannins present in lentils is the discoloration they cause in the seeds (Nozzolillo and De Bezada 1984).

High levels of saponins reduce the biological availability of micronutrients, but positive effects have also been reported (Thompson 1993). Oligosaccharides and hemagglutinin are substances responsible for indigestion induced after consumption (Fleming 1981).

The heat treatment they usually undergo when cooked at home and in the food industry inactivates the inhibitors more or less completely (Rathod and Annapure 2016), so they are not expected to cause harm to the human body. As a result of the varying thermal stability of the inhibitors, constant and careful monitoring of raw materials and products is required, especially when new products are introduced and new processes are applied (Belitz et al. 2005).

In conclusion, lentils are an important protein and carbohydrate food and are a valuable source of nutrients and microelements. It is also a common and fast-cooking legume in many developing countries. The total production of legumes in India (which is the leader producer country) has remained stable over the past two decades, while the population has grown dramatically over the same (Singh and Singh 2014). As a result, the per capita availability of legumes has decreased significantly. Efforts are currently being made to develop higher-yielding varieties of legumes in order to increase their production and availability per capita, while improving their nutritional value. In addition to yield and productivity, the nutritional and organoleptic characteristics of legumes are very important from consumer's standpoint. In terms of nutritional value, protein, amino acids, calcium, iron, and vitamins are of great importance in legumes due to their complementary effects on cereals (Jarpa-Parra 2018). Comparatively shorter cooking time and lower antioxidants make lentils more suitable for human consumption than other legumes. Breeders follow cross-breeding strategies to upgrade the chemical and nutritional value of the available genetic resources (Solanki et al. 1999).

# 9 Classical Breeding Approaches

Classical breeding is based on genetic diversity to apply suitable methodologies and produce new crop varieties with desirable characteristics. This valuable genetic variability is found either in local populations or is produced through the combination of genes by crossing of selected parents. The appropriate stewardship and breeding strategy are determined by the source genetic material, as well as by practical and economic factors, such as the duration of the program, the space required for the installation of the experiments, the cost of the labor, etc.

The main breeding methods used to manage genetic variability and develop improved lentil genotypes are similar to those utilized to breed other self-pollinated crops with specific modifications and are presented as follows:

1. **Bulk Population Breeding (BPB)**. In BPB all the plants of one generation are harvested in bulk and randomly sampled to form the next generation (Stefaniak and McPhee 2015). This inbreeding procedure is repeated until $F_4$ or $F_5$ generation and then single-plant selection is applied. When high level of homozygosity is reached, evaluation trials are conducted and breeder seed of the promising lines is developed. The main advantage of the method is its ease of application, little record keeping, low cost, and ample seed for multilocation trials. The disadvantage of BPB is the effect of natural selection in favor of more competitive and not always more efficient genotypes. For this reason, bulk population breeding is popular for traits such as seed size or quality when these traits are considered fixed (Muehlbauer and McPhee 2004).
2. **Pedigree Selection (PS)**. In PS, individual lentil plants are selected from $F_2$ and subsequent generations, their progenies are grown in separate lines, and a record of all parent-offspring relationship is maintained (pedigree record) so as each progeny in every generation can be tracked back to the $F_2$ plant from which it is originated (Muehlbauer and Slinkard 1982). The progeny of one $F_3$ plant constitutes a family and in subsequent generations, selection is applied between and within families until the genotypes have reached a high level of homozygosity ($F_5$-$F_6$). After $F_7$-$F_8$ the superior lines are tested in replicated yield trials at several locations. The main advantage of PS in lentil is the rapid elimination of less

desirable genotypes and the option to select in each generation, whereas the disadvantage is that record keeping is slow, tedious, time-consuming, and expensive. In addition, gene recombination is possible to be severely restricted when selection in early generations is limited to quality traits (i.e., simply inherited, strongly heritable traits) (Muehlbauer et al. 1988).

3. **Single Seed Descent (SDS)**. SSD is performed in lentil as a method of advancing lentil populations to higher homozygosity and consists of harvesting and sowing only one pod from each single plant to form the next generation (Haddad and Muehlbauer 1981). When homozygosity is achieved, multilocation trials are conducted. A disadvantage of the method is the possibility that some seeds will not germinate and therefore the genetic background of some crosses will be lost. The advantage of the method is the avoidance of keeping a large number of lines, its cost-effectiveness over the pedigree method, less space for the experimentation, and that it could be applied under controlled conditions (glasshouse, growth chamber) since no selection is made until homozygosity has been achieved. Haddad and Muehlbauer (1981) evaluated the efficiency of BPB in comparison to the SSD method in lentil and observed that the lines selected by BPB were taller and formed pods higher. In contrast, shorter and erect cultivars were not selected. The above researchers attributed the observations to the effect of natural selection favoring the selection of more competitive genotypes. Losses from the original lentil population ranged from 75% in BPB to 30% by the SSD method after four generations of selection. Based on the aforementioned and given that Martin et al. (1978) found that the efficiency of the SSD method is affected when the initial population losses exceeded 50%, Haddad and Muehlbauer (1981) concluded that the SSD method was more effective than the BPB method to improve lentil. Recently, the SSD method has been performed in lentil to enhance breeding process through the application of extended light period under controlled conditions (Mobini et al. 2016). Idrissi et al. (2019) and Idrissi (2020) studied a breeding protocol based on the SSD method under extended photoperiod and concluded that after optimization of light characteristics it could be proposed as a simple, rapid, and efficient breeding method for accelerated genetic gain in lentil.
4. **Bulk Pedigree Selection (BPS)**. Bulk pedigree selection (BPS) is a modification of the Bulk Population Breeding (BPB) method applied in lentil breeding (Stefaniak and McPhee 2015). Bulk Pedigree Selection (BPS) differs (1) in the generation that selection is performed (single plants are selected within $F_{2:3}$ families) and (2) the best plants within selected families are advanced in bulk to a desired level of homozygosity (usually $F_6$ or $F_7$). Advantages of BPS are that selection of a plant in early generations can ensure that a desired allele will be present in that family and will be advanced through the program. Disadvantage to this method is that the breeder must be able to identify the phenotype for selection in the $F_2$-derived lines using relatively few families (~50).
5. **Pure Line Selection (PLS)**. PLS comprises the following four steps: (1) selection of single plants that fit better to the target phenotype, (2) inbreeding until reaching the suitable homozygosity level, (3) seed increasement, and (4)

multilocation replicated trials across 2 or 3 years. The results of the evaluation trials could be used to select desirable parents to be employed for breeding populations or even to predict varieties' performance for commercial use (Stefaniak and McPhee 2015). The aforementioned researchers pointed out that in most lentil PLS programs, primary attention was paid to yield and plant architecture for mechanized harvesting, while secondary selection was focused on tolerance to biotic factors, maturity, and seed traits.

PLS is the most popular method to exploit genetic variation of landraces. Lentil landraces are naturally evolving populations of a predominantly self-pollinated crop and therefore are considered as a mixture of fairly homozygous plants (Tokatlidis et al. 2010; Kargiotidou et al. 2014). Landraces are surprisingly heterogeneous and include individuals that meet many breeding goals without the need of hybridization. Vlachostergios et al. (2018a) applied PLS in three lentil landraces under ultra-low plant density and found that the second-generation sister lines had mean grain yields by 8%, 10%, and 20% higher compared to their respective ancestors under the farming density across five environments. The aforementioned researchers suggested this methodology as an efficient tool that allows the breeder to exploit the natural genetic variability within lentil landraces and develop in short time pure-line varieties adaptable to a wide range of environments.

The ICARDA lentil breeding program is a modification of the BPB method (Erskine et al. 2016). The theoretical basis of the program was that heterogeneous populations tend to be more stable than homogeneous populations through the preservation of variability within selected varieties (Erskine 1997). The lentil breeding program is based on a bulk population method with single-plant selection at $F_4$ to develop $F_5$ progeny. More analytically, the next phases comprise the following steps: selection among $F_5$ progenies for uniformity, agronomic traits, and Fusarium wilt reaction, evaluation in preliminary screening nursery of $F_6$ progeny for yield, agronomic traits, and F. wilt reaction, preliminary yield trials in $F_7$ at three contrasting locations, advanced yield trials in $F_8$ at three locations under rainfed and irrigated conditions, seed increase of selected lines in $F_9$ and their incorporation into international nurseries in $F_{10}$ generation (Erskine et al. 2016). Inbred lines and/or segregating populations are then distributed internationally. The program follows a decentralized breeding strategy to develop cultivars for short season environments of southerly latitude countries and winter-hardy cultivars for the highlands. Also, the lentil breeding program has adopted Participatory Varietal Selection (PVS) to enhance the adoption of improved varieties (Witcombe et al. 1996; Stefaniak and McPhee 2015). This economical way of improvement enabled the creation of improved populations with adaptability to a wide range of environments and helped the rural development of many developing areas.

In Canada individual plants are selected from the $F_1$ generation after multiple crossings or in the $F_2$ generation after simple crosses and continue with pedigree modifications. Since most of the imported lentil germplasm was unadapted to the long-day environment of Canada, application of multi-parent crosses has become a standard practice to dilute the genetic contribution of unadapted parents (Muehlbauer et al. 2009; Stefaniak and McPhee 2015). In addition, the BPS method was adopted by the Canadian breeding program and resulted in the development of improved varieties for tolerance to ascochyta blight, anthracnose, and seed characters (Vandenberg et al. 2002a, b, 2005).

Lentil breeding in Australia is a part of the Pulse Breeding Australia (PBA) initiated in the 1990s (Erskine and Sarker 2004). Bulk population method is utilized along with single seed selection at $F_2$ or $F_3$ generation, while mass selection is preferred for morphological seed traits by hand picking, using sieves or gravity tables and color sorters (Materne and McNeil 2007). Screening for ascochyta blight is done in early generation via natural selection for resistant plants in the field, whereas infected seeds with characteristic ascochyta blight blemishes are eliminated using a color sorter (Materne and McNeil 2007). Nowadays, numerous cultivars have been released with significant traits like disease resistance (clvs. *Nipper*, *Boomer*), optimum flowering response (clvs. *Nugget*, *Digger*), adaptation to low rainfall (clvs. *PBA Flash*, *PBA Bounty*, *PBA Blitz*), imidazolinone tolerance (clv. *PBA Hurricane XT*), seed size (clv. *PBA Jumbo*), improved grain quality, and others, while the majority are red lentils that are exported throughout the world (Materne et al. 2002; Inder et al. 2008).

Little is known about the Greek lentil breeding program despite the fact that Greece was among the first countries in Europe to systematically apply lentil breeding. The establishment of the Research Station of Forage and Pulse Crops in 1933 has contributed to the development of the first improved varieties (clv. *Pelasgia* in 1939) and the evaluation of local populations for valuable agronomic and nutrient traits. Similarly, to other countries the first released lentil varieties were selections from existing local landraces (Solanki 2001; Kumar et al. 2005); however, very soon crosses between imported germplasm and local lentils have resulted in the development of a wide spectrum of improved lines with significant traits evaluated in a multilocational network. Due to the strategy of systematic cross-breeding between parents of diverse origin with desired traits, the phenomenon of genetic bottleneck (Erskine et al. 1998) was not observed in Greek germplasm. Thus gradually, older varieties with inferior agronomic or quality traits were substituted by the modern ones. The methodology applied was either PLS in the case of local populations or BPS. After 2000, lentil breeding is conducted according to the principles of honeycomb selection under nil competition (Iliadis et al. 2003; Vlachostergios et al. 2011, 2018a; Vlachostergios and Roupakias 2017; Ninou et al. 2019). Nowadays, national lentil breeding is implemented as a self-funded program by the Institute of Industrial and Forage crops. The main varieties cultivated in Greece are also registered in the European Variety Catalogue and are crossbred cultivars with the exception of the new ultra-early flowering cultivar ELPIDA that was produced via PLS under nil competition within germplasm originated from ICARDA (Vlachostergios et al. 2018b).

## Key Issues in Classical Lentil Breeding

### The Optimum Plant Density in Single-Plant Selection

One of the major issues discussed by the lentil breeders was the optimum plant density, expressed by interplant distance, for the selection of individual lentil plants that meet the improvement goals. Initial questioning on suitable plant density in lentil selection was posed by Erskine et al. (1990). The abovementioned researchers investigated the effect of interplant distance and concluded that the plant density of the environment for plant selection can be low enough to avoid interplant entanglement by tendrils, allowing a focus in plant selection on characters with a higher heritability than yield. Formerly, Fasoulas (1973) advocated that selection of individual plants should be applied at the critical interplant distance, where the range of phenotypic expression is enlarged and the negative effect of the competition on selection efficiency is eliminated (Fig. 3). Later, a negative relation between yield capacity and competitive ability has been observed (Kyriakou and Fasoulas 1985; Fasoula and Fasoula 1997; Benaragama et al. 2014); therefore, high yielders could fully express their yielding capacity at the nil competition regime (Chatzoglou and Tokatlidis 2012; Ninou et al. 2014; Fischer et al. 2019; Fischer 2020). Vlachostergios et al. (2011) tested this hypothesis in lentil by investigating the effect of two plant

**Fig. 3** Lentil plants grown under nil competition in a honeycomb design maximize phenotypic differentiation of heterogeneous population

**Fig. 4** Different rate of infected lentil plants with fusarium wilt screened under three interplant distances (Vlachostergios and Roupakias 2017)

densities (12.8 plants $m^{-2}$/30 cm interplant distance and 3.2 plants $m^{-2}$/60 cm interplant distance) on the efficiency of single-plant selection and concluded that single-plant selection at the lower plant density could be applied in lentil breeding programs with considerable effectiveness.

The interpretation of this finding is based on the observation that heterogeneous genetic materials exhibit plant-to-plant variation and comprise plants with high or low competitive ability considered as "strong" or "weak competitors," respectively (Tokatlidis and Vlachostergios 2016). In order to mitigate the intra-species competition and the interplant differences, varieties should comprise genotypes of the "weak competitor" (Sedgley 1991; Pan et al. 2003). Therefore, when selection is applied at high plant density (dense stand), the presence of "strong competitor" ideotypes hinders the recognition of genotypes representing the opposite features, i.e., a "weak competitor" ideotype (Tokatlidis 2017). This is the main reason why the condition of selection under nil competition permits identification of the "weak competitors" with high single-plant grain yield.

Furthermore, it is worth noticing that lentil selection under low plant density resulted in the production of virus-free plants and facilitated the identification of single plants with tolerance to other biotic stresses. Hence, Kargiotidou et al. (2014), Kargiotidou et al. (2015), and Kargiotidou et al. (2016) found that selection at ultra-low density of the highest-yielding plants constituted an effective way to improve the health status of the seeds produced and resulted in high-yielding and potentially virus-tolerant pure-line varieties, while Vlachostergios and Roupakias (2017) examined the effectiveness of screening lentil single plants in a wilt sick plot under three plant densities (interplant distances: 15, 30, and 60 cm) and found that screening under low plant density reinforces the identification of lentil plants with various levels of resistance to fusarium wilt (Fig. 4).

## Conventional Methods for Multiple Trait Breeding

The modern lentil breeding programs aim at making progress by pyramiding traits that increase yield potential and reliability, tolerance to biotic/abiotic stresses, harvestability, improved grain quality, and general adaptation to more marginal pedo-climatic environments for lentil. The conventional breeding by using suitable methodologies to integrate multiple traits into a single genotype is considered feasible, although challenging to implement.

Multiple trait breeding depends on the germplasm under consideration and more specifically on the genetic background and genetic relationships among the target traits. Target traits are inherited independently or associated in a favorable or unfavorable manner. Two types of trait association are recognized: (1) genetic association (additive, dominance effects) and (2) nongenetic (environmental) association. A successful multiple trait breeding program prerequires a correlation among the traits near 1.0 and heritability of the second trait near 0.9 (Rana et al. 2019). A vast knowledge of the association degree and heritability between the target traits is considered of crucial importance to breeding program effectiveness. Nowadays, ample data exist on the degree of correlation among significant agronomic and quality traits (Rana et al. 2019) which could be proven a strong tool to breeders for systematically organizing target traits pyramiding. Traits with favorable correlation and additive genetic effects are in the foreground as the selection for one trait will cause an indirect change in the mean of the second trait (Bernardo 2001).

Three basic strategies including tandem selection, independent culling, and selection index can be utilized to simultaneously breed or select for multiple traits:

**Tandem selection** is the method in which the target traits are improved in succession (serial improvement). Thus, selection is performed for the first trait, until a desired genetic level is reached. Afterwards, a next breeding cycle begins for the second, third, and following traits. Tandem selection is usually performed when correlation does not exist between the target traits and is generally considered as the least efficient method.

**Independent culling or** truncation selection involves simultaneous but independent culling for each trait. Independent culling in one generation follows a specified order, whereas the level of culling for each trait represents the selection level. A key point is that breeders must apply less strict culling for the first trait to ensure sufficient variability for the traits in subsequent culling.

**Index selection** is a method that allows multiple features of interest to be selected. The breeder assigns a weighting scheme to each character based on their economic importance and the potential for improvement, creating a univariate character (index) that is then selected (Céron-Rojas and Crossa 2018). It is considered as the most efficient method for crop improvement, given the concept of aggregate genetic value (Bernardo 2001; Hallauer et al. 2010).

## 10 Induced Genetic Diversity

### *Induction of Genetic Variability*

Plant breeding is a function of heritability, exploitation or production of genetic diversity, and application of effective selection schemes, techniques, and methodologies (Fig. 5).

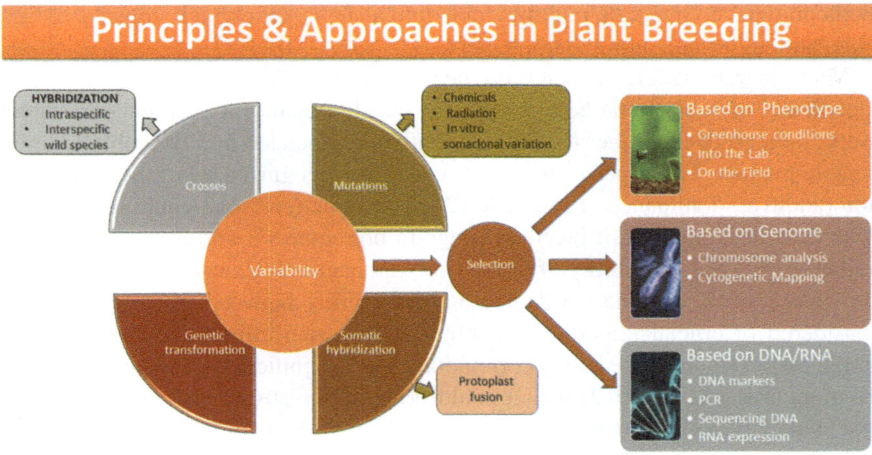


**Fig. 5** Graphical representation of principles and approaches in plant breeding

The success of a breeding program in a plant species depends upon the present nature and magnitude of genetic variability in local populations, landraces, traditional cultivars, and wild relative species that give the possibility of producing new genetic variation for desirable traits through hybridization. The higher levels of physical genetic variability could help in choosing the diverse parents for meaningful hybridization giving consequently greater chances for new combinations and attaining better success for valuable genetic variation (Gautam et al. 2013). Specifically, hybridization provides the opportunity to increase genetic variability, allowing the introgression of important alleles for yield components, quality traits as well as resistance or tolerance to biotic and abiotic stresses (Ocampo et al. 2000).

Preserving and exploiting genetic diversity is of paramount concern to lentil breeders but unfortunately, cultivated lentil germplasm collections contain relatively low amounts of genetic variation (Muench et al. 1991; Alvarez et al. 1997; Ford et al. 1997; Ferguson 2000; Durán 2002). Several studies have been conducted to assess the variability in germplasm (genetic stocks) regarding yield and its components, biotic and abiotic stresses, and nutritional parameters (Rajput and Sarwar 1989; El Attar 1991; Pandey et al. 1992; Esmail et al. 1994; Kumar et al. 1995; Chauhan and Singh 1998; Kumar et al. 1999; Chakraborty and Haque 2000; Solanki and Sharma 2001a, b; Rathi et al. 2002; Hamdi et al. 2003; Mishra et al. 2005; Poonam 2006, etc.).

## *Induction of New Genetic Variation Through Wide Hybridization*

Natural agro-biodiversity stored in genebanks can be used to expand the diversity in crops. These collections are a vital source for discovering useful genes/alleles, which serve as a cornerstone for any prebreeding program. Substantial improvement in lentil productivity could not be achieved due to the loss of invaluable alleles for higher productivity, low extent of genetic variation in lentil cultivars, and narrow genetic base of the local varieties (Ferguson 2000; Durán 2002). Therefore, the key to increase lentil yield is through widening the available genetic base (Erskine and Saxena 1993). Wide hybridization is expected to widen the genetic base of the cultivated gene pool (Singh et al. 2014b, 2018b) and disintegrate the yield plateau (Gorim and Vandenberg 2018). The wild relatives and exotic lines generate new variability through recombination breeding approaches. The ultimate goal of wide hybridization is to transfer useful genes from alien species into cultivated species. For this reason, extensive prebreeding efforts have been made worldwide using wild species, which show large genetic diversity for morphological and seed traits and carry useful alien genes for improving yield, quality, and stress resistance. Gupta and Sharma (2006) reported genetic variability for yield attributes and biotic and abiotic stresses among 70 accessions of four wild species/subspecies (*L. culinaris* ssp. *orientalis*, *L. odemensis*, *L. ervoides*, and *L. nigricans*). As a result, donors for resistance to powdery mildew (*L. culinaris* ssp. *orientalis*, *L. nigricans*), rust, and wilt resistance have been found in all the above four species and also in *L. nigricans* for drought tolerance (Gupta and Sharma 2006; Gautam et al. 2013). Sources of resistance to lentil rust, the major foliar disease, vascular wilt, the most important soil-borne disease (induced by *Fusarium oxysporum*), and ascochyta blight (induced by *Ascochyta Fabae*) have been identified in cultivated lentil (Bayaa et al. 1997; Erskine et al. 1994). Resistance to vascular wilt and ascochyta blight has also been found in *L. culinaris* subsp. *orientalis* (Bayaa et al. 1997).

Furthermore, resistance to anthracnose is found in *Lens ervoides* germplasm and is exploited in Canada by introgressing resistance genes into cultivated backgrounds (Fiala 2006; Tullu et al. 2006a, b; Fiala et al. 2009). This successful use of *Lens ervoides* holds promise as a source of genes for resistance to other diseases and possibly for plant habit, biomass production, and other important agronomic and marketing traits. Crossing of cv "Eston" (*L. culinaris*) with PI72815 and L01-827 (*L. ervoides*) was successful with the aid of embryo rescue. Production of hybrid seeds followed by $F_2$–$F_7$ seeds led to the development of two recombinant inbred populations (RILs) with varying degrees of sterility. However, among these RILs, several lines were observed as transgressive segregants for various agronomic traits including an 8% increase in seed size. These lines could be useful as pre-breeding material in lentil breeding programs (Tullu et al. 2011a, b).

In India, a network project in collaboration between the Indian Council of Agricultural Research (ICAR) and ICARDA funded by the Department of Agriculture and Cooperation (DAC), Government of India, has been launched in 2010 for introgression of alien genes from unadapted germplasm and wild species in order to break the yield barriers in lentil. This project led to the introgression of alien genes from wild species into the cultivated species using knowledge of cross-compatibility between wild and cultivated species.

Recently, Singh et al. (2013) made successful crosses between cultivated (*Lens culinaris* ssp. *orientalis*) and wild lentils (*L. culinaris* ssp. *orientalis*, *odemensis*, *lamottei* and *ervoides*) while effects of species groups, day length, and temperature on crossability were clearly visible and a high level of heterosis was observed in $F_1$

crosses for the important traits studied. With regard to inter-subspecific hybrids of lentil, it has been reported that the domesticated lentil *Lens culinaris* ssp. *culinaris* is readily crossable with *Lens culinaris* ssp. *orientalis* (Muehlbauer et al. 1989; Vandenberg and Slinkard 1989; Vaillancourt and Slinkard 1992), although the fertility of the hybrids depends on the chromosome arrangement of the wild parent (Ladizinsky et al. 1984).

## Mutation Breeding

During the last decades, breeding efforts were made to increase productivity of lentils using conventional breeding methods to develop varieties with high and stable yield, resistance to main lentil diseases, high protein content, and good culinary quality. However, the current productivity of lentils is still well below its potential (Kumar et al. 2016a). Narrow genetic base and limited genetic diversity are the major research constraints that affect the efficacy of classic breeding methods in lentil (Tabti et al. 2018). Also creating genetic variation with classical breeding methods, for example through hybridization, is a difficult practice in lentil. The small-sized fragile flowers make the emasculation process difficult (Rana and Solanki 2015). Moreover, the self-fertile nature of the crop imposes further limitations to hybridization programs (Tabti et al. 2018).

Alternatively, mutation breeding seems to be a feasible and sustainable technique to broaden narrow genetic base and create a new technical gene pool for a number of desirable traits with economic importance in lentil (Amin et al. 2015). Mutations are highly useful to create variability, when a desired trait may not be available in the existing germplasm and when suitable screening methods are available that can be adopted to evaluate large mutagenized populations (Kumar et al. 2016a, b). Induced mutagenesis creates a new allelic combination in the desirable traits within a short period of time without disturbing the basic genomic structure of the plant; thereby a sustainable acceleration in creating variation towards desirable direction is possible with very limited risk and resource (Amin et al. 2015).

Mutation breeding programs have become an interesting tool for the development of different plant types with specific seed size and grain quality, as well as nonshattering, high yielding, disease resistant, and early maturing plants (Shahwar et al. 2017). Research applied with lentil mutagenesis has shown that macrosperma cultivars are more prone to mutagenesis than the microsperma genotypes (Kumar et al. 2016a).

Usually, mutants are usually induced by different types of chemical mutagens and ionizing radiations. In the FAO/IAEA Mutant Variety Database (2020) out of total 458 released mutant varieties of legumes, only 18 mutant lentil varieties have been approved since 1981 till 2017, from Bangladesh, India, Pakistan, Moldova, and Bulgaria. Most of them have been developed through radiation (gamma rays) and only three through chemical mutagen (EMS and extract of Datura seeds) (FAO/IAEA MVD 2020). Gamma rays are the most used mutagen to change gene(s) in lentil, due to their easy application (Toker et al. 2007). Most of the improved characteristics/traits concern high yield, early maturity, resistance to major diseases, high protein content as well as morphological characteristics which lead to higher productivity (Table 1).

The FAO/IAEA Mutant Variety Database or MVD collects information on plant mutant varieties (cultivars) released officially or commercially worldwide. Data on the mutagen and dose used, the characters improved, and agronomic data if available are among the information provided (FAO/IAEA MVD 2020).

The purpose of the database is to demonstrate the significance of mutation breeding as an efficient tool for preserving and enhancing global food security, to serve as a platform for breeders to showcase their varieties to a global audience, and to stimulate germplasm transfer for cultivation, breeding, or genomics studies.

In the context of MVD a mutant variety is a new plant variety that is bred through:

1. Direct use of a mutant line that is developed through physical and chemical mutagenesis, or somaclonal variation and activation of endogenous transposable elements
2. Indirect use of a mutant line/lines, which is/are used as a parental variety/varieties in cross-breeding (cross between mutant lines or with a commercial variety/varieties)
3. Use of wild species' genes translocated into plant genomes through irradiation-facilitated translocations, e.g., genes of wheat wild relative species (FAO/IAEA MVD 2020).

## *Chemical Mutagenesis*

Chemical mutagenesis provides an inexpensive and straightforward way to generate a high density of novel diversity in the genomes of plants (Jankowicz-Cieslak and Till 2016) on a small and/or large scale (Amin et al. 2015).

Chemical mutagenesis of lentil still needs to be exploited and standardized to develop reproducible protocols for mutation breeding (Amin et al. 2015). Chemical mutagens therefore in many studies result in more frequent mutations (Yeong and Kim 2019) and possess significant ability to improve qualitative and quantitative traits in plants (Shahwar et al. 2017).

The most commonly used chemical mutagen in plants is ethyl methanesulfonate (EMS) which has shown to induce primarily single base point mutations (Jankowicz-Cieslak and Till 2016). Afterwards EMS mutagenized plant populations can be screened phenotypically (forward genetics), or mutations in genes can be identified in advance of phenotypic characterization (reverse genetics) (Jankowicz-Cieslak and Till 2016). Other chemical mutagens being used in lentil are methyl methanesulfonate (MMS), dimethyl sulfoxide (DMSO), hydrazine hydrate (HZ), sodium azide (SA), caffeine (caffeine), and others.

**Table 1** Some records of mutant lentil (*Lens culinaris* Medik.) varieties in FAO/IAEA Mutant Variety Database (FAO/IAEA MVD 2020)

| Variety name | Country | Registration year (local/ natl.) | Short description |
|---|---|---|---|
| S-256 (Ranjan) | India | 1981 | The mutant variety S-256 was developed by **γ-irradiation.** Main improved attributes of mutant variety are high yield and spreading type. |
| Rajendra Masoor 1 | India | 1996 | The mutant variety Rajendra Masoor 1 was developed by **irradiation with gamma rays (100 Gy)**. Main improved attributes of mutant variety are tolerance to low temperatures and earliness. |
| NIAB MASOOR-2006 | Pakistan | 2006 | The mutant variety NIAB MASOOR-2006 was developed by **irradiation with gamma rays (200 Gy)**. Main improved attributes of mutant variety are higher number of pods, resistance to lodging, and resistance to blight and rust. This mutant variety covers 90% of the lentil growing area in Pakistan. |
| Binamasur-1 | Bangladesh | 2001 | The mutant variety Binamasur-1 was developed by using **a chemical extract from Datura seeds** which was applied as mutagen on wet seeds of a local genotype L-5. Main improved traits of mutant variety Binamasur-1 are high yield, tolerant to rust and blight, black seed coat. |
| Binamasur-2 | Bangladesh | 2005 | The mutant variety Binamasur-2 was developed by **Gamma radiation** of dry seeds into a local genotype Utfala using **200 Gy**. Main improved traits are high yield, early maturity, and tolerant to rust and blight. |
| Binamasur-3 | Bangladesh | 2005 | The mutant variety Binamasur-3 was originated after **soak pretreatment** onto seeds of a local genotype L-5 using a **chemical mutagen (0.5% EMS)**. Main improved traits are high yield, early maturity, rust, and blight tolerance. |
| Mutant 17 MM | Bulgaria | 1999/1999 | The mutant variety Mutant 17 MM was officially approved in 1999. Main improved attributes of mutant variety are vigorous growth habit, large leaflet, pods and seeds, resistance to anthracnose, Stemphylium and viruses, high yield, drought tolerance, and improved cooking quality. |
| Zornitsa | Bulgaria | 2000/2000 | The mutant variety Zornitsa was developed by treatment with **chemical mutagen 0.1% EMS**. Main improved traits are high yield, high protein content (28.7%), good culinary and organoleptic quality, resistance to anthracnose, viruses, and ascochyta blight. |

(continued)

**Table 1** (continued)

| Variety name | Country | Registration year (local/ natl.) | Short description |
|---|---|---|---|
| Djudje | Bulgaria | 2000/2000 | The mutant variety Djudje was developed by irradiation of seeds with **gamma rays (30 Gy)**. Main improved traits are high yield, dwarf bushy habit, suitable for mechanized harvesting, nonshattering, resistance to fusarium and botrytis, high protein content (27.9%), good culinary and organoleptic quality. |
| Elitsa | Bulgaria | 2001/2001 | The mutant variety Elitsa was developed by irradiation of seeds of Tadjikskaya 95 **with gamma rays (40 Gy)**. Main improved traits are high yield (34.4%) and resistance to major diseases. |
| Verzuie | Moldova, Republic | 2004/2007 | The mutant variety Verzuie was developed by irradiation with **gamma rays (250 Gy)**. Main improved attributes of mutant variety are drought resistance, vegetative period (98 days), proteins—26.7%, oils—1.5%, fructose—0.17%, glucose—0.08%, saharose—1.23%, starch—45.30%, cellulose—7.16%. |
| Aurie | Moldova, Republic | 2005/2008 | The mutant variety Aurie was developed by irradiation with **gamma rays (250 Gy)**. Main improved attributes of mutant variety are drought resistance, high yields, early maturity, and high protein content. |
| NIAB MASOOR 2002 | Pakistan | 2002 | The mutant variety NIAB MASOOR 2002 was developed by **irradiation**. Main improved attributes of mutant variety are erect growth habit, early maturity (120 days), black seed coat color, high grain yield, diseases resistance, and synchronous pod maturity. |
| Binamasur-9 | Bangladesh | 2014 | The high-yielding mutant variety was developed through **gamma irradiation** onto a popular lentil cultivar, BARI masur-4. Out of 12 M4 plants, the mutant line LM-132-7 derived from **200 Gy** of gamma rays was selected for earliness and high yield potential. |
| Binamasur-8 | Bangladesh | 2006/2014 | The high-yielding mutant variety was developed through gamma irradiation of the popular lentil cultivar, BARI masur-4. Out of 10 M4 plants, the mutant line LM-75-4 derived from **200 Gy of gamma rays** was selected for earliness and high yield potential. |

(continued)

**Table 1** (continued)

| Variety name | Country | Registration year (local/natl.) | Short description |
|---|---|---|---|
| Binamasur-6 | Bangladesh | 2002/2011 | The high-yielding mutant variety was developed through **gamma irradiation** of the popular lentil cultivar, BARI masur-4. Out of 15 M4 plants, the mutant line LM-115-7 derived from **250 Gy** of gamma rays was selected for earliness and high yield potential. |
| Binamasur-5 | Bangladesh | 2002/2011 | The high-yielding mutant variety was developed through **gamma irradiation** of the popular lentil cultivar, BARI masur-4. Out of 15 M4 plants, the mutant line LM-79-5 derived from **200 Gy** of gamma rays was selected for earliness and high yield potential. |
| Binamasur-11 | Bangladesh | 2006/2017 | The high-yielding mutant variety was developed through **gamma irradiation** of the popular lentil cultivar, BARI masur-4. Out of 15 M4 plants, the mutant line LM-156-1 derived from **200 Gy** of gamma rays was selected for earliness and high yield potential. Mutant LM-156-1 produced the highest seed yield ranged 2200–2400 while the check variety, BARI masur-5, produced 2100 kg/ha. |

## *Radiation*

Ionizing radiation is a widely used physical agent to treat mainly the seeds or other somatic plant tissues to create heritable mutations and develop new cultivars or genetic material for functional genomic studies (Yeong and Kim 2019). Stadler in 1928 (Stadler 1928a, b) first published the results of mutations induced by irradiation in crops, like barley and maize. Radiation induces a high proportion of mutations, which substantially affect gene functions (e.g., the deletion of a complete gene), thus resulting in more loss-of-function mutations related to target traits with fewer mutations per genome (Yeong and Kim 2019). X-rays, gamma rays, neutrons, and ion beams have been applied for systematic characterizations and utilization into plant species in lentil (Shu et al. 2012; Tanaka et al. 2010). Among them, gamma rays are most commonly used to create variation in lentil (Singh et al. 2011). Gamma rays cause chromosomal rearrangements, deletions, and hence produce new sources of genetic variation in plants (Bhat et al. 2007). The application of gamma rays is also found easy because it demands only two steps: preparation of seeds and mutagen treatment (Toker et al. 2007). Gamma rays have been successfully used to develop lentil cultivars for resistance to main lentil diseases, high protein content, and culinary quality traits (Tomlekova 2010).

## *Induction of Genetic Variability Through In Vitro Culture Techniques*

For making genetic improvement by generating new variability, useful genes identified in the primary gene pool have readily been used for the production of new varieties. However, the frequency of occurrence of these useful genes has been observed much more in the species belonging to secondary and tertiary gene pools (Collard et al. 2001; Tullu et al. 2006a, b). Sometimes, strong crossability barriers limit utilization of wild germplasm for lentil improvement. Problems like reproductive isolation, embryo or endosperm abortion, hybrid sterility, and limited levels of genetic recombination are significant obstacles for introgression of alien genes from wild germplasm at a larger extent in lentil breeding programs. All these barriers can occur before or after the fertilization of male with female gametes because the pollen tube does not reach the ovary, or the male gametes do not fuse with female gametes (Chowdhury and Chowdhury 1983). Therefore, more efforts are required on deployment of novel techniques or biotechnological methods for using these gene pools to the improvement of lentils.

Tissue culture techniques may be utilized conveniently to overcome incompatibility barriers through in vitro ovule culture and embryo rescue or by fusion of protoplasts from vegetative cells for interspecific, intergeneric, and interfamiliar group (Rao and Chadha 1986). As an example, Fratini and Ruiz (2006) suggested that hybrids between *L. culinaris* ssp. and *L. nigricans*, *L. ervoides*, and *L. odemensis* developed through embryo rescue and could be viable with a rate of 3–9%. The first interspecific lentil hybrid using embryo tissue culture was obtained by crossing *L. culinaris* with *L. ervoides* (Ladizinsky et al. 1985). In Canada a breeding program was initiated to introgress the anthracnose genes from *L. ervoides* into the high-yielding cultivars of *L. culinaris* (Fiala et al. 2009; Tullu et al. 2013) and also has been noted the success in the cross between *L. culinaris* and *L. tomentosus* (Tullu et al. 2011a, b).

For the induction of somaclonal variation, optimistic results were given from embryo-derived callus cultures (Saxena and King 1987; Polanco et al. 1988), nodal explants (Polanco et al. 1988; Singh and Raghuvanshi 1989), intact seedlings (Malik and Saxena 1992), immature seeds (Polanco and Ruiz 2001), cotyledonary nodes (Gulati et al. 2001; Talib et al. 2003), and decapitated embryos (Sarker et al. 2003). There are only a few reports of rooting from in vitro regenerated shoots in lentil, suggested that these explants are not ideal for *Agrobacterium*-based genetic transformation, since it is difficult for the *Agrobacterium* to access the meristematic parts during the co-cultivation process (Sarker et al. 2003). Furthermore, protoplasts were isolated from lentil (*Lens culinaris*) roots and their suitability as a transient expression system was investigated (MacCarrone et al. 1995). After transfecting the protoplasts with the β-glucuronidase (GUS) gene by either electroporation or polyethylene glycol (PEG), the specific activity of the reporter enzyme and the cell viability were determined. Electroporation was more effective than PEG treatment as transfection procedure and its efficiency was affected by the plasmid length (MacCarrone et al. 2000).

Genetic engineering has high potential to improve tolerance of lentil against biotic and abiotic stresses. Genetic engineering may also have a role in eliminating antinutritional factors and improving the nutritional quality of lentil. Lentil can also be employed as a host to produce drugs using gene transformation. However, for genetic transformation a successful in vitro plant regeneration *protocol* is necessary when *Agrobacterium tumefaciens* method will be applied. In this case the percentages of successive events depend on the determination of critical regeneration parameters such as explant source, genotype, and the composition of medium used (Ahmad et al. 1997). Lentil shoots have been regenerated from shoot apices (Williams and McHughen 1986; Singh and Raghuvanshi 1989), nodal explants (Polanco et al. 1988; Fratini and Ruiz 2003), intact seedlings (Malik and Saxena 1992), cotyledonary nodes (Gulati et al. 2001; Sarker et al. 2003; Chhabra et al. 2008), epicotyls (Omran et al. 2008), decapitated embryos (Sarker et al. 2003), and immature seeds (Polanco and Ruiz 2001). Among all these explants, the higher frequency of shoot regeneration was observed from cotyledonary nodes (Warkentin and McHughen 1993; Gulati et al. 2001). Transient expression of the reporter gene GUS was achieved by electroporation, particle bombardment, and by *Agrobacterium* transfer method. The first successful work was reported by Barton et al. (1997), using pCGP1258 plasmid construct on four lentil genotypes. Öktem et al. (1999) reported the transient and stable chimeric transgene expression on cotyledonary lentil nodes using particle bombardment. Gulati et al. (2002) reported regeneration of the first fertile transgenic lentil plants on MS medium with 4.4μM benzyladenine (BA), 5.2μM gibberellic acid (GA3), and chlorsulfuron (5 nM) followed by micrografting and transplantation into the soil.

## 11 Genomic Resources

Lentil, among edible plants, carries one relatively large genome with size of over 4 Gbp (in comparison human genome is ~3 Gb while among major edible plants only cereals have larger genomes at the range of 5–6 Gb). This makes the effort of genome sequencing and assembly a daunting task. Recently, a group led by Kirstin Bett of the University of Saskatchewan in Canada released a draft assembly of the *L. culinaris* (cv. CDC Redberry) genome (Bett et al. 2016; http://knowpulse.usask.ca). However, gene duplications, chromosomal rearrangements, and large amounts of repetitive elements in addition to the large size of the genome made the assembly of the seven chromosomes very difficult with the total assembled bases representing approximately half of the 4.3 Gb lentil genome (Bett et al. 2016).

Based on the CDC Redberry draft genome, the Bett's group developed an exome capture array using multiple sources of transcript resources. The probes were designed based on 16 wild lentils and 22 cultivar accessions of lentil, and the data set was used to assess the interspecific relationships within the genus *Lens* (Ogutcen et al. 2018).

The application of "omics" (genomics, transcriptomics, proteomics, and metabolomics) and bioinformatics tools is essential for an integrated approach in understanding the molecular basis of plant response to biotic and abiotic stresses and the underlying mechanisms of tolerance. Transcriptome datasets in lentil supporting functionally associated gene-based marker discovery for assessment of genetic diversity, linkage mapping, and trait dissection have been produced. The expressed transcriptomes of lentil in various tissues and genotypes were determined to examine expression changes during different developmental stages, in different tissues, and upon exposure to toxic compounds (Kaur et al. 2011; Verma et al. 2013; Sudheesh et al. 2016a, b), to develop molecular markers for breeding purposes (Sharpe et al. 2013; Kaur et al. 2014), and examine the transcriptional changes that occur during a pathogen attack (Khorramdelazad et al. 2018).

The genetic variation of lentils has attracted much attention as a subject for genetic analysis during the last decades, using available marker technologies. Ferguson and Robertson (1996) could evaluate the genetic diversity of more than 400 species, subspecies, and landraces of lentils using allozyme markers. Ford et al. (1997) used RAPD molecular markers to study the genetic distance of lentils originating from targeted breeding. All types of molecular markers have been widely used for genetic differentiation studies in lentil varieties, such as RFLP (Havey and Muehlbauer 1989), AFLP (Fiocchetti et al. 2009), RAPD (Sultana and Ghafoor 2008), SSR (Gaad et al. 2017; Idrissi et al. 2017; Yadav et al. 2016), ISSR (Baloch et al. 2015; Fikiru et al. 2011; Sonnante and Pignone 2007), expressed sequence tag (EST)-derived simple sequence repeats (SSRs) (Dikshit et al. 2015; Kumar et al. 2018), and SNP (Lombardi et al. 2014; Basheer-Salimia et al. 2015). Recently, an exome capture array which represents the coding fraction of the genome was developed and applied to evaluate variation in lentil species (Ogutcen et al. 2018). Estimation of genetic variation in landraces is important in species where seeds are the primary product, where adulteration is often when the market price is high, and consequently the nutritional and commercial value can be compromised (Bosmali et al. 2012; Torricelli et al. 2012). Many researchers evaluated the genetic variation of lentils around the world using a multitude of morphological and molecular markers. Toklu et al. (2009) used a combination of ISSR and AFLP markers for the estimation of the genetic variation among varieties. El-Nahas et al. (2011) analyzed proteins and applied ISSR markers in exotic varieties of lentils, while Alghamdi et al. (2014) used SRAP and AFLP markers and combined the results with several morphological/agronomical and nutritional characters. In studies of landrace variation, Torricelli et al. (2012) examined genetic diversity in local lentil varieties of the Santo Stefano di Sessanio area, Abruzzo Region, in Italy using a combination of AFLP, agro-morphological traits, and imaging analysis for their characterization. Fikiru et al. (2011) used ISSR and morphological markers to distinguish Ethiopian lentil varieties. Bosmali et al. (2012) particularly focused on the 'Eglouvis' lentils aiming to facilitate discrimination of bulk samples of the landrace from other commercial varieties. Most of the studies estimated high genetic diversity among different landraces. However, data on variation within a landrace are limited and when available show inconsistent results. Recently, Tsanakas et al. (2018) used morphological and molecular markers to estimate genetic diversity within the Greek lentil landrace 'Eglouvis' that is cultivated continuously at the Lefkada island for more than 400 years. Results suggested that although morphological diversity based on seed characteristics is high, the landrace is genetically homogeneous and at the same time genetically distinct from cultivated commercial varieties.

## 12 Modern Breeding Approaches

The recent application of next generation sequencing and genotyping by sequencing technologies has facilitated the lentil genome sequencing project and large discovery of genome-wide single nucleotide polymorphism (SNP) markers (Kumar et al. 2015; Pavan et al. 2019). Following the advent of next-generation sequencing technologies, SNP markers have been used in lentil to describe genetic variation of germplasm collections and link-specific markers to phenotypic traits, including seed quality, disease resistance, and micronutrient concentration (Fedoruk et al. 2013; Lombardi et al. 2014; Khazaei et al. 2016, 2017, 2018; Sudheesh et al. 2016a, b). Genotyping by sequencing (GBS), based on the sequencing of reduced-representation genomic libraries obtained by restriction enzyme digestion, is one of the most effective methods for SNP discovery and genotyping (Elshire et al. 2011). GBS protocols have been successfully applied for the study of genetic diversity in a collection of wild and cultivated *Lens* spp. germplasm (Wong et al. 2015) and for the identification of DNA polymorphisms in lentil through key links between population structure, geographical origin, and phenotypic and molecular variables (Pavan et al. 2019). Other changes leading to the plant adaptation include mechanisms which affect gene expression and trait phenotypes but do not alter the DNA (Springer and Schmitz 2017). Taking into consideration the phenotypic alterations as a result of epigenetics variation (Mousavi-Derazmahalleh et al. 2019), abiotic-methylome variation will be useful in epi-breeding programs towards the development of climate smart lentil cultivars (Harkess 2018; Kumar et al. 2019). Metabolomics analyses of genotypes with different levels of stress tolerance provide complementing evidence for better understanding the role of different metabolites in the acclimation to adverse environments (Muscolo et al. 2015). Metabolic adjustments in response to the adverse conditions are transient and depend on the type and severity of the stress and are organ- and genotype-specific, indicating that stress tolerance varies based on the developmental stage of plants (Muscolo et al. 2015; Skliros et al. 2018). Comparison of the metabolic profiles of lentil accessions with different tolerance to salinity and drought stresses (Sidari et al. 2008; Muscolo et al. 2015; Skliros et al. 2018) underlining that may overlap, or be specific, in metabolic adjustments under different conditions.

Lentil efficiently improves soil nitrogen levels and carbon sequestration, and in general ameliorates soil quality (Ram et al. 2018; Liu et al. 2019), providing sustainability in crop production systems. Lentil has adapted in various and adverse agricultural environments, while it has low requirements for fertilization (Skliros et al. 2018).

QTL mapping for disease resistance in lentil using molecular markers has identified several resistance genes for major lentil pathogens. The dominant locus *Fw* conferring resistance to *Fusarium oxysporum* has been tagged with the RAPD marker OPK-15900 (Eujayl et al. 1998) and with SSR and AFLP markers (Hamwieh et al. 2005; Choudhary and Kumar 2016). Genetic resistance to ascochyta blight has been linked to several loci (Sudheesh et al. 2016a, b; Ford et al. 1999; Ye et al. 2003; Rodda et al. 2017) connected to molecular markers and tested in linkage mapping to validate QTLs for resistance (Sharpe et al. 2013; Kaur et al. 2014; Sudheesh et al. 2016a, b; Muehlbauer et al. 2006; Chowdhury et al. 2001; Nguyen et al. 2001; Tar'an et al. 2003). Bhadauria et al. (2017) reported three QTLs significantly associated with resistance to *S. botryosum* revealed by composite interval mapping and Cao et al. (2019) observed two QTLs for resistance to stemphylium blight and one gene with unknown function as candidate susceptibility gene. QTL analysis highlighted the linkage group 6 as a significant location for both ascochyta and anthracnose resistance (Tar'an et al. 2003; Tullu et al. 2003) while Bhadauria et al. (2017) discovered five QTLs significantly associated with resistance to anthracnose at the same chromosomal region where three QTLs associated with resistance to *S. botryosum* were found. QTLs connected to resistance to rust due to *Uromyces viciae-fabae* have been reported by Saha et al. (2010) and Dikshit et al. (2016). Marker-assisted selection (MAS) can speed up breeding programs and is deemed necessary to achieve durable resistance. At the present efforts to utilize mapped QTLs have been marginally successful, mainly due to loose linkage of identified markers to resistance genes. Also, virus-resistant lentil varieties have been identified through screening of different genotypes (Kumari et al. 2009). No successful marker-assisted selection for resistance to virus diseases has been reported to date.

Regarding abiotic stresses, QTLs have been identified for tolerance to several adverse conditions. QTLs related to winter survival, frost tolerance, and yield have been characterized (Kahraman et al. 2004a; Barrios et al. 2017). QTLs for heat tolerance were reported (Singh et al. 2017b). Kaur et al. (2014) identified QTLs for boron tolerance. SSR markers linked to the *Sdt* locus conferring drought tolerance in lentil have been reported (Singh et al. 2016a, b). A genomic region harboring numerous QTLs related to root and shoot characteristics associated with drought tolerance has been identified (Idrissi et al. 2016). Further, a linkage map with 291 SSR markers and 75 QTLs for drought tolerance and yield has been provided for an intraspecific RIL mapping population (cited in Gupta et al. 2019).

Research on the genetic control and linkage of SSR markers for drought tolerance in global collections of lentil accessions with different adaptation to drought conditions has led to the identification of different genes/QTLs controlling root traits; recently linked molecular markers were associated with the *Sdt* locus that controls seedling survival drought tolerance in lentil (Singh et al. 2016a, b, c) and a QTL-hotspot genomic region controlling a number of root and shoot characteristics linked to drought tolerance (Idrissi et al. 2016). In lentil under heat stress several

QTLs have thus far been identified for seedling survival and pod setting traits (Singh et al. 2017b), days to maturity, pods/plant, secondary branches/plant, 100 seed weight, yield/plant, and reproductive duration (Kumar et al. 2018). Recently, an NGS-based transcriptome approach in lentil has revealed candidate genes expressed under heat stress conditions related to physiological and pollen phenotypes, cell wall, and secondary metabolism in lentil (Singh et al. 2019).

Recently, the PCR-based markers are being replaced by such systems amenable to high-throughput automation, as the DNA chip-based markers, particularly SNPs. Discovery of SNPs in lentil has progressed utilizing NGS technologies, and SNP linkage maps harboring thousands of loci have been constructed allowing for future high-density linkage map construction and association mapping (Sharpe et al. 2013; Yilmaz Temel et al. 2015). Further recognition of markers closely linked to QTLs could provide further opportunities to dissect the candidate genes and develop molecular markers for MAS and pyramiding of potentially beneficial genes to improve lentil varieties for stress tolerance. The future of lentil crop improvement, driven by climate change and population growth, is the development of high-yielding and stable genotypes under diverse agro-ecological conditions. Molecular-linked functional physiology for developing plants with improved tolerance coupled with higher sustainable yields should be the aim in order to combat the adverse effects of abiotic stresses. In such endeavor, an integrated systems biology approach could advance breeding of lentil towards climate change ready crops.

## 13 Phenotyping Approaches Including High-Throughput Phenotyping

Phenotyping of lentil includes morphological observations that need to be recorded using lentil descriptors (www.icarda.org). Several descriptors require only visual inspection (seedling stem pigmentation, flower color, cotyledon color, leaf pubescence, pod dehiscence, stipule shape, seed shape at maturity, ground color of testa at maturity, pattern of testa), while others are subject of suitable measurement (leaflet size, tendril length, days to 50% flowering, plant height, number of branches per plant, number of peduncles per plant, pedicle length, number of pods per plant, number of seeds per plant, days to maturity, seed yield per plant, biological yield per plant, 100-seed weight). Accuracy in visual observation is a matter of uncertainty; therefore imaging technologies and analytical software have been developed. Artificial vision and image analysis are important for many civil, industrial, and scientific applications. The three main application areas are studies on object and shape recognition (machine vision), classification or cataloguing (grading), and use for the check and choice (tools of measures and evaluation) (Venora et al. 2007). Applications in agriculture include inspection of fruits, vegetables, grains, as well as remote sensing useful in determining the vegetation indices, canopy measurement, land mapping, etc. with greater accuracies (Vibhute and Bodhe 2012; Pandurng and Lomte 2015).

Image analysis in lentils is routinely used for classification into color grades by inspectors following guidelines established by Canadian Grain Commission (Venora et al. 2007). Such official imaging systems generally are based on expensive hardware involving color video cameras, limiting their use in breeding applications. Nevertheless, convenient imaging systems based on low-cost flatbed scanners have been applied successfully for color grading and shape parameters estimation in lentils (Shahin and Symons 2001). Using image captured by a flatbed scanner, Venora et al. (2007) were able to measure seed size, shape, and mean color on individual seeds, and applying each seed imaging to Linear Discriminant Analysis (Classifier) were able to identify five Sicilian landraces with accuracy reaching 97.1%. In another study, Tsanakas et al. (2018) classified seed samples of a landrace to different color categories and assessed seed shape characteristics using a commercial photograph processing software. Automation of image acquisition and processing might enable future integration of high-throughput seed characteristics analysis in lentil breeding programs. Towards this goal, a portable imaging system (BELT) supported by image acquisition and analysis software (phenoSEED) was created for small-seed optical analysis and demonstrated using lentils its ability to extract quantifiable traits of seed color, size and shape, as well as lentil seed coat patterning, characteristics that were traditionally estimated by visual inspection amenable to operator biases (Halcro et al. 2020). Image-based phenotyping approaches may prove especially important for difficult to estimate features such as root characteristics. Roots having an underground establishment are difficult to observe and require time and labor-intensive efforts for their study. High-throughput root phenotyping using imaging platforms may provide a useful alternative. Utilizing such a facility Idrissi and Draye (2019) followed root growth and branching dynamics at seedling stage in order to determine genetic control of drought tolerance in lentil. Preliminary results indicate high heritability and wide genetic variation among the studied genetic material for root biomass, shoot biomass, and root-shoot ratio.

Also, image-based phenotyping approaches were used to evaluate *Aphanomyces euteiches* resistance in lentil genotypes in greenhouse using digital Red-Green-Blue/RGB and hyperspectral imaging and in the field using unmanned aerial system-based multispectral imaging. Results demonstrated moderate to strong correlations between RGB, multispectral, and hyperspectral derived features extracted from lentil shoots/roots and visual scores (Marzougui et al. 2019).

High-throughput methods have been used to study plant responses to various types of biotic and abiotic stresses (drought, heat, salinity, nutrient starving, UV light) and their application in lentil research will be immensely important. Imaging techniques with different spectrum such as visible, infrared, and hyperspectral imaging, integrated in high-throughput phenotyping in the field, are now available for acquisition of both morphological and physiological parameters in one shot. Three-dimensional models of individual plants, tray, plot, or even entire field can be generated to calculate all plant parameters needed for breeding research. These innovations, although requiring expensive facilities, could be of immense importance to accelerate progress in lentil breeding research and provide new robust varieties to benefit low-input agriculture and poor farmers of semiarid areas.

## 14 Conclusions and Future Prospects

Lentil is an important crop, an integral part of the modern agriculture utilized in low- as well as high-input cropping systems, in rotation with cereals and able to improve soil and maintain production in challenging environments. It has enhanced stress tolerance and stable yielding ability. However, several enemies and diseases threaten productivity and have been the subject of intense breeding efforts.

Classical breeding, successful as it has been to provide cultivars with high yield and stress tolerance, has also narrowed the genetic basis of cultivated lentil varieties. Molecular breeding techniques such as MAS or transgenic technologies have limited application to lentil, thus far.

A large number (approximating 10,000 accessions in the EURISCO database) of lentil genotypes including landraces, old and obsolete varieties as well as species crop wild relatives are conserved ex situ in Genebanks throughout the world. This wealth of genetic material, analyzed with molecular tools and evaluated for valuable genes, may prove important for expanding the genetic basis of new lentil varieties and improve significant traits.

This effort can be supported and accelerated by incorporating new technologies, specifically next generation sequencing and phenomics in the breeding schemes. Recognition of numerous markers/QTLs and associations with specific phenotypes will enable breeders to better utilize genetic resources and improve lentil varieties. New genomic selection methodologies could be implemented using large sets of genome-spanning markers to potentially capture all the QTL underlying a trait. These tools are expected to have a profound impact on lentil breeding and produce new varieties designed to be high yielding and stable under different cultivation systems and environmental conditions.

**Acknowledgments** Research on lentil diversity and breeding in the labs of the authors has been co-financed by the European Union and Greek national funds through the Operational Program Competitiveness, Entrepreneurship and Innovation, under the call RESEARCH–CREATE–INNOVATE (project code: T1EDK-04633).

**Author Contribution** Authors contributed equally according to their expertise and names are in alphabetical order after ANP who coordinated the effort, with students and postdocs first and senior scientists following. PVM with EK and AG wrote Sects. 2–5 on origin, diversity of wild relatives and progenitors, genetic resources, and in situ and ex situ collections of lentil germplasm; INO and ES wrote Sect. 6 on crop improvement, climate resilience, and herbicide resistance and also contributed to Sects. 11 and 12; AL wrote Sect. 7 on disease and insect pest resistance; DG and MAS wrote Sect. 8 on end use and nutritional quality; DV wrote Sect. 9 on classical breeding approaches; AM with IDA and SDK wrote Sect. 10 on induced genetic diversity; ANP wrote Sect. 11 on genomic resources, Sect. 12 on modern breeding approaches, and Sect. 13 on phenotyping.

## References


Abbas A (1995) Variation in some cultural and physiological characters and host/pathogen interaction of Fusarium oxysporum f. sp. lentis and inheritance of resistance to lentil wilt in Syria. PhD Thesis, Faculty of Agriculture, University of Aleppo, Aleppo, Syria

Abu-Shakara S, Tannous RI (1981) Nutritional value and quality of lentils. In Webb C, Hawtin G (Eds) Lentils. Commonwealth Agricultural Bureaux, Slough, UK, pp 191–202

Ahmad M, McNeil DL, Fautrier AG (1997) Phylogenetic relationships in Lens species and parentage determination of their interspecific hybrids using RAPD markers. Euphytica 94:101–110

Alghamdi SS, Khan AM, Ammar MH et al (2014) Phenological, nutritional and molecular diversity assessment among 35 introduced lentil (*Lens culinaris* Medik.) genotypes grown in Saudi Arabia. Int J Mol Sci 15:277–295

Alonso R, Aguirre A, Marzo F (2000) Effects of extrusion and traditional processing methods on antinutrients and in vitro digestibility of protein and starch in faba and kidney beans. Food Chem 68(2):159–165. https://doi.org/10.1016/S0308-8146(99)00169-7

Al-Quraan NA, Al-Smadi ML, Swaleh AF (2015) GABA metabolism and ROS induction in lentil (*Lens culinaris* medik) plants by synthetic 1,2,3thiadiazole compounds. J Plant Interact 10(1):185–194. https://doi.org/10.1080/17429145.2015.1056262

Alvarez MT, García P, Pérez de la Vega M (1997) RAPD polymorphism in Spanish lentil landraces and cultivars. J Genet Breed 51:91–96

Amin R, Laskar RA, Khan S (2015) Assessment of genetic response and character association for yield and yield component in lentil (*Lens culinaris* L.) population development through chemical mutagenesis. Cogent Food Agric 1:1–15

Arjenaki FG, Dehaghi MA, Jabbari R (2011) Effects of priming on seed germination of marigold (Calendula officinalis). Adv Environ Biol 5(2 spec. issue):276–280

Babayeva S, Akparov Z, Damania A et al (2014) Genetic diversity for drought tolerance in lentils from Central Asia and the Caucasus: CACLentil. Albanian J Agric Sci 13:1–8. https://doi.org/10.31046/proceedings.2018.47

Bakr MA, Ahmed F (1992) Development of stemphylium blight of lentil and its chemical control. Bangladesh J Plant Pathol 8:39–40

Baloch FS, Derya M, Andeden EE et al (2015) Inter-primer binding site retrotransposon and inter-simple sequence repeat diversity among wild Lens species. Biochem Syst Ecol 58:162–168

Bandeoğlu E, Eyidoğan F, Yücel M et al (2004) Antioxidant responses of shoots and roots of lentil to NaCl-salinity stress. Plant Growth Regul 42:69–77. https://doi.org/10.1023/B:GROW.0000014891.35427.7b

Barrios A, Caminero C, García P et al (2017) Deep Super-SAGE transcriptomic analysis of cold acclimation in lentil (*Lens culinaris* Medik.). BMC Plant Biol 17:1–15. https://doi.org/10.1186/s12870-017-1057-8

Barton JE, Klyne A, Tennakoon D et al (1997) Development of a system for gene transfer to lentils. In: International Food Legume Research Conference III B Program and Posters. Adelaide, Australia (Abstract), p 85

Barulina H (1930) Lentil of the U.S.S.R. and of other countries. Bull Appl Bot Genet Plant Breed 40:265–304

Basheer-Salimia R, Camilli B, Scacchi S et al (2015) Assessment of genetic diversity in lentils (*Lens culinaris* Medik.) based on SNPs. Genet Mol Res 14(2):5870–5878. https://doi.org/10.4238/2015.June.1.4

Bayaa B, Erskine W, Singh A (1997) Screening lentil for resistance to fusarium wilt: methodology and sources of resistance. Euphytica 98:69–74

Belitz H-D, Grosch W, Schieberle P (2005) Food chemistry. Springer, Berlin; Heidelberg; New York, NY, pp 1209–1218. ISBN: 3-540-40817-7

Benaragama D, Rossnagel BG, Shirtliffe SJ (2014) Breeding for competitive and high-yielding crop cultivars. Crop Sci 54:1015–1025. https://doi.org/10.2135/cropsci2013.04.0223

Benlioğlu B, Adak MS (2019) Importance of crop wild relatives and landraces genetic resources in plant breeding programmes. J Exp Agric Int 37(3):1–8

Bernardo R (2001) What if we knew all the genes for a quantitative trait in hybrid crops? Crop Sci 41:1–4

Bett K et al (2016) OP06: the lentil genome – from the sequencer to the field. In: PAG XXIV: Plant and Animal Genomics Conference, San Diego, California. https://mel.cgiar.org/reporting/download/hash/gJFbVXUc

Bhadauria V, Bett KE, Zhou T et al (2013) Identification of *Lens culinaris* defense genes responsive to the anthracnose pathogen *Colletotrichum truncatum*. BMC Genet 14:31. https://doi.org/10.1186/1471-2156-14-31

Bhadauria V, Ramsay L, Bett KE et al (2017) QTL mapping reveals genetic determinants of fungal disease resistance in the wild lentil species *Lens ervoides*. Sci Rep 7:3231

Bhandari K et al (2016) Heat stress at reproductive stage disrupts leaf carbohydrate metabolism, impairs reproductive function, and severely reduces seed yield in lentil. J Crop Improv 30(2):118–151. https://doi.org/10.1080/15427528.2015.1134744

Bhat RS, Upadhyaya NM, Chaudhury A (2007) Chemical and irradiation-induced mutants and tilling. In: Upadhyaya NM (ed) Rice functional genomics: challenges, progress and prospects. Springer, New York, NY, pp 148–180

Bolibok-Brągoszewska H, Rakoczy-Trojanowska M (2015) Molecular marker based assessment of genetic diversity in rye. In: Ahuja MR, Jain SM (eds) Genetic diversity and erosion in plants. Springer, New York, NY, p 110

Bosmali I, Ganopoulos I, Madesis P et al (2012) Microsatellite and DNA-barcode regions typing combined with High Resolution Melting (HRM) analysis for food forensic uses: a case study on lentils (*Lens culinaris*). Food Res Int 46(1):141–147. https://doi.org/10.1016/j.foodres.2011.12.013

Bouhssini EI, Sarker M, Eriskine W et al (2008) First sources of resistance to Sitona weevil (*Sitona crinitus* Herbst) in wild Lens species. Genet Resour Crop Evol 55:1–4

Buchwaldt L, Anderson KL, Morrall RAA et al (2004) Identification of lentil germ plasm resistant to *Colletotrichum truncatum* and characterization of two pathogen races. Phytopathology 94:236–243

Buchwaldt L, Shaikh R, Adam J et al (2013) Recessive and dominant genes confer resistance to *Colletotrichum truncatum* in cultivated lentil. Can J Plant Pathol 35(2):222–231. https://doi.org/10.1080/07060661.2013.768296

Buchwaldt L, Dzananovic E, Durkin J (2018) Lentil anthracnose: epidemiology, fungicide decision support system, resistance and pathogen races. Can J Plant Pathol 40:189–198. https://doi.org/10.1080/07060661.2018.1441185

Cao Z, Li L, Kapoor K et al (2019) Using a transcriptome sequencing approach to explore candidate resistance genes against stemphylium blight in the wild lentil species *Lens ervoides*. BMC Plant Biol 19:399

Ceccobelli S, Ciancaleoni S, Lancioni H et al (2019) Genetic distinctiveness of a protected geographic indication lentil landrace from the Umbria region, Italy, over 20 years. Genet Resour Crop Evol 66:1483–1493

Céron-Rojas JJ, Crossa J (2018) Linear selection indices in modern plant breeding. Springer, New York, NY

Chakraborty M, Haque MF (2000) Genetic variability and component analysis in lentil. J Res Birsa Agric Univ 12(2):199–204

Chatzoglou T, Tokatlidis IS (2012) Decision on germplasm choice to apply breeding within a local population of common vetch is affected by crowding. Span J Agric Res 10(3):752–755

Chauhan MP, Singh IS (1998) Genetic variability, heritability and expected genetic advance for seed yield and other characters over two years in lentil. Lens Newsl 25(1):3–6

Chauhan BS, Matloob A, Mahajan G et al (2017) Emerging challenges and opportunities for education and research in weed science. Front Plant Sci 8:1–13. https://doi.org/10.3389/fpls.2017.01537

Chen W, Sharma H, Muehlbauer FJ (eds) (2011) Compendium of chickpea and lentil diseases and pests. APS Press, St Paul, MN, p 165

Chhabra G, Chaudhary D, Varma M et al (2008) TDZ-induced direct shoot organogenesis and somatic embryogenesis on cotyledonary node explants of lentil (*Lens culinaris* Medik.). Physiol Mol Biol Plants 14(4):347–353

Chopra R, Aparna, Saini R (2012) Use of sonication and vacuum infiltration for Agrobacterium - mediated transformation of an Indian lentil (*Lens culinaris* Medik.) cultivar. Sci Hortic 143:127–134. https://doi.org/10.1016/j.scienta.2012.06.019

Choudhary AK, Kumar S (2016) Genetic improvement for fusarium wilt resistance in lentil. In: Singh AK, Bhakta N, Sangale UR, Manibhushan, Sundaram PK, Kumar S, Yasin JK (eds) Scientific lentil production: Indian perspectives. Society for Upliftment of Rural Economy, Varanasi, pp 59–72

Chowdhury RK, Chowdhury JB (1983) Compatibility between *Vigna radiata* (L.) Wilczek and Vigna umbellata Thumb. Ohwi and Ohashi. Genet Agraria 37:257–266

Chowdhury MA, Andrahennadi CP, Slinkard AE et al (2001) RAPD and SCAR markers for resistance to Ascochyta blight in lentil. Euphytica 118:331–337. https://doi.org/10.1023/A:1017581817201

Cokkizgin A, Shtaya MJY (2013) Lentil: origin, cultivation techniques, utilization and advances in transformation. Agric Sci 1(1):55–62. https://doi.org/10.12735/as.v1i1p55

Collard BCY, Ades PK, Pang ECK et al (2001) Prospecting for sources of resistance to Ascochyta blight in wild Cicer species. Australas Plant Pathol 30:271–276

Dadu RHR, Ford R, Sambasivam P et al (2017) A novel *Lens orientalis* resistance source to the recently evolved highly aggressive Australian *Ascochyta lentis* isolates. Front Plant Sci 8:1038. https://doi.org/10.3389/fpls.2017.01038

Dalgetty DD, Baik BK (2006) Fortification of bread with hulls and cotyledon fibers isolated from peas, lentils, and chickpeas. Cereal Chem 83(3):269–274. https://doi.org/10.1094/CC-83-0269

Damm U, O'Connell RJ, Groenewald JZ, Crous PW (2014) The Colletotrichum destructivum species complex - hemibiotrophic pathogens of forage and field crops. Stud Mycol 79:49–84. https://doi.org/10.1016/j.simyco.2014.09.003. PMID: 25492986; PMCID: PMC4255528

Das R, Nath R, Dikshit HK (2017) Host resistance of lentil genotypes against Stemphylium blight caused by *Stemphylium botryosum* Wallr. in lower gangetic alluvial zone of West Bengal, India. J Mycopathol Res 55:169–172

Das A, Dutta S, Jash S et al (2019) Current knowledge on pathogenicity and management of *Stemphylium botryosum* in lentils (*Lens culinaris* ssp. *culinaris* Medik). Pathogens 8:225. https://doi.org/10.3390/pathogens8040225

Davis KR (1981) Effect of processing on composition an Tetrahy-mena relative nutritive value on green and yellow peas, lentils and white pea beans. Cereal Chem 58:454–460

Davidson J, Smetham G, Russ MH et al (2016) Changes in aggressiveness of the *Ascochyta lentis* population in southern Australia. Front Plant Sci 7:393. https://doi.org/10.3389/fpls.2016.00393

De Almeida Costa GE, Da Silva Q-MK, Pissini Machado Reis SM et al (2006) Chemical composition, dietary fibre and resistant starch contents of raw and cooked pea, common bean, chickpea and lentil legumes. Food Chem 94(3):327–330. https://doi.org/10.1016/j.foodchem.2004.11.020

Delahunty A, Nuttall J, Nicolas M et al (2018) Response of lentil to high temperature under variable water supply and carbon dioxide enrichment. Crop Past Sci 69:1103–1112. https://doi.org/10.1071/CP18004

Dhindsa KS, Sood DR, Chaudhary MS (1985) Nutritional evaluation of some varieties of lentil (*Lens esculenta*). Ind J Nutr Diet 22:187–189

Dikshit HK, Singh A, Singh D et al (2015) Genetic diversity in lens species revealed by EST and genomic simple sequence repeat analysis. PLoS One 10(9):e0138101

Dikshit HK, Singh A, Singh D et al (2016) Tagging and mapping of SSR marker for rust resistance gene in lentil (*Lens culinaris* Medikus subsp. culinaris). Indian J Exp Biol 54:394–399

Dubey K, Mishra A (2019). Identification of lentil genotype(s) resistant to the Fusarium oxysporum f. sp. Lentis. Int J Agri Plant Sci 1(3):31–33
Duke JA (1981) Handbook of legumes of world economic importance. Plenum Press, New York, USA, 345 p. (online ISBN: 978-1-4684-8151-8)
Duranti M, Gius C (1997) Legume seeds: protein content and nutritional value. Field Crops Res 53(1/3):31–45
Durán Y (2002) Utilización de marcadores moleculares en Lens (Miller): variabilidad y mapa genético. PhD thesis Universidad de León, León, Spain
El Attar AH (1991) Genetic variability among some exotic lentil germplasm in Egypt. Bull Fac Agric Univ Cairo 42(3):993–1000
El-Monem A, Sharaf M (2008) Tolerance of five genotypes of lentil to NaCl-salinity stress. N Y Sci J 1:70–80
El-Nahas AI, El-Shazly HH, Ahmed SM et al (2011) Molecular and biochemical markers in some lentil (*Lens culinaris* Medik.) genotypes. Ann Agric Sci 56(2):105–112. https://doi.org/10.1016/j.aoas.2011.11.001
Elshire RJ, Glaubitz JC, Sun Q et al (2011) A robust, simple genotyping-by-sequencing (GBS) approach for high diversity species. PLoS One 6:1–10. https://doi.org/10.1371/journal.pone.0019379
Erskine W (1997) Lessons for breeders from landraces of lentil. Euphytica 93:107–112
Erskine W, Sarker A (2004) Lentil/breeding. In: Wrigley C, Corke H, Walker C (eds) Encyclopedia of grain science. Elsevier, Amsterdam, pp 142–150
Erskine W, Saxena MC (1993) Breeding lentil at ICARDA for southern latitudes. In: Erskine W, Saxena MC, Roush S (eds) Lentils in South Asia. International Center for Agricultural Research in the Dry Areas (ICARDA), Aleppo
Erskine W, Isawi J, Masoud K (1990) Single plant selection for yield in lentil. Euphytica 48:113–116
Erskine W, Tufail M, Russell A et al (1994) Current and future strategies in breeding lentil for resistance to biotic and abiotic stresses. Euphytica 73:127–135
Erskine W, Chandra S, Chaudhry M et al (1998) A bottleneck in lentil: widening its genetic base in South Asia. Euphytica 101:207–211
Erskine W, Muehlbauer FJ, Saker A et al (2009) The lentil: botany, production and uses. CAB International, Wallingford, p 456
Erskine W, Sarker A, Kumar S (2011) Crops that feed the world 3. Investing in lentil improvement toward a food secure world. Food Secur 3(2):127–139. https://doi.org/10.1007/s12571-011-0124-5
Erskine W, Sarker A, Kumar S (2015) Lentil: breeding. In: Encyclopedia of food grains, 2nd edn. Elsevier Inc, Amsterdam, pp 317–324. https://doi.org/10.1016/B978-0-12-394437-5.00210-2
Erskine W, Sarker A, Kumar S (2016) Lentil: breeding. In: Wrigley C, Corke H, Seetharaman K, Faubion J (eds) Encyclopedia of food grains, 2nd edn. Elsevier Ltd, Academic Press, Oxford, pp 317–324
Esmail AM, Mohamed AA, Hamdi A et al (1994) Genetic variability and heritability for agronomic traits in segregating populations of lentil (*Lens culinaris* Medik.). Ann Agric Sci Moshtohor 32(3):1107–1118
Eujayl I, Baum M, Powell W et al (1998) A genetic linkage map of lentil (*Lens* sp.) based on RAPD and AFLP markers using recombinant inbred lines. Theor Appl Genet 97(1–2):83–89. https://doi.org/10.1007/s001220050869
Eujayl I, Erskine W, Bayaa B et al (2006) Fusarium vascular wilt in lentil: inheritance and identification of DNA markers for resistance. Plant Breed 117:497–499
Fahad S, Bajwa AA, Nazir U et al (2017) Crop production under drought and heat stress: plant responses and management options. Front Plant Sci 8:1–16. https://doi.org/10.3389/fpls.2017.01147

Farooq M, Gogoi N, Hussain M et al (2017) Effects, tolerance mechanisms and management of salt stress in grain legumes. Plant Physiol Biochem 118:199–217. https://doi.org/10.1016/j.plaphy.2017.06.020

FAOSTAT (2019) Available at: http://www.fao.org/faostat/en/#data/QC

Fasoula DA, Fasoula VA (1997) Competitive ability and plant breeding. Plant Breed Rev 14:89–138

Fasoulas A (1973) A new approach to breeding superior yielding varieties. Dept. of Genetics and Plants Breeding Pub. No. 3. Aristotelian University of Thessaloniki, Thessaloniki

Fedoruk LK, Johnson EN, Shirtliffe SJ (2011) The critical period of weed control for lentil in western Canada. Weed Sci 59(4):517–526. https://doi.org/10.1614/ws-d-11-00051.1

Fedoruk MJ, Vandenberg A, Bett KE (2013) Quantitative trait loci analysis of seed quality characteristics in lentil using single nucleotide polymorphism markers. Plant Genome 6:plantgenome2013.05.0012. https://doi.org/10.3835/plantgenome2013.05.0012

Ferguson M (2000) Lens spp: conserved resources, priorities for collection and future prospects. In: Knight R (ed) Linking research and marketing opportunities for pulses in the 21st Century. Proceedings of the Third International Food Legumes Research Conference, Current plant science and biotechnology in agriculture, vol 34. Kluwer Academic Publishers, Dordrecht; Boston, MA; London, pp 613–620

Ferguson ME, Robertson LD (1996) Genetic diversity and taxonomic relationships within the genus Lens as revealed by allozyme polymorphism. Euphytica 91:163–172

Fernández-Aparicio M, Sillero JC, Pérez-De-Luque A et al (2008) Identification of sources of resistance to crenate broomrape (Orobanche crenata) in Spanish lentil (*Lens culinaris*) germplasm. Weed Res 48:85–94. https://doi.org/10.1111/j.1365-3180.2008.00604.x

Fernández-Aparicio M, Sillero JC, Rubiales D (2009) Resistance to broomrape in wild lentils (*Lens spp.*). Plant Breed 128:266–270. https://doi.org/10.1111/j.1439-0523.2008.01559.x

Fernández-Aparicio M, Reboud X, Gibot-Leclerc S (2016) Broomrape weeds. Underground mechanisms of parasitism and associated strategies for their control: a review. Front Plant Sci 7:135. https://doi.org/10.3389/fpls.2016.00135

Fiala JV (2006) Transferring resistance to Colletotrichum truncatum from wild lentil species to cultivated lentil species (*Lens culinaris* subsp *culinaris*). Dissertation, University of Saskatchewan

Fiala JV, Tullu A, Banniza S et al (2009) Interspecies transfer of resistance to anthracnose in lentil (*Lens culinaris* Medic). Crop Sci 49:825–8305

Fikiru E, Tesfaye K, Bekele E (2011) Morphological and molecular variation in Ethiopian lentil (*Lens culinaris* Medikus) varieties. Int J Genet Mol Biol 3(4):60–67

Fiocchetti F, Laddomada B, Roselli M et al (2009) Fingerprinting of three typical macrosperma Italian lentil (*Lens culinaris* Medik.) landraces using fluorescence-based AFLP markers. Sci Hortic 121:383–387

Fischer RA (2020) Breeding wheat for increased potential yield: contrasting ideas from Donald and Fasoulas, and the case for early generation selection under nil competition. Field Crop Res 252:107782. https://doi.org/10.1016/j.fcr.2020.107782

Fischer RA, Moreno Ramos OH, Ortiz Monasterio I et al (2019) Yield response to plant density, row spacing and raised beds in low latitude spring wheat with ample soil resources: an update. Field Crop Res 232:95–105. https://doi.org/10.1016/j.fcr.2018.12.011

Fleming SE (1981) A study of relationships between flatus potential and carbohydrate distribution in legume seeds. J Food Sci 46:794–798

Flint-Hamilton KB (1999) Legumes in Ancient Greece and Rome: food, medicine, or poison? Hesperia 68:371. https://doi.org/10.2307/148493

Ford R, Pang ECK, Taylor PWJ (1997) Diversity analysis and species identification in Lens using PCR generated markers. Euphytica 96:247–255

Ford R, Pang ECK, Taylor PWJ (1999) Genetics of resistance to Ascochyta blight (*Ascochyta lentis*) of lentil and the identification of closely linked RAPD markers. Theor Appl Genet 98:93–98. https://doi.org/10.1007/s001220051044

Foti C, Khah E, Pavli O (2018) Response of lentil genotypes under PEG-induced drought stress: effect on germination and growth. Plant 6:75. https://doi.org/10.11648/j.plant.20180604.12

Foti C, Khah EM, Pavli O (2019) Germination profiling of lentil genotypes subjected to salinity stress. Plant Biol 21:480–486. https://doi.org/10.1111/plb.12714

Fratini R, Ruiz ML (2003) A rooting procedure for lentil (*Lens culinaris* Medik) and other hypogeous legumes (Pea, chickpea and Lathyrus) based on plant polarity. Plant Cell Rep 21:726–732

Fratini R, Ruiz ML (2006) Interspecific hybridization in the genus Lens applying in vitro embryo rescue. Euphytica 150:271–280

Gaad D, Laouar M, Udupa SM et al (2017) Diversity study of Algerian accessions of lentil (*Lens culinaris* Medik.) by using microsatellite markers. Res Crop 18(4):722–727. https://doi.org/10.5958/2348-7542.2017.00119.x

Gahoonia TS, Ali O, Sarker A et al (2005) Root traits, nutrient uptake, multi-location grain yield and benefit-cost ratio of two lentil (*Lens culinaris* Medikus.) varieties. Plant Soil 272:153–161. https://doi.org/10.1007/s11104-004-4573-x

Gahoonia TS, Ali O, Sarker A et al (2006) Genetic variation in root traits and nutrient acquisition of lentil genotypes. J Plant Nutr 29:643–655. https://doi.org/10.1080/01904160600564378

Gaur PM, Samineni S, Krishnamurthy L et al (2015) High temperature tolerance in grain legumes. Legum Perspect 6–7:23

Gautam NK, Singh M, Khan Z et al (2013) Assessment of lentil germplasm with respect to agronomic performance and major biotic stress. Legum Res 36(3):214–219

Gepts P, Beavis WD, Brummer EC et al (2005) Legumes as a model plant family. Genomics for Food and Feed Report of the Cross-Legume Advances through Genomics Conference. Plant Physiol 137:1228–1235

Ghanem ME, Kibbou F, Guiguitant J et al (2017) Opportunities to improve the seasonal dynamics of water use in lentil (*Lens culinaris* Medik.) to enhance yield increase in water-limited environments. Chem Biol Technol Agric 4:1–6. https://doi.org/10.1186/s40538-017-0103-y

Gorim LY, Vandenberg A (2018) Can wild lentil genotypes help improve water use and transpiration efficiency in cultivated lentil? Plant Genet Res 16:459–468. https://doi.org/10.1017/S1479262117000399

Gulati A, Schryer P, McHughen A (2001) Regeneration and micrografting of lentil shoots. In Vitro Cell Dev Biol Plant 37:798–802

Gulati A, Schryer P, McHughen A (2002) Production of fertile transgenic lentil (*Lens culinaris* Medik) plants using particle bombardment. In Vitro Cell Dev Biol Plant 38(4):316–324. https://doi.org/10.1079/IVP2002303

Gupta D, Sharma SK (2006) Evaluation of wild Lens taxa for agro-morphological traits, fungal diseases and moisture stress in northwestern Indian hills. Genet Resour Crop Evol 53:1233–1241

Gupta D, Sharma SK (2007) Widening the gene pool of cultivated lentils through introgression of alien chromatin from wild Lens subspecies. Plant Breed 126:58–61. https://doi.org/10.1111/j.1439-0523.2007.01318.x

Gupta DS, McPhee K, Kumar S (2017) Development of molecular markers for iron metabolism related genes in lentil and their expression analysis under excess iron stress. Front Plant Sci 8:1–8. https://doi.org/10.3389/fpls.2017.00579

Gupta D, Harinath R, Dadu R et al (2019) Toward climate-resilient lentils: challenges and opportunities. In: Genomic designing of climate-smart pulse crops. Springer International Publishing, Cham, pp 165–234. https://doi.org/10.1007/978-3-319-96932-9_4

Haddad NI, Muehlbauer FJ (1981) Comparison of random bulk population and single seed descent methods for Lentil breeding. Euphytica 30:643–651

Hahn DH, Rooney LW, Earp CF (1984) Tannins and phenols of sorghum. Cereal Foods World 29:776–779

Halcro K, McNabb K, Lockinger A et al (2020) The BELT and phenoSEED platforms: shape and colour phenotyping of seed samples. Plant Methods 16(1):1–13. https://doi.org/10.1186/s13007-020-00591-8

Hallauer AR, Carena MJ, Filho JB (2010) Quantitative genetics in maize breeding. Springer, New York, NY; Dordrecht; Heidelberg; London

Hamdi A, El-Ghareib AA, Shafey SA, Ibrahim MAM (2003) Genetic variability, heritability and expected genetic advance for earliness and seed yield from selection in lentil. Egypt J Agric Res 81(1):125–138

Hamwieh A, Udapa SM, Choumane W et al (2005) A genetic linkage map of lentil based on microsatellite and AFLP markers and localization of Fusarium vascular wilt resistance. Theor Appl Genet 110:669–677

Hansen J, Renfrew JM (1978) Paleolithic-Neolithic seed remains at Franchthi cave, Greece. Nature 71:349–352

Harkess A (2018) Handling the heat: methylome variation underlying heat tolerance in cotton. Plant Cell 30:1947–1948. https://doi.org/10.1105/tpc.18.00698

Havey MJ, Muehlbauer FJ (1989) Variability for restriction fragment lengths and phylogenies in lentil. Theor Appl Genet 77:839–843

Hawtin GC, Singh KB, Saxena MC (1980) Some recent developments in the understanding and improvement of Cicer and Lens. In: Summerfield RJ, Bunting AH (eds) Advances in legume science. Royal Botanical Garden, London, pp 613–624

Hiremani NS, Dubey SC (2018) Race profiling of Fusarium oxysporum f. sp. lentis causing wilt in lentil. Crop Prot 108:23–30. https://doi.org/10.1016/j.cropro.2018.02.010

Hobson K, Armstrong R, Nicolas M et al (2006) Response of lentil (*Lens culinaris*) germplasm to high concentrations of soil boron. Euphytica 151:371–382. https://doi.org/10.1007/s10681-006-9159-7

Hossain MI, Islam MM, Wahed MA et al (2009) Lentil-based high protein diet is comparable to animal-based diet in respect to nitrogen absorption and nitrogen balance in malnourished children recovering from shigellosis. Asia Pac J Clin Nutr 18(1):8–14

Hulse JH (1990) Nature, composition and utilization of grain legumes. Uses of tropical grain legumes. In: Proceedings of a consultants' meeting, ICRISAT, pp 11–27

IAEA (2020) Mutant Variety Database. http://mvd.iaea.org. Accessed 18 Mar 2020

Ibrahim HM (2011) Heat stress in food legumes: evaluation of membrane thermostability methodology and use of infra-red thermometry. Euphytica 180(1):99–105. https://doi.org/10.1007/s10681-011-0443-9

Idrissi O (2020) Application of extended photoperiod in lentil: towards accelerated genetic gain in breeding for rapid improved variety development. Morrocan J Agric Sci 1(1):14–19

Idrissi O, Draye X (2019) High-throughput phenotyping for drought tolerance-related root traits in lentil: steps ahead for the development of efficient screening protocol for climate change resilient varieties. In: International Conference Climate Resilient Agriculture: Ways of Adaptation, Rabat, 2019

Idrissi O, Houasli C, Udupa SM et al (2015) Genetic variability for root and shoot traits in a lentil (*Lens culinaris* Medik.) recombinant inbred line population and their association with drought tolerance. Euphytica 204:693–709. https://doi.org/10.1007/s10681-015-1373-8

Idrissi O, Udupa SM, De Keyser E et al (2016) Identification of quantitative trait loci controlling root and shoot traits associated with drought tolerance in a lentil (*Lens culinaris* medik.) recombinant inbred line population. Front Plant Sci 7:1174. https://doi.org/10.3389/fpls.2016.01174

Idrissi O, Piergiovanni AR, Toklu F et al (2017) Molecular variance and population structure of lentil (*Lens culinaris* Medik.) landraces from Mediterranean countries as revealed by simple sequence repeat DNA markers: implications for conservation and use. Plant Genet Resour Charact Util 16:249–259. https://doi.org/10.1017/S1479262117000260

Idrissi O, Piergiovanni A, Faruk T et al (2018) Molecular variance and population structure of lentil (*Lens culinaris* Medik.) landraces from Mediterranean countries as revealed by simple sequence repeat DNA markers: implications for conservation and use. Plant Genet Res 16(3):249–259. https://doi.org/10.1017/S1479262117000260

Idrissi O, Sahri A, Udupa S et al (2019) Single seed descent under extended photoperiod as a simple, rapid and efficient breeding method for accelerated genetic gain in lentil. In: Proceedings of the 3rd International Legume Society Conference, Poznan, Poland, 2019

Iliadis CG, Roupakias DG, Goulas CK (2003) Effectiveness of honeycomb selection for yield superiority at three interplant distances: a field simulation study using chickpea (*Cicer arietinum* L.) inbred lines. Euphytica 133:299–311

Inder P, Materne M, Taylor PWJ et al (2008) Genotyping elite genotypes within the Australian lentil breeding program with lentil-specific sequenced tagged microsatellite site (STMS) markers. Aust J Agric Res 59:222–225

Jankowicz-Cieslak J, Till BJ (2016) Chemical mutagenesis of seed and vegetatively propagated plants using EMS. Curr Protoc Plant Biol 1:617–635

Jarpa-Parra M (2018) Lentil protein: a review of functional properties and food application. An overview of lentil protein functionality. Int J Food Sci Technol 53:892. https://doi.org/10.1111/ijfs.13685

Kafi M, Goldani M, Shariat Jafari MH (2012) Effectiveness of nutrient management in managing saline agro-ecosystems: a case study of *Lens culinaris* Medik. Pak J Bot 44:269–274

Kahraman A, Kusmenoglu I, Aydin N et al (2004a) QTL mapping of winter hardiness genes in lentil. Crop Sci 44:13–22. https://doi.org/10.2135/cropsci2004.1300

Kahraman A, Kusmenoglu I, Aydin N et al (2004b) Genetics of winter hardiness in 10 lentil recombinant inbred line populations. Crop Sci 44:5–12. https://doi.org/10.2135/cropsci2004.5000

Kahraman A, Demirel U, Ozden M et al (2010) Mapping of QTLs for leaf area and the association with winter hardiness in fall-sown lentil. Afr J Biotechnol 9(50):8515–8519. https://doi.org/10.5897/AJB10.572

Kamboj RK, Pandey MP, Chaube HS (1990) Inheritance of resistance to Fusarium wilt in Indian lentil germplasm (Lens culinaris Medic.). Euphytica 50:113–117

Kant P, Materne M, Rodda MS et al. (2017) Screening lentil germplasm for stemphylium blight resistance. Australasian Plant Pathol 46:129–136. https://doi.org/10.1007/s13313-017-0469-0

Karagounis D, Katsavou E, Vlachostergios D et al (2011) A comparative study on physicochemical and sensory properties in lentil varieties cultivated under conventional and organic growing conditions. In: 25th General Congress of HSHS Horticulture in Europe, Cyprus

Kargiotidou A, Chatzivassiliou E, Sinapidou E et al (2014) Selection at ultra-low density highlights plants escaping virus infection and leads towards high-performing pure-line cultivars in lentil. J Agric Sci 152:749–758

Kargiotidou A, Tzantarmas C, Chatzivassiliou E et al (2015) Seed propagation at low density facilitates the selection of healthy plants to produce seeds with a reduced virus load in a lentil landrace. Seed Sci Technol 43:31–39

Kargiotidou A, Vlachostergios DN, Tzantarmas C et al (2016) Addressing huge spatial heterogeneity induced by virus infections in lentil breeding trials. J Biol Res (Thessaloniki) 23:2. https://doi.org/10.1186/s40709-016-0039-6

Katerji N, Van Hoorn JW, Hamdy A et al (2001) Response of two varieties of lentil to soil salinity. Agric Water Manag 47:179–190. https://doi.org/10.1016/S0378-3774(00)00109-8

Katerji N, Van Hoorn JW, Hamdy A et al (2003) Salinity effect on crop development and yield, analysis of salt tolerance according to several classification methods. Agric Water Manag 62(1):37–66. https://doi.org/10.1016/S0378-3774(03)00005-2

Kaur S, Cogan NOI, Pembleton LW et al (2011) Transcriptome sequencing of lentil based on second-generation technology permits large-scale unigene assembly and SSR marker discovery. BMC Genomics 12:265. https://doi.org/10.1186/1471-2164-12-265

Kaur S, Cogan NOI, Stephens A et al (2014) EST-SNP discovery and fine-resolution genetic mapping in lentil (*Lens culinaris* Medik.) enables candidate gene selection for boron tolerance. Theor Appl Genet 127:703–713. https://doi.org/10.1007/s00122-013-2252-0

Kay DE (1979) Food legumes TPI crop and product digest, vol 3. Tropical Development and Research Institute (TPI), London

Khan A, Khan S (2011) Induced variation in quantitative traits due to chemical mutagen (hydrazine hydrate) treatment in lentil (*Lens culinaris* Medik.). Ind Stream Res J 1(VII)

Khatib F, Koudsieh S, Ghazal B et al (2007) Developing herbicide resistant lentil (*Lens culinaris* Medikus subsp. *culinaris*) through *Agrobacterium*-mediated transformation. Arab J Plant Prot 25(2):185–192

Khazaei H, Caron CT, Fedoruk M et al (2016) Genetic diversity of cultivated lentil (*Lens culinaris* Medik.) and its relation to the world's agro-ecological zones. Front Plant Sci 7:1–7. https://doi.org/10.3389/fpls.2016.01093

Khazaei H, Podder R, Caron CT et al (2017) Marker-trait association analysis of iron and zinc concentration in lentil (*Lens culinaris* Medik.) seeds. Plant Genome 10(2). https://doi.org/10.3835/plantgenome2017.02.0007

Khazaei H, Fedoruk M, Caron CT et al (2018) Single nucleotide polymorphism markers associated with seed quality characteristics of cultivated lentil. Plant Genome 11:1–7. https://doi.org/10.3835/plantgenome2017.06.0051

Khorramdelazad M et al (2018) Transcriptome profiling of lentil (*Lens culinaris*) through the first 24 hours of Ascochyta lentis infection reveals key defence response genes. BMC Genomics 19(1):1–21. https://doi.org/10.1186/s12864-018-4488-1

Knuckles BE, Kuzmicky DD, Betschart AA (1982) HPLC analysis of phytic acid in selected foods and biological samples. J Food Sci 49:1257–1258

Krohn RM, Raqib R, Akhtar E et al (2016) A high-selenium lentil dietary intervention in Bangladesh to counteract arsenic toxicity: study protocol for a randomized controlled trial. Trials 17:218. https://doi.org/10.1186/s13063-016-1344-y

Kumar A, Singh DP, Singh BB (1995) Genetic variability of yield and its components in lentil (*Lens culinaris* Medik.). Ind J Puls Res 8(1):60–66

Kumar R, Kumar D, Kumar S (1999) Genetic variability in lentil (*Lens culinaris* Medik.). Ann Agric Res 4(1):75–77

Kumar R, Mishra SK, Sharma B (2001) Genetics of rust resistance in lentil (*Lens culinaris* Medik.). Ind J Genet Plant Breed 61:238–241

Kumar R, Sharma SK, Luthra OP et al (2005) Phenotypic stability of lentil genotypes under different environments. Ann Biol 21(2):155–158

Kumar J, Basu PS, Srivastava E et al (2012) Phenotyping of traits imparting drought tolerance in lentil. Crop Past Sci 63(6):547–554. https://doi.org/10.1071/CP12168

Kumar SK, Barpete S, Kumar J et al (2013) Global lentil production: constraints and strategies. SATSA Mukhapatra Annu Tech Issue 17:1–13

Kumar S, Rajendran K, Kumar J et al (2015) Current knowledge in lentil genomics and its application for crop improvement. Front Plant Sci 6:1–13. https://doi.org/10.3389/fpls.2015.00078

Kumar J, Gupta S, Gupta P et al (2016a) Breeding strategies to improve lentil for diverse agro-ecological environments. Ind J Genet Plant Breed 76:530–549. https://doi.org/10.5958/0975-6906.2016.00071.7

Kumar J, Kant R, Basu PS et al (2016b) Heat tolerance in lentil under field conditions. Legum Genom Genet 7:1–11. https://doi.org/10.5376/lgg.2016.07.0001

Kumar J, Basu PS, Gupta S et al (2018) Physiological and molecular characterisation for high temperature stress in *Lens culinaris*. Funct Plant Biol 45(4):474–487. https://doi.org/10.1071/FP17211

Kumar J, Choudhary AK, Gupta DS et al (2019) Towards exploitation of adaptive traits for climate-resilient smart pulses. Int J Mol Sci 20(12):2971. https://doi.org/10.3390/ijms20122971

Kumari J, Ahmad R, Chandra S et al (2009) Determination of morpholological attributes imparting resistance against aphids (*Aphis craccivora* Koch) in lentil (*Lens culinaris* Medik). Arch Phytopathol Plant Protect 42:52–57. https://doi.org/10.1080/03235400600940806

Kyriakou DT, Fasoulas AC (1985) Effects of competition and selection pressure on yield response in winter rye (*Secale cereale* L.). Euphytica 34:883–895

Ladizinsky G (1979) The origin of lentil and its wild genepool. Euphytica 28:179–187

Ladizinsky G, Braun D, Goshen D et al (1984) The biological species of the genus Lens L. [*Lens nigricans*]. Bot Gaz 145:253–261

Ladizinsky G, Cohen D, Muehlbauer FJ (1985) Hybridization in the genus Lens by means of embryo culture. Theor Appl Genet 70:97–101

Laserna-Ruiz I, De-Los-Mozos-Pascual M, Santana MO et al (2012) Screening and selection of lentil (*Lens* Miller) germplasm resistant to seed bruchids (*Bruchus* spp.). Euphytica 188:153–162

Liener IE, Kakade ML (1980) Protease inhibitors. In: Liener IE (ed) Toxic constituents of plant foodstuffs. Academic Press, London

Lira Junior MDA, Lima AST, Arruda JRF et al (2005) Effect of root temperature on nodule development of bean, lentil and pea. Soil Biol Biochem 37:235–239. https://doi.org/10.1016/j.soilbio.2004.07.032

Liu K, Blackshaw RE, Johnson EN et al (2019) Lentil enhances the productivity and stability of oilseed-cereal cropping systems across different environments. Eur J Agron 105:24–31. https://doi.org/10.1016/j.eja.2019.02.005

Ljuština M, Mikić A (2010) Archaeological evidence of the domestication of lentil (*Lens culinaris*) and its distribution in Europe. J Lentil Res 4:26–29

Lombardi M, Materne M, Cogan NOI et al (2014) Assessment of genetic variation within a global collection of lentil (*Lens culinaris* Medik.) cultivars and landraces using SNP markers. BMC Genet 15:1–10. https://doi.org/10.1186/s12863-014-0150-3

MacCarrone M, Veldinka GA, Finazzi A et al (1995) Lentil root protoplasts: a transient expression system suitable for coelectroporation of monoclonal antibodies and plasmid molecules. Biochim Biophys Acta 1243(1):136–142

MacCarrone M, Van Zadelhoff G, Veldink GA et al (2000) Early activation of lipoxygenase in lentil (*Lens culinaris*) root protoplasts by oxidative stress induces programmed cell death. Eur J Biochem 267(16):5078–5084

Malik KA, Saxena PK (1992) Thidiazuron induces high frequency shoot regeneration in intact seedlings of pea (*Pisum sativum*), chickpea (*Cicer arietinum*) and lentil (*Lens culinaris*). Aust J Plant Physiol 19:731–740

Martin RJ, Wilcox JR, LaViolette FA (1978) Variability in soybean progenies developed by single seed descent at two plant populations. Crop Sci 18:359–363

Marzougui A, Ma Y, Zhang C et al (2019) Advanced imaging for quantitative evaluation of aphanomyces root rot resistance in lentil. Front Plant Sci 10:1–16. https://doi.org/10.3389/fpls.2019.00383

Materne M, McNeil DL (2007) Breeding methods and achievements. In: Yadav S, McNeil D, Stevenson P (eds) Lentil. An ancient crop for modern times. Springer, Dordrecht, pp 241–255

Materne MA, McMurray L, Nitschke S et al (2002) The future of Australian lentil production. In: Brouwer JB (ed) Proceedings of Lentil Focus. Horsham, Victoria, 2002

Mbasani-Mansi J, Briache FZ, Ennami M et al (2019a) Resistance of Moroccan lentil genotypes to *Orobanche crenata* infestation. J Crop Improv 33:306–326. https://doi.org/10.1080/15427528.2019.1581866

Mbasani-Mansi J, Ennami M, Briache FZ et al (2019b) Characterization of genetic diversity and population structure of Moroccan lentil cultivars and landraces using molecular markers. Physiol Mol Biol Plants 25(4):965–974

McLean LA, Sosulski FW, Youngs CG (1974) Effects of nitrogen and moisture on yield and protein in field peas. Can J Plant Sci 54:301–305

Mishra SK, Sarker A, Singh BB et al (2005) Slow rusting and its potential donors for resistance in lentil (*Lens culinaris* Medik.). Ind J Genet Plant Breed 65(4):319–320

Mishra BK, Srivastava JP, Lal JP et al (2016) Physiological and biochemical adaptations in lentil genotypes under drought stress. Russ J Plant Physiol 63(5):695–708. https://doi.org/10.1134/S1021443716040117

Mobini SH, Lulsdorf M, Warkentin TD et al (2016) Low red: far-red light ratio causes faster in vitro flowering in lentil. Can J Plant Sci 96:908–918

Mohammadi N, Goltapeh EM, Babaie-Ahari A et al (2011) The genetic diversity of Iranian isolates causing fusarium wilt of Lentil. J Agric Technol 7:63–72

Mondal D, Bhattacharyya PK, Das R (2017) Disease reaction of lentil genotypes against Stemphylium blight caused by *Stemphylium botryosum* Wallr. in west Bengal. J Agroecol Nat Res Manag 4:149–152

Mousavi-Derazmahalleh M, Bayer PE, Hane JK et al (2019) Adapting legume crops to climate change using genomic approaches. Plant Cell Environ 42:6–19. https://doi.org/10.1111/pce.13203

Muehlbauer FJ (1991) Use of introduced germplasm in cool season food legume cultivar development. In: Shands HL, Wiesner LE (eds) Use of plant introductions in cultivar development (Part 2). Crop Science Society America, Madison, WI, pp 49–73

Muehlbauer FJ, McPhee KE (2004) Registration of 'Pennell' lentil. Crop Sci 44:1488–1488

Muehlbauer F, Slinkard A (1982) Genetics and breeding methodology. In: Webb C, Hartwin GC (eds) Lentils. Commonwealth Agricultural Bureau, Farnham Royal, pp 69–90

Muehlbauer FJ, Cubero JL, Summerfield RJ (1985) Lentil (*Lens culinaris* Medic.). In: Summerfield RJ, Roberts EH (eds) Grain legume crops. Collins, London, pp 266–311

Muehlbauer F, Redden R, Nassib A et al (1988) Population improvement in pulse crops: an assessment of methods and techniques. In: Summerfield RJ (ed) World crops: cool season food legumes. Kluwer Academic Publishers, Dordrecht, pp 941–966

Muehlbauer FJ, Weeden NF, Hoffman DL (1989) Inheritance and linkage relationships of morphological and isozyme loci in lentil (Lens Miller). J Hered 80:298–303

Muehlbauer FJ, Cho S, Sarker A et al (2006) Application of biotechnology in breeding lentil for resistance to biotic and abiotic stress. Euphytica 147:149–165. https://doi.org/10.1007/s10681-006-7108-0

Muehlbauer FJ, Mihov M, Vandenberg A et al (2009) Improvement in developed countries. In: Yadav SS, McNeil DL, Stevenson PC (eds) Lentil: an ancient crop for modern times. Springer, Berlin, pp 137–154

Muench DG, Slinkard AE, Scoles GJ (1991) Determination of genetic variation and taxonomy in lentil (Lens Miller) species by chloroplast DNA polymorphism. Euphytica 56:213–218

Muscolo A, Sidari M, Anastasi U et al (2014) Effect of PEG-induced drought stress on seed germination of four lentil genotypes. J Plant Interact 9:354–363. https://doi.org/10.1080/17429145.2013.835880

Muscolo A, Junker A, Klukas C et al (2015) Phenotypic and metabolic responses to drought and salinity of four contrasting lentil accessions. J Exp Bot 66:5467–5480. https://doi.org/10.1093/jxb/erv208

Mwakutuya E, Banniza S (2010) Influence of temperature and wetness periods on the development of stemphylium blight on lentil. Plant Dis 94:1219–1224

Nadeem M, Li J, Yahya M et al (2019) Research progress and perspective on drought stress in legumes: a review. Int J Mol Sci 20(10):2541. https://doi.org/10.3390/ijms20102541

Naivikul O, D'Apollonia BL (1979) Carbohydrates of legume flours compared with wheat flour: II. Starch. Cereal Chem 56:24–28

Nene YL (2006) Indian pulses through the millennia. Asian Agric Hist 10(3):179–202

Nguyen TT, Taylor PWJ, Brouwer JB et al (2001) A novel source of resistance in lentil (*Lens culinaris* ssp. culinaris) to Ascochyta blight caused by A. lentis. Aust Plant Path 30:211–215

Ninou E, Mylonas I, Tsivelikas A et al (2014) Wheat landraces are better qualified as potential gene pools at ultraspaced rather than densely grown conditions. Sci World J 2014:957472. https://doi.org/10.1155/2014/957472

Ninou E, Papathanasiou F, Vlachostergios DN et al (2019) Intense breeding within lentil landraces for high-yielding pure lines sustained the seed quality characteristics. Agriculture 9:175. https://doi.org/10.3390/agriculture9080175

Nourollahi K, Madahjalali M (2017) Analysis of population genetic structure of Iranian Fusarium oxysporum f. sp. lentis isolates using microsatellite markers. Aust Plant Pathol 46:35–42

Nozzolillo C, De Bezada GM (1984) Browning of lentil seeds, concomitant loss of viability and the possible role of soluble tannins in both phenomena. Can J Plant Sci 64:815–824

Ocampo B, Conicella C, Moss JP (2000) Wide crossing: opportunities and progress. In: Knight R (ed) Linking research and marketing opportunities for pulses in the 21st Century, 3rd International Food Legumes Research Conference, Current plant science and biotechnology in agriculture, vol 34. Kluwer Academic Publishers, Dordrecht; Boston, MA; London, pp 411–419

Ogutcen E, Ramsay L, Bishop von Wettberg E et al (2018) Capturing variation in Lens (Fabaceae): development and utility of an exome capture array for lentil. Appl Plant Sci 6(7):e1165. https://doi.org/10.1002/aps3.1165

Öktem HA, Mahmoudian M, Eyidooan F et al (1999) GUS gene delivery and expression in lentil cotyledonary nodes using particle bombardment. Lens Newsl 26:3–6

Öktem HA, Eyidodan F, Demirba D et al (2008) Antioxidant responses of lentil to cold and drought stress. J Plant Biochem Biotechnol 17:15–21. https://doi.org/10.1007/BF03263254

Omran VG, Bagheri A, Moshtaghi N (2008) Direct in vitro regeneration of lentil (*Lens culinaris* Medik.). Pak J Biol Sci 11(18):2237–2242

Ouji A, El-bok S, Mouelhi M et al (2015) Effect of salinity stress on germination of five tunisian lentil (*Lens culinaris* L.) genotypes. Eur Sci J 11:63–75

Pan XY, Wang GX, Yang HM et al (2003) Effect of water deficits on within-plot variability in growth and grain yield of spring wheat in northwest China. Field Crop Res 80:195–205

Pandey A, Singh DP, Singh BB (1992) Evaluation of indigenous germplasm for yield and yield components in lentil (*Lens culinaris* Medik.). N. D. University of Agriculture & Technology, Faizabad, India. Res Bull 1:45

Pandurng JA, Lomte SS (2015) Digital image processing applications in agriculture: a survey. IJARCSSE 5(3):622–624

Papakosta – Tasopoulou D (2005) Legumes for fruits and hay. Sinhroni Paedia Press, Thessaloniki, Greece (ISBN: 978-960-357-067-2)

Pavan S, Bardaro N, Fanelli V et al (2019) Genotyping by sequencing of cultivated lentil (*Lens culinaris* Medik.) highlights population structure in the Mediterranean gene pool associated with geographic patterns and phenotypic variables. Front Genet 10:1–9. https://doi.org/10.3389/fgene.2019.00872

Petterson D, Sipsas S, Mackintosh JB (1997) The chemical composition and nutritive value of Australian pulses. Grains Research and Development Corporation, Kingston, ACT

Podder R, Banniza S, Vandenberg A (2013) Screening of wild and cultivated lentil germplasm for resistance to stemphylium blight. Plant Genet Res 11:26–35. https://doi.org/10.1017/S1479262112000329

Polanco MC, Ruiz ML (2001) Factors that affect plant regeneration from in vitro culture of immature seeds in four lentil (*Lens culinaris* Medik.) cultivars. Plant Cell Tissue Org Cult 66:133–139

Polanco MC, Pelaez MI, Ruiz ML (1988) Factors affecting callus and shoot formation from in vitro cultures of *Lens culinaris* Medik. Plant Cell Tissue Organ Cult 15:175–182

Poonam S (2006) Characterization of lentil germplasm through morphological and metric traits and molecular markers. Ph.D. Thesis, Bundelkhand University, Jhansi, India

Pouralibaba HR, Rubiales D, Fondevilla S (2016) Identification of pathotypes in Fusarium oxysporum f.sp. lentis. Eur J Plant Pathol 144:539–549

Rajput M, Sarwar G (1989) Genetic variability, correlation studies and their implications in selection of high yielding genotypes in lentil. Lens Newsl 16(2):5–8

Ram S, Ghosh AK, Nema AK et al (2018) Influence of long term tillage, organic and inorganic fertilization on primary nutrient and S in rice - lentil cropping sequence under dry land ecosystem. Int J Curr Microbiol App Sci 7:2511–2522. https://doi.org/10.20546/ijcmas.2018.704.287

Rameshwaran P, Qadir M, Ragab R et al (2016) Tolerance of faba bean, chickpea and lentil to salinity: accessions' salinity response functions. Irrig Drain 65:49–60. https://doi.org/10.1002/ird.1922

Rana A, Solanki IS (2015) Ethyl methane sulphonate induced genetic variability and heritability in macrosperma and microsperma lentils. J Environ Biol 36:1119–1123

Rana M, Sood A, Hussain W et al (2019) Gene pyramiding and multiple character breeding. In: Singh M (ed) Lentils: potential resources for enhancing genetic gains. Academic Press, London, pp 83–124

Rao PS, Chadha MS (1986) Protoplast culture of some economically important plants. In: Nuclear techniques and in vitro culture for plant improvement. Proceedings of a symposium, Vienna, pp 493–496

Rathi AS, Sindhu JS, Singh VS (2002) Variability, heritability and genetic advance in lentil. Legum Res 25(2):113–116

Rathod RP, Annapure US (2016) Effect of extrusion process on antinutritional factors and protein and starch digestibility of lentil splits. LWT Food Sci Technol 66:114. https://doi.org/10.1016/j.lwt.2015.10.028

Raza A, Razzaq A, Mehmood SS et al (2019) Impact of climate change on crops adaptation and strategies to tackle its outcome: a review. Plants 8(2):34. https://doi.org/10.3390/plants8020034

Razzak MA, Islam MA, Rahman MH et al (2018) Screening of lentil germplasm against stemphylium blight by observing disease reaction in three different stages. Malay J Halal Res 1(2):15–18

Reddy NR, Sathe SK, Pierson MD (1988) Removal of phytate from great northern beans (Phaseolus vulgaris L.) and its combined density fraction. J Food Sci 53:107

Reddy NR, Pierson MD, Sathe SK, Salunkhe DK (1984) Chemical, nutritional and physiological aspects of dry bean carbohydrates—A review. Food Chem 13(1):25–68

Rispail N, Dita MA, González-Verdejo C et al (2007) Plant resistance to parasitic plants: molecular approaches to an old foe: research review. New Phytol 173:703–712. https://doi.org/10.1111/j.1469-8137.2007.01980.x

Rizwan M, Aslam M, Asghar MJ et al (2017) Pre-breeding of lentil (*Lens culinaris* Medik.) for herbicide resistance through seed mutagenesis. PLoS One 12:1–15. https://doi.org/10.1371/journal.pone.0171846

Rodda MS, Davidson J, Javid M et al (2017) Molecular breeding for ascochyta blight resistance in lentil. Current progress and future directions. Front Plant Sci 8:1136. https://doi.org/10.3389/fpls.2017.01136

Rogers D, McGuire P (2015) Genetic erosion: context is key. In: Ahuja MR, Mohain Jain S (eds) Genetic diversity and erosion in plants. Indicators and prevention, vol 1. Springer, Cham, pp 1–24

Rubiales D, Mónica FA (2012) Innovations in parasitic weeds management in legume crops. A review. Agron Sustain Dev 32:433–449. https://doi.org/10.1007/s13593-011-0045-x

Rubiales D, Pérez-de-Luque A, Fernández-Aparico M et al (2006) Screening techniques and sources of resistance against parasitic weeds in grain legumes. Euphytica 147:187–199. https://doi.org/10.1007/s10681-006-7399-1

Rubiales D, Flores F, Emeran AA et al (2014) Identification and multi-environment validation of resistance against broomrapes (Orobanche crenata and Orobanche foetida) in faba bean (Vicia faba). Food Crop Res 166:58–65. https://doi.org/10.1016/j.fcr.2014.06.010

Rutter JB (2017) Aegean prehistoric archaeology. http://www.dartmouth.edu/~prehistory/aegean/?page_id=107

Saha GC, Sarker A, Chen WD et al (2010) Identification of markers associated with genes for rust resistance in *Lens culinaris* Medik. Euphytica 175:261–265

Salunkhe DK, Kadam SS (1989) Handbook of world food legumes, nutritional chemistry, processing technology and utilization. CRC Press, Boca Raton, FL

Sambasivam P, Taylor PWJ, Ford R (2017) Pathogenic variation and virulence related responses of Ascochyta lentis on lentil. Eur J Plant Pathol 147:265–277. https://doi.org/10.1007/s10658-016-0999-2

Sandhu JS, Singh S (2007) History and origin. In: Yadav S, McNeil DL, Stevenson PC (eds) Lentil: an ancient crop for modern times. Springer, Dordrecht, pp 1–10

Sari D, Mutlu N, Toker C (2016) Assessment of resistance gene analog, peroxidase, and WRKY gene polymorphisms in the genus lens miller. Turk J Bot 40(2):121–129. https://doi.org/10.3906/bot-1501-41

Sarker A, Erskine W (2006) Recent progress in the ancient lentil. J Agric Sci 144:19–29

Sarker RH, Mustafa BM, Biswas A et al (2003) In vitro regeneration in lentil (*Lens culinaris* Medik.). Plant Tissue Cult 13:155–163

Sarker A, Erskine W, Singh M (2005) Variation in root and shoot traits and their relationship in drought tolerance in lentil. Genet Resour Crop Evol 52:87–95

Saxena PK, King J (1987) Morphogenesis in lentil: plant regeneration from callus cultures of *Lens culinaris* Medik. via somatic embryogenesis. Plant Sci 52:223–227

Sedgley RH (1991) An appraisal of the Donald ideotype after 21 years. Field Crop Res 26:93–112

Sehgal A, Sita K, Kumar J et al (2017) Effects of drought, heat and their interaction on the growth, yield and photosynthetic function of lentil (*Lens culinaris* medikus) genotypes varying in heat and drought sensitivity. Front Plant Sci 8:1776. https://doi.org/10.3389/fpls.2017.01776

Shahin MA, Symons SJ (2001) A machine vision system for grading lentils. Can Biosyst Eng/Le Genie des biosystems au Canada 43:77–714

Shahwar D, Ansari M, Mohsin T et al (2017) Evaluation of high yielding mutant of lentil developed through caffeine of an exotic germplasm. Int J Plant Breed Genet 11:55–62. https://doi.org/10.3923/ijpbg.2017.55.62

Sharma SR, Singh S, Aggarwal N et al (2018) Genetic variation for tolerance to post-emergence herbicide, imazethapyr in lentil (*Lens culinaris* Medik.). Arch Agron Soil Sci 64:1818–1830. https://doi.org/10.1080/03650340.2018.1463519

Sharpe AG, Ramsay L, Sanderson LA et al (2013) Ancient orphan crop joins the modern era: gene-based SNP discovery and mapping in lentil. BMC Genomics 14:192. https://doi.org/10.1186/1471-2164-14-192

Shrestha R, Turner NC, Siddique KHM et al (2006) Physiological and seed yield responses to water deficits among lentil genotypes from diverse origins. Aust J Agric Res 57:903–915. https://doi.org/10.1071/AR05204

Shu QY, Forster BP, Nakagawa H (2012) Plant mutation breeding and biotechnology. CAB International, Wallingford

Sidari M, Santonoceto C, Anastasi U et al (2008) Variations in four genotypes of lentil under NaCl-salinity stress. Am J Agric Biol Sci 3:410–416. https://doi.org/10.3844/ajabssp.2008.410.416

Singh RK, Raghuvanshi SS (1989) Plantlet regeneration from nodal segment and soot tip-derived explants of lentil. Lens Newsl 16:33

Singh U, Singh B (1992a) Tropical grain legumes as important human foods. Econ Bot 46:310–321

Singh JP, Singh IS (1992b) Genetics of rust resistance in lentil (*Lens culinaris*). Indian J Agric Sci 62:337–338

Singh KM, Singh AK (2014) Lentil in India: an overview. SSRN. https://doi.org/10.2139/ssrn.2510906

Singh M, Sardana S, Sharma SK (2011) Genetic resources of lentil and its utilization in India. Plant Genet Res 9:30–37

Singh M, Rana MK, Kumar K et al (2013) Broadening the genetic base of lentil cultivars through inter-sub-specific and interspecific crosses of Lens taxa. Plant Breed 132:667–675

Singh G, Kaur H, Khanna V (2014a) Short communication Weed management in lentil with post-emergence herbicides. Ind J Weed Sci 46:187–189

Singh M, Bisht IS, Kumar S et al (2014b) Global wild annual Lens collection: a potential resource for lentil genetic base broadening and yield enhancement. PLoS One 9:e107781

Singh D, Singh CK, Taunk J, Tomar RSS (2016a) Genetic analysis and molecular mapping of seedling survival drought tolerance gene in lentil (*Lens culinaris* Medikus). Mol Breed 36(58). https://doi.org/10.1007/s11032-016-0474-y

Singh D, Singh CK, Tomar RSS et al (2016b) Exploring genetic diversity for heat tolerance among lentil (*Lens culinaris* Medik.) genotypes of variant habitats by simple sequence repeat markers. Plant Breed 135:215–223. https://doi.org/10.1111/pbr.12341

Singh D, Singh CK, Tomar RSS et al (2016c) Molecular assortment of lens species with different adaptations to drought conditions using SSR markers. PLoS One 11:e0147213. https://doi.org/10.1371/journal.pone.0147213
Singh D, Singh CK, Kumari S et al (2017a) Discerning morpho-anatomical, physiological and molecular multiformity in cultivated and wild genotypes of lentil with reconciliation to salinity stress. PLoS One 12(12):e0190462
Singh D, Singh CK, Tomar RSS, Pal M (2017b) Genetics and molecular mapping of heat tolerance for seedling survival and pod set in lentil. Crop Sci 57:3059–3067. https://doi.org/10.2135/cropsci2017.05.0284
Singh M, Rana JC, Singh B et al (2017c) Comparative agronomic performance and reaction to Fusarium wilt of *Lens culinaris* × L. orientalis and L. culinaris × L. ervoides derivatives. Front Plant Sci 8:1162. https://doi.org/10.3389/fpls.2017.01162
Singh D, Singh CK, Singh YP et al (2018a) Evaluation of cultivated and wild genotypes of Lens species under alkalinity stress and their molecular collocation using microsatellite markers. PLoS One 13:1–25. https://doi.org/10.1371/journal.pone.0199933
Singh M, Sharma SK, Singh B et al (2018b) Widening the genetic base of cultivated gene pool following introgression from wild Lens taxa. Plant Breed 137(4):470–485
Singh D, Singh CK, Taunk J et al (2019) Genome wide transcriptome analysis reveals vital role of heat responsive genes in regulatory mechanisms of lentil (*Lens culinaris* Medikus). Sci Rep 9:1–19. https://doi.org/10.1038/s41598-019-49496-0
Singh M, Kumar S, Basandrai AK et al (2020) Evaluation and identification of wild lentil accessions for enhancing genetic gains of cultivated varieties. PLoS One 15(3):e0229554. https://doi.org/10.1371/journal.pone.0229554
Sinha RP, Yadav BP (1989) Inheritance of resistance to rust in lentil. Lens Newsl 16:41
Sita K, Sehgal A, Hanumantharao B et al (2017a) Food legumes and rising temperatures: effects, adaptive functional mechanisms specific to reproductive growth stage and strategies to improve heat tolerance. Front Plant Sci 8:1–30. https://doi.org/10.3389/fpls.2017.01658
Sita K, Sehgal A, Kumar J et al (2017b) Identification of high-temperature tolerant lentil (*Lens culinaris* Medik.) genotypes through leaf and pollen traits. Front Plant Sci 8:1–27. https://doi.org/10.3389/fpls.2017.00744
Sita K, Sehgal A, Bhandari K et al (2018) Impact of heat stress during seed filling on seed quality and seed yield in lentil (*Lens culinaris* Medikus) genotypes. J Sci Food Agric 98:5134–5141. https://doi.org/10.1002/jsfa.9054
Skliros D, Kalloniati C, Karalias G et al (2018) Global metabolomics analysis reveals distinctive tolerance mechanisms in different plant organs of lentil (*Lens culinaris*) upon salinity stress. Plant Soil 429:451–468. https://doi.org/10.1007/s11104-018-3691-9
Slinkard A, Vandenberg A, Holm F (2015) Lentil plants having increased resistance to imidazolinone herbicides. US Patent 7,232,942 B2, 30 May 2007
Sneep J (1977) Selection for yield in early generations of self-fertilizing crops. Euphytica 26:27–30
Solanki IS (2001) Stability of seed yield and its component characters in lentil (*Lens culinaris*). Indian J Agric Sci 71(6):414–416
Solanki IS, Sharma B (2001a) N-nitroso-N-ethyl Urea induced genetic variability for quantitative characters in lentil (*Lens culinaris* Medik.). Natl J of Plant Improv 3:102–106
Solanki IS, Sharma B (2001b) Early generation selection of polygenic mutations in lentil. Ind J Genet Plant Breed 6:330–334
Solanki IS, Kapoor AC, Singh U (1999) Nutritional parameters and yield evaluation of newly developed genotypes of lentil (*Lens culinaris* Medik.). Plant Foods Hum Nutr 54(1):79–87. https://doi.org/10.1023/A:1008182302900
Sonnante G, Pignone D (2007) The major Italian landraces of lentil (*Lens culinaris* Medik.): their molecular diversity and possible origin. Genet Resour Crop Evol 54(5):1023–1031. https://doi.org/10.1007/s10722-006-9153-x
Sosulski FW, Garratt MO, Slinkard AE (1976) Functional properties of ten legume flours. Can Inst Food Sci Technol J 9:66–69

Springer NM, Schmitz RJ (2017) Exploiting induced and natural epigenetic variation for crop improvement. Nat Rev Genet 18:563–575. https://doi.org/10.1038/nrg.2017.45
Srivastava RP, Vasishtha H (2013) Dietary fiber, protein and lectin contents of lentils (*Lens culinaris*) on soaking and cooking. Curre Adv Agric Sci 5(2):238–241
Stadler LJ (1928a) Genetic effects of X-rays in maize. Proc Natl Acad Sci U S A 14:69–75
Stadler LJ (1928b) Mutations in barley induced by X-rays and radium. Science 68:186–187
Stefaniak TR, McPhee KE (2015) Lentil. In: De Ron A (ed) Grain legumes, Handbook of plant breeding, vol 10. Springer, New York, NY, pp 111–140
Stevenson PC, Dhillon MK, Sharma HC et al (2007) Insect pests of lentil and their management. In: Yadav S, McNeil DL, Stevenson PC (eds) Lentil: an ancient crop for modern times. Springer, Dordrecht, pp 331–348
Sudheesh S, Verma P, Forster JW et al (2016a) Generation and characterisation of a reference transcriptome for lentil (*Lens culinaris* Medik.). Int J Mol Sci 17:1887. https://doi.org/10.3390/ijms17111887
Sudheesh S, Rodda MS, Davidson J et al (2016b) SNP-based linkage mapping for validation of QTLs for resistance to Ascochyta blight in lentil. Front Plant Sci 7:1604. https://doi.org/10.3389/fpls.2016.01604
Sultana T, Ghafoor A (2008) Genetic diversity in *ex-situ* conserved *Lens culinaris* for botanical descriptors, biochemical and molecular markers and identification of landraces from indigenous genetic resources of Pakistan. J Integr Plant Biol 50(4):484–490. https://doi.org/10.1111/j.1744-7909.2007.00632.x
Sultani MI, Gill MA, Anwar MM et al (2007) Evaluation of soil physical properties as influenced by various green manuring legumes and phosphorus fertilization under rain fed conditions. Int J Environ Sci Technol 4:109–118. https://doi.org/10.1007/BF03325968
Suschetet M (1975) Influence of tannic acid on the hepatic content of vitamin A in rats fed a vitamin A-containing diet or vitamin A-deficient diet. C. R. Seances SOC. Biol Ses Fil 169:970–978
Tabekhia M, Luh BS (1980) Effect of germination cooking and canning on phosphorus and phytate retention in dry beans. J Food Sci 45:406–408
Tabti D, Laouar M, Rajendran K et al (2018) Identification of desirable mutants in quantitative traits of lentil at early (M) generation. J Environ Biol 39:137–142
Talib BS, Safarnejad A, Lahooti M et al (2003) Effect of genotype, explant and plant regulator in tissue culture of lentil (*Lens culinaris* M.). In: The Third National Conference of Biotechnology, Islamic Republic of Iran, pp 336–339
Talukdar D (2013) Comparative morpho-physiological and biochemical responses of lentil and grass pea genotypes under water stress. J Natl Sci Biol Med 4:396–402. https://doi.org/10.4103/0976-9668.116983
Tanaka A, Shikazono N, Hase Y (2010) Studies on biological effects of ion beams on lethality, molecular nature of mutation, mutation rate, and spectrum of mutation phenotype for mutation breeding in higher plants. J Radiat Res 233:223–233
Tar'an B, Buchwaldt L, Tullu A et al (2003) Using molecular markers to pyramid genes for resistance to Ascochyta blight and anthracnose in lentil (*Lens culinaris* Medik). Euphytica 134:223–230. https://doi.org/10.1023/B:EUPH.0000003913.39616.fd
Taylor P, Lindbeck K, Chen W et al (2007) Lentil diseases. In: Yadav S, McNeil DL, Stevenson PC (eds) Lentil: an ancient crop for modern times. Springer, Dordrecht, pp 291–313
Teja KC, Duary B, Dash S et al (2017) Post-emergence application of imazethapyr for weed management in lentil. SATSA Mukhapatra Annual Tech Issue 21:183–188
Thavarajah D, Ruszkowski J, Vandenberg A (2008) High potential for selenium biofortification of lentils (*Lens culinaris* L.). J Agric Food Chem 56(22):10747–10753
Thavarajah D, Thavarajah P, Sarker A et al (2009a) Lentils (*Lens culinaris* medikus subspecies culinaris): a whole food for increased iron and zinc intake. J Agric Food Chem 57(12):5413–5419
Thavarajah P, Thavarajah D, Vandenberg A (2009b) Low phytic acid lentils (*Lens culinaris* L.): a potential solution for increased micronutrient bioavailability. J Agric Food Chem 57(19):9044–9049

Thompson LU (1993) Potential health benefits and problems associated with antinutrients in foods. Food Res Int 26:131–149
Tiwari N, Ahmed S, Sarker A (2018) Fusarium wilt: a killer disease of lentil. In: Tulin A (ed) Fusarium—plant diseases, pathogen diversity, genetic diversity, resistance and molecular markers. IntechOpen, Rijeka, pp 119–138
Tokatlidis IS (2017) Crop adaptation to density to optimise grain yield: breeding implications. Euphytica 213:92. https://doi.org/10.1007/s10681-017-1874-8
Tokatlidis I, Vlachostergios D (2016) Sustainable stewardship of the landrace diversity. Diversity 8:29. https://doi.org/10.3390/d8040029
Tokatlidis IS, Papadopoulos II, Baxevanos D, Koutita O (2010) GxE effects on single-plant selection at low density for yield and stability in climbing dry bean. Crop Sci 50:775–783
Toker C, Yadav SS, Solanki IS (2007) Mutation breeding. In: Yadav S, McNeil DL, Stevenson PC (eds) Lentil: an ancient crop for modern times. Springer, Dordrecht, pp 209–224
Toklu F, Biçer B, Karaköy T (2009) Agro-morphological characterization of the Turkish lentil landraces. Afric J Biotechnol 8(17):4121–4127. https://doi.org/10.4314/AJB.V8I17.62138
Tomlekova NB (2010) Induced mutagenesis for crop improvement in Bulgaria. Plant Mutat Rep 2(2):4–27
Torricelli R, Silveri DD, Ferradini N et al (2012) Characterization of the lentil landrace Santo Stefano di Sessanio from Abruzzo, Italy. Genet Resour Crop Evol 59:261–276
Tripathi HS (2015) Advances in diseases of lentil. In: Awasthi LP (ed) Recent advances in the diagnosis and management of plant diseases. Springer, New Delhi, pp 55–68
Tsanakas GF, Mylona PV, Koura K et al (2018) Genetic diversity analysis of the Greek lentil (*Lens culinaris*) landrace "Eglouvis" using morphological and molecular markers. Plant Genet Res 16(5):469–477. https://doi.org/10.1017/S1479262118000096
Tullu A, Buchwaldt L, Warkentin T et al (2003) Genetics of resistance to anthracnose and identification of AFLP and RAPD markers linked to the resistance gene in PI 320937 germplasm of lentil (*Lens culinaris* Medikus). Theor Appl Genet 106:428–434
Tullu A, Buchwaldt L, Lulsdorf M et al (2006a) Sources of resistance to anthracnose Colletotrichum truncatum in wild Lens species. Genet Resour Crop Evol 53:111–119
Tullu A, Tar'an B, Breitkreutz C et al (2006b) A quantitative-trait locus for resistance to ascochyta blight (Ascochyta lentis) maps close to a gene for resistance to anthracnose (Colletotrichum truncatum) in lentil. Can J Plant Pathol 28:588–595
Tullu A, Banniza S, Bett K et al (2011a) A walk on the wild side: exploiting wild species for improving cultivated lentil. Grain Legum 56:13–14
Tullu A, Diederichsen A, Suvorova G et al (2011b) Genetic and genomic resources of lentil: status, use and prospects. Plant Genet Resour Charact Util 9:19–29. https://doi.org/10.1017/S1479262110000353
Tullu A, Bett K, Banniza S et al (2013) Widening the genetic base of cultivated lentil through hybridization of *Lens culinaris* "Eston" and L. ervoides accession IG 72815. Can J Plant Sci 93:1037–1047
Vaillancourt RE, Slinkard AE (1992) Inheritance of new genetic markers in lentil (Lens Miller). Euphytica 64:227–236
Van Hoorn JW, Katerji N, Hamdy A et al (2001) Effect of salinity on yield and nitrogen uptake of four grain legumes and on biological nitrogen contribution from the soil. Agric Water Manag 51:87–98. https://doi.org/10.1016/S0378-3774(01)00114-7
Van Oss H, Aron Y, Ladizinsky G (1997) Chloroplast DNA variation and evolution in the genus Lens Mill. Theor Appl Genet 94:452–457
Vandenberg A, Slinkard AE (1989) Inheritance of four new quantitative genes in lentil. J Hered 80:320–322
Vandenberg A, Kiehn FA, Vera C et al (2002a) CDC vantage lentil. Can J Plant Sci 82:109–110
Vandenberg A, Kiehn FA, Vera C et al (2002b) CDC robin lentil. Can J Plant Sci 82:111–112
Vandenberg A, Vera C, Buchwaldt L et al (2005) CDC plato lentil. Can J Plant Sci 85:161–162

Venora G, Grillo O, Shahin MA et al (2007) Identification of Sicilian landraces and Canadian cultivars of lentil using an image analysis system. Food Res Int 40(1):161–166. https://doi.org/10.1016/j.foodres.2006.09.001
Verma P, Shah N, Bhatia S (2013) Development of an expressed gene catalogue and molecular markers from the de novo assembly of short sequence reads of the lentil (*Lens culinaris* Medik.) transcriptome. Plant Biotechnol J 11:894–905. https://doi.org/10.1111/pbi.12082
Vibhute A, Bodhe SK (2012) Applications of image processing in agriculture: a survey. Int J Comput Appl 52(2):34–39. https://doi.org/10.5120/8176-1495
Vlachostergios DN, Roupakias DG (2017) Screening under low plant density reinforces the identification of lentil plants with resistance to fusarium wilt. Crop Sci 57:1285–1294. https://doi.org/10.2135/cropsci2015.10.0626
Vlachostergios DN, Lithourgidis AS, Roupakias DG (2011) Effectiveness of single plant selection at low density under organic environment. A field study with lentil. Crop Sci 51:41–51. https://doi.org/10.2135/cropsci2010.03.0137
Vlachostergios DN, Tzantarmas C, Kargiotidou A et al (2018a) Single-plant selection within lentil landraces at ultra-low density: a short-time tool to breed high yielding and stable varieties across divergent environments. Euphytica 214:58. https://doi.org/10.1007/s10681-018-2139-x
Vlachostergios DN, Lithourgidis AS, Baxevanos D et al (2018b) Evaluation of lentil varieties and farming system effect on seed damage and yield loss due to bruchid infestation. Crop Past Sci 69:387–394. https://doi.org/10.1071/CP17309
Wang N, Daun JK (2003) Quality of Western Canadian pulse crops. http://grainscanada.gc.ca
Wang N, Daun JK (2006) Effects of variety and crude protein content on nutrients and antinutrients in lentils (*Lens culinaris*). Food Chem 95:493–502. https://doi.org/10.1016/j.foodchem.2005.02.001
Warkentin TD, McHughen A (1993) Regeneration from lentil cotyledonary nodes and potential of this explant for transformation by Agrobacterium tumefaciens. Lens Newsl 20:26–28
Weiss E, Zohary D (2011) The neolithic southwest Asian founder crops their biology and archaeobotany. Curr Anthropol 52(4):237–254. https://doi.org/10.1086/658367
Westphal E (1974) Pulses in Ethiopia, their taxonomy and agricultural significance. Agricultural research reports 815. Centre for Agricultural Publishing and Documentation, Wageningen, p 263
Williams DJ, McHughen A (1986) Plant regeneration of the legume *Lens culinaris* Medik (lentil) in vitro. Plant Cell Tiss Org 7:149–153
Williams JT, Sanchez AMC, Jackson MT (1974) Studies on lentils and their variation. I. The taxonomy of the species. SABRAO J 6:133–145
Williams PC, Bhatty RS, Deshpande SS, Hussein LA, Savage GP (1994) Improving nutritional quality of cool season food legumes. In: Expanding the production and use of cool season food legumes. Muchbauer FJ, Kaizer WJ (Eds) Kluwer Academic Publishers, Dordrecht, The Netherlands, 113–129
Wiraguna E, Malik AI, Erskine W (2017) Waterlogging tolerance in lentil (*Lens culinaris* Medik. subsp. culinaris) germplasm associated with geographic origin. Genet Resour Crop Evol 64:579–586. https://doi.org/10.1007/s10722-016-0385-0
Witcombe JR, Joshi A, Joshi KD et al (1996) Farmer participatory crop improvement. I. Varietal selection and breeding methods and their impact on biodiversity. Exp Agric 32:445–460. https://doi.org/10.1017/S0014479700001526
Wong MML, Gujaria-Verma N, Ramsay L et al (2015) Classification and characterization of species within the genus lens using genotyping-by-sequencing (GBS). PLoS One 10:1–16. https://doi.org/10.1371/journal.pone.0122025
Yadav NK, Ghimire SK, Shakya SM et al (2016) Genetic diversity analysis of lentil (*Lens culinaris* L.) germplasm using DNA based SSR markers. Am J Food Sci Health 2(3):18–24
Yadav NK, Ghimire SK, Shrestha SM et al (2017) Source of resistant against fusarium wilt and stemphylium blight in lentil (*Lens culinaris* Medikus). Int J Appl Sci Biotechnol 5:102–107
Yau SK, Erskine W (2000) Diversity of boron-toxicity tolerance in lentil growth and yield. Genet Resour Crop Evol 47:55–62. https://doi.org/10.1023/A:1008733106108

Ye G, McNeil DL, Hill GD (2002) Breeding for resistance to lentil Ascochyta blight. Plant Breed 121:185–191. https://doi.org/10.1046/j.1439-0523.2002.00705.x

Ye G, McNeil DL, Hill GD (2003) Inheritance of foliar resistance to Ascochyta blight in lentil (*Lens culinaris*). New Zeal J Crop Hort 31:77–83. https://doi.org/10.1080/01140671.2003.9514239

Yeong J, Kim J-B (2019) Frequency and spectrum of radiation-induced mutations revealed by whole-genome sequencing analyses of plants. Quant Beam Sci 3(7):1–13. https://doi.org/10.3390/qubs3020007

Yilmaz Temel H, Gol D, Betul Hilal K et al (2015) Single nucleotide polymorphism discovery through illumina-based transcriptome sequencing and mapping in lentil. Turk J Agric For 39(3):470–488. https://doi.org/10.3906/tar-1409-70

Zhang H, Pala M, Oweis T et al (2000) Water use and water-use efficiency of chickpea and lentil in a Mediterranean environment. Aust J Agric Res 51:295–304. https://doi.org/10.17700/jai.2015.6.1

# Part V
# Fruits and Nuts

# Date Palm Genetic Resources for Breeding

**Mehfuz Hasan, Hasan M. Abdullah, Abu Sayeed Md. Hasibuzzaman, and Mir Aszad Ali**

**Abstract** The date palm tree offers a wide range of benefits. It contributes significantly to environmental balance—it thrives in severe climatic conditions. It is a known source of several useful medical and nutritional chemical compounds. In recent times, environmental and non-environmental factors like unregulated commercial exploitations, extensive genetic and soil erosion, and pests and disease infestations threaten the existence of the date palm cultivars. Traditional breeding of date palm takes a long time due to its extended life cycle and difficulty in identifying the sex as the sex becomes prominent a few years after planting. However, with the advancement of tissue culture and other biotechnology and molecular breeding, date palm researches have been accelerated. Both in vitro and in situ conversation options have been explored as a way to conserve the existing cultivars. Traditional propagation via offshoots is simply not enough to meet the present-day date palm tree demands. Conversely, micropropagation has shown promising signs as a better large-scale propagation method for cultivating date palm trees. In vitro regeneration involving different methods has been explored. This came with noticeable shortcomings like browning, latent contamination, extended in vitro cycle, ex vitro acclimatization and transplanting, and somaclonal variation. Further research into these issues has resulted in improved micropropagation systems of various date palm cultivars. The genetic conformity is retained in regenerated plantlets via molecular characterization—an essential requirement for commercial production. More work is required to perfect the micropropagation of exclusive date palm varieties. This will facilitate the commercial production of date palm utilizing tissue culture laboratories worldwide.

**Keywords** Allelic richness · Characterization · Conservation · Genetic diversity · Geographical Information System

M. Hasan (✉) · A. S. M. Hasibuzzaman
Department of Genetics and Plant Breeding, Bangabandhu Sheikh Mujibur Rahman Agricultural University, Gazipur, Bangladesh

H. M. Abdullah
Department of Agroforestry and Environment, Bangabandhu Sheikh Mujibur Rahman Agricultural University, Gazipur, Bangladesh

M. A. Ali
International Potato Center, Dhaka, Bangladesh



P. M. Priyadarshan, S. M. Jain (eds.), *Cash Crops*,
https://doi.org/10.1007/978-3-030-74926-2_12

# 1 Introduction

Date palm (tree of life) is a prominent member of the *Arecaceae* family, with a pronounced prominence in North and Middle East Africa. The last century has seen date palm agriculture penetrate the American, Sothern African, and Australian soils (Tengberg 2014). The uniqueness of the distribution is attributed to the unusual hot climate in these areas, which is needed for fruit setting and adequate pollination of the date palm. The cultivation record of date palm for the last thousands of years revealed the emergence of multiple date palm varieties with high-quality nutritional value via selection procedures, in different parts of the world. Date fruits are known for their vital medicinal and nutritional contents. They also play a prevalent role in a religious setting and are a common edible fruit for everyone. The nutritional composition of dates varies based on the maturation and dehydration stage. For instance, the carbohydrate component ranges from 44% to 88%, protein 2.3–5.6%, fat 0.2–9.3%, and essential minerals and salts, vitamins, and dietary fiber constituting 6.4–11.5% of the entire nutritional value. The seed encompasses 7.7–9.7% oil, while the flesh involves 0.2–0.5% oil. The seed accounts for 5.6–14.2% of the total weight of the fruit (Ali et al. 2012; Farag 2016).

Carbohydrates are the main component of dates. Dates contain maltose, mannose, glucose, fructose, and other sugars. Sucrose and other non-reducing sugars are also present, accounting for >80% dry matter. Vitamins like folic acid, ascorbic acid (C), niacin, riboflavin (B), thiamine (B), and β-carotene are all present in various proportions. While carotenoids are also present, the content depends on the cultivar and stage of ripeness—stored dates or dates in their final ripening stages have less carotenoid contents. Up to 15 essential minerals, all in considerable amounts, are seen in dates. These are selenium, fluorine, iron, copper, magnesium, zinc, sodium, potassium, and phosphorus. These minerals each account for 0.1–100 mg per 100 g dry matter of date palm (Amadou 2016; Farag 2016).

Eating dates comes with different medicinal effects. For instance, the tannins in the fruit make it an excellent astringent remedy for intestine-related issues. Date fruit formulations, including pastes, syrups, decoctions, and infusions, have helped relieve symptoms of colds, cough, and sore throat. Dates contain anti-carcinogenic agents (Al Alawi et al. 2020). It contains phenolic compounds—known sources of anti-fungal, -bacterial, and -viral characteristics. This is why they are highly regarded in preventing chronic inflammations and treating some diseases. Rich compositions of antioxidants, selenium, and dietary fibers are present in date fruit and its by-products. The scientifically backed anti-carcinogenic characteristics of phenolics are replicated in dates. Studies suggest that phenolics influence the development of malignant tumors at different levels. Phenolics like ferulic and caffeic acids—present in high concentrations in dates—prevent the proliferation of skin tumors.

Finally, a balanced mix of date-based and protein-rich food items, like fish and milk products, results in a highly beneficial diet. Natural fibers are abundant in dates. It is an excellent replacement for refined sugar. The anti-mutagenic and antioxidant properties are great for relieving heart disease symptoms. But, the recent overexploitation of date palms has harmed its diversity. The present environmental situations continuously alter the production and application patterns of date fruits across various regions of the earth. Date palm production faces a wide array of challenges, including drought, salinity, disease/pest infestation, and inadequate harvest/postharvest practices. Vegetative offshoot is an excellent method of propagating date palm. However, the heterozygous nature of date palm means the seeds are not fit for propagating commercially elite cultivars—they result in off-type propagules. Only 20–30 offshoots are obtained from a date palm tree throughout its existence. This translates to a very low on-field persistence rate and a high risk of biotic factors transmittal. These are some of the reasons the increasing industrial demand for date palm propagules is difficult to meet.

## 2 Date Palm Husbandry

A date garden offers up to an average economic life of 50 years. However, 150 years is possible in some cases, as evident in a few date palms existing for several hundred years (Chao and Krueger 2007). The date palm's biology means it requires a couple of unusual features other perennial crops can do without. For instance, a few cultural practices involve the crown of the tree. This becomes a problem in very tall old trees where the crown is practically unreachable. Pollination, pruning, covering, harvesting, and bunch-tie down are all important cultural practices occurring on the tree crown. There are two options—climbing the tree up to the crown or using mechanical lifts. The latter is found in industrialized or advanced date palm-production regions and the former in a few underdeveloped date-producing countries.

The natural pollination process in date relies on wind and sometimes insects. Artificial pollination is ideal for commercial plantation, where the male date palm trees supply the required pollen. This option dates back several centuries in the Middle East and North Africa. Successful pollination of up to 80% of the female flowers provides a proper fruit set (Awad 2011). The exact fruit set percentage varies across cultivars. There are unclear cases of partial or zero compatibility between various male and female cultivars. The fertilization is such that one fertilized ovule per flower results in one fruit (Fig. 1).

**Fig. 1** Various types of date palm fruits. (Hegazy 2020a)

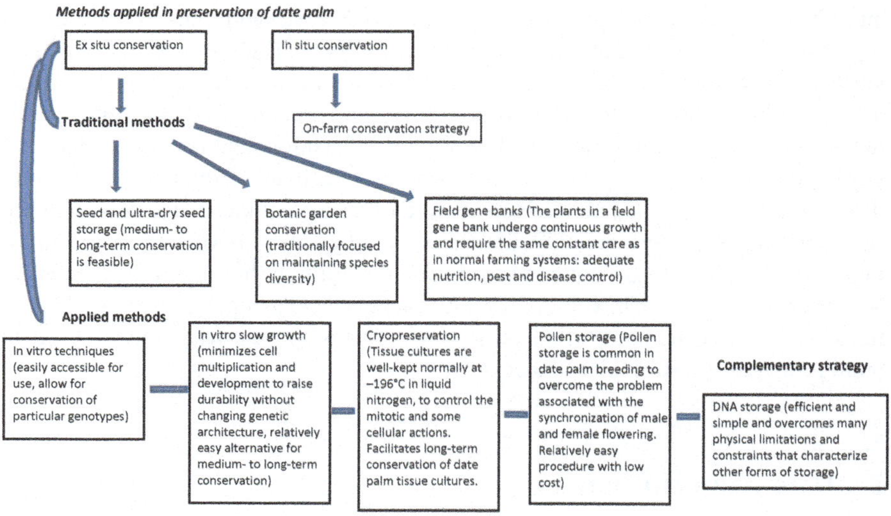


**Fig. 2** Preservation methods applied in different crops

## 3 Conservation of Genetic Diversity

***Methods of Conservation*** Two different conservation methods are known for the preservation of crop genetic resources: in situ conservation and ex situ conservation. Figure 2 shows the details of the conservation methods.

### *In Situ Conservation*

When farmers cultivate and manage a wide range of populations in the agroecosystems where a crop originates, it is called an on-farm in situ conservation. While it is possible to sustain the breeding efforts in the long run, it depends on the long-term availability of the genetic variation, which can be nurtured in farmer's fields. On-farm conservation systems are built on different objectives, including (1) conservation of the evolution process and adaptation of crops to their environments, (2) conservation diversity at multiple levels including ecosystem, species, and within species, (3) regulate the influence and access of the farmer to crop genetic resources, (4) making farmers an integral part of the national plant genetic resources conservation system, and (5) making lives easier for farmers who are low on resources. It is impossible to properly maintain the inherent constituents of date palm without proper conservation of its varieties and cultivars in their natural surroundings and habitats. Date palms thrive best in areas with extended and very hot summers, where the humidity and rainfall are low. This explains their Middle East origin—the arid subtropical deserts of the Middle East are known for their oases and river alleys (Krueger 2011).

The diversity of date palm cultivars is extensive. Each growing country has different cultivars, resulting from a constant selection designed to achieve improved quality and crop yield. This strategy has recorded up to 5000 various cultivars worldwide, with notable differences in ripening time, size, shape, flavor, and color (Jaradat and Zaid 2004). There is currently a disturbing reduction in the range of date palm groves of the oasis due to social and economic changes. These, alongside soil salinization and desertification, put the natural date palm environment under constant threats. The Arabian Gulf has recorded many saltwater invasions that led to the complete loss of date palm groves. These developments may lead to irreversible loss of genetic diversity, resulting from loss of local varieties with unique genetic compositions.

The maintenance of date palm genetic diversity is largely dependent on traditional oases. Fixing the negative characteristics and maintaining their positive aspects are necessary to support traditional oases. Better agricultural practices, disease and salinity, and provision of adequate irrigation systems are viable ways of optimizing the natural habitats of date palms for maximum cultivation. It is also important to keep track of the amount and distribution of genetic diversity on-farm. This involves the collection of metrics like the pattern of farmer-named varieties and their pattern of occurrence and populations' genetic structure.

## *Ex Situ Conservation*

When farmers attempt to conserve their plant materials exterior their usual habitation, it is called ex situ conservation. Common methods of ex situ conservation include botanical gardens, cryopreservation or in vitro preservation, collection farms, gene banks, and seed and DNA storage (Abul-Soad et al. 2017). All of these preservation methods come with advantages and disadvantages. The complementary approach may be an alternative to meeting the requirements of conservation strategies (Bekheet and Taha 2013).

### Date Palm Field Gene Bank

One way to move genotypes from their natural environment to areas where they are not adapted to is through field gene bank conservation. The collection of samples on the field is by both random and non-random sampling methods. Non-random sampling helps isolate samples with distinct morphological characters while excluding those with disease resistance and other physiological features. A collection simply means collecting samples from populations in natural habitats and fields. These samples are meant to be conserved, with the collection information well-documented.

The collection data comprises information like date and title of the expedition, plant identification information, collector details, collection date, site, and material status.

Field gene bank management comes with a couple of challenges, notably environmental disasters and diseases. For instance, pest's infestation is possible at every level of crop growth. A combination of proper phytosanitary practices and prevention procedures can help to prevent and control the infestation. Environmental disasters and hazards like haze, thunderstorms, strong winds, drought, and flooding can disrupt the established plants in the gene banks. The date palm field gene bank's accessions are best created with several listed items, i.e., two or three plants of each type are recommended for a start.

Preservation of wild date palm in field gene banks is often hampered by their different reproductive systems (Jaradat 2011). Hence, it requires special care in field gene bank. To preserve both plant and wild date palms, it is important to have information about the structure and distribution of genetic variation in date palms (Jain 2011; Chaluvadi et al. 2019).

Most of the date palm-producing countries like Saudi Arabia (King Faisal University), Tunisia (Centre de Recherches Phoénicicoles, INRA, Degache), Kuwait (Kuwait Institute for Scientific Research, KISR), and Iraq (southern Iraq) have field gene bank for date palm conservation. In addition to these major date palm-producing countries, the USA (USDA-ARS National Germplasm Repository for Citrus and Dates, NCGRCD), Nigeria (Dutse, Jigawa state), Pakistan (Shah Abdul Latif University), and India (Central Institute for Arid Horticulture, CIAH) have also established date palm field gene banks (Ataga et al. 2012; Abul-Soad et al. 2017). These gene banks have a wide variety of local and foreign germplasm which will help the farmers to select the varieties suitable for the local environment.

## In Vitro Conservation of Date Palm

Over time, various advancements in plant biotechnology have complemented the traditional methods employed in conserving plant genetic resources. These advancements include in vitro preservation methods, designed to facilitate the preservation of plant tissue cultures. The targeted storage duration determines the suitable in vitro preservation method. The in vitro storage of germplasm is of two types. The first type is the slow growth storage, where the culture medium or temperature requirements are tweaked. The second type is called cryopreservation, storing germplasm at an extremely low-temperature (−196 °C) in liquid nitrogen. The explants in storage cultures can come from any part of the plant. However, parts like embryos, axillary buds, and apical meristems have shown impressive outputs. There are two aspects of in vitro gene bank—the in vitro base active bank and the in vitro base gene bank. In in vitro active gene bank, the accessions are made available for immediate distribution and multiplication. The in vitro base gene bank is suited for long-term preservation, with the germplasm hardly available to users.

The traditional germplasm conservation methods will not work on date palm for two reasons—it is commercially propagated via vegetative offshoots, and it is a heterozygous and dioecious fruit tree. Tissue culture techniques better handle the acquisition, reproduction, and stowage of the date palm materials (El-Bahr et al. 2016; Bekheet 2017). In this case, the germplasm materials—different types of explants—are all preserved in vitro. This helps to reduce the need for space. Likewise, the explants are disease-free. Hence, the exchange of germplasm (both regionally and internationally) requires only simple quarantine measures. All of these point to the fact that it is possible to effectively scale-up the production systems of date palm micropropagation to achieve seamless germplasm exchange and in vitro preservation.

#### Preservation by Slow Growth

The slow growth preservation method involves a significant reduction in cell division and growth of active germplasm. This, in turn, leads to increased longevity and zero genetic modifications. The clonal plant material in slow growth procedures may stay up to 5 years under tissue culture conditions. Regular sub-culturing procedures may occur during this period, based on the species being conserved. Slow growth utilizes a few tissue culture techniques for germplasm preservation. One of these is storage under low temperature, which results in delayed cell elongation and division and cell membrane thickening caused by the continuous deposit of unsaturated lipids on the cell membrane. With slow growth, date palm can be preserved for short- and mid-term (Bekheet et al. 2001). They demonstrated a case study, where different date palm explants of Egypt origin were subjected to preserve at 5 °C in dark conditions. Bekheet et al. (2005) also noted the artificial seed preservation method of date palm germplasm. Shoot-tip explants sheathed in sodium alginate casings for a year produced somatic embryos via an in vitro approach. Fki et al. (2011) also came up with practical ways to produce synthetic seeds of date palm from somatic embryogenesis.

#### Cryopreservation

Cryopreservation is the stowage of plant materials in liquefied nitrogen at a temperature of −196 °C. These conditions slow down or stop the biological decomposition and metabolic processes of the preserved material, leaving it genetically stable. Not all specimens can survive these storage conditions, with the deciding factors ranging from genotype and physiological conditions to pre- and post-freezing tweaks. Conversely, as advantageous as cryopreservation is, the preserved specimen's membrane structure is at risk of thawing and freezing injuries. This may, in turn, affect the function and ultimately reduce the survival percentages of the germplasm.

Bekheet et al. (2006) designed a practical cryopreservation method for date palm. They cryopreserved undifferentiated nodular tissue cultures using freezing means. This led to the regeneration of the plantlets. The preculture medium of the cryopreservation process had various types of sugars as osmotic agents, with sucrose emerging as the best option. Sucrose-containing date palm tissue cultures showed the highest survival rate. Alansi et al. (2019), conversely, successfully cryopreserved date palm embryogenic callus using dehydrated-encapsulation, vitrification, and vitrification-encapsulation methods. Al-Bahrany and Al-Khayri (2012a, b) improved on the cryopreservation procedures of cell suspension of date palm cv. Cells cryoprotected in 10% DMSO, and 0.75 M sucrose recorded the highest embryos, utmost callus development, and maximum colony formation. In this case, the cryopreservation entailed procedures like preculturing in media with vitrification compounds (or osmotically), treating with cold shock and stowing at −196 °C, defrosting and post-defrosting measures, and growth repossession.

## Date Palm DNA Banking

The creation of PCR means it is now possible to amplify certain oligonucleotides or genes from the complete set of genomic DNA. Subsequently, we now have an international network of DNA repositories where genomic DNA is stored. DNA storage is more affordable and accessible, especially for research and future references. This preservation method may be the go-to preservation option for potentially extinct wild species (Hasan et al. 2020) and species that struggle with vegetative and other conventional preservation methods. The genetic diversity of these suitable species is preserved, as well as their population. However, the preservation is short-term, at least, until there are effective, long-term methods. The only downside of DNA preservation is its inability to regenerate entire plants.

DNA sequencing technology and reduction have continued to evolve. Likewise, sequencing reagents are now cheaper and more available than ever. Both the developments have contributed significantly to the advancement in genome analysis of date palm. Researchers have engaged in highly rated DNA sequencing programs centered on identifying the genes expressed during the crucial development stages of the date palm. The findings from one of these studies showed that there are 18 chromosome pairs in the date palm genome. The estimated genome size of the date palm, obtained from flow cytometry, was put at 1.2 billion base pairs (Zhang et al. 2012). Based on these premises, Yang et al. (2010) documented the entire sequence of the date palm chloroplast genome from an elite cultivar Khalas. This documentation was based on pyrosequencing. Tested and proven isolation and sequencing methods, handling and preservation methods, and the collection materials involved in the date palm DNA have all found vital applications in DNA banking. Subsequently, it is considered as a reliable conservation method for this crop.

### GIS in Date Palm Conservation

Date palm thrives in harsh conditions where traditional cropping fails. The extensive climatic-physical adaptability is one of the major reasons date palm is cultivated in arid regions of Australia, America, Africa, and Asia. While the value and relevance of date palm are high, there is little research and scientific studies. In comparison, other members of the family, such as oil palm, rank high in this aspect. The problem of inadequate research on date palm extends to the geospatial analysis, which is scarce and basic when available.

Geospatial data has continued to find more agricultural applications, especially with the need for spatial and temporal dimensions in the problem-solving process. Agriculture, now more than ever, relies on data. Almost all farming decisions, ranging from sowing to harvest, rely on data to answer the vital questions of how, when, and where.

Geospatial data is becoming popular in agricultural applications due to its spatial and temporal dimension of understanding a problem and for delivering solutions. In recent times, agriculture is data-driven. Farming is data intensive. Farming decisions from sowing to harvest widely depend on the conditions, which need to answer when, where, and how (Abdullah and Rahman 2015). In changing climate and limited resource conditions, near-real-time/real-time data are precious. Advances in sensors and the increasing power of computing and algorithms made those geospatial data usable for farming. Geospatial data can be obtained from sensors mounted in satellite, aircraft, drone, or even ground platforms (Abdullah et al. 2011).

Further, those data can be generated from extensive field observation in combination with a global positioning system (GPS) data and sensor-based imagery. According to need and purpose, geospatial data could be vector and raster formats. Different sensors (multispectral, hyperspectral, thermal, LiDAR) have their opportunities and challenges in agricultural applications. In the case of the date palm, geospatial data has definite scopes for conservation, breeding, facilitating agronomic practices, stress monitoring, and even crop yield monitoring and forecasting.

Date palm is a commercial crop and planted in a large area suitable for plantation. Utilizing the strength of GIS, and considering edaphic and agroclimatic factors, land suitability map for date palm can be determined (El-Hassanin et al. 2020). Counting date palm trees in a large plantation is often critical and required for the different operations which can be done using remote sensing (Mansour and Chockalingam 2020). Geospatial data can quantify the efficacy and need assessment of agro-inputs (irrigation, fertilizer, herbicides, pesticides) in an orchard. Further, pest infestation (Al-Kindi et al. 2017) is a challenge for successful crop harvest, so monitoring the health, biotic and abiotic (Abdullah 2014; Hasan and Abdullah 2015; Al-Rahbi et al. 2019) stress can be quantified using remote sensing tools. For successful date palm breeding, like many other crops, especially tree crop, requires wider genetic-based which can be ensured by proper genetic resource conservation. Geospatial tools are playing a critical role in determining which genetic resource to conserve and where to conserve based on allelic richness mapping (Huda et al. 2019). Further, tree crop breeding requires a long time to have the desired variety, so it demands a large amount of investment. Incorporation of remote sensing-based field phenotyping could play an essential role in this regard.

### Allelic Richness via GIS–Molecular Combined Approach: Wild Date Palm in South-Central Coastal Bangladesh as a Case Study

The molecular components of various populations can be interpolated using Geographical Information System (GIS). GIS also comes handy in setting up a grid-centered clustering for allelic discernibility (Jahan et al. 2020). The easiest way to better understand how plant genetic resources are distributed is by demonstrating GIS-centered outputs on maps. In this study, the possible integration of SSR marker data into the GIS was explored. When properly incorporated into the GIS system, the molecular data provide an unobstructed view and a clearer understanding of the spatial pattern of the allelic richness of PGR in an area of interest. The local plant diversity is also better conserved by leveraging this combined GIS–molecular analysis.

In terms of materials, this study relied on leaf samples of 38 date palm genotypes. These materials were sourced from the south-central region of Bangladesh. The collection was done at 5-km interludes and in a consistent lattice. In terms of genetic diversity, the differences in DNA base pair composition was observed at each locus and documented as absent or present. Integrating molecular data into the GIS system via the DIVA-GIS helped to estimate the allelic diversity accurately.

The geographic frequencies and distributions revealed different types of alleles. In terms of conservation, the highest value was attributed to the locally common alleles. These alleles were repeatedly found within a small area—a reliable metric for identifying zones of unique and high diversity.

The allelic richness of wild date palm in Bangladesh's south-central coast is depicted in Fig. 3. The sampling errors in the data were removed using the rarefaction technique. From the results, Bangladesh's south-central coastal areas of Dumki, Doshmina, Kalapara, and Kukata all showed the highest allelic richness.

As seen in the map, there is also the estimation and visualization of rare alleles to confirm the conservation areas (Fig. 4). Only the Doshmina area of south-central coastal Bangladesh had all the rare alleles.

The genetic similarity of the cell composition is the basis for the clustering of the subunits. There are eight various colors, each representing a cluster of sampled date palm from Bangladesh's south-central coastal area (Fig. 5). The same colors denote similar genotypes.

The GIS–Molecular combined analysis, in this case, resulted in the visualization of the allelic richness of the wild date palm germplasm collected in south-central coastal regions of Bangladesh. Breeders and conservationists will be able to leverage this combined analysis to easily obtain the visual representations of the allelic richness of species at a specific location. These findings come handy in planning new and improving on existing conservation and breeding methodologies that fix the problem of genetic abrasion of crops.

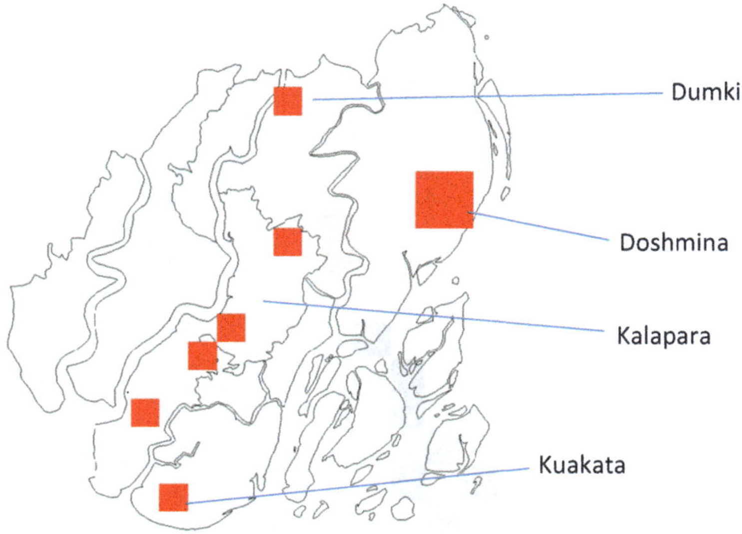


**Fig. 3** Visualization of the allelic richness of wild date palm sourced from Bangladesh's south-central coastal area

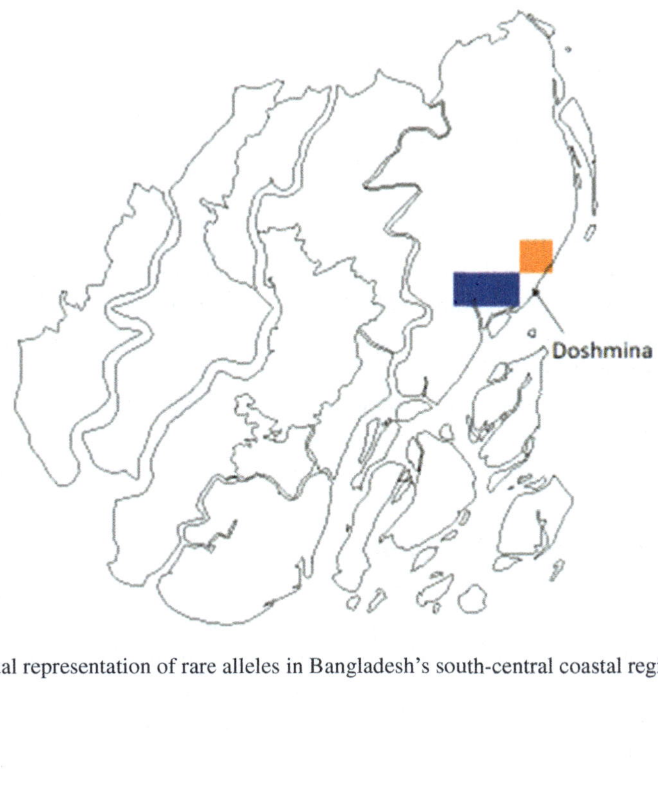


**Fig. 4** Visual representation of rare alleles in Bangladesh's south-central coastal region

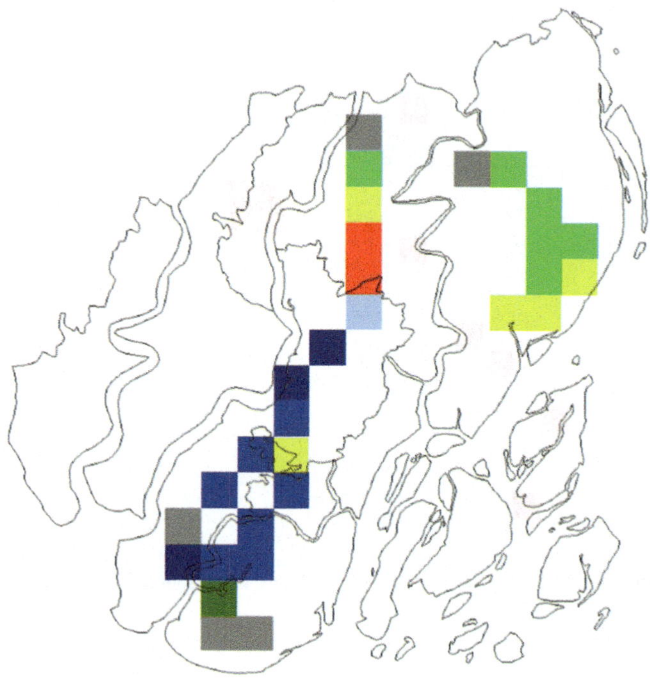

**Fig. 5** Variations in the diversity of raster cells

# 4 Breeding of Date Palm

The traditional date palm breeding follows the crosses and backcrosses methodology (El Hadrami and El Hadram 2009). However, it takes more time—an average of 30 years is needed to conclude three backcrosses—because of the extended lifecycle of the date palm (El Hadrami et al. 2011; Al-Khayri et al. 2018). Another downside of the traditional breeding methodology is gender identification–genders only become prominent after fruiting, i.e., 5–7 years after planting. However, with the adoption of tissue culture techniques and other biotechnologies, the processes are now faster but not perfect (Al-Khayri et al. 2018).

## *Utilization of Genetic Variation*

Germplasm characterization and genetic variation identifications are both knowledge-intensive areas of general crop improvement. While the breeding objectives may differ, new or improved elite cultivars can only be achieved by investigating this genetic variation. Breeders rely on techniques like induced mutagenesis, artificial, and traditional hybridizations to produce new material. Through the various technological advancements, the date palms' genome, metabolome, and proteome are now better understood (El Rabey et al. 2016; Khalil et al. 2017; Hazzouri et al. 2019). Further, there are more ways of expanding the creation and identification of genetic variation.

There are both qualitative and quantitative genetic variation of traits. Date palm breeders rely on a few parameters and statistical criteria to determine the genetic distinction present in the recognized crosses relying on a wide range of morphological traits. The data served as the backbone of the chromosome linkage maps. However, with SNP, SSR, AFLP, RFLP, SCAR, and other polymorphic molecular markers, the approach is better and faster (Saker et al. 2006; Khanam et al. 2012; Sabir et al. 2014). The arrival of QTL means breeders can now do a proper correlation of genetic variability with trait distinctness, utilizing molecular markers sequestering in an offspring. The previous data gathered from the date palm sequencing suggested that the crop genome size is short relative to other monocotyledonous and perennial species (Mathew et al. 2014; Gros-Balthazard et al. 2018). This is expected to speed up the understanding of how the genetics and inheritability of this species work. Furthermore, it is now possible to acquire fresh evidence on the evolutionary features for the crop and several new crop species due to the recent advancements in genomics.

## *Varietal Characterization of Date Palm*

DNA-based molecular data, morphological attributes, biochemical data, agronomic traits, and pedigree data are some of the bases on which population, germplasm accessions, or breeding line variation can be explored. This, in turn, helps to introduce desired genes from a different source into present crop cultivars. It also helps to achieve maximum diversity for selection and to identify the varieties that need to be protected. Cultivar identification of female date palm trees is made using morphological characters. Sex, on the other hand, is identified by the fruit traits. Since male plants are usually seed grown, they are distinct in appearance from their female parents. This makes their identification challenging and forces farmers to rely on experience to identify them.

Dissemination of commercial cultivars of the date palm is mainly via offshoots sourced from oasis close to lower Eastern Arabia and Mesopotamia (Jaradat 2011). A great deal of similarity is usually observed among cultivars propagated by cultivars. The human and natural assortment has created adapted cultivars. Fruit characteristics and how well the plant fares under abiotic and biotic pressures are the bases for creating a human collection. Natural selection is also common to noncommercial cultivars propagated via seeds. The variations in reproductive traits of date palm cultivars mean the fruit features remain the best means of identification (Hammadi et al. 2009; Sakr et al. 2010).

The continuous interaction between the environment and genotypes leads to continuous change in traits. This is why it is important to update the plant's morphological properties often. This documented information can serve as vital knowledge for improving crops and creating techniques and equipment that can pack clean, sort, and grade fruits, post-harvest. Technological and physiological studies of date palm also depend heavily on the fruit chemical composition. The farming practices adopted, and the climatic conditions, affect the value of the fruit, which varies with the type of variety.

Breeders and farmers use the morphological and biochemical features when assessing diversity and characterizations of the germplasm. However, they are not entirely reliable for two reasons: (1) they are few, and (2) the growth stage of the plants alongside the environmental factors influences the markers. Genetic characterization, on the other hand, does not suffer these shortcomings. It relies on gene sequencing, which can be individually incorporated at any stage of the plant growth, without being influenced by environmental factors. The best way to determine the diversity level and possible relationship among cultivars will be to combine molecular, biochemical, and morphological characterization into a single approach.

## Gene-Based Characterization

Productivity traits in date palm trees take several years to be expressed, due to their long generation cycle. This is one of the biggest challenges facing date palm tree breeding. Hence, an efficient tool that can speed up breeding will go a long way in improving date palm and other tree crops. For example, determining the sex of the clone, pest and disease resistance, and other traits early enough during the developmental stages will go a long way in helping breeders identify and choose elite trees with valuable traits quickly and efficiently. DNA-based marker techniques are now used in detecting genetic variability in different crop germplasm and those resulting from somaclonal variation and mutation (Sweety et al. 2017; Bhowal et al. 2019; Hasibuzzaman et al. 2020; Jahan et al. 2020). There are several studies on the use of molecular markers to understand the genetic relationship and diversity in date palm (Chaluvadi et al. 2019; Huda et al. 2019; Mohamoud et al. 2019). These markers include Amplified Fragment Length Polymorphic (AFLP), Randomly Amplified Polymorphic DNAs (RAPDs), Simple Sequence Repeat (SSR), and Single-Nucleotide Polymorphism (SNP).

Plant research now prefers gene-targeted markers to random DNA markers because the former is cheaper and faster for estimating genome sequence. This trend is expected to contribute significantly to the development of more such gene-based makers. Gene-based markers will find wide applications across the board, including gene cloning, molecular breeding, and in the mapping of qualitative trait loci (QTL) of date palm. Zhao et al. (2012) used genomic databases to analyze EST sequences (28,889) to develop gene-based markers. They succeeded in designing 4967 EST-SSR primers. These markers will help to identify variant differences through diversity studies, QTL mapping, or other molecular methods in date palm varietal developmental programs.

The Expressed Sequence Tags (EST)-derived SSRs or generic SSRs are considered an important genetic marker type when mapping candidate genes of date palm. This is because they are functional markers that can serve as a putative function. When distributed on the genetic map, generic SSRs can depict genomic gene distribution. Therefore, EST-SSRs have found important implications in the construction of high-density linkage maps. Some EST-SSRs with phenotypic links are employed in marker-assisted breeding programs. While genomic SSRs are not transferrable among similar genera and species, generic SSR markers can be transferred.

## *Sex Determination of Date Palm*

Every plant that reproduces sexually undergoes various developmental processes in their life cycle. Sex determination is one of these processes, notable for its economic importance. The breeding and cultivation technique adopted for any commercial crop depends on the sexual phenotypes. Date palm is known for its extended juvenile phase—a challenge to breeding progenies. Molecular tools have shown great potentials to make it easier to distinguish male and female seedlings right from the early developmental stages (Fig. 6).

A lot of research has gone into possible ways of identifying the sex of date palm plants with little or no success. This led to an unexpected female plant population in the field, and consequently, uncertainty in the crop's production potential. Female plants are expected to outnumber the males. Therefore, this is practically impossible to ascertain the plants' sex before the flowering stage or the reproductive age. This takes about 5–10 years to identify the sex of the plants (Bekheet and Hanafy 2011; Kharb and Mitra 2017). Another challenge of the breeding of this crop is the extended juvenile and dioecism phases. The genetic diversity is also low, making it impossible to devise simple and effective differentiation techniques to identify genders prior to the first flowering. If available, an effective means of identifying the sex of this crop would go a long way in boosting the efficiency and profitability of

**Fig. 6** Date palm flowers: (**a**) Male flowers (Singh n.d.); (**b**) Female flowers (Roy n.d.).

seed-based cultivation. Biochemical and molecular markers and other various identification techniques have been employed to simplify the foundation for sex recognition in date palm. Such techniques have been successfully applied in similar circumstances involving other plants (Khierallah et al. 2011; Jaskani et al. 2016).

Cherif et al. (2013) studied 107 cultivars sourced from various geographical locations to identify the sex-linked loci for male and female plants. With the aid of simple sequence repeat markers, they documented three linked loci heterozygous solely in male plants. This served as viable evidence to back up the presence of an XY chromosome organization, including a non-recombining XY-akin state in the genome. Having confirmed the presence of an XY chromosome scheme, it was logical to conclude that it is also possible to discover the Y chromosome-related molecular marker for recognizing males. Recently, the *SRY* gene has been identified as a sex-linked gene in mammals. Findings showed that the male has a functional *SRY* while the female has none. This *SRY* molecular factor has proven to be a reliable sex determination methodology in mammals and papaya. Polymerase chain reaction (PCR) is one of the established molecular applications for sex identification. This is a laboratory-based sterile method that synthesizes certain DNA sequences enzymatically and can be applied to distinguish. The sex identification through the PCR confirms the absence or presence of the *SRY* gene. When present, it is a male, and when absent, a female.

Primers were devised to aim at a part of the *SRY* gene, leading to a 355-bp male-confined PCR product. Subjecting this fragment to a normal PCR procedure amplified it, making the sex identification process faster in date palms. To further determine how specific or applicable the primers were, the researchers applied a DNA from an entirely different female and male date palm tree. Further, a new gene was discovered to be specifically amplified on the Y chromosome. This chromosome is homologous with the Y-chromosome in both human and papaya. Some other researchers described the existence of Y chromosome, carrying distinct repeat sequences in different species (Yamato et al. 2007; Yano et al. 2020). This discovery confirmed the similarities between the *SRY* gene of human and date palm trees. A plant is said to be a male once it contains the *SRY* gene. All these discoveries have contributed to the collection of facts and information on molecular events that help with sex determination.

## *Advancements in Biotechnologies*

Biotechnologies have taken center stage at some point in traditional and non-traditional breeding methodologies of the date palm. The most-utilized application is the tissue culture employed in regenerating and transforming plants. There has been significant progress in this regard, especially on the identification of condition factors and stimuli affecting the totipotency and plasticity of the date palm tissue.

## Organogenesis

The basic internal side of young leaves found in the offshoots is the basis of organogenic cultures. Induction is carried out in the dark, so that tissue browning and collection of phenolics are limited. This may last for several months, provided the conditions remain stable. The process is dependent upon different factors like the genotype, the components of the culture medium, and the collection time of explants and its' in vitro introduction (Loutfi and El Hadrami 2005; Sidky et al. 2007; Al-Khateeb 2008; Aslam and Khan 2009). Tissue cultures may exhibit physiological disorders, including hyperhydricity, although the specific regulatory factors remain largely unknown (Fig. 7).

Usually Murashige and Skoog (1962) medium is applied for date palm organogenesis with different supplements (Table 1). According to preliminary studies, the rapid development of date palm explants may be due to the presence of a notable amount of ammonium nitrate (Shehata et al. 2014; Amine Mazri et al. 2016). The offshoots usually become elongated once the shoot buds are transferred to media containing a large amount of auxin to cytokinin ratio. Regeneration is due after 1 or 2 years of induction, and the regenerated plantlets are moved to the greenhouse, where they are screened. The last stages of the in vitro culture favor roots formation.

## Somatic Embryogenesis

Somatic embryogenesis is another way to propagate date palm. There are two ways to do this: (1) via organogenesis, i.e., the origination of plantlets from multiplied bud clusters while skipping the callus stage, and (2) producing embryos from embryogenic callus before germinating them to create complete plantlets. Somatic embryogenesis has found multiple applications in the propagation of palms at commercial scales. However, it requires a long period, sometimes 2–3 years, to yield

**Fig. 7** Organogenesis of date palm. (Hegazy 2020b)

**Table 1** Direct organogenesis of date palm applied by different scientists

| Cultivars | Explant | Media supplement | Reference |
|---|---|---|---|
| Siwy | Shoot tip | MS + 1 mg/L NAA, 1 mg/L NOA, 2.5 mg/L BA, and 2.5 mg/L 2iP | Sidky (2017) |
| Maktoom | Inflorescence, shoot tip | MS + 2 mg/L NAA, 4 mg/L BAP | Khierallah et al. (2017) |
| Shaishee | Inflorescence | MS + 100 mg/L 2,4-D, 3 mg/L kinetin, 3 mg/L 2iP | Al-Ali et al. (2017) |
| Zaghlool | Shoot tip | MS + 2 mg/L 2ip, 1 mg/L NAA, 5 mg/L silver nitrate | Bekheet (2013) |
| Maktoom | Shoot tip | MS + 2.0 mg/L 2ip, 1.0 mg/L BA, 1.0 mg/L NAA and 1.0 mg/L NOA | Khierallah and bader (2006) |
| Barhee | Shoot tip | MS + 2iP (0, 1, 2, 3, 4, 6, 8 and 12 mg/L), 2iP (1.5, 3.0, and 4.5 mg/L) and NAA (0.2, 0.5, 1.0, and 2.0 mg/L) | Jazinizadeh et al. (2015) |
| Hillawi | Shoot tip | MS + 1 mg/L 6-benzyladenine (BA) and different concentrations (0.1–2.0 mg/L) of thidiazuran (TDZ), or free of BA and TDZ | Al-Mayahi (2014) |
| Mejhoul | Shoot tip | MS + 14.2 μM IAA, 13.4 μM NAA and 0.5 μM 2iP, and 3 months on MS/2 medium supplemented with 1.1 μM IAA, 1.1 μM NAA, 0.5 μM 2iP, 2.2 μM BA and 0.4 μM kinetin | Mazri et al. (2018) |
| Halawy | Shoot tip | MS/2 + activated charcoal (1.5 and 3.0 g/L), PVP (1.0 and 2.0 g/L), thiourea (250.0 and 500.0 mg/L), NAA (0.0, 0.1 and 1.0 mg/L), 2-iP and TDZ (1.5 and 3.0 mg/L) | Bhan et al. (2013) |
| Zaghlol | Shoot tip | MS + BA or KIN at 1.0 mg/L, NAA at 0.0, 1.0, 2.0 and 4.0 mg/L in the presence of 1.0 mg/L NOA | Baghdady (2018) |
| Medjool, Mazafati | Shoot tip | MS + 1–2iP: 1 mg/L, BAP: 1 mg/L, 2–2iP: 0.5 mg/L, BAP: 0.5 mg/L, 3–2iP: 0.2 mg/L, BAP: 0.2 mg/L, 4–2iP: 0.1 mg/L, BAP: 0.1 mg/L | Rad et al. (2015) |

acclimatized plants. To fix this, breeders have adopted different approaches to improve the developmental stages of somatic embryos.

Studies have shown that the addition of polyethylene glycol to the tissue culture media, alongside sorbitol and growth retardants, enhanced embryo multiplication and plantlet survival and rooting in the acclimatization phase. Date palm somatic embryogenesis also showed significant improvement when partial desiccation techniques were combined with medium modification, coconut water, thiamine, and biotin (Al-Khayri 2001; Al-Khateeb and Ghazzawy 2015; El-Bahr et al. 2016). The most viable way to make the process faster and cost-efficient in commercial terms is to devise a simple, yet effective and fast method to improve the development of culture steps in date palm tissue culture. In vitro date palm callus culture recorded significant improvement in growth and development when the salinity levels were reduced. Conversely, when increased, the callus development and total quantity of somatic embryos reduced. Further, the somatic embryogenesis was enhanced through the application of seawater at entire phases of in vitro processes (Yaish and Kumar 2015).

## Haploid Breeding of Date Palm

Production of doubled haploids is becoming an imperative tool in plant breeding. This technique has been successfully utilizing for different crops (Hasan et al. 2001; Pauk et al. 2004; Hooghvorst et al. 2020). The research on haploid breeding and its positive response for date palm is scanty. Globular embryoids have been formed from uninucleate microspores, as part of the cell division process under different conditions. One of the most effective conditions for generating embryoids is the combination of cold treatment with two auxins and one cytokinin. There was no significant progress in assessing various treatments and exogenous factors, except for the development of the fragile calli that did not generate into plantlets. These studies were faced with different challenges, notably the short-time flowering period that makes limitation of uninucleate microspores. Also, the anthers only lasted a few weeks after their culture, with most dying after turning brown. According to Chaibi et al. (2002), the ideal stage to treat anthers with a heat shock at 37–38 °C is just before the cell culture. They used a modified MS medium containing 2,4-D, 2-isopentenylaminopurin (2-iP), and activated charcoal to preclude callus browning, thus allowing more numbers of microspore divisions. There have been a few recovery attempts on date palm unfertilized ovules, but most of these ovules suffered necrosis and browning as a result of their small sizes. While culturing these carpels increased their size, they only survived long enough to form callus and roots when activated charcoal was introduced.

## Protoplast Culture of Date Palm

It may be possible to develop resistant date palm cultivars using somatic hybridization by protoplast fusion, considering its impressive success in the genetic improvement of other species. With protoplast fusion, it is possible to combine the cytoplasmic and nuclear genomes of related cultivars or species. The combination can be partially or completely, but always at the intergeneric and interspecific levels. It solves the problem of natural sex incompatibility. A couple of factors drive protoplast isolation, including the culture environments, material source, genotype, and the kind and concentrations of PGR used in preparing the media.

Chabane et al. (2010) investigated the factors that affect the regeneration of date palm plantlets by isolating protoplasts. The young leaves of offshoot and callus showed better performance in protoplast isolation when cellulase RS, pectolyase, and hemicellulose were used in combination.

The concentration of plant growth regulators (PGRs) plays an important role in every development stage of the protoplast. Researchers are using different PGRs in different combinations. Assani et al. (2011) found 2,4-D and 2iP as a good combination of PGRs for cell proliferation and calli induction.

In general, only a few reports are available on protoplast isolation and culture. So, there is a need to create an effective regeneration system from date palm protoplasts. This is expected to kickstart the phase of applying somatic hybridization to the development of new cultivars, with decent fruit quality and biotic and abiotic stress resistance.

#### Cryopreservation

For cryopreservation to be successful, variables like specimen size, sample water content, thawing, and freezing rates, and the type and concentration of cryoprotectant must be optimized. It is also important to optimize cryostorage methods to ensure that cryopreserved plant tissues survived better. This can be achieved by introducing pretreatments like air drying, preculture of abscisic acid-supplemented medium, and cold acclimation. In the same vein, research on date palm tissues cryopreservation showed a higher post-cryopreservation survival rate. This was achieved by shelving other cryoprotection procedures for a preculturing media fortified with high amounts of sucrose (Fki et al. 2017). Further, the vitrification procedure works perfectly without slow controlled freezing. It also favors cryopreservation of tissues via direct transfer to liquid nitrogen (Al-Khayri et al. 2018; Solliman et al. 2019).

Researchers applied cold hardening and different types of sugars to improve cryopreservation of date palm explants (Bekheet et al. 2006; Fki et al. 2017). In many cases sucrose is seen as the best for persistence of cryopreserved date palm materials (Bekheet et al. 2006; Al-Bahrany and Al-Khayri 2012a, b).

Although extensive work on the cryopreservation status of date palm has not been published, some works demonstrate the successful protocol of cryopreservation of this crop. Researchers are using a variety of cryoprotocols like standard vitrification, droplet vitrification, and encapsulation vitrification for date palm (Bekheet et al. 2006; Fki et al. 2017; Alansi et al. 2019).

## 5 Conclusions and Prospects

Over the past several years, the date palm has been pivotal to the livelihood of inhabitants of hot and barren deserts: several old sculptures and medieval civilizations exclusively documented these. In recent times, date palm is now cultivated in non-date palm-cultivating countries, and credit goes to the tissue culture technology. But, there is the need to maintain and conserve the rich diversity of date palm using in situ or ex situ methodologies. Date palms are constantly at the risk of pest and pathogen infestation and a wide range of abiotic factors that continuously adversely affect the ecosystems. There has been a significant loss of cultivars and genetic diversity erosion, caused by the replacement of the poor yielding cultivars with few top commercial cultivars when creating a date palm orchard. Several generations of breeders have ensured the adequate preservation of the date palm diversity via in situ or on-farm conservation methods, with the aid of traditional knowledge. However, such moves are practically useless nowadays, especially with the lack of assistance from government organizations that may motivate or pique growers' interests. This inadequacy has led to an even more increased interest in modern conservation methods. One of these modern conservation methods is the in vitro germplasm conservation, known for its sluggish development of cultures. Another is cryopreservation technique, which works effectively with little space. Several dates producing countries have adopted the concept of date palm collection farms. While this is a good move, it does not completely solve the problem of global genetic diversity of date palm.

## References


Abdullah HM (2014) Standardized precipitation evapotranspiration index (SPEI) based drought assessment in Bangladesh. In: Proceedings of 5th International Conference on Environmental Aspects of Bangladesh [ICEAB 2014], vol 23, pp 40–42

Abdullah HM, Rahman MM (2015) Initiating rain water harvest technology for climate change induced drought resilient agriculture: scopes and challenges in Bangladesh. J Agric Environ Int Dev 109(2):189–208

Abdullah HM, Akiyama T, Shibayama M, Awaya Y (2011) Estimation and validation of biomass of a mountainous agroecosystem by means of sampling, spectral data and QuickBird satellite image. Int J Sust Dev World 18(5):384–392

Abul-Soad AA, Jain SM, Jatoi MA (2017) Biodiversity and conservation of date palm. In: Biodiversity and conservation of woody plants. Springer, Cham, pp 313–353

Al Alawi R, Alhamdani MSS, Hoheisel JD, Baqi Y (2020) Antifibrotic and tumor microenvironment modulating effect of date palm fruit (*Phoenix dactylifera* L.) extracts in pancreatic cancer. Biomed Pharmacother 121:109522

Al-Ali AM, Ko CY, Al-Sulaiman SA, Al-Otaibi SO, Abd Ulmoneem H, Ammar MH (2017) Indirect somatic embryogenesis from mature inflorescence explants of date palm. In: Date palm biotechnology protocols, vol I. Humana Press, New York, NY, pp 89–97

Alansi S, Al-Qurainy F, Nadeem M, Khan S, Tarroum M, Alshameri A, Gaafar ARZ (2019) Cryopreservation: a tool to conserve date palm in Saudi Arabia. Saudi J Biol Sci 26(7):1896–1902

Al-Bahrany AM, Al-Khayri JM (2012a) In vitro responses of date palm cell suspensions under osmotic stress induced by sodium, potassium and calcium salts at different exposure durations. Am J Plant Physiol 7(3):120–134

Al-Bahrany AM, Al-Khayri JM (2012b) Optimizing in vitro cryopreservation of date palm (Phoenix dactylifera L.). Biotechnology 11(2):59–66

Ali A, Waly M, Essa MM, Devarajan S (2012) Nutritional and medicinal value of date fruit. In: Dates: production, processing, food, and medicinal values. CRC Press, New York, NY, pp 361–375

Al-Khateeb A (2008) Comparison effects of sucrose and date palm syrup on somatic embryogenesis of date palm (Phoenix dactylifera L.). Am J Biochem Biotechnol 4(1):19–23

Al-Khateeb AA, Ghazzawy HS (2015) Influence of polyethylene glycol and gibberelline on somatic embryogenesis of date palm (*Phoenix dactylifera*) cv Sukkary. J Global Biosci 4(6):2446–2451

Al-Khayri JM (2012) Determination of the date palm cell suspension growth curve, optimum plating efficiency, and influence of liquid medium on somatic embryogenesis. Emir J Food Agric 24(5)

Al-Khayri JM, Naik PM, Jain SM, Johnson DV (2018) Advances in date palm (Phoenix dactylifera L.) breeding. In: Advances in plant breeding strategies: fruits. Springer, Cham, pp 727–771

Al-Kindi KM, Kwan P, Andrew NR, Welch M (2017) Modelling spatiotemporal patterns of Dubas bug infestations on date palms in northern Oman: a geographical information system case study. Crop Prot 93:113–121

Al-Mayahi AMW (2014) Thidiazuron-induced in vitro bud organogenesis of the date palm (*Phoenix dactylifera* L.) cv. Hillawi. Afr J Biotechnol 13(35)

Al-Rahbi S, Al-Mulla YA, Jayasuriya H, Choudri B (2019) Analysis of true-color images from unmanned aerial vehicles to assess salinity stress on date palm. J Appl Remote Sens 13(3):034514

Amadou I (2016) Date fruits: nutritional composition of dates (*Balanites aegyptiaca Delile* and *Phoenix dactylifera* L.). In: Nutritional composition of fruit cultivars. Academic Press, New York, NY, pp 215–233

Amine Mazri M, Meziani R, El Fadile J, Ezzinbi AE (2016) Optimization of medium composition for in vitro shoot proliferation and growth of date palm cv. Mejhoul. 3 Biotech 6:1–11

Aslam J, Khan SA (2009) In vitro micropropagation of ‘Khalas’ date palm (Phoenix dactylifera L.), an important fruit plant. J Fruit Ornam Plant Res 17(1):15–27

Assani A, Chabane D, Shittu H, Bouguedoura N (2011) Date palm cell and protoplast culture. In: Date palm biotechnology. Springer, Dordrecht, pp 605–629

Ataga CD, Mohammed AH, Yusuf AO (2012) Status of date palm (Phoenix dactlylifera L.) genetic resources in Nigeria. Int J Life Sci Pharm Res 2(2):L46–L51

Awad MA (2011) Pollination of date palm (*Phoenix dactylifera* L.) cv. ‘Lulu’ TM with pollen grains-water suspension. J King Abdulaziz Univ Meteor, Env Arid Land Agric Sci 22(1):91–101

Baghdady GA (2018) Micro propagation of date palm (*Phoenix dactylifera* L.) var. Zaghlol via direct organogenesis. Ann Agric Sci 56(4):127–136

Bekheet S (2013) Direct organogenesis of date palm (*Phoenix dactylifera* L.) for propagation of true-to-type plants. Sci Agric 4(3):85–92

Bekheet SA (2017) In vitro conservation of date palm tissue cultures. In: Date palm biotechnology protocols, vol II. Humana Press, New York, NY, pp 15–24

Bekheet SA, Hanafy MS (2011) Towards sex determination of date palm. In: Date palm biotechnology. Springer, Dordrecht, pp 551–566

Bekheet SA, Taha HS (2013) Complementary strategy for conservation of date palm germplasm. Glob J Biodivers Sci Manag 3(1):96–107

Bekheet SA, Saker MM, Taha HS, Moursy HA (2001) Plant regeneration via somatic embryogenesis in date palm (*Phoenix dactylifera* L.). Arab J Biotechnol 4(1):111–118

Bekheet SA, Taha HS, El-Bahr MK (2005) Preservation of date palm cultures using encapsulated somatic embryos. Arab J Biotechnol 8:319–328

Bekheet SA, Taha HS, Solliman ME, Hassan NA (2006) Cryopreservation of date palm (*Phoenix dactylifera* L.) cultured in vitro. In III International Date Palm Conference. Acta Hortic 736:283–291

Bhan C, Sivalingam PN, Singh D, Sharma MK (2013) Studies on in-vitro organogenesis in date palm. Ind J Hortic 70(4):475–479

Bhowal RR, Hossain MM, Kayesh E, Hasan M (2019) Morphological and molecular characterization of tropical strawberry. Plant Tissue Cult Biotechnol 29(2):267–276

Chabane D, Bouguedoura N, Assani A (2010) Importance of protoplast culture in the genetic improvement of date palm (Phoenix dactylifera L.). In IV International Date Palm Conference. Acta Hortic 882:185–192

Chaibi N, Ben Abdallah A, Harzallah H, Lepoivre P (2002) Potentialités androgénétiques du palmier dattier Phoenix dactylifera L. et culture in vitro d’anthères. Biotechnol Agron Soc Environ 6(4):201–207

Chaluvadi SR, Young P, Thompson K, Bahri BA, Gajera B, Narayanan S et al (2019) Phoenix phylogeny, and analysis of genetic variation in a diverse collection of date palm (Phoenix dactylifera) and related species. Plant Div 41(5):330–339

Chao CT, Krueger RR (2007) The date palm (*Phoenix dactylifera* L.): overview of biology, uses, and cultivation. HortScience 42(5):1077–1082

Cherif E, Zehdi S, Castillo K, Chabrillange N, Abdoulkader S, Pintaud JC et al (2013) Male-specific DNA markers provide genetic evidence of an XY chromosome system, a recombination arrest and allow the tracing of paternal lineages in date palm. New Phytol 197(2):409–415

El Hadrami I, El Hadram A (2009) Breeding date palm. In: Breeding plantation tree crops: tropical species. Springer, New York, NY, pp 191–216

El Hadrami A, Daayf F, El Hadrami I (2011) Date palm genetics and breeding. In: Date palm biotechnology. Springer, Dordrecht, pp 479–512

El Rabey HA, Al-Malki AL, Abulnaja KO (2016) Proteome analysis of date palm (Phoenix dactylifera L.) under severe drought and salt stress. Int J Genomics 2016:7840759

El-Bahr MK, Abd El-Hamid A, Matter MA, Shaltout A, Bekheet SA, El-Ashry AA (2016) In vitro conservation of embryogenic cultures of date palm using osmotic mediated growth agents. J Genet Eng Biotechnol 14(2):363–370

El-Hassanin AS, Abd El Hady AA, Ali RR, Maksoud KA, Oda MMM (2020) Land resources assessment of Siwa oasis, western desert, Egypt. Plant Arch 20(1):3084–3093

Farag KM (2016) Date palm, a wealth of healthy food. In: Encyclopedia of food and health (FOHE). Chapter, 00215. Elsevier, Amsterdam

Fki L, Bouaziz N, Kriaa W, Benjemaa-Masmoudi R, Gargouri-Bouzid R, Rival A, Drira N (2011) Multiple bud cultures of 'Barhee' date palm (*Phoenix dactylifera*) and physiological status of regenerated plants. J Plant Physiol 168(14):1694–1700

Fki L, Chkir O, Kriaa W, Nasri A, Baklouti E, Masmoudi RB et al (2017) In vitro cryopreservation of date palm caulogenic meristems. In: Date palm biotechnology protocols, vol II. Humana Press, New York, NY, pp 39–48

Gros-Balthazard M, Hazzouri KM, Flowers JM (2018) Genomic insights into date palm origins. Genes 9(10):502

Hammadi H, Mokhtar REJELI, Mokhtar E, Ali F (2009) New approach for the morphological identification of date palm (Phoenix dactylifera L.) cultivars from Tunisia. Pak J Bot 41(6):2671–2681

Hasan M, Abdullah HM (2015) Plant genetic resources and traditional knowledge: emerging needs for conservation. In: Plant genetic resources and traditional knowledge for food security. Springer, Singapore, pp 105–120

Hasan M, Pauk J, Jankulosky L, Mihály R, Tóth-Lőkös K, Falusi J et al (2001) Comparison of in vitro androgenic response and F 1 seed production in wheat (*Triticum aestivum* L.). Cereal Res Commun 29(3–4):297–304

Hasan M, Hasibuzzaman ASM, Abdullah HM, Kallol MMH (2020) Genetic and genomic resources and their exploitation for unlocking genetic potential from the wild relatives. In: Rediscovery of genetic and genomic resources for future food security. Springer, Singapore, pp 193–210

Hasibuzzaman ASM, Islam AA, Miah MG, Hasan M (2020) Phylogeographic diversity and population structure of *Carica papaya* L. revealed through nuclear microsatellites. Braz J Bot 43:147–154

Hazzouri KM, Gros-Balthazard M, Flowers JM, Copetti D, Lemansour A, Lebrun M et al (2019) Genome-wide association mapping of date palm fruit traits. Nat Commun 10(1):1–14

Hegazy A (2020a) Various types of date palm fruits. Image by Adel Hegazy. University of Sadat City, Egypt.

Hegazy A (2020b) Organogenesis of date palm. Image by Adel Hegazy. University of Sadat City, Egyp

Hooghvorst I, Torrico O, Hooghvorst S, Nogués S (2020) In situ parthenogenetic doubled haploid production in melon "Piel de Sapo" for breeding purposes. Front Plant Sci 11:378

Huda MN, Hasan M, Abdullah HM, Sarker U (2019) Spatial distribution and genetic diversity of wild date palm (*Phoenix sylvestris*) growing in coastal Bangladesh. Tree Genet Genomes 15(1):3

Jahan T, Ali MA, Raihan MS, Rahman MM, Abdullah HM, Huda MN, Hasan M (2020) Indigenous Mota-named coarse rice germplasm is distinct from fine-grained rice collected from south-central coastal Bangladesh as compared with the morphological descriptors and molecular markers. Braz J Bot 43(4):933–945

Jain SM (2011) Prospects of in vitro conservation of date palm genetic diversity for sustainable production. Emir J Food Agric 23:110–119

Jaradat AA (2011) Biodiversity of date palm. In: Encyclopedia of life support systems: land use, land cover and soil sciences. Eolss Publishers, Oxford, p 31

Jaradat AA, Zaid A (2004) Quality traits of date palm fruits in a center of origin and center of diversity. Food Agric Environ 2:208–217

Jaskani MJ, Awan FS, Ahmad S, Khan IA (2016) Development of molecular method for sex identification in date palm (*Phoenix dactylifera* L.) plantlets using novel sex-linked microsatellite markers. 3 Biotech 6(1):22

Jazinizadeh E, Zarghami R, Majd A, Iranbakhsh A, Tajaddod G (2015) In vitro production of date palm (*Phoenix dactylifera* L.) cv. 'Barhee' plantlets through direct organogenesis. Biol Forum Res Trend 7(2):566

Khalil MN, Fekry MI, Farag MA (2017) Metabolome based volatiles profiling in 13 date palm fruit varieties from Egypt via SPME GC–MS and chemometrics. Food Chem 217:171–181

Khanam S, Sham A, Bennetzen JL, Aly MA (2012) Analysis of molecular marker-based characterization and genetic variation in date palm (Phoenix dactylifera L.). Aust J Crop Sci 6(8):1236

Kharb P, Mitra C (2017) Early sex identification in date palm by male-specific sequence-characterized amplified region (SCAR) markers. In: Date palm biotechnology protocols, vol II. Humana Press, New York, pp 199–207

Khierallah HS, Bader SM (2006) Micropropagation of date palm (Phoenix dactylifera L.) var. Maktoom through direct organogenesis. In III International Date Palm Conference. Acta Hortic 736:213–224

Khierallah HS, Bader SM, Baum M, Hamwieh A (2011) Genetic diversity of Iraqi date palms revealed by microsatellite polymorphism. J Am Soc Hortic Sci 136(4):282–287

Khierallah HS, Bader SM, Al-Khafaji MA (2017) NAA-induced direct organogenesis from female immature inflorescence explants of date palm. In: Date palm biotechnology protocols, vol I. Humana Press, New York, NY, pp 17–25

Krueger RR (2011) Date palm germplasm. In: Date palm biotechnology. Springer, Dordrecht, pp 313–336

Loutfi K, El Hadrami I (2005) 4.3 Phoenix dactylifera date palm. In: Biotechnology of fruit and nut crops, vol 29. CABI, Wallingford, p 144

Mansour S, Chockalingam J (2020) Diagnostically counting palm date trees in Al-Ahssa Governorate of Saudi Arabia: an integrated GIS and remote sensing processing of IKONOS imagery. Spat Inf Res 28:579–588

Mathew LS, Spannagl M, Al-Malki A, George B, Torres MF, Al-Dous EK et al (2014) A first genetic map of date palm (*Phoenix dactylifera*) reveals long-range genome structure conservation in the palms. BMC Genomics 15(1):1–10

Mazri MA, Meziani R, Belkoura I, Mokhless B, Nour S (2018) A combined pathway of organogenesis and somatic embryogenesis for an efficient large-scale propagation in date palm (*Phoenix dactylifera* L.) cv. Mejhoul. 3 Biotech 8(4):215

Mohamoud YA, Mathew LS, Torres MF, Younuskunju S, Krueger R, Suhre K, Malek JA (2019) Novel subpopulations in date palm (*Phoenix dactylifera*) identified by population-wide organellar genome sequencing. BMC Genomics 20(1):498

Murashige T, Skoog F (1962) A revised medium for rapid growth and bio assays with tobacco tissue cultures. Physiol Plant 15(3):473–497

Pauk J, Hasan MS, Puolimatka M, Lantos C, Mihály R, Mesterhazy A et al (2004) Microspore and anther culture improvements for wheat breeding. In: In vitro application in crop improvement. CRC Press, New York, NY, pp 151–172

Rad MR, Zarghami R, Hassani H, Zakizadeh H (2015) Comparison of vegetative buds formation in two date palm cultivars, Medjool and Mazafati through direct organogenesis. Int J Farm Alli Sci 4(6):549–553

Roy C (n.d.) The female flowers of date palm covered in spathe borne in a cluster (Online image). http://www.dreamstime.com

Sabir JS, Arasappan D, Bahieldin A, Abo-Aba S, Bafeel S, Zari TA et al (2014) Whole mitochondrial and plastid genome SNP analysis of nine date palm cultivars reveals plastid heteroplasmy and close phylogenetic relationships among cultivars. PLoS One 9(4):e94158

Saker MM, Adawy SS, Mohamed AA, El-Itriby HA (2006) Monitoring of cultivar identity in tissue culture-derived date palms using RAPD and AFLP analysis. Biol Plant 50(2):198–204

Sakr MM, Zeid IA, Hassan AE, Baz AGIO, Hassan WM (2010) Identification of some date palm (Phoenix dactylifera) cultivars by fruit characters. Indian J Sci Technol 3(3):338

Shehata WF, Aldaej MI, Alturki SM, Ghazzawy HS (2014) Effect of ammonium nitrate on antioxidants production of date palm (*Phoenix dactylifera* L.) in vitro. Biotechnology 13(3):116–125

Sidky R (2017) Optimized direct organogenesis from shoot tip explants of date palm. In: Date palm biotechnology protocols, vol I. Humana Press, New York, NY, pp 37–45

Sidky RA, Zaid ZE, El-Bana A (2007) Optimized protocol for in vitro rooting of date palm (Phoenix dactylifera L.). In: Proceedings fourth international symposium on date palm, King Faisal University, Al Hassa, Saudi Arabia, pp 454–467

Singh AK (n.d.) Tiny white male dates flower with pollen (Online image). http://www.dreamstime.com

Solliman MED, Al-Khateeb AA, Mohasseb HAA, Aldaej M, Al-Khateeb SA (2019) Towards development of new technique for cryopreservation of date palm (Phoenix dactylifera L.) in vitro grown cultures. Fresenius Environ Bull 28(8):6256–6263

Sweety MA, Hossain MM, Hoque MA, Hasan M, Ivy NA, Islam MN et al (2017) Molecular characterization of tropical strawberry genotypes. PTCB 27(1):33–39

Tengberg M (2014) Date palm: origins and development. In: Encyclopedia of global archaeology. Springer, New York, NY, pp 2033–2035

Yaish MW, Kumar PP (2015) Salt tolerance research in date palm tree (Phoenix dactylifera L.), past, present, and future perspectives. Front Plant Sci 6:348

Yamato KT, Ishizaki K, Fujisawa M, Okada S, Nakayama S, Fujishita M et al (2007) Gene organization of the liverwort Y chromosome reveals distinct sex chromosome evolution in a haploid system. PNAS 104(15):6472–6477

Yang M, Zhang X, Liu G, Yin Y, Che K, Yun Q et al (2010) The complete chloroplast genome sequence of date palm (*Phoenix dactylifera* L.). PLoS One 5(9):e12762

Yano Y, Chiba T, Asahara H (2020) Analysis of the mouse Y chromosome by single-molecule sequencing with Y chromosome enrichment. Front Genet 11:406

Zhang G, Pan L, Yin Y, Liu W, Huang D, Zhang T et al (2012) Large-scale collection and annotation of gene models for date palm (*Phoenix dactylifera*, L.). Plant Mol Biol 79(6):521–536

Zhao Y, Williams R, Prakash CS, He G (2012) Identification and characterization of gene-based SSR markers in date palm (*Phoenix dactylifera* L.). BMC Plant Biol 12(1):1–8

# Genetic Diversity for Breeding Tomato

**Dariusz Kulus**

**Abstract** Tomato (*Solanum lycopersicum* L.) is the second most important vegetable crop on the market. It was primarily domesticated in Latin America. Italy and Spain are considered secondary centres of its diversification. In the past few decades, the tomato was subjected to intensive breeding programmes for the improvement of desirable traits. Next-generation sequencing-based mutation mapping approaches have currently enormous potential to accelerate breeding in this species. However, despite the high morphological diversity, the cultivated tomato suffers from serious genetic erosion and loss of genetic variability, due to the domestication syndrome. The genetic diversity of the species could be supported with wild tomato relatives and landraces. They constitute an important basis for both breeding and genetic research. Their utilization on a food production scale is limited because of the introduction of high-yielding modern cultivars. This makes conservation of the *Lycopersicon* section gene pool a necessity, both in situ and ex situ. Tomato germplasm collections can be stored either in seed banks (at low or freezing temperature) or at cryogenics (liquid nitrogen). In order to optimize the management of germplasm collections, efforts have been made to generate subsets of individuals that represent the diversity conserved in the whole collection. The development of highly parallel genome-wide assays has made the study of genetic polymorphism the fastest and most reliable method for the analysis of variability in germplasm collections. The aim of this chapter is to summarize the information on genetic diversity, conservation strategies, and breeding progress in tomato.

**Keywords** Biodiversity · Gene bank · Landraces · Molecular markers · Somaclonal variation

---

D. Kulus (✉)
Laboratory of Ornamental Plants and Vegetable Crops, Department of Environmental Bases of Agriculture and Horticulture, Faculty of Agriculture and Biotechnology, UTP University of Science and Technology, Bernardyńska, Bydgoszcz, Poland



P. M. Priyadarshan, S. M. Jain (eds.), *Cash Crops*,
https://doi.org/10.1007/978-3-030-74926-2_13

# 1 Origin

The cultivated tomato was domesticated in Mesoamerican maize fields probably from *S. lycopersicum* var. *cerasiforme* (wild cherry), which spread as a weed from the Andean region to Vera Cruz and Puebla areas of México, where it is still found (Carelli et al. 2006; Bai and Lindhout 2007; Giovannoni 2018). However, it is Peru which is considered to be the centre of diversity for wild relatives of tomato (Gerszberg et al. 2015). This New World plant was referenced in Europe at the beginning of the sixteenth century. According to Corrado et al. (2014), at that time its potential as nourishment was restrained by the suspicion of the presence of harmful alkaloids in the fruit causing 'brain fever' (similarly to Mandrake and Belladonna) and, therefore, it was treated more as a botanical curiosity cultivated as an ornamental plant. On the other hand, Matthioli wrote in the mid-sixteenth century: 'One can find in our time another species in Italy segmented like a rose hip and made in segments, first of colour green and when it is mature of colour gold the which one eats in the same way (as the eggplant)' which means that by then tomato was appreciated in Southern Europe. It also appears in a shopping list of a hospital de la Sangre in Seville, Spain, in 1608 (www.mkcc.rhawn.com). Nevertheless, the value of tomato as an edible product was commonly recognized in the seventeenth century. Back then the cultivation of the species rapidly spread through the Old World, revolutionizing European eating habits. From Spain and Italy, it quickly spread to France, where it was known as the pomme d'amour and thought to have aphrodisiac properties (Goodwin and Goodwin 1984; Mazzucato et al. 2008). It was reintroduced into North America in the eighteenth century when an eccentric New Jersey Colonel Robert Gibbon Johnson brought the tomato back from a trip overseas. By consuming an entire basket of tomatoes in front of a crowd of spectators, he proved the fruits are not harmful. Still, its cultivation in the U.S. was rather rare until the twentieth century (Bhowmik et al. 2012).

# 2 Economic Value

Nowadays, tomato is the seventh among the most important crops in production/consumption in the world following maize, rice, wheat, potatoes, soybeans, and cassava. It is grown on the surface of a few million hectares at a great latitude range (Foolad 2007; Fentik 2017). In 2009 there were 153 million tons of tomatoes produced worldwide and in 2011, 160 million tons. Currently, the global tomato production is around 130 million tons, of which 88 million are destined for the fresh market and 42 million are processed. The greatest tomato producers include China, the EU, India, the USA, Turkey, Egypt, and Italy. They account for 70% of world production (www.eurofresh-distribution.com). The species is of a great economic value reaching billions of dollars (Van Eck 2018).

Tomatoes are used both as fresh fruits and in the form of various processed products (juices, sauces, chutneys, soups, canned tomatoes, concentrates, and powders), being an important part of a diverse and balanced diet. Its fleshy berry-type fruits consist mainly of water (approximately 95% of fresh weight) with moderate nutritional value. However, due to its high consumption level, it is an important source of mineral salts (e.g. iron, magnesium, phosphorus, and potassium), as well as vitamins/antioxidants (A, B1, B2, B5, C, and E), bioflavonoids and carotenoids; with lycopene accounting for up to 90% of the total carotenoids (Flores et al. 2017). Therefore, sometimes it is referred to as 'poor man's orange' (Devi et al. 2008). Bhowmik et al. (2012) have reported that tomato consumption can decrease the risk of human diseases, such as cancer (e.g. breast, cervical, colorectal, oesophageal, lung, oral, pancreatic, prostate, stomach), osteoporosis, and cardiovascular problems. It is also protective against neurodegenerative diseases and helps to replace skin cells and in sunburns (Bhowmik et al. 2012; Capel et al. 2017). Moreover, tomato is a model crop to study fleshy fruit development and plant-pathogen interactions (Adhikari et al. 2020).

## 3 Systematics and Diversity

The cultivated tomato is a member of the nightshade family—Solanaceae, division Magnoliophyta, class Magnoliopsida, subclass Asteridae, order Solanales, and suborder Solanineae (Foolad 2007; Anderson et al. 2010). Several studies on the classification of tomato have been performed over the years. Originally, Carl Linnaeus placed the tomato in the *Solanum* genus under the name *S. lycopersicum* (in 1753). A year later, Philip Miller moved it to its own genus, naming it *Lycopersicum esculentum* (to avoid the infamy of the nightshade family). However, the use of genome mapping and morphological information eventually reintroduced tomato in the *Solanum* genus in the *Lycopersicon* section (Foolad 2007).

The *Lycopersicon* section is one of the smallest within the Solanaceae family. Robertson and Labate (2006) distinguished 3 to 4 major groupings by genetic markers within this small monophyletic clade. Overtime, solely *S. lycopersicum* L. was domesticated for large-scale food production. *Solanum pimpinellifolium* L., with small (~1 cm diameter) red fruit, is the closest relative of the cultivated tomato; there is barely a 0.6% nucleotide divergence between both the above-mentioned tomatoes (Gerszberg et al. 2015). This species is also locally grown for consumption, although its significance on the market is negligible (Zorzoli et al. 2007; Giovannoni 2018). Other wild species of tomato include: *S. arcanum* Peralta, Knapp & Spooner, *S. cheesmaniae* (Riley) Fosburg, *S. chilense* (Dunal) Reiche, *S. chmielewskii* (C. M. Rick et al.) D. M. Spooner et al., *S. corneliomulleri* J. F. Macbr., *S. galapagense* S. C. Darwin & Peralta, *S. habrochaites* S. Knapp & D. M. Spooner, *S. huaylasense* Peralta, Knapp & Spooner, *S. jugandifolium* Dun.,

*S. lycopersicoides* Dun., *S. neorickii* (C. M. Rick) Spooner et al., *S. ochranthum* Dun., *S. pennellii* Correll, *S. peruvianum* L., *S. sitiens* (Anderson et al. 2010).

The tomato germplasm can be classified into the following categories: modern cultivars, obsolete cultivars, commercial breeding stocks, genetic stocks (monosomics, alien addition lines), landraces, heirloom varieties, and wild forms of cultivated species (Grout and Crisp 1995). The first classification of tomato, based on morphology, was performed by Luckwill (1943). Currently, more than 10,000 morphologically different tomato cultivars and forms exist (Gerszberg et al. 2015), which is more than any other vegetable (Foolad 2007). The typical turnover time of commercial cultivars is about 5 years (Bai and Lindhout 2007). Despite this great morphological variability of tomato (resulting often from the agrotechnics and local climate), its genetic diversity is highly limited due to bottlenecking during domestication which took place in Europe 400 years ago, and further intensive selection for desirable traits (Fig. 1; Bhattarai et al. 2016). Domestication has favoured traits related to productivity at the expense of other economically important characteristics such as stress resistance, organoleptic properties, and diversity as a whole (Gonias et al. 2019). To overcome this problem, the genetic diversity of the cultivated tomato could be supported by wild tomato relatives and landraces.

**Fig. 1** Tomato fruits show wide morphological diversity but genes involved in various plant developmental processes (*ACS2*, *COP1*, *CYC*-B, *RIN*, *MSH2*, *NAC-NOR*, *PHOT1*, *PHYA*, *PHYB*, and *PSY1*) screened for SNP polymorphism by EcoTILLING show low genomic diversity (average frequency of polymorphism across the genes is 0.038/Kb) as described by Mohan et al. (2016) in 127 tomato accessions from 20 countries

# 4 Genetic Diversity in Landraces and Heirloom Varieties

Tomato landraces and heirlooms are classified as open-pollinated, non-hybrid cultivars (Fullana-Pericàs et al. 2020). They include old introductions and locally developed cultivars, evolved under natural and low-input agricultural conditions. They originated as a result of isolation from other populations of the species (Mazzucato et al. 2008). Landraces represent dynamic populations of cultivated plant species with certain eco-geographical origins, adapted to the climatic conditions of the area of cultivation, and traditional management and use (Casañas et al. 2017). Landraces and heirlooms present a greater genetic variability (heterozygosity) for quantitative and qualitative characteristics than the modern cultivars produced since the 1990s (Mazzucato et al. 2008). Landraces usually have a higher number of rare alleles and a lower number of private alleles when compared with contemporary cultivars (Corrado et al. 2013, 2014) For example, Labate et al. (2009) extensively surveyed SNP (single nucleotide polymorphism) variability among tomato landraces by analysing 50 gene fragments (23.2 kb) per plant in a 31 plant diversity panel. Most of the *loci* (80%) were polymorphic, although Carelli et al. (2006) found that the frequency of rare alleles was similar in Brazilian commercial and landrace tomato accessions. This high level of genetic variability within landraces is related to their plasticity, i.e. ability to adapt to the natural and cultural environment of agriculture (Hassan et al. 2013). According to Lázaro (2018), consumers positively respond to heirloom tomatoes, especially in terms of their total soluble solids content and some morphological traits.

Unfortunately, despite landraces having several traits of interest, such as biotic stress resistance, superb flavour, yield stability, and great value for sustainable farming, this important gene pool has not been explored exhaustively. The introduction of pure lines and hybrids and the evolution of highly mechanized farming systems have a negative impact on the large-scale cultivation of tomato landraces (Henareh et al. 2015). The cultivation of tomato landraces is currently limited to private gardens for personal consumption and in small-size farms for local markets, although they are valued in some parts of Italy; e.g. San Marzano, Corbarino, Sorrento; East Anatolian region of Turkey; and North-West of Iran. Limited information about the origin and the relationship of landraces, along with the phenotyping costs for specific traits, is the main factor limiting their utilization (Corrado et al. 2014). This is quite disturbing, as those genetic resources, once lost, are extremely difficult to reconstitute (Coste et al. 2015).

## 5 In Situ and Ex Situ Collections

Due to the highly reduced level of genetic diversity in the cultivated gene pool of tomato and the progressing phenomenon of genetic erosion, it is necessary to maintain germplasm collections, accessible to researchers and plant breeders. Two main conservation strategies can be implied for this purpose: in situ and ex situ.

In situ conservation is the on-site conservation of plant genetic resources in their natural surroundings or in the surroundings where they have developed their distinctive properties (on-farm). According to Corrado et al. (2014) in situ conservation can be effective for the preservation of *S. lycopersicum* genetic resources. By those means, it is possible to study and include the important role of environmental factors and the possibility to evolve under natural conditions through crossing with wild or weedy relatives (Tripp and Heide 1996). The drawbacks of in situ protection include significant expenditure and labour, as it requires large areas, nursing, and agrotechnology. Moreover, it is limited by climatic conditions and threatened with loss due to biotic stresses. Therefore, ex situ (off-site) conservation of tomato is also performed.

Approximately 83,000 accessions of the wild and cultivated species of *Lycopersicon* section are kept in germplasm banks located in over 120 countries, although the exact number is difficult to estimate (Bauchet and Causse 2012). The most important collections are located at the CM Rick Tomato Genetics Resource Center (www.tgrc.ucdavis.edu), U.S., the United States Department of Agriculture (www.usda.gov), and laboratories of the International Board for Plant Genetic Resources (www.cgspace.cgiar.org) network (Sharifova et al. 2013). Other vital facilities include Universidade Estadual do Norte Fluminense in Rio de Janeiro, Brazil, The Mexican National Genetic Resources Center (CNRG), and the World Vegetable Center (WVC) in Tainan, Taiwan (Gonçalves et al. 2008). The European Cooperative Programme for Plant Genetic Resources tomato database maintains more than 20,000 accessions of several tomato species (Todorovska et al. 2014). The tomato collection of the European Solanaceae database is composed of approximately 7000 domesticated and wild lines of *S. lycopersicum* (www.eu-sol.wur.nl) obtained from public and private collections. Large collections of tomato germplasm are also located in the Netherlands (BGARD), Russia (VIR), Peru (DHUNA), Japan (NIAS), Cuba (INIFAT), and Norway (Svalbard Global Seed Vault) (Foolad 2007). Other countries such as Bulgaria, Canada, China, Colombia, France, Germany, Hungary, the Philippines, and Spain conserve mainly *S. lycopersicum* genetic resources (Robertson and Labate 2006).

Traditional gene banks collect mainly dried seed samples, which are stored in plastic or glass vessels at reduced temperature (4–5 °C) and relative humidity (20%). Accessions must be characterized and evaluated to determine their genetic diversity, e.g. by using the Ward-modified location model (Gonçalves et al. 2009). Seeds of tomato are quite suitable for storage under such conditions; however, maintaining seed collections requires their intermittent regeneration and viability testing resulting in the material loss (Kulus 2018a). Banking of vegetative explants under in vitro tissue culture conditions is a valuable complementary approach. Such collections are not threatened with pest activity or climatic conditions, and they maintain true-to-type clones without the risk of variation occurrence, typical for generative reproduction by seeds. Moreover, plants obtained in such conditions are often of superior quality. An even more effective strategy of long-term storage of biological material is based on cryopreservation, i.e. storage of tissues at the cryogenic temperature of liquid nitrogen (−196 °C). A few up-to-date reports on the cryostorage of tomato pollen, seeds, and shoot tips were successfully developed, although this approach is not yet implemented on a large scale (Kulus 2019).

## 6 Genomic Resources and Application of Molecular Markers

Studies on genetic distances are necessary for the proper organization of gene banks and in selecting parental combinations that produce progenies with maximum genetic diversity. There are several methods of investigating variation within and between plant lines, including various types of markers, e.g. phenotypic, biochemical, and molecular. Genetic markers such as AFLP (Amplified Fragment Length Polymorphism), ISSR (Inter Simple Sequence Repeat), RFLP (Restriction Fragments Length Polymorphism), RAPD (Random Amplification of Polymorphic DNA), SRAP (Sequence-Related Amplified Polymorphism), SSR (microsatellites or Simple Sequence Repeats/short tandem repeats), SSAP (Sequence-Specific Amplification Polymorphism), STMS (Sequence Tagged Microsatellite Site) are recognized as the most reliable and fast method for fingerprinting and molecular characterization of germplasm. Primer sets available in public databases can be utilized to investigate the genetic background of species. However, the narrow genetic base of tomato cultivars (less than 10% of the total genetic diversity in the *Lycopersicon* gene pool is found in *S. lycopersicum*) makes it difficult to distinguish them via many classical genetic markers (Zhou et al. 2015). For example, the mean genetic distance detected by RFLP markers among nine cultivars was only 0.003 (Miller and Tanksley 1990). Fortunately, the development of specific DNA fragment-based markers, i.e. $(GATA)_4$, $(CCTA)_4$, and $(GGAT)_4$ oligonucleotide motifs (complementary to simple-sequence repeats), enabled the identification of tomato cultivars with high levels of monomorphism that are otherwise difficult to distinguish (García-Martínez et al. 2013).

Also SSRs, due to a great discrimination power, can detect specific genetic relationships in tomato (El-Awady et al. 2012; Ohyama et al. 2017). Their main advantage is the co-dominant type of inheritance, which allows to easily distinguish heterozygous individuals from homozygous ones. SSRs also do not require knowledge of the amplified sequences or high-throughput technologies. In the study conducted by He et al. (2003), about 50% of 129 obtained SSR bands were polymorphic among a geographically broad sample of 19 tomato cultivars with the Polymorphic Information Content (PIC) ranging from 0.09 to 0.67. Todorovska et al. (2014) successfully applied SSR markers to cover almost all 12 tomato chromosomes with a minimum of seven SSRs per chromosome, located near the quantitative trait loci (QTLs). The core set of SSR markers is also a useful tool for cultivar identification and seed purity testing (Phan et al. 2016). Unfortunately, microsatellite development is expensive and laborious, so other markers are also used.

Currently, SNPs are very useful in the molecular characterization of tomato germplasm (Kulus 2018b). According to The 100 Tomato Genome Sequencing Consortium (2014), in wild tomato accessions, the number of SNPs exceeds 10MM, i.e. 20-fold higher than in most of the modern cultivars. Labate et al. (2009) extensively surveyed SNP variability among tomato landraces by analysing 50 gene fragments (23.2 kb) per plant in a 31 plant diversity panel. Most *loci* (80%) were polymorphic. On the other hand, in a sample of four *S. lycopersicum* cultivars (two heirloom and two modern types), only one SNP was detected by Nesbitt and Tanksley (2002) in 7 kb of sequence. Nonetheless, SNPs provide a vast amount of information for sequence variation among individuals.

Insertion-deletions (InDels) are another frequent choice of molecular markers for genetic diversity analysis, due to high reproducibility, co-dominance, and their multi-allelic nature (Causse et al. 2013; Jin et al. 2019). Many InDel markers are available for tomato research. In 2014, Yang et al. developed 145,695 InDels, in which 2272 markers were polymorphic in 22 tomato cultivars including wild species, processing, fresh market, and cherry tomatoes. According to Phan et al. (2016), InDel markers also could be a useful tool for cultivar identification.

In the past 15 years, resources for genetic studies in tomato were provided through genome sequencing projects (Todorovska et al. 2014). Sequencing technologies are constantly developing, e.g. next-generation sequencing (NGS), providing unparalleled ability to determine genome diversity across entire clades, at both the structural and genotype level, by rapid scoring of several thousand markers in parallel (Celik et al. 2017). Sequence-based genotyping provides a cost-efficient alternative to arrays for studying genetic variation (Koboldt et al. 2013). Plastid genome sequencing can also be a great source of information for phylogenetic and genetic engineering studies in the *Solanum* genus. Wu (2016) assembled eight tomato chloroplast (cp) genomes using NGS. A novelty described by Raime and Remm (2018) is based on the identification of taxon-specific k-mers from cp genome that is applicable for the fast detection of plant taxa directly from raw sequencing reads without aligning, mapping, or assembling the reads.

## 7 Breeding Strategies

Tomato is one of the better-known crops from the genetic point of view, as it is self-fertilized, has a relatively short life cycle, and good morphology markers (Buiatti and Morpurgo 1990). The creation of new tomato cultivars for improved characteristics started over 200 years ago in Europe (Foolad 2007). Breeding aims in this crop have gone through four major phases: breeding for yield in the 1970s, for shelf-life in the 1980s, for flavour in the 1990s, and nutritional value (such as antioxidants: lycopene and vitamin C content) at present (Bai and Lindhout 2007). In the twentieth century, controlled hybridization allowed crossing between cultivated and wild tomato compartments (Bauchet and Causse 2012). Due to the tremendous progress in biotechnology, currently, also more sophisticated methods such as marker-assisted or marker-aided selection (MAS), induced mutation breeding, high-throughput phenotyping, genome-wide association mapping, and genomic selection are utilized for breeding advances. Consequently, the genetic diversity of tomato, which was very low during the 1960s, is now eight times higher when compared to that dip (Schouten et al. 2019).

## *Heterosis Breeding*

Heterosis is the improved function of any biological quality in a hybrid offspring. It is also known as hybrid vigour or outbreeding enhancement. Heterosis (superiority of the F1 hybrid over its parental mean) and heterobeltiosis (superiority of the F1 hybrid over its better parent) are widely used with the *Lycopersicum* section. Most of the tomato species are intercrossable (Bauchet and Causse 2012), apart from *S. peruvianum* and *S. chilense* (Beddows et al. 2017), which allows heterosis breeding studies. It is an important tool that can facilitate yield enhancement from 30% to even 400% and help enrich many other desirable quantitative and qualitative traits (Afzal and Jindal 2016). Isolation of pure lines from the segregating generation of heterotic F1 is an attractive approach to improve the processing quality and dual disease tolerance, e.g. against tomato leaf curl virus (ToLCV) and bacterial wilt (BW) diseases (Acharya et al. 2018). Moreover, heterosis may manifest in tomato in the form of greater vigour, faster growth and development, and earliness in maturity (Afzal and Jindal 2016). Mukherjee et al. (2019) found that commercial cherry tomato hybrids with high antioxidant activity and tolerance to leaf curl virus can be developed with the involvement of a single parent tolerant to this disease. Kumar et al. (2019) reported that the cross NDTVR60 × Floradade manifested positive desirable heterosis over the best parent for lycopene (60.22%) and Selection7 × Floradade (25.12%) for ascorbic acid contents. The cross NDTVR60 × Floradade exhibited the highest positive heterosis over the best parent for titrable acidity (31.37%), whereas cross CO3 × Azad T5 (30.65%) exhibited the significantly highest positive heterosis over the best parent for total soluble solids. On the other hand, cross combination DT2 × Azad T5 (56.82%) exhibited positive heterobeltiosis for pericarp thickness. Several other successful studies based on heterosis breeding were recently published (Kumar et al. 2017; Tamta and Singh 2017; Salim et al. 2019).

## *Induced Genetic Diversity*

*Mutation breeding*. Mutagenesis is a process of generating mutations either spontaneously or by mutagens. Induced mutagenesis is among the most effective strategies for trait improvement in plants. Due to relatively low cost and social acceptance, it is an attractive alternative to genetic transformation. In this technique, seeds, pollen, or in vitro-derived explants are treated with chemical or physical mutagens. The former ones include the use of alkylating agents, especially ethyl methanesulfonate (EMS). Radiation-induced mutants were developed primarily using gamma rays, X-rays, and fast neutron mutagenesis. Mutation breeding has been applied to crop species since the 1930s (Ahloowalia 2004). Tomato is a good example of the successful utilization of mutations affecting important genes for plant breeding. Among such major genes one can mention self-pruning (*sp*), ovate (*o*), uniform fruit (*u*), and jointless2 (*j2*), whose recessive alleles are broadly used in the tomato industry to produce compact plants (*sp*) with elongated fruits (*o*) lacking green shoulders (*u*), and which are also more adequate for mechanical harvesting (*j2*) (Pino-Nunes et al. 2009). Approximately 6000 Micro-Tom mutant lines were generated using gamma-rays irradiation (Matsukura et al. 2007). EMS mutagenesis was employed to develop resistance against potyvirus, a widespread destructive virus for tomato (Piron et al. 2010). Classical chemical and physical mutagenesis approaches, however, induce random mutations in the genome, which lead to several non-target mutations. This problem may be overcome by newly developed mutagenesis approaches based on genetic engineering tools, which are more specific in altering the target gene. Chaudhary et al. (2019) reviewed several novel next-generation sequencing-based mutation mapping approaches, i.e. MutMap (mapping-by-sequencing), MutChromeSeq (helpful to identify the candidate genes in the shortest time), and whole-genome sequencing-based mapping, which have enormous potential to accelerate the mutation breeding in tomato.

*Somaclonal variation* is another method of inducing genetic diversity. Somaclonal variation is the variation found in plants produced by in vitro tissue culture. It is particularly common in individuals regenerated from callus (Larkin and Scowcroft 1981). The variations can be genotypic or phenotypic (in the latter case of either genetic or epigenetic origin). The advantages of somaclonal variation include the creation of additional genetic variability relatively rapidly and without sophisticated technology. Increasing the concentration of plant growth regulators in the culture medium is the easiest way to induce this variation. Ploidy-level somaclones were regenerated from leaf, cotyledon, and hypocotyl explants of tomato ‘Moneymaker’. Neither the explant source nor the duration of the tissue culture period significantly affected the mutation frequency or spectrum. Somaclonal variation was also used to produce plants resistant to biotic (*Alternaria solani*, *Fusarium oxysporum*, *Phytophthora infestans*, TMV pathogens) and abiotic stresses (e.g. high salinity, water stress, cadmium, and aluminium) (Buiatti and Morpurgo 1990; Rzepka-Plevneš et al. 2010). The efficiency of somaclonal variation, however, is limited. With tomato ‘Moneymaker’, mutant frequencies were higher after EMS treatment than those generated through tissue culture (van den Bulk et al. 1990). Other limitations of somaclonal variability include randomness and instability.

## *Modern Breeding Methods*

Application of modern biotechnological tools, e.g. haploid induction, somatic hybridization, embryo rescue, marker-assisted selection, and genetic transformation, provided good results in breeding of biotic and abiotic stress resistant *S. lycopersicum* plants with improved characteristics, such as fruit abscission, size, flavour, texture, colour, and storage ability (Zorzoli et al. 2007; Kwon et al. 2009; Uluisik et al. 2016; Fentik 2017).

Protoplast fusion, i.e. somatic cell hybridization, can be used to produce hybrids of tomato species that cannot be obtained by sexual hybridization. Genes from *Solanum* and *Nicotiana* species have been introgressed into the cultivated tomato to obtain novel cytoplasm-nucleus combinations (cybrids) and to extend the genetic variation available for breeding (Wolters et al. 1994). However, despite the improvement of protoplast culture techniques, the problem with fusions between distantly related species is the reduced fertility of the hybrids and the limited homologous recombination between chromosomes of the parental species. This indicates the importance of both cytoplasm-nucleus and nucleus-nucleus interactions in somatic incongruity. Some of these problems can be overcome by partial genome transfer from donor to recipient through microprotoplast (+) protoplast fusion, and the production of monosomic or disomic chromosome addition lines, when embryo rescue is performed (Wolters et al. 1994). Alternatively, genetic transformation can be used.

van Roekel et al. (1993) found that *Agrobacterium* strains harbouring an L,L-succinamopine-type helper plasmid yielded significantly higher transformation frequencies (averaging 9%) than those with octopine- or nopaline-type helper plasmids. A delayed ripening tomato, Flavr-savr, with a reduced expression of a cell wall polygalacturonase, produced by an antisense RNA construction, was the first transgenic plant released on the market (Kramer and Redenbaugh 1994). The creation of genetic linkage maps made it possible to study the chromosomal locations of QTL genes for improving yield and other complex traits in tomato (Frary et al. 2000). The reverse genetics tool—targeting induced local lesions in genomes (TILLING)—has been developed to efficiently identify single base mutations (also called point mutations) in a target gene (Kurowska et al. 2011). Finally, the *S. arcanum, S. habrochaites, S. lycopersicum, S. pennellii,* and *S. pimpinellifolium* genomes have been sequenced and made publicly available (The 100 Tomato Genome Sequencing Consortium 2012; 2014). This opened further breeding possibilities. Sequencing technologies (both DNA-seq and RNA-seq), genome-wide association studies (GWAS), genomic selection, multi-omic analysis, and recent CRISPR/Cas9 techniques are now utilized for tomato improvement (Sacco et al. 2015; Shirasawa et al. 2016; Zhu et al. 2018).

GWAS have become a valuable method in quantitative trait dissection, i.e. fruit weight, flavour volatiles, earliness, and plant growth. To date, GWAS have been performed in tomato on panels of heirloom and wild accessions (Zhang et al. 2016; Bauchet et al. 2017). Site-directed mutagenesis using genetic approaches also can provide a wealth of resources for crop breeding, as well as for biological research (Ito et al. 2015). CRISPR/Cas9 system, i.e. the so-called molecular scissors that make cuts at specific locations in a plant genome, can induce precise mutation in tomato plants, in the first and later generations (Pan et al. 2016). CRISPR/Cas9-induced Targeted Mutagenesis and Gene Replacement were used to obtain long-shelf-life tomato lines, parthenocarpic plants, and others (Ueta et al. 2017; Wang et al. 2017; Yu et al. 2017). Wang et al. (2019) used CRISPR/Cas9-mediated mutagenesis to knock out the encoding genes and to study the function of transcription factors (TF) in tomato fruit ripening. Transcription Activator-like Effector Nucleases (TALENs) is another commonly used approach of targeted mutagenesis in gene editing (Chaudhary et al. 2019). A review of recent discoveries in tomato molecular breeding was published by Kulus (2018b).

## 8 Conclusions and Prospects

Tomato origin is from the Andean regions of South America. During the migration to Mesoamerica, where it was domesticated, the first bottlenecking in tomato diversity took place. In the sixteenth century, Spanish conquistadors shifted the already limited tomato genetic resources to Europe, from which it was spread all around the world (including a re-introduction to America in the eighteenth century). Consequently, despite the great progress in breeding and diversity of morphological traits, dramatic genetic erosion occurred (Barrantes et al. 2016). Moreover, the results for increasing tolerance in tomato so far obtained both with conventional and biotechnological breeding approaches have been rather inadequate due to the complexity of responses to various biotic stresses (Gonias et al. 2019). Therefore, genetic diversity studies in elite tomato germplasm are among major issues with an enormous impact on the effective management of genetic resources (Todorovska et al. 2014). Strengthening the conservation and popularization of landraces and tomato wild relatives is of key importance, as often they have superior traits compared with contemporary cultivars. However, the management, maintenance, and characterization of all accessions in a traditional seed bank collection are economically demanding. Those problems can be overcome by utilizing modern tissue-culture- and molecular-based approaches. Germplasm can be cryo-conserved in liquid nitrogen for a long time, without the need for periodical regeneration. High-resolution molecular profiling, such as the analysis of SNPs and InDels, can be used to extract core subsets, dodging the need to phenotypically characterize and maintain the entire collection (Corrado et al. 2014). Additionally, modern molecular breeding approaches and heterosis can be used to increase the variability within the *Lycopersicon* section.

## References


Acharya B, Dutta S, Dutta S, Chattopadhyay A (2018) Breeding tomato for simultaneous improvement of processing quality, fruit yield, and dual disease tolerance. Int J Veg Sci 24(5):407–423

Adhikari P, Adhikari TB, Louws FJ, Panthee DR (2020) Advances and challenges in bacterial spot resistance breeding in tomato (*Solanum lycopersicum* L.). Int J Mol Sci 21(5):1734

Afzal K, Jindal K (2016) Exploiting yield potential in tomato (*Solanum lycopersicum* L.) through heterosis breeding. Plant Gene Trait 7(8):1–7

Ahloowalia BS (2004) Global impact of mutation-derived varieties. Euphytica 135:187–204

Anderson LK, Covey PA, Larsen LR, Bedinger P, Stack SM (2010) Structural differences in chromosomes distinguish species in the tomato clade. Cytogenet Genome Res 129(1–3):24–34

Bai Y, Lindhout P (2007) Domestication and breeding of tomatoes: what have we gained and what can we gain in the future? Ann Bot 100(5):1085–1094

Barrantes W, López-Casado G, García-Martínez S, Alonso A, Rubio F, Ruiz JJ, Fernández-Muñoz R, Granell A, Monforte AJ (2016) Exploring new alleles involved in tomato fruit quality in an introgression line library of *Solanum pimpinellifolium*. Front Plant Sci 7:1172

Bauchet G, Causse M (2012) Genetic diversity in tomato (*Solanum lycopersicum*) and its wild relatives. In: Caliskan M (ed) Genetic diversity in plants. InTech, Rijeka, pp 133–162

Bauchet G, Grenier S, Samson N, Bonnet J, Grivet L, Causse M (2017) Use of modern tomato breeding germplasm for deciphering the genetic control of agronomical traits by genome wide association study. Theor Appl Genet 130(5):875–889

Beddows I, Reddy A, Kloesges T, Rose LE (2017) Population genomics in wild tomatoes—the interplay of divergence and admixture. Genome Biol Evol 9(11):3023–3038

Bhattarai K, Louws FJ, Williamson JD, Panthee DR (2016) Diversity analysis of tomato genotypes based on morphological traits with commercial breeding significance for fresh market production in eastern USA. Aust J Crop Sci 10(8):1098–1103

Bhowmik D, Sampath Kumar KP, Paswan S, Srivastava S (2012) Tomato—a natural medicine and its health benefits. J Pharmacog Phytochem 1(1):24–36

Buiatti M, Morpurgo R (1990) Somaclonal variation in tomato. In: Bajaj YPS (ed) Somaclonal variation in crop improvement I. Biotechnology in agriculture and forestry, vol 11. Springer, Heidelberg, pp 400–415

Capel C, Yuste-Lisbona FJ, López-Casado G, Trinidad A, Heredia A, Cuartera J, Fernández-Muñoz R, Lozano R, Capel J (2017) QTL mapping of fruit mineral contents provides new chances for molecular breeding of tomato nutritional traits. Theor Appl Genet 130:903

Carelli BP, Gerald LTS, Grazziotin FG, Echeverrigaray S (2006) Genetic diversity among Brazilian cultivars and landraces of tomato *Lycopersicon esculentum* mill. revealed by RAPD markers. Gent Res Crop Evol 53:395–400

Casañas F, Simó J, Casals J, Prohens J (2017) Toward an evolved concept of landrace. Front Plant Sci 8:145

Causse M, Desplat N, Pascual L, Le Paslier M-C, Sauvage C, Bauchet G, Bérard A, Bounon R, Tchoumakov M, Brunel D, Bouchet J-P (2013) Whole genome resequencing in tomato reveals variation associated with introgression and breeding events. BMC Genomics 14:791

Celik I, Gurbuz N, Uncu AT, Frary A, Doganlar S (2017) Genome-wide SNP discovery and QTL mapping for fruit quality traits in inbred backcross lines (IBLs) of *Solanum pimpinellifolium* using genotyping by sequencing. BMC Genomics 18:1

Chaudhary J, Alisha A, Bhatt V, Chandanshive S, Kumar N, Mir Z, Kumar A, Yadav SK, Shivaraj SM, Sonah H, Deshmukh R (2019) Mutation breeding in tomato: advances, applicability and challenges. Plan Theory 8:128

Corrado G, Piffanelli P, Caramante M, Coppola M, Rao R (2013) SNP genotyping reveals genetic diversity between cultivated landraces and contemporary varieties of tomato. BMC Genomics 314:835

Corrado G, Caramante M, Piffanelli P, Rao R (2014) Genetic diversity in Italian tomato landraces: implications for the development of a core collection. Sci Hortic 18:138–144

Coste A, Şuteu D, Bacila I, Deliu C, Valimareanu S, Halmagyi A (2015) Genetic integrity assessment of cryopreserved tomato (*Lycopersicon esculentum* mill.) genotypes. Turk J Biol 39:638–648

Devi R, Dhaliwal MS, Kaur A, Gosal SS (2008) Effect of growth regulators on *in vitro* morphogenic response of tomato. Indian J Biotechnol 7:526–530

El-Awady MAM, El-Tarras AAE, Hassan MM (2012) Genetic diversity and DNA fingerprint study in tomato (*Solanum lycopersicum* L) cultivars grown in Egypt using simple sequence repeats (SSR) markers. Afr J Biotechnol 11(96):16233–16240

Fentik DA (2017) Review on genetics and breeding of tomato (*Lycopersicon esculentum* mill). Adv Crop Sci Tech 5(5):306

Flores P, Sánchez E, Fenoli J, Heltin P (2017) Genotypic variability of carotenoids in traditional tomato cultivars. Food Res Int 100(3):510–516

Foolad MR (2007) Genome mapping and molecular breeding of tomato. Int J Plant Genomics 2007:64358

Frary A, Nesbitt TC, Frary A, Grandillo S, van der Knaap E, Cong B, Liu J, Meller J, Elber R, Alpert KB, Tanksley SD (2000) fw2.2: a quantitative trait locus key to the evolution of tomato fruit size. Science 289(5476):85–88

Fullana-Pericàs M, Conesa MA, Ribas-Carbó M, Galmés J (2020) The use of a tomato landrace as rootstock improves the response of commercial tomato under water deficit conditions. Agronomy 10(5):748. https://doi.org/10.3390/agronomy10050748

García-Martínez S, Corrado G, Ruiz JJ, Rao R (2013) Diversity and structure of a sample of traditional Italian and Spanish tomato accessions. Genet Resour Crop Evol 60(2):789–798

Gerszberg A, Hnatuszko-Konka K, Kowalczyk T, Kononowicz AK (2015) Tomato (*Solanum lycopersicum* L.) in the service of biotechnology. Plant Cell Tiss Org Cult 120:881–902

Giovannoni J (2018) Tomato multiomics reveals consequences of crop domestication and improvement. Cell 172(1–2):6–8

Gonçalves LSA, Rodrigues R, Amaral Júnior AT, Karasawa M, Sudré CP (2008) Comparison of multivariate statistical algorithms to cluster tomato heirloom accessions. Genet Mol Res 7(4):1289–1297

Gonçalves LSA, Rodrigues R, Amaral Júnior AT, Karasawa M, Sudré CP (2009) Heirloom tomato gene bank: assessing genetic divergence based on morphological, agronomic and molecular data using a ward-modified location model. Genet Mol Res 8(1):364–374

Gonias ED, Ganopoulos I, Mellidou I, Bibi AC, Kalivas A, Mylona PV, Osanthanunkul M, Tsaftaris A, Madesis P, Doulis AG (2019) Exploring genetic diversity of tomato (*Solanum lycopersicum* L.) germplasm of genebank collection employing SSR and SCAR markers. Genet Resour Crop Evol 66:1295–1309

Goodwin JS, Goodwin JM (1984) The tomato effect—rejection of highly efficacious therapies. JAMA 251(18):2387–2390

Grout BWW, Crisp PC (1995) Cryopreservation of germplasm of tomato. Biotechnol Agric Forestry 32:371–380

Hassan NA, Mostafa S, Twfik A (2013) Assessment of genetic diversity of tomato (*Lycopersicon esculentum* L.) germplasm using molecular markers (RAPD and ISSR). Egypt J Genet Cytol 42:163–182

He C, Poysa V, Yu K (2003) Development and characterization of simple sequence repeat (SSR) markers and their use in determining relationships among *Lycopersicon esculentum* cultivars. Theor Appl Genet 106(2):363–373

Henareh M, Dursun A, Mandoulakani BA (2015) Genetic diversity in tomato landraces collected from Turkey and Iran revealed by morphological characters. Acta Sci Pol Hort Cult 12(2):87–96

Ito Y, Nishizawa-Yokoi A, Endo M, Mikami M, Toki S (2015) CRISPR/Cas9-mediated mutagenesis of the RIN *locus* that regulates tomato fruit ripening. Biochem Biophys Res Commun 467(1):76–82

Jin L, Zhao L, Wang Y, Zhou R, Song L, Xu L, Cui X, Li R, Yu W, Zhao T (2019) Genetic diversity of 324 cultivated tomato germplasm resources using agronomic traits and InDel markers. Euphytica 215:69

Koboldt DC, Steinberg KM, Larson DE, Wilson RK, Mardis ER (2013) The next-generation sequencing revolution and its impact on genomics. Cell 155(1):27–38

Kramer MG, Redenbaugh K (1994) Commercialization of a tomato with an antisense polygalacturonase gene: the FLAVR SAVR™ tomato story. Euphytica 79(3):293–297

Kurowska M, Daszkowska-Golec A, Gruszka D, Marzec M, Szurman M, Szarejko I, Maluszynski M (2011) TILLING: a shortcut in functional genomics. J Appl Genet 52(4):371–390. https://doi.org/10.1007/s13353-011-0061-1

Kulus D (2018a) Genetic resources and selected conservation methods of tomato. J Appl Bot Food Qual 91:135–144

Kulus D (2018b) Molecular breeding of tomato—a mini review of recent discoveries. Nauka Przyroda Technologie 12(1):65–72

Kulus D (2019) Managing plant genetic resources using low and ultra-low temperature storage: a case study of tomato. Biodivers Conserv 28(5):1003–1027

Kumar P, Choudhary R, Jat BL (2017) Heterosis breeding in tomato for yield and quality contributing trait. Asian J Bio Sci 12(2):259–279

Kumar R, Srivastava K, Kumar V, Saroj SK, Sharma SK, Singh RK (2019) Heterosis analysis in tomato (*Solanum lycopersicum* L.) for *Lycopene*, TSS, titrable acidity and ascorbic acid. Plant Breed 10(4):1547–1553

Kwon Y-S, Park S-G, Yi S-I (2009) Assessment of genetic variation among commercial tomato (*Solanum lycopersicum* L) varieties using SSR markers and morphological characteristics. Genes Genomics 31(1):1–10

Labate JAL, Robertson D, Baldo AM (2009) Multilocus sequence data reveal extensive departures from equilibrium in domesticated tomato (*Solanum lycopersicum* L.). Heredity 103:257–267

Larkin PJ, Scowcroft WR (1981) Somaclonal variation—a novel source of variability from cell cultures for plant improvement. Theoret Appl Genet 60:197–214

Lázaro A (2018) Tomato landraces: an analysis of diversity and preferences. Plant Gent Resour 16(4):315–324

Luckwill LC (1943) The genus *Lycopersicon*: an historical, biological, and taxonomical survey of the wild and cultivated tomatoes. Aberdeen Univ Stud 120:1–44

Matsukura C, Yamaguchi I, Inamura M, Ban Y, Kobayashi Y, Yin Y-G, Saito T, Kuwata C, Imanishi S, Nishimura S (2007) Generation of gamma irradiation-induced mutant lines of the miniature tomato (*Solanum lycopersicum* L.) cultivar 'micro-tom'. Plant Biotechnol 24:39–44

Mazzucato A, Papa R, Bitocchi E, Mosconi P, Nanni L, Negri V, Picarella ME, Siligato F, Soressi GP, Tiranti B, Veronesi F (2008) Genetic diversity, structure and marker-trait associations in a collection of Italian tomato (*Solanum lycopersicum* L.) landraces. Theor Appl Genet 116(5):657–669

Miller JC, Tanksley SD (1990) RFLP analysis of phylogenetic relationships and genetic variation in the genus *Lycopersicon*. Theor Appl Genet 80:437–448

Mohan V, Gupta S, Thomas S, Mickey H, Charakana C, Chauhan VS, Sharma K, Kumar R, Tyagi K, Sarma S, Gupta SK, Kilambi HV, Nongmaithem S, Kumari A, Gupta P, Sreeelakshmi Y, Sharma R (2016) Tomato fruits show wide phenomic diversity but fruit developmental genes show low genomic diversity. PLoS One 11(4):e0152907

Mukherjee D, Maurya PK, Banerjee S, Bhattacharjee T, Chatterjee S, Chatterjee S, Mandal AK, Maji A, Chattopadhyay A (2019) Breeding cherry tomato grown under open field conditions for simultaneous improvement in yield, nutritional quality, and leaf curl virus disease tolerance. Int J Veg Sci 26(3):211–248

Nesbitt TC, Tanksley SD (2002) Comparative sequencing in the genus *Lycopersicon*: implications for the evolution of fruit size in the domestication of cultivated tomatoes. Genetics 162:365–379

Ohyama A, Shirasawa K, Matsunaga H, Negoro S, Miyatake K, Yamaguchi H, Nunome T, Iwata H, Fukuoka H, Hayashi T (2017) Bayesian QTL mapping using genome-wide SSR markers

and segregating population derived from a cross of two commercial F1 hybrids of tomato. Theor Appl Genet 130(8):1601–1616
Pan C, Ye L, Qin L, Liu X, He Y, Wang J, Chen L, Lu G (2016) CRISPR/Cas9-mediated efficient and heritable targeted mutagenesis in tomato plants in the first and later generations. Sci Rep 6:24765
Phan TN, Kim MK, Sung-Chur S (2016) Genetic variations of F1 tomato cultivars revealed by a core set of SSR and Indel markers. Sci Hortic 212:155–161
Pino-Nunes LE, Figueira AV, Figueira O, Neto AT, Zsögön A, Piotto FA, Silva JA, Bernardi WF, Peres LEP (2009) Induced mutagenesis and natural genetic variation in tomato 'micro-tom'. Acta Hortic 821:63–71
Piron F, Nicolaï M, Minoïa S, Piednoir E, Moretti A, Salgues A, Zamir D, Caranta C, Bendahmane A (2010) An induced mutation in tomato eIF4E leads to immunity to two potyviruses. PLoS One 5:e11313
Raime K, Remm M (2018) Method for the identification of taxon-specific k-mers from chloroplast genome: a case study on tomato plant (*Solanum lycopersicum*). Front Plant Sci 9:6
Robertson LD, Labate JA (2006) Genetic resources of tomato (*Lycopersicon esculentum* mill.) and wild relatives. In: Razdan MK, Mattoo AK (eds) Genetic improvement of Solanaceae crops, vol. 2: tomato. Taylor & Francis, Boca Raton, pp 285–313
Rzepka-Plevneš D, Kulpa D, Palka E, Wisniewska M (2010) Somaclonal variability in callus culture of *Lycopersicon hirsutum* f. typicum and *Lycopersicon chilense*. Acta Sci Pol Hort Cult 4(9):63–73
Sacco A, Ruggieri V, Parisi M, Festa G, Rigano MM, Picarella ME, Mazzucato A, Barone A (2015) Exploring a tomato landraces collection for fruit-related traits by the aid of a high-throughput genomic platform. PLoS One 10(9):e0137139
Salim MMRS, Rashid MH, Ali R, Akter L (2019) Studies on character improvement in tomato (*Solanum lycopersicum* L.) by heterosis. Asian Plant Res J 2(3):1–12
Schouten HJ, Tikunov Y, Verkerke W, Finkers R, Bovy A, Bai Y, Visser RGF (2019) Breeding has increased the diversity of cultivated tomato in the Netherlands. Front Plant Sci 10:1606
Sharifova S, Mehdiyeva S, Theodorikas K, Roubos K (2013) Assessment of genetic diversity in cultivated tomato (*Solanum lycopersicum* L.) genotypes using RAPD primers. J Hortic Res 21(1):83–89
Shirasawa K, Hirakawa H, Isobe S (2016) Analytical workflow of double-digest restriction site-associated DNA sequencing based on empirical and in silico optimization in tomato. DNA Res 23(2):145–153
Tamta S, Singh JP (2017) Heterosis in tomato for growth and yield traits. Int J Veg Sci 24(2):169–179
The 100 Tomato Genome Sequencing Consortium (2014) Exploring genetic variation in the tomato (*Solanum* section *Lycopersicon*) clade by whole-genome sequencing. Plant J 80(1):136–148
The Tomato Genome Consortium (2012) The tomato genome sequence provides insights into fleshy fruit evolution. Nature 485:635–641
Todorovska E, Ivanova A, Ganeva D, Pevicharova G, Molle E, Bojinov B, Radkova M, Danailov Z (2014) Assessment of genetic variation in Bulgarian tomato (*Solanum lycopersicum* L.) genotypes, using fluorescent SSR genotyping platform. Biotechnol Biotechnol Equip 28(1):68–76
Tripp R, Heide W (1996) The erosion of crop genetic diversity: challenges, strategies and uncertainties. Nat Res Perspect 7:1–10
Ueta R, Abe C, Watanabe T, Sugano SS, Ishihara R, Ezura H, Osakabe Y, Osakabe K (2017) Rapid breeding of parthenocarpic tomato plants using CRISPR/Cas9. Sci Rep 7:507
Uluisik S, Chapman NH, Smith R, Poole M, Adams G, Gillis RB, Besong TMD, Sheldon J, Stiegelmeyer S, Perez L, Samsulrizal N, Wang D, Fisk ID, Yang N, Baxter C, Rickett D, Fray R, Blanco-Ulate B, Powell ALT, Harding SE, Craigon J, Rose JKC, Fich EA, Sun L, Domozych DS, Fraser PD, Tucker GA, Grierson D, Seymour GB (2016) Genetic improvement of tomato by targeted control of fruit softening. Nat Biotechnol 34:950–952

van den Bulk RW, Löffler HJ, Lindhout WH, Koornneef M (1990) Somaclonal variation in tomato: effect of explant source and a comparison with chemical mutagenesis. Theor Appl Genet 80(6):817–825

Van Eck J (2018) Genome editing and plant transformation of solanaceous food crops. Curr Opin Biotechnol 49:35–41

van Roekel JS, Damm B, Melchers LS, Hoekema A (1993) Factors influencing transformation frequency of tomato (*Lycopersicon esculentum*). Plant Cell Rep 12(11):644–647

Wang L, Chen L, Li R, Zhao R, Yang M, Sheng J, Shen L (2017) Reduced drought tolerance by CRISPR/Cas9-mediated SlMAPK3 mutagenesis in tomato plants. J Agric Food Chem 65(39):8674–8682

Wang R, Tavano ECDR, Lammers M, Martinelli AP, Angenent GC, de Maagd RA (2019) Re-evaluation of transcription factor function in tomato fruit development and ripening with CRISPR/Cas9-mutagenesis. Sci Rep 9(1):1696

Wolters A, Jacobsen E, O'Connell M, Bonnema G, Sree RK, de Jong H, Schoenmakers H, Wijbrandi J, Koornneef M (1994) Somatic hybridization as a tool for tomato breeding. Euphytica 79:265–277

Wu X (2016) The completed eight chloroplast genomes of tomato from *Solanum* genus. Mitochondrial DNA A DNA Mapp Seq Anal 27(6):4155–4157

Yang J, Wang Y, Shen H, Yang W (2014) *In silico* identification and experimental validation of insertion–deletion polymorphisms in tomato genome. DNA Res 21:429–438

Yu Q-H, Wang B, Li N, Tang Y, Yang S, Yang T (2017) CRISPR/Cas9-induced targeted mutagenesis and gene replacement to generate long-shelf fife tomato lines. Sci Rep 7:11874

Zhang H, Mittal N, Leamy LJ, Barazani O, Song B-H (2016) Back into the wild—apply untapped genetic diversity of wild relatives for crop improvement. Evol Appl 10(1):5–24

Zhou R, Wu Z, Cao X, Jiang FL (2015) Genetic diversity of cultivated and wild tomatoes revealed by morphological traits and SSR markers. Genet Mol Res 14(4):13868–13879

Zhu G, Wang S, Huang Z, Zhang S, Liao Q, Zhang C, Lin T, Qin M, Peng M, Yang C, Cao X, Han X, Wang X, van der Knaap E, Zhang Z, Cui X, Klee H, Fernie AR, Luo J, Huang S (2018) Rewiring of the fruit metabolome in tomato breeding. Cell 172(1–2):249–261

Zorzoli R, Pratta PR, Rodriguez GR, Picardi LA (2007) Advances in biotechnology: tomato as a plant model system. Funct Plant Sci Biotechnol 1(1):146–159

# Genetic Diversity of Cashew

**Charles Konan Kouakou, Jacky Amenan Konan, Dabé Doga, and Ahouly Boris Roméo Kouadio**

**Abstract** Cashew (*Anacardium occidentale* L.) is a multipurpose tropical edible tree nut crop. Originated from northeastern Brazil, cashew tree was disseminated to current growing areas. Due to the importance of the plant, several studies have been carried out on it. Previous researches have acknowledged the applications of cashew nuts and apples and molecular advances in genomics of cashew. The present review summarizes the current knowledge of the morphological and genetic diversity, the breeding approaches, the conservation strategies, the diseases and insects pests resistance status and the cashew management facing the climate change.

**Keywords** Cashew, Genetic resources, Diversity, Breeding

## 1 Origin and Domestication

Cashew is from northeastern Brazil, and its name cajou is from the dialect of an Indian tribal dialect 'aca-yú' meaning 'yellow apple'. Dissemination of the cashew was spontaneous and natural in the South American continent. However, its presence in other continents is attributed to human intervention (Lautié et al. 2001). The tree was introduced by the Portuguese in Mozambique and India between 1563 and 1578, then to Southeast Asia. It spread to Africa in the second half of the fifteenth century, first on the east coast and then on the west (Lacroix 2003). Since the introduction of the modern cashew into Africa and Asia continents about five centuries ago, the crop has spread widely, and these new areas have become the centre of diversity of cashew today (Aliyu 2012). Cashew was introduced in Côte d'Ivoire in 1951 by the Ivorian State as plant of reforestation of the savannicole area and to reduce soil erosion, deforestation and bush lights (Lebailly et al. 2012). It was in

Ahouly Boris Roméo Kouadio was formerly at the CNRA.

C. K. Kouakou (✉) · J. A. Konan · D. Doga · A. B. R. Kouadio
Centre National de Recherche Agronomique (CNRA), Abidjan, Côte d'Ivoire



P. M. Priyadarshan, S. M. Jain (eds.), *Cash Crops*,
https://doi.org/10.1007/978-3-030-74926-2_14

1970 that cashew became rent farming and that the forest plantations have been transformed into orchards of cashew (Lebailly et al. 2012). Starting in 1980, the Brazilian variety 'Jumbo', with large nuts, was introduced in Côte d'Ivoire by SODEFEL (Fruit and Vegetables Development Company) to provide producers with more efficient planting equipment. A field experimental and a seed orchard of plant material originating in Brazil were set up in Badikaha near Ferkéssedougou. In 1985, the Lataha (Korhogo) research station established the first collection of 54 seeds based on various origins including nuts descending from the 'Jumbo' variety originating in Brazil (Djaha et al. 2014). From 1992 to 2001, more than 120,000 hectares of plantations was established from the seeds. From 2001, the Research Station of Lataha decided to stop delivering these nuts to the producers and began its elite tree selection programme to provide grafted plants with improved yield (Lebailly et al. 2012; Adopo 2018).

Cashew was introduced into Nigeria by the Portuguese traders around the sixteenth century. It was first planted in Agege, Lagos State. From there, it spread to a few other parts of the country through transfer of nuts by man. For over 400 years after introduction, cashew trees were exploited mainly for apple; no commercial value was attached to the nuts (Aliyu 2012). The first commercial cashew planting in Nigeria was in the mid-1950 at Ogbe, Oji, Udi, Mbala, Iwo, Eruwa and Upper Ogun. These plantations were established with introduced Indian cashew varieties. Progress in the cashew industry then was low due to general neglect and poor management of the plantations. With the involvement of private entrepreneurs, Federal and State Governments, and affluent farmers, more nuts were obtained in 1978, 1980 and 1982 from India, Tanzania, Mozambique and Brazil to broaden the cashew genetic base of the country. Today, cashew cultivation has spread to almost all the states of Nigeria with increased processing, shipping and exporting activities (Adeigbe et al. 2015).

About the presence of cashew in India, literature commonly mentions Portuguese introduction of cashew nut into India from Brazil in sixteenth century. However, archaeological depiction of cashew fruit and plant in ancient monuments of India and its mention in ancient writings indicate its ancient pre-Columbian presence, negating common belief about Portuguese introduction. These evidences suggest that cashew was domesticated parallelly in India after phyto-geographic spread in the early phases of geological and geographical evolution of earth from super continental times onwards; similar to that proposed for its presence in Europe, or got introduced from the East coast of Africa through biogeographic links between South America, Africa and Asia (Singh 2018).

## 2 Diversity, Wild Relatives and Progenitors

The research on cashew began in 1980 with the introduction of plant material of Brazilian origin. This variety put into a seed orchard helped provide the quality seed producer. The first collection of cashew was installed at the fruit-growing research station of the National Center for Agricultural Research (CNRA), located in

Lataha /Korhogo, in northern Côte d'Ivoire since 1985. This collection composed of 54 trees from Côte d'Ivoire (descendants of the Brazilian variety Jumbo: 47 accessions), Australia (2 accessions), Benin (3 accessions) and Brazil (5 accessions) (Djaha et al. 2014). This collection has enabled the direct creation of more than 2920 ha to orchard by seed. Externally, this means more than 120,000 hectares of plantations were established from 1992 to 2001 from the seed provided by research. The morphological characterization of trees in the cashew collection attests to one of large variability within accessions. The variability observed was due to factors such as the difference in geographical origin of the accessions, and adaptation to the new agro-ecological conditions (Djaha et al. 2014). The variability encountered in cashew in Côte d'Ivoire is presented in Figs. 1, 2, 3, 4 and 5.

**Fig. 1** (**a**, **b**) Cashew trees bearing inflorescences

**Fig. 2** (**a–d**) Cashew tree inflorescences

## 3 Genetic Diversity in Landraces

Even if the cashew is grown over six million hectares in more than 30 countries of the world (Savadi et al. 2020), much remains to be done in the breeding of the cashew tree. Different landraces have been adopted for different ecological zones, and these possess differing characteristics of crop growth, seed sizes, quality and yields (Olubode et al. 2018). Doing so, the majority of the samples analysed in diversity studies are still landrace samples. For example, Chabi et al. (2015) evaluated the agro-morphological diversity of the cashew varieties for their better management and utilization in crop improvement. For this purpose, 120 cashew trees

**Fig. 2** (continued)

sampled from the three major production areas were characterized based on 36 (14 qualitative and 22 quantitative) agro-morphological descriptors. PCA results revealed that the parameters related to apple, nuts, kernel and inflorescence were the most important parameters that heavily contribute to overall diversity within the studied cashew varieties. Based on HCA of the quantitative variables, four phenotypic classes were distinguished within the sampled cashew trees. Significant correlations (positive or negative) were detected between variables such as the length and weight of the apple on the one hand and the weight, length and width of the nut and inflorescence on the other hand. This preliminary work suggested the existence of an important genetic variability among Beninese cashew accessions that could be used in a breeding.

Djaha et al. (2014) characterized the morphological diversity of cashew trees in the collection of the National Center for Agronomic Research (CNRA) of Côte

a

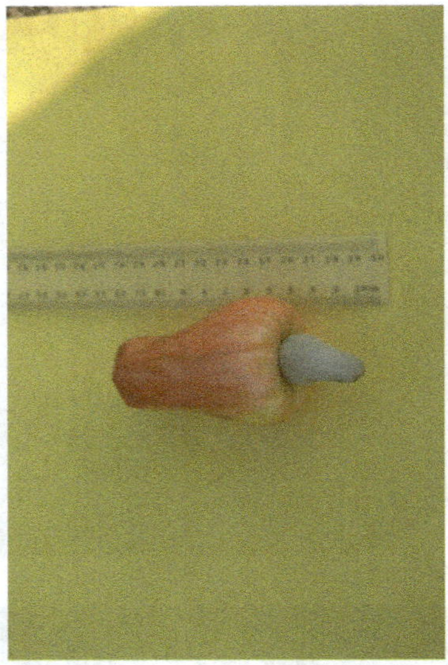

b

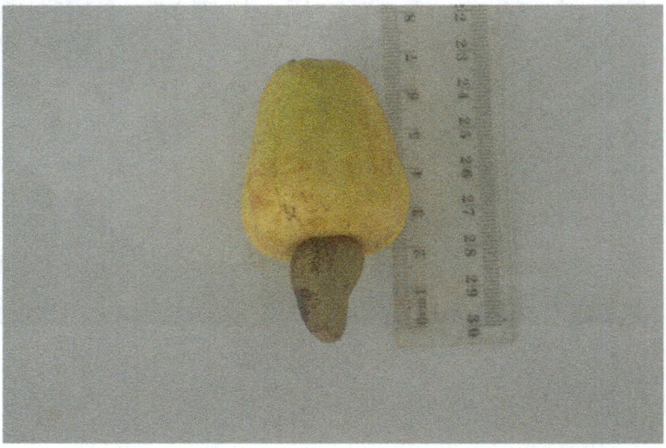

**Fig. 3** (**a-f**) Morphological variability of cashew fruits

d'Ivoire. Fifty-seven cashew accessions of various origins were assessed using four quantitative parameters and one qualitative parameter. Four significantly distinct classes of cashew trees have been identified: The first consists of large trees with an average number of carpenter branches. The second contains medium cashew trees with few carpenter branches. The third contains trees with a high bole, medium wingspan and few carpenter branches. The fourth class consists of trees of small stature, with many carpenter branches. The total height, the circumference and the

c

d

**Fig. 3** (continued)

diameter of the canopy were the main discriminating parameters that can serve as the basic criteria for identifying cashew trees in Côte d'Ivoire.

Wallis et al. (2017) characterized the cashew trees present in the municipality of Parakou, in North Benin. They chose 30 accessions based on their apple colour in a plantation containing the most varieties of cashew tree that can be found in the town. On each accession, 10 fully mature fruits were randomly collected. The physical characteristics (length, apple circumference, apple weight, nut weight and length and fruit weight) were evaluated. It appears that the characteristics of the fruit vary depending on the colour of the apple. The categorization made it possible to distinguish three groups of fruits which differed in the weight of the whole fruit

e

f

**Fig. 3** (continued)

(apple with nuts) and the length of the apple. The group consisting mainly of fruits bearing pale yellow apples and fruit bearing red apples exhibited the best characteristics. Knowledge of these superior traits is fundamental to any tree improvement programme in Parakou.

Molecular analysis using 13 SSR markers revealed among 172 accessions that cashew grown in Côte d'Ivoire belong to six different genetic groups. This diversity forms a set of 'gene pool', in which adequate characters can be developed to ensure continuous process improving species in their own environment (Kouakou et al. 2020). Various factors may explain the variability observed in this study. First location, the mode of reproduction of the cashew is mainly allogamy. Therefore, cross fertilization promotes natural genetic brewing leading to diversity important genetics within cashew (Tiendrebeogo et al. 2016). In addition, the plant material used

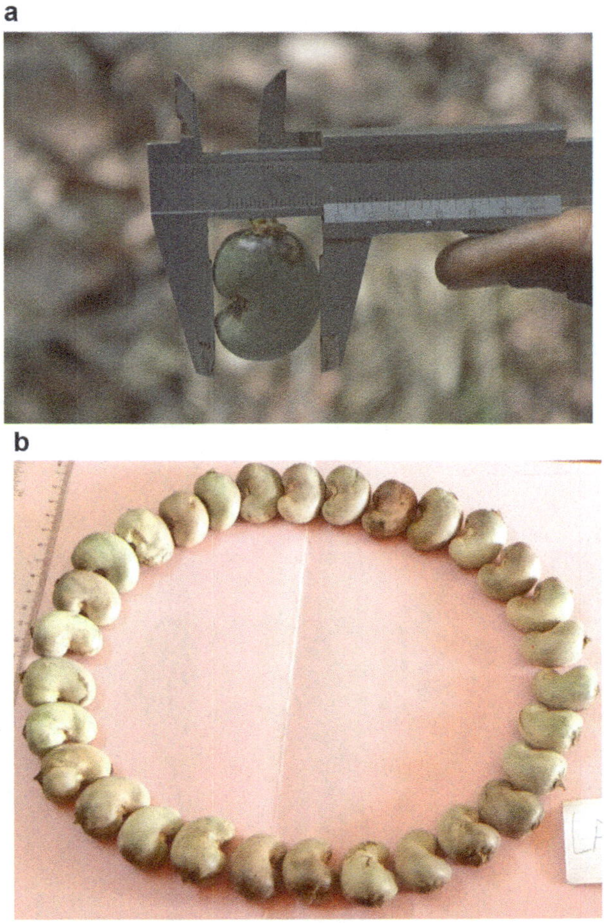

Fig. 4 (**a**, **b**) Cashew nut

also contributes to observed variability. Most orchards of cashew in Côte d'Ivoire are derived from plant material consisting of 'all coming' nuts, highly heterogeneous, not selected material (Djaha et al. 2014). N'djolossè et al. (2019) assessed the agro-morphological diversity of preselected cashew mother trees for facilitating their further selection in cashew plantations in Benin. Thirty-one variables including 12 quantitative and 19 qualitative were measured. Statistical tools were performed for the assessment of the 394 cashew mother trees preselected in 23 administrative districts originating from the four major cashew nut production regions in Benin: Atacora-Donga, Borgou, Zou-Collines and Plateau. A factorial analysis on mixed data and a hierarchical cluster analysis (HCA) were carried out to group the individuals. Finally, a morpho-metric characterization of the different

**Fig. 5** (**a**–**d**) Variability of cashew leaves

groups based on the 31 morpho-metric quantitative discriminant descriptors showed that 15 were found to be the most discriminating (5 quantitative and 10 qualitative descriptors) of the morphometric groups ($P < 0.001$). The leaf vein crossing, hermaphrodite flower rate, apples unit weight, abnormal flower number and nut basis shape showed significant positive or negative correlations. These results suggest the existence of an important phenotypic variability among Beninese cashew

c

d

**Fig. 5** (continued)

accessions that could be used in cashew tree breeding programme being implemented in Benin.

Since better productivity is always sought in cashew cultivation, its traditional varieties could end up being replaced by selected improved varieties such as early dwarf varieties (Castro et al. 2011; Uaciquete et al. 2013) and giant type (Bhat et al. 2010).

## 4 Ex Situ and In Situ Collections

The ex situ cashew collection is the CNRA-Lataha's cashew collection currently comprised of 36 cashew accessions. Eight of the 36 were selected and were installed in 2010 in three first wooden parks: Madinani zone (Zone Northwest), Tanda (East Zone) and Lataha/Ferké (East Zone). The three wooden parks have a capacity of 5 ha each. From 2016 to 2017, new wood parks from 1 to 1.5 ha were set up in different localities: Badikaha, Séguéla, Toumodi, Bouna and Béoumi. These different wooden parks contain genotypes from their own agro-ecological zones, and each contains 9–18 new genotypes plus the three genotypes already released (Adopo 2018). The ex situ conservation remains the most important and widespread means of conservation of Phytogenetic resources. The main objective is to keep as many people as possible genetic traits potentially usable in plant selection (Hunter and Heywood 2011).

In situ conservation refers to the maintenance of local varieties or cultivars in the environment in which they have developed their distinctive properties, as well as the maintenance of their pollinating insects, living organisms of the soil and other elements of biodiversity (Hunter and Heywood 2011). The objectives are to maintain the processes of evolution, to adapt crops in their environment and to maintain or increase control of farmers on their genetic resources (Del Castillo et al. 2008). It is important to adopt a complementary approach to conservation as these methods do not have the same objectives and do not retain the same quantity and quality of diversity. In situ conservation will be chosen to maintain a level of diversity dynamic adapted to socio-environmental changes. While conservation ex situ will be recommended for a fixed collection of useful genes that are easily accessible to breeders (Del Castillo et al. 2008).

## 5 Climate Resilience and Herbicide Resistance

The production of cashew nuts finds itself facing climate variability. They constitute a major constraint of production of cashew nuts. The main climatic factors affecting the productivity of the cashew are the water deficit, the poor distribution of the rains, the strong winds and harmattan severity. These factors cause the fall of flowers and fruit. Harmattan duration reduces flowering and causes flowers to fall after having them dried (Tchétangni et al. 2016). Excess rains can have a negative impact on the production of nuts. They entered a good vegetative development of the tree but flowering is less good, the fruits ripen badly or rot (Houenou 2008). The cashew was introduced to Africa and particularly to Côte d'Ivoire as a tree Reforestation (Lebailly et al. 2012). The culture of the cashew allows to reforest the agricultural fallow and to protect them against degradation agents, including erosion (Bassett 2017). It is a culture that can adapt to external stress with minimal effect on its productivity. In Côte d'Ivoire, culture is growing more to areas previously considered unsuitable for the cashew (Adopo 2018). Although adapting genotypes to change climate is included in cashew varietal improvement programmes (Yabi et al. 2013). Adaptive techniques should therefore be used to make up for effects of climatic variations as the majority of the orchards of cashew in Côte d'Ivoire has been set up from non-selected plant material (500 kg/hectare versus at least 1000–1500 kg/ha in India and Brazil (Gniré 2017)). Different strategies for adaptation to climate change can be used in the cashew plantations. The ploughing followed by the mowing of the weeds at the end of the season rainfall reduces soil evapotranspiration: ploughing creates cashed spaces to retain water, and mottled weeds are humus that can be exploited by cashew plants. Thinning and pruning followed by maintenance is a strategy climate change adaptation as they contribute to reducing competition in nutrients and photosynthetic. In Benin, a study proved that mulching and the organic amendment were a good strategy for adapting to change climate (Bello et al. 2017). The mulching and one amendment applied significantly induces a higher number of hermaphrodite flowers, longer apples.

In addition, the production of grafted plants would mitigate the effects climate change. Cashew plantations made from grafted plants have a better vegetative development than direct seedlings (Balogoun et al. 2014). Indeed, the grafted plants develop a good rooting which would lead to a great exploration of land, good nutrition and good yield.

## 6 Disease and Pest Resistance

Cashew is known as a tree resistant to its adaptability and tolerance to various kinds of biotic factors. However, it is subject to insect attacks and abiotic diseases (Lautié et al. 2001). Cashew production is severely penalized by infestations of several types of insect pests that negatively affect yields (up to 80% losses) and the quality of nuts (Vayssières et al. 2017). Various insect pests of the cashew orchard have been identified as the main insects in the order of the Coleoptera. *Analeptes trifasciata* and *Apate terebrans* are predominant and most damaging. The damage identified is significant and diverse, concerning all aerial parts of the plant (N'Depo et al. 2017). These include foliar cutting and perforation, leaf silt attacks, shear followed by slaughter of branches, branches and sometimes young cashew in shear orchards at the beginning of the dry season (Kra et al. 2017). There is also the manufacture of orifices and galleries in young branches, trunks and branches cashew followed by the release of wood sawdust at the end of the 'living' orifices, the pudding and leaf deformation under sap suction Hétéroptera and Homoptera (N'Depo et al. 2017). Producers should always burn the branches they crush in the plantations. This practice is the only one that would eliminate not only the eggs but also larvae contained in the sectioned parts (Kra et al. 2017). According to Akessé et al. (2020), the best control technique against *Analeptes trifasciata* or *Diastocera trifasciata* is the mechanical wrestling that will consist of removing and burning branches between January and March, when adults are absent from orchards (Akessé et al. 2020).

Onzo et al. (2018) found that cashew orchards in Benin that were subjected to systematic collection burning of dead wood were less attacked by the driller. To prevent pest to destroy more branches and/or stems of cashew tree, the use of control Mechanical will consist of killing the pest with a machete or a hoe (Kra et al. 2017). A more widespread and long-term application of this practice will reduce remarkably the pressure of pests in orchards (Onzo et al. 2018).

In addition to the mechanical wrestling which is the most used method by producers followed by chemical control is biological control. The Natural Enemies of Pest Insects observed are the ants *Oecophylla longinoda*, *Camponotus* sp., *Dorylus* sp., *Isoptera xylophages* and *Iphiaulax* sp. (Hymenoptera, Braconidae). These species attack only the vulnerable stages such as eggs, larvae, nymphs (Akessé et al. 2018). Recent studies have highlighted that *O. longinoda* was an effective biological control agent against these pests in Benin, Ghana and Tanzania (Vayssières et al. 2017). However, the intensity of their activities (nest construction with young leaves or on branches) could more or less affect orchard productivity (N'Depo et al. 2017).

## 7 End-Use Quality and Nutritional Quality

The culture of the cashew is of considerable socio-economic importance economically and nutritionally (Gniré 2017). The cashew almond is characterized by a richness of essential fatty acids (48.3% of total weight) (Lautié et al. 2001). Matter fat in cashew almond consists of 79.7% unsaturated fatty acids, 20.1% of saturated fatty acids and 0.2% of trans fatty acids (Rico et al. 2016). The cashew almond is also rich in protein and has an acid composition amino acids similar to soybean seed (Lautié et al. 2001). The total content of proteins is estimated at 21.3 g/100 g. Vitamin E is the most abundant vitamin with an average contribution of 5.80 mg/100 g. Vitamins A, B1, B2 and B3 are present with magnesium and potassium, respectively, 23 times and 2 times more than in orange (Lautié et al. 2001). Potassium with an average value of 6225 mg/g is the most important mineral followed by phosphorus 502.5 mg/g and potassium 248 mg/g. The most amino acid present is glutamic acid with 4.60 ± 0.23 g/100 g followed by arginine (2.22 ± 0.13 g/100 g), aspartic acid (1.89 ± 0.05 g/100 g), leucine (1.47 ± 0.09 g/100 g) and valine (1.12 ± 0.05 g/100 g). The average energy content is estimated at 2525 kJ/100 g (Rico et al. 2016). From the false fruit that accompanies the nut, it can be extracted 85% of very sweet juice often astringent, can be directly eaten raw. The juice contains about 10% sugars (sugars) Reducers in the majority, which is an important contribution of energy for man. The juice extract can be pasteurized, fermented or distilled into single or alcoholic beverages. There is also the possibility of drying the fruit or turning it into jam (Lebailly et al. 2012). The apple also contains a relatively large proportion of compounds polyphenolic (tannins), some of which are responsible for its characteristic astringence. Its rich polyphenols give it antioxidant properties (Hêdiblè et al. 2017). In addition, the cashew apple has a vitamin C content between 200–300 mg × 100 $g^{-1}$ fresh material, about 5 times that of orange (Lautié et al. 2001). The juice cashew apple can therefore be a good alternative to supplementation daily vitamin C for humans (Adou et al. 2012).

## 8 Traditional Breeding Approaches

Long-term cashew breeding programmes have been developed for the improvement of useful traits, such as apple and nut yield and quality, dwarfing traits, resistances to diseases and pests, drought tolerance. Until now, the breeding of cashew has been largely dependent on traditional methods of trait selection using phenotypic parameters. The good planting material is a very important input in crop production because it determines the upper limits on yield and the ultimate productivity of other inputs. The crop yield is also largely dependent on the number of female flowers, length of panicle, fruit set and fruit loss. Objectives of crop improvement in cashew generally include development of new high-yielding commercial varieties with characters such as desired tree size (dwarf / semi dwarf canopy), bold nut size (>8 g) with higher shelling percentage (>28%) and higher kernel grade (180 to 210 W), bigger and juicy apple, resistance / tolerance to biotic stress (pests and diseases) and abiotic stress (Adeigbe et al. 2015). Wide crossing of parents in a breeding programme is necessary for introgression of some economic traits with composite features in hybrid varieties. There are reports of developed clones in Tanzania (Masawe et al. 1998) and improved hybrids in Australia (Blaikie et al. 2002). High-yielding dwarf genotype was reported (Ashante et al. 2002) to have been developed in Brazil; the variety gave good yields, precocious fruiting and bold nuts. In Nigeria, a low rate of fertilization and lack of progression to maturity was obtained in previous hybridization trials (Aliyu 2008). Poor fertilization may be due to incompatibility (Aliyu and Awopetu 2007a), leading to reduced fruit set. From records, fruit-set in cashew can be as low as between 0.7 and 4.1% (Aliyu 2008). One reason for difficulty of hybrid generation in cashew may be due to differences in ploidy level of parents. Cytological examination of some Nigeria cashew populations revealed a diploid and haploid chromosomes of $2n = 42$ and $n = 21$, respectively (Aliyu and Awopetu 2007b), further screening for ploidy levels and particularly of parent before hybridization may be useful in promoting proper match of chromosome for higher fruit set. Identification of good parental lines is important in enhancing the effectiveness of hybridization programme. Furthermore, adoption of polyploid breeding in cashew might open opportunity for development of improved varieties and widening of the genetic base of the crop (Adeigbe et al. 2015).

# 9 Genomic Resources

## *Molecular Markers*

Genetic diversity in crop species is traditionally analysed based on differences in morphological (quantitative and qualitative) traits/markers. Due to the ease of usage, morphological markers are continued to be used at the first step as descriptors in most of the germplasm characterization and classification programmes. However, traits are often affected by the environmental factors and are limited in number, which makes it difficult to unequivocally characterize germplasms. Molecular markers exhibit high polymorphism and occur at high frequencies than morphological markers and also allow genetic characterizations even at the seedling stage. Hence, along with morphological descriptors, use of molecular markers would allow better characterization of germplasm (Savadi et al. 2020). The cashew plant has a high content of phenol and tannin which hinder the extraction of quality genomic DNA from leaf tissues for molecular studies. Rout et al. (2002) established a method of plant DNA extraction from high tannin content leaf tissues of cashew for molecular studies. Until recently, few molecular markers and genomic resources were available in cashew. DNA markers such as RAPD, ISSR and/or AFLP, SSR, SNP markers have been used to analyse genetic diversity in germplasm accessions from different regions of the world in cashew.

Dhanaraj et al. (2002) used random amplified polymorphic DNA (RAPD) markers to estimate the diversity among 90 cashew accessions from the National Cashew Gene Bank of India. The dendrogram constructed using Ward's method, squared Euclidean distance confirmed that the diversity of Indian cashew collections can be considered to be ''moderate' to 'high'. These authors identified core collection based on the study which represents the same diversity as the entire population, giving data for a more efficient germplasm management of cashew in India.

Archak et al. (2003a) reported DNA profiling of 24 selections and 11 hybrids with increased yield and excellent nut characters. They used both random amplified polymorphic DNA (RAPD) and inter simple sequence repeats (ISSR) markers to identify genotypes for a breeding programme. A total of 94 DNA fragments were generated which differentiated all the varieties. There was no correlation between the relationships based on molecular data and the pedigree of the varieties. Differences in the average similarity coefficients between selections and hybrids was very low, indicating hybridization. These authors also analysed 19 cashew accessions from India with 50 random primers, 12 ISSR primers and 6 AFLP primer pairs to compare the efficiency and utility of these techniques for detecting variation in cashew germplasm. In their study, each marker system could discriminate between all of the accessions, albeit with varied efficiency of polymorphism detection. AFLP exhibited maximum discrimination efficiency with a genotype index of 1. The utility of each molecular marker technique, expressed as marker index, was estimated as a function of average band informativeness and effective multiplex ratio. Marker index was calculated to be more than 10 times higher in AFLP than in RAPD and ISSR (Archak et al. 2003b).

Samal et al. (2003, 2004) noted the genetic relationships of 25 varieties of cashew, collected from India on the basis of phenotypic characters and RAPD markers. They found 80 distinct DNA fragments with a range of 0.2–3.0 kb by using 11 selected random decamer primers. Cluster analysis clearly showed 25 varieties of cashew grouped into two major clusters based on similarity indices. They reported that Ullal3 and Dhana showed the highest similarity indices (87%) among the 20 varieties. They also indicated that Vengurla2 has very close similarity (85%) with variety Vridhachalam3. Genotypes Vengurla2 and Vengurla3 were not grouped into a single cluster but Vengurla4 has 82% similarity to Vengurla3. The analysis of genetic relationships in cashew using morphological traits and RAPD banding data can be useful for plant improvement, descriptions of new varieties and also for assessment of variety purity in plant certification programmes.

In 2005, Syed et al. isolated two cashew LTR sequences, Tao1 and Tao2, by means of unidirectional PCR from conserved regions in the *RNAseH* gene of cashew Ty1-copia LTR retrotransposons. Both the LTRs were successfully used to develop SSAP marker systems using several individuals of an F1 population. The efficiencies of SSAP and AFLP technologies were compared using the same selective bases in SSAP and AFLP experiments. In their analysis, SSAP appears to give a comparable number but higher proportion of polymorphic markers compared to AFLP (Syed et al. 2005).

Croxford et al. (2006) used an automated, high-throughput system to isolate cashew microsatellites from a non-enriched genomic library blotted onto membranes at high density for screening. Sixty-five sequences contained a microsatellite array, of which 21 proved polymorphic among a closely related seed garden population of 49 genotypes. Twelve markers were suitable for multiplex analysis. Of these, 10 amplified in all three related tropical tree species tested: *Anacardium microcarpum*, *Anacardium pumilum* and *Anacardium nanum*. The markers developed have provided a valuable resource for future breeding efforts and map-based applications in cashew.

Aliyu and Awopetu (2007a) reported the pattern of diversity among 59 cashew accession breeding populations, assessed using proteinisozyme markers. All the accessions grouped into six clusters, indicating moderate diversity among Nigerian cashew collections. The clustering pattern was depended on the ecogeographical origin of the accessions. Closer genetic affinity was observed between Indian and local clonal populations.

In 2007, the first genetic map of cashew was constructed using an F1 mapping population of 85 individuals from a cross between CP 1001 (dwarf commercial clone) and CP 96 (giant genotype). That mapping population was used to generate two linkage genetic maps comprising of 205 genetic markers (194 AFLP and 11 SSR markers). The female map (CP 1001) contains 122 markers over 19 linkage groups and the male map (CP 96) comprises 120 markers assembled over 23 linkage groups. The total map distance of the female map is 1050.7 cM, representing around 68% genome coverage, whereas the male map spans 944.7 cM (64% coverage). The average map distance between markers is 8.6 cM in the female map and 7.9 cM in the male map. Homology between the two maps was established between

13 linkage groups of the female map and 14 of the male map using 46 bridging markers that include 11 SSR markers (Cavalcanti and Wilkinson 2007). These maps represent a platform from which to identify loci controlling economically important traits in this crop.

By investigating the level of genetic variation and genetic structure of cashew populations collected from different geographical regions of India, Archak et al. (2009) attempted to construct the introduction and diversification events of cashew in India. A total of 91 individuals from four populations were analysed using AFLP markers and morphometric data. AFLP analysis based on 354 polymorphic loci revealed Indian cashew to have low but relatively substantial genetic diversity for an introduced species (HE = 0.262 and IS = 0.404). Twenty-seven qualitative and quantitative traits also revealed the existence of considerable morphometric variation (24% around the mean values). Bayesian cluster analysis based on AFLP data did not indicate the existence of definite population differentiation. Morphometric analysis allocated 12% variation among all four populations, whereas AFLP variation observed in cashew individuals was entirely within populations. The results support the possibility of cashew having been introduced into India repeatedly over a period of time but at a single location (west coast) (Archak et al. 2009).

Asolkar et al. (2011) studied the diversity pattern in 51 half-sib genotypes with five maternal parents (accessions namely Tiswadi-3, Mutant, 8A, RL-2 and 15A) using RAPD marker employing 10 random primers. The objective of the study was to understand the extent of genetic similarity, the open pollinated seedlings would have with their maternal parents. Based on the DNA data profile, Jaccard's similarity coefficient values were generated for the interpretation of results. Random primers namely OPN-3, OPN-6, OPO-14, OPR-15 and OPS-11 were found to be the most efficient primers. In general, all ten siblings of RL-2 exhibited more than 58% of genetic makeup similar to their maternal parent. Pair-wise comparison among siblings of RL-2 maternal parent indicated a relative genetic similarity in the range of 0.43 (between RL-2/16 and RL-2/22; between RL-2/20 and RL-2/22) and 0.9455 (between RL-2/16 and RL-2/4). Out of eleven half-sib seedlings of 15A genotype, four half siblings (15A/2, 15A/8, 15A/11 and 15A/13) showed more than 73% of similarity with their maternal parent. As regards the siblings of Tiswadi-3 maternal parent, diversity skewed around 66%, indicating moderate to low diversity among the sibling genotypes of Tiswadi-3 accession. On the other hand, the half-sib seedlings 8A/3, 8A/5 and 8A/11 showed similarity coefficient values of more than 0.69 compared with maternal parent. Similarly, the maternal parent genotype (Mutant/P) contributed more than 50% of its genetic makeup to nine half-sib seedlings with Jaccard's similarity coefficient values in the range of 0.508–0.666 (Asolkar et al. 2011). This information would be of practical significance in understanding the breeding behaviour of cashew genotypes.

In 2012, one hundred and eighty-seven (187) cashew accessions from different populations were evaluated using 10 simple sequence repeats (SSR) markers in Nigeria. An overview of the phenogram derived from the molecular data showed that eight (8) of the ten (10) microsatellite loci were polymorphic and could be efficiently used for characterization and management of cashew germplasm. Preliminary data from the study showed significant level of redundancy (homogenous group) within the Nigerian cashew germplasm and the need to develop a core collection to reduce the cost of management of cashew germplasm. The data also depicts narrow genetic base of the Nigerian cashew germplasm (Aliyu 2012). Apparently, future efforts at enhancing genetic base of Nigeria cashew germplasm should focus on collaborations and extensive linkages with relevant research institutions from Asia and South America.

The genome-wide selection (GWS) was proposed as a way to increase efficiency and accelerate the genetic improvement, emphasizing the simultaneous prediction of the genetic effects of a large number of DNA genetic markers scattered throughout the genome of an organism, in order to capture the effects of all loci and explain the genetic variation of a quantitative character. Cavalcanti et al. (2012a) applied the principle of GWS in the cashew breeding, simultaneously estimating the effects of 238 markers assessed in 74 individuals from a full-sib family to explain the large percentage of total genotypic variation of the character kernel weight and increasing the efficiency of cashew breeding. It was found that the predictive power and accuracy are nearly maximized in the analysis with 70 markers to the greatest effect. The increase in the number of markers does not increase linearly the accuracy of the method GWS RR-BLUP. The 70 markers with the greatest effects captured 74% of the total genotypic variation and provided a selective high accuracy (86%) of the selection for kernel weight, while the five markers with the greatest effects captured only 19% of the total genotypic variation and provided a selective accuracy of only 44%. Thus assisted selection (MAS) based on a few (five) markers provided significant effects efficiency much lower than GWS. The predicted genomic genetic values in the population of cross-validation had a good approximation of the observed phenotype values, with a correlation of 0.79. These results indicate that the simultaneous estimation of the effects of the markers according to the concept of GWS is an interesting alternative to enhance the efficiency of cashew breeding.

Dasmohapatra et al. (2014) studied the molecular and agromorphological assessment of Indian cashew genotypes and found wide genetic variation with regard to nut yield, nut weight, shelling percentage, height and trunk girth of the potentially superior genotypes. A wide variation was noted with regard to fruit quality, fruit colour, nut weight, shelling percentage, and nut yield and apple weight. Maximum and minimum plant height was observed in variety Chintamani1 and Ullal1, respectively, with a 0.65 CD value. Twenty RAPD primers and 14 ISSR primers were used to detect the genetic variability among and between the genotypes. One hundred eighty-eight polymorphic bands and 31 monomorphic bands were observed by using both RAPD and ISSR primers. Twenty RAPD primers yielded 19 monomorphic and 84 polymorphic bands with percent of polymorphism being 81.55%. Of a total of 116 ISSR bands generated by using 14 ISSR primers, 104 bands (89.65%) were found to be polymorphic. Cumulative data generated from these two markers precisely arranged genotypes into 14 clusters.

Sethi et al. (2016) used 15 RAPD primers to study the genetic diversity of 20 cashew hybrids and their 8 parents. They found 1742 alleles with an average of 62.2 alleles per genotype, with 87.8% polymorphism. On average 7.66 bands were produced with a range of 4–15 fragments per primer. They observed that hybrids H6 and D19 maintained a high level of average genetic dissimilarity with the rest of the tested genotypes. The hybrids comprising the most divergent genetic group showed significantly aboveaverage productivity, and these could enrich the cashew gene pool.

Thimmappaiah et al. (2016) used different DNA markers (RAPD, ISSR, SSR) to detect the genetic similarity among 40 cashew genotypes in India. They found 71.8% polymorphism in RAPD, 87.5% in the case of ISSR and 93.3% in terms of SSR markers, which indicated high genetic variation existing among the tested genotypes. PIC (polymorphic information content) and MI (marker index) indicated that SSR and ISSR markers are highly informative. The average genetic similarity was 0.53, 0.63 and 0.76, respectively, which revealed predominance of high genetic similarity rather than diversity. The genetic similarity coefficient with combined markers showed a range of 0.46–0.86, with an average similarity of 0.66, also indicating narrow genetic distances and low diversity occurring between the varieties. The high genetic similarity (0.86) found between V7 and BPP5 indicates their close relationship and low genetic divergence. The low similarity (0.46) observed between Ullal1 and Jhargram1 indicates high genetic divergence with these varieties. Both RAPD and ISSR markers detected high genetic similarity (0.95) between genotypes Goal 1/6 and VRI3 and also high genetic diversity between Jhargram1 and Ullal1, and Jhargram1 with Chintamani1. (Thimmappaiah et al. 2016). The unique markers may be useful to identify genotypes for breeding programmes.

Jena et al. (2016) reported an elucidation of genetic diversity among 12 promising Indian cashew nut cultivars, including 9 local selections and 3 hybrids on the basis of morphometric and yield parameters as well as random amplified polymorphic DNA (RAPD) markers. Dice dissimilarity coefficient was used to discriminate cultivars with respect to gross phenotypic characters such as plant growth, flowering, fruit and nut characters. DNA from these cultivars were subjected to polymerase chain reaction (PCR) using 35 decamer oligonucleotide primers of which 15 primers were able to generate informative polymorphism (average 79.187%). The unweighted pair group method using arithmetic average (UPGMA) dendrogram segregated the cashew cultivars into two main clusters at 0.53 Jaccard's similarity coefficient. Five different primers were identified with the ability to discriminate eight cultivars based on a unique band pattern. Although both morphometric descriptors and RAPD markers proved their effectiveness for distinguishing the different cultivars in cashew germplasm of India, the superiority of the latter was explicit being more informative as to genotypic discrimination and establishing genetic relationships.

Bhadra et al. (2019) characterized six cashew accessions from Bangladesh using random amplified polymorphic DNA (RAPD) markers. Using four random primers, a total of 33 reproducible bands were obtained, out of which 11 were monomorphic and 22 were polymorphic. On average 74.12% polymorphism was observed. Cluster analysis revealed two main distinct groups. The genetic distance between the groups was found low and varied from 0.002 to 0.0308. The low genetic distance, which is unusual for this out crossing long-lived tree species, indicates the probability of having common ancestry among the germplasm or may be due to the use of a narrow range of populations for the investigation.

In 2019, Dos Santos et al. investigated whether wild coastal and domesticated populations of cashew could be distinguished genetically using inter-simple repeat molecular markers (ISSRs). In total, 94 polymorphic loci from five primers were used to characterize genetic diversity, structure and differentiation in four wild resting populations and four domesticated ones from eight localities in Piauí state (30 individuals per population). Genetic diversity was greater overall in wild (%P: 57.2%, I: 0.24, He: 0.15) than domesticated populations (%P: 49.5%, I: 0.19, He: 0.12). Significant structure was observed among the eight populations (between-population variance 22%, ΦPT = 0.217, $P \geq 0.001$), but only weak distinctions between wild and domesticated groups. Cluster and principal coordinate analyses showed marked genetic disparity in populations. No correlation of genetic and geographical interpopulation distance was found (Mantel test, $r = 0.02032$, $P = 0.4436$). Bayesian analysis found an eight-group optimal model ($\Delta K = 50.2$, $K = 8$), which mostly corresponded to sampled populations. Wild populations show strong genetic heterogeneity within a small geographical area despite probable gene flow between them. Within-population genetic diversity of wild plants varied considerably and was lower where extractive activities by local people are most intense (Labino population) (Dos Santos et al. 2019). The study underlines the importance of wild populations as in situ genetic reserves and the urgent need for further studies to support their conservation.

Hawerroth et al. (2019) evaluated the use of RAPD and ISSR molecular markers to determine the genetic variability among cashew genotypes and to indicate possible promising crosses based on cashew genetic variability and phenotypic performance. Ten hybrids derived from the crosses CCP 76 × BGC 589, CCP 76 × BRS 226, CCP 76 × HAC 276-1, CCP 76 × Embrapa 51, CCP 76 × BRS 253, CCP 76 × HAC222-4 and BRS226 × Embrapa 51 and their parents were assessed at the molecular level. The hybrids were evaluated for nut yield, mean nut weight, bored nuts and powdery mildew on nuts (scale 0–4). The RAPD and ISSR markers were efficient in the determination of the genetic variability among cashew genotypes, allowing for the grouping of 21 clusters. Associated with the phenotypic characterization of cashew nut for yield, weight and health, the used markers can efficiently identify possible combinations with higher genetic variability and higher probability of developing transgressive genotypes in segregating populations.

Kouakou et al. (2020) evaluated the genetic diversity of high-yielding cashew cultivars for better use in breeding programmes in Côte d'Ivoire. Using 13 SSR markers, they identified the first two introduced populations of cashew in Côte d'Ivoire. The average allelic richness was 3.56 (±1.45) alleles per locus, the fixation index (FIS) indicated an overall heterozygosity deficit of 0.332 (±0.076), and the average population differentiation (FST) was 0.014 (±0.004). Much of the total genetic variability occurred at the intra-population level (98.6%), compared to only 1.4% variability attributable to differences between populations. Average value of gene flow was 22.528. This study has shown that the gene flow within Ivorian cashew populations maintains high intra-population genetic diversity. This flow rate reflects a long-term exploitable genetic variability for use in selection and conservation.

## *Genome Sequencing*

In 2007, aceto-orcein squashing of meristematic tissues (root tips and flower buds) of some genotypes has been analysed for the number, structure and behaviour of chromosome in cashew. The study revealed that most genotypes had a diploid and haploid chromosomes of $2n = 42$ and $n = 21$, respectively. The karyotypes were mostly symmetric, composed mainly of metacentric pairs and several submetacentrics (Aliyu and Awopetu 2007b). There were similarity in the morphology, number and behaviour of the chromosomes among genotypes from different sources of introduction, which could imply common progenitors and slow mutation rate (chromosomal) among the existing cashew.

In 2014, flow cytometric analysis of 43 cashew accessions from the Nigerian germplasm was carried out to determine the relative genome size, intraspecific variation and ploidy status of the species using *Solanum lycopersicum* cv. Stupicke as an internal standard reference. The mean genome size was significantly different ($P < 0.01$) across the 43 cashew accessions, suggesting intra-specific variation (about 3.6% coefficient of variation) for absolute nuclear DNA content in this species, but with no evidence of endopolyploidy or endoreduplication. Flow cytometric analysis showed that absolute genome size of cashew is about 0.857 pg/2C (419 Mb/1C). However, there was a slight gradual increment in genome size from south to north gradient across sampled populations. The data also showed significant correlation ($r = 0.368$, $P < 0.01$) between nuclear DNA content (genome size) and phenotypic trait, nut weight. The results obtained confirmed relatively small genome size in this tree species, almost twice genome size of diploid *Arabidopsis thaliana* but slightly lower than mango, *Mangifera indica*, a close relative in *Anacardiaceae* (Aliyu 2014).

Crop breeding has long relied on cycles of phenotypic selection and crossing, which generate superior genotypes through genetic recombination. When genome sequences are available, all genes and genetic variants contributing to agronomics traits can be identified and changes made during breeding processes can be assessed at the genotype level. Genetic improvement of cashew by traditional breeding is slow and unpredictable due to the long juvenile phase, high degree of heterozygosity, lack of juvenile-mature traits correlations and large size of the mature plant. Additionally, most of the yield and agronomic traits are genetically complex which complicate its breeding. Recently, the next-generation sequencing (NGS) and high-throughput genotyping technologies have expedited the pace of development of genomic tools and resources for genomics-orphan crops like cashew (Limera et al. 2017). The genomics advancements allow designing novel molecular breeding technologies with the potential for enhancing genetic gains and accelerating crop improvement, which is of utmost importance in the breeding of long juvenile species like cashew.

do Amaral et al. (2017) performed a genetic study of several canopies of *Anacardium occidentale* L. specimen called 'Cashew King' in order to assess whether all canopies belong to the same plant. The DNA analysis was made from automated sequencing by capillary electrophoresis equipment ABI3500 Genetic Analyzer (Applied Biosystems) and alignment of nucleotide sequences generated with reference sequences deposited in GenBank. According to the analysis, it was shown that all canopies belong to the same plant. Thus, the 'Cashew King' can be considered as a single specimen *A. occidentale* occupying an area of 8834 $m^2$.

Rabah et al. (2017) completed plastome sequence of ten species including the orchard tree cashew. In the latter, they reported the insertion of an ~6.7 kb fragment of mitochondrial DNA into the plastome IR. BLASTn analyses returned high identity hits to mitogenome sequences including an intact ccmB open reading frame. Using three plastome markers for five species of Anacardium, they generated a phylogeny to investigate the distribution and timing of the insertion. Four species share the insertion, suggesting that this event occurred <20 million years ago in a single clade in the genus.

Because of the ready availability of genomic data for breeders today, genomics plays an increasingly important role in all aspects of crop breeding, such as quantitative trait loci (QTL) mapping and genome-wide association studies (GWAS), where genomic sequencing of crop populations can allow gene-level resolution of agronomic variation (Hu et al. 2018). Mzena et al. (2018) developed single-nucleotide polymorphic (SNPs) markers and genetic linkage map for identifying quantitative traits loci (QTL) associated with yield, nut quality and plant size in cashew. In their study, genotyping by sequencing (GBS) technique, tagged as DArTSeq, was used to identify SNPs from reduced complexity cashew genome of F1 and F2 as well as ATA19/250 (male parent) and Cook05 (female parent). The reduced libraries were sequenced using Hiseq 2500 illumina sequencer, besides, SNPs call was done by using DArTsoft v.7.4.7 (DArT P/L, Australia). The study identified 6364 high-quality SNPs, of which 57.6% were transition and 42.4% were transversion. The average call rate, reproducibility and polymorphic information content (PIC) of the SNPs was 99%, 98% and 0.31, respectively. Among the 6364 markers, 1009 were codominant markers of which 761 (75%) were assigned to 21 linkage groups of total length 2330.6 cM with an average distance of 3.3 cM between markers. Their study revealed the first SNPs linkage map which could be used in identifying QTL for efficient breeding programme through marker-assisted selection in cashew.

Bastiaan-Net et al. (2020) investigated the presence of pathogenesis-related protein 10 (*PR10*) genes in cashew nut. Using RNA-seq analysis, they were able to identify several PR10-like transcripts in cashew nut and clone six putative PR10 genes. In addition, PR10 protein expression in raw cashew nuts was confirmed by immunoblotting and liquid chromatography–mass spectrometry (LC–MS/MS) analyses. An in silico allergenicity assessment suggested that all identified cashew

PR10 proteins were potentially allergenic and may represent three different isoallergens.

In 2014, Dario Grattapaglia Research scientist at EMBRAPA, the Brazilian Agricultural Research Corporation launched the cashew genome project. A key motivation was the almost complete absence of genomic resources, including a reference genome, for cashew. In 2020, Grattapaglia and Silva-Junior have generated a chromosome-scale assembly of the ~488 Mbp genome of the Brazilian Agricultural Research Corporation (EMBRAPA) dwarf clonal variety BRS226 and an Axiom 16KSNP-chip from whole-genome resequencing data of 24 diverse cashew accessions (https://phytozome-next.jgi.doe.gov/info/Aoccidentale_v0_9). Using a combination of long-read sequencing from Pacific Biosciences, Chicago and Hi-C chromatin proximity ligation from Dovetail Genomics, they assembled 377.95 Mb (77.4%) of the cashew tree genome into 581 scaffolds, with a scaffold N50/L50 of 17.05 Mb/10 and 90% of the genome contained in 20 scaffolds. The genome assembly was annotated with 41,197 gene models using transcript evidence from RNA-Seq approaches. Repetitive sequences comprised 43% of the cashew tree genome. Additional genome coverage is being pursued by assembling the sequencing data into the correctly phased diploid genome by combining long and short reads with SNP and pedigree data (https://phytozome-next.jgi.doe.gov/info/Aoccidentale_v0_9).

## 10 Induced Genetic Diversity (Mutation Breeding, Soma Clonal Variation, Etc)

In commercial production, the uniformity of individual plants within a clone population is a major advantage of clonal cultivars. However, genetic variations do occur in undifferentiated cells, isolated protoplasts, calli, tissues and morphological traits of in vitro raised plants (Currais et al. 2013). The general term 'somaclonal variation' is used for plant variants derived from any form of cell or tissue cultures. From the point of commercial micropropagation, these somaclonal variations may be considered obstructive and worthless; since, such variations may lead to loss of genetic fidelity. However, plant cell and tissue cultures render increased genetic variability comparatively faster and without applying a sophisticated technology. This technology holds ample scope in crop improvement of horticultural crops, which are largely propagated vegetatively, partly, due to reasons like longer juvenile phase as in perennial fruit crops, occasional inbreeding depression, self and cross incompatibility, narrow genetic base especially in ornamentals, etc. Further, somaclonal variations require less space and time for screening of desirable traits in vitro unlike cross seedlings of perennial crops, which require a great deal of land area and time. Somaclones may itself have numerous applications in plant breeding and genetic improvements and recovery of such novel variants can be enhanced by applying suitable in vitro selection pressure (Krishna et al. 2016). To date, numerous studies on cashew have treated with in vitro tissues culture, but none of them had reported the occurrence of somaclonal variation in cultivated tissues.

In 2002, Ananthakrishnan et al. induced in vitro adventious formation from cotyledon explants of cashew. Excised cotyledons from mature embryos of cashew developed adventitious shoots from the proximal end when cultured on Murashige and Skoog's (MS) medium supplemented optimally with 22.2 mM BAP and 3% sucrose. Initiation occurred only at the proximal end of the cotyledon. Differentiation of roots occurred on the same medium after the initiation of leaves. Histology revealed that there was no vascular connection between regenerated shoots and roots. Isolated shoots rooted after treatment on half-strength MS liquid medium containing IBA (122.6 mM) for 72 h. Shoots were transferred to half-strength MS solid medium containing IBA (4.9 mM) and IAA (5.7 mM). About 55% of in vitro raised shoots developed roots. This procedure was first applicable to the three commercial high-yielding cultivars of cashew in India (Ananthakrishnan et al. 2002).

In Nigeria, efforts have been made in the application of tissue culture technique for mass propagation of cashew (Aliyu 2005). The recalcitrant nature of cashew, abnormal development of calli from explants and browning of explants were some of the limiting factors of the tissue culture success observed in cashew in Nigeria. Browning of explants has been reduced through frequent transfers, addition of activated charcoal and dark treatments. Furthermore, use of explants from in vitro germinated seedlings or fungicidal treated young flush in MS (Murashige and Skog) basal salt medium supplemented with cytokinins was found to improve the in vitro success rate of cashew significantly (Aliyu 2005). However, culture to full plant and successful transfer to field has not been achieved by that author.

In 2016, Sija et al. established an efficient in vitro shoot proliferation protocol through excised shoot tip and nodal segment of *Cashew*. The effect of different growth regulators alone or in combination on multiple shoot production from different explants of *cashew* was studied. Shoot proliferation was obtained from excised shoot tips, and nodal segments from in vitro grown seedlings of cashew were cultured on MS medium supplemented with NAA alone or in combination with BAP and Kinetin. Maximum number of shoots (3–5 per explants) was obtained from excised nodal segment (Sija et al. 2016).

In Côte d'Ivoire, Nanti et al. (2018) identified the type of seeds suitable for in vitro germination and plantlets development. Seeds without shell with entire cotyledons (SWSEC), seeds without shells with half cotyledons (SWSHC) and the embryonic axis (EA) were placed on the basal medium of Murashige and Skoog (MS) without any plant growth regulators. The final rate of germination ranged from 75 to 95%, the time to have 50% germination and the average germination time were statistically identical with the three types of seeds. Higher plants (7 cm) with larger numbers of leaves, which are more vigorous were obtained with SWSHC. These authors have shown that seeds without shells with half cotyledons are more suitable to provide seedling used as explant sources for the micropropagation of the cashew tree.

In 2020, Nanti et al. propagated elite plants of cashew be used as planting material under in vitro conditions. Shoot tip and basal part explants derived from vitro plants of 16-day-old were transferred onto Murashige and Skoog (MS) medium containing different concentrations of cytokinins. After 1 month of culture, the induced shoots were placed onto different strengths of MS medium with various concentrations of sucrose and auxin. The highest number of buds (9) was recorded with the basal explants on medium supplemented with Thidiazuron (TDZ) at 0.01 mg/L. The highest shoots (3 cm) were obtained with these same explants on a medium without growth regulators. A½ MS with 60 g/L of sucrose and 5 mg/L of IBA induced the highest rooting percentage (72%) and number of roots (4 roots) in a short time (16 days) (Nanti et al. 2020).

## 11 Modern Breeding Approaches: Marker-Assisted Selection, Genome-Wide Association Mapping and Genomic Selection

An important way to increase productivity in any crop plant is to first select desirable genotypes from the existing variations and to use the superior types in breeding programmes. Until recently, variability studies based on morphological characters coupled with classical hybridization formed the bases for initiating crop improvement programmes in cashew. However, such approaches would not be precise in accomplishing the breeding objectives for enhanced production and productivity, since the conventional methods involve higher costs and long time. The advent of DNA markers has opened a new path for understanding crop diversity, phylogenetic relationship and synteny.

Syed et al. (2005) isolated two LTR sequences from the cashew genome and used them successfully to develop SSAP marker systems. These were shown to have higher levels of polymorphism than amplified fragment length polymorphic markers for this species.

QTLs for yieldrelated traits such as nut weight, male and hermaphrodite flowers have been identified by Cavalcanti et al. (2012b). The traits were identified in 71 FI genotypes of the cross CCP 1001 × CP 96; 3 QTLs were detected for nut weight, 4 for male flowers and 4 for hermaphrodite flowers. They also suggested that QTLs accounted for 3.8–13% of the total phenotypic variance and had phenotypic effects of 31.8–34.2%. The effective potential for markerassisted selection of the QTL hf2f and hf3 m is great and the percentage of phenotypic variation higher than that of the others.

Santos et al. (2010) reported the identification of MAS (markerassisted selection) and QTLs (quantitative trait loci) which are of great interest in breeding programmes with regard to fruit quality. Physicochemical characteristics including vitamin C contents, phenolics, total titrable acidity, oligomeric and total soluble solids were analysed in the mapped cashew population. They were detected by QTL intervals and multiple QTL mapping with high phenotypic variation in the segregating FI generation for ail traits, and also identified 18 QTLs linked with fruit quality (dos Santos et al. 2011). QTLs are promising for markerassisted selection since they have the greatest phenotypic effects and contribution to phenotypic variation. The first genetic map of cashew was published by Cavalcanti and Wilkinson (2007). These authors produced an F1 mapping population of 85 individuals from a cross between CP 1001 (dwarf commercial clone) and CP 96 (giant genotype) and used it to generate two linkage genetic maps comprising of 205 genetic markers (194 AFLP and 11 SSR markers). The female map (CP 1001) contained 122 markers over 19 linkage groups and the male map (CP 96) comprised 120 markers assembled over 23 linkage groups. The total map distance of the female map was 1050.7 cM, representing around 68% genome coverage, whereas the male map spans 944.7 cM (64% coverage). The average map distance between markers was 8.6 cM in the female map and 7.9 cM in the male map. Homology between the two maps was established between 13 linkage groups of the female map and 14 of the male map using 46 bridging markers that include 11 SSR markers (Cavalcanti and Wilkinson 2007). These maps represent a platform from which to identify loci controlling economically important traits in this crop.

The genome-wide selection (GWS) was proposed as a way to increase efficiency and accelerate the genetic improvement, emphasizing the simultaneous prediction of the genetic effects of a large number of DNA genetic markers scattered throughout the genome of an organism, in order to capture the effects of all loci and explain the genetic variation of a quantitative character. In 2012, Cavalcanti et al. applied the principle of GWS in the cashew breeding, simultaneously estimating the effects of 238 markers assessed in 74 individuals from a full-sib family to explain the large percentage of total genotypic variation of the character kernel weight and increasing the efficiency of cashew breeding. It was found that the predictive power and accuracy are nearly maximized in the analysis with 70 markers to the greatest effect. The increase in the number of markers does not increase linearly the accuracy of the method GWS RR-BLUP. The 70 markers with the greatest effects capture 74% of the total genotypic variation and provide a selective high accuracy (86%) of the selection for kernel weight, while the five markers with the greatest effects capture only 19% of the total genotypic variation and provide a selective accuracy of only 44%. Thus, marker-assisted selection (MAS) based on a few (five) markers provide significant effects efficiency much lower than GWS. The predicted genomic genetic values in the population of cross-validation had a good approximation of the observed phenotype values, with a correlation of 0.79 showing that the simultaneous estimation of the effects of the markers according to the concept of GWS is an interesting alternative to enhance the efficiency of cashew breeding.

## 12 Phenotyping Approaches Including High-Throughput Phenotyping

Crop improvement based on plant breeding needs accurate phenotyping (Mir et al. 2019). Thus, there is an increasing interest from research institutions to develop systems for high-throughput plant phenotyping both at greenhouse and at field scale (Roitsch et al. 2019).

Samal et al. (2003) used morphological characters such as number of laterals, flowering laterals, length of panicle, sex ratio, umber of nut per panicle, nut weight, shelling percentage and yield per plant were used to describe the genetic relationships of 20 varieties of cashew. Results obtained for the phenotypic characters based on similarity coefficient were divided into four clusters with 70% similarity. By means of similarity coefficients (SG), cluster I was found to consist of 12 varieties. Cluster II consisted of a single variety, NRCC-1, cluster III consisted of six varieties and cluster IV had only one variety, ridhachalam-2.

In 2016 the prediction of cashew apple quality by non-destructive analysis using a portable (100 g), near infrared spectrophotometer was investigated. The quality parameters of interest were: soluble solids content (SSC), titratable acidity (TA), SSC/TA ratio and ascorbic acid (AA). The results have shown that that highly portable spectrometer can be used to monitor and control cashew apple postharvest quality, and may constitute an important tool for genome wide selection in cashew breeding programmes (Ribeiro et al. 2016).

Adeigbe et al. (2016) investigated the diversity in the cashew nut sizes in commercial cultivation in Nigeria in relation to kernel and shell ratios. Six categories were examined for their nut, shell and kernel properties. Analysis of variance was employed to detect variability among the group for each trait while principal component analysis was used to identify distinguishing traits and to group the different sizes based on similarities. Nut weight ranged from 2.95 to 23.20 g while kernel weight ranged from 0.04 to 6.88 g. Two groups evolved from the classification: the larger and the smaller nut groups. Kernel weight was a function of the nut girth ($r = 0.93$) and nut weight ($r = 0.87$). Nut weight, nut length, nut girth and nut width, respectively, had the highest broad sense heritability of 99.25, 99.01, 98.82 and 98.74. In that study, Extra-large and small nuts were best in shelling percentages of 32.46% and 37.29% respectively; however, the former had bigger and bolder kernel. The 'Jumbo' had the highest mean nut weight (17.86 g), much of it was attributed to the shell compared to extra-large nuts. Identification and characterization of the six categories for kernel and nut properties is of paramount practical significance in direct selection of superior genotype for subsequent use in cashew improvement programmes (Adeigbe et al. 2016).

In 2018, some cashew apples have been phenotyped by near infrared spectroscopy for quality traits such as firmness, pH, total soluble sugar (SSC), soluble solid (SS), titratable acidity (TA), SS/AT ratio (favour), vitamin C (VC), total carotenoids (TC), total flavonoids (TF), total polyphenols (TP) and total antioxidant activity (TAA). A set of results obtained with a near infrared spectrometer using the 830–2500 nm range showed good prediction of the quality traits cited above in the ripe early dwarf cashew clones by partial least squares (PLS) algorithm. The intact cashew apples spectra resulted in good predictions of firmness (R2 = 0.92; RMSEP = 0.71), pH (R2 = 0.84; RMSEP = 0.17), SSC (R2 = 0.86; RMSEP = 0.99), SS (R2 = 0.90; RMSEP = 0.70), TA (R2 = 0.96; RMSEP = 0.055), favour (R2 = 0.87; RMSEP = 7.5), VC (R2 = 0.92; RMSEP = 38), TC (R2 = 0.97; RMSEP = 0.089), TF (R2 = 0.95; RMSEP = 0.86), TP (R2 = 0.94; RMSEP = 27) and TAA (R2 = 0.93; RMSEP = 34). The NIR methodology is expected to be an alternative for cashew germplasm banks as a tool to support the database of images spectra for a rapid and robust phenotyping (Samamad et al. 2018).

Alves et al. (2019) developed an advanced fast phenotyping tool to explore the cashew apple compositions from different genotypes, based on a portable near-infrared (MicroNIR) spectroscopy. This was in addition to associating the variability of the respective cashew apple pulps with the genotypes by ultra-performance liquid chromatography (UPLC), coupled with high-resolution mass spectrometry (HRMS). The UPLC-HRMS analysis is characterized by specific bioactive compounds, such as the derivatives of hydroxybutanoic acid, galloyl and flavonoids. Furthermore, both techniques allowed the identification of a group of accessions, which presented similarities among the chemical profiling. However, to improve the understanding of cashew chemical and physical variability, further variables related to the cashew apple composition, such as edaphoclimatic conditions, should be considered for future studies. These approaches lead to the conclusion that these two tools are useful for the maintenance of BAG-Caju (Cashew Germplasm Bank) and for the cashew-breeding programme (Alves et al. 2019).

de Andrade et al. (2019) investigated whether wild populations on coastal dunes in Piauí, Brazil differed from non-wild populations. Ten populations were sampled, and 21 morphological variables related to leaf, petiole, inflorescence and fruit were measured. Variation within and between populations was investigated with univariate and multivariate statistics. Dune populations were mostly more similar to one another than to domesticated ones. There was significant correlation between inter-population geographical distance and morphological dissimilarity. Classification methods showed 96.4% successful assignment to the dune category and 86%–100% to dune populations individually, but dune and non-dune populations overlap morphologically. Dune populations had shorter, broader leaves, shorter drupes and fewer secondary veins. Non-dune coastal populations showed strongest similarity to dune populations. Populations distant from the coast were most divergent. The population from the 'cerrado' region was most distinct, with thicker leaf blades and narrower petioles. The dune populations are recognized as the 'restinga ecotype' of *A. occidentale* (de Andrade et al. 2019).

## 13 Conclusions and Perspectives

Cashew is a tree nut crop of significant socioeconomic importance. Increasing demand for cashew nut in the confectionery and industrial trade is due to increasing per capita cashew consumption. However, this growing demand may not be met if nothing is done. Indeed, current yields are low due to the use of low-yielding varieties, losses due to pests and abiotic stresses and the poor management of the crop. Furthermore, the conventional breeding in cashew that has been used to date is slow. Recent developments in genomics of cashew would provide opportunities for designing novel molecular tools and breeding strategies that improve speed up the development of superior cultivars.

## References

Adeigbe OO, Olasupo FO, Adewale BD, Muyiwa AA (2015) A review on cashew research and production in Nigeria in the last four decades. Sci Res Essays 10:196–209

Adeigbe OO, Adewale BD, Muyiwa AA et al (2016) Quantitative descriptors of cashew nut categories in Nigeria: providing indices for superior nut selection. J Agric Biol Sci 11:142–148

Adopo AN (2018) Status of high yielding cashew varieties and their adoption in Cote d'Ivoire. In: Cashew technology handbook 2018, p 180

Adou M, Tetchi FA, Gbané M, Kouassi KN, Amani NG (2012) Physico-chemical characterization of cashew apple juice (*Anacardium occidentale*, L.) from Yamoussoukro (Côte d'Ivoire). Innov Rom Food Biotechnol 11:32–43

Akessé EN, Ouali-N'Goran S-WM, N'Dépo OR, Tano K (2018) Population fluctuation of Diastocera trifasciata (Fabricius, 1775) (Coleoptera: Cerambycidae), cashew branches girdler in the Brobo locality (central Côte d'Ivoire). J Entomol Zool Stud 6:1–8

Akessé EN, N'goran S-WMO, Minhibo YM, Koffi KM, Koné D (2020) Efficacité d'une lutte mécanique associée au biopesticide Neco 50 EC dans le contrôle des adultes de *Diastocera trifasciata* (Coleoptera: Cerambycidae), ciseleur des branches d'anacardier en Côte d'Ivoire. Int J Biol Chem Sci 14:1038–1051

Aliyu OM (2005) Application of tissue culture to cashew (*Anacardium occidentale* L.) breeding: an appraisal. Afr J Biotechnol 4:1485–1489

Aliyu OM (2008) Compatibility and fruit-set in cashew (*Anacardium occidentale* L.). Euphytica 160:25–33

Aliyu OM (2012) Genetic diversity of Nigerian cashew germplasm. In: Caliskan M (ed) Genetic diversity in plants. InTech, Rijeka, p 23

Aliyu OM (2014) Analysis of absolute nuclear DNA content reveals a small genome and intra-specific variation in cashew (*Anacardium occidentale* L.), Anacardiaceae. Silvae Genet 63:285–292

Aliyu OM, Awopetu JA (2007a) Multivariate analysis of cashew (*Anacardium occidentale* L.) Germplasm in Nigeria. Silvae Genet 56:170–179

Aliyu OM, Awopetu JA (2007b) Chromosome studies in cashew (*Anacardium occidentale L.*). Afr J Biotechnol 6:131–136

Alves F, Silva L et al (2019) Metabolomic variability of different genotypes of cashew by LC-Ms and correlation with near-infrared spectroscopy as a tool for fast phenotyping. Metabolites 9:121

Ananthakrishnan G, Ravikumar R, Girija S, Ganapathi A (2002) In vitro adventitious shoot formation from cotyledon explants of cashew (*Anacardium occidentale* L.). Sci Hortic 93:343–355

Archak S, Gaikwad AB, Gautam D, Rao EVVB, Swamy KRM, Karihaloo JL (2003a) DNA fingerprinting of Indian cashew (*Anacardium occidentale* L.) varieties using RAPD and ISSR techniques. Euphytica 130:397–404

Archak S, Gaikwad AB, Gautam D, Rao EVVB, Swamy KRM, Karihaloo JL (2003b) Comparative assessment of DNA fingerprinting techniques (RAPD, ISSR and AFLP) for genetic analysis of cashew (*Anacardium occidentale* L.) accessions of India. Genome 46:362–369

Archak S, Gaikwad AB, Swamy KRM, Karihaloo JL (2009) Genetic analysis and historical perspective of cashew ( *Anacardium occidentale* L.) introduction into India (F Belzile, Ed.). Genome 52:222–230

Ashante AK, Barneit R, Caligari PD (2002) Graft studies on cashew genotypes. Ghana J Agric Sci 35:33–39

Asolkar T, Desai AR, Singh NP (2011) Molecular analysis of cashew genotypes and their half-sib progeny using RAPD marker. Biotechnol Bioinform Bioeng 1:495–504

Balogoun I, Saïdou A, Ahoton EL et al (2014) CARACTERISATION DES SYSTEMES DE PRODUCTION A BASE D'ANACARDIER DANS LES PRINCIPALES ZONES DE CULTURE AU BENIN. Agron Afr 26:9–22

Bassett TJ (2017) The cashew boom in the cotton basin of northern Côte d'Ivoire. Afrique Contemporaine 263–264:59–83

Bastiaan-Net S, Pina-Pérez MC, Dekkers BJW et al (2020) Identification and in silico bioinformatics analysis of PR10 proteins in cashew nut. Protein Sci 29:1581–1595

Bello DO, Ahoton LE, Saidu A et al (2017) Climate change and cashew (*Anacardium occidentale* L.) productivity in Benin (West Africa): perceptions and endogenous measures of adaptation. Int J Biol Chem Sci 11:924–946

Bhadra T, Obaidullah A, Sultana MS, Ahmed M, Islam M (2019) Genetic diversity analysis in cashew (*Anacardium occidentale* L.) germplasm using RAPD marker. J Bangladesh Agric Univ 17:461–465

Bhat MG, Nagaraja KV, Rupa TR (2010) Cashew research in India. J Hortic Sci 5:1–16

Blaikie SJ, O'Farrell P, Muller W et al (2002) Assessment and selection of new cashew hybrids: a report for the Rural Industries Research and Development Corporation. Rural Industries Research and Development Corporation, Barton, ACT

Castro ACR, Bordallo PN, Cavacanti JJV, Barros LM (2011) Brazilian cashew germplasm bank. Acta Hortic 918:857–861

Cavalcanti JJV, Wilkinson MJ (2007) The first genetic maps of cashew (*Anacardium occidentale* L.). Euphytica 157:131–143

Cavalcanti JJV, de Resende MDV, dos Santos FHC, Pinheiro CR (2012a) Simultaneous prediction of the effects of molecular markers and genome wide selection in cashew. Rev Bras Frutic 34:840–846

Cavalcanti JJV, dos Santos FHC, da Silva FP, Pinheiro CR (2012b) QTL detection of yield-related traits of cashew. Crop Breed Appl Biotechnol 12:60–66

Chabi SK, Adoukonou-Sagbadja H, Ahoton LE et al (2015) Morphological characterization and agronomic performances of cashew (*Anacardium occidentale* L.) accessions from Benin. J Agric Crop Res 3:27–40

Croxford AE, Robson M, Wilkinson MJ (2006) Characterization and PCR multiplexing of polymorphic microsatellite loci in cashew (*Anacardium occidentale* L.) and their cross-species utilization. Mol Ecol Notes 6:249–251

Currais L, Loureiro J, Santos C, Canhoto JM (2013) Ploidy stability in embryogenic cultures and regenerated plantlets of tamarillo. Plant Cell Tiss Org Cult 114:149–159

Dasmohapatra R, Rath S, Pradhan B, Rout GR (2014) Molecular and agromorphological assessment of cashew (*Anacardium occidentale* L.) genotypes of India. J Appl Hortic 16:215–221

de Andrade IM, de Nascimento JD, de Sousa MV, dos Santos JO, Mayo SJ (2019) A morphometric study of the restinga ecotype of *Anacardium occidentale* (Anacardiaceae): wild coastal cashew populations from Piauí, Northeast Brazil. Feddes Repertorium 130:89–116

Del Castillo C, Mahy G, Winkel T (2008) La quinoa en Bolivie: une culture ancestrale devenue culture de rente "bio-équitable". BASE 12:421–435

Dhanaraj AL, Rao EVVB, Swamy KRM, Bhat MG, Prasad DT, Sondur SN (2002) Using RAPDs to assess the diversity in Indian cashew (*Anacardium occidentale* L.) germplasm. J Hortic Sci Biotech 77:41–47

Djaha AJ-B, N'Da HA, Koffi KE, Adopo AN (2014) Diversité morphologique des accessions d'anacardier (*Anacardium occidentale* L.) Introduits en Côte d'Ivoire. Revue Ivoirienne de Sciences et Technologies 23:244–258

do Amaral FPM, de Sa GH, Filgueiras LA et al (2017) Genetics analysis of the biggest cashew tree in the world. Genet Mol Res 16:1–7

dos Santos FHC, Cavalcanti JJV, da Silva FP (2011) QTL detection for physicochemical characteristics of cashew apple. Crop Breed Appl Biotechnol 11:17–26

Dos Santos JO, Mayo SJ, Bittencourt CB, de Andrade IM (2019) Genetic diversity in wild populations of the Restinga ecotype of the cashew (*Anacardium occidentale*) in coastal Piauí, Brazil. Plant Syst Evol 305:913–924

Grattapaglia and Silva-Junior (2020) at Phytozome 13. (https://phytozome-next.jgi.doe.gov/info/Aoccidentale_v0_9)

Gniré OM (2017) Les déterminants de l'adoption de certaines bonnes pratiques culturales avant récolte de la noix de cajou en Côte d'Ivoire. J Econ Financ 8:8–15

Hawerroth MC, do Bordallo PN, Oliveira LCP, Vale EH, Vidal Neto F das C, Melo DS (2019) Genetic variability among cashew hybrids and prediction of superior combinations based on agronomic performance. Pesq Agrop Brasileira 54:1–13

Hêdiblè LG, Adjou ES, Tchobo FP, Agbangnan P, Soumanou MM (2017) Caractérisation physico-chimique et morphologique de trois morphotypes de pommes d'anacarde (*Anacardium occidental* L.) pour leur utilisation dans la production d'alcool alimentaire et de boissons spiritueuses. J Appl Biosci 116:11546–11556

Houenou B (2008) Evaluation des conditions de mise en oeuvre des normes de production issues des directives CEE/ONU dans la production des noix brutes d'anacarde à Kouandé. Atacora, Bénin

Hu H, Scheben A, Edwards D (2018) Advances in integrating genomics and bioinformatics in the plant breeding pipeline. Agriculture 8:1–18

Hunter D, Heywood V (2011) Espèces sauvages apparentées à des plantes cultivées: Manuel de conservation in situ. CTA/Bioversity International

Jena RC, Samal KC, Pal A, Das BK, Chand PK (2016) Genetic diversity among some promising Indian local selections and hybrids of cashew nut based on morphometric and molecular markers. Int J Fruit Sci 16:69–93

Kouakou CK, Adopo AN, Djaha AJ-B et al (2020) Genetic characterization of promising high-yielding cashew (*Anacardium occidentale* L.) cultivars from Côte d'Ivoire. Biotechnol Agron Soc Environ 24:46–58

Kra KD, Kwadjo KE, Douan BG, Kouame KL, Ouattara KV, Doumbia M (2017) Évaluation des dégâts de *Analeptes trifasciata* (Coleoptera: Cerambycidae) sur les anacardiers dans les régions du Béré et de l'Iffou (Côte d'Ivoire). J Appl Biosci 112:10969–10977

Krishna H, Alizadeh M, Singh D et al (2016) Somaclonal variations and their applications in horticultural crops improvement. 3 Biotech 6:1–18

Lacroix E (2003) Les Anacardiers, les Noix de Cajou et la Filière Anacarde À Bassila et Au Bénin: plantations et Production Agricole en Afrique. Independently Published

Lautié E, Dornier M, de Souza Filho M, Reynes M (2001) Les produits de l'anacardier: caractéristiques, voies de valorisation et marchés. Fruits 56:235–248

Lebailly P, Lynn S, Seri H (2012) Etude pour la préparation d'une stratégie pour le développement de la filière anacarde en Côte d'Ivoire. AGRER Consortium 92

Limera C, Sabbadini S, Sweet JB, Mezzetti B (2017) New biotechnological tools for the genetic improvement of major Woody fruit species. Front Plant Sci 8:1–16

Masawe PAL, Cundall EP, Caligari PDS, Naliendele ARI (1998) Performance in South-Eastern Tanzania of local cashew germplasm and selected clones in: dares Salam. BioHybrids International, Tanzania, pp 146–150

Mir RR, Reynolds M, Pinto F, Khan MA, Bhat MA (2019) High-throughput phenotyping for crop improvement in the genomics era. Plant Sci 282:60–72

Mzena GP, Kusolwa PM, Rwegasira GM, Yao NK (2018) Discovery of novel single nucleotide polymorphic (SNP) markers for genetic mapping of cashew (*Anacardium occidentale*. L). Int J Agric Environ Biores 3:186–196

N'Depo OR, Cherif M, Johnson F et al (2017) Inventaire des insectes ravageurs du verger anacardier dans les régions de Bounkani, Gontougo et Indénie-Djablun au Nord-Est en Côte d'Ivoire. Afrique Sci 13:333–343

N'djolossè K, Adoukonou-Sagbadja H, Gbèmavo CDSJ et al (2019) Agro-morphological characterization of preselected cashew (*Anacardium occidentale* L.) mother trees in Benin farmer's plantations. J Agric Environ Int Dev (JAEID) 113:17–34

Nanti BTJ-I, Soumahoro BA, Gnamien YG et al (2018) In vitro seeds germination and seedling growth of cashew (*Anacardium occidentale* L.). Agron Afr 30:271–278

Nanti BTJ-I, Gnamien YG, Kone T, Soumahoro BA, Kone M (2020) In vitro regeneration of *Anacardium occidentale* from shoot tip and basal part. Int J Environ Agric Biotechnol 5:621–630

Olubode OO, Joseph-Adekunle TT, Hammed LA, Olaiya AO (2018) Evaluation of production practices and yield enhancing techniques on productivity of cashew (Anacardium occidentale L.). Fruits 73:75–100

Onzo A, Biaou JT, Agboton C (2018) Efficacité du ramassage et du brûlage systématiques des bois morts dans la lutte contre le foreur de bois, *Apate terebrans*, dans les anacarderaies du Nord-Bénin. J Appl Biosci 121:12168–12180

Rabah SO, Lee C, Hajrah NH et al (2017) Plastome sequencing of ten nonmodel crop species uncovers a large insertion of mitochondrial DNA in cashew. Plant Genome 10:1–14

Ribeiro LPD, da Silva APM, de Lima AA, de Oliveira SE, Rinnan Å, Pasquini C (2016) Non-destructive determination of quality traits of cashew apples (*Anacardium occidentale*, L.) using a portable near infrared spectrophotometer. J Near Infrared Spectrosc 24:77–82

Rico R, Bulló M, Salas-Salvadó J (2016) Nutritional composition of raw fresh cashew (*Anacardium occidentale* L.) kernels from different origin. Food Sci Nutr 4:329–338

Roitsch T, Cabrera-Bosquet L, Fournier A et al (2019) Review: new sensors and data-driven approaches—a path to next generation phenomics. Plant Sci 282:2–10

Rout GR, Samal S, Nayak S, Nanda RM, Lenka PC, Das P (2002) An alternative method of plant DNA extraction of cashew (*Anacardium occidentale* L.) for randomly amplified polymorphic DNA (RAPD) analysis. Gartenbauwissenschaft 67:114–118

Samal S, Rout GR, Lenka PC (2003) Analysis of genetic relationships between populations of cashew (*Anacardium occidentale* L.) by using morphological characterisation and RAPD markers. Plant Soil Environment 49:176–182

Samal S, Lenka PC, Nanda RM, Nayak S, Rout GR, Das P (2004) Genetic relatedness in cashew (*Anacardium occidentale* L.) germplasm collections as determined by randomly amplified polymorphic DNA. Genet Resour Crop Evol 51:161–166

Samamad NTI, Ribeiro LPD, de Almeida Lopes MM, Puschmann R, de Oliveira Silva E (2018) Near infrared spectroscopy, a suitable tool for fast phenotyping—the case of cashew genetic improvement. Sci Hortic 238:363–368

Santos FHC, Cavalcanti JJV, Silva FP (2010) Detection of quantitative trait loci for physical traits of cashew apple. Crop Breed Appl Biotechnol 10:101–109

Savadi S, Muralidhara BM, Preethi P (2020) Advances in genomics of cashew tree: molecular tools and strategies for accelerated breeding. Tree Genet Genomes 16:1–15

Sethi K, Tripathy SK, Lenka PC (2016) Assessment of genetic diversity and identification of elite cashew hybrids. Adv Life Sci 16:6200–6205

Sija SL, Potty VP, Santhoshlal PS (2016) In vitro shoot proliferation from excised shoot tip and nodal segment of *Anacardium occidentale* L. Int J Curr Microbiol Appl Sci 5:635–642

Singh A (2018) Early history of crop presence/introduction in India: III. *Anacardium occidentale* L., cashew nut. Asian Agric Hist 22:1–6

Syed NH, Sureshsundar S, Wilkinson MJ, Bhau BS, Cavalcanti JJV, Flavell AJ (2005) Ty1-copia retrotransposon-based SSAP marker development in cashew (*Anacardium occidentale* L.). Theor Appl Genet 110:1195–1202

Tchétangni YA, Assogbadjo AE, Houéhanou T, Bello DO (2016) Perception Paysanne Des Effets Du Changement Climatique Sur La Production Des Noix D'anacardier (*Anacardium occidentale* L.) Dans La Commune De Savalou Au Bénin. Eur Sci J 12:220–239

Thimmappaiah SD, Mohana G, Adiga JD, Bhat P (2016) Fingerprinting of released varieties of cashew based on DNA markers. Vegetos 29:1–7

Tiendrebeogo FK, Sawadogo N, Nanema RK, Traore ER, Zongo J-D, Sawadogo M (2016) Assessment of genetic diversity of *Jatropha curcas* L. in BURKINA FASO. Int J Innov Appl Stud 16:11

Uaciquete A, Korsten L, Van der Waals JE (2013) A search for anthracnose resistant cashew cultivars in Mozambique. Crop Prot 50:6–11

Vayssières J-F, Anato F, Sinzogan A et al (2017) African farmers have amazing allies in their cashew. In: Colloque International d'Échanges Scientifiques sur l'Anacarde, pp 143–149

Wallis NZ, Bagnan MA, Akossou AYJ, Kanlindogbe CB (2017) Caractérisation morphologique d'une collection de fruits d'anacardier provenant de la commune de Parakou (Bénin). Int J Biol Chem Sci 10:2413

Yabi I, Yabi Biaou F, Dadegnon S (2013) Diversité des espèces végétales au sein des agro-forêts à base d'anacardier dans la commune de Savalou au Benin. Int J Biol Chem Sci 7:696–706

# Part VI
# Spices

# Genetic Resources of Small Cardamom

**K. Mary Mathew, Reshma Ranjanan, Varghese C. Rithin, S. K. Bhat, and A. B. Remashree**

**Abstract** *Elettaria cardamomum* Maton or small cardamom is one of the most valued spices worldwide for its versatile flavour and a pleasant aroma. The dried mature fruit capsule which forms the spice is also referred to as the 'Queen of Spices'. The cardamom plant, a shade-loving monocot, has a rhizomatous herbaceous perennial habit and is indigenous to the moist evergreen forests of the Western Ghats of south India. India, formerly the foremost producer, consumer and exporter of small cardamom, is second to Guatemala which fortuitously is currently the largest producer of cardamom. India has reported substantial diversity in small cardamom and is considered to be the centre of origin of *E. cardamomum*. The distribution of abundant genetic variability of the species in the Western Ghats substantiates the hypothesis that it is the major centre of genetic diversity of small cardamom. However, more impetus has to be given in broadening the genetic base, developing new varieties with wide adaptability, and strategies to reduce crop vulnerability by promoting genetic diversity rather than monocultures, which would certainly help sustainable production. During the past years, selection, multiplication, hybridization, evaluation, conservation and in vitro multiplication have all been done in cardamom, but not much introduction of plant materials from other zones have been carried out except those brought to repositories, probably due to its niche specificity. Much work has been done on evolving genotypes with high yield potential and superior capsule quality with relatively lesser importance to resistance or tolerance to biotic and abiotic stress. Concerted efforts are required for creating a detailed database for small cardamom. Molecular characterization and next-generation sequencing (NGS) technologies would help introgression of new alleles from unutilized germplasm accessions, related wild species, and exotic relatives which would be vital for further improvement and enhancement of genetic base of cultivated cardamom.

**Keywords** *Elettaria cardamomum* · Genetic resources · Genetic diversity · Small cardamom

---

K. M. Mathew (✉) · R. Ranjanan · V. C. Rithin · A. B. Remashree
Indian Cardamom Research Institute, Spices Board, (Government of India), Idukki, Kerala, India

S. K. Bhat
ICRI, RRS, Saklespur, Karnataka, India



P. M. Priyadarshan, S. M. Jain (eds.), *Cash Crops*,
https://doi.org/10.1007/978-3-030-74926-2_15

# 1 Introduction

Small cardamom (*Elettaria cardamomum* Maton) is a rhizomatous herbaceous perennial belonging to the family *Zingiberaceae* under the order Zingiberales. The cardamom of commerce is the dried fruit capsule of the plant (Fig. 1), which has a versatile flavour and aroma due to which it is used in sweets and savouries alike universally. It is popularly referred to as the Queen of Spices and is also named as the Grains of Paradise. It is also traditionally known as green cardamom, small cardamom, Malabar cardamom or true cardamom. The combination of the major components of essential oil, viz. α-terpinyl acetate and 1, 8-cineole in cardamom, is responsible for its characteristic fragrance and flavour (Guenther 1975).

Cardamom plant is a large herbaceous perennial monocot growing to about 2–5 m in height and has a subterranean rhizomatous stem with pseudostem (tillers) made of leaf sheaths. The inflorescence is a racemose panicle directly emerging from the base of the pseudostem and possesses a long cane-like peduncle having nodes, the axil from which flowers are borne on cincinnus (Madhusoodanan et al. 2002). Panicles emerge from the base of a tiller. Usually, flowering in cardamom commences with the onset of monsoon and is spread over a period of 6 months, from May to October.

India accounted for 70% of the world production of cardamom in the 1970s, which has come down to 41% (George 2007), and still lower at 26.2% in 2018, while Guatemala contributed to around 30.5% in 2018 of the present world production (Thomas et al. 2019). In India, during 1987–1988, cultivation was under an area of 1,05,000 ha, with a production of 3200 tons (Ravindran 2002). However, the area has reduced to 69,970 ha with a production of 18,000 tons in 2014–2015 (https://www.indianspices.com). Production in Guatemala has gone up to 33,000 metric tons during the crop year 2017–2018 (https://www.statista.com/

**Fig. 1** Dried fruit capsules of cardamom

statistics/964678/guatemala-cardamom-harvest-volume/), whereas India's estimated production is 11,230 metric tons from an area of 69,993 ha during 2019–2020 (https://www.indianspices.com). Indiscriminate deforestation that led to climate change and in turn to drought, changed rainfall patterns and vicissitudes in temperature, increased incidence of diseases and pests and changes in soil pH due to indiscriminate use of pesticides. All these have all led to drastic changes in the habitat of cardamom. This inevitably resulted in the search for alternate crops leading to crop diversification and decline in area under cardamom, which otherwise had been a major source of income. Despite several limiting factors, it is noteworthy that cardamom productivity in India is on the increase due to the selection of high-yielding clones, better crop management through integrated pest, disease and nutrient management. Most of the produce have an extensive domestic market in India, including industrial consumption, whereas Guatemala exports the whole produce (Nair 2011; Indhushree and Kuruvila 2019).

The importance of improving conservation and use of the genetic diversity of crop plants has been widely recognized recently. In cardamom, broadening the genetic base, developing new varieties with wide adaptability and strategies to reduce crop vulnerability by promoting genetic diversity rather than monocultures would certainly help to sustain production. During the past years, selection, multiplication, hybridization, evaluation, conservation and in vitro multiplication have all been done in cardamom. However, not much introduction of plant materials from other zones have been carried out except those brought to repositories, probably due to its niche specificity. Much work has been done on evolving genotypes with high yield potential and superior capsule quality with lesser importance to resistance or tolerance to major pests and diseases, as well as to drought. The reported availability of genetic diversity in abundance in the Western Ghats (Kuruvilla and Madhusoodanan 1988; Madhusoodanan et al. 1994) for a monotypic species such as cardamom makes it inspiring enough to explore and exploit these for improvement of the crop.

## 2 History and Trade

The history of spices is as old as human civilization, and that of cardamom is not different as it was highly valued even from the Vedic period (about 3000 BC). Mentioned by Susruta (eighth century BC) as Ela in Sanskrit, cardamom has been used in India from ancient times (Ridley 1912). Cardamom finds mention in ancient Ayurvedic compendium Charaka Samhita, written between the second century BC and second century AD, as an ingredient in various herbal remedies. The Greek physician Dioscorides (40–90 AD), an author of the legendary Materia Medica mentioned cardamom in his work. Cardamom was widely used to aid digestion, which could have been the reason why the Greeks and Romans imported large quantities of spices from India and became one of the popular oriental spices in Greek and Roman cuisine. Cardamom was listed among the spices liable for duty in Alexandria in 176–180 AD. It was mentioned by Idrisi as a product of Ceylon at about AD 1154 (Charles 2013) and was probably a traded spice in Europe long before that, though there is no definite record of it. Barbosa, the Portuguese traveller, mentioned it as a product of Malabar Coast in 1514. Most of the Indian names of the spice were derived from the Sanskrit word Ela, such as Elaichi (Hindi and Bengali), Yelaki (Kanarese) and Eletari (Tamil, name of the Genus *Elettaria*) (Khan 1990). Cardamom was a very important product of commercial trade in British India, and buyers were mainly the Muslim merchants and the best lot reserved for export as Aleppey green (Watt 1872). Materialistic evidence for cardamom trade was unearthed from Quseir-al-Qadim (formerly Myos Hormos), an ancient Red-Sea port, in archaeological excavation as capsule of cardamom. It is dated to around the eleventh century AD (Van der Veen and Morales 2015).

A summary of cardamom-growing practices in South India was given by Watt (1872). The system of cardamom collection from naturally growing plants was the major practice in India until 1803, but demand shoot up in later years, which naturally led to organized cultivation and the initiation of large-scale plantations in India and Sri Lanka (Ridley 1912). Domestication thus started only in the nineteenth century when the European settlers began large-scale clearing of forests to establish plantations of tea and cardamom (Ravindran 1999). According to Van Sloten (1990), the loss of plant genetic resources is mainly due to environmental degradation, monoculturing, and urban development. However, the wild populations of cardamom also diminished as a result of the large-scale devastation of its natural habitats. Natural populations of small cardamom presently exist only in the Silent Valley forests of the Western Ghats (Ravindran 1999).

The major cardamom exporting countries are Guatemala, India and Indonesia (Fig. 2). Cardamom is also grown in Sri Lanka, Papua New Guinea, Tanzania, El-Salvador, Vietnam, Laos and Cambodia. In India, cardamom is grown mainly in three southern states, viz. Kerala, Karnataka and Tamil Nadu (Nair 2011).

| Cardamom-growing states in India | Districts/places | Production (%) |
|---|---|---|
| Kerala | Idukki, Palakkadu, Waynad | 60 |
| Tamil Nadu | Pulney, Kodai hills | 31 |
| Karnataka | Coorg, Chickmagalur, and Hassan | 9 |

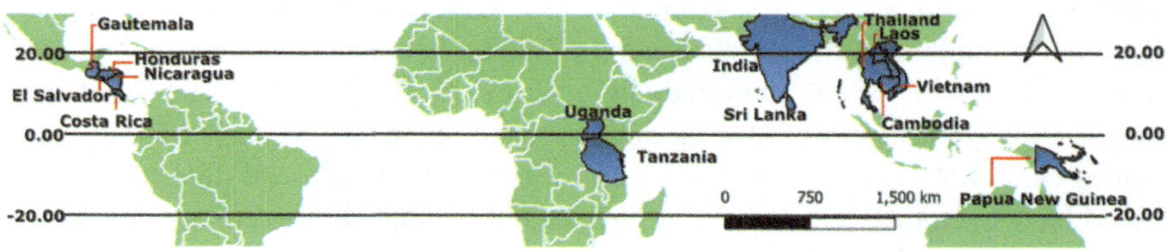


Fig. 2 Cardamom-growing countries

Lawrence (1978) reported that cardamom was introduced in Guatemala in the early 1920s from Sri Lanka or India with the help of a German coffee planter Oscar Majus Kloeffer and was planted in the vicinity of Coban city in the Department of Alta Verapaz in the north central part of Guatemala. After the Second World War, the cardamom production in Guatemala diversified, and it became the top cardamom producer in the world. The natives do not relish the taste of cardamom, and so almost all of the cardamom produced is exported. However, exported Indian cardamom was found superior to that of Sri Lankan and Guatemalan cardamoms for physical and biochemical parameters tested (Kizhakkayil et al. 2006; Nair 2011).

In Tanzania, cardamom was introduced by German immigrants during the 1890s and is being grown in certain localities like Amani and East Usambaras (Lawrence 1978) and cultivation picked up in certain localities in Papua New Guinea, under virgin forests. However, Newmark, in 2002, reported that conversion of natural forest for cardamom cultivation had been one of the major obstacles threatening the East Usambaran forest and its flora.

## 3 Origin and Distribution

*Elettaria cardamomum* Maton, a shade-loving monocot, is indigenous to the moist evergreen forests of Western Ghats of South India (Purseglove et al. 1981; Ravindran 2002). The cardamom-growing region of South India lies within 8′ and 30′ N latitudes and 75′ and 78′30’ E longitudes under the forest canopy (Fig. 3), where an annual rainfall of 1500–4000 mm and a temperature range of 10–35 °C prevail (Anonymous 1976).

**Fig. 3** Cardamom habitat

The genus *Elettaria* is distributed in India, Sri Lanka, Malaysia and Indonesia (Holttum 1950; Willis 1967; Mabberley 1987). Mabberley (1987) reported six species in this genus. The major centre of diversity for the genus *Elettaria* is the Sarawak (Malaysia) and Borneo region, from where eight species have been reported (Sakai and Nagamasu 2000). Small cardamom *(E. cardamomum)* and the Malaysian *Elettaria* (*E. ensal*) were considered to be representations of parallel development from different points of origin in the *Alpinia* stock (Holttum 1950). Recent studies based on molecular phylogeny reveal that *Elettaria* is endemic in India and Sri Lanka. In India, the genus *Elettaria* is represented by only one species *E. cardamomum*, and it is the only economically important member under the genus (Ravindran 2002). Nevertheless, India has reported substantial genetic diversity in small cardamom (Mayne 1951; Abraham and Thulasidas 1958). The distribution of abundant genetic variability of the species in the Western Ghats substantiates the hypothesis that it is the major centre of genetic diversity and therefore the centre of origin of *E. cardamomum.*

# 4 Climatic Conditions

Cardamom is a pseophyte (shade loving) cultivated at an optimum altitude of 600–1500 m above MSL with an annual rainfall of 1500–5750 mm and a temperature range of 17–25 °C (Anonymous 1982; Spices Board 2009). It is a climate-sensitive crop normally grown as rain-fed under forest cover in cardamom hills (Murugan et al. 2007). Viral disease like *katte* mosaic is more severe in summer than in the rainy season. The cold climate also results in poor or no setting of capsules. In South India, all cardamom-growing areas lie between 700 m and 1300 m above mean sea level (MSL) and could adapt even up to 1500 m but with poor growth and yield. Cardamom culture conditions in Guatemala are more favourable than those in India. Guatemala's altitude ranges from 90 to 1200 m above MSL, and rainfall varies from 2000 to 5000 mm. High rainfall and the absence of drought substantiate cardamom yield in Guatemala. In India, most of the cardamom varieties are location specific and very sensitive to elevation, and therefore, erroneous selection of cultivar and location can severely affect the yield and growth. In India, the variety Malabar is highly productive at lower elevations of 500–700 m MSL. Varieties Mysore and Vazhuka exhibit higher yield at higher elevations (Abraham and Thulasidas 1958). The plants are sensitive to moisture stress, and it acts as one of the crucial confining factors in cardamom production as was reported by Subbarao and Korikanthimath (1983, 1991). The devastation of forest habitats changed the weather patterns of the Western Ghats, resulting in persistent drought in the cardamom-cultivating regions leading to a sharp decline in cardamom yield. General climatic conditions of cardamom-growing countries is given in Table 1.

**Table 1** General climatic conditions of cardamom-growing countries

| Country | General climate |
|---|---|
| India | Tropical monsoon type with winter (November to January), summer with hot, dry weather (though normally with high atmospheric humidity) (March to May), southwest monsoon season with hot, wet weather (June to September) and post-monsoon or northeast monsoon season with cool, dry weather (October to December). Most of the rainfall brought by the southwest monsoon. Because of the geographical diversity of India, including elevation, type of relief and proximity to bodies of water, regional climate conditions in the extreme north, east and west vary from the general conditions given here. Cardamom thrives in its natural habitat in the Western Ghats |
| Guatemala | Located within the tropics and with elevations ranging between sea level and more than 13,000 ft (4000 m). Experiences a diversity of climate. Year-round temperatures hover around 60–70 °F (16–21 °C). Winters (November–April) tend to be the warmest months of the year. The dry season (verano) usually lasts from November until May, while the rainy season (invierno) typically runs from May through November. An average of 40–80 in. (1000–2000 mm) of precipitation is received in southern and eastern Guatemala, but this is doubled in areas located nearer the Caribbean shoreline. Severe tropical storms, especially during September and October, often deluge the country with damaging floods |
| Costa Rica | The climate is tropical and subtropical. Dry season (December to April); rainy season (May to November); cooler in highlands. However, the country has many microclimates depending on elevation, rainfall, topography, and the geography of each particular region. Costa Rica's seasons are defined by how much rainfall during a particular period. The location receiving the most rain is the Caribbean slopes of the Cordillera Central mountains, with an annual rainfall of over 5000 mm (196.9 in.). Humidity is also higher on the Caribbean side than on the Pacific side. The mean annual temperature on the coastal lowlands is around 27 °C (81 °F), 20 °C (68 °F) in the main populated areas of the cordillera central, and below 10 °C (50 °F) on the summits of the highest mountains |
| Sri Lanka | Tropical and consists of very distinctive dry and wet seasons. The average temperature ranges from 28 to 32 °C, which may differ due to global weather conditions as a whole. The coldest months, according to the mean monthly temperature are December and January while the warmest months are April and August. Tropical monsoon; northeast monsoon (December to March); southwest monsoon (June to October). Most parts of the country receive an average annual rainfall of more than 50 in. (1270 mm). However, regional differences in the amount of rain, seasonality, variability and effectiveness have formed the basis of a distinction in Sri Lanka between a wet zone and a dry zone. Droughts that persist for more than three months are common in some region |
| El Salvador | Tropical but is moderated by elevation in the interior; in general, it is warm rather than hot, varying between the high 50s and low 70s F (about 15 and 23 °C). Heavy rains, known as the temporales, fall in the winter season, from May to October. The dry summer season lasts from November to April. There is considerable climatic variation in different regions. Annual precipitation on the Pacific lowlands averages about 65 to 70 in. (about 1700 mm); on the southern and northern mountain ranges, at elevations between 2000 and 3500 ft (600 and 1060 m), the average is between 70 and 100 in. (1800 and 2500 mm); the higher mountains receive a little more. Annual precipitation recorded in the deeper valleys and surrounding plateau-like areas is between about 45 and 60 in. (1100 and 1500 mm) |

(continued)

**Table 1** (continued)

| Country | General climate |
|---|---|
| Vietnam | Steamy, tropical climate. It is cooler to the north and gets fairly cold in the winter. July, August and September are the hottest months. Monsoon rains hit the north and south of Vietnam from May to August, while Central Vietnam experiences heavy rain and flooding between October and December. Central Vietnam usually remains warm, with a typical temperature of 65 °F (18 °C) to 85 °F (29 °C). Warm, dry season (mid-October to mid-March). North Vietnam, like China, has two basic seasons: a cold, humid winter from November to April and warm, wet summer for the remainder of the year. The northern provinces share the climate of the north, while the southern provinces share the tropical weather of the south. South Vietnam is relatively warm. The period between December and April is drier and is characterized by the winds of the northeast monsoon and, in the south, by high temperatures |
| Laos | Typical tropical monsoon (wet-dry) climate of the region, though the mountains provide some variations in temperature. During the rainy season (May to October), the winds of the southwest monsoon deposit an average rainfall of 50 to 90 in. (1300 to 2300 mm), with totals reaching some 160 in. (4100 mm) on the Bolivian plateau. The dry season (November to April) is dominated by the northeast monsoon. Minimum temperatures average between 60 and 70 °F (16 and 21 °C) in the cool months of December through February, increasing to highs of more than 90 °F (32 °C) in March and April, just before the start of the rains. In the wet season, the average temperature is 80 °F (27 °C) |
| Thailand | Tropical; rainy, warm, cloudy southwest monsoon (mid-May to September); dry, cool northeast monsoon (November to mid- March); southern isthmus always hot and humid. Between November and February, the winds reverse direction, and the northeast monsoon brings cool, relatively dry air in a south-westerly flow to create cooler temperatures for much of the country. Stagnant air in March and April produces a distinct hot-and-dry inter-monsoonal period. Uplands cause local variations in the general weather patterns, especially on the peninsula: Ranong on the west coast receives approximately 160 in. (4000 mm) of precipitation annually, while Hua Hin on the east coast receives less than 40 in. (1000 mm). Nationwide, temperatures are relatively steady throughout the year, averaging between 77 and 84 °F (25 and 29 °C). The cooler, drier air of the northeast monsoon produces frequent morning fogs that generally dissipate by midday in the north and northeast regions. Humidity is extremely high during the rainy season |
| Cambodia | Climate is governed by the monsoon winds, which define two major seasons. From mid-May to early October, the southwest monsoon brings heavy rains and high humidity. From early November to mid-March, the northeast monsoon brings variable cloudiness, infrequent precipitation and lower humidity. The weather between these seasons is transitional. Maximum temperatures are high throughout the year, ranging from about 82 to 83 °F (28 °C) in January, the coolest month, to about 95 °F (35 °C) in April. Annual precipitation varies considerably throughout the country, from more than 200 in. (5000 mm) on the seaward slopes of the southwestern highlands to about 50–55 in. (1270–1400 mm) in the central lowland region. Three-fourths of the annual rainfall occurs during the months of the southwest monsoon |

(continued)

**Table 1** (continued)

| Country | General climate |
|---|---|
| Honduras | Tropical with cooler, more temperate weather in the mountains with temperature ranging from 16 °C (61 °F) to 20 °C (68 °F). The dry season is from November to April, and the wet season runs from May to October. In the northern and eastern coastal and alluvial plains and on adjacent mountains, mean annual precipitation ranges from 70 to 110 in. (1800 to 2800 mm) or more, with a less rainy season from March to June; these areas occasionally have summer hurricanes that are accompanied by heavy rains. Pacific plains and mountain slopes get 60–80 in. (1500–2000 mm) of rain annually, but from December to April receive little or no rain. Interior sheltered mountain basins and valleys receive 40–70 in. (1000–1800 mm) annually |
| Nicaragua | A tropical climate with two seasons: dry from January to June and rainy season from July to November. The temperatures vary roughly between 54 °F at night and 68 °F in the daytime (12–20 °C). Temperature can reach 100 °F in May (38 °C). The rainfall here is around 1439 mm\|56.7 in. per year |
| Papua & New Guinea | Tropical, with the coastal plains averaging a temperature of 28 °C, the inland and mountain areas averaging 26 °C, and the higher mountain regions, 23 °C. The area's relative humidity is quite high and ranges between 70% and 90%. The extreme variations in rainfall are linked with the monsoons. Generally speaking, there is a dry season (June to September) and a rainy season (December to March). Western and northern parts of Papua New Guinea experience the most precipitation. Rainfall typically ranges from 2000 to 4000 mm (80 to 160 in.) per year, with some higher peaks on the most exposed slopes, where it reaches as high as 7/8 m (23/26 ft) per year, while it drops to 1000/1500 mm (40/60 in.) on the southern coast, overlooking the Gulf of Papua and the Coral Sea |
| Uganda | Tropical weather is perfect all year round with temperatures between 16 and 28 °C for most of the year. Being crossed by the equator, its climate is mitigated by an altitude of 900–1800 m. There is little fluctuation in temperature because there is no real winter or summer. The dry season is experienced from December to February and June to July. The wet season is from September to November and March to May. The northeast, which is more subjected to drought, however, has one continuous wet season from March to November and dry season from December to February |

Sources: www.britannica.com; www.worldatlas.com; www.anywhere.com; www.climatestotravel.com

## 5 Diversity, Including Crop Wild Relatives and Progenitors

Cardamom is highly location-specific with three domesticated natural cultivars, namely Malabar, Mysore and Vazhuka, based on the nature of panicles (Sastri 1952). Var. Malabar is characterized by prostrate panicles; var. Mysore by erect panicles and var. Vazhukka by semi-erect panicles. Var. Vazhukka is considered a natural hybrid between var. Malabar and var. Mysore (Madhusoodanan et al. 2002). Cultivar Malabar is suited to Karnataka, while cultivar Mysore is cultivated mainly in Kerala and parts of Tamil Nadu. Cultivar Vazhukka is mostly cultivated in Kerala. In India, other types of cultivars of cardamom are also identified. Generally, they can be considered as ecotypes of var. *Mysore, Malabar* or *Vazhukka*. Some of them are *Munjarabad, Bijapar, Kannielam, Makaratelam, Thara* and *Nadan* (Table 2).

**Table 2** Major characteristic features of the three varieties of cardamom

| Sl. no. | Parameters | Malabar | Mysore | Vazhukka |
|---|---|---|---|---|
| 1 | Adaptability | Low elevation (600–1000 m msl) | High elevation (900–1200 m msl) | High elevation (900–1200 m msl) |
| 2 | Tolerance to drought | Withstands long dry spell (4–6 months) | Needs well-distributed rain | Needs well-distributed rain |
| 3 | Plant stature | Dwarf (2–3 m) | Tall (3–5 m) | Tall (3–5 m) |
| 4 | Leaf | Short petiole | Long petiole | Long petiole |
| 5 | Panicle | Prostrate | Erect | Semi erect |
| 6 | Bearing nature | Early, short span of flowering | Late, long span of flowering | Late, long-flowering span |
| 7 | Capsule colour | Pale, golden yellow | Green | Green |

Source: Sudharsan et al. (1991)

**Fig. 4** Variety Malabar

## *Var. Malabar*

Malabar type, also called Alleppey green cardamom, is the best grade available in the world. Plants of Malabar variety are medium-sized and attain 2–3 m height. The leaves are pubescent or glabrous. Panicles are prostrate in nature. Fruits are globose, round or ovoid and lightly ribbed. This type is better suited at 600–900 m elevation and is commonly cultivated in Kerala, Tamil Nadu and Karnataka at lower elevation (Fig. 4).

## *Var. Mysore*

Plants are of robust growth and attain 3–4 m in height. The leaves are lanceolate and glabrous. Panicles are erect, and the capsules are bold, longer, fusiform and dark green. It is considered to be more suitable for higher elevations than the Malabar cardamom (Fig. 5).

## *Var. Vazhukka*

This is considered to be a natural cross between Malabar and Mysore varieties. Plants are robust. Leaves are dark green coloured, with oblong lanceolate or ovate shaped. Panicles are semi-erect with round to oblong capsules (Fig. 6).

**Fig. 5** Variety Mysore

**Fig. 6** Variety Vazhukka

## *Other Varieties*

Based on some morphological characters, two more varieties were recognized, viz. var. *mysorensis* and var. *laxiflora* (Madhusoodanan et al. 2002).

***Var. mysorensis***: Plants are tall and robust with either glabrous or pubescent leaves. It has flexuous panicles and flowers are produced in short racemes. The capsules are bold and distinctly three angled. It is common throughout India and mainly cultivated in Karnataka near Mysore.

***Var. laxiflora:*** Plants are short and less robust compared to *var. mysorensis*. This variety has glabrous leaves and flexuous decumbent panicles with oblong–oblong fusiform capsule.

## *The Sri Lankan Wild Cardamom (*E. ensal *(Gaertn) Abeyw)*

Sri Lankan cardamom is known in different names, viz. *Ceylon* (Owen 1901) *E. cardamomum* var. *majus* (Ridley 1912), *E. cardamomum* var. *major* (in Rees Cyclop., 39, 1819),*E. ensal* (Abheywickrama 1959). Burtt (1980) reported that there is not much variability to segregate Sri Lankan cardamom into a new species. But biochemical evidences supported the distinctness of Sri Lankan wild cardamom (Bernhard 1971; Rajapakse 1979) (Table 3).

**Table 3** Cardamom nomenclature

| Author | Cardamom nomenclature | Plant characteristics |
|---|---|---|
| Ridley (1912) | *E. cardamomum* var. majus | Plant with shorter stems with broader leaves, fruit oblong and long narrower than Malabar type, dried fruit is greyish brown in colour, seeds large sized, less aromatic. It is found in Ceylon |
| | *E. cardamomum* var. minus | Narrower and less firm leaves, long globose fruits greyish yellow or buff in colour. This is confined to South India |
| Owen (1901) | *Ceylon* | Plant base has a pink colour |
| | *Malabar* | Plant is green and base of the arial stem is whitish |
| | *Mysore* | Plant robust, panicles are borne perpendicularly from the bulbs and the fruit grows in clusters of five or seven and is better suited for higher elevations |
| Rees Cyclop (1819) | *E. cardamomum* var. *major* | Wild type, indigenous to Ceylon, oblong and long cardamom |
| | *E. cardamomum var. minor* | Smaller cultivated varieties, like Malabar and Mysore cardamom |

### *Allied Species and Their Occurrence*

Sakai and Nagamasu (2000) have described eight species under the genus *Elettaria* (*E. stolonifera, E. kapitensis, E. surculosa, E. linearicrista, E. longipilosa, E. brachycalyx* and *E. rubida*). Poulsen et al. (2018) conducted molecular phylogenetic studies using ITS, At103, plastid matK, ndhF and trnL-F sequences, which resulted in the re-allocation of several species from the genus *Elettaria* into a new genus named as *Sulettaria* and a few under the genus *Alpinia*. They also reported that *Elettaria* were endemic to India and Sri Lanka, and all other species from Peninsular Malaysia, Sumatra and Borneo were *Sulettaria*. Presently, the genus *Elettaria* consists of only two species *E. cardamomum* (L.) Maton and *E. ensal* (Gaertn.) Abeyw.

## 6 Released Varieties Including Selections and Hybrids

Establishment of germplasm repositories, evaluation of accessions and identification have eventually resulted in release of improved varieties of cardamom (Thomas 2006) (Fig. 7). The Indian Cardamom Research Institute (Spices Board, Government of India), released eight high-yielding cardamom selections/hybrids as a series, viz. ICRI-1 to ICRI-8. Out of these ICRI-1, ICRI-2, ICRI-5, ICRI-6, ICRI-7 were for Kerala zone, ICRI-3 and ICRI-8 for Karnataka zone and ICRI-4 for Tamil Nadu zone.

The Indian Institute of Spices Research (IISR-ICAR), Kozhikode, Kerala and Cardamom Research Centre at Appangala in Coorg district (Karnataka), Zonal Agriculture and Horticulture Research Station at Mudigere (University of

**Fig. 7** A few released varieties of cardamom. (**a**) ICRI 1 (selection), (**b**) ICRI 3 (selection), (**c**) ICRI 5 (hybrid), (**d**) ICRI 7 (hybrid)

Agricultural Sciences and Horticulture, Bangalore) and Cardamom Research Station at Pampadumpara, Idukki under the Kerala Agricultural University also have released improved varieties (Table 4). Besides these, several farmers have also conducted selection of superior clumps from populations which resulted in the farmer's selection Njallani Green Gold which is the most popular cultivated variety in Kerala (Fig. 8). Variety releases were through germplasm selection and less frequently hybridization (Zachariah et al. 1998; Kuruvilla et al. 2002, 2006; Radhakrishnan et al. 2004, 2006; Hrideek et al. 2011, 2015). ICRI 5 and ICRI 7 are the only man-made hybrids in cardamom.

## *Genetic Diversity in Landraces*

Elite cardamom landraces native to specific agro-climatic zones provide immense potential in terms of exploitation of genetic resources for the benefit of farmers in the form of superior planting material and for researchers as unique raw material for selection. Screening of such landraces could lead to the selection of promising genotypes and in turn new releases in cardamom. Till date, only a few hybrids (Kuruvilla et al. 2006) have been released in cardamom and hence hybridization and selection of promising genotypes from the resultant segregating populations can also be used as important tools in cardamom improvement. Koshy (2002) reported elite landraces from Idukki District of Kerala. Hrideek et al. (2015) pointed out that even though about 15 improved varieties have been released in India, many of the

**Table 4** Released varieties of cardamom

| S. no. | Variety | Centre which developed | Year of release | Pedigree/parentage | *Av. yield (kg/ha) | Dry recovery | Oil (%) | 1,8-Cineol % | α-Terpenyl acetate % | Area of adaptability | Salient features |
|---|---|---|---|---|---|---|---|---|---|---|---|
| 1. | ICRI 1 | ICRI (Spices Board), Myladumpara, Kerala | 1992 | Selection from Chakkupalam collection, Malabar type | 325 | 22.90 | 8.70 | 29 | 38 | South zone of Idukki, Kerala | An early maturing type globose, round and extra bold dark green capsules; medium-sized panicle with profusely flowering, early maturing type |
| 2. | ICRI 2 | ICRI (Spices Board), Myladumpara, Kerala | 1992 | Clonal selection from germplasm, Mysore type | 375 | 22.50 | 6.67 | 29 | 36 | Parts of Kerala and Tamil Nadu | Performs well under high altitude and irrigated condition, medium long panicles, oblong bold and parrot green capsules, tolerant to azhukal disease |

(continued)

**Table 4** (continued)

| S. no. | Variety | Centre which developed | Year of release | Pedigree/parentage | *Av. yield (kg/ha) | Dry recovery | Oil (%) | 1,8-Cineol % | α-Terpenyl acetate % | Area of adaptability | Salient features |
|---|---|---|---|---|---|---|---|---|---|---|---|
| 3. | ICRI 3 | ICRI (Spices Board), Myladumpara, Kerala | 1994 | Selection from Malabar type | 440 | 22.00 | 6.60 | 54 | 24 | Cardamom growing tracts of Karnataka | Early maturing long pubescent leaves, tolerant to rhizome rot disease, oblong, bold parrot green capsules suitable for hill zone of Karnataka |
| 4. | ICRI 4 | ICRI (Spices Board), Myladumpara, Kerala | 1997 | Clonal selection from lower pulleys, a Malabar type | 455 | 22.76 | 6.40 | – | – | Adapted to lower Palani hills, Tamil Nadu | Early maturity, medium-sized panicles, globose bold capsules. Suitable for low rainfall areas, relatively tolerant to rhizome rot and capsule borer |

| | | | | | | | | | | | |
|---|---|---|---|---|---|---|---|---|---|---|---|
| 5. | ICRI 5 | ICRI (Spices Board), Myladumpara, Kerala | 2006 | Hybrid between MCC 260 × MCC 49 | 1543 | 23.15 | 7.13 | – | – | Kerala and parts of Tamil Nadu | First hybrid variety, early maturity, moderately tolerant to drought, high yield under intensive management, capsule size 68%; more than 70 mm |
| 6. | ICRI6 (MCC-73) | ICRI (Spices Board), Myladumpara, Kerala | 2006 | Selection from the germplasm (acc. MCC-73 | 1200 | 19.00 | 7.33 | – | – | Kerala and parts of Tamil Nadu | High yield, medium maturity, relatively tolerant to drought, high percentage of bold capsules, capsule size 71%; more than 7 mm |
| 7. | ICRI 7 | ICRI (Spices Board), Myladumpara, Kerala | 2010 | Hybrid | – | 22.00 | 8.84 | – | – | Suitable for Wayanad, Kerala. | Suitable for Wayanad, Kerala, semi-erect panicles Angular bold capsules, oleoresin 7.99% |

(continued)

**Table 4** (continued)

| S. no. | Variety | Centre which developed | Year of release | Pedigree/parentage | *Av. yield (kg/ ha) | Dry recovery | Oil (%) | 1,8-Cineol % | α-Terpenyl acetate % | Area of adaptability | Salient features |
|---|---|---|---|---|---|---|---|---|---|---|---|
| 8. | ICRI 8 | ICRI (spices Board), Myladumpara, Kerala | 2014 | A selection from Sakleshpur with oval, bold and pale green capsules | 655 | – | 7.10 | – | – | Suitable for cardamom growing tracts of Karnataka region | Suitable for cardamom growing tracts of Karnataka region |
| 9. | IISR Kodagu Suvasini (CCS-1) | Indian Institute of Spices Research, Calicut, Kerala | 1997 | Selection from OP progeny of CL-37 from RRS Mudigere,Malabar type | 745 | 22 | 8.70 | 42 | 37 | Suitable for Kodagu, Hassan and Chikmagalur | Early maturing, suitable for high density planting, long panicle, tolerant to rhizome, rot, thrips, shoot/ panicle/capsule borer |
| 10. | IISR Avinash (RR-1) | Indian Institute of Spices Research, Calicut, Kerala | 2001 | A selection from OP progeny of CCS-1, a Malabar type | 847 | 20.8 | 6.7 | 30.4 | 35.5 | Suited for hotspots of rhizome rot prone areas | Dark green capsules. Tolerant to rhizome rot, and shoot/ panicle/capsule borer |

| | | | | | | | | | | | |
|---|---|---|---|---|---|---|---|---|---|---|---|
| 11. | IISR Vijetha (NKE-12) | Indian Institute of Spices Research, Calicut, Kerala | 2001 | Clonal selection from field resistant plants for katte, a Malabar type | 643 | 22 | 7.9 | 45 | 23.4 | Suitable for Kodagu, Hassan, Chikmagalur and north Waynad; suitable for katte-prone areas. Adapted to moderate rainfall and moderate to high shade areas | Resistant to katte virus, bold capsules. Field tolerant to thrips and borer |
| 12. | Mudigere 1 | Regional Research Station, UAS, Mudigere. Karnataka | 1984 | Clonal selection from Malabar type | 300 | 20 | 8 | 36 | 42 | Traditional cardamom growing Malanad areas of Karnataka, under rainfed conditions | Erect and compact panicle, suitable for high-density planting, moderately tolerant to thrips, hairy caterpillar and white grubs, pubescent leaf. Short panicle, pale green, oval bold capsule 2 |

(continued)

**Table 4** (continued)

| S. no. | Variety | Centre which developed | Year of release | Pedigree/parentage | *Av. yield (kg/ha) | Dry recovery | Oil (%) | 1,8-Cineol % | α-Terpenyl acetate % | Area of adaptability | Salient features |
|---|---|---|---|---|---|---|---|---|---|---|---|
| 13. | Mudigere 2 | Regional Research Station, UAS, Mudigere. Karnataka | 1994/1996 | Clonal selection from open pollination of Malabar type | 475 | – | 8 | 45 | 38 | Traditional cardamom growing Malanad areas of Karnataka, under rainfed conditions | Early maturing variety, suitable for high density planting, round/oval bold capsules |
| 14. | PV 1 | Cardamom Research Station, KAU, Pampadumpara, Idukki, Kerala | 1991 | A selection from Walayar collection, Malabar type | 260 | 19.9 | 6.8 | 33 | 46 | All cardamom growing tracts of Kerala and parts of Tamil Nadu | An early maturing type, short panicle, elongated slightly ribbed light green capsules, long, bold capsule |

| | | | | | | | | | | | |
|---|---|---|---|---|---|---|---|---|---|---|---|
| 15. | PV 2 | Cardamom Research Station, KAU, Pampadumpara, Idukki, Kerala | 2001 | Selection from OP seedlings of PV-1,a Malabar type | 982 | 23.8 | 10.45 | – | – | All cardamom growing tracts of Kerala and parts of Tamil Nadu | Early maturing, lengthy panicle, long bold capsules, high dry recovery percentage, field tolerant to stem borer and thrips, suitable for elevation range of 1000–1200 m above MSL 5 |
| 16. | PV 3 | Cardamom Research Station, KAU, Pampadumpara, Idukki, Kerala | 2018 | Malabar type | 416 | 18.5 | – | – | – | Suitable for partial shade (50–60%) in tropical evergreen forests of the state | Drought tolerant and moderately tolerant to capsule borer. Ellipsoid capsule; number of capsules 342/ panicle |

(continued)

**Table 4** (continued)

| S. no. | Variety | Centre which developed | Year of release | Pedigree/parentage | *Av. yield (kg/ha) | Dry recovery | Oil (%) | 1,8-Cineol % | α-Terpenyl acetate % | Area of adaptability | Salient features |
|---|---|---|---|---|---|---|---|---|---|---|---|
| 17. | PV 5 | Cardamom Research Station, KAU, Pampadumpara, Idukki, Kerala | 2018 | Vazhukka type | 594 | 18 | – | – | – | Suitable for cardamom growing tracts of Kerala | Tolerant to thrips. Ovoid capsule, number of capsules 153/panicle |
| 18. | Njallani | Farmers selection popularly known as Njallani green gold | 1990s | Selection from Vazhukka type | 1600 | – | 9.01 | – | – | All cardamom growing tracts of Kerala and parts of Tamil Nadu | Non-pubescent, semi-erect, globose extra bold and dark green capsules, high yielder |

Courtesy: Cardamom cultivation practices Spices Board 2009, AICRPS; http://aicrps.res.in/sites/default/files/inline-files/Varieties.pdf, www.kau.in

**Fig. 8** Njallani, farmer's selection

farmers of the traditional cardamom tracts use farmer selected landraces as planting material and hence the importance of such landraces is not questionable. This study on variability, performance and adaptability to analyse the extent of genetic variability in these landraces revealed significant genotypic differences between the landraces. The study claimed to have generated a considerable quantum of information on the variability among ten elite landraces of cardamom and the heritability, genetic advance, genetic divergence and association of characters and helped to identify the lead characters to be considered while practising selection in cardamom (Table 3). Another study by Akhila et al. 2017) characterized and evaluated the variability and performance of the four new landrace of cardamom in comparison with the released variety ICRI-2 and the popular landrace Njallani Green Gold with regard to growth, yield and quality characters which revealed superiority over the popular landrace. Njallani Green Gold had received the first National Grassroots Innovation Awards in 2001 from the Government of India. (http://nif.org.in/innovation/New_cardamom_variety_-Njallani/84). Incidentally, some of the landraces, namely Elarajan, Thiruthali, Panikulangara, Wonder cardamom have also been registered as farmers' varieties with PPV and FRA (https://spices.icar.gov.in/sites/default/files/farmers-advisory/Farmers_varieties.pdf). Diversity of landraces should be further analysed through molecular markers for ascertaining uniqueness and have to be conserved in germplasm for further breeding programmes (Table 5).

**Table 5** Elite landraces of small cardamom

| Sl. no. | Selection | Plant type | Average yield (kg/ plant) | Oil (%) | Distinguishing characteristics |
|---|---|---|---|---|---|
| 1 | Panikulangara 1 | Vazhukka | 4.05 | 8.79 | The plants have oval-shaped and green-coloured capsules. This plant performs well under rainfed conditions |
| 2 | Panikulangara 2 | Vazhukka | 4.82 | 9.08 | The plants are bushy in nature with vigorous growth. Branched panicles are also seen in this type. The capsules are oblong and deep green in colour |
| 3 | Njallani Green Gold | Vazhukka | 3.96 | 9.04 | The plants are robust with tall tillers with purple coloured swollen base. It has simple unbranched inflorescence. Capsules are extra bold, dark green and thick skinned |
| 4 | Vali Green Gold | Vazhukka | 4.81 | 9.85 | The plants are robust in nature and the rhizome is fleshy and very stout with the nodes and internodes crowded. Panicles are unbranched and capsules are extra bold and dark green |
| 5 | Palakkudi | Malabar | 4.48 | 8.49 | Plants are robust in nature and rhizomes stout and fleshy. Each tiller has 2–3 unbranched panicles. Capsules are round bold and deep green in colour |
| 6 | PNS Vaigai | Vazhukka | 5.04 | 8.45 | Plants are robust in nature with tall tillers. The dark green coloured non-pubescent leaves and the prominent ligules with light pale green colour are the striking features of this landrace. It has simple unbranched inflorescence. Capsules are extra bold with parrot green colour and thick skin |
| 7 | Wonder cardamom | Vazhukka | 3.51 | 8.99 | Plants are robust and bushy with very long shoots and deep green foliage. Branched panicles are also seen in this plant. The capsules are extra bold in size and deep green in colour |
| 8 | Ela Rani 1 | Vazhukka | 4.08 | 8.69 | Plants are robust and bushy in nature with very stout pseudostem base. Capsules are round oblong and pale green in colour |
| 9 | Ela Rani 2 | Vazhukka | 4.00 | 8.94 | Plants show only average growth and size with average sized tiller base. Capsules are round, slightly oblong and pale green in colour |
| 10 | Ela Rani 3 | Vazhukka | 4.17 | 8.61 | Plants are robust and bushy in nature with vigorous growth and strong tiller base. Capsules are medium-sized, round oblong in shape with light green colour |

(continued)

**Table 5** (continued)

| Sl. no. | Selection | Plant type | Average yield (kg/ plant) | Oil (%) | Distinguishing characteristics |
|---|---|---|---|---|---|
| 11 | Pulari | Malabar | 2.33 | 7.00 | Plants robust with lesser number of bearing tillers, larger leaves, long panicles, moderate yielder, more seeds per capsule |
| 12 | Pappalu | Malabar | 4.18 | 8.00 | Plants tall and robust, long panicles, high yielder |
| 13 | Kalarikkal Bold | Vazhukka | 3.36 | 9.00 | Plants tall and robust with a high number of bearing tillers, high volatile oil content |
| 14 | Arjun | Mysore | 1.89 | 6.56 | Plants robust with higher number of bearing tillers, moderate yielder |
| 15 | Thiruthali | Malabar | 4.5 | 5.8 | The plant high yielding and resistant to stem borers, root grubs disease. Capsules are medium bold and parrot green coloured |

Source: Hrideek et al. (2011) and Akhila et al. (2017)

# 7 Genetic Resources of Cardamom and Its Conservation

## *Cardamom Germplasm*

Cardamom is cross-pollinated and is propagated mostly by seeds, and hence, the existing variability would be fairly high. The first step in breeding programme is to assemble and evaluate a comprehensive germplasm collection (Cameron 1983). Therefore, collection, conservation and evaluation of genetic resources is of paramount importance in cardamom, like in any other agronomic crop. However, as suggested by Paterniani et al. (2000), these resources collected are to be used in breeding programs and not just to be conserved, as plant genetic resource conservation is much broader than just collection and storage. Characterization and evaluation, an integral part of genetic resource conservation, have been relegated to a second level because of the great effort made to increase the number of accessions. As suggested by Cameron (1983), breeding programs should begin only after appropriate germplasm characterization.

Two surveys were conducted in the 1950s in the cardamom-growing areas of India, which were the first of their kind. The former recorded genetic resources and wild populations and the latter sought to understand the geographical distribution and the environmental impact on cardamom (Abraham and Thulasidas 1958). The conservation of cardamom genetic resources started only in the early 1960s, and six research organizations in the country have conducted exploration for germplasm conservation. The total available accessions at present in these research stations were 1350 (Madhusoodanan et al. 1998, 1999), which has considerably increased

**Table 6** Germplasm collection of cardamom in India

| Sl. no. | Field gene bank | Total no. of germplasm collection | Wild and allied taxa |
|---|---|---|---|
| 1. | IISR Regional Station (Indian Council of Agricultural Research) Appangala, Karnataka | 416 | 13 |
| 2. | Indian Cardamom Research Institute (Spices Board) Myladumpara, Idukki, Kerala | 563 | 12 |
| | Regional Research Station (Spice Board) Saklespur, Karnataka | 244 | 10 |
| 3. | Cardamom Research Centre (Kerala Agriculture Research), Pampadumpara, Idukki, Kerala | 77 | 14 |
| 4. | Zonal Agriculture and Horticulture Research Station, UAHS, Mudigere, Chickmagalur, Karnataka | 236 | 7 |

since then. The Indian Cardamom Research Institute (ICRI, Spices Board, Ministry of Commerce, Government of India) has a National Germplasm Conservatory at its station at Myladumpara, Idukki (Kerala), with the largest germplasm collection (Table 6). The Indian Institute of Spices Research (IISR-ICAR) and its Research centre at Appangala in Coorg district (Karnataka) are the other major conservation centres. ICRI Regional Research Station, Sakleshpur (Karnataka), Zonal Agriculture and Horticulture Research Station at Mudigere (University of Agricultural Sciences and Horticulture, Bangalore) and Cardamom Research Station at Pampadumpara, Idukki under the Kerala Agricultural University also maintain ex situ cardamom germplasm.

## *In Situ and Ex Situ Conservation in Cardamom*

In situ conservation and ex-situ conservation are the two basic conservation strategies. In situ conservation is considered as the primary conservation strategy and ex-situ also plays a supportive role to achieve its conservation targets, as is mentioned in Article 8 of the Convention on Biological Diversity (CBD). But in the case of cardamom in situ conservation actually does not exist except for some naturally surviving genotypes found in protected forest areas of Silent Valley Biosphere Reserve (Ravindran 1999). Ex-situ conservation of cardamom is maintained by four research organizations as mentioned in Sect. 7.1, in the form of field gene banks and is being characterized and evaluated based on IPGRI descriptor (1994). However, due to biotic and abiotic factors, ex-situ conservation always pose a certain amount of risk.

## *In Vitro Conservation*

In vitro conservation is an integrated conservation strategy and is mainly used as an alternative to conserve the vegetative parts of plants under artificially controlled conditions. However, this technique requires specialized laboratories and skilled workers, whereas conservation in field gene banks require a larger area, labour workforce and is generally threatened by natural calamities and edaphic factors. Very few literature is available on in vitro conservation techniques on cardamom. Nirmal Babu et al. (1999) and Geetha et al. (1995), in their studies, found that slow growth technique was ideal for in vitro storage of cardamom. The medium used was that of MS (1962) at half-strength salts devoid of growth regulators but with 10 mg/L sucrose and mannitol as an osmoregulator. Tyagi et al. (2009) developed a single step method for in vitro conservation in cardamom on MS (1962) half-strength medium containing the growth regulator BAP (5μM). However, presently, no information is available on in vitro conservation practices or on scaled up protocols for long-term culture and retrieval.

## *Cryo-Conservation*

Cryo-conservation at ultra-low temperature was attempted in cardamom by suspending seeds in cryovials in the vapour phase of liquid nitrogen (−150°C) by slow freezing and also by direct immersion in liquid nitrogen (−196°C) by fast freezing (Choudhary and Chandel 1995). The seeds were successfully cryo-preserved, and 80% germination was observed when tested after one year of storage in liquid nitrogen. Refinement and utilization of such technologies have not further been reported in cardamom.

## *Collection and Cataloguing of Germplasm Data*

The International Board of Plant Genetic Resources (IBPGR) is aimed at collection and conservation of plant genetic resources and provides essential support for collection, conservation and utilization of plant genetic resources throughout the world. In 1994, a comprehensive plant descriptor for cardamom was published by the International Plant Genetic Resources Institute (IPGRI) in Rome, Italy. The cardamom germplasm is being characterized and evaluated based on these IPGRI descriptors (1994). Besides, National Bureau of Plant Genetic Resources (NBPGR), India, has compiled data of different crops in the form of crop catalogues based on descriptors and DUS guidelines (PPV and FRA 2009). Passport data sheets were used for recording all details, viz. scientific name of crop, crop collector name and cultivar name, geographical coordinates, habitat, data on pests and pathogens, soil characteristics and other special attributes. Each accession was assigned a specific IC (indigenous collection) number for further evaluation and maintenance.

## 8 Studies on Genetic Variability

Studies on variability and selection in cardamom were first reported in the early 1970s and 1980s for yield characters (Pattanshetti et al. 1973; Parameswar and Haralappa 1980; George et al. 1981). Documentations during 1980s were based on an old descriptor (Dandin et al. 1981), and Sudharsan et al. (1991) formulated a key for the identification of various types. Since 1994, with the introduction of cardamom descriptor (IPGRI 1994), several exploration missions have been carried out, which resulted in a good amount of documentation of genetic resources of cardamom (Sudharsan et al. 1988, 1991; Madhusoodanan et al. 1994). The germplasm accessions have been conserved ex-situ in germplasm repositories maintained as field gene banks and are used for characterization and evaluation. In situ conservation of cardamom genetic resources is uncommon, but a sizable population of cardamom plants, in its natural setting, exists in protected forest areas, especially in the Silent Valley Biosphere Reserve in Kerala (Nair 2011). Accessions with distinct morphological marker characters such as compound panicle types (Figs. 9 and 10), terminal panicle bearing, narrow leaf types, pink pseudostem types (Fig. 11), dark green and parrot green capsule colour variants, capsule shape variants such as round bold or globose, oval and ellipsoid types (Fig. 12), flower labellum variability

**Fig. 9** Variability in panicle morphology showing simple, compound and multiple branching racemes

**Fig. 10** Compound panicle type

**Fig. 11** Variability in pseudo stem colour—pink and normal green tillers

**Fig. 12** Variability in capsule shape—globose, oval, ellipsoid

**Fig. 13** Variability in flower labellum

(Fig. 13) and high-yielding biotic stress-tolerant types are conserved in the repositories besides others. The very structure of the cardamom flower is predisposed for entomophily as the stigma is positioned above the anther.

Madhusoodanan et al. (1994) reported that phenotypic variants exist naturally due to cross pollination which is facilitated by honey bees (*Apis cerana indica* and *Apis dorsata*). Kuriakose et al. (2009) reported that no reproductive barrier existed between wild and cultivated populations of cardamom. Nirmal Babu et al. (2012) have also pointed out that all the varieties and races of cardamom are inter–crossable, and the observed variations are probably due to natural crossing. Being cross-pollinated, natural populations of cardamom exhibit variability for quantitative and qualitative characters, precocity of bearing, panicle emergence, length, position and branching of panicles, number of panicles per tiller, size of tiller, number of capsules per raceme, size, shape and colour of mature capsules, essential oil content and resistance to diseases (Prasath and Venugopal 2004).

Many variations in morphological, chemical and yield characters were recorded by Zachariah and Lukose (1992) and Zachariah et al. (1998) and from different regions (Krishnamurthy et al. 1989). Several documents have been published on morphological variability in cardamom (Sastri 1952; Kuruvilla et al. 1992, 2006; Zachariah et al. 1998; Padmini et al. 1999; Madhusoodanan et al. 1994, 2002; Radhakrishnan et al. 2006, 2010; Sasikumar et al. 2012; Sherin 2015). Many have been based on the qualitative and quantitative characters specified for cardamom characterization (IPGRI 1994) and DUS guidelines (PPV and FRA 2009), viz. plant habit, panicle type, leaf characters, number of bearing tillers per clump, number of panicles per plant, panicle length, capsule morphology, etc. to name a few.

Variability within the three natural cultivars was extensively studied by George et al. (1981), Padmini et al. (1999), Prasath and Venugopal (2004). Padmini et al.

(1999) observed that, in general, the varieties Vazhukka and Mysore were more robust than Malabar. Likewise, number of tillers were also more in Vazhukka and Mysore than that in Malabar. Earlier reports of Sudharsan et al. (1991) have also listed the distinctive features of the three natural varieties as the plant stature being tall, medium or of dwarf nature (Table 2). Anisha (2016), however, contradicted that these qualitative characters did not have any close association in particular to any of the three varieties Malabar, Mysore and Vazhuka, revealing close affinities between the varieties. No truly diagnostic morphological features could also be defined to assign an individual cardamom plant or a population of the released varieties to Malabar, Vazhukka or Mysore with certainty before panicle emergence.

The highest variability was reported in panicle characters among the different characters of cardamom (Anonymous 1958, 1987). George et al. (1981) recorded high panicle variability in wild and cultivated cardamom and reported multiple branched panicles, basal and terminal panicles in certain accessions. Panicle length was also reported to be highly variable, especially among the three natural cultivars (George et al. 1981). Padmini et al. (2000) observed that compound panicle types mostly had Vazhukka type of inflorescence and that proximal branching was more prevalent in compound panicle types. Prasath and Venugopal (2009) also reported prominent variation in panicle morphology with compound panicle and branched cincinnus in panicle besides the simple non-branching types.

Radhakrishnan et al. (2006) evaluated 90 cardamom accessions for available genetic diversity in the accessions wherein wide diversity for growth and yield attributes was observed. Hrideek et al. (2011) studied the genetic divergence of some elite landraces of small cardamom cultivated in the cardamom tracts of Idukki district of Kerala. Hrideek et al. (2015) studied the phenotypic and genotypic variance, coefficient of variation, heritability and genetic advance to analyse the extent of variation in each character in elite landraces. Anjali et al. (2016), in their studies on intraspecific variation in cardamom, reported that the genetic variability in different cardamom accessions occurred due to slight insertion or deletion of nucleotide sequences. They observed variation in genome size in the samples studied and revealed that the distribution of the genetic diversity and differences in genome size had no correlation to the type of the genotypes studied such as wild, feral, released variety or cultivar.

Kuruvilla et al. (2000) reported that though cardamom genotypes varied in yield and morphological attributes, variability was not significant in quality aspects like colour and shape of the capsules. Nevertheless, cardamom capsules are globose or ovoid or narrowly ellipsoid to elongate in shape, trilocular, and on maturity attain colour ranging from pale green to dark green (Ravindran 2002). Capsule shapes such as elongated, oblong, oval bold, round bold, oblong bold, extra bold, and pale green to dark green and parrot green coloured capsules have been recorded in different released varieties of cardamom (Madhusoodanan et al. 2002).

Pertaining to quality aspects, the most commercially desirable component in cardamom is its volatile oil. In general, the oil content in cardamom capsules varies from 6.5 to 10.5% (Prasath and Venugopal 2004). Table 4 illustrates the variability in oil content in the released varieties of cardamom. The major chemical constituents which impart a sweet flavour to the oil are terpinyl acetate, linalyl acetate and linalool. Kizhakkayil et al. (2006) reported that Indian cardamom was superior to Guatemalan cardamom in physical and biochemical parameters as well as higher quantities of α-terpinyl acetate and 1,8-cineole, which impart aroma and flavour to cardamom. Variety wise, 'Mysore' possesses the mildly herbaceous, sweet spicy flavour (Govindarajan et al. 1982) with more terpinyl acetate than 'Malabar', while the latter possesses more 1,8-cineole, which imparts a harsh camphor-like note to the oil (Sarath Kumara et al. 1985). Zachariah (2002) reported that capsules of Malabar variety possessed the maximum flavonoid content (0.159%), followed by Vazhukka (0.137%), Guatemala (0.132%) and Mysore (0.090%). Manoj et al. (2018) studied the variability in oil profile from the three different zones of cardamom (A, B and C) in the Cardamom Hill Reserve (CHR) rather than in varieties. However, the maximum essential oil and its constituents α-terpinyl acetate and 1,8-cineole were recorded from zone A, where cultivation of Njallani predominated.

## 9 Plant Propagation and Cultivation Practices

Cardamom is a long-lived perennial plant with a life span of more than 15 years, and it can be propagated both by vegetative methods and from seeds. Cardamom has bisexual flowers, and it is self-compatible (Kuriakose et al. 2009). Cross-pollination is most common in cardamom because self-pollination is hindered due to the slight protrusion of the stigma above the stamens (Parthasarathy and Prasath 2012). Since cardamom is a naturally cross-pollinated crop, seedling population is highly heterogeneous and hence vegetative propagation is mostly adopted in case of elite clones. Vegetative propagation is mostly done by sucker (rhizome propagules) multiplication or less frequently by micro propagated plants. Disease- and pest-free suckers selected from large clumps of growing plants are separated into small clumps and planted. A planting unit in cardamom usually consists of a portion of the rhizome, one old tiller and a small developing shoot. Efficient methods have been devised for attaining high rate of rhizome multiplication of selected high-yielding clumps (Spices Board 2009).

## 10 Traditional Breeding

The rich genetic diversity in cardamom (Mayne 1951; Abraham and Thulasidas 1958) and the considerable variation under cultivation (Nirmal Babu et al. 2012) have led researchers to use techniques such as selection, hybridization, mutation and polyploid breeding for genetic modification and upgradation in this crop. Crop improvement studies in cardamom have been reported from other cardamom-growing countries such as Sri Lanka (Melgode 1938), Guatemala (Rubido 1967),

Tanzania (Rijekbusch and Allen 1971), Papua New Guinea (PNG) (Stone and Willmer 1989) and Columbia (Tamayo and Afanador 2007). Though Guatemala is the top producer, support for research investment on cardamom is poor (Bentley 2016), and therefore, crop improvement studies are literally nil in the recent past.

## Selection

The primary requisite of selection is the availability of variability. Traditionally, cardamom cultivar identification is based on morphological traits. In cardamom, after identifying a variable population, individuals that are the best performers for the desired traits, such as high yielding or tolerant lines, are chosen, and the rest of the population is rejected. The progeny of the selected individuals are screened for the desired characters until a more or less identical plant population is attained, which is time consuming due to the perennial nature of cardamom. Initial collections for desirable traits were made from planters' fields as well as from in situ wild habitats based on definite parameters. These collections are then subjected to primary evaluation trials (PET) followed by comparative yield trials (CYT) and multilocation trials (MLT) under different agro-ecological conditions, the later one due to location specificity of genotypes.

Studies on selection and genetic variability in cardamom started in the 1970s (Pattanshetti et al. 1973). George et al. (1981) reported that high panicle variability exists in wild and cultivated variety that could be used in selection and breeding of cardamom. The documentation of several reproductive characters by Dandin et al. (1981) and the first model descriptor for cardamom and later taxonomic key for varietal identification (Sudharsan et al. 1991; IPGRI 1994) were fundamental in selection of different genotypes for crop improvement in cardamom.

Korikanthimath and Mulge (1998) carried out studies on 49 accessions of small cardamom for growth and panicle characters. These accessions varied among each other for number of tillers, bearing tillers, panicles per plant and branches per panicle. Characters such as number of bearing tillers, panicles per plant and panicle length were found to have a positive correlation with yield in cardamom (Korikanthimath et al. 2000; Backiyarani et al. 2002; Radhakrishnan 2003). Korikanthimath et al. (1999) carried out a selection programme in the seedling progenies and found that six lines were very early flowering types and poor yielders.

Miniraj et al. (2000) evaluated 72 accessions of cardamom for assessing vegetative and economic characters and susceptibility to insect pests. They reported that multiple branching panicle and compound panicle could be used in breeding programmes in view of their higher yield potential. Radhakrishnan et al. (2004) evaluated 17 characters, out of which six factors were identified for having an influence on growth and yield of cardamom. The characters with maximum factor loadings in each group included bearing tillers per clump, seeds per capsule, internodal length, racemes per panicle, leaf breadth and capsules (dry) per kilogram.

Documented reports have indeed mentioned the superiority of the variety Malabar with distinct features such as a relatively higher number of panicles (Padmini et al. 2000), which has a direct correlation to yield; relative tolerance to low rainfall (Sudharsan et al. 1991), high flavonoid content (Zachariah 2002) which has antioxidant and anti-inflammatory properties; and early maturing trait and suitability for high-density planting (Madhusoodanan et al. 2002). Moreover, it has been reported that most of the resistant genotypes reported in cardamom are of Malabar type (Madhusoodanan et al. 2002).

Systematic study of germplasm accessions in India during the 1980s resulted in the identification and release of some elite clonal selections of cardamom (Madhusoodanan et al. 2002). Regional Horticultural Research and Extension Centre of the University of Agricultural and Horticultural Sciences, Chikmagalur, Karnataka, was the first research organization to release a cardamom variety Mudigere 1 in 1984. Mudigere 1 is a cardamom selection from Malabar type, suitable for high-density planting due to its compact nature and is also tolerant to hairy caterpillars and white grubs. Followed by these, several cardamom varieties were released by different research organizations (Spices Board 2009). Selections of these varieties were done on the basis of better yield, disease resistance, early maturing and suitability to different geographical areas. PV1 is an early maturing variety released by the Cardamom Research Station, Kerala Agricultural University, Pampadumpara in the early 1990s. In 1992, the Indian Cardamom Research Institute, Myladumpara, Kerala, released two varieties ICRI 1 and ICRI 2. ICRI 2 is tolerant to azhukal disease and has erect panicles. ICRI 3 is a rhizome rot tolerant variety released by ICRI, Myladumpara, in 1994. ICRI 5 is the first hybrid variety in cardamom, which was released by ICRI, Myladumpara, in the year 2006. ICRI 7, another hybrid variety suitable for Wayanad region, was released in the year 2010. Indian Institute of Spices Research, Calicut, released varieties through selection of which IISR Kudagu Suvasini from Appangala station was found to be early maturing and tolerant to thrips and rot. IISR Avinash was found suitable for planting in rhizome rot-prone areas. Vijetha, another variety by IISR with resistance to katte virus, was released in 2001. KAU Pampadumpara further released PV2 in 2001, PV3 and PV 5 in 2018, the former two Malabar types and the latter Vazhukka. Apart from these research organizations, progressive farmers have also carried out selection of superior varieties from populations. Njallani green gold (vazhukka) is one such farmer's selection from Idukki district, having high yield potential and is presently the most popular one. Though the yield potential of several of the released varieties was above 1000 kg/ha, Njallani outperformed all the improved varieties, even though notable exceptions were found with a few varieties yielding as much as the landrace (Table 3). Till date, only a few hybrids have been released in cardamom and hence segregating populations from hybridization can be used in cardamom improvement.

## *Selection for Biotic and Abiotic Stress Tolerance*

At IISR Cardamom Research Centre at Appangala (Karnataka), collections of Natural Katte Escapes (NKE) lines were subjected to artificial inoculation through the use of insect vectors and further screened. Some of these resistant lines were also high yielding, comparable to the released selections, in both yield and quality (IISR 1997). For further studies and upgradation, the NKE lines were used in a diallele-crossing programme (IISR 1998). IISR Vijetha (NKE12) line is resistant to cardamom mosaic or katte virus (Sasikumar et al. 2012). Extensive exploration for natural resistance, which is the most economical, feasible and sustainable strategy to manage diseases, was successfully adopted in the identification and release of the rhizome rot resistant variety IISR Avinash (Venugopal et al. 2006). ICRI-2 selection developed by Indian Cardamom Research Institute is also tolerant to Azhukal or rot. ICRI-6 has also been found to be moderately tolerant to rot. PV-2 developed by Kerala Agricultural University exhibits 23% tolerance to capsule rot. Senthil Kumar et al. (2018) studied resistance sources for diseases like rhizome rot and leaf blight in addition to morphological and yield parameters of 70 genotypes of small cardamom. They categorized 29 genotypes under highly resistant to rhizome rot disease with percent disease index ranging from 0.0 to 5.0%. Selection for drought tolerance yielded more or less successful results from KAU Pampadumpara in 2018 with the release of PV3, which was also moderately tolerant to capsule borer and PV 5 variety was tolerant to thrips (http://www.kau.in/basic-page/varieties-released).

## *Hybridization*

The perennial, cross-pollinated and heterozygous nature of cardamom calls for a longer breeding cycle through conventional means to achieve homozygous lines. Thus, exploitation of direct heterosis by means of crossing two homozygous types is remote. As such, hybridization of varied types is the only strategy to evolve new and divergent cultivars. Hybridization studies involving inter-generic and inter-varietal techniques were used. Inter-varietal and inter-cultivar hybridizations have been carried out for producing high-yielding progenies. The inter-generic hybridization in cardamom was mainly tried with the intention of transferring disease resistance (Krishnamurthy et al. 1989; Madhusoodanan et al. 1990). Studies involving inter-generic crosses involving *Amomum, Alpinia, Hedychium* and *Aframomum* resulted in sterile progenies (Krishnamurthy et al. 1989; Madhusoodanan et al. 1990) except in a cross with *Alpinia neutans* that yielded fruits (Parameswar 1977). Polycross progeny selection studies were conducted by Chandrappa et al. (1998), with clonal selections of *Malabar*-type cardamom and its open pollinated seedlings. It was found that selections from polycross seed was very effective for the production of high-yielding progenies in cardamom and also stated that more than 3000 plants are required for a successful selection. Krishnamurthy et al. (1989) conducted a study on diallele cross of six genotypes and found that the hybrid progenies were more vigorous than its parental lines. Another study on evaluation of F1 hybrids from inter-varietal crosses in cardamom by Madhusoodanan et al. (1998, 1999) led to the isolation of a few high-yielding heterotic recombinants for cultivation under moderate management practices. The released hybrid ICRI 5 resulted from a cross between ICRI 1 and Njallani and ICRI 7 from PV1 × MCC 12, a white flowered accession moderately tolerant to rot and a high yielder (Anonymous 2009).

### *Polyploidy*

Sudharsan (1987, 1989) successfully induced polyploids ($2n = 4x = 96$) in cardamom by treating with 0.5% aqueous solution of colchicine in the sprouting seeds. The polyploid lines exhibited gigantism, traits related to drought tolerance and reasonably good fertility. However, tetraploids were inferior to their diploid counterparts in all traits that governed yield (Anonymous 1986).

### *Mutation Breeding*

Mutagens like x-rays, gamma rays, ethyl methane sulphonate and maleic hydrazide were used for generating desirable mutants in cardamom. Though *katte* virus tolerance was reported (Bavappa 1986), the studies did not lead to any effective, desirable mutant in cardamom. There have been reports on sterility as early as the 1960s (Pillai and Santha Kumari 1965). The absence of macro-mutation in $M_1$ generation and its progeny was reported in studies by ICRI (Anonymous 1987).

## 11 Biotechnological Approaches

### *Micropropagation*

Cardamom is one of the first plantation crops where commercialization of micropropagation has been achieved (Nadgauda et al. 1983; Vatsya et al. 1987; Rao et al. 2000). Micropropagation of small cardamom was standardized by many research groups (Rao et al. 1982; Nadgauda et al. 1983; Kumar et al. 1985; Reghunath 1989; Lukose et al. 1993; Nirmal Babu et al. 1997; Pradip Kumar et al. 1997; Rao et al. 2000). The various approaches used were callus regeneration, adventitious bud proliferation and enhanced axillary branching. Reghunath and Bajaj (1992) provided a detailed account of micropropagation in cardamom using various explants such as shoot primordium, inflorescence primordium, immature inflorescence segments and immature capsules. Priyadarshan et al. (1992) used shoot primordia as explants and got the best results with MS medium fortified with IAA, BAP, kinetin, calcium pantothenate, biotin and coconut water. Micropropagation system through increased axillary branching leading to large-scale production of disease-free homogenous plantlets was achieved (Vatsya et al. 1987; Priyadarshan et al. 1992). Rao et al. (1994) stated that uniformity in plant population and thereby uniformity in yield can be attained through micropropagation in cardamom due to the high variability in seedling population of cardamom. Nirmal Babu et al. (1997) reported micropropagation of cardamom from rhizome bits with vegetative buds. They used MS medium supplemented with 1 mg/L BAP and 0.5 mg/L NAA for shoot multiplication and rooting in one step. Scaled-up cost-efficient protocol for large-scale propagation of cardamom through micropropagation from rhizome buds was reported by Rao et al. (2000), and the protocols were transferred to Biotech firms in the country for large-scale propagation (Figs. 14, 15 and 16). Recently, Malhotra et al. (2020) claimed to have developed a reliable method for producing genetically stable micropropagated plants in cardamom from shoot tips with reliability tests using ISSR markers.

During 1989–1992, a total of 1.85 lakhs of tissue culture (TC) plants were raised in 100 ha of farmers' fields for evaluating their performance against their seedling counterparts under a scheme implemented by Spices Board and The Department of

**Fig. 14** Cardamom TC primary culture

**Fig. 15** Cardamom TC multiplication

**Fig. 16** Primary hardened plants

Biotechnology, Government of India (Anonymous 1996). The results showed that tissue culture plants under high management exhibited a 25–36% increase in yield over their seedling counterparts (Anonymous 1996). In this study, however, suckers were not included. TC cardamom project plan was the first-ever large-scale field trials of tissue culture plants in India. However, Sudharsan et al. (1997) reported an overall variability of 4.5% in tissue cultured plants as against 3.0% in open-pollinated seedling counterparts for a specified set of characters. The reasons attributed for variations were adventitious bud formation during micropropagation, genetic instability of adventitious meristem and tissue culture-induced disorganization of meristems which are not established till date. Chandrappa et al. (1997) tested eight tissue cultured cardamom selections against their suckers along with two local checkouts of which three TC clones were found promising in yield-contributing characters. In spite of the success of tissue culture plan projects, farmers preferred suckers, possibly due to a less significant awareness about TC plants. The major reasons could be initial establishment constraints of TC plants in moderately managed plantations in contrast to sucker multiplication that gave yields in a relatively shorter duration and higher field establishment costs of TC plants. But the need and importance still exist because suckers are one of the major sources of cardamom mosaic virus (katte) inoculum that cause severe damages to cardamom plantations. The production of virus free planting material can be achieved through scaling up of meristem cultures in cardamom or through mandatory testing of micropropagated plants using molecular markers.

Other systems of tissue culture, such as somatic embryogenesis (SE), anther and protoplast culture, have been reported in cardamom to a very limited extent. Somatic embryogenesis, which is the development of embryos from somatic cells, has been reported in cardamom (Manohari et al. 2008) from rhizome callus followed by successful plantlet regeneration and hardening. The highest frequency of embryogenic calli (68%) and plantlets (86%) were obtained from MS medium containing 4.4μM BAP and 0.5μM NAA. Further, shoot development and rooting were obtained in 13.2μM BAP and 0.5μM NAA. This method if scaled up through bioreactors, could be used as an alternative to conventional micropropagation for large-scale production in cardamom as somatic embryos offer many advantages due to its bipolar nature where shoots and roots are formed simultaneously.

There have been only a very few reports on another culture, protoplast culture and development of synthetic seeds in cardamom. In an attempt to produce haploids in cardamom, which would have been a breakthrough in producing pure lines, Ravindran (2002) reported callus induction and proliferation from anther cultures but failed to develop whole plantlets. Successful isolation of protoplasts from cell suspensions and leaf tissues (mesophyll) in cardamom to microcalli stage was achieved by Geetha et al. (2000). The technique would have been immensely useful in somatic hybridization of cardamom with wild genera for incorporation of desirable characters. Though viability of protoplasts was reported to be 75% in mesophyll and 40% in cell suspensions, no reports on further plantlet formation are available. Production of synthetic seeds, as a new propagation method, by encapsulation of cardamom buds was also reported in cardamom (Ganapathi et al. 1994;

Sajina et al. 1997). Synthetic seeds were generated by encapsulating somatic embryos or shoot tips in hydrogel sodium alginate and then dipped in calcium chloride or calcium nitrate to make the beads firm and permeable. Ganapathi et al. (1994) encapsulated shoot tips in 3% alginate and then further cultured it on MS medium. However, in further studies, encapsulated shoot tips from in vitro grown plantlets of cardamom gave only a 60% success rate (Ravindran et al. 2004; George, 2007). Synthetic seeds mimic natural seeds in structure and though offer huge possibilities for propagation, storage, in vitro conservation and germplasm exchange, the challenges remain un-attempted in cardamom to make it more amenable for farmers and stakeholders.

## 12 Modern Breeding Approaches

### *Genetic Transformation*

Genetic transformation, the innovative technique that has become the most debated one globally, has been more or less widely applied to quite a large number of crop plants. Plant genetic transformation leads to generation of transgenic plants, which express genes of foreign origin that are integrated into the plants and are commonly called genetically modified plants. It mainly aims to improve agricultural production and crop protection by transferring genes between gene pools and unrelated organisms artificially, in contrast to conventional breeding, as only the gene of interest for desired traits are being transferred to the recipient plants. Genetic transformation if applied in cardamom, could lead to significant upgradation of pest and disease control strategies, eventually augmenting crop production.

Nirmal Babu et al. (1998) were the first to attempt particle bombardment of cardamom callus using glucuronidase (*GUS* gene), the enzyme in the bacterium of *E. coli* which is one of the popular and common plant reporter systems. They confirmed transient expression of *GUS* gene in the bombarded callus tissues of cardamom. Also, some preliminary study was conducted on the transformation of cardamom using the biolistic process to study the optimum conditions for gene delivery and the efficiency of the plasmid vector pAHC 25 and promoter Ubi-1 (maize ubiquitin) followed by gene expression studies in callus. In a recent study, Jebasingh et al. (2017) successfully introduced RNA-dependent RNA polymerase gene (NIb) of cardamom mosaic virus (CdMV) into cardamom embryogenic callus by particle bombardment in an attempt to produce plants resistant to *katte* disease caused by the virus. The method was based on resistance derived from the pathogen. The transformants were proved with the presence of NIb and the marker genes Green Fluorescent Protein genes (*gfp*) by PCR, Southern hybridization and GFP expression. The variety selected for the studies was the popular Njallani green gold, and the evaluation of these NIb transgenic plants is reportedly in progress.

## *Molecular Marker Studies*

Morphological characterization and germplasm selection and hybridization have been carried out in cardamom (Zachariah et al. 1998; Kuruvilla et al. 2002; Radhakrishnan et al. 2004, 2006; Hrideek et al. 2011, 2015; Akhila et al. 2017) that has led to several varietal releases though less frequently hybrids. Incidentally, reports on molecular marker studies on cardamom have been fewer (Radhakrishnan and Mohanan 2005; Nirmal Babu et al. 2012; Sherin et al. 2013, 2014; Cyriac et al. 2015; Anjali et al. 2015, 2016) and remain to be assessed for its contribution to prebreeding efforts. Molecular markers have been infrequently used in estimating the relationships among genotypes to analyse genetic diversity in cardamom. Random amplified polymorphic DNA (RAPD), restriction fragment length polymorphism (RFLP), inter simple sequence repeats (ISSR) and simple sequence repeat (SSR) are a few of them (Radhakrishnan and Mohanan 2005; Nirmal Babu et al. 2012; Cyriac et al. 2015). Nirmal Babu et al. (2012) employed RAPD, RFLP-PCR and ISSR polymorphism techniques for genetic diversity analysis and phylogenetic relationship in small cardamom and related genera. They studied about 11 related species representing five major tribes, viz. *Amomum, Aframomum, Alpinia, Hedychium* and *Elettaria* and 96 accessions of cardamom germplasm representing the range of genetic variability available in different Cardamom Research Centres. The study helped to discriminate all the genotypes and to ascertain that no duplicates existed in germplasm collections studied. Furthermore, they reported that the genetic variation expressed by molecular markers (RAPD) was relatively low among cultivated cardamom though most of them were morphological variants which was expected from *Elettaria* a monotypic genus in India. Molecular characterization of selected cardamom genotypes using AFLP markers was done in Columbia (Tamayo and Afanador 2007). The study included 15 accessions comprising nine improved varieties and one wild species, among others, and significantly different groupings were observed. Thomas et al. (2006) performed molecular profiling of Indian, Sri Lankan and Guatemalan exported cardamoms by RAPD and ISSR primers for profiling and reported that in spite of a lack of genetic polymorphism among the tested samples, they showed variation in quality parameters.

Studies on genetic diversity of 50 germplasm accessions using ISSRs and microsatellite markers revealed high polymorphism (Sherin et al. 2014). A variety-specific marker was generated for Malabar variety which has the potential for testing seedling progenies in breeding populations (Sherin et al. 2013). Anjali et al. (2015) developed EST-SSR markers to assess genetic diversity in cardamom, suggesting their usefulness for quantitative trait mapping and marker-assisted selection besides genetic diversity and phylogenetic studies in cardamom. Genetic diversity studies using ISSR markers were conducted on a total of 13 released varieties of cardamom, including selections and hybrids from different research stations in South India along with a Farmer's Selection, Njallani (Anisha 2016). The study, however, revealed a relatively high level of genetic redundancy among the tested varieties with the exception of a few, which could be an indicator of the narrowing genetic base in the released varieties. Earlier, Prasath and Venugopal (2004) reported that cluster analysis using morphological markers had demonstrated that there was no definite clustering of accessions for Malabar, Mysore and Vazhukka type of panicles, as accessions of three cultivar groups often grouped in the same cluster, ascertaining the common ancestry between the three varieties.

As regards released varieties of cardamom are concerned, Anisha (2016) reported that in spite of the availability of a good number of germplasm accessions, efforts for exploiting variability from these germplasm resources has been limited. The study reported that not much genetic enhancement is reflected in the released varieties and that it would be logical to assume that the genetic base of cardamom in terms of released varieties might have become narrower due to selection consequential to domestication and, more recently, commercialization. Some of the recent studies have also concentrated more on yield performance of newer landraces brought out by farmers (Akhila et al. 2017) as would be in any such crop, but not much efforts for hybridization of varieties or for the introduction of new alleles such as for biotic and abiotic stresses have been noticed. Nirmal Babu et al. (2012) suggested that hybridization rather than selection should be the preferred method of improvement to broaden the genetic base of cultivated cardamom. As recently documented, natural populations of cardamom were replaced by 'selections' which narrowed down the genetic base and thus the evolutionary potential of cardamom (Anjali et al. 2016).

Variability can be analysed using molecular marker data, which would enable categorization of varieties at the genome level and could be used to complement conventional morphological characterization in identifying populations and for studying the pattern of diversity. The ability to characterize genetic variability with DNA-based technologies has revolutionized germplasm characterization, and DNA markers provide much clear insight into genetic variation at molecular level (Rafalski et al. 1996). The rapid advancement of sequencing technologies such as next-generation sequencing technology makes possible the sequencing of the whole genome and transcriptome of a species and helps to study the expression profile of entire genes. Genome-wide expression studies help breeders with an understanding of the molecular basis of complex traits and can be extensively utilized for crop genome upgradation. Anjali et al. (2017) reported that to date, no information is available about the regulatory roles of miRNAs in response to drought stress in cardamom. In their studies, they identified 20 novel microRNAs from cardamom samples collected from natural forest areas. They suggested that 17 differentially expressed miRNAs discovered under drought stress could have been involved in various biological processes to improve tolerance to water stress. This study has been the first report of drought-responsive miRNAs and their targets in cardamom. Nadiya et al. (2018, 2019) studied miRNA profiling of cultivar and wild cardamom genotypes using Ion Proton sequencer and identified conserved and novel miRNAs also its differential expression analysis revealed that most miRNAs identified were highly expressed in cultivars and, furthermore, lower expression of miR169 and higher expression of miR529 in wild cardamom proved evidence that wild genotypes have stronger drought stress tolerance and floral development than cultivars. Cardamom plants are vulnerable to various biotic and abiotic stresses. So for identifying molecular mechanisms underpinning host plants, cardamom mosaic virus (CdMV) (Khan et al. 2019) and *Phytophthora meadi* infected leaf transcriptome genomic resources were developed (Mathew et al. 2019) and are available in NCBI. (https://www.ncbi.nlm.nih.gov/bioproject/?term=elettaria%20cardamom). Capsule rot caused by *Phytophthora meadii* has been one of the major destructive diseases in cardamom plantations. A limited number of transcriptome-based sequencing studies are available in cardamom so far, and genetic map or whole genome-wide expression studies have not been reported. Mathew et al. (2019) reported transcriptomes of capsule rot-affected cardamom plants, which revealed differential gene expressions in infected and healthy plants. This finding has yet to be utilized in future breeding programmes in marker-assisted selection (MAS) in cardamom as, transcriptomes are valuable resources for ecological, genetic and evolutionary studies apart from giving evidence for genes resistant/ tolerant to diseases and pests. Markers that associate at high frequency with the gene of interest such as productivity, biotic and abiotic stress tolerance and quality have to be identified from DNA/RNA sequence variations and utilized for MAS in cardamom breeding programmes which is otherwise a cumbersome task due to the perennial nature of the plant.

## 13 Conclusions and Future Prospects

India has the largest germplasm collection of small cardamom in its ex situ germplasm repositories. However, only a limited contribution from that has been made in the development of varieties which points out that diversity assessment using molecular markers is a prerequisite for the evaluation of germplasm which can also help to minimize pre-breeding attempts. Despite being a prestigious Indian spice renowned worldwide, research on cardamom has not been given much impetus it deserves. Likewise, being monotypic also has not prompted thorough research on geographical origin, crop improvement, genetic enhancement, analysis of intervarietal diversity, hybridization, breeding for resistance, heterosis, geographical diversity, to name a few. Concerted efforts are also required for creating a detailed database for small cardamom. Molecular characterization and NGS technologies would help introgression of new alleles from unutilized germplasm accessions, related wild species, and exotic relatives, which would be vital for further improvement and enhancement of the genetic base of cultivated cardamom. Specific molecular markers that could reveal genes governing resistance to biotic and abiotic stresses have to be generated and used to exploit germplasm resources for use in resistance breeding and Marker Assisted Selection (MAS) for attaining sustainable crop production. Documented evidences (Nair 2011; Jyoti et al. 2019) reveal that though the area under cultivation of cardamom has come down significantly, production has not declined proportionally, which certainly gives hope for regaining the monopoly of top producer of quality cardamom that India had formerly enjoyed for a considerable period of time.

**Acknowledgements** The authors are grateful to Dr. S. K. Bhat, Scientist, ICRI, Saklespur, for providing relevant images for the manuscript and Shri Basil K. Eldose, ICRI, Myladumpara, for the assistance received for data compilation and processing of the manuscript.

## References

Abheywickrama BA (1959) A provisional checklist of the flowering plants of Ceylon. Ceylon J Sci (Brol Sea) 2:119–240

Abraham P, Thulasidas G (1958) South Indian cardamom and their agricultural value. In: Technical Bulletin, vol 79. ICAR, New Delhi, pp 1–27

Akhila PR, Radhakrishnan VV, Pradipkumar K, Mohanan KV (2017) Assessment of variability and performance of new landraces of small cardamom (*Elettaria cardamomum* Maton). J Plant Crop 45(1):43–48

Anisha CS (2016) Agro-morphological and molecular characterization of some released varieties of small cardamom (*Elettaria cardamomum* Maton). PhD thesis, Andhra University, Visakhapatnam

Anjali N, Dharan SS, Nadiya F, Sabu KK (2015) Development of Est-ssr markers to assess genetic diversity in elettaria cardamomum maton. Int J Appl Sci Biotechnol 3(2):188–192

Anjali N, Ganga KM, Nadiya F, Shefeek S, Sabu KK (2016) Intraspecific variations in cardamom (*Elettaria cardamomum* Maton): assessment of genomic diversity by flow cytometry, cytological studies and ISSR analysis. Springer Plus 5:1560

Anjali N, Nadiya F, Thomas J, Sabu KK (2017) Discovery of microRNAs in cardamom (Elettaria cardamomum Maton) under drought stress. Hindawi Dataset Papers in Science: 9507485

Anonymous (1958) South Indian cardamoms—their evolution and natural relationships. In: Technical bulletin, vol 57. ICAR, New Delhi, pp 1–27

Anonymous (1976) Tissue cultured cardamom: product plan. Final report, Spices Board, India

Anonymous (1982) Condiments and spices. In: Hand book of agriculture. ICAR, New Delhi

Anonymous (1986) Annual report, ICRI. Spices Board, Myladumpara, India

Anonymous (1987) Annual Report, ICRI. Spices Board, Myladumpara, India

Anonymous (1996) Annual Report, ICRI. Spices Board India, Kochi, Kerala

Anonymous (2009) Annual Report. Spices Board, Kochi, India

Backiyarani S, Kurian PS, Joseph Rajkumar A, Murugan M (2002) Evaluation of high yielding accessions of small cardamom (*Elettaria cardamomum* Maton) for suitability in the high ranges of Idukki District. J Spices Aromatic Crops 11(2):93–96

Bavappa KVA (1986) Research at CPCRI. In: Technical Bulletin. CPCRI, Kasaragod, India, p 34

Bentley JW (2016) Cardamom—a crop cannot run away from its problems. Outlook Pest Manag 27:4–6. https://doi.org/10.1564/v27_feb_02

Bernhard RA (1971) Terpenoids of cardamom oil and their comparative distribution among varieties. Phytochemistry 10:177–184

Burtt BL (1980) Cardamoms and other Zingiberaceae in Hortus Malabaricus. In: Manilal KS (ed) Botany and history of Hortus Malabaricus. Oxford and IBH, New Delhi, pp 139–148

Cameron DF (1983) To breed or not to breed. In: McIvor JG, Bray RA (eds) Genetic resources of forage plants. CSIRO, East Melbourne, pp 238–250

Chandrappa HM, Shadakshari YG, Sudharsan MR, Raju B (1997) Preliminary yield trial of tissue cultured cardamom selections. In: Edison S, Ramana KV, Sasikumar B, Nirmal Babu K, Eapen SJ (eds) Biotechnology of spices, medicinal and aromatic plants. Indian Society of Spices, Calicut, India, pp 102–105

Chandrappa HM, Shadakshari YG, Dushyandhakumar BM, Edison S, Shivashankar KT (1998) Breeding studies in cardamom (Elettaria cardamomum Maton.). In: Mathew NM, Jacob CK (eds) Developments in plantation crops research. Allied, New Delhi, pp 20–27

Charles DJ (2013) Antioxidant properties of spices, herbs and other sources. Springer, Berlin

Choudhary R, Chandel KPS (1995) Studies on germination and cryopreservation of cardamom (*Elettaria cardamomum* Maton.) seeds. Seed Sci Biotechnol 23(1):235–240

Cyriac A, Paul R, Prasath D, Deepesh PV, Nirmal Babu K, Parthasarathy VA (2015) Transferability of ginger, turmeric and large cardamom SSR primers to small cardamom (*Elettaria cardamomum* Maton). J Trop Agric 53(2):107–115

Dandin SB, Madhusoodanan KJ, George KV (1981) Cardamom descriptor. In: Proc. IVth Symp. Plant. Crops (Placrosym-IV). CPCRI, Kasargode, pp 401–406

Ganapathi TR, Bapat VA, Rao PS (1994) In vitro development of encapsulated shoot tips of cardamom. Biotech Tech 8:239–244

Geetha SP, Manjula C, Sajina A (1995) In vitro conservation of genetic resources of spices. In: Proc. Kerala Sci. Congress, Palakkad, pp 12–16

Geetha SP, Nirmal Babu K, Rema J, Ravindran PN, Peter KV (2000) Isolation of protoplasts from cardamom (*Elettaria cardamomum* Maton.) and ginger (*Zingiber officinale* Rosc.). J Spices Aromat Crops 9:23–30

George Y (2007) Studies on Cryopreservation of Spices Genetic Resources, Ph. D Thesis, University of Calicut, Kerala, India

George KV, Dandin SB, Madhusoodanan KJ, Koshy J (1981) Natural variations in the yield parameters of cardamom (*Elettaria cardamomum*). In: Visveshawara S (ed) Proc. PLACROSYM 1981. ISPS, Kasaragod, India, pp 216–231

Govindarajan VS, Shanthi N, Raghuveer KG, Lewis YS (1982) Cardamom production, technology chemistry and quality. Crit Rev Food Sci Nutr 16:326

Guenther E (1975) The cardamom oils. In: Robort EK (ed) The essential oils, vol 5. Van Nostrand, New York, pp 85–106

Holttum RE (1950) The Zingiberaceae of Malay peninsula. Gard Bull 13:236–239

Hrideek TK, Radhakrishnan VV, Mohanan KV, Kuruvilla KM, Madhusoodanan KJ, Thomas J (2011) Genetic divergence in some elite landraces of small cardamom (*Elettaria cardamomum* Maton). J Plant Crops 39(1):201–202

Hrideek TK, Radhakrishnan VV, Kuruvilla KM, Mohanan KV (2015) A study on variability of elite landraces of small cardamom (*Elettaria cardamomum* Maton). Indian J Plant Genet Resour 28(3):311–316. https://doi.org/10.5958/0976-1926.2015.00041.8

IISR (NRCS) (1997–1998) Annual report (1997–1998). Indian Institute of Spices Research, Calicut, India

Indhushree A, Kuruvila A (2019) Performance of small cardamom export from India. J Trop Agric 57(2):122–131

IPGRI (1994) Descriptors for cardamom (*Elettaria cardamomum* Maton). International Plant Genetic Resources Institute, Rome, pp 27–41

Jebasingh T, Backiyarani S, Manohari C, Somanath A, Usha R (2017) Transformation of cardamom with the RNA dependent RNA polymerase gene of cardamom mosaic virus. J Appl Biotechnol Bioeng 3(3):00066. https://doi.org/10.15406/jabb.2017.03.00066

Jyoti BB, Bhat ARS, Hanabar S, Megha J, Rijoy T, Pavitra NL (2019) Instability analysis of productivity and production of cardamom. Int J Curr Microbiol Appl Sci 8(07):1500–1503. https://doi.org/10.20546/ijcmas.2019.807.178

Khan MT (1990) Spices in Indian economy. Academic Foundation, New Delhi, pp 25–30

Khan A, George KJ, Jasrotia RS, Aravind S, Angadi UB, Iquebal MA, Manju KP, Jaiswal S, Umadevi P, Rai A, Kumar D (2019) Plant virus interaction mechanism and associated pathways in mosaic disease of small cardamom (*Elettaria cardamomum* Maton) by RNA-Seq approach. Genomics 112(2):2041–2051

Kizhakkayil J, Thomas E, Zacharia TJ, Syamkumar S, Sasikumar BA (2006) Comparative quality appraisal of exported cardamoms of India, Sri Lanka and Guatemela. Nat Prod Radiance 5:361–365. (Presently Ind J Natural Prod Resour)

Korikanthimath VS, Mulge R (1998) Assessment of elite cardamom lines for dry matter distribution and harvest index. J Med Aromatic Plants 20:28–31

Korikanthimath VS, Mathew MJ, Mulge R, Hegde R, Hosmani M (1999) Studies on selection of elite lines of CL-37 seedling progenies of cardamom. J Spices Aromat Crops 8:57–61
Korikanthimath VS, Prasath D, Mohammed Sayed AA (2000) Evaluation of cardamom accessions in relation to yield, recovery and oil content. Indian J Horticult 57(3):273–275
Koshy J (2002) Elite landraces in cardamom fiom Idukki district. Spice India 15(8):2–4
Krishnamurthy K, Khan MM, Avadhani KK, Venkatesh J, Siddaramaiah AL, Chakravarthy AK, Gurumurthy BR (1989) Three decades of cardamom research at Regional Research Station, Mudigere. In: Technical bulletin no. 2. University of Agricultural Science, Bangalore, p 94
Kumar KB, Kumar PP, Balachandran SM, Iyer RD (1985) Development of clonal plantlets in cardamom (Elettaria cardamomum). J Plant Crops 13:31–34
Kuriakose G, Sinu PA, Shivanna KR (2009) Domestication of cardamom (Elettaria cardamomum) in Western Ghats, India: divergence in productive traits and a shift in major pollinators. Ann Bot 103(5):727–733
Kuruvilla KM, Sudharsan MR, Madhusoodanan KJ, Priyadharshan PM, Radhakrishnan VV, Naidu R (1992) Phenology of tiller and panicle in cardamom, *Elettaria cardamomum* (Maton). J Plant Crops 20(Suppl):162–165
Kuruvilla KM, Madhusoodanan KJ (1988) Effective pollination for better fruit set in cardamom. Spice India 1(6):19–21
Kuruvilla KM, Pradip Kumar K, Radhakrishnan VV, Vadivel V, Madhusoodanan KJ (2000) Evaluation of cardamom genotypes for yield and quality. In: Proc PLACROSYM-XIV, pp 267–270
Kuruvilla KM, Pradip Kumar K, Radhakrishanan VV, Vadivel V, Madhusoodanan KJ (2002) Evaluation of cardamom genotypes for yield and quality. In: Proceedings of the Plantation Crops and Development in the New Millennium (PLACROSYM – XIV-2000), 267–270
Kuruvilla KM, Madhusoodanan KJ, Vadivel V, Radhakrishnan VV, Patil DV, Thomas J (2006) Hybrid cardamom 'MHC'-26 with high yield and quality capsule traits. J Plant Crops 34:208–211
Lawrence BM (1978) Major tropical spices—cardamom (*Elettaria cardamomum*). In: von Bruchhausen F (ed) Hagers Handbuch der pharmazeutischen Praxis. Allured Publishers, Wheaton, pp 105–155
Lukose R, Saji KV, Venugopal MN, Korikanthimath VS (1993) Comparative field performance of micropropagated plants of cardamom. (Elettaria cardamomum). Indian J Agric Sci 63:417–418
Mabberley DJ (1987) The Plant Book. Cambridge University Press, Cambridge
Madhusoodanan KJ, Sudharsan MR, Priyadarshan PM, Radhakrishnan VV (1990) Small cardamom—botany and crop improvement. In: Cardamom Production Technology. ICRI, Spices Board, pp 7–13
Madhusoodanan KJ, Kuruvilla KM, Priyadarshan PM (1994) Cardamom improvement. In: Chadha KL, Rethinam P (eds) Advances in horticulture. Plantation crops, vol 9. Malhotra Publication House, New Delhi, pp 121–129
Madhusoodanan KJ, Kuruvilla KM, Potty SN (1998) Cardamom hybrids for higher yield and better quality capsule. Spice India 11(3):6–7
Madhusoodanan KJ, Radakrishnan VV, Kuruvilla KM (1999) Genetic resources and diversity in cardamom. In: Sasikumar B, Krishnamoorthy B, Rema J, Ravindran PN, Peter KV (eds) Biodiversity, conservation and utilization of spices, medicinal and aromatic plants. IISR, Calicut, pp 68–72
Madhusoodanan KJ, Pradip Kumar K, Ravindran PN (2002) Botany, crop improvement and biotechnology of cardamom. In: Ravindran PN, Madhusoodanan KJ (eds) Cardamom—the genus *Elettaria*. Taylor and & Francis Inc., London, pp 11–68
Malhotra EV, Kamalapriya M, Bansal S et al (2020) Improved protocol for micropropagation of genetically uniform plants of commercially important cardamom (*Elettaria cardamomum* Maton). In Vitro Cell Dev Biol Plant. https://doi.org/10.1007/s11627-020-10131-7
Manohari C, Backiarani S, Jebasingh T, Somanath A, Usha R (2008) Efficient plant regeneration in small cardamom (*Elettaria cardamomum* Maton), through somatic embryogenesis. Indian J Biotechnol 7:407–409

Manoj O, Xavier R, Vadiraj BA, Kumar PM, Remashree AB (2018) Quality appraisal of small cardamom (*Elettaria cardamomum* Maton) sourced from a, B and C zones of CHR in Idukki District of Kerala, India. J Essen Oil Bearing Plants 21(5):1315–1326. https://doi.org/10.1080/0972060X.2018.1538821

Mathew KM, Reshma R, Geethu M, Rithin V, Sabu KK, Nadiya F, Noushad MA, Dharan SS, Rao YS, Remashree AB (2019) Data on small cardamom transcriptome associated with capsule rot disease. Data in Brief 27:104625

Mayne WW (1951) Report on cardamom cultivation in South India. In: ICAR Technical Bulletin. ICAR, p 50

Melgode E (1938) Cardamom I Ceylon-part 1. Trop Agric 91:325–328

Miniraj N, Murugan M, Joseph CR (2000) Evaluation of cardamom (*Elettaria cardamomum* Maton) germplasm. J Spices Aromatic Crops 9(1):55–56

Murugan M, Backiyarani S, Josephrajkumar A, Hiremath MB, Shetty PK (2007) Yield of small cardamom (Elettaria cardamomum M) variety PV1 as influenced by levels of nutrients and neem cake under rain fed condition in southern Western Ghats, India. Caspian J Environ Sci 5(2):19–25

Nadgauda R, Mascarenhas AF, Madhusoodanan KJ (1983) Clonal multiplication of cardamom (E. cardamomum Maton.) by tissue culture. J Plant Crops 11:60–64

Nadiya F, Anjali N, Thomas J, Gangaprasad A, Sabu KK (2018) Genome-wide differential expression profiling in wild and cultivar genotypes of cardamom reveals regulation of key pathways in plant growth and development. Agri Gene 8:18–27

Nadiya F, Anjali N, Thomas J, Gangaprasad A, Sabu KK (2019) Deep sequencing identified potential miRNAs involved in defence response, stress and plant growth characteristics of wild genotypes of cardamom. Plant Biol (Stuttg) 21(1):3–14. https://doi.org/10.1111/plb.12888

Nair KPP (2011) The agronomy and economy of cardamom (*Elettaria cardamomum* M.): the "queen of spices". In: Agronomy and economy of black pepper and cardamom; the king and queen of spices, pp 109–366. https://doi.org/10.1016/C2011-0-04238-2

Nirmal Babu K, Ravindran PN, Peter KV (1997) Protocols for micropropagation of spices and aromatic crops. Indian Institute of Spices Research, Calicut, India

Nirmal Babu K, Geetha SP, Minoo D, Ravindran PN, Peter KV (1999) In vitro conservation of cardamom (Elettaria cardamomum Maton) germplasm. Plant Gene Res News Lett 11(9):41–45

Nirmal Babu K, Minoo D, Geetha SP, Samsudeen K, Rema J, Ravindran PN, Peter KV (1998) Plant biotechnology—it's role in improvement of spices. Indian J Agric Sci 68(8):533–547

Nirmal Babu K, Jayakumar VN, Divakaran M, Venugopal MN, Sudharsan MR, Radhakrishnan VV, Backiyarani S, Narayanaswami MM, Peter KV, Parthasarathy VA (2012) Genetic diversity and phylogenetic relationships among small cardamom (Elettaria cardamomum Maton.) cultivars and related genera using DNA markers. Int J Innov Hortic 1(1):47–56

Owen TC (1901) Notes on cardamom cultivation, Colombo

Padmini K, Venugopal MN, Korikanthimath VS (1999) Biodiversity and conservation of cardamom (*Elettaria cardamomum* Maton). In: Sasikumar B, Krishnamoorthy B, Rema J, Ravindran PN, Peter KV (eds) Biodiversity, conservation and utilization of spices, medicinal and aromatic plants. IISR, Calicut, pp 73–78

Padmini K, Venugopal MN, Sasikumar B (2000) Performance of hybrids, open pollinated progenies and inbreds of cardamom (E. cardamomum) under nursery conditions. Indian Journal of Agricultural Sciences 70:550–551

Parameswar NS (1977) Intergeneric hybridization between cardamom and related genera. Curr Sci 6:10

Parameswar NS, Haralappa HS (1980) Studies on some new high yielding cardamom clones. Curr Res 9(12):212–213

Parthasarathy VA, Prasath D (2012) Cardamom. In: 2nd (ed) Handbook of herbs and spices. Elsevier, London, pp 131–170

Paterniani MEAGZ, Sawazaki E, Dudienas C, Duarte AP, Gallo PB (2000) Diallel crossing among maize lines with emphasis on resistance to foliar diseases. Genet Mol Biol 23:381–385

Pattanshetti HV, Prasad ABN, Kololgi SD (1973) Selection of prepotent mother lines for obtaining precocious bearers in cardamom. Curr Res 2(1):1
Pillai PK, Santha Kumari S (1965) Studies on the effect of growth regulators on fruit setting in cardamom. Agric Res J Kerala 3:5–15
Poulsen AD, Mathisen BH, Newman MF, Ardiyani M, Lofthus Ø, Bjora C (2018) Sulettaria: a new ginger genus disjunct from Elettaria cardamomum. Taxon 67:725–738
PPV & FRA (2009) Distinctiveness, uniformity and stability guidelines. Protection of plant varieties and farmers rights authority, Ministry of Agriculture, Government of India
Prasath D, Venugopal MN (2004) Genetic diversity and conservation of cardamom (*Elettaria cardamomum* Maton) in India. Plant Genet Resour Newsl 138:55–60
Prasath D, Venugopal MN (2009) Compound inflorescence cardamom (*Elettaria cardamomum* (L.) Maton) in India. Genet Resour Crop Evol 56:749–753
Pradip Kumar, Mary Mathew K, Rao YS, Madhusoodanan KJ, Potty SN (1997) Rapid propagation of cardamom through in vitro techniques Proc. of Kerala Science Congress (9th), Ernakulam, Kerala
Priyadarshan PM, Kuruvilla KM, Madhusodanan KJ, Naidu R (1992) Effect of various media protocols and genotype specificity on micropropagation of cardamom (*Elettaria cardamomum* Maton.). In: Subba Rao NS, Rajagopalan C, Ramakrishna SV (eds) New trends in biotechnology. Oxford and IBH Publications, New Delhi, pp 109–117
Purseglove JW, Brown EG, Green CL, Robbins SRJ (1981) Spices, vol 2. Longman, New York
Radhakrishnan VV (2003) Studies on variability, genetic divergence and crop improvement in cardamom (*Elettaria cardamomum* Maton). PhD thesis, University of Calicut, Kerala, India
Radhakrishnan VV, Mohanan KV (2005) Molecular characterization of some elite genotypes of cardamom (*Elettaria cardamomum* Maton). Indian J Genet Plant Breed 65(3):227–228
Radhakrishnan VV, Priya PM, Madhusoodanan KJ, Kuruvilla KM, Thomas J (2004) Factor analysis in cardamom (Elettaria cardamomum Maton). J Spices Aromatic Crops 13(1):37–39
Radhakrishnan VV, Mohanan KV, Priya PM (2006) Genetic divergence in cardamom (Elettaria cardamomum Maton). J Plant Crops 34:149–151
Radhakrishnan VV, Madhusoodanan KJ, Reji K, Kuruvilla KM, Mohanan KV, Thomas J (2010) High yielding cardamom clones for Wayanad. Gregor Mendel Found J 1:23–26
Rafalski JA, Morgante M, Powell J, Tinggey S (1996) Generating and using DNA markers in plants. In: Birren B, Lai E (eds) Analysis of non-mammalian genomes: a practical guide. Academic Press, Boca Raton, pp 75–134
Rajapakse LS (1979) G.L.C. study of the essential oil of wild cardamom oil of Sri Lanka. J Sci Food Agric 30:521–527
Rao NSK, Narayana Swamy S, Chacko EK, Doreswamy ME (1982) Regeneration of plantlets from callus of Elettaria cardamomum Maton. Proc Indiana Acad Sci 91(B):37–41
Rao YS, Mary Mathew K, Pradip Kumar K, Madusoodanan KJ, Naidu R (1994) Tissue culture studies on Spice Crops. Indian Review of Life Sciences 14:129–155 (Review)
Rao YS, Mary Mathew K, Pradip Kumar K, Madhusoodanan KJ, Potty SN, George CK (2000) Practical manual for micropropagation of some selected spices. Indian Cardamom Research Institute, Spices Board, Cochin, p 60
Ravindran PN (1999) Genetic diversity of major spices and their conservation in India. In: Biodiversity, conservation and utilization of spices and aromatic plants. Indian Institute of Spices Research, Calicut, pp 16–44
Ravindran PN (2002) Introduction. In: Ravindran PN, Madhusoodanan KJ (eds) Cardamom—the genus *Elettaria*. Taylor and Francis, London, pp 1–10
Ravindran PN, Nirmal Babu K, Saji KV, Geetha SP, Praveen K, Yamuna G (2004) Conservation of Spices genetic resources in in vitro gene banks. ICAR Project report. Indian Institute of Spices Research, Calicut, Kerala, India, pp 81
Reghunath BR (1989) In vitro studies on the propagation of cardamom (*Elettaria cardamomum* Maton). PhD Thesis. Kerala Agricultural University, Trichur, India

Reghunath BR, Bajaj YPS (1992) Micropropagation of cardamom. In: Bajaj YPS (ed) Biotechnology in agriculture and forestry: high-tech and micropropagation III. Springer-Verlag, Berlin, pp 175–198

Ridley HN (1912) Spices. Macmillan, London, pp 326–327

Rijekbusch PAH, Allen DJ (1971) Cardamom in Tanzania. Acta Hort 21:144–150

Rubido JF (1967) Prospects of cardamom growing Gautemala. Revista Cafeteria Gautemala 71:10–13

Sajina A, Minoo D, Geetha P, Samsudeen K, Rema J, Nirmal Babu K, Ravindran PN, Peter KV (1997) Production of synthetic seeds in few spice crops. In: Edison S, Ramana KV, Sasikumar B, Nirmal Babu K, Eapen SJ (eds) Biotechnology of spices, medicinal and aromatic plants. Indian Society for Spices, Calicut, India, pp 65–69

Sakai S, Nagamasu H (2000) Systematic studies of Bornean Zingiberaceae II, Elettaria of Sarawak. Edinb J Bot 57:227–243

Sarath Kumara SJ, Packiyajothy EV, Jansz ER (1985) Some studies on the effect of maturity and storage on the chlorophyll content and essential oils of the cardamom fruit (*Elettaria cardamomum*). J Sci Food Agric 36(6):491–498

Sasikumar B, Dinesh R, Anandaraj M (2012) Sadaram—summary of advancement achieved in research and management cardamom. Indian Institute of Spices Research, pp 1–138

Sastri BN (1952) The wealth of India—raw materials, D-E, pp 150–160

Senthil Kumar R, Aravind S, Biju CN, Ankegowda SJ (2018) Characterization of cardamom genotypes 143 characterization of genotypes of small cardamom (*Elettaria cardamomum* Maton) for yield parameters and disease resistance. J Plant Crop 46(2):143–146

Sherin J (2015) Molecular characterization of small cardamom (*Elettaria cardamomum* Maton) germplasm and development of variety specific scar markers. PhD thesis, Andhra University, Visakhapatnam

Sherin J, Mary MK, Rao YS, Kuruvilla KM, Sudharsan MR (2013) Identification of variety-specific ISSR markers in small cardamom (*Elettaria cardamomum* Maton). J Plant Crop 41(2):233–237

Sherin J, Anisha CS, Mary Mathew K, Aniel Kumar O, Rao YS (2014) ISSR markers as a tool for assessing genetic diversity in small cardamom. Int J Trop Agric 32(3–4):465–467

Spices Board (2009) Cultivation practices for cardamom, *Elettaria cardamomum* Maton, p 45

Stone GN, Willmer PG (1989) Pollination of cardamom in Papua New Guinea. J Agric Res 28(4):228

Subbarao G, Korikanthimath VS (1983) The influence of rainfall on the yield of cardamom (*Elettaria cardamomum* Maton) in Coorg dist. J Plant Crops 11(1):68–69

Sudharsan MR (1987) Colchicine—induced tetraploids in cardamom (*Elettaria cardamomum* Maton). Curr Sci 56(1):36–37

Sudharsan MR, Madhusoodanan KJ, Jagadeesan P (1989) Evaluation of germplasm in cardamom. J Plant Crops 16(Suppl):331–334

Sudharshan MR, Kuruvilla KM, Madhusoodanan KJ (1988) Periodicity phenomenon in cardamom and its importance in plantation management. Spice India 1(6):14–20

Sudharsan MR, Kuruvilla KM, Madhusoodanan KJ (1991) A key to the identification of types in cardamom. J Plant Crops 18:52–55

Sudharsan MR, Bhatt SS, Narayanaswamy M (1997) Variability in the tissue cultured cardamom plants. In: Edison S, Ramana KV, Sasikumar B, Nirmal Babu K, Eapen SJ (eds) Biotechnology of spices, medicinal and aromatic plants. Indian Society for Spices, Calicut, pp 98–101

Tamayo AC, Afanador L (2007) Molecular characterization of selected cardamom [*Elettaria cardamomum* L. (Matón) (*Zingiberaceae*)] genotypes using AFLP markers. Tierra Tropical 3(2):199–216

Thomas J (2006) Re-crowning the queen of spices- cardamom. In: Thomas J, Hrideek TK, Thomas J, Kuruvilla KM (eds) Plantation crops research—an overview. Indian Cardamom Research Institute, Spices Board, Myladumpara, Kerala, India, pp 1–2

Thomas E, Kizhakkayil J, Zachariah TJ, Syamkumar S, Sasikumar B (2006) Comparative quality characterization and molecular profiling of Indian, Sri Lankan and Guatemalan cardamoms. J Food Agric Environ 4(2):129–133

Thomas L, Rajeev P, Sanil PC (2019) Trade competitiveness and export performance of Indian cardamom. J Spices Arom Crops 28(1):34–42

Tyagi RK, Goswami R, Sanayaima R, Singh R, Tandon R, Agrawal A (2009) Micropropagation and slow growth conservation of cardamom (*Elettaria cardamomum* Maton). In Vitro Cell Dev Biol Plant 45:721–729

Van der Veen M, Morales J (2015) The Roman and Islamic spice trade: new archaeological evidence. J Ethnopharmacol 167:54–63

Van Sloten DH (1990) IBPGR and the challenges of the1990's: a personal point of view. Diversity 6:36–39

Vatsya B, Duiesh K, Kundapurkar AR, Bhaskaran S (1987) Large scale plant formation of cardamom (Elettaria cardamomum) by shoot tip cultures. Plant Physiol Biochem 14:14–19

Venugopal MN, Prasath D, Mulge R (2006) IISR Avinash—a rhizome rot resistant and high yielding variety of cardamom (Elettaria cardamomum Maton). J Spices Aromat Crops 15(1):14–18

Watt G (1872) A dictionary of the economic products of India, vol II. Today and Tomorrows Pub, New Delhi

Willis JC (1967) A dictionary of the flowering plants and ferns—7th edition (revised by H.K. Airey Shaw). Cambridge University Press, Cambridge

Zachariah TJ (2002) Chemistry of cardamom. In: Ravindran PN, Madhusoodanan KJ (eds) Cardamom—the genus *Elettaria*. Taylor and Francis, New York

Zachariah TJ, Lukose R (1992) Aroma profile of selected germplasm accessions in cardamom (*Elettaria cardamomum* Maton). J Plant Crops 20(suppl):310–312

Zachariah TJ, Mulge R, Venugopal MN (1998) Quality of cardamom from different accessions. In: Mathew NM, Jacob CK (eds) Developments in plantation crops research. Allied Pub, India, pp 337–340

# Index




P. M. Priyadarshan, S. M. Jain (eds.), *Cash Crops*,
https://doi.org/10.1007/978-3-030-74926-2

**D**

## K



## L



## M

## N



## O

T



U

The manufacturer's authorised representative in the EU is Springer Nature Customer Service Centre GmbH, Europaplatz 3, 69115 Heidelberg, Germany. If you have any concerns regarding our products, please contact ProductSafety@springernature.com

Printed and bound by CPI Group (UK) Ltd, Croydon, CR0 4YY
22/09/2026
02221048-0015